GROWTH HORMONE

Stephen Harvey
Colin G. Scanes
William H. Daughaday

CRC Press
Boca Raton Ann Arbor London Tokyo

Library of Congress Cataloging-in-Publication Data

Harvey, Stephen, 1951–
Growth hormone / Stephen Harvey, C.G. Scanes, William Daughaday.
p. cm.
Includes bibliographical references and index.
ISBN 0-8493-8697-7
1. Somatotropin. I. Scanes, C. G. II. Daughaday, William H., 1918– III. Title.
[DNLM: 1. Hepatitis B virus. QW 710 G289h]
QP572.S6H37 1994 1995
612.6—dc20
DNLM/DLC
for Library of Congress 94-19378
CIP

Direct all inquiries to CRC Press, Inc., 2000 Corporate Blvd., N.W., Boca Raton, Florida 33431.

International Standard Book Number 0-8493-8697-7
Library of Congress Card Number 94-19378
Printed in the United States of America 1 2 3 4 5 6 7 8 9 0
Printed on acid-free paper

For our families, colleagues and students.

PREFACE

At the beginning of this century the critical requirement of the anterior pituitary gland for growth was first recognized. The occurrence of a growth hormone (GH)-like substance in this tissue was subsequently demonstrated in the 1920s by the ability of pituitary extracts to restore growth in hypophysectomized animals and to promote growth and induce gigantism in pituitary-intact laboratory species. These initial observations resulted in the development of a sensitive bioassay which facilitated the eventual isolation, purification, and characterization of bovine GH from ox anterior pituitary glands. Since then GH has been purified from many vertebrate species and its physiochemical and biological properties have been extensively studied. Its gene has now been isolated, cloned, and sequenced and utilized by recombinantion technologies for the transgenic overproduction of GH and the synthesis of GH for numerous clinical and agricultural applications.

These advances in GH research have been extensively documented in a vast body of literature that appears to expand exponentially. The aim of this book was thus to comprehensively review recent literature on the chemistry, synthesis, regulation, and actions of pituitary GH. This information has been compiled in thirty chapters that address basic, clinical, and applied aspects of GH research.

Undaunted by Ernst Knobil's and Julane Hotchkiss's insight in earlier review of this field ("Anyone undertaking a compilation, evaluation and summation of the contributions which have been made to the nature and actions of the pituitary growth hormone does so at the peril of his own clarity of thought, to say nothing of that of his readers." *Growth Hormone. Ann. Rev. Physiol.* **26**:47–74; 1964), we hope this monograph provides a clarification of recent literature and will be a source of reference for those who practice or teach GH physiology.

S. Harvey
C. G. Scanes
W. H. Daughaday

March 1994

Stephen Harvey, Ph.D. is a Professor in the Department of Physiology and in the Division of Endocrinology, Department of Medicine in the Faculty of Medicine at the University of Alberta, Edmonton.

Dr. Harvey graduated from the University of Leeds with a B.Sc. Honors degree in Zoology (1974) and a Ph.D. from the Department of Animal Physiology and Nutrition (1977). After postdoctoral training in the Department of Zoology, University of Hull, Dr. Harvey became a Research Lecturer in the Wolfson Institute, University of Hull from 1981 to 1986. Dr. Harvey then joined the faculty at the University of Alberta, serving as an Associate Professor until 1990.

Dr. Harvey is a member of the Endocrine Society, the Society for Endocrinology (U.K.), the European Neuroendocrine Association, the European Society for Comparative Endocrinology, and the British Neuroendocrine Group. In 1980 he received the Heller Memorial Prize, awarded to a leading endocrinologist under 35, from the Society for Endocrinology. In 1986 he was elected as a Fellow in the Institute of Biology for his contributions to biological research.

Dr. Harvey is the author or co-author of more than 200 original research papers and 20 review articles. His research has been funded by the Science and Engineering Research Council (U.K.), the Natural Sciences and Engineering Research Council, the Medical Research Council, the Alberta Heritage Foundation for Medical Research, and by private industry. His research has focused on the comparative regulation of growth hormone secretion and on the mechanism of growth hormone action.

Colin G. Scanes, Ph.D., D.Sc. is Professor II and Chairman of the Department of Animal Sciences, Rutgers – The State University of New Jersey.

Dr. Scanes graduated in 1969 from Hull University (U.K.) with a B.Sc. degree (first class honors) in Biological Chemistry and Zoology and obtained his Ph.D. in 1972 from the University of Wales (U.K.). In 1985 he received a D.Sc. from Hull University for published research.

Dr. Scanes is a member of the American Association for the Advancement of Science; American Institute of Nutrition; American Physiological Society; American Society of Zoologists; American Society of American Science; Endocrine Society; Poultry Science Association; Society for Environmental Toxicology and Chemistry; Society for Experimental Biology and Medicine; and Society for the Study of Reproduction.

Among other awards, he has received the Poultry Science Association Merck Award for Research and the Rutgers University Research Award. He has been previously a Program Manager in the United States Department of Agriculture Competitive Grants Program, a member of the National Research Council animal growth panel and a member of the Office of Technology Assessment Review of Animal Biotechnology.

Dr. Scanes has presented over 40 invited lectures at International Symposia and over 100 invited Seminars/Presentations at universities and meetings. He has edited 7 books and published over 220 research papers. His current research interests include the hormonal control of growth and metabolism in agricultural animals and the relationship between toxic metals and growth factors.

William H. Daughaday, M.D. is the Irene E. and Michael M. Karl Professor of Metabolism in Medicine, Emeritus, Washington University School of Medicine, Saint Louis, Missouri.

Dr. Daughaday graduated from Harvard College in 1940 with an A.B. degree (Magna cum laude) and from Harvard Medical School in 1943.

Dr. Daughaday is a member of the American Society for Clinical Investigation, the Association of American Physicians, the Endocrine Society (President 1971–1972), American Diabetes Association, the National Academy of Sciences and the American Academy of Arts and Sciences. He is the recipient of the Fred Conrad Koch Award of the Endocrine Society, the Joslin Medal of the Massachusetts Affiliate, American Diabetes Association, and an Honorary Member of the Japan Endocrine Society. He has delivered ten named lectureships.

Dr. Daughaday is a former Editor of the Journal of Laboratory and Clinical Medicine, the Journal of Clinical Endocrinology and Metabolism and an Associate Editor of the Journal of Clinical Investigation. He is a former member of the American Board of Internal Medicine, the Endocrine Study Section, National Institutes of Health. He served on the Board of Scientific Counselors of the National Institute of Arthritis and Metabolic Diseases, and the Advisory Council, National Institute of Diabetes, Digestive and Kidney Diseases.

His major research interest has been clinical disorders of pituitary growth hormone and insulin-like growth factors. He is the author of over 300 medical and scientific papers.

CONTRIBUTORS

Robert M. Campbell
Hoffmann-LaRoche
Nutley, New Jersey

Kerry L. Hull
Department of Physiology
University of Alberta
Edmonton, Alberta, Canada

CONTENTS

Chapter 1

Growth Hormone: Chemistry

Colin G. Scanes and Robert M. Campbell

CONTENTS

I. INTRODUCTION

Growth hormone (GH) is, in its major form, a 22-kDa protein. It is apparent, on the basis of homology of the sequences of amino acid residues in cDNA, that there is a GH-prolactin family of proteins. This family has two major trunks, GH and prolactin, which diverged prior to the evolution of the vertebrates (this is discussed in more detail later in this chapter). In addition, proteins with sequences somewhat intermediate between GH and prolactin are found in fish. These are referred to as *somatolactins*. Thus, the GH-prolactin family might be viewed as a trichotomy of three trunks with GH, prolactin, and somatolactins (for details see Chapter 2). There are further branching points in the family. For instance, the chorionic somatomammotropins (CS) of primates show close homology to GH and are thought to have diverged or branched (by gene duplication and point mutation) from the GH trunk during primate evolution. In addition, proteins with marked sequence homology to prolactin have been identified. These include the placental lactogens (PLs) and proliferins. These are presumed to be derived by gene duplication of the prolactin gene followed by divergence due to point mutation.

0-8493-8697-7/95/$0.00+$.50

II. GROWTH HORMONE: STRUCTURE

A. AMINO ACID SEQUENCE

Newly synthesized GH is a protein with about 191 amino acid residues; the number of residues varying slightly with GH from different species. Figure 1[1–66] shows amino acid sequences for GH from 38 different species representing the major vertebrate classes (mammals, birds, reptiles, amphibians, bony fish, and cartilaginous fish). Figure 2[67–115] shows the amino acid sequences for prolactin (the second member of the GH-prolactin family of hormones). For purposes of comparison, the primary structures of the variant form of human GH (hGHV) and hCS are also included with GH in Figure 1; hCS being thought to have evolved by gene duplication from GH.[116] Moreover, rodent and ungulate placental lactogens (PLs) and proliferins are included in the prolactin chart (Figure 2), as these show marked homologies to prolactin. Fish somatolactins are included in the prolactin chart (Figure 2) simply for convenience, it not being clear yet whether somatolactins are ancestrally derived from either the GH or prolactin genes. Growth hormone, prolactin, and PL display varying biological activities and yet retain homologous amino acid sequences.[117] It is apparent that the differences between these molecules confer subtle, but significant, changes in secondary structure and associated biological activity.

B. CONFORMATION AND THREE-DIMENSIONAL STRUCTURE

Growth hormone has two disulfide bridges that link, respectively, large and small peptide loops (Figure 3, top).[118] There are four cysteine residues present in all GHs from as evolutionarily diverse species as human and shark (Figure 1). Thus, the schematic representation for monomeric GH in Figure 3 (top) is applicable to all vertebrate GHs, which are characterized by large and small loops of amino acid residues.

The conformation of GH is about 50% α-helical. Chou-Fasman calculations and circular dichroism spectra indicate that vertebrate GH, prolactin, and PL are approximately 45–55% α-helical. Of these hormones, few detailed conformational data are available (i.e., two-dimensional nuclear magnetic resonance [NMR] or X-ray crystallography), with the exception of porcine and human GH.

Figure 4[119] shows the three-dimensional structure of human GH adapted from the x-ray structure of porcine GH.[120] Crystalline porcine GH is composed of four antiparallel α-helices, extending from residues 7–34, 75–87, 106–127 and 152–183.[120] These four helices exhibit distinct amphiphilic character, with helix 4 being primarily hydrophobic (see Figures 4 and 6). In agreement with earlier circular dichroism and Chou-Fasman calculations, 85 of 191 or ~45% of the porcine GH residues are involved in the four major helices. Other smaller α-helical regions, noncolinear with the major helices, are found between residues 53 and 58 (~1.5 turns) and between 89 and 96 (~2 turns[120]), bringing the total helical content to ~52%.

C. STRUCTURAL REQUIREMENTS FOR RECEPTOR BINDING

It would seem to be axiomatic that GH radioreceptor activity should reflect biological activity. Thus, it is not surprising that a radioreceptor assay was employed to examine the GH-binding activity of mutated GH analogs or the effects of monoclonal antibodies on specific epitopes on the GH molecule. Wells and colleagues at Genentech (South San Francisco, CA)[121] initially employed the ingenious approach of substituting various sequences from prolactin into GH to create chimeric molecules as prolactin does not bind to the GH receptor. On the basis of these studies, together with use of monoclonal antibodies, it was concluded that three determinants on the hGH molecule (loop residues 54–74; the central to C-terminal region of helix 4; and also, somewhat, the N-terminal region of helix 1) were involved in binding of GH (hGH) to its receptor. In addition, a series of GH analogs were engineered with single amino acid residue substitutions.[122] Table 1 shows a series of substituted GH analogs; some with reduced GH radioreceptor activities. These substitutions throw considerable light on the active site of GH for binding to the GH receptor (GHR). For instance, they confirm the critical importance of specific amino acid residues in helix 4, the loop residues 54 to 74, and, to a lesser extent, helix 1 to GH binding. Moreover, if as few as 8 amino acid residues in these same regions of prolactin are replaced with the residue found in GH, then the association constant for binding to the GH receptor for the mutated prolactin analog is increased over 10,000-fold.[123] These data provided strong evidence for one region in the GH molecule (helix 4, loop residues 54 to 74 and perhaps also helix 1) (see Figure 4) being involved in GH binding to the GH receptor.

The three-dimensional structure of the complex of hGH and the extracellular domain of the hGHR (the hGH-binding protein) has been determined by X-ray crystallography.[124] The complex was made up of one molecule of hGH and two molecules of the GHR. There were, in fact, two distinct sites on GH that bound to the GHR (see Figure 4). As might be expected on the basis of mutagenesis studies, site 1 included the central to C-terminal region of helix 4 together with loop 54–74 with salt and hydrogen bonds between Lys-41, Gln-46, Pro-61, Arg-167, Lys-168, Asp-171, Thr-175 and Arg-178 and the GHR.[124] In addition, there was a second site that included helix 3 and helix 1 (with bonds between residues Asn-12, Arg-16, Arg-19, and the GHR). Thus, GH and its receptor form a heterotrimer ("sandwich"), there being also binding (and specific salt and hydrogen bonds) between the two GHR molecules.[124] The presently accepted view is that dimerization of the two GHR molecules is necessary for GH to exert its biological effect.[125] This is summarized in Figure 5.

The scanning mutagenesis/radioreceptor studies largely ignored helix 3 of GH. This is probably because peptides equivalent to helix 3 of GH exhibit little radioreceptor activity.[126] However, peptides equivalent to helix 3 in both hGH and bovine GH (bGH) have significant growth-promoting activity; respectively 1.6% of hGH (Table 2) and 5% of bGH[127] on a weight basis. The role of α-helix 3 in GH action is considered below.

D. THE ROLE OF HELIX 3

The three-dimensional structure of GH includes four antiparallel α-helical regions.[120] Innovative studies by John J. Kopchick and colleagues have provided evidence that α-helix 3 plays a pivotal role in GH action. When represented as a "helical wheel," employing an Edmundson wheel projection (Figure 6), it was evident that α-helix 3 of bGH, between residues 109 and 126, was largely amphiphilic. However, three amino acid residues were inconsistent with an idealized amphiphilic helix. To investigate whether bGH with an idealized amphiphilic α-helix 3 would have enhanced biological activity, a mutated bGH was engineered with Glu-117 substituted to leucine (E117L), Gly-119 to arginine (G119R), and Ala-122 to aspartate (A122D).[128] Transgenic mice expressing this modified bGH did not have enhanced growth rates; in fact, both growth and circulating concentrations of IGF-I were depressed.[128,129] The modified GH did, however, inhibit binding of ^{125}I-bGH to liver membranes.[128,129] This suggested that the modified bGH was a functional antagonist for GH.[129] This substituted form of bGH also antagonizes other actions of GH, including adipocyte differentiation[130] and the insulin-like effect of stimulating glucose oxidation[130] and lipolysis by either rat[130] or chicken[131] adipose tissue.

Analogs of bGH or hGH have been generated with single substitutions to examine further which amino acid residues may be critical to GH action. Expression of bGH G119R in transgenic mice is associated with dwarfism,[132] due presumably to the GH antagonist activity of the GH analog. Similarly, the human GH analog (G120R) inhibits binding of GH to hGH binding protein but lacks GH biological activity (proliferation of FDC-P1 cells).[125] It is probably not coincidental that glycine at position 119 is consistently found in GH throughout the vertebrates (see Figure 1). Nicoll and colleagues[117] argued strongly for the view that when identical amino acid residues are found across

Figure 1 Amino acid sequences of growth hormone (GH) from various vertebrate and hypothetical ancestral species. Sequences were aligned using theoretical deletions to maximize homology. Standard single-letter abbreviations are used for amino acids (IUPAC-IUB Commission on Biochemical Nomenclature): A, Ala; C, Cys; D, Asp; E, Glu; F, Phe; G, Gly; H, His; I, Ile; K, Lys; L, Leu; M, Met; N, Asn; P, Pro; Q, Gln; R, Arg; S, Ser; T, Thr; V, Val; W, Trp; Y, Tyr. (–) Gap/deletion. (For the convenience of the reader, the classification ["relationships"] of the species is included as a footnote.)

GH Species	Ref. #	[1] [10] [20] [30] [40] [50] [60] [70] [80] [90] [100]
human	1,2	-FPTIPLSRLFDNAMLRAHRLHQLAFDTYQEFEEAYIPKEQKYSFLQNPQTSLCFSESIPTPSNREETQQKSNLELLRISLLLIQSWLEPVQFLRSVFANS
human variant	3	-FPTIPLSRLFDNAMLRARRLYQLAYDTYQEFEEAYILKEQKYSFLQNPQTSLCFSESIPTPSNRVKTQQKSNLELLRISLLLIQSWLEPVQLLRSVFANS
human CS	4-6	-VQTVPLSRLFDHAMLQAHRAHQLAIDTYQEPGGTYIPKDQKYSFLHDSQTSFCFSDSIPTPSNMEETQQKSNLELLRISLLLIESWLEPVRFLRSMFANN
monkey, rhesus	7	-FPTIPLSRLFDNAMLRAHRLHQLAFDTYQEFEEAYIPKEQKYSFLQNPQTSLCFSESIPTPSNREETQQKSNLQLLRISLLLIQSWLEPVQFLRSVFANS
rat	8	-FPAMPLSSLFANAVLRAQHLHQLAADTYKEFERAYIPEGQRYS-IQNAQAAFCFSETIPAPTGKEEAQQRTDMELLRFSLLLIQSWLGPVQFLSRIFTNS
mouse	9	-FPAMPLSSLFSNAVLRAQHLHQLAADTYKEFERAYIPEGQRYS-IQNAQAAFCFSETIPAPTGKEEAQQRTDMELLRFSLLLIQSWLGPVQFLSRIFTNS
hamster	10	-FPAMPLSSLFANAVLRAQHLHQLAADTYKEFERAYIPEGQRYS-IQNAQTAFCFSETIPAPTGKEEAQQRSDMELLRFSLLLIQSWLGPVQFLSRIFTNS
whale, fin	11	-FPAMPLSSLFANAVLRAQHLHQLAADTYKEFERAYIPEGQRYS-IQNAQAAFCFSETIPAPTGKNEAQQRSDVELLRLSLLLIQSWLGPLQFLSRVFTNS
whale, sei	12	-FPAMPLSSLFANAVLRAQHLHELAADTYKEFERAYIPEGQRY-FLQNAQSTGCFSEVIPTPANKDEAQQRSDVELLRFSLLLIQSWLGPVQFLEKAYANE
fox	13	-FPAMPLSSLFANAVLRAQHLHQLAADTYKEFERAYIPEGQRYS-IQNAQAAFCFSETIPAPTGKDEAQQRSDVELLRFSLVLIQSWLGPLQFLSRVFTNS
mink	14	AFPAMPLSSLFANAVLRAQHLHQLAADTYKDFERAYIPEGQRYS-IQNAQAAFCFSETIPAPTGKDEAQQRSDMELLRFSLLLIQSWLGPVQFLSRVFTNS
cattle	15,16	AFPAMSLSGLFANAVLRAQHLHQLAADTFKEFERTYIPEGQRYS-IQNTQVAFCFSETIPAPTGKNEAQQKSDLELLRISLLLIQSWLGPLQFLSRVFTNS
sheep	17,18	AFPAMSLSGLFANAVLRAQHLHQLAADTFKEFERTYIPEGQRYS-IQNTQVAFCFSETIPAPTGKNEAQQKSDLELLRISLLLIQSWLGPLQFLSRVFTNS
goat	19	AFPAMSLSGLFANAVLRAQHLHQLAADTFKEFERTYIPEGQRYS-IQNTQVAFCFSETIPAPTGKNEAQQKSDLELLRISLLLIQSWLGPLQFLSRVFTNS
pig	16	-FPAMPLSSLFANAVLRAQHLHQLAADTYKEFERAYIPEGQRYS-IQNAQAAFCFSETIPAPTGKDEAQQRSDVELLRFSLLLIQSWLGPVQFLSRVFTNS
alpaca	20	-FPAMPLSSLFANAVLRAQHLHQLAADTYKEFERTYIPEGQRYS-IQNAQAAFCFSETIPAPTGKDEAQQRSDVELLRFSLLLIQSWLGPVQFLSRVFTNS
horse	21	-FPAMPLSSLFANAVLRAQHLHQLAADTYKEFERAYIPEGQRYS-IQNAQAAFCFSETIPAPTGKDEAQQRSDMELLRFSLLLIQSWLGPVQLLSRVFTNS
elephant	22	-FPAMPLSSLFANAVLRAQHLHQLAADTYKEFERAYIPEGQRYS-IQNAQAAFCFSETIPAPTGKDEAQQRSDVELLRFSLLLIQSWLGPVQFLSRVFTNS
ancestral mammal		-FPAMPLSSLFANAVLRAQHLHQLAADTYKEFERAYIPEGQRYS-IQNAQAAFCFSETIPAPTGKDEAQQRSDLELLRFSLLLIQSWLGPVQFLSRVFTNS
duck	23	TFPAMPLSNLFANAVLRAQHLHLLAAETYKEFERSYIPEDQRHT-NKNSQ-AFCYSETIPAPTGKDDAQQKSDMELLRFSLVLIQSWLTPVQYLSKVFTNN
chicken	24,25	TFPAMPLSNLFANAVLRAQHLHLLAAETYKEFERTYIPEDQRYT-NKNSQAAFCYSETIPAPTGKDDAQQKSDMELLRFSLVLIQSWLTPVQYLSKVFTNN
turkey	26	TFPTMPLSNLFTNAVLRAQHLHLLAAETYKEFERTYIPEDQRYT-NKNSQAAFCYSETIPAPTGKDDAQQKSDMELLRFSLVLIQSWLTPMQYLSKVFTNN
sea turtle	27	AFPAMPLSSLFANAVLRAQHLHLLAADTYKEFERTYIPEEQRHS-NKISQSASCYSETIPAPTGKDDAEQKSDMELLRFSLILIQSWLNPVQFLSRVFTNS
ancestral amniote		-FPAMPLSSLFANAVLRAQHLHQLAADTYKEFERAYIPEGQRYS-IQNAQAAFCFSETIPAPTGKDEAQQRSDMELLRFSLLLIQSWLGPVQFLSRVFTNS
xenopus I	28	-???FCYSETMPYPTDKDNTHQKSDMELLRFSQTLIQSWLNPVQALSKVFSNN
bullfrog (from cDNA)	29	-FPQMSLSNLFTNAVIRAQHLHQMVADTYRDYERTYIPEDQRFK-QTLLISVYCYSETIPAPTDKDNTHQKSDIDLLRFSLTLLQSWMTPIQIVNRVFGNN
bullfrog (from protein)	30	-FPQMSLSNLFTNAVIRAQHLHQMVADTYRDYERTYIPEDQRFS-NKHSYSVYCYSETIPAPTDKDNTHQKSDIELLRFSLLLLQSWMNPIQIVNRVFGNN
bullfrog (from PCR)	31	-FPQMSLSNLFTNAVIRAQHLHQMVADTYRDYERTYIPEDQRFS-NKHSYSVYCYSETIPAPTDKDNTHQKSDIDLLRFSLTLLQSWMTPIQIVNRVFGNN
bonito	32	-QPITESQRLFSIAVSRVQNLHLLAQRLFSDFESSLQTQEQRQL-NKIFLQDFCNSDYIISPIDKHETQRSSVLKLLSISYRLVESWEFPSRSLSG-----
yellow tail	33	-QPITDSQHLFSIAVSRIQNLHLLAQRLFSNFESTLQTEDQRQL-NKIFLQDFCNSDYIISPIDKHETQRSSVLKLLSISYRLVESWEFSSRFLSGGGS--
tuna	34,35	-QPITDSQRLFSIAVSRVQHLHLLAQRLFSDFESSLQTEEQRQL-NKIFLQDFCNSDYIISPIDKHETQRSSVLKLLSISYRLVESWEFPSRSLSG-GS--
snook (l. calcarifer)	36	-QQITDSQRLFSIAVSRVQHLHLLAQRRFSEFESSLQTEEQRHV-NKIFLQDFCNSDYIISPIDKHETQRSSVLKLLSISYRLVESWEFSSRSLSG-GS--
tilapia	37	-QQITDSQRLFSIAVNRVTHLHIIAQRLFSDFESSLQTEEQRQL-NKIFLQDFCNSDYIISPIDKHETQRSSVLKLLSISYGLVESWEFPSRSLSG-GS-S
flounder	38,39	-QPITENQRLFSIAVGRVQYLHLVAKKLFSDFENSLQLEDQRLL-NKIASKEFCHSDNFLSPIDKHETQGSSVQKLLSVSYRLIESWEFFSRFL--VAS--
sea bream, red	40	-QPITDGQRLFSIAVSRVQHLHLLAQRLFSDFESSLQTEEQLKL-NKIFP-DFCNSDYIISPIDKHETQRSSVLKLLSISYRLVESWEFPSRSLSG-GS--
sea bream, black	36	-QPITDGQRLFSIAVSRVQHLHLLAQRLFSDFESSLQTEEQRQL-NKIFLQDFCNSDYIISPIDKHETQRSSVLKLLSISYRLVESWEFPSRSLAG-GS--
sea bream, gilthead	41	-QPITDGQRLFSSAVSRVQHLHLLAQRLFSDFESSLQTEEERQL-NKIFLQDFCNSDYIISPIDKHETQRSSVLKLLSISYRLVESWEFPSRSLSG-GS--
porgy, yellowfin	42	-QPITDGQRLFSIAVSRVQHLHLLAQRLFSDFESSLQTEEQRQL-NKIFLQDFCNSDYIISPIDKHETQRSSVLKLLSISYRLVESWEFPSRSLAG-GS--
hard tail	43	-QPIPNNQHLFSMAVSRIHHLHLRAQRLFANFESSLQSDDQRQL-NKIFLQDFCNSDYIISPIDKHETQRSSVLKLLLISKQLVESWEISSHFLPG-G---
cod	44	-HPLIDSQRLFSIAVNRIQHLHMLAQRIFSELESSLQIEEQRQL-NKIFLQDFCNSDSIISPIDKHETQRSSVLRLLTVSYRLIESWEFPSQSLPG-GS-V
carp, common	45-47	----SDNQRLFNNAVIRVQHLHQLAAKMINDFEDSLLPEERRQL-SKIFPLSFCNSDYIEAPTGKDETQKSSMLKLLRISFRLIESWEFPSQTLSGTVSNS
carp, bighead/grass/silver	48,49	----SENQRLFNNAVIRVQHLHQLAAKMINDFEDNLLPEERRQL-SKIFPLSFCNSDSIEAPTGKDETQKSSMLKLLRISFRLIESWEFPSQTLSGAVSNS
salmon I, chum	50,51	----IENQRLFNIAVSRVQHLHLLAQKMFNDFDGTLLPDERRQL-NKIFLLDFCNSDSIVSPVDKHETQKSSVLKLLHISFRLIESWEYPSQTL--IISNS
salmon II, chum	52	----MENQRLFNIVVNRVQHLHLLAQKMFNDFEGTLLSDERRQL-NKIFLLDFCNSDSIVSPIDKQETQKSSVLKLLHISFRLIESWEYPSQTL--TISNS
salmon, coho	53,54	----IENQRLFNIAVSRVQHLHLLAQKMFNDFDGTLLPDERRQL-NKIFLLDFCNSDSIVSPVDKHETQKSSVLKLLHISFRLIESWEYPSQTL--IISNS
salmon, atlantic	55,56	----MENQRLFNIAVNRVQHLHLMAQKMFNDFEGTLLPDERRQL-NKIFLLDFCNSDSIVSPIDKLETQKSSVLKLLHISFRLIESWEYPSQTL--TISNS
trout I	57,58	----IENQRLFNIAVSRVQHLHLLAQKMFNDFDGTLLPDERRQL-NKIFLLDFCNSDSIVSPVDKHETQKSSVLKLLHISFRLIESWEYPSQTL--IISNS
trout II	57,58	----MENQRLFNIAVNRVQHLHLLAQKMFNDFEGTLLPDERRQL-NKIFLLDFCNSDSIVSPIDKQETQKSSVLKLLHISFRLIESWEYPSQTL--IISNS
catfish	59	----FESQRLFSIAVIRVQHLHLLAAKMMDDFEEALLPEERKQL-SKIFPLFFCNSDSIEAPAGKDEAQKSSVLKLLHTSYRLIESWEFPSRNL-------
catfish (p. pangasius)	60	----FENQRLFNNAVIRVQHLHQLAAKMMDDFEEALLPEERKQL-SKIFPLSFCNSDSIEAPAGKDETQKSSVLKLLHTSYRLIESWEFPSKNL-------
eel	61-63	VEP-ISLYNLFTSAVNRAQHLHTLAAEIYKEFERSIPPEAHRQL-SKTSPLAGCYSDSIPTPTGKDETQEKSDGYLLRISSALIQSWVYPLKTLSDAFSNS
sturgeon	64	-YPMIPLSSLFTNAVLRAQYLHQLAADIYKDFERTYMPNEQRHS-SKNSPSAFCYSETIPAPTGKDEAQQRSDVELLQFSLALIQSWISPLQSLSRVFTNS
ancestral boney fish		-QPI?ENQRLFSIAVSRVQHLHLLAQRLFSDFESSLQPEEQRQL-NKIFL?DFCNSD?IISPIDKHETQRSSVLKLLSISYRLIESWEFPSRTLSG-GSNS
shark	65	-YPLLPLSDLFAKAVHRAQHLHAVAAETTKDFERKYIPEEQRHS-NKSSPSAFCQSETIPAPTGKEDAQQRSDRELLLYSLLLIQSWLNPIQNLS------
ancestral vertebrate		-FPAMPLSRLFANAVLRAQHLHQLAADTYKEFERTLIPEEQRYS-NKN?QAAFCFSETIPAPTGKDEAQQSSDLELLRISLLLIQSWLGPSQFLSRVFTNS

GH Species	Ref. #	[110] [120] [130] [140] [150] [160] [170] [180] [190]
human	1,2	LVYGASDSNVYDLLKDLEEGIQTLMGRLEDG--SPRTG-QIFKQ---TYSKFDTNSHNDDALLKNYGLLYCFRKDMDKVETFLRIV--QCR-SVEGSCGF
human variant	3	LVYGASDSNVYRHLKDLEEGIQTLMWRLEDG--SPRTG-QIFNQ---SYSKFDTKSHNDDALLKNYGLLYCFRKDMDKVETFLRIV--QCR-SVEGSCGF
human CS	4-6	LVYDTSDSDDYHLLKDLEEGIQTLMGRLEDG--SRRTG-QILKQ---TYSKFDTNSHNHDALLKNYGLLYCFRKDMDKVETFLRMV--QCR-SVEGSCGF
monkey, rhesus	7	LVYGTSYSDVYDLLKDLEEGIQTLMGRLEDG--SSRTG-QIFKQ---TYSKFDTNSHNDDALLKNYGLLYCFRKDMDKIETFLRIV--QCR-SVEGSCGF
rat	8	LMFGTSD-RVYEKLKDLEEGIQALMQELEDG--SPRIG-QILKQ---TYDKFDANMRSDDALLKNYGLLSCFKKDLHKAETYLRVM--KCRRFAESSCAF
mouse	9	LMFGTSD-RVYEKLKDLEEGIQALMQELEDG--SPRVG-QILKQ---TYDKFDANMRSDDALLKNYGLLSCFKKDLHKAETYLRVM--KCRRFVESSCAF
hamster	10	LMFGTSD-RVYEKLKDLEEGIQALMQELEDG--SPRVG-QILKQ---TYDKFDTNMRSDDALLKNYGLLSCFKKDLHKAETYLRVM--KCRRFVESSCAF
whale, fin	11	LVFGTSD-RVYEKLKDLEEGIQALMRELEDG--SPRAG-QILKQ---TYDKFDTNMRSDDALLKNYGLLSCFKKDLHKAETYLRVM--KCRRFVESSCAF
whale, sei	12	LVFGTSD-RVYEKLKDLEEGIQALMRELEDG--SPRAG-QILKQ---TYDKFDTNMRSDDALLKNYGLLSCFKKDLHKAETYLRVM--KCRRFVESSCAF
fox	13	LVFGTSD-RVYEKLKDLEEGIQALMRELEDG--SPRAG-QILKQ---TYDKFDTNLRSDDALLKNYGLLSCFKKDLHKAETYLRVM--KCRRFVESSCAF
mink	14	LVFGTSD-RVYEKLKDLEEGIQALMRELEDG--SPRAGP-ILKQ---TYDKFDTNLRSDDALLKNYGLLSCFKKDLHKAETYLRVM--KCRRFVESSCAF
cattle	15,16	LVFGTSD-RVYEKLKDLEEGILALMRELEDG--TPRAG-QILKQ---TYDKFDTNMRSDDALLKNYGLLSCFRKDLHKTETYLRVM--KCRRFGEASCAF
sheep	17,18	LVFGTSD-RVYEKLKDLEEGILALMRELEDV--TPRAG-QILKQ---TYDKFDTNMRSDDALLKNYGLLSCFRKDLHKTETYLRVM--KCRRFGEASCAF
goat	19	LVFGTSD-RVYEKLKDLEEGILALMRELEDV--TPRAG-QILKQ---TYDKFDTNMRSDDALLKNYGLLSCFRKDLHKTETYLRVM--KCRRFGEASCAF
pig	16	LVFGTSD-RVYEKLKDLEEGIQALMRELEDG--SPRAG-QILKQ---TYDKFDTNLRSDDALLKNYGLLSCFKKDLHKAETYLRVM--KCRRFVESSCAF
alpaca	20	LVFGTSD-RVYEKLKDLEEGIQALMRELEDG--SPRAG-QILRQ---TYDKFDTNLRSDDALLKNYGLLSCFKKDLHKAETYLRVM--KCRRFVESSCAF
horse	21	LVFGTSD-RVYEKLRDLEEGIQALMRELEDG--SPRAG-QILKQ---TYDKFDTNLRSDDALLKNYGLLSCFKKDLHKAETYLRVM--KCRRFVESSCAF
elephant	22	LVFGTSD-RVYEKLKDLEEGIQALMRELEDG--SPRPG-QVLKQ---TYDKFDTNMRSDDALLKNYGLLSCFKKDLHKAETYLRVM--KCRRFVESSCAF
ancestral mammal		LVFGTSD-RVYEKLKDLEEGIQALMRELEDG--SPRAG-QILKQ---TYDKFDTNMRSDDALLKNYGLLSCFKKDLHKAETYLRVM--KCRRFVESSCAF
duck	23	LVFGTSD-RVFEKLKDLEEGIQALMRELEDR--SPR-GPQLLK---PTYDKFDIHLRNEDALLKNYGLLSCFKKDLHKVETYLKVM--KCRRFGESNCTI
chicken	24,25	LVFGTSD-RVFEKLKDLEEGIQALMRELEDR--SPR-GPQLLR---PTYDKFDIHLRNEDALLKNYGLLSCFKKDLHKVETYLKVM--KCRRFGESNCTI
turkey	26	LVFGTSD-RVFEKLKDLEEGIQALMRELEDR--SPR-GPQLLR---PTYDRFDIHLRSEDALLKNYGLLSCFKKDLHKVETYLKVM--KCRRFGESNCNI
sea turtle	27	LVFGTSD-RVYEKLRDLEEGIQALMRELEDG--SLR-GFQVLR---PTYDKFDINLRNEDALLKNYGLLSCFKKDLHKVETYLKLM--KCRRFGESNCTI
ancestral amniote		LVFGTSD-RVYEKLKDLEEGIQALMRELEDG--SPRAG-QILKQ---TYDKFDTNMRSDDALLKNYGLLSCFKKDLHKAETYLRVM--KCRRFVESSCAF
xenopus I	28	LVFGSSD--VYERLKYLEEGIQALMRELEDG--SFRSFP-VLR---PLYERFDINLRSDEALVKVYGLLSCFKKDMHKVETYLKVM--KCRRFVESNCTI
bullfrog	29-31	QVFGNID-RVYDRLRDLDEGLHILIRELDDGNVRNYG---VL---TFTYDKFDVNLRSEEGRAKNYGLLSCFKKDMHKVETYLKVM--KCRRFVESNCTF
bonito	32	-AQRN---QISEKLSDLKMGIQLLIRANQDGAEMFADSS-AL-QLAP-YGNYYQSLGGDESLRRNYELLACFKKDMHKVETYLMVA--KCRLSPEANCTL
yellow tail	33	-ALRN---QISPRLSELKTGIQLLITANQDGAEMFSDVS-AL-QLAP-YGNFYQSLGGEELLRRNYELLACFKKDMHKVETYLTVA--KCRLSPEANCTL
tuna	34,35	-APRN---QISPKLSELKTGIHLLIRANQDGAEMFADSS-AL-QLAP-YGNYYQSLGADESLRRSYELLACFKKDMHKVETYLTVA--KCRLSPEANCTL
snook (l. calcarifer)	36	-APRD---QISPKLSELKTGILLLIRANQDGAEMFSDSS-AL-QLAP-YGNYYQSLGADESLRRTYELLACFKKDMHKVETYLTVA--KCRLSPEANCTL
tilapia	37	L--RN---QISPRLSELKTGILLLIRANQDEAENYPDTD-TH-QHAP-YGNYYQSLGGNESLRQTYELLACFKKDMHKVETYLTVA--KCRLSPEANCTL
flounder	38,39	FAVRT---QVTSKLSELKMGLLKLIEANQDGAGGFSESS-VL-QLTP-YGN--------------SELFACFKKDMHKVETYLTVA--KCRLFPEANCTL
sea bream, red	40	-APRN---QISPKLSELKMGIHLLIRANEDGAEIFPDSS-AL-QLAP-YGNYYQSLGADESLRRTYELLACFKKDMHKVETYLTVA--KCRLSPEANCTL
sea bream, black	36	-APRN---QISPKLSELKTGIHLLIRANEDGAELFPDSS-AL-QLAP-YGDYYHSPGTDESLRRTYELLACFKKDMHKVETYLTVA--KCRLSPEANCTL
sea bream, gilthead	41	-APRN---QISPKLSELKTGIHLLIRANEDGAEIFPDSS-AL-QLAP-YGNYYQSLGTDESLRRTYELLACFKKDMHKVETYLTVA--KCRLSPEANCTL
porgy, yellowfin	42	-APRN---QISPKLSELKTGIHLLIRANEDGAELFPDSS-AL-QLAP-YGDYYQSPGTDESLRRTYELLACFKKDMHKVETYLTVA--KCRLSPEANCTL
hard tail	43	LAERS---QISSRLAELREEIQMLITTNQEGAEVFSDSS-TL-PLAPPFGNFFQTQGGDELQRRSYELLACFKKDMHKVETYLTVA--KCRLSTEANCTL
cod	44	L--RN---QISPKLSELKNGIHILIRTSQ-GAG-DALVE-AD-PM-SPYGGYYQALGGDGSLRKSYEMLACFKKDMHKVETYLTVA--KCRLSPEANCTL
carp, common	45-47	LTVGNPN-QITEKLADLKMGISVLIKGCLDGQPNMDDND-SL-PL-P-FEDFYLTMG-ENNLRESFRLLACFKKDMHKVETYLRVA--NCRRSLDSNCTL
carp, bighead/grass/silver	48,49	LTVGNPN-QITEKLADLKVGISVLIKGCLDGQPNMDDND-SL-PL-P-FEDFYLTMG-ESSLRESFRLLACFKKDMHKVETYLRVA--NCRRSLDSNCTL
salmon I, chum	50,51	LMVRNAN-QISEKLSDLKVGINLLITGSQDGVLSLDDND-SQ-QL-PPYGNYYQNLGGDGNVRRNYELLACFKKDMHKVETYLTVA--KCRKSLEANCTL
salmon II, chum	52	LMVRNSN-QISEKLSDLKVGINLLIEGSQEGVLSLDDND-SQ-HL-PPYGNYYQNLGGDGNVRRNYELLACFKKDMHKVETYLTVA--KCRKSLEANCTL
salmon, coho	53,54	LLVGNAN-QISEKLSDLKVGINLLIMGSQDGLLSLDDND-SQ-QL-PRYGNYYQNPGGDGNVRRNYELLACFKKDMHKVETYLTVA--KCRKSLEANCTL
salmon, atlantic	55,56	LMVRNSN-QISEKLSDLKVGINLLIKGSQDGVLSLDDND-SQ-QL-PPYGNYYQNLGGDGNVRRNYELLACFKKDMHKVETYLTVA--KCRKSLEANCTL
trout I	57,58	LMVRNAN-QISEKLSDLKVGINLLITGNQDGVLSLDDND-SQ-QL-PPYGNYYQNLGGAGNVRRNYELLACFKKDMHKVETYLTVA--KCRKSLEANCTL
trout II	57,58	LMVRNSN-QISEKLSDLKVGINLLIKGSQDLALSLDDND-SQ-HL-PPYGNYYQNLGGAGNVRRNYELLACFKKDMHKVETYLTVA--KCRKSLEANCTL
catfish	59	---GNPN-GISEKLADLKMGIGYLIEGCVDGQTGLDEED-S---LAPPFEDFYQTL-SEGNLRKSFRLLSCFKKDMHKVETYLSVA--KCRRSLDSNCTL
catfish (p. pangasius)	60	---GNPN-HISEKLADLKMGIGVLIEGCLDGQTSLDEND-S---LAPPFEDFYQTL-SEGNLRKSFRLLSCFKKDMHKVETYLSVA--KCRRSLDSNCTL
eel	61-63	LMFGTSD-GIFDKLEDLNKGINELMKVVGDGGIYIEDVR-NL-RY-ENFDVHLRN--DAG-LMKNYGLLACFKKDMHKVETYLTVKVTKCRRFVESNCTL
sturgeon	64	LVFLTSD-RVFEKLKDLEEGIVALMRDLGEGG--FG-SSTLLK-LT--YDKFDVNLRNDDALFKNYGLLSCFKKDMHKVETYLKVM--KCRRFVESNCTL
ancestral boney fish		L?VRN-N-QISEKLSDLK?GINLLIRGNQDGAE?FDDNS-SL-QLAPPYGNYYQNLGGDGNLRRNYELLACFKKDMHKVETYLTVA--KCRLSPEANCTL
shark	65	-AFRTSD-RVYDKLRDLEEGIFALMKTLEDGGSSQGFA--WLK-FS--YERFDGNL-SEEALMKNYGLLACFKKDMHKVETYLKVM--NCKRFAESNCTV
ancestral vertebrate		LVFGTSD-RVYEKLKDLEEGIQALMRELEDG--SPRDG-QILKQL-PTYDKFDQNLRSDDALLKNYGLLSCFKKDMHKVETYLRVM--KCRRFVESNCTL

Figure 1 (continued)

Classification of vertebrates (classification of fish from (66)).

- **Phylum Chordata**
  - **Subphylum Vertebrata**
    - Division Agnatha (jawless vertebrates; e.g. hagfish, lamprey)
    - Division Gnathostomata (jawed vertebrates)
      - Class Chondrychthyes (cartilaginous fish)
        - Subclass Holocephali (e.g. rabbitfish, ratfish)
        - Subclass Elasmobranchi
          - Order Squaliformes (e.g. most sharks[G], dogfish)
          - Order Rajiformes (e.g. skates, rays)
          - Order Torpediniformes (e.g. electric ray, torpedo)
      - Class Osteichthyes (boney fish)
        - Subclass Crossopterygii (lobefish, ancestor to tetrapods; e.g. coelocanth)
        - Subclass Dipnoi (lungfish[P])
        - Subclass Actinopterygii (higher boney fish; e.g., rayfin fish)
          - Order Polypteriformes (e.g. birchirs, reedfish)
          - Order Acipenseriformes (e.g. sturgeons[G], paddlefish)
          - Order Amiiformes (e.g. bowfin)
          - Order Lepisosteiformes (e.g. gars)
          - Order Anguilliformes (e.g. eels[G])
          - Order Clupeiformes (e.g. herring)
          - Order Salmoniformes
            - Suborder Salmonoidei
              - Family Salmonidae (e.g. trout[G,P], salmon[G,P])
            - Suborder Esocoidei (e.g. pike)
          - Order Cypriniformes
            - Suborder Cyprinoidei (e.g. carp[G,P], suckers)
          - Order Siluriformes (e.g. catfish[G])
          - Order Perciformes (e.g. sea bass, sunfish, perch)
            - Suborder Percoidei (e.g. tilapia[G,P], sea bream[G], yellowtail[G], snook[G], porgy[G], hard tail[G])
            - Suborder Scombroidei (e.g. mackerel, bonito[G], tuna[G])
            - Suborder Gobioidei (e.g. gobies)
          - Order Pleuronectiformes
            - Suborder Pleuronectoidei (e.g. flounder[G])
            - Suborder Soleoidei (e.g. sole)
          - Order Sporpaeniformes (e.g. sculpin)
          - Order Gadiformes (e.g. cod[G], hake)
          - Order Lophiiformes (e.g. anglerfish)
      - Class Amphibia (e.g. frogs, toads)
        - Order Anura
          - Suborder Archeobatrachia (e.g., xenopus[G,P])
          - Suborder Neobatrachia (e.g., bullfrog[G,P])
      - Class Reptilia
        - Order Squamata (e.g. lizards, snakes)
        - Order Chelonia (e.g. turtles[G,P])
        - Order Crocodilia (e.g. alligators[P], crocodiles[P])
      - Class Aves
        - Order Galliformes (e.g. chickens[G,P], turkeys[G,P])
        - Order Anseriformes (e.g. ducks[G])
      - Class Mammalia
        - Subclass Eutheria (placental mammals)
          - Order Cetacea (e.g. whales[G,P])
          - Order Rodentia (e.g. rats[G,L,P], mice[G,L,P], coypu, guinea pig, hamster[G,L,P])
          - Order Artiodactyla (e.g. goat[G], pigs[G,P], sheep[G,L,P], cows[G,L,P], camel[P], alpaca[G])
          - Order Perissodactyla (e.g. horses[G,P])
          - Order Proboscidea (e.g. elephants[G,P])
          - Order Carnivore (e.g. cats, dogs, bears, fox[G], mink[G,P])
          - Order Primates (e.g. monkeys[G], apes, human[G,L,P])

Superscripts indicate where primary sequences of hormones have been reported:
G = Growth Hormone
L = Placental lactogen
P = Prolactin

Footnote to Figure 1

Species	Ref. #	Sequence ([1] … [10] … [20] … [30] … [40] … [50] … [60] … [70] … [80] … [90])
human PRL	67	LPIC-PGGAARC-QVTLRDLFDRAVVLSHYIHNLSSEMFSEFDKRYTH------------GRG---FIT--KAINSCHTSSLATPEDKEQAQQMNQKDFLS-LIVSILR--SWNEPL
rat PRL	68	LPVC-SGG--DC-QTPLPELFDRVVMLSHYIHTLYTDMFIEFDKQYVQ------------DRE---FIA--KAINDCPTSSLATPEDKEQAQKVPPEVLLN-LILSLVH--SWNDPL
mouse PRL	9,69	LPIC-SAG--DC-QTSLRELFDRVVILSHYIHTLYTDMFIEFDKQYVQ------------DRE---EMV--KVINDCPTSSLATPEDKEQALKVPPEVLLN-LILSLVQ--SSSDPL
hamster PRL	10	LPIC-PGG--NC-QMPLQELFDRVIMLSHYIYMLSADMFIELDKQYAQ------------DHE---FIA--KAISDCPTSSLATPEGKEEAQQVPPEVLLN-LILSLVH--SWNDPL
fin whale PRL	70	IPIC-PSGAVNC-QVSLRDLFDRAVILSHYIHNLSSEMFNEFDKRYAQ------------GRG---FIT--KAINSCHTSSLQTPEDKEQAQQIHHEVLVS-LILGVLR--SWNDPL
mink PRL	71	????-???????-??????????????LSHYIHNLSSEMFNEFDKRYAQ------------GRG---FIT--KAINSCHTSSLSTPEDKEQAQQIHHEDLLN-LILRVLR--SWNDPL
cattle PRL	72,73	TPVC-PNGPGNC-QVSLRDLFDRAVMVSHYIHDLSSEMFNEFDKRYAQ------------GKG---FIT--MAINSCHTSSLPTPEDKEQAQQTHHEVLHS-LILGLLR--SWNDPL
sheep PRL	74	TPVC-PNGPGDC-QVSLRDLFDRAVMVSHYIHNLSSEMENEFDKRYAQ------------GKG---FIT--MAINSCHTSSLPTPEDKEQAQQTHHEVLMS-LILGLLR--SWNDPL
pig PRL	75	LPEC-PSGAVNV-QVSLRDLFDRAVILSHYIHNLSSEMFNEFDKRYAQ------------GRG---FIT--KAINSCHTSSLSTPEDKEQAQQIHHEVLIN-LILRVLR--SWNDPL
camel PRL	76	LPIC-PSGAVNC-QVSLRDLFDRAVILSHYIHNLSSEMFNEFDKRYAQ------------GRG---FMT--KAINSCHTSSLSTPEDKEQAQQIHHEDLLN-LVLRVLR--SWNDPL
horse PRL	77	LPIC-PSGAVNC-QVSLRELFDRAVILSHYIHNLSSEMFNEFDKRYAQ------------GRG---FVT--KAINSCHTSSLSTPEDKEQAQQIHHEDLLN-LILRVLK--SWNDPL
elephant PRL	78	IPVC-PRGSVRC-QVSLPDLFDRAVMLSHYIHSLSSDMFHEFNKQYAL------------GRG---FIP--RAINSCHTSSISTPEDKDQAQQTHHEVLMD-LILGLLR--SWNDPL
ancestral mammal PRL		LPIC-PSGAVNC-QVSLRDLFDRAVILSHYIHNLSSEMFNEFDKRYAQ------------GRG---FIT--KAINSCHTSSLSTPEDKEQAQQIHHEVLLN-LILGLLR--SWNDPL
chicken PRL	79,80	LPIC-PIGSVNC-QVSLGELFDRAVKLSHYIHYLSSEIFNEFDERYAQ------------GRG---FIT--KAVNGCHTSSLTTPEDKEQAQQIHHEDLLN-LVVGVLR--SWNDPL
turkey PRL	81	LPIC-SSGSVNC-GVSLGELFDRAVRLSHYIHFLSSEIFNEFDERYAQ------------GRG---FIT--KAVNGCHTSSLTTPEDKEQTQQIHHEELLN-LILGVLR--SWNDPL
sea turtle PRL	82	LPIC-PSGSVGC-QVSLENLFDRAVKLSHYIHSLSSEMFNEFDERYAQ------------GRG---FLT--KAINGCHTSSLTTPEDKEQAQQIHHEDLLN-LVLGVLR--SWNDPL
crocodile PRL	83	LPIC-PSGSVNC-QVSLGELFDRAVKLSHYIHFLSSEMFNEFDERYAQ------------GRG---FIT--KAVNGCHTASLTTPEDKEQAQQIHHEDLLN-LVLGVLR--SWNDPL
alligator PRL	83	LPIC-PSGSVNC-QVSLGELFDRAVKLSHYIHFLSSEMFNEFDERYAQ------------GRG---FIT--KAVNGCHTASLTTPEDKEQAQQIHHEDLLN-LVLGVLR--SWNDPL
ancestral amniote PRL		LPIC-PSGAVNC-QVSLRDLFDRAVKLSHYIH?LSSEMFNEFDKRYAQ------------GRG---FIT--KAINSCHTSSLTTPEDKEQAQQIHHEVLLN-LILGVLR--SWNDPL
xenopus PRL	84	LPIC-PDGGTNC-QMSTGALFDKAVKLSHYIHSLSSEMFNEFDERFTP------------SRR---FLA--KSIMSCHTSSLNTPEDREQAQQIQHEDLLN-LVMRVLK--SWNDPL
bullfrog PRL	85	QPIC-PNGGTNC-QIPTSALFDRAVKLSHYIHSLSSEMFNEFDERFTP------------GRR---FLA--KSGISCHTSSLNTPEDKEQARQIQHEDLLN-LVLKVLR--SWNDPL
lungfish PRL	60	LPIC-ANGSTNCHAIPLDDLFEFVVKLAHRIHSLTSDMFNEFDERYAQ------------GR----GFIS-RAINNCHTSSLTTPEAKENAQKFHHDDLLR-LVMKVLR SWNDPL
tilapia PRL-188	86,87	VPINELLERASQHSDKLHSLSTTLTQELDSHFPPI------------GR----VIMP-RP-AMCHTSSLQTPIDKDQALQVSESDLMS-LARSLLQ--AWSDPL
tilapia PRL-177	86,87	VPINDLIYRASQQSDKLHALSTMLTQELGSEAFPI------------DR----VLA-------CHTSSLQTPTDKEQALQVSESDLLS-LARSLLQ--AWSDPL
common carp PRL	88	VGLNDLLERASELSDKLHSLSTSLTNDLDSHFPPV------------GR----VMMP-RP-SMCHTSSLQVPNDKDQALKVPEDPLLS-LARSLLL--AWSDPL
bighead carp PRL	89	VGLNDLLERASQLSDKLHSLSTSLTNDLDSHFPPV------------GR----VMMP-RP-SMCHISSLQIPNDKDQALKVPEDELLS-LAHSLLL--AWSDPL
silver carp PRL	90	VGLNDLLERASQLSDKLHSLSTSLTNDLDSHFPPV------------GR----VMMP-RP-SMCHTSSLQIPNDKDQALKVPEDELLS-LAHSLLL--AWSDPL
chum salmon PRL	91,92	IGLSDLMERASQRSDKLHSLSTSLTKDLDSHFPPM------------GR----VMMP-RP-SMCHTSSLQTPKDKEQALKVSENELIS-LARSLLL--AWNDPL
chinook salmon PRL	93	IGLSDLMERASQRSDKLHSLSTSLTKDLDSHFPPM------------GR----VMMP-RP-SMCHTSSLQTPKDKEQALKVSENELIS-LARYLLL--AWNDPL
trout PRL	94	IGLSDLMERASQRSDKLHSLSTSLTKDLDSHFPPM------------GR----VMMP-RP-SMCHTSSLQTPKDKEGALKVSENELIS-LARSLLL--AWNDPL
ancestral boney fish PRL		IGLSDLMERASQRSDKLHSLSTSLTKDLDSHFPPM------------GR----VMMP-RP-SMCHTSSLQTPKDKEQALKVSENELIS-LARSLLL--AWNDPL
rat PL-I	95	--KP----TA-IVSTDDLYHRLVEQSHNTFIMAADVYREFDINFAK------------RSW----MKD-RILPLCHTASIHPTENLEEVHEMKTEDFLN-SIINVSV--SWKEPL
mouse PL-I	96	--KP----TA-MVPTEDLYTRLAELLHNTFILAADVYREFDLDFFD------------KTW----ITD-RTLPLCHTASIHTPENREEVHETKTEDLLK-AMINVSI--SWKEPL
sheep PL	97,98	QHPPYCRNQPG-KC-QIPLQSLFDRATTVANYNSKLAGEMVNRFDEQYGQ------------GIN----SES-KVIN-CHTSSITTPNSKAEAINTEDKILFK-LVISLLH--SWDEPL
cattle PL-Ala/Val	99,100	DYAPYCKNQPG-NC-RIPLQSLFERATLVASNNYRLAREMFNEFNKQFGE------------GKN---F-TS-KVINSCHTEFMTTPNNKEAAANTEDEALLR-LVISLLH--SWDEPL
cattle PRP-I	101	KSCPS-CGPDVFVSL-RKSFTDRFMNAASLSHDFYNLSTIMFNEFDEKYAQ------------GKLYY-INVT-K---SCHTNSFHAPEERDIVQQTNIEDLSK-WTLVLLY--SWNNPL
cattle PRP-II	102	ISCPS-CGPDMFVSL-QKSLIDVFINAASLSHDFHNLSTIMFNEFDEKYAQ------------GKLYY-INAT-K---SCHTNSFHTPEERDKAQQMNNEDLSK-WTLVLLY--SWNNPL
cattle PRP-III	102	SCPS-CCPDVFDIP-LESLTHLFLNASRLSHDIVNHTTIMFHEFDEKYAQ------------NQPYT-INAT-K---SCHTNSLHTPQEREKALRMNNEDLSK-WILMLLY--SWHRPL
cattle PRP-IV	99,100	ISCPS-CGPDMFVSL-QKSLIDVFINAASLSHDFHNLSTIMFNEFDEKYAQ------------GKLYY-INAT-K---SCHTNSFHTPEERDKAQQMNNEDLSK-WTLVLLY--SWNNPL
cattle PRP-V	103	ILCPSLC-PDGDDVC-RASLIDLFVHASVLSTGMYNTSVKMFTDFDEQYAQ------------GKEYF-INVSDK----CHTNSLHLPEDMQHNPTDEQKGLIM-WILMLLY--SWQRPL
cattle PRP-VI	104	NSCPSCC-PDVSDIP-LDLLRELFLNATLLSQSILKHSRIMLNEFDEKYAQ------------GKLYY-LTAT-K---SCHTNSLHTTEDMDKAEKIDNEDLSK-WILMLLY--SWHRPL
rat PLP-A	105	TASAMRAK-LLNVHN-YTSYGDTWNQAIKISQDMNQYISDLSTHVKIFYAQ------------GRG-F-ERRTTR----CHTSSLSSPENKEQAQQFQLEVLLG-LSHSLLQ--AWLNPL
rat PLP-B	106	ASVPRHA-SGAGRG-EMSLHGLLDHAIILAHNVTELIAEMNSVFLEDVLYKP----------GRW-FPERDLTA----CHRSPFTIAVSKEGTQQR-LGVFL--VKEMIGMLETWTFSL
rat PLP-C	107	ASIPACM-VEDGGC-WDPLREAFNSATQRAETLRNLSDQLYVEFFQNQFSSRQFADL---SSQLIRKDETVLKAGTYCHSNR-AKPKSRGVNIDI--EEYLKMSINFCGF---MDQPL
rat PL-II	108	AP---------NY-RMSTGSLYQRVVELSHYTHDLASKVFIEFDMKF--------------GRTVW-THNLM--LSPCHTAAIPTPENSEQVHQAKSEDLLK-VSITILQ--AWQEPL
mouse PL-II	109	LP---------NY-RLPTESLYQRVIVVSHNAHDLASKAFMEFEMKF--------------GRTAW-TYGLM--LSPCHTAAILTPENSEQVHQTTSEDLLK-VSITILQ--AWEEPL
hamster PL-II	110	SA----------SP-RLSTRNLYQRVVELSHCTHDLASKVFTDFNMKF--------------GKSIC-RQKLM--LYTCHTSSIPTPENREQVHQTNSEDLLK-VTISVLQ--AWEEPV
mouse PLF	111	SFPMCAMRNGRC--FMSFEDTFELAGSLSHNISIEVSELFTEFEKHYSNVS------GLRDKS------PMR----CNTSFLPTPENKEQARLTHYSALLK-SGAMILD--AWESPL
mouse PRP	112	SSAPINA-SEAVL---SDLKDLFDNATVLSGEMSKLGVIMRKEFFMNSFSSETFNKIILDLHKST--ENIT-KAFNSCHTVPINVPETVEDVRKTSFEEFLKMVLHMLL---AWKEPL
flounder SL	113	PLDCKEEQG-SLSRCP-SISQEKLLDRVIQHAELIYRVSEESCSMFEEMFVPFP----------LR---L-QR-NQAGYACITKALPIPSSKSEIQQISDTWLLH-SVLMLVQ--SWIEPL
cod SL	114	PVDCREEQAGSS-QCP-TISQEKLLDRVIQHTELIYRVSEESCSMFEDMFVPFP----------VR---L-QR-NQAGNTCITKDFPIPTSKNELQQISDTWLLH-SVLMLVQ--SWIEPL
salmon SL	115	PLECKDEQG-SIILCA-SISKEKLLDRVIQHAELIYRVSEESCTLFEEMFVPFP-------- MRS----QR-NQAGYTCATKAFPIPGSKSEIQQISDKWLLH-SVLILVQ--SWIEPL

Figure 2 Amino acid sequences of prolactin (PRL), placental lactogens (PL), prolactin-like proteins (PLP), prolactin-related cDNA (PRC), soma-tolactins (SL), proliferins (PLF) and proliferin-related protein (PRP) from various vertebrate and hypothetical ancestral species. Cattle PL and PLP-I differ only in their putative signal sequences (residue-2, Ala/Val) and have also been referred to as PL-Ala and PL-Val, respectively. Mature cattle PLP-II differs from cattle PRC-I by only one amino acid; however, their putative signal sequences differ greatly. Sequences were aligned using theoretical deletions to maximize homology. See Figure 1 for single-letter abbreviations used for amino acids. (–) Gap/deletion.

Species	Ref. #	[100] [110] [120] [130] [140] [150] [160] [170] [180] [190]
human PRL	67	YHLVTEVRGMQEA-PE--ALLSKAVEIEEQT--KRLLRGMELIVS-QVHPET---KENEIYPVWSGLPSLQMADEES-RLSAYYNLLHCLRRDSHKIDNYLKLLKCRII--HNNNC
rat PRL	68	FQLITGLGGIHEA-PD--AIISRAKEIEEQN--KRLLEGIEKIIG-QAYPEA---KGNEIYLVWSQLPSLQGVDEES-KDLAFYNNIRCLRRDSHKVDNYLKFLRCQIV--HKNNC
mouse PRL	9,69	FQLITGVGGIQEA-PE--YILSRAKEIEEQN--KQLLEGVEKIIS-QAYPEA---KGNGIYFVWSQLPSLQGVDEES-KILSLRNTIRCLRRDSHKVDNFLKVLRCQIA--HQNNC
hamster PRL	10	FQLVTEVDGIHEAS-D--AIISRAKEIGEQN--KRLLEGIEKILG-QAYPEA---KGNEIYSVWSQFPSLQGVDEES-RDLAIYNKVRCLRRDSHKVDNYLKLLRCRVV--HNNNC
fin whale PRL	70	YHLVTEVRGMQEA-PD--AILSRAIQEEEEN--KRLLEGMEKIVG-QVHPGV---KENEVYSVWSGLPSLQMADEDT-RLFAFYNLLHCLRRDSHKIDSYLKLLKCRII--YNSNC
mink PRL	71	YHLVSEVRGMQEA-PD--SILSRAIEIEEQN--RRLLEGMEKIVG-QVHPGV---RENEVYSVWSGLPSLQMADEDS-RLFAFYNLLHCLRRDSHKIDNYLKLLKCRIV--YDSNC
cattle PRL	72,73	YHLVTEVRGMKGA-PD--AILSRAIEIEEEN--KRLIEGMEMIFG-QVIPGA---KETEPYPVWSGLPSLQTKDEDA-RYSAFYNLLHCLRRDSSKIDTYLKLLNCRII--YNNNC
sheep PRL	74	YHLVTEVRGMKGV-PD--AILSRAIEIEEEN--KRLLEGMENIFG-QVIPGA---KETEPYPVWSGLPSLQTKDEDA-RHSAFYNLLHCLRRDSSKIDTYLKLLNCRII--YNNNC
pig PRL	75	YHLVTEVRGMQEA-PE--AILSRAIQEEEEN--KRLLEGMEKIVG-QVHPGI---KENEVYSVWSGLPSLQMADEDT-RLFAFYNLLHCLRRDSHKIDNYLKLLKCRII--YNSNC
camel PRL	76	YHLVTEVRGMQEA-PD--AILSRAIEIEEQN--KRLLEGMEKIVG-QVHPGV---KENEIYSVWSGLPSLQMADEDT-RLFAFYNLLHCLRRDSHKIDNYLKLLKCRII--YDSNC
horse PRL	77	YHLVSEVRGMQEA-PE--AILSKAIEIEEQN--RRLLEGMEKIVG-QVQPRI---KENEVYSVWSGLPSLQMADEDS-RLFAFYNLLHCLRRDSHKIDNYLKLLKCRIV--YNSNC
elephant PRL	78	DHLASEVHSLPKA-PS--ALLTKATEVKEEN--QRLLEGIEKIVD-QVHPGA---KENKAYSVWSGLPSLQTTDEDA-RLFAFYNLFRCLRRDSHKIDSYLKLLKCRIV--YNNNC
ancestral mammal PRL		YHLVTEVRGMQEA-PE--AILSRAIEIEEEN--KRLLEGMEKIVG-QVHPGA---KENEIYSVWSGLPSLQMADEDS-RLFAFYNLLHCLRRDSHKIDNYLKLLKCRII--YNNNC
chicken PRL	79,80	LHLASEVQRIKEA-PD--TILWKAVEIEEQN--KRLLEGMEKIVG-RVHSGH---AGNEIYSHSDGLPSLQLADEDS-RLFAFYNLLHCHRRDSHKIDNYLKVLKCRLI--HDSNC
turkey PRL	81	IHLASEVQRIKEA-PD--TILWKAVEIEEQN--KRLLEGMEKIVG-RIHSGD---AGNEVFSQWDGLPSLQLADEDS-RLFAFYNLLHCLRRDSHKIDNYLKVLKCRLI--HDNNC
sea turtle PRL	82	LHLVSEVQSIKEA-PD--TIL-KAVEIEEQD--KRLLEGMEKIVG-QVHPGE---IENEVYSPWSGLPSLQQVDEDS-RLFAFYNLLHCLRRDSHKIDNYLKLLKCRLI--HDNDC
crocodile PRL	83	LHLVTEVQRIKEA-PD--TILWKAVEIEEQN--KRLLEGMEKIIG-RVQPGD---TGNEVYSRWSGLPSLQLADEDS-RLFAFYNLLHCGRRDSHKIDNYLKLLKCRLI--HDSNC
alligator PRL	83	LHLVTEVQRIKEA-PD--TILWKAVEIEEQN--KRLLEGMEKVIG-RVQPGD---TGNEVYSRWSGLPSLQLADEDS-RLFAFYNLLHCGRRDSHKIDNYLKLLKCRLI--HDSNC
ancestral amniote PRL		LHLVTEVRGMKEA-PD--AILSKAIEIEEQN--KRLLEGMEKIVG-QVHPGA---KENEVYSVWSGLPSLQ?ADEDS-RLFAFYNLLHCLRRDSHKIDNYLKLLKCRII--HNNNC
xenopus PRL	84	LHMVGEVQDIREA-PD--TILWKTVEVEEQT--KRLLEGMEKIVG-RIHPFD---LENDVNSLWSGPPAAQSADENS-RLFGFYNLLHCLRRDSHLIDNYLKLLKCRLI--HDSNC
bullfrog PRL	85	VHMVSEVQDIREA-PD--TILWKTVEVEEQT--KRLLEGMERIIG-RIQPGD---LENEIYSPWPGPASIPG-DENS-RLFAFYNLLHCLRRDSHKIDNYLKLLKCRLI--HEGNC
lungfish PRL	60	LQLVSEVQGIGEA-PG--TILWKVTEVEDQT--KQLIEGMEKILS-RMHPNG---LDNEVLSLWPMPGAM-HAGDGS-KLFALYNLLHCFRRDSFKIDSYLKLLRCRLF--HEGGC
tilapia PRL-188	86,87	VVLSSSASTLPH--PAQSSIFNKIQEMQQYS--KSLKDGLD-VLSSKMGSPA-----QAITSLPYRGGTNLGHDKIT-KLINFNFLLSCLRRDSHKIDSFLKVLRCRAAKMQPEMC
tilapia PRL-177	86,87	EVLSSSTNVLPY-S-AQSTLSKTIQKMQEHS--KDLKDGLD-ILSSKMGPAA-----QTITSLPFIETNEIGQDKIT-K------LLSCFRRDSHKIDSFLKVLRCRAANMQPQVC
common carp PRL	88	ALLSSEASSLAH--PERNTIDSKTKELQENI--NSLGAGLEHVF-NKMDSTS-----DNLSSLPFYTNSL-GEDKTS-RLVNFHFLLSCFRRDSHKIDSFLKVLRCRAAKKRPEMC
bighead carp PRL	89	ALLSSEASSLAH--PERNTINSKTKELQDNI--NSLGAGLERVV-HKMGSSS-----DNLSSLPFYSNSL-GQDKTS-RLVNFHFLLSCFRRDSHKIDSFLKVLRCRAAKKRPEMC
silver carp PRL	90	ALLSSKASSLAH--PERNTINSKTKELQDNI--NSLVPGLEHVV-HKMGSSS-----DNLSSLPFYSNSL-GQDKTS-RLVNFHFLLSCFRRDSHKIDSFLKVLRCRAAKKRPEMC
chum salmon PRL	91,92	LLLSSEAPTCPH--PSNGDISSKIRELQDYS--KSLGDGLD-IMVNKMGPSS-----QYISSIPFKGGDL-GNDKTS-RLINFHFLMSCFRRDSHKIDSFLKVLRCRATKMRPETC
chinook salmon PRL	93	LLLSSEAPTLPHT-PSNGDISSKIRELQDYS--KSLGDGLD-IMVNKMGPSS-----QYISSIPFKGGDL-GNDKTSPRLINFHFLMSCFRRDSHKIDSFLKVLRCRATNMRPETC
trout PRL	94	LLLSSEAPTLPH--PSNGDISSKIRELQDYS--KSLGDGLD-IMVNKMGPSS-----QYISSIPFKGGDL-GNDKTS-RLINFHFLMSCFRRDSHKIDSFLKVLRCRATKMRPEAC
ancestral boney fish PRL		LLLSSEAPTLPH--PS--NGDISSKIRELQDYSKSLGDGLD-IMVNKMGPSS-----QYISSIPFKGGDL-GNDKTS-RLINFHFLMSCFRRDSHKIDSFLKVLRCRATKMRPE?C
rat PL-I	95	KHLVVCSDCSSGAS-V--SMGKKAVDMKDK--NLIIILEGLQTLY-NRTQAKVEENFENFDYPAWSGLKDLDSSDEEH-HLFAICNLCRCVKRDIHKIDTYLKVLRCRVV--FKNECGVSTF
mouse PL-I	96	KHLVSALTALPGAS-E--SMGKKAADIKGR--NLVIILEGLQTIY-NRSQANIEEN-ENFDYPAWSGLEELQSPNEDT-HLFAVYNLCRCIKRDIHKIDSYIKVLRCRVV--FQNEC
sheep PL	97,98	HHAVTELANSKGTSPA---LLTKAQEIKEK--AKVLVDGVEVIQ-KRIHPG--E-K-NEPYPVWSEQSSLTSQDENV-RRVAFYRLFHCLHRDSSKIYTYLRILKCR----LT-SCET
cattle PL-Ala/Val	99,100	HQAVTELLHRNGASPD---ILARAKEIEDK--TKVLLEGVEMIQ-KRVHPG--E-KKNEPYPVWSEKSSLTADDEDV-RQTAFYRMFHCLHRDSSKISTYINLLKCR----FT-PC
cattle PRP-I	101	HHLVTELQHMKELSNA---FLSSATRFENM--SEKLQAFIER-QFSKIIVPVLNTMIQARSS-WTGLPSLMSSAEDR-RHSEFYNLFYCLRRDSRKVDMYIKILTCR-T--HK-TC
cattle PRP-II	102	YYLLLELRNMKNLSEA--VI-SSAMEIENM--SEKLQAFIES-QFRKIIVPV-LKMIHEVSNTWSRFSSMTFSDEDR-SISEYYNLFYCLRRDSRKVDMYIKILTCR-T--RK-TC
cattle PRP-III	102	YLLVKDLQSMKEVS-D--TILSSAKENMRK--IEELQAFIER-QFSQVIYPVIRT-IFKARIYWSGLASLVSNDEDV-RHSAFYKLFMCLYRDSRKLDMYTEILACRIT----NTC
cattle PRP-IV	99,100	YYLLLELRNMKNLSEA--VI-SSAMEIENM--SEKLQAFIES-QFRKIIVPV-LKMIHEVSDTWSRFSSMTFSDEDR-SISEYYNLFYCLRRDSRKVDMYIKILTCR-T--RK-TC
cattle PRP-V	103	YQLVTDLRSMKEVSN---TILSSARENVKK--LKELQALIER-PFSQVIFTARRKM-YIARIYWFGLRSLLSSNEDR-RHSAFYSLFFCLRRDTRKLDIYTKFVACRLI--YKK-C
cattle PRP-VI	104	NHLVTDMRGMIEDS-D--TILSNAKENVKK--VEELQAFIER-QFCQIIIPVMRKML-ENNIYWSGLPSLLSSDEDV-RHSAFYSLFHCLHRDTLEIDIYTKILARRMI----NT-
rat PLP-A	105	HHL-WEAMCDRLGSTP--PTLYKALMLKES--NIKLLDAIKNIAKKGNFE--INQKAN--YTAWSELGFLQSPNRDT-RYFAFYNLFHCLKKDSNNVEMYLKLLKCRLI--RSK-C
rat PLP-B	106	YHIANEMSHMEEP-PD--EIISRAKNIEEK--IKELMDVLKGIL-NKIQPG--SPQ-NERFPMWNELAYLRSPDEER-RHFAFTNLFQCLLQDSRKFDSKVRLLKCRLI--YNRDC
rat PLP-C	107	FHLVIELSAMEGV-PE--TILSKAKDLEEN--NRQLLDDLRWIL-TKVFPTA-EIKEEFPS--WDYLSFLKSSNKNH-KFLAIFNLSSCLDYDTQVHYTLSQILNC-LI--TGKDC
rat PL-II	108	KHIVAAVATLPDGS-D--TLLSRTKELEER--IQGLLEGLETIL-SRVQPG----AVGSDYTFWSEWSDLQSSDKST-KNGVLSVLYRCMRRDTHKVDNFLKVLKCRDI--YNNNC
mouse PL-II	109	KHMVAAVAALPHV-PD--TLLSRTKELEER--IQGLLEGLKIIF-NRVYPG----AVASDYTFWSAWSDLQSSDEST-KNSALRTLWRCVRRDTHKVDNYLKVLKCRDV--HNNNC
hamster PL-II	110	KHMVAAVAALPGTS-D--AMLSRAKELEER--VLGLLEGLKIIL-NRIHPG----AVENDYTFWSGWSDLQSSDEAT-RNIAFYTMGRCLRRDTHKVDNYLKVLKCRDI--HNNNC
mouse PLF	111	DDLVSELSTIKNV-PD--IIISKATDIKKK--INAVRNGV-NALMSTMLQNGDEEKKN---PAWF----LQSDNEDA-RIHSLYGMISCLDNDFKKVDIYLNVLKCYML--KIDNC
mouse PRP	112	KHLVTELSALPEC-PY--RILSKAEAIEAK--NKDLLEYIIRII-SKVNP-A--IKENEDYPTWSDLDSLKSADKET-QFFALYMFSFCLRIDLETVDFLVNFLKCLLL--YDDVCYSGF
flounder SL	113	VYLQTTLDRYDNA-PD--MLLNKTKWVSDK--LISLEQGV-VVLIRKMLDEGMLTATYNEQGLFQY--DAQPDMLES-VMR-DYTLLSCFKKDAHKMEIFLKLLKCRQT--DKYNCA
cod SL	114	VYLQTTLDRYDDV-PD--VLLNKTKWMSEK--LISLEQGV-VVLIRKMLDGAILNSSYNEYSAVQL--DVQPEVLES-ILR-DYNVLCCFKKDAHKIETILKLLKCRQI--DKYNCALY
salmon SL	115	VYLQTTLDRYDDA-PD--TLLKKTKWVSEK--LLSLEQGV-VVLIRKMLDDDMLTTSYYEQGVAPY--ALQPEVLES-VLR-DYTLLSCFKKDAHKMETFLKLLKCRQT--DKYSCFLH

Figure 2 (continued)

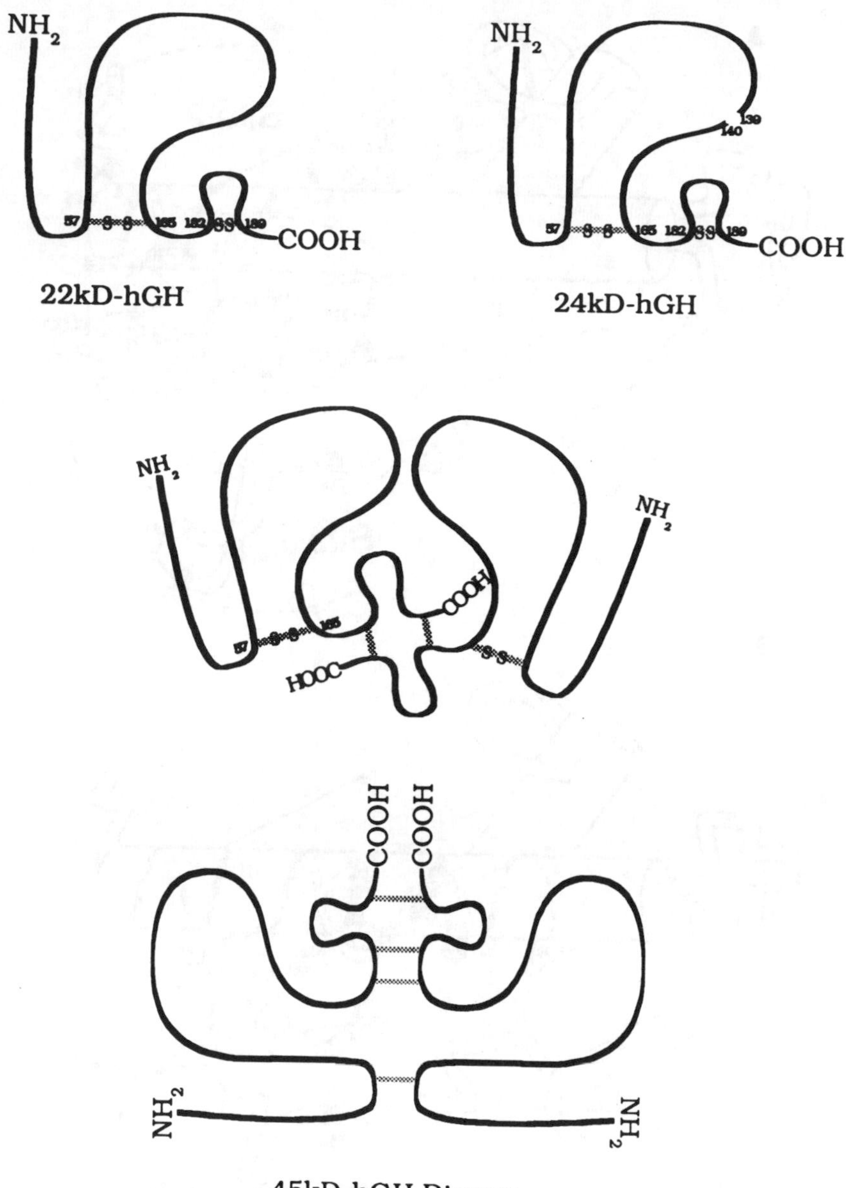

Figure 3 Schematic representation of hGH mass/size variants: 22-kDa hGH, the predominant monomeric form; 24-kDa hGH, produced by cleavage between residues 139 and 140 in the large loop; and 45-kDa hGH dimers (two possible conformations are depicted). (Adapted from Charrier and Martal.[118])

species separated by long periods of evolutionary divergence, then it is likely that these are of importance to the action of GH.

Amino acid residues in helix 3 are part of binding site 2 on GH.[124] The observations that the substituted GH analogs are GH antagonists are consistent with the prevailing view of GH action. It is thought that a GHR binds first to binding site 1 of GH. A second GHR then binds to binding site 2 of GH prior to GHR dimerization and GH action[125] (see Figure 5).

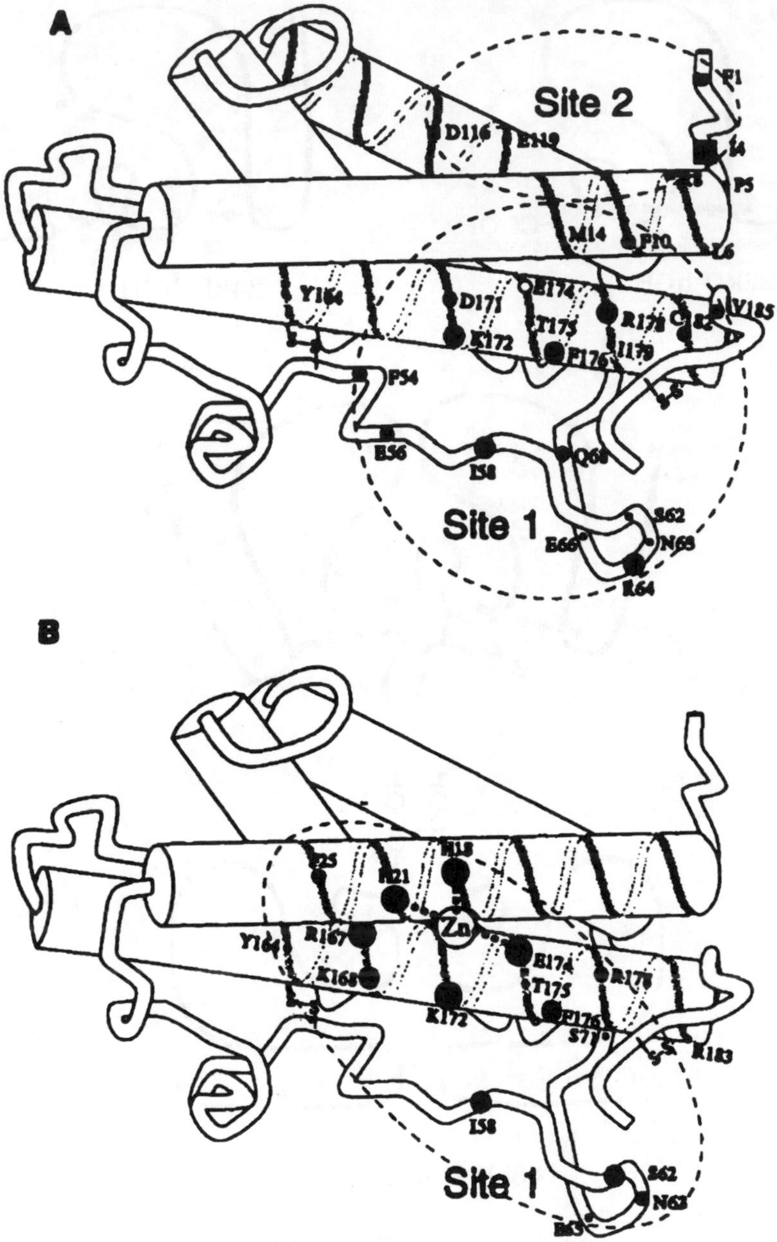

Figure 4 Three-dimensional structure of human GH, showing two binding sites. (From Cunningham, B. C., Ultsch, M., DeVos, A. M., et al., *Science*, 254, 821–825, 1991. Copyright ©1991 by the American Association for the Advancement of Science. Reproduced with permission).

III. BIOLOGICAL ACTIVITY

The model of GH action involves binding sequentially to two GHR molecules prior to GHR dimerization (Figure 5).[125] However, it may be questioned whether this model applies to all biological activities of GH. If GH were exerting all its activities via the same

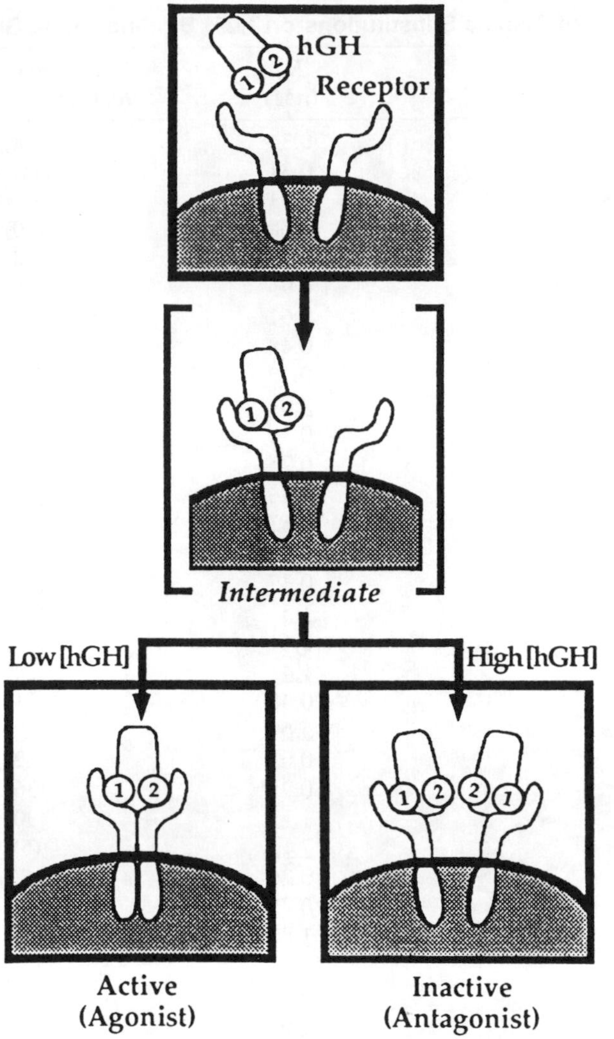

Figure 5 Schematic representation the possible mechanism of GH action, with GH binding to GH receptors (GHRs) and GHR dimerizations. (From Fuh, G., Cunningham, B. C., Fukunaga, R., et al., *Science*, 256, 1677–1678, 1992. Copyright ©1992 by the American Association for the Advancement of Science. Reproduced with permission.)

system, then we would expect the same relative potencies for GH isoforms, variants, fragments, analogs, and GH preparations from different species in the various biological assays systems. Examples of the potencies of such GH forms are given in Tables 2[133–145] and 3.[146–163] One caveat should be introduced for the interpretation of results in *in vivo* bioassays; and that is, that differences in biological activity may reflect metabolic clearance rate and not inherent activity. Nonetheless, there are marked differences in the relative activity of at least some GH preparations.

The diabetogenic activity of GH *in vivo* in *ob/ob* mice appears to have much less strict, or certainly different, structural requirements for GH to exert its effects than the growth (somatogenic) assays (Tables 2 and 3). The antilipolytic activity of GH on chicken adipose tissue *in vitro* is observed with GH preparations that have little or no lipolytic activity

Table 1 **Effect of Alanine Substitutions on hGH Binding to the GHBP**

Mutation	K_d (n*M*)	Percentage Activity (K_d [Wild] × 100kDa)
hGH wild type	0.34	(100)
P2A	0.31	(110)
T3A	0.31	(110)
I4A	0.68	(50)
P5A	0.71	(48)
L6A	0.95	(36)
S7A	0.61	(56)
R8A	0.48	(71)
L9A	0.32	(106)
F10A	2.0	(17)
N12A	0.40	(85)
M14A	0.75	(45)
L15A	0.44	(77)
R16A	0.51	(67)
H18A	0.24	(142)
R19A	0.37	(92)
F54A	1.5	(23)
S55A	0.41	(83)
E56A	1.4	(24)
S57A	0.48	(71)
I58A	5.6	(6)
P59A	0.65	(52)
S62A	0.95	(36)
N63A	1.12	(30)
R64A	7.11	(5)
E65A	0.20	(170)
E66A	0.71	(48)
Q68A	1.8	(19)
Q69A	0.31	(110)
K70A	0.82	(41)
S71A	0.68	(50)
L73A	0.24	(142)
R167A	0.26	(131)
K168A	0.37	(92)
D171A	2.4	(14)
K172A	4.6	(7)
E174A	0.075	(453)
T176A	5.4	(6)
I179A	0.92	(37)
V180A	0.34	(100)
Q181A	0.54	(63)
C182A	1.9	(18)
R183A	0.71	(48)
S194A	0.31	(110)
V185A	1.5	(23)
E186A	0.27	(126)
G187A	0.61	(56)

Table 1 **Effect of Alanine Substitutions on hGH Binding to the GHBP (continued)**

Mutation	K_d (n*M*)	Percentage Activity (K_d [Wild] × 100kDa)
S188A	0.24	(142)
F191A	0.47	(72)

Source: From Cunningham and Wells.[122]

Table 2 **hGH Isoforms, Variants, and Fragments: Possible Examples Where Some But Not All GH Biological Activities Are Retained**

	Biological Activities (Potency per Microgram Relative to hGH)		
	Growth Promoting[a]	Insulin-Like[b]	Diabetogenic[c]
Isoforms/variants			
hGH	100	100	100
20K hGH	97[133]	<5[134]	—
20K met-hGH	100[135]	~20[135]	~100[135]
Dimer hGH	10,[136] 22[137]	—	—
Fragments			
hGH	100	100	100
$des_{1-17}hGH$	3[138]	<4[138]	~20[138]
$des_{1-43}hGH$[d]	<5[139]	—	>100[139]
hGH_{95-131}(Dc2)	1.6[140]	<1[140]	~50[141]
hGH_{1-134T}	3.6[140,141]	2.2[140,141]	~50[140,141]
Reduced and *S*-carboxy methylated hGH (CAM hGH)	~1[142]	<1[142]	~50[142]
hGH (cleaved 134–155)	61[141]	19[141]	—
Da1- a noncovalent complex of hGH_{1-134} and $hGH_{141-191}$ (formed by plasmin treatment of reduced and *S*-carbamido-methyl hGH[e])	84[140]	~10–16[143,144]	100[141]
$des_{1-8,\ 134-145}hGH$[f]	70[145]	<50[145]	<50[145]

[a] Rat tibia test

[b] [^{14}C]Glucose oxidation by rat adipose tissue *in vitro.*

[c] Glucose tolerance test in chronically GH-treated *ob/ob* mice.

[d] Diabetogenic effect in yellow obese mice. Note also that the preparation shows lipolytic activity (as indicated by an increase in circulating free fatty acids 5 h following its administration to fasted, hypophysectomized rats) but does not appear to have insulin-like activity, as indicated by the lack of change in circulating concentrations of either glucose or free fatty acids (1 h administration to fasted, hypophysectomized rats).[139]

[e] Lipolytic potencies, >100[(143,144)]; leucine oxidation activity, ~100[(144)]; antilipolytic, ~100.[(143)]

[f] Antilipolytic potency, <25; diabetogenic activity determined in dogs.[(145)]

(Table 3). It is difficult to reconcile these results with the prevailing model (Figure 5) of GH binding to two GHRs that then dimerize. It may be that posttranslational modification of GHR influences the requirements for GH binding, and/or binding of either binding site or of binding site 2 of GH to a single GHR may be sufficient for some biological activity.

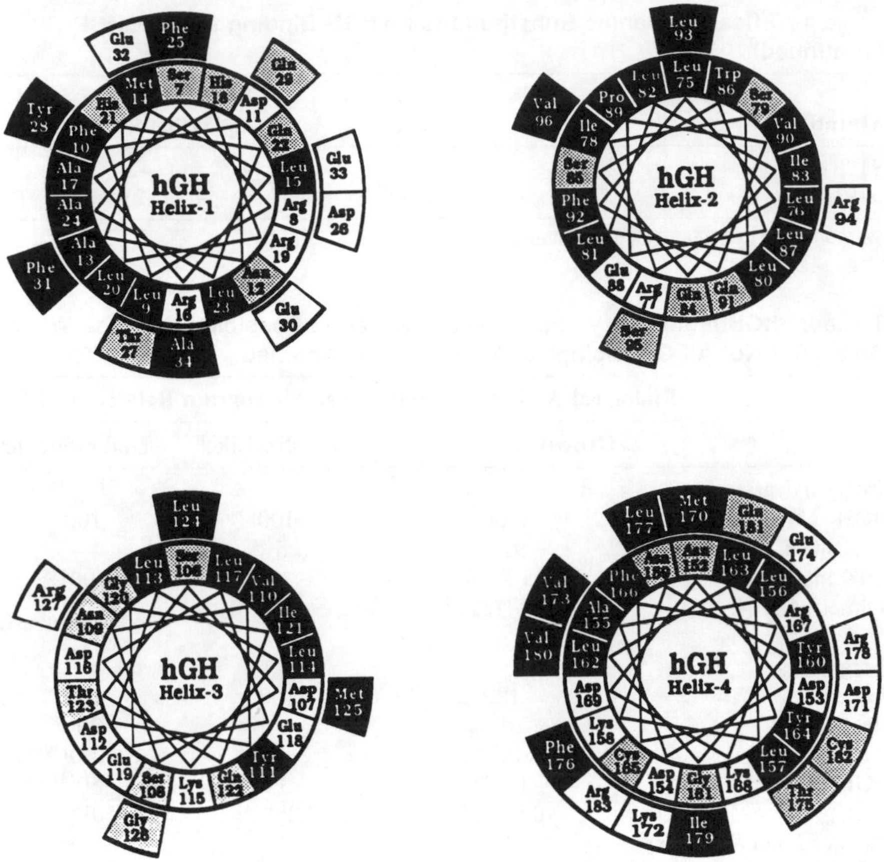

Figure 6 Edmundson wheel projection of human GH, assuming an α-helical conformation (turn frequency about four residues) based on porcine GH crystallographic studies.[120] As per Figure 1, bovine GH differs at helix 3 by seven substitutions (Arg^{109}, Glu^{112}, Lys^{113}, Leu^{122}, Ala^{123}, Arg^{126} and Glu^{127}) plus one theoretical deletion at Ser^{106}. Hydrophobic residues are black, neutral residues are shaded, and charged residues are white.

IV. ISOHORMONES AND VARIANTS

A. INTRODUCTION

Growth hormone heterogeneity is receiving increasing attention (reviewed by Nicoll et al.,[117] Lewis et al.,[164] Lewis,[165] and Baumann[166]). Following the definitions employed in the review by Baumann,[166] the term *isohormone* will be used for GH or GH-like proteins with different primary sequences (due to the presence of several GH genes or to differential splicing of the GH mRNA). *Variant* will be the term employed for posttranslational changes of GH, including aggregation/oligomerization, proteolytic cleavage, reduction, glycosylation, deamidation, and phosphorylation. This section considers first the different isoforms and variants prior to a general discussion of the biological activities and clearance rates.

B. DIFFERENT GENE PRODUCTS

In humans, there are two GH genes, *hGH-N* and *hGH-V*, each with respective proteins hGH-N and hGH-V. These proteins show differences in 13 amino acid residues (see

Table 3 **Relative Activities of Vertebrate Growth Hormones in Various Biological Assays**

	Mammalian Biological Assays			Avian Bioassays	
	Growth (Rat Tibia Test)	Insulin-Like [^{14}C] Glucose Oxidation by Rat Adipose Tissue *in Vitro*[a]	Diabetogenic (Glucose Tolerance Test (*in Vitro*) in *ob/ob* Mice)	Lipolytic (*In Vitro*)	Antilipolytic (*In Vitro*)
Mammalian GH					
22-kDa hGH, bGH	100	100	100	100[146,147]	100[147,148]
M8[b]	0[b]	<1[130]	—	~1[131]	100[131]
hCS	<1[149–151]	<1[151]	—	<1[147]	<1[147]
Avian GH					
Chicken GH	Active[152]	—	—	~100[146]	~100[148]
Ostrich GH	Active[155]	1.2[153]	~10[153]	—	—
Reptile GH					
Sea turtle GH	12[154,155]	1.1[153]	~5[153]	<1[156]	67[156]
Snapping turtle GH	24[154,155]	<1.0[153]	~10[153]	—	—
Amphibian GH					
Bullfrog GH	115[154,157]	4.5[153]	~10[153]	<1[156]	22[156]
Bony Fish GH					
Sturgeon GH	19[158]	5.9[153]	~5[153]	<1[156]	103[156]
Tilapia GH	~5NP [159]	<1.0[153]	~10[153]	—	—
Salmon GH[c]	Active[160]	—	—	<1[156]	609[156 c]

Table 3 **Relative Activities of Vertebrate Growth Hormones in Various Biological Assays (continued)**

	Mammalian Biological Assays			Avian Bioassays	
	Growth (Rat Tibia Test)	**Insulin-Like [^{14}C] Glucose Oxidation by Rat Adipose Tissue *in Vitro*[a]**	**Diabetogenic (Glucose Tolerance Test (*in Vitro*) in *ob/ob* Mice)**	**Lipolytic (Insulin-Like)**	**Antilipolytic (*In Vitro*)**
Elasmobranch GH					
Blue shark GH	~15[161]	—	—	<1[156]	78[156]
Tapeworm					
Pleurocoid growth factor[d]	Active[162]	Active[163]	Inactive[162]	—	—

Note: —, No data available; NP, nonparallel response.

[a] Potency calculated from Ref. 153.

[b] M8 is a bGH analog with Glu-117, Gly-119, and Ala-122 replaced by Leu-117, Gly-119, and Ala-122; when expressed in transgenic mice it reduced growth.[128] M8 also exerted antagonist activity in insulin-like and lipolytic assays using rat adipose tissue.[136]

[c] Other fish, e.g., bonito and yellow tail, also show very high antilipolytic activity.[156]

[d] The pleurocoid growth factor is also inactive in the lipolytic assay using rat adipose *in vitro*.[163]

Figure 1). The *h*GH-N gene is expressed in the pituitary gland with "normal" hGH-N released. The *hGH-V* gene is expressed in the placenta[167,168] with the predominant form of GH in the circulation of women in the third trimester being hGH-V.[169] There is little evidence for two forms of GH in other species except in salmonid fish, in which two forms of GH have been isolated (in both trout and chum salmon) (Figure 1). These may represent different genes or microheterogeneity in different populations of these fish. There is one well-characterized example of microheterogeneity. In bGH, the amino acid residue as position 127 is either valine or leucine[170,171] with bGH in an individual pituitary gland being all Val-127 or all Leu-127 or half Val-127 and half Leu-127. Thus, there appears to be allelic polymorphism of GH in cattle.[172]

C. DIFFERENT SPLICING OF GH mRNA

There is one excellent example of the *GH* RNA being spliced into two different mRNAs. This is the case with the *hGH-N* gene transcript. The resulting proteins are, respectively, 22-kDa hGH with 191 amino acid residues (Figure 1) and 20-kDa hGH with a deletion of 15 amino acid residues (residues 32 to 46).[173–176] The 20-kDa hGH represents 5 to 10% of monomeric hGH in the pituitary gland.[166] This different splicing pattern is not observed with the transcript of *hGH-V* gene.[177] Additional alternative splicing products of *hGH-N* pre-mRNA are possible although the putative protein products have not been characterized. The existence of different splicing of GH RNA from species other than human has not been established as yet.

D. DIFFERENT PROCESSING OF PROGROWTH HORMONE

In view of the proteolytic transformation of pro-GH (alternatively referred to as pre-GH) to GH itself, it is perhaps not surprising that some N-terminal heterogeneity of GH has been observed in some species. For example, the N terminal of bGH or ovine GH (oGH) has either phenylalanine or an extra alanine.[178–182] These result from differential processing of the precursor pro-GH.[183]

E. POSTTRANSLATIONAL VARIANTS

After synthesis, GH may undergo posttranslational modification to yield a series of GH variants. These modifications include dimerization/oligomerization, deamidation, proteolytic cleavage with or without reduction, glycosylation, phosphorylation, and acetylation, and are considered below.

1. Dimer and Oligomeric Forms

Growth hormone can form dimers and other oligomers, trimers, tetramers, and even pentamers.[165,184] These have been estimated to represent approximately 40% of immunoreactive GH secreted into the plasma in humans.[166] This estimate is, of course, based on the assumption that oligomeric forms of GH each have the same inherent immunoreactivity. Dimeric GH is thought to be two 22-kDa GH monomers linked either noncovalently or by disulfide bridges[166] (see Figure 3 for a diagrammatic representation). However, at least some dimeric GH remains, on the basis of studies in the chicken, after both sodium dodecyl sulfate (SDS) and vigorous 2-mercaptoethanol treatment (disrupting both noncovalent and disulfide bonds) (C. Aramburo and C. G. Scanes, unpublished observations). Similarly in humans some secreted GH is nondisulfide but covalently linked GH.[166] The nature of these covalently linked GH dimers is not known.

Relatively little is known of the three-dimensional structures or the physiological significance of oligomeric GHs. However, dimeric (44-kDa) hGH has been isolated and its biological activity determined. This is considered below (Section IV.E.7). In view of the proportion of oligomeric GH secreted and the reduced clearance rate of dimeric hGH and presumably other oligomeric forms of GH (see Section IV.E.8), it is likely that a considerable percentage of the circulating concentration of GH is, in fact, oligomeric as opposed to monomeric GH.

2. Deamidated GH

Two forms of deamidated GH have been characterized in the human pituitary gland (Asp-152 hGH and Glu-137 hGH).[185,186] These are also found in the circulation.[187] Deamidated GH has also been demonstrated in cattle.[188] Deamidation of GH does not appear to affect its biological activity in rat growth bioassays.[185]

3. Cleaved GH/GH Fragments

Two distinct types of cleaved GH and GH fragments have been detected or indicated in pituitary tissue[189–191] with proteolytic cleavage occurring either in the region of amino acid residues 133 to 146 (in hGH) or between residues 43 and 44 in hGH.[117,166,192] It is apparent that proteolytic "nicking" between residues 133 and 146 (at trypsin-like enzyme-susceptible sites—namely Arg-134, Lys-140, and Lys-145 in hGH) would open the large loop (disulfide linked) of amino acid residues (see Figure 3 for a schematic representation).[158,166] This results in a larger apparent molecular size of approximately 24 kDa.[186,193] Following reduction of the disulfide bonds, two fragments of 14 and 8 kDa, respectively, would be generated.[193] Such proteolytic cleavage has been reported in pituitary tissue,[189–191] with exogenous proteolytic enzymes (see, e.g., Ref. 193), and with plasmalemma from GH target tissues (e.g., liver).[194] Proteolytic cleavage has been reported with hGH, bGH, and rat GH (rGH).[165,166,189–191,193,194]

There is considerable controversy as to whether proteolytic cleavage of the large loop of GH does occur *de novo* or whether this represents an experimental artifact.[117,167] It is possible that there is postmortem cleavage of GH in the pituitary gland or that purified GH becomes cleaved during isolation by contaminating proteases. Furthermore, cleavage of GH may be part of an intrasomatotroph degradation pathway for GH. There is evidence that cleaved hGH is not secreted.[195] However, cleaved GH has been detected in the circulation.[196] This cleaved GH may be the result of target tissue action.[194] Irrespective of whether cleaved GH exists in the pituitary and/or circulation, chemically modified, proteolytically cleaved GH preparations have been useful in examining the biological role(s) of GH (see above). The biological activity and metabolism clearance rate of cleaved GH are discussed below.

Human pituitary glands contain a fragment of GH, hGH_{1-43}, which has been isolated and characterized.[192] The biological activity of hGH_{1-43} is considered elsewhere (see Section IV.E.7, below). Assuming 22-kDa hGH is cleaved to produce hGH_{1-43}, we could expect a second fragment—hGH_{44-191}. This has not been definitively identified although synthetic hGH_{44-191} shows interesting biological activities.[226]

4. Glycosylated GH

Glycosylated form(s) of GH have been characterized or identified in a number of species including humans[197] with hGH-V,[198] porcine GH,[199] murine GH,[200] rat GH,[201] and chicken GH.[202] The structure of glycosylated GH (and particularly that of the oligosaccharide itself) remains to be elucidated.

There is little to suggest that glycosylated GH has different biological effects compared to nonglycosylated GH. There is, however, evidence that in the case of prolactin, glycosylation may influence some biological activities. Two glycosylated forms of human prolactin have been isolated. One had full prolactin activity in the pigeon crop sac assay whereas the other had only 24% of the potency of the reference preparation.[203] Similarly, glycosylated human prolactin was reported to have lower radioreceptor and biological activity (Nb2 cell assay) than nonglycosylated prolactin.[204] Glycosylated ovine prolactin also has been reported to have reduced activity with potencies of 22 and 24% that of ovine prolactin in, respectively, a radioreceptor and the Nb2 rat lymphoma cell assay.[205] Glycosylated prolactin has also been reported to have lost insulin-like action.[206]

Glycosylated GH represents less than 1% of pituitary GH.[166,199] The extent to which glycosylated GH is secreted, and whether its synthesis/secretion is under differential control relative to nonglycosylated GH monomer, remains to be determined, As with prolactin, there is evidence that the ratio of glycosylated to nonglycosylated forms of GH in the pituitary/plasma change during pregnancy[199] and fetal development.[207]

5. Phosphorylated GH

There is evidence that a phosphorylated form(s) of GH exists. Phosphate has been found in native rat, ovine, and chicken GH preparations.[208–210] Moreover, human GH can be phosphorylated by a tyrosine kinase[211] and chicken GH can be phosphorylated by either protein kinase A or C.[210,212] Incubation of salmon GH with alkaline phosphatase results in a shift in charge,[213] indicating the presence of a phosphorylated GH.

6. Acetylated GH

The N-terminal amino group of a small proportion of hGH is acetylated with an acetyl group.[166] This does not appear to influence the biological activity of the GH. Acetylation does affect the pI and, hence, electrophoretic pattern. Thus, acetylated hGH is also called "fast" GH.[166]

7. Biological Activity

If there are both isoforms and variants of GH, it is reasonable to consider why such forms exist. Do GH isoforms and/or variants have the same potencies, and do they exhibit similar activities as indicated by different biological assays?

There is, in fact, relatively little information on the biological activities of many of the forms of GH. An exception to this is the situation with some isoforms and size variants of hGH, the biological activities of some of which are included in Table 2. The hGH-V form exhibits growth stimulating (somatogenic) activity when tested at a single dose but appears to have much reduced lactogenic activity.[214] The 20-kDa isoform of hGH has both full growth promoting (rats, Table 2) and diabetogic activities (mice, Table 2)[215] in *in vivo* assays. In contrast, the 20-kDa form of hGH has low insulin-like activity *in vitro* (Table 2). This reduced insulin-like activity with rat adipose tissue correlates well with the low radioreceptor activity of 20-kDa hGH.[216,217] It may be questioned whether 20-kDa hGH is active at all (or at least acts via the GHR) in humans, owing to its inherent lack of activity in human GHBP radioreceptor assays.[218] There is, however, evidence of a GHBP that binds the 20-kDa hGH.[219] The disproportionately high activity of 20-kDa hGH in the *in vivo* growth-promoting and diabetogenic assays may be due to the reduced metabolic clearance rate (MCR) for this isoform (see Table 4).

In contrast, despite its slow clearance rate (Table 4) and high radioreceptor activity,[220] dimeric hGH shows low biological activity in *in vitro* biological assays (Table 2). Certain cleaved hGH forms appear to retain full GH activity.[166] If GH inhibition of binding of hGH to rabbit liver membranes is subjected to Scatchard analysis, a two-site model is preferred with the intact hGH binding to both sites but cleaved hGH binding only to the high-affinity site.[221] This might suggest different activities for cleaved hGH although there is not evidence for this as yet.

Another possible size variant of hGH is hGH_{1-43}. This represents about 1% of hGH in the pituitary,[166] but may still prove to be an isolation artifact. This variant has no detectible growth-promoting activity but shows insulin-potentiating effects in hypophysectomized rodents.[222,223]

Table 4 **Metabolic Clearance Rates of Human GH Isoforms/Variants**

Isoform/Variant	Metabolic Clearance Rate ± SEM	
	ml/min • 100 g (Body Weight)	Liters/M^2 • day
MCR determined in rats[1]		
hGH_{22K} monomer	1.40 ± 0.22[a]	
hGH_{20K} monomer	0.73 ± 0.22[a]	
hGH_{22K} dimer	0.27 ± 0.02[c]	
hGH_{20K} dimer	0.38 ± 0.06[c]	
MCR determined in humans[2]		
hGH		108 ± 8.1[a]
Cleaved hGH (C or hGH 1)		68 ± 12.3[b]
Cleaved hGH (D or hGH 2)		56 ± 4.9[b]

Note: Different superscript letter indicates difference $p < 0.05$.

[1] Data from Baumann et al.[225]
[2] Data from Baumann.[228]

8. Clearance Rates

The biological activity of isoforms and variants of GH *in vivo* depends on the inherent activity of the isoform/variant and also the circulating concentration. The latter, in turn, reflects the rate of clearance. Some isoforms/variants have been demonstrated to have markedly different metabolic clearance rates. For instance, the monomeric 20-kDa hGH is cleared more slowly than the 22-kDa[224] (Table 4). Moreover, dimeric hGH (either the dimeric 20-kDa or dimeric 22-kDa form) is cleared at an even slower rate[225] (Table 4). Glycosylation may also reduce the MCR loss[226,227] as does proteolytic cleavage.[228]

The relative proportion of the different isoforms/variants of a hormone in the circulation will reflect the secretion rate (SR) of hormone and the MCR.

$$\text{Hormone concentration (ng/ml)} = \text{SR (ng/min)/MCR (ml/min)}$$

Thus large differences in MCR (e.g., 5.2-fold between the monomeric and dimeric forms of 22-kDa hGH) will be expected to change the relative proportions of the isoforms/variants in the circulation.

9. Problems with Detection

It is intuitively obvious that different forms of GH may not be equally immunoreactive. Individual epitopes may be missing as in the case of 20-kDa hGH with its omitted sequence of its amino acid residues or cleaved GH. Alternatively, antigenic epitopes may be masked by dimerization/oligomerization or by the presence of a GH-binding protein. Examples of GH isoforms/variants not retaining immunoreactivity include some radioimmunoassays not detecting 20-kDa hGH or high molecular weight forms of hGH.[229,230]

REFERENCES

1. **Niall, H. D.,** *Nature (Lond.)*, 230, 90, 1971.
2. **Martial, J. A., Baxter, J. D., and Goodman, H. M.,** *Science*, 205, 602, 1979.
3. **Seeburg, P. H.,** *DNA*, 1, 239, 1982.
4. **Brewley, T. A. and Li, C. H.,** *Experientia*, 27, 1368, 1971.
5. **Niall, H. D., Hogan, M. L., Sater, R., et al.,** *Proc. Natl. Acad. Sci. U.S.A.*, 68, 866, 1971.

6. **Shine, J., Seeburg, P. H., Martial, J. A., et al.,** *Nature (Lond.)*, 270, 494, 1977.
7. **Li, C. H., Chung, D., Lahm, H.-W., et al.,** *Arch. Biochem. Biophys.*, 245, 287, 1986.
8. **Seeburg, P. H., Shine, J., Martial, J. A., et al.,** *Nature (Lond.)*, 270, 486, 1977.
9. **Linzer, D. I. H. and Talamantes, F.,** *J. Biol. Chem.*, 260, 9574, 1985.
10. **Southard, J. N., Sanchez-Jimenez, F., Campbell, G. T.,** *Endocrinology*, 129, 2965, 1991.
11. **Tsubokawa, M. and Kawauchi, H.,** *Int. J. Peptide Protein Res.*, 25, 297, 1985.
12. **Pankov, Y. A., Bulatov, A. A., and Osipova, T. A.,** *Int. J. Peptide Protein Res.*, 20, 396, 1982.
13. **Li, C. H., Izdebski, J., and Chung, D.,** *J. Peptide Protein Res.*, 33, 70, 1989.
14. **Shoji, K., Ohara, E., Watahiki, M., et al.,** *Nucleic Acids Res.*, 18, 6424, 1990.
15. **Graf, L. and Li, C. H.,** *Biochem. Biophys. Res. Commun.*, 56, 168, 1974.
16. **Seeburg, P. H., Sias, S., Adelman, J., et al.,** *DNA*, 2, 37, 1983.
17. **Li, C. H., Gordon, D., and Knorr, J.,** *Arch. Biochem. Biophys.*, 156, 493, 1973.
18. **Orian, J. M., O'Mahoney, J. V., and Brandon, M. R.,** *Nucleic Acids Res.*, 16, 9046, 1988.
19. **Yamano, Y., Oyabayashi, K., Okuno, M., et al.,** *FEBS Lett.*, 228, 301, 1988.
20. **Bonino, M. D. D.-J., DeNue, I. A., Ore, R., et al.,** *Int. J. Peptide Protein Res.*, 38, 193, 1991.
21. **Zakin, M. M., Pokus, E., Langton, A. A., et al.,** *Int. J. Peptide Protein Res.*, 8, 435, 1976.
22. **Hulmes, J. D., Meidel, M. C., Li, C. H., et al.,** *Int. J. Peptide Protein Res.*, 33, 368, 1989.
23. **Chen, H.-T., Pan, F.-M. and Chang, W.-C.,** *Biochem. Biophys. Acta,* 949, 247, 1988.
24. **Souza, L. M., Boone, T. C., Murdock, D., et al.,** *J. Exp. Zool.*, 232, 465, 1984.
25. **Lamb, I. C., Galehouse, D. M., and Foster, D. N.,** *Nucleic Acids Res.*, 16, 9339, 1988.
26. **Foster, D. N., Kim, S. U., Enyeart, J. J., et al.,** *Biochem. Biophys. Res. Commun.*, 173, 967, 1990.
27. **Yasuda, A., Yamaguchi, K., Papkoff, H., et al.,** *Gen. Comp. Endocrinol.*, 73, 242, 1989.
28. **Martens, G. J. M., Groenen, P. J. T. A., et al.,** *Nucleic Acids Res.*, 17, 3974, 1989.
29. **Pan, F.-M. and Chang, W. C.,** *Biochim. Biophys. Acta*, 950, 238, 1988.
30. **Kobayashi, T., Kikuyama, S., Yasuda, A., et al.,** *Gen. Comp. Endocrinol.*, 73, 417, 1989.
31. **Takahashi, N., Kikuyama, S., Gen, K., et al.,** *J. Mol. Endocrinol.*, 9, 283, 1992.
32. **Noso, T., Yasuda, A., Kawazoe, I., et al.,** *Int. J. Peptide Protein Res.*, 32, 579, 1988.
33. **Watahiki, M., Tanaka, M., Masuda, N., et al.,** *Gen. Comp. Endocrinol.*, 70, 401, 1988.
34. **Sato, N., Watanabe, K., Murata K., et al.,** *Biochim. Biophys. Acta*, 949, 352, 1988.
35. **Kariyas, Y., Sato, N., Kawazoe, I., et al.,** *Agric. Biol. Chem.*, 53, 1679, 1989.
36. **Knibb, W., Robins, A. J., Crocker, L. A., et al.,** *DNA Seq.*, 2, 121, 1992.
37. **Rentier-Delrue, F., Swennen, D., Philippart, J. C., et al.,** *DNA*, 8, 271, 1989.
38. **Momota, H., Kosugi, R., Chugai, H., et al.,** *Nucleic Acids Res.*, 16, 10362, 1988.
39. **Watahiki, M., Yamamoto, M., Yamakawa, M., et al.,** *J. Biol. Chem.*, 264, 312, 1989.
40. **Momota, H., Kosugi, R., Hiramatsu, H., et al.,** *Nucleic Acids Res.*, 16, 3107, 1988.
41. **Funkenstein, B., Chen, T. T., Powers, D. A., et al.,** *Gene*, 103, 243, 1991.
42. **Tsai, H.-J., Lin, K.-L., and Chen, T. T.,** *Comp. Biochem. Physiol.*, 104B, 803, 1993.
43. **Yamakawa, M., Watahiki, M., Kamioka, Y., et al.,** *Biochim. Biophys. Acta*, 1087, 247, 1990.
44. **Rand-Weaver, M., Noso, T., and Kawauchi, H.,** *Gen. Comp. Endocrinol.*, 81, 39, 1991.
45. **Chao, S.-C., Pan, F.-M., and Chang, W.-C.,** *Biochim. Biophys. Acta*, 1007, 233, 1989.
46. **Koren, Y., Sarid, S., Ber, R., et al.,** *Gene*, 77, 309, 1989.
47. **Chiou, C.-S., Chen, H.-T., and Chang, W.-C.,** *Biochim. Biophys. Acta*, 1087, 91, 1990.
48. **Chang, Y. S., Liu, C. S., Huang, F. L., et al.,** *Gen. Comp. Endocrinol.*, 87, 385, 1992.
49. **Ho, W. K. K., Tsang, W. H., and Dias, N. P.,** *Biochem. Biophys. Res. Commun.*, 161, 1230, 1989.
50. **Sekine, S., Mizukami, T., Nishi, T., et al.,** *Proc. Natl. Acad. Sci. U.S.A.*, 82, 4306, 1985.
51. **Sekine, S., Mizukami, T., Saito, A., et al.,** *Biochim. Biophys. Acta*, 1009, 117, 1989.
52. **Kawauchi, H., Moriyama, S., Yasuda, A., et al.,** *Arch. Biochem. Biophys.*, 244, 542, 1986.
53. **Nicoll, C. G., Steiny, S. S., King, D. S., et al.,** *Gen. Comp. Endocrinol.*, 68, 387, 1987.
54. **Gonzalez-Villasenor, L. I., Zhang, P., Chen, T. T., et al.,** *Gene*, 65, 239, 1988.
55. **Johansen, B., Johansen, O. C., and Valla, S.,** *Gene*, 77, 317, 1989.
56. **Lorens, J., Nerland, A. H., Male, R., et al.,** *Nucleic Acids Res.*, 17, 2352, 1989.
57. **Agellon, L. B. and Chen, T. T.,** *DNA*, 5, 463, 1986.
58. **Agellon, L. B., Davies, S. L., Chen, T. T., et al.,** *Proc. Natl. Acad. Sci. U.S.A.*, 85, 5136, 1988.
59. **Rand-Weaver, M., Kawauchi, H., and Ono, M.,** In: *The Endocrinology of Growth, Development and Metabolism in Vertebrates* (Eds. M. P. Schreibman, C. G. Scanes, and P. K. T. Pang). Academic Press, San Diego, 1993, pp. 13–42.
60. **Lemaire, C. and Panyim, S.,** *GenBank Accession* #M63713, 1991.
61. **Yamaguchi, K., Yasuda, A., Kishida, M., et al.,** *Gen. Comp. Endocrinol.*, 66, 447, 1987.

62. **Kishida, M., Hirano, T., Kubota, J., et al.,** *Gen. Comp. Endocrinol.*, 65, 478, 1987.
63. **Saito, A., Sekine, S., Komatsu, Y., et al.,** *Gene*, 73, 545, 1988.
64. **Yasuda, A., Yamaguchi, K., Noso, T., et al.,** *Biochim. Biophys. Acta*, 1120, 297, 1992.
65. **Yamaguchi, K., Yasuda, A., Lewis, U. J., et al.,** *Gen. Comp. Endocrinol.*, 73, 252, 1989.
66. **Lagler, K. F., Bardach, J. E., Muller, R. R., et al.,** John Wiley & Sons, New York, 1977.
67. **Cooke, N. E., Colt, D., Shine, J., et al.,** *J. Biol. Chem.*, 256, 4007, 1981.
68. **Cooke, N. E., Colt, D., Weiner, R. I., et al.,** *J. Biol. Chem.*, 225, 6502, 1980.
69. **Kohmoto, K., Tsunasawa, S., and Sakiyama, F.,** *Eur. J. Biochem.*, 138, 227, 1984.
70. **Tsubokawa, M., Muramoto, K., and Kawauchi, H.,** *Int. J. Peptide Protein Res.*, 25, 442, 1985.
71. **Bondar, A. A., Golovin, S. J., and Mertvetsov, N. P.,** GenBank Accession #X63235, 1991.
72. **Sasavage, N. L., Nilson, J. H., Horowitz, S., et al.,** *J. Biol. Chem.*, 257, 678, 1982.
73. **Miller, W. L., Thirion, J. P., and Martial, J. A.,** *Endocrinology*, 107, 851, 1980.
74. **Li, C. H., Dixon, J. S., Lo, T. B., et al.,** *Arch. Biochem. Biophys.*, 141, 705, 1970.
75. **Li, C. H.,** *Int. J. Peptide Protein Res.*, 8, 205, 1976.
76. **Martinat, N., Hoet, J.-C., Nespoulous, C., et al.,** *Biochim. Biophys. Acta*, 1077, 339, 1991.
77. **Lehrman, S. R., Lahm, H. W., Meidel, M. C., et al.,** *Int. J. Peptide Protein Res.*, 31, 544, 1988.
78. **Li, C. H., Oosthuizen, M. M. J. and Chung, D.,** *Int. J. Peptide Protein Res.*, 33, 67, 1989.
79. **Hanks, M. C., Alonzi, J. A., Sharp, P. J., et al.,** *J. Mol. Endocrinol.*, 2, 21, 1989.
80. **Watahiki, M., Tanaka, M., Masuda, N., et al.,** *J. Biol. Chem.*, 264, 5535, 1989.
81. **Karatzas, C. N., Zadwormy, D., and Kuhnlein, U.,** *Nucleic Acids Res.*, 18, 3071, 1990.
82. **Yasuda, A., Kawauchi, H., and Papkoff, H.,** *Gen. Comp. Endocrinol.*, 80, 363, 1990.
83. **Noso, T., Swanson, P., Lance, V. A., et al.,** *Int. J. Peptide Protein Res.*, 39, 250, 1992.
84. **Buckbinder, L. and Brown, D. D.,** *Proc. Natl. Acad. Sci. U.S.A.*, 90, 3820, 1993.
85. **Takahashi, N., Yoshihama, K., Kikuyama, S., et al.,** *J. Mol. Endocrinol.*, 5, 281, 1990.
86. **Yamaguchi, K., Specker, J. L., King, D. S., et al.,** *J. Biol. Chem.*, 263, 9113, 1988.
87. **Rentier-Delrue, F., Swennen, D., Prunet, P., et al.,** *DNA*, 8, 261, 1989.
88. **Yasuda, A., Miyajima, K., Kawauchi, H., et al.,** *Gen. Comp. Endocrinol.*, 66, 280, 1987.
89. **Chang, Y. S., Huang, F. L., Mao, M., et al.,** *GenBank Accession* #X61049, 1991.
90. **Chang, Y. S., Huang, F. L., Mao, M., et al.,** *GenBank Accession* #X61052, 1991.
91. **Yasuda, A., Itoh, H., and Kawauchi, H.,** *Arch. Biochem. Biophys.*, 244, 528, 1986.
92. **Kuwana, Y., Kuga, T., Sekine, S., et al.,** *Agric. Biol. Chem.*, 52, 1033, 1988.
93. **Song, S., Trinh, K. Y., Ilew, C. L., et al.,** *Eur. J. Biochem.*, 172, 279, 1988.
94. **Mercier, L., Rentier-Delrue, F., Swennen, D., et al.,** *DNA*, 8, 119, 1989.
95. **Robertson, M. C., Croze, F., Schroedter, I. C., et al.,** *Endocrinology*, 127, 702, 1990.
96. **Colosi, P., Talamantes, F., and Linzer, D. I. H.,** *Mol. Endocrinol.*, 1, 767, 1987.
97. **Colosi, P., Thordarson, G., Hellmiss, R., et al.,** *Mol. Endocrinol.*, 3, 1462, 1989.
98. **Warren, W. C., Liang, R., Krivi, G. G., et al.,** *J. Endocrinol.*, 126, 141, 1990.
99. **Schuler, L. A., Shimomura, K., Kessler, M. A., et al.,** *Biochemistry*, 27, 8443, 1988.
100. **Krivi, G. G., Hauser, S. D., Stafford, J. M., et al.,** *Proc. 71st Annu. Mtg. Endocrine Soc.*, Abstr., 1523, 1989.
101. **Schuler, L. A. and Hurley, W. L.,** *Proc. Natl. Acad. Sci. U.S.A.*, 84, 5650, 1987.
102. **Kessler, M. A., Milosavljevic, M., Zieler, C. G., et al.,** *Biochemistry*, 28, 5154, 1989.
103. **Tanaka, M., Yamakawa, M., Watahiki, M., et al.,** *Biochim. Biophys. Acta*, 1008, 193, 1989.
104. **Deb, S., Roby, K. F., Faria, T. N., et al.,** *J. Biol. Chem.*, 266, 23027, 1991.
105. **Duckworth, M. L., Peden, L. M., and Friesen, H. G.,** *J. Biol. Chem.*, 261, 10879, 1986.
106. **Tanaka, M., Minoura, H., Ushiro, H., et al.,** *Biochim. Biophys. Acta*, 1088, 385, 1991.
107. **Duckworth, M. L., Lynn, M., Peden, L. M., et al.,** *Mol. Endocrinol.*, 2, 912, 1988.
108. **Duckworth, M. L., Kirk, K. L., and Friesen, H. G.,** *J. Biol. Chem.*, 261, 10871, 1986.
109. **Jackson, L. L., Colosi, P., Talamantes, F., et al.,** *Proc. Natl. Acad. Sci. U.S.A.*, 83, 8496, 1986.
110. **Southard, J. N., Do, L., Smith, W. C., et al.,** *Mol. Endocrinol.*, 3, 1710, 1989.
111. **Linzer, D. I. H. and Nathans, D.,** *Proc. Natl. Acad. Sci. U.S.A.*, 81, 4255, 1984.
112. **Linzer, D. I. H. and Nathans, D.,** *EMBO J.*, 4, 1419, 1985.
113. **Ono, M., Takayama, Y., Rand-Weaver, M., et al.,** *Proc. Natl. Acad. Sci. U.S.A.*, 87, 4330, 1990.
114. **Rand-Weaver, M., Noso, T., Muramoto, K., et al.,** *Biochemistry*, 30, 1509, 1991.
115. **Takayama, Y., Rand-Weaver, M., Kawauchi, J., et al.,** *Mol. Endocrinol.*, 5, 778, 1991.
116. **Miller, W. L. and Eberhardt, N. L.,** *Endocr. Rev.*, 4, 97, 1983.
117. **Nicoll, C. S., Mayere, G. L., and Russell, S. M.,** *Endocr. Rev.*, 7, 169, 1986.
118. **Charrier, J. and Martal, J.,** *Reprod. Nutr. Dev.*, 28, 857, 1988.
119. **Cunningham, R. C., Ultsch, M., DeVos, A. M., et al.,** *Science*, 254, 821, 1991.

120. **Abdel-Meguid, S. S., Shieh, H. S., Smith, W. W., et al.**, *Proc. Natl. Acad. Sci. U.S.A.*, 84, 6434, 1987.
121. **Cunningham, B. C., Thurani, P., Ng, P., et al.**, *Science*, 243, 1330, 1989.
122. **Cunningham, B. C. and Wells, J. A.**, *Science*, 244, 1081, 1989.
123. **Cunningham, B. C., Henner, D. J., and Wells, J. A.**, *Science*, 247, 1461, 1990.
124. **DeVos, A. A., Ultsch, M., and Kossiakoff, A. A.**, *Science*, 255, 306, 1992.
125. **Fuh, G., Cunningham, B. C., Fukunaga, R., et al.**, *Science*, 256, 1677, 1992.
126. **Donner, D. B., Nakayama, K., Tani, S., et al.**, *J. Biol. Chem.*, 253, 6717, 1978.
127. **Hara, K., Chen, C.-J. H., and Sonenberg, M.**, *Biochemistry*, 17, 550, 1978.
128. **Chen, W. Y., Wight, D. C., Wagner, T. E., et al.**, *Proc. Natl. Acad. Sci. U.S.A.*, 87, 5061, 1990.
129. **Chen, W. Y., White, M. E., Wagner, T. E., et al.**, *Endocrinology*, 129, 1402, 1991.
130. **Okada, S., Chen, W. Y., Wiehl, P., et al.**, *Endocrinology*, 130, 2284, 1992.
131. **Campbell, R. M., Chen, W. Y., Wiehl, P., et al.**, *Proc. Soc. Exp. Biol. Med.*, 203, 311, 1993.
132. **Chen, W. Y., Wight, D. C., Mehta, B. V., et al.**, *Mol. Endocrinol.*, 5, 1845, 1991.
133. **Smal, J., Closset, J., Hennen, G., et al.**, *J. Biol. Chem.*, 262, 11071, 1987.
134. **Frigeri, L. G., Peterson, S. M., and Lewis, U. J.**, *Biochem. Biophys. Res. Commun.*, 91, 778, 1979.
135. **Kostyo, J. L., Cameron, C. M., Olson, K. C., et al.**, *Proc. Natl. Acad. Sci. U.S.A.*, 82, 4250, 1985.
136. **Lewis, U. J., Peterson, S. M., Bonewald, L. F., et al.**, *J. Biol. Chem.*, 252, 3697, 1977.
137. **Becker, G. W., Bowsher, R. B., MacKeller, W. C., et al.**, *Biotech. Appl. Biochem.*, 9, 478, 1987.
138. **Towns, R., Kostyo, J. L., Vogel, T., et al.**, *Endocrinology*, 130, 1225, 1992.
139. **Lewis, U. J., Lewis, L. J., Salem, M. A. M., et al.**, *Mol. Cell. Endocrinol.*, 78, 45, 1991.
140. **Reagan, C. R., Kostyo, J. L., Mills, J. B., et al.**, *Endocrinology*, 102, 1377, 1978.
141. **Reagan, C. R.**, *Diabetes*, 27, 883, 1978.
142. **Cameron, C. M., Kostyo, J. L., Rillema, R. A., et al.**, *Am. J. Physiol.*, 247, E639, 1984.
143. **Goodman, H. M. and Kostyo, J. L.**, *Endocrinology*, 108, 553, 1981.
144. **Chipkin, S. R., Szecowka, J., Tai, L.-R., et al.**, *Endocrinology*, 125, 450, 1989.
145. **Becker, G. and Shaar, C. J.**, United States Patent, 5,079,345, 1992.
146. **Campbell, R. M. and Scanes, C. G.**, *Proc. Soc. Exp. Biol. Med.*, 180, 513, 1985.
147. **Campbell, R. M., Kostyo, J. L., and Scanes, C. G.**, *Proc. Soc. Exp. Biol. Med.*, 193, 269, 1990.
148. **Campbell, R. M. and Scanes, C. G.**, *Proc. Soc. Exp. Biol. Med.*, 184, 456, 1987.
149. **Friesen, H.**, *Endocrinology*, 76, 369, 1965.
150. **Florini, J. R., Tonelli, G., Brener, C. B., et al.**, *Endocrinology*, 79, 692, 1966.
151. **Turtle, J. R. and Kipnis, D. M.**, *Biochim. Biophys. Acta*, 144, 583, 1967.
152. **Harvey, S. and Scanes, C. G.**, *J. Endocrinol.*, 73, 321, 1977.
153. **Cameron, C. M., Kostyo, J. L., and Papkoff, H.**, *Endocrinology*, 116, 1501, 1985.
154. **Farmer, S. W., Papkoff, H., and Hayashida, T.**, *Endocrinology*, 99, 692, 1986.
155. **Farmer, S. W. and Papkoff, H.**, In: *Hormones and Evolution* (Ed. E. J. W. Barrington). Academic Press, New York, 1979, pp. 525–559.
156. **Campbell, R. M., Kawauchi, H., Lewis, U. J., et al.**, *Proc. Soc. Exp. Biol. Med.*, 197, 409, 1991.
157. **Farmer, S. W., Licht, P., and Papkoff, H.**, *Endocrinology*, 101, 1145, 1977.
158. **Farmer, S. W., Hayashida, T., Papkoff, H., et al.**, *Endocrinology*, 108, 377, 1981.
159. **Farmer, S. W., Papkoff, H., Hayashida, T., et al.**, *Gen. Comp. Endocrinol.*, 30, 91, 1976.
160. **Wagner, G. F., Fargher, R. C., Brown, J. C., et al.**, *Gen. Comp. Endocrinol.*, 60, 27, 1985.
161. **Hayashida, T. and Lewis, U. J.**, *Gen. Comp. Endocrinol.*, 36, 530, 1978.
162. **Salem, M. A. M. and Phares, C. K.**, *Proc. Soc. Exp. Biol. Med.*, 191, 187, 1989.
163. **Salem, M. A. M. and Phares, C. K.**, *Proc. Soc. Exp. Biol. Med.*, 190, 203, 1989.
164. **Lewis, U. J., Singh, R. N. P., Tutwiler, G. F., et al.**, *Recent Prog. Horm. Res.*, 36, 477, 1980.
165. **Lewis, U. J.**, *Annu. Rev. Physiol.*, 46, 33, 1984.
166. **Baumann, G.**, *Endocr. Rev.*, 12, 424, 1991.
167. **Frankenne, F., Rentier-Delrue, F., Scippo, M. L., et al.**, *Endocrinol. Metab.*, 64, 635, 1987.
168. **Liebhaber, S. A., Urbanek, M., Ray, J., et al.**, *J. Clin. Invest.*, 83, 1985, 1989.
169. **Frankenne, F., Closset, J., Gomez, F., et al.**, *J. Clin. Endocrinol. Metab.*, 66, 1171, 1988.
170. **Fellows, R. E. and Rogol, A. D.**, *J. Biol. Chem.*, 244, 1567, 1969.
171. **Fernandez, H., Daurant, S. T., Pena, C., et al.**, *FEBS Lett.*, 18, 53, 1971.
172. **Seavey, B. K., Singh, R. N. P., Lewis, U. J., et al.**, *Biophys. Res. Commun.*, 43, 189, 1971.
173. **Lewis, U. J., Dunn, J. T., Bonewald, L. F., et al.**, *J. Biol. Chem.*, 253, 2679, 1978.

174. **Lewis, U. J., Bonewald, L. F., and Lewis, L. J.,** *Biochem. Biophys. Res. Commun.*, 92, 511, 1980.
175. **DeNoto, F. M., Moore, D. D., and Goodman, H. M.,** *Nucleic Acids Res.*, 9, 3719, 1981.
176. **Estes, P. A., Cooke, N. E., and Liebhaber, S. A.,** *J. Biol. Chem.*, 267, 14903, 1992.
177. **Cooke, N. E., Ray, J., Watson, M. A., et al.,** *J. Clin. Invest.*, 82, 270, 1988.
178. **Li, V. H. and Ash, L.,** *J. Biol. Chem.*, 203, 419, 1953.
179. **Wallis, M.,** *FEBS Lett.*, 3, 118, 1969.
180. **Wallis, M.,** *Biochim. Biophys. Acta*, 310, 388, 1983.
181. **Wallis, M. and Davies, R. V.,** In: *Growth Hormone and Related Peptides* (Eds. A. Pecile and E. E. Miller). Excepta Medica, Amsterdam, 1976, pp. 1–3.
182. **Pena, C., Paladini, A. C., Dellacha, J. M., et al.,** *Eur. J. Biochem.*, 17, 27, 1970.
183. **Pena, C., Paladini, A. C., Dellacha, J. M., et al.,** *Biochim. Biophys. Acta*, 194, 320, 1969.
184. **Stolar, M. W., Amburn, K., and Baumann, G.,** *J. Clin. Endocrinol. Metab.*, 59, 212, 1984.
185. **Lewis, U. J., Singh, R. N. P., Bonewald, L. F., et al.,** *Endocrinology*, 104, 1256, 1979.
186. **Lewis, U. J., Singh, R. N. P., Bonewald, L. F., et al.,** *J. Biol. Chem.*, 256, 11645, 1981.
187. **Baumann, G., MacCart, J. G., and Amburn, K.,** *J. Clin. Endocrinol. Metab.*, 56, 946, 1983.
188. **Secchi, C., Biondi, P. A., Negri, A., et al.,** *Int. J. Peptide Protein Res.*, 28, 298, 1986.
189. **Chrambach, A., Yadley, R. A., Ben-David, M., et al.,** *Endocrinology*, 93, 848, 1973.
190. **Yadley, R. A., Rodbard, D., and Chrambach, A.,** *Endocrinology*, 93, 866, 1973.
191. **Singh, R. N. P., Seavey, B. K., Rice, V. P., et al.,** *Endocrinology*, 94, 883, 1974.
192. **Singh, R. N. P., Seavey, B. K., Lewis, L. J., et al.,** *J. Protein Chem.*, 2, 525, 1983.
193. **Maciag, T., Forand, R., Ilsley, S., et al.,** *J. Biol. Chem.*, 255, 6064, 1989.
194. **Schepper, J. M., Hughes, E. F., Postel-Vinay, M.-C., et al.,** *J. Biol. Chem.*, 259, 12945, 1984.
195. **Baumann, G. and MacCart, J.,** *J. Clin. Endocrinol. Metab.*, 55, 611, 1982.
196. **Chawla, R. K., Parks, J. S., and Rudman, D.,** *Annu. Rev. Med.*, 34, 519, 1983.
197. **Sinha, Y. N. and Lewis, U. J.,** *Biochem. Biophys. Res. Commun.*, 140, 491, 1986.
198. **Ray, J., Jones, B. K., Liebhaber, S. A., et al.,** *Endocrinology*, 125, 566, 1989.
199. **Sinha, Y. N., Klemcke, H. G., Maurer, R. R., et al.,** *Endocrinology*, 127, 410, 1990.
200. **Sinha, Y. N. and Jacobsen, B. P.,** *Biochem. Biophys. Res. Commun.*, 145, 1368, 1987.
201. **Bollenger, F., Velkeniers, B., Hooghe-Peters, E., et al.,** *J. Endocrinol.*, 120, 201, 1989.
202. **Berhman, L. R., Lens, P., Decuypere, E., et al.,** *Gen. Comp. Endocrinol.*, 68, 408, 1987.
203. **Lewis, U. J., Singh, R. N. P., and Lewis, L. J.,** *Endocrinology*, 124, 1558, 1989.
204. **Pellegrini, I., Gunz, G., Ronin, C., et al.,** *Endocrinology*, 122, 2667, 1988.
205. **Markoff, E., Sigel, M. B., Lacour, N., et al.,** *Endocrinology*, 123, 1303, 1988.
206. **Frigeri, L. G., Lewis, L. J., Teguh, K., et al.,** *Biochem. Biophys. Res. Commun.*, 134, 764, 1987.
207. **Sinha, Y. N., Klemcke, H. G., Maurer, R. R., et al.,** *Proc. Soc. Exp. Biol. Med.*, 194, 293, 1990.
208. **Liberti, J. P. and Joshi, G. S.,** *Biochem. Biophys. Res. Commun.*, 137, 806, 1986.
209. **Liberti, J. P., Antoni, B. A., and Chlebowski, J. F.,** *Biochem. Biophys. Res. Commun.*, 128, 713, 1985.
210. **Aramburo, C., Montiel, J. L., Proudman, J. A., et al.,** *J. Mol. Endocrinol.*, 8, 183, 1992.
211. **Baldwin, G. S., Grego, B., Hearn, M. T. W., et al.,** *Proc. Natl. Acad. Sci. U.S.A.*, 80, 5276, 1983.
212. **Aramburo, C., Donoghue, D., Montiel, J. L., et al.,** *Life Sci.*, 47, 947, 1990.
213. **Skibeli, V., Andersen, U., and Gaitvik, K. M.,** *Gen. Comp. Endocrinol.*, 80, 333, 1990.
214. **MacLeod, J. N., Worsley, I., Ray, J., et al.,** *Endocrinology*, 128, 1298, 1991.
215. **Ader, M., Agajanian, T., Finegood, D. T., et al.,** *Endocrinology*, 120, 725, 1987.
216. **Siegel, M. B., Thorpe, N. A., Kobrin, M. S., et al.,** *Endocrinology*, 108, 1600, 1981.
217. **Closset, J., Smal, J., Gomez, F., et al.,** *Biochem. J.*, 214, 885, 1983.
218. **McCarter, J., Shaw, M. A., Winer, L., et al.,** *Mol. Cell. Endocrinol.*, 73, 11, 1990.
219. **Baumann, G. and Shaw, M. A.,** *J. Clin. Endocrinol. Metab.*, 70, 215, 1990.
220. **Brostedt, P., Luthman, M., Wide, L., et al.,** *Acta Endocrinol.*, 122, 241, 1990.
221. **Ingram, R. T., Afshari, N., and Nicoll, C. S.,** *Endocrinology*, 130, 3085, 1992.
222. **Frigeri, L. G., Teguh, C., Ling, N., et al.,** *Endocrinology*, 122, 2940, 1988.
223. **Salem, M. A. M.,** *Endocrinology*, 123, 1565, 1988.
224. **Baumann, G., Stolar, M. W., and Buchanan, T. A.,** *Endocrinology*, 117, 1309, 1985.
225. **Baumann, G., Stolar, M. W., and Buchanan, T. A.,** *Endocrinology*, 119, 1497, 1986.
226. **Sinha, Y. N., DePaulo, L. V., Lewis, U. J., et al.,** *Proc. 73rd Annu. Mtg. Endocrine Soc.*, 214, 1991.
227. **Berghman, L. R., Buyse, J., Huybrechts, L. M., et al.,** 1994, in preparation.
228. **Baumann, G.,** *J. Clin. Endocrinol. Metab.*, 49, 495, 1979.
229. **Celniker, A. C., Chen, A. B., Wert, R. M., et al.,** *J. Clin. Endocrinol. Metab.*, 68, 469, 1989.
230. **Luthman, M., Jonsdottir, I., Skoog, B., et al.,** *Acta Endocrinol.*, 123, 317, 1990.

Chapter 2

Growth Hormone: Evolution

Colin G. Scanes and Robert M. Campbell

CONTENTS

I. EVOLUTIONARY ASPECTS OF THE GROWTH HORMONE AND PROLACTIN GENE FAMILY

At least in mammals, genes encoding growth hormone (GH) and prolactin (PRL) are not located on the same chromosome. The prolactin gene is located, respectively, on chromosome 6 in humans,[1] 17 in the rat,[2] and 13 in the mouse[3] whereas the *GH* gene is on chromosome 17 in humans,[4–7] 10 in the rat, [2] and 11 in the mouse.[3] It is not known when the ancestral genes for GH and prolactin became dispersed onto two chromosomes.

Genes encoding prolactin-related proteins, as indicated by the degree of homology of amino acid residue sequence, appear to be confined to the same chromosome as prolactin. For instance, the genes encoding placental lactogen (PL), proliferin and proliferin-related proteins, and prolactin have been mapped to chromosome 13 in the mouse.[3] In an analogous manner, genes encoding GH-like proteins appear to be located on the same chromosome as GH. In humans, genes encoding (GH) and human chorionic somatomammotropin (hCS) show close homologies and are found in a cluster of five genes (for GH and GH-related products) on the long arm of chromosome 17[6,7] (see Figure 1).

II. EVOLUTION OF THE GROWTH HORMONE AND PROLACTIN FAMILY OF HORMONES

A. INTRODUCTION

On the basis of homologies between sequences of GH and prolactin, it is hypothesized that both protein hormones were derived from a single ancestral proto-GH/prolactin that underwent gene duplication. This is illustrated in Figure 2.[8] Moreover, following the

0-8493-8697-7/95/$0.00+$.50

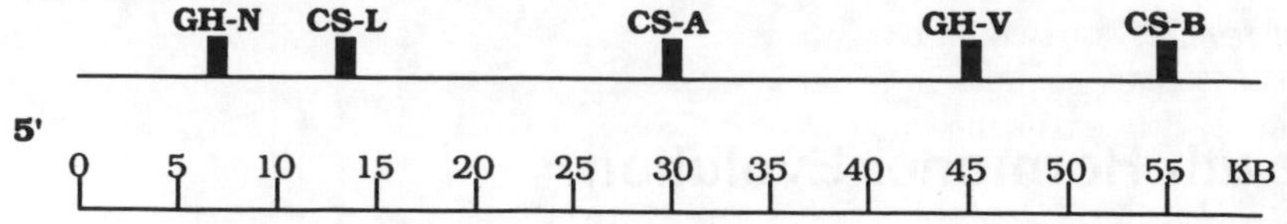

Figure 1 The human *GH/CS* gene cluster (redrawn from Ref. 8). *GH-N*, normal pituitary GH; *GH-V*, placental GH variant; *CS-A*, *CS-B*, and *CS-C*, three genes for chorionic somatomammotropin (*CS-L* appears to be nonfunctional; *CS-A* and *CS-B* are functional). *CS-A* and *CS-B* are expressed, respectively, to a major and minor extent by the placenta.

evolution of the major groups of placental mammals, there is evidence of multiple gene duplications for both the prolactin gene (giving rise to the rodent and ungulate PLs, proliferins, and other prolactin-like proteins) and the *GH* gene (giving rise, e.g., to primate CS)[9,10] (Figure 2).

Before addressing the questions concerning when the ancestral proto-GH/prolactin originated and when the proto-GH/prolactin gene underwent gene duplication, it is necessary to consider a series of questions related to whether GH and prolactin (or GH- and prolactin-like proteins) are found throughout the vertebrates and even in invertebrates.

1. Are GH and Prolactin Found throughout the Vertebrates?

Both GH and prolactin have been isolated from examples of all vertebrate groups except members of the Division Agnatha. The primary structures of mammalian, avian, reptilian, amphibian, fish, and elasmobranch GHs are shown in Chapter 1 (Figure 1), and structures for prolactin are presented in Chapter 1 (Figure 2). Thus, there is strong evidence that the GH/prolactin gene duplication (Figure 2) occurred prior to the divergence of the chondrichthyes and the osteichthyes (Chapter 1, Figure 1;[11] also see review[12]) approximately 430 million years ago. In addition, there would appear to be separate prolactins and GH among members of the Division Agnatha, which has been separate from the rest of the vertebrates (Division Gnathostomata) for approximately 500 million

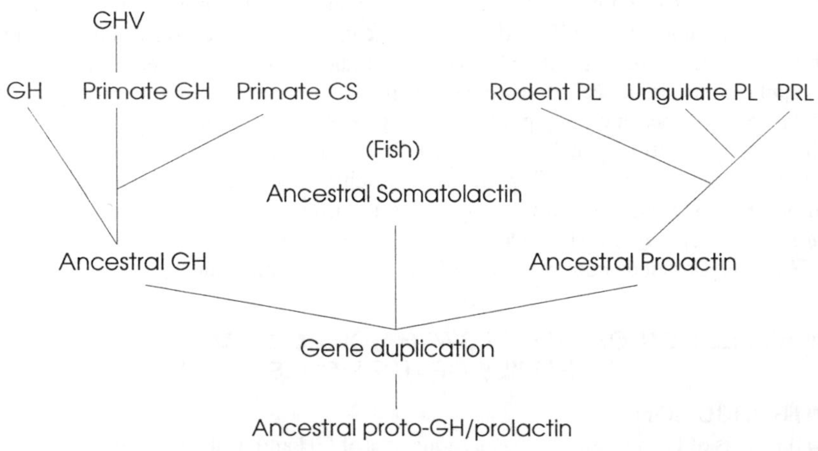

Figure 2 Hypothetical phylogenetic tree for the GH-prolactin family of hormones. Abbreviations: GH, growth hormone; GHV, placental growth hormone variant; CS, chorionic somatomammotropin; PL, placental lactogen; PRL, prolactin.

years.[11,12] This is based on immunocytochemical studies of the lamprey pituitary gland using antisera to mammalian GH and prolactin. Cross-reactivity to these antisera against GH and prolactin was indeed observed but in distinctly different pituitary cells.[13]

2. Is GH and/or Prolactin Found in Invertebrates?

To date, the evidence for invertebrate GH and/or prolactin rests largely on immunochemical studies with antisera against vertebrate, largely mammalian, GH or prolactin. Cross-reactivity to GH or prolactin has been observed but whether this is specifically related to invertebrate analogs of GH or prolactin is open to question. A polypeptide(s) cross-reacting with human, bovine, and chicken GH monoclonal antibodies has been demonstrated in insect (*Locusta migratoria*) brain.[14,15] Moreover, with respect to invertebrates, prolactin-like immunoreactivity has been detected in at least five insect orders: (1) Blattodea (*Leucophaea maderae*),[16] (2) Orthoptera (*L. migratoria*),[14,16] (3) Diptera (*Sarcophaga bullata*),[16] (4) Coleoptera (*Leptinotarsa decemlineata*),[17] and (5) Hymenoptera (*Apis mellifica*).[18,19] An immunoreactive, prolactin-like substance has been reported even in different microorganisms.[20]

In some insects studied, GH- and prolactin-like immunoreactivities are observed in different cells.[15] This is suggestive of separate GH and prolactins in the arthropods. Hence, this might indicate that the gene duplication of the ancestral proto-GH/prolactin gene occurred prior to the separation of the protostomian phyla (annelids, arthropods, and mollusks) and deuterostome stock (which gave rise to the chordates and echinoderms) in the pre-Cambrian era (i.e., prior to 590 million years ago). An abstract reported the identification of a GH homolog in a mollusk, the abalone,[21] with considerable similarity to vertebrate GH and presumably distinct from prolactin. This provides further evidence for the possibility that the common ancestor to mollusks and vertebrates processed a GH-like protein. It is anticipated that when the sequences of putative invertebrate GH and prolactin homologs are established, then it might be possible to reach firm conclusions on when the gene duplication divergence of GH and prolactin occurred.

B. EVOLUTION OF GROWTH HORMONE AND PROLACTIN

1. Origin of Growth Hormone

The origin of GH is impossible to determine definitively, as hormones obviously do not leave fossil records. Some individuals have speculated on the origin of GH. It has been suggested that the ancestral proto-GH prolactin molecule was, like insulin, derived from an ancestral serine protease.[22] Critical analysis of primary sequences of GH preparations from all the major vertebrates demonstrates many amino acid residues and entire sequences that have been retained throughout vertebrate evolution (see Figure 1). However, these regions of GH show little similarity to proposed ancestral insulin[12] or to the proposed ancestral serine protease.[22]

It has also been suggested that the proto-GH/prolactin molecule was derived from smaller peptides[23] that were enlarged, perhaps by exon duplication. The evidence for this was based on the observation that regions of supposed internal homology were identified in both GH and prolactin.[23] This contention has been questioned by Nicoll and colleagues, as the same but low degrees of internal homology can be found at a number of alternative sites.[10] Moreover, the supposed regions of internal homology do not necessarily correspond to the conserved sequences or to the sequences in the putative ancestral forms of GH or prolactin (Chapter 1, Figures 1 and 2).

In view of the evidence for GH- and prolactin-like proteins in arthropods and mollusks (see above), it is reasonable to suggest that the proto-GH/prolactin existed at least at the branching point of the ancestors of the arthropods and mollusks (the protostomian branch) and that of the vertebrates (the deuterostome branch) in the pre-cambrian era. By the Cambrian period (590 to 500 million years ago), species that can be clearly identified with presently existing phyla are present in the fossil record. The origin of the metazoa is thought to have occurred approximately 1 billion years ago. Thus, at some time between 1 billion and 590 million years ago, the common ancestors of mollusks and vertebrates existed. An intermediate date might be suggested, although an earlier time is perhaps more probable. Thus, a conservative estimate of the evolutionary age of ancestral proto-GH prolactin is approximately 700 to 800 million years. It is also possible to envisage the presence of the ancestral proto-GH/prolactin even preceding the evolution of the multicellular organisms. Certainly, there is evidence for some peptide hormones in present-day protozoa.[20]

2. Molecular Evolution

Several mechanisms are envisaged for the evolution of the structure of proteins, including hormones. The most discussed are gene duplication and/or point mutation, but, in addition, exon deletion, substitution, and rearrangement are also possible. There are two schools of thought on the incorporation of changes due to point mutation into a population and thence its descendants. One view is that the incorporation of nondeleterious changes is a neutral phenomenon (the "neutralist" position). Alternatively, mutation can be viewed as either deleterious or advantageous; the former are selected against whereas the latter are selected for (the "selectionist" position). There is evidence in support of both schools of thought on the evolution of GH and prolactin.

Growth hormone appears to have a relatively consistent rate of evolutionary change. For instance, Kawauchi and colleagues estimated the rate of evolutionary change to be 9.6 UEP (unit evolutionary period) in most tetrapods (excluding primates) and a slightly different rate (6.3 UEP) in fish.[24,25] The UEP is defined as "the time in millions of years for a 1% of sequence divergence between two related proteins."[24] The rate of evolutionary change of both GH and PRL has been calculated on the basis of the hypothetical structures of ancestral mammalian and amniote GH and prolactins (see Chapter 1, Figures 1 and 2). These changes, as shown in Table 1, generally agree with those of Yasuda and colleagues.[24] Virtually all amniote GHs (except primate GH) show a low rate of evolutionary change (Table 1). This conservatism is consistent with strong selection pressure against the survival of animals with mutations in the *GH* gene and also suggests that most such changes to GH are deleterious. The lack of conservative substitutions implies that the structure of GH is critical to its role, but not necessarily for its receptor binding and signal transduction.

It might be noted that the rate of evolutionary change for prolactin appears slightly higher than that of GH (in mammals for GH, 1.06 ± 1.38 [S.D.] Paulings; for prolactin, 2.29 ± 1.59 Paulings; in amniotes for GH, 0.85 ± 1.20 Paulings; for prolactin, 1.97 ± 1.63 Paulings) (see Table 1). However, many of the differences in prolactins are due to apparently conservative substitutions.

The primary structures of known vertebrate GHs exhibit considerable homology within the four major α-helices (see Chapter 1, Figures 1 and 4), suggesting that the general three-dimensional conformation has also been highly conserved throughout vertebrate evolution. Abdel-Meguid and co-workers[26] have suggested that the invariant residues comprising the helices are required for maintenance of structural integrity and are not involved in species specificity. It may be expected that prolactin, and PL also possess conformations similar, but not identical, to GH. For instance, Chou-Fasman calculations predict that residues 160 to 183 (corresponding to porcine GH helix 4) of chicken prolactin[27,28] are not α-helical.

The cysteine disulfide bonds are largely conserved throughout the GH/prolactin family. All known vertebrate GHs contain two disulfide bonds between Cys-57 and Cys-165 and between Cys-182 and Cys-189 (see Chapter 1, Figure 1). All tetrapod prolactins sequenced contain three disulfide bonds in comparable locations, whereas teleost prolactins

Table 1 **Relationship between Higher Vertebrate GH, PRL, and Hypothetical Ancestral Forms with Calculated Rate of Evolutionary Change**

Hormone[a]	Comparator: Hypothetical Ancestral Form	% Difference[b]	Divergence Time (years)[c]	Evolutionary Change UEP	Rate (Paulings)
Human	Ancestral mammalian GH	30.0	70 million	2.3	4.30
Rat GH	Ancestral mammalian GH	5.2	70 million	13.5	0.74
Whale GH	Ancestral mammalian GH	9.8	70 million	7.1	1.40
Fox GH	Ancestral mammalian GH	2.6	70 million	26.9	0.37
Cattle GH	Ancestral mammalian GH	6.7	70 million	10.4	0.96
Pig GH	Ancestral mammalian GH	2.6	70 million	26.9	0.37
Horse GH	Ancestral mammalian GH	1.3	70 million	53.8	0.19
Elephant GH	Ancestral mammalian GH	1.0	70 million	70.0	0.14
Duck GH	Ancestral amniote GH	9.6	310 million	32.3	0.30
Chicken GH	Ancestral amniote GH	7.7	310 million	40.3	0.25
Sea turtle GH	Ancestral amniote GH	11.1	310 million	27.9	0.36
Human PRL	Ancestral mammalian PRL	14.6	70 million	4.8	2.10
Rat PRL	Ancestral mammalian PRL	30.4	70 million	2.3	4.30
Whale PRL	Ancestral mammalian PRL	5.3	70 million	13.2	0.76
Elephant PRL	Ancestral mammalian PRL	24.4	70 million years	2.9	3.50
Chicken PRL	Ancestral amniote PRL	10.8	310 million years	28.7	0.35

Note: UEP = 1% change in 1 million years; 1 Pauling = 1×10^{-9} changes per site per year.

[a] Only one example for order is usually used.

[b] Based on Chapter 1, Figures 1 and 2. Where no consensus can be reached on the ancestral form, a difference is assumed. This assumption will produce a somewhat higher rate of evolutionary change.

[c] Based on Benton.[11]

(like GH) contain only two, lacking the N-terminal cysteine residues (Cys-4 and Cys-11 deleted; see Chapter 1, Figure 2). Teleost somatolactins also contain three disulfide bonds in positions analogous (i.e., accounting for theoretical deletions/substitutions) to those in tetrapod prolactin.

The disulfide cross-linking would be expected to be essential for maintenance of secondary structure and biological activity, having been retained through multiple evolutionary divergences. However, evidence from various laboratories suggests that the N-terminal and C-terminal disulfides are not required: (1) human GH, human CS, and teleost prolactin, which do not have the N-terminal disulfide, display significant activity in prolactin bioassays; (2) reduced ovine prolactin retains mammotropic activity; (3) C-terminal disulfide bond reduction does not eliminate bovine GH or ovine prolactin activities; and (4) reduced/carboxymethylated human GH is active in diabetogenic[29] and chicken lipolytic assays.[30] The situation is not straightforward, as reduced/carboxymethylated human GH, reduced/aminoethylated porcine GH, and completely reduced (all three disulfides) ovine PRL are biologically inactive (reviewed in Ref. 31).

Due to the size and complexity of GH, PRL, and PL molecules, it is difficult to isolate the residue(s) that endows each hormone with its unique set of biological properties. The use of naturally occurring hormones from different species yields a limited amount of information, as the mutated residues are not subject to the control of the investigator. Multiple residue changes in different regions of the molecule confound the interpretation of structure-activity studies. Only by difference analysis can certain inferences be made. For instance, to ascertain the structural determinants of GH necessary for lipolytic effects, a sequence database consisting of naturally occurring GH homologs (hCS and human, bovine, chicken, sea turtle, bullfrog, shark, salmon, bonito, yellow tail, tuna, and sturgeon GH) was constructed and cross-referenced with *in vitro* activity (also see Chapter 1, Table 3), that is, correlating conserved amino acid residues with activity and eliminating nonconserved residues that did not affect activity.[30,32] In this case it was proposed that Pro-132, found in all lipolytic GH species but nonconservatively substituted in nonlipolytic species, was a major determinant of lipolytic activity. Proline possesses a cyclic side-chain and low conformational flexibility often associated with α-helix disruption, and so could impart particularly profound effects to the three-dimensional structure of GH.

Techniques (e.g., site-specific mutagenesis; transgenic animals) have become available that provide some insight into the discrete differences that delineate GH-, prolactin-, CS-, and PL-like activities. Recombinantly generated bovine prolactin variants with amino acid substitutions or deletions have been examined in the Nb2 cell lactogen assay. In this assay, independent nonconservative substitutions at residue 20 (Asp → Ala), 27 (His → Gln), 28 (Tyr → Arg, Ala), 65 (Thr → Ala), 93 (Asp → Gly) or 151 (Ser → Asp), or deletion of Phe-50 or Ser-114 did not dramatically affect lactogenic activity, whereas deletion of Tyr-28, Trp-91, Arg-125, or Ser-179 virtually eliminated the activity.[33,34] These observations suggest that residues 20, 27, 28, 65, 93, and 151 are not critical for lactogenic activity. The authors speculated that the relative inactivity of the deletion mutants was due to alterations in the α-helical conformation (i.e., spatial constraints of helices 2 through 4), not a requirement for single amino acids involved in receptor binding. Site-specific mutagenesis of human GH (hGH) has shown that residues His-18, His-21, and Phe-25 (helix 1), Ile-58, Ser-62, and Asn-63 (loop region), and Arg-167, Lys-168, Lys-172, Glu-174, and Phe-176 (helix 4) are major determinants in prolactin receptor binding.[122] Residues His-18, Glu-174 and His-188 are further postulated to comprise a Zn^{2+}-binding site that modulates the degree of hGH binding to the prolactin receptor.[35] However, zinc is not required for hGH binding to the hGH receptor.

3. Evolutionary Relationships

There are marked homologies between GH and prolactin. In view of the homologies between GH and prolactin, it has been proposed that GH and prolactin are derived from a common ancestral gene that underwent gene duplication.[9] This divergence has been estimated to be approximately 350 million years ago,[9] between 460 and 640 (mean, 550) million years ago,[25] and 800 million years[10]; these being calculated from amino acid residue differences[9] and UEP.[36] Is it possible to reconcile these differences and come up with a reasonable and acceptable estimate for the gene duplication? The mean rate of evolutionary change for GH and prolactin for those mammals considered in Table 1 is 19.0 million years with a standard deviation of 21.0 million years. A similar estimate for evolutionary changes in GH and prolactin from the hypothetical ancestral amniote forms is 22.1 ± 19.2 (SD). Estimates of divergence can be made on the basis of either the ancestral mammalian or amniote GHs and prolactins (Chapter 1, Figures 1 and 2). Assuming a constant rate of evolutionary change based on the 77% difference between hypothetical ancestral mammalian GH and prolactin, the divergence would be $(77 \times 9)/2 + 70$ million years, the latter term representing the time when mammalian stock diverged from a common ancestor.[11] This is equal to 800 million years but with a standard deviation of 807 million years! The estimate based on amniote GHs and prolactins is 1138 ± 720 million years. With the wide standard deviation, the values are clearly not satisfactory.

An alternative approach is to compare the homologies between prolactin and GH in present species and ancestral forms. One would imagine that the more ancient the ancestor, the closer the GH and prolactin should be. However, this is not the case (human prolactin and GH are 20% identical, pig prolactin and GH are 24% identical, ancestral mammalian prolactins and GH are 23% identical, and ancestral amniote prolactin and GH are 25% identical). Some tentative conclusions can be reached. First, the existence of separate GHs and prolactins in bony fish and tetrapods would indicate that the gene duplication resulting in GH and prolactin occurred before the branching points leading to these groups (405 million years ago)[11] or that gene duplication may have occurred more than once. Second, the presumption that the rate of evolutionary change is consistent over geological time is unreasonable, as there are significant differences even within the mammals (Table 1). Moreover, it can readily be envisaged (using a selectionist argument) that selection pressure led to a faster rate change in GH than in prolactin early in its evolution. Alternatively, the view that neutral changes in sequences would be more acceptable early in the evolution of the two hormones could be consistent with a neutralist position. Thus, the divergence of GH from prolactin in all probability occurred much more than 405 million years ago. Little more can be definitively stated unless sequences of GH and prolactin or an intermediate product can be characterized in representative phyla that diverged from the chordate stock during the pre-Cambrian period.

4. Evolution of Fish Growth Hormone

Examination of the amino acid sequences for fish GHs (Chapter 1, Figure 1) leads to a number of conclusions. First, it is apparent that eel, sturgeon, and shark GH show greater homology with tetrapod GH than with bony fish. This is consistent with a low rate of evolutionary change for GH through the cartilaginous fish, the tetrapods, and the ancestral stock for the bony fish (before the divergence of eel, sturgeon, and shark). Second, within the higher bony fish (Subclass Actinopterygii), the structural relationships of GH are consistent within the phylogenetic tree[37] and classification with the single exception of eel GH. For instance, there is closer homology within the salmonids than with GH from different actinopterygiian orders. Third, the marked difference between the structure of eel GH and that of other bony fish leads to the conclusion that eels diverged earlier from the major bony fish stock.[25] As the sequences of more vertebrate GHs are reported, it will be increasingly possible to make conclusions on the changes to the hormone during evolution.

III. EVOLUTION OF PRIMATE GROWTH HORMONE

Primate GHs show marked differences compared to GH in other mammals. These include the existence of several genes for GH and GH-like proteins, the rapidity of the evolutionary change in the amino acid sequence of GH, and the unusual biological activites of at least hGH (with both GH and prolactin biological activity) and hCS (with no GH but still retaining prolactin biological activity). In humans, there are five *GH/GH-like* genes (see Figure 1). A similar situation exists in the rhesus monkey with a *GH*, a *GH* variant, and three *CS* genes.[38] This appears to be unlike the situation in other mammals, although there is evidence for multiple *GH* genes in rats.[39] It would be tempting to conclude, on the basis of Occam's razor, that a five-gene *GH* family represents an ancestral situation in primates. However, it is argued that this is not the case.

The evolution of the five *GH* and *GH*-like genes (Figure 1) in humans might be viewed essentially as a three-step process[8,23,40,41] of gene duplication overlaid by a random process of (effectively) neutral point mutations. The initial gene duplication would result in two genes: one encoding GH and the other, ultimately, for hCS.[23,40,41] The duplication is thought to be the result of combining of two *GH* genes and 5′ flanking regions through unequal crossing over during meiosis.[8] With the sequencing of the five rhesus monkey GHs (GH and GH variants) and of CS-1, -2, and -3,[38] it is possible to compare sequences among other primate GHs and CSs. If the gene duplication giving rise to human *CS* and *GH* occurred prior to the common ancestor of humans and rhesus monkeys, then it would be predicted that at least some of the amino acid residues that differ between hGH and hCS should also differ between rhesus (rh) GH and rhCS. Furthermore, where the differences occur the specific amino acid residue should be the same in GH as in monkey GH or hCS and monkey CS (all forms). This previous requirement is met for residues 12, 16, 100, and 179 of hCS (see Figure 1 for numbering system). Moreover, hCS differs from hGH at residues 52 and 109 but is identical to both monkey CS and GH, suggesting that hGH underwent changes at these positions after the divergence of the monkey and human ancestral stocks. Thus, the gene duplication giving rise to GH and CS probably occurred prior to the human: old world monkey divergence time of 25 to 27 million years ago.[42]

The second stage in the evolution of the human GH and CS series involves the duplication of both the ancestral *GH* and *CS* genes to give four genes.[8] In the third stage, duplication of one of the *CS* genes occurred.[8] The presence of similar restriction fragments in humans, chimpanzees, and gorillas but not in orangutan led to the speculation as to when the third gene duplication occurred.[8] It is suggested that it occurred subsequent to the evolutionary divergence of the orangutans from the ancestors of humans, chimpanzees and gorillas (~9 million years ago) but prior to the trichotomy of the ancestors of humans, chimpanzees, and gorillas (~5 million years ago).[8] Comparisons of the differences between the sequences of hGH with those in rhesus monkey GH and monkey GH variant form show no obvious similarities. It is suggested, therefore, that there have been independent gene duplications leading to hGHV and rhGHV. Moreover, in view of the great similarities between monkey CS-1, -2, and -3,[38] gene duplications giving rise to these probably occurred relatively recently in evolutionary history.

The situation with primate GH is an exception to the picture of consistently slow rates of change for amniote GH. There is close homology between monkey and human GH, but relatively high divergence between these primate GHs and other mammalian GHs or the hypothetical ancestral mammalian GH (Chapter 1, Figure 1; 30% difference). The similarity between monkey and human GH indicates that the primate/"other mammal" structural divergence occurred prior to the divergence of monkeys and humans. The differences between GH in primate and other mammals is discussed below. The mean rate of evolutionary changes for mammalian GHs, excluding primates, was 0.50 ± 0.45 (SD) Paulings (1 Pauling = 10^{-9} replacements per site per year) (Table 1). However, the rate for human GH was some 7.2-fold higher (Table 1). The reason for this is not clear but speculations on this are considered below: it is suggested that the propensity for primate GH to undergo gene duplication would reduce the selective constraints against retaining mutations. Moreover, the selective advantage in primate GH acquiring prolactin activity may provide a rationale for rapid change. Experimentally, it is possible with relatively few point mutations to give prolactin GH-like activity.[43] The converse case (GH → prolactin) is obviously possible and presumably occurred during primate evolution. It is easy to envisage a situation in vertebrate evolution whereby a random mutation in the *GH* gene engendered prolactin activity in the translated product. Normally, this would be disadvantageous and hence selected against, an occurrence consistent with both the selectionist and neutralist positions. However, the situation in primate evolution differed for some reason. It may be that the specific mutations that caused GH to acquire lactogenic activity were advantageous and subject to positive selection pressure. A possible scenario is for primate GH gaining lactogenic activity prior to the evolution of a separate placental lactogen. This GH might function to promote mammary development and the initiation of lactation. A placental site for expression might also be possible as is the case for prolactin gene-derived placental lactogens. With gene duplication, the precursor for CS would have arisen with both GH and prolactin activities. Random mutations to this proto-*CS* gene would occur. Some mutations would reduce GH activity (these for GH itself would be disadvantageous and hence selected against). For CS, the loss of GH activity could be viewed as advantageous; removing the possible adverse effects of excess GH such as acromegaly during pregnancy. It is not known, as yet, whether rhCS-1, -2 and -3 possess both GH and/or prolactin biological activity. When this is established it may be possible to delineate as to the evolutionary point at which primate CS lost its GH-like biological activity.

The lack of GH activity in hCS is thought to be due to specific amino acid residues that impede binding to the GH receptor.[36] In addition, there may have been selection pressure on the primate GH receptor for high-affinity binding of GH. This may have led to changes in the specificity of binding with other mammalian GHs, so that other mammalian GHs were no longer capable of binding to the primate GH receptor.[44,45] There would also be continuous selection pressure on both the GH and PRL receptor gene throughout vertebrate evolution, such that the translation products retained their specifity for GH and PRL, respectively.

IV. EVOLUTIONARY AND COMPARATIVE ASPECTS OF GROWTH HORMONE MULTIPLE BIOLOGICAL ACTIVITY

A. INTRODUCTION

The most widely used biological assay for GH is the rat tibia assay, in which the response is a thickening or widening of the epiphyseal cartilage plate in hypophysectomized rats. There is information on the activities of GH from different species from different vertebrate groups (see Chapter 1, Table 3). Reptilian and amphibian GHs are reported to be active in the rat tibia test.[46–48] Similarly, distinct GH activities have been observed with GH from an elasmobranch shark[49] and a primitive bony fish, the sturgeon.[50] Although teleost GH is generally thought to be inactive,[10] in some cases distinct GH bioactivity is observable, albeit at high doses (e.g., with tilapia GH[51] or salmon GH[52]). There is a single report of the biological activities of lower vertebrate GHs in rodent bioassays with end points reflecting the metabolic actions of GH.[53] Lower vertebrate GH preparations appear to have reduced biological activities (Chapter 1, Table 3). Potencies were not reported, but considerably higher doses of lower vertebrate GHs were required to elicit a response compared to mammalian GH. For instance, sea turtle, bullfrog, and sturgeon GH were active in both systems, but doses at least 10-fold that of mammalian GH were required to evoke a comparable response.[53] Published reports are consistent with poikilotherm GHs having reduced activity in rodent bioassays. The low biological activity of bony fish GH is consistent with their poor ability to compete with mammalian GH for mammalian GH receptors. For example, sturgeon GH was reported to be <0.4% as active as bovine or human GH in inhibiting ^{125}I-labeled bovine GH binding to rabbit liver membranes.[54] The relatively low activities of poikilotherm GHs in mammalian GH assays is explicable in terms of the degree of homologies between the sequences of GH in lower vertebrates and in mammals (30 to 50%) (Chapter 1, Figure 1).

Among the mammals, there are several peculiarities concerning GH with respect to the species specificity of the human GH receptor/growth response and the lactogen effect of human GH. It is widely held that non-primate GH is not active in promoting growth in humans, on the basis of clinical growth trials.[45] Moreover, bovine GH has poor, if any, ability to inhibit binding of ^{125}I-labeled human GH to human liver membranes.[44] It is also interesting to note that human GH, together presumably with other primate GHs, is unique in its lactogen activity. Human GH is active in both the prolactin bioassays and prolactin radioreceptor assays (reviewed in Ref. 10).

The ability of GH, derived from a wide spectrum of vertebrates, to influence avian adipose tissue has been reported using two end points: (1) stimulation of basal lipolysis as indicated by glycerol release from chicken adipose explants *in vitro* (a lipolytic response) and (2) inhibition of glucagon-stimulated lipolysis *in vitro* (an antilipolytic response). These data are summarized in Chapter 1, Table 3. Mammalian and avian GH evoked similar lipolytic and antilipolytic responses. As was the case in the rat tibia assay, hCS did not evoke either lipolytic or antilipolytic responses.[30] The situation with lower vertebrate GHs is perplexing. Glucagon-stimulated lipolysis was inhibited by all lower vertebrate GH preparations examined (possessing 37 to 88% amino acid sequence homology with chicken GH), but little or no lipolytic activity was observed (see Chapter 1, Table 3).[32] The absence of lipolytic activity by reptilian GH was unexpected in view of the similarity (88% sequence identity) between chicken and sea turtle GH (Chapter 1, Figure 1). In contrast, rat bioassay studies have provided some evidence that there are different structural requirements for the different actions of GH and, by implication, different receptor/effector mechanisms.

The specificity of the avian liver GH receptor has been examined by a number of workers. As might be expected avian and mammalian GH preparations show similar abilities to inhibit binding of ^{125}I-labeled GH to chicken liver membranes.[55,56] However, somewhat surprisingly, ovine prolactin shows strong activity, inhibiting binding albeit in a nonparallel manner.

Growth of lower vertebrates is stimulated by mammalian GH. Mammalian GH stimulates growth in hypophysectomized birds,[57] reptiles,[58–60] amphibians,[61–66] and fish.[67–74] The ability of mammalian GH to stimulate growth in fish does not completely comply with observations concerning the binding of mammalian GH to fish liver membranes. *Tilapia* and bovine GH have similar potencies in stimulating growth in both *Tilapia* and salmon.[75] However, bovine GH is much less effective than *Tilapia* GH in inhibiting binding of ^{125}I-labeled *Tilapia* GH to liver membranes.[75] Moreover, no specific binding of ^{125}I-labeled bovine GH was found with liver membranes from various species of fish.[54]

B. ANCESTRAL ROLE FOR GH

A prime question is the following: What is the most ancient of function of GH? The naming of the hormone—growth hormone or somatotrop(h)in—does not *a priori* define either its major function or its ancestral function. We should not, however, capriciously exclude from consideration the possibility that the ancestral role for GH is growth. In all groups of vertebrates examined, GH can stimulate growth. This might represent a "self-fulfilling prophesy" as investigators tend to look for (and find) growth effects of GH that they expect. Other actions that are viewed as unlikely will be less likely to be investigated. Not only is GH involved in growth regulation in vertebrates, but there is evidence for an analogous role in invertebrates. A homolog of vertebrate GH has been reported in a mollusk (abalone).[21] In addition, growth-promoting effects of vertebrate GH have been reported in mollusks and arthropods. Trout GH stimulates growth in oysters (T. T. Chen and K. T. Paynter, personal communication). The presence of immunoreactivity resembling hGH in arthropods led Charmantier and colleagues[77] to examine and subsequently to demonstrate that hGH increases growth of lobsters. This growth-promoting effect of GH in existing invertebrates and vertebrates suggests that a GH stimulated growth in the common ancestor of the deuterostomes and protostomes. This argues for an extremely ancient role (>500 million years) for GH. It is not clear whether this role precedes or follows the gene duplication and divergence between GH and PRL.

C. SIMILARITIES AND DIFFERENCES: EVOLUTIONARY CONSIDERATIONS

If we compare the list of generalized functions of GH with that of prolactin, a large element of overlap exists. Both GH and prolactin influence growth and development, lipid metabolism, osmoregulation, reproduction, and adrenal steroidogenesis. In most cases, while GH and prolactin influence these functions, their roles and exact effects differ. Moreover, caution should be taken in particular with effects of human GH, which has substantial PRL activity, and with ovine PRL, which exerts some GH-like effects (including conversion of $T_4 \rightarrow T_3$ in fish[77,78] and binding to the chicken GH receptor).[55,56] In general, the use of heterologous hormone preparations may lead to problems in interpretation of data. Finally, it should be stressed that the name *prolactin* with its connotations pertaining to mammary function might lead to confusion with respect to the role of the hormone in most vertebrates, in which lactation is not present.

The essential concept of this section is that both GH and PRL have a range of activities, sometimes overlapping, influencing different aspects of animal physiology depending on the developmental and/or physiological state. Frequently, the effects of the hormones are analogous, but far from identical. Examples of this can be drawn from the five areas listed above. If we first consider growth, both GH and prolactin influence amphibian development. Prior to metamorphosis, both GH and prolactin can stimulate growth.[61,62] Prolactin is arguably more effective in stimulating body weight, tail length and tail height, but limb growth appears to be stimulated only by GH.[67,69] GH has been demonstrated to exert a strong growth-promoting effect following metamorphosis in frogs,[63] whereas prolactin has much less effect.[62] In higher vertebrates, GH is the major pituitary hormone stimulating growth (body weight, length or height), but prolactin has been reported to have some effect in a few isolated cases (e.g., lizards[58]) and dwarf mice.[80,81]

Both GH and prolactin influence reproduction, lipid metabolism, and adrenocortical functioning. In some cases, the effects are similar, as with the effects of salmon GH and prolactin on androgen synthesis by *Fundulus* testes[82] and in reducing liver lipid concentrations in Coho salmon parr.[83] The effects of GH and prolactin may differ between species, with milk production (as distinct from initiation of lactogenesis) being predominantly under the control of prolactin in some species (e.g., humans) but under GH control in others (e.g., cattle). In other cases, the effect is similar but the mechanism differs. For instance, corticosterone production is increased by GH,[84,85] due to increased biosynthetic capacity, whereas prolactin will also tend to evoke a similar overall effect but by decreasing the catabolism of glucocorticoids via 5α-reductase.[86,87] The effects of GH and prolactin may be totally different. For example, only prolactin stimulates the crop sac gland in pigeons and doves, the corpus luteum in rats, and the initiation of milk production in many mammals. Conversely, GH has lipolytic, antilipolytic (i.e., inhibition of glucagon- or catecholamine-stimulated lipolysis), antilipogenic, and diabetogenic effects that are not seen with prolactin. The overlap and/or differences in the effects of GH and PRL are not fully established, but may perhaps reflect vestigial functions or lack of receptor specificity.

REFERENCES

1. **Owenbach, D., Rutter, W. J., Cooke, N. E., et al.,** *Science*, 212, 815, 1981.
2. **Cooke, N. E., Szpirer, C., and Levan, G.,** *Endocrinology*, 119, 2451, 1986.
3. **Jackson-Grusby, L. L., Pravtcheva, D., Ruddle, F. H., et al.,** *Endocrinology*, 122, 2462, 1988.
4. **Seeburg, P. H.,** *DNA*, 239, 1982.
5. **Overbach, D., Rutter, W. J., Martial, J. A., et al.,** *Science*, 209, 289, 1980.
6. **George, D. L., Phillips, J., Francke, U., et al.,** *Hum. Genet.*, 57, 138, 1981.
7. **Harper, M. E., Barrera-Saldana, H. A., and Saunders, G. F.,** *Ann. J. Hum. Genet.*, 34, 227, 1982.
8. **Parks, J. S., Kassels, M., McKean, M. C., et al.,** Pythagora Press (Rome-Milan) and Springer-Verlag (Berlin-Heidelberg), 1989.
9. **Miller, W. L. and Eberhardt, N. L.,** *Endocr. Rev.*, 4, 97, 1983.
10. **Nicoll, C. S., Mayere, G. L., and Russell, S. M.,** *Endocr. Rev.*, 7, 169, 1986.
11. **Benton, M. J.,** *J. Mol. Evol.*, 30, 409, 1990.
12. **Scanes, C. G. and Campbell, R. M.,** in:*The Endocrinology of Growth, Development and Metabolism in Vertebrates* (Eds., M. P. Schreibman, C. G. Scanes, and P. K. T. Pang), Academic Press, San Diego, 1993, p.559.
13. **Wright, G. M.,** *Gen. Comp. Endocrinol.*, 55, 269, 1984.
14. **Vanden Broeck, J., Cardoen, J., Neyts, J., et al.,** *Comp. Biochem. Physiol.*, 97A, 35, 1990.
15. **Swinnen, K., Verhaert, P., and DeLoof, A.,** *Gen. Comp. Endocrinol.*, 66, 47, 1985.
16. **Hansen, G. N., Hansen, B. L., and Scharrer, B.,** *Cell Tiss. Res.*, 252, 557, 1988.
17. **Veenstra, J. A., Romberg-Privee, H. M., Schooneveld, H., et al.,** *Histochemistry*, 82, 9, 1985.
18. **Schmid, K. P., Maier, V., Obert, B., et al.,** *Histochemistry*, 91, 469, 1989.
19. **Schmid, K. P., Maier, V., and Pfeiffer, E. F.,** *Horm. Metab. Res.*, 22, 413, 1990.
20. **Ghione, M. and Dell'Orto, P.,** *Microbiologica*, 6, 315, 1983.
21. **Moriyama, S., Atsuta, S., Kobayashi, M., et al.,** *Endocrinology*, 242, 1990.
22. **Adelson, J. L.,** *Nature (Lond.)*, 229, 321, 1971.
23. **Niall, H.D., Hogan, M.L., Saver, R., et al.,** *Proc. Natl. Acad. Sci. U.S.A.*, 68, 866, 1971.
24. **Yasuda, A., Yamaguchi, K., Papkoff, H., et al.,** *Gen. Comp. Endocrinol.*, 73, 242, 1989.
25. **Kawauchi, H. and Yasuda, A.,** in: *Advances in Growth Hormone and Growth Factor Research* (Eds., E. E. Muller D. Cochia, and V. Locatelli). Pythagora Press, Rome, 1989, pp. 51–68.
26. **Abdel-Meguid, S. S., Shieh, H.-S., Smith, et al.,** *Proc. Natl. Acad. Sci. U.S.A.*, 84, 6434, 1987.
27. **Hanks, M. C., Alonzi, J. A., Sharp, P. J., et al.,** *J. Mol. Endocrinol.*, 2, 21, 1989.
28. **Watahiki, M., Tanaka, M., Masuda, N., et al.,** *J. Biol. Chem.*, 264, 5535, 1989.
29. **Cameron, C. M., Kostyo, J. L., Rillema, R. A., et al.,** *Am. J. Physiol.*, 247, E639, 1984.
30. **Campbell, R. M., Kostyo, J. L., and Scanes, C. G.,** *Proc. Soc. Exp. Biol. Med.*, 193, 269, 1990.
31. **Nicoll, C. G., Tarpey, J. F., Mayer, G. L., et al.,** *Am. Zool.*, 26, 965, 1986.
32. **Campbell, R. M., Kawauchi, H., Lewis, U. J., et al.,** *Proc. Soc. Exp. Biol. Med.*, 197, 409, 1991.
33. **Luck, D. N., Gout, P. W., Beer, C. T., et al.,** *Mol. Endocrinol.*, 3, 822, 1989.
34. **Luck, D. N., Gout, P. W., Kelsay, K., et al.,** *Mol. Endocrinol.*, 4, 1011, 1990.
35. **Cunningham, B. C. and Wells, J. A.,** *Science*, 244, 1081, 1989.
36. **Russell, S. M. and Nicoll, C. S.,** in: *Progress in Comparative Endocrinology* (Eds. A. Epple, C. G. Scanes, and M. H. Stetson). John Wiley & Sons, New York, 1990, pp. 168–173.
37. **Bernardi, G., D'Onofrio, G., Caccio, S., et al.,** *J. Mol. Evol.*, 37, 644, 1993.
38. **Kawauchi, H. and Yasuda, A.,** In: *Advances in Growth Hormone and Growth Factor Research* (Eds. E. E. Muller D. Cochia, and V. Locatelli). Pythagora Press, Rome, 1989, pp. 51–68.
39. **Golos, T. G., Darning, M., Fisher, J.M., et al.,** *Endocrinology*, 133, 1744, 1993.
40. **Ogilvie, S., Buhi, W. C., Olson, J. A., et al.,** *Endocrinology*, 126, 3271, 1990.

41. **Barsh, G. S., Seeburg, P. H., and Gelinas, R. E.,** *Nucleic Acids Res.*, 11, 3939, 1983.
42. **Hirt, H., Kimelman, J., Birnbaum, M. J., et al.,** *DNA*, 6, 59, 1987.
43. **Li, W. H. and Tanimura, M.,** *Nature* (*Lond.*) 326, 93.
44. **Cunningham, B. C., Henner, D. J., and Wells, J. A.,** *Science*, 247, 1461, 1990.
45. **Carr, D. and Friesen, H. G.,** *J. Clin. Endocrinol. Metab.*, 116, 1501, 1976.
46. **Juskevich, J. C. and Guyer, C. G.,** *Science*, 249, 875, 1990.
47. **Farmer, S. W., Papkoff, H., and Hayashida, T.,** *Endocrinology*, 99, 692, 1986.
48. **Farmer, S. W., and Papkoff, H.,** in: *Hormones and Evolution* (Ed. E. J. W. Barrington). Academic Press, New York, 1984, p. 525.
49. **Farmer, S. W., Licht, P., and Papkoff, H.,** *Endocrinology*, 101, 1145, 1977.
50. **Hayashida, T. and Lewis, U. J.,** *Gen. Comp. Endocrinol.*, 36, 530, 1978.
51. **Farmer, S. W., Hayashida, T., Papkoff, H., et al.,** *Endocrinology*, 108, 377, 1981.
52. **Farmer, S. W., Papkoff, H., Hayashida, T., et al.,** *Gen. Comp. Endocrinol.*, 30, 91, 1976.
53. **Wagner, G. F., Fargher, R. C., Brown, J. C., et al.,** *Gen. Comp. Endocrinol.*, 60, 27, 1985.
54. **Cameron, C. M., Kostyo, J. L., and Papkoff, H.,** *Endocrinology*, 116, 1501, 1985.
55. **Tarpey, J. F. and Nicoll, C. S.,** *Gen. Comp. Endocrinol.*, 60, 39, 1985.
56. **Leung, F. C., Taylor, J. E., Steelman, S. L., et al.,** *Gen. Comp. Endocrinol.*, 56, 389, 1984.
57. **Krishman, K. A., Proudman, J. A., and Bahr, J. M.,** *Mol. Cell. Endocrinol.*, 66, 125, 1989.
58. **King, D. B. and Scanes, C. G.,** *Proc. Soc. Exp. Biol. Med.*, 182, 201, 1986.
59. **Licht, P. and Hoyer, H.,** *Gen. Comp. Endocrinol.*, 11, 338, 1986.
60. **Nichols, C. W.,** *Gen. Comp. Endocrinol.*, 21, 219, 1973.
61. **Owens, D. W., Hendrickson, R., and Endres, D. B.,** *Gen. Comp. Endocrinol.*, 38, 53, 1979.
62. **Enemar, A., Essuik, B., and Klang, R.,** *Gen. Comp. Endocrinol.*, 11, 328, 1968.
63. **Zipser, R. D., Licht, P., and Bern, H. A.,** *Gen. Comp. Endocrinol.*, 13, 382, 1969.
64. **Brown, P. S. and Frye, B. E.,** *Gen. Comp. Endocrinol.*, 13, 139, 1969.
65. **Brown, P. S. and Brown, S. C.,** *J. Exp. Zool.*, 178, 29, 1971.
66. **Cohen, D. C., Greenberg, J. A., Licht, P., et al.,** *Gen. Comp. Endocrinol.*, 18, 384, 1972.
67. **Delidow, B. C.,** *J. Exp. Zool.*, 249, 279, 1989.
68. **Pickford, G. E.,** *Endocrinology*, 55, 274, 1954.
69. **Higgs, D. A., Donaldson, E. M., Dye, H. M., et al.,** *Gen. Comp. Endocrinol.*, 27, 240, 1975.
70. **Higgs, D. A., Fagerlund, U. H. M., McBride, J. R., et al.,** *Can. J. Zool.*, 55, 1048, 1977.
71. **Kayes, T.,** *Gen. Comp. Endocrinol.*, 33, 382, 1977.
72. **Inui, Y., Miwa, S., and Ishioka, H.,** *Gen. Comp. Endocrinol.*, 59, 287, 1985.
73. **Schute, P. M., Down, N. E., and Souza, L. M.,** *Aquaculture*, 76, 145, 1989.
74. **Sakata, S., Noso, T., Moriyama, S., et al.,** *Gen. Comp. Endocrinol.*, 89, 396, 1993.
75. **Clark, W. C., Farmer, S. W., and Hartwell, K. M.,** *Gen. Comp. Endocrinol.*, 33, 174, 1977.
76. **Fryer, J. N.,** *Gen. Comp. Endocrinol.*, 39, 123, 1979.
77. **Charmantier, G., Charmantier-Daures, M., and Aiken, D. E.,** *C.R. Acad. Sci. (Paris)*, 308, 21, 1989.
78. **deLuze, A. and LeLoup, J.,** *Gen. Comp. Endocrinol.*, 56, 308, 1989.
79. **deLuze, A., LeLoup, J., Papkoff, H., et al.,** *Gen. Comp. Endocrinol.*, 73, 186, 1989.
80. **Berman, R., Bern, H. A., Nicoll, C. S., et al.,** *J. Exp. Zool.*, 156, 353, 1964.
81. **Wallis, M. and Dew, J. A.,** *J. Endocrinol.*, 56, 235, 1973.
82. **Smeets, T. and van Buul-Offers, S.,** *Growth*, 47, 160, 1983.
83. **Singh, H., Griffith, R. W., Takahashi, A., et al.,** *Gen. Comp. Endocrinol.*, 72, 144, 1988.
84. **Sheridan, M. A.,** *Gen. Comp. Endocrinol.*, 64, 220, 1986.
85. **Young, C.,** *Gen. Comp. Endocrinol.*, 71, 85, 1988.
86. **Carsia, R. V., Weber, H., King, D. B., et al.,** *Endocrinology*, 117, 928, 1985.
87. **Witorsch, R. J. and Kitay, J. I.,** *Endocrinology*, 91, 764, 1972.
88. **Carsia, R. V., Scanes, C. G., and Malamed, S.,** *Endocrinology*, 115, 2464, 1984.

Chapter 3

Growth Hormone-Secreting Cells

Stephen Harvey

CONTENTS

I. INTRODUCTION

Growth hormone (GH) has been isolated from the pituitary glands of species belonging to all vertebrate taxa. The physiochemical and biochemical characteristics of this molecule have been largely conserved throughout vertebrate evolution and in all groups is essential for normal growth and development. GH and GH-like peptides are also produced by "pharyngeal" pituitary glands and by the pars tuberalis,[1] the brain,[2] placenta,[3] extrapituitary tumors,[1,4] and the immune system (Chapter 24). In this chapter only the pituitary cells secreting GH are reviewed.

II. GH SECRETING CELLS

A. SOMATOTROPH MORPHOLOGY

In most species, acidophil GH-secreting cells (somatotrophs) are the most abundant pituitary cell type.[5] The morphology of GH-secreting cells in human,[6,7] monkey,[8] pig,[9] cow,[10,11] sheep,[12] goat,[13] rat,[14-16] mouse,[17] musk shrew,[18] guinea pig,[19] hedgehog,[20] European ferret,[21] housebats,[22] Afghan pika,[23] quail,[24] chicken,[25-27] frog,[28,29] molly,[30,31] stickleback,[32] goldfish,[33,34] *Tilapia*,[35] and *Gambusia*[36] pituitary glands have been described. These cells are round,[37] oval, or triangular in shape and have characteristic ultrastructural features.[38] These cells typically have a large nucleus with heterochromatin and prominent nucleoli and numerous nuclear pores. Golgi complexes are well developed, especially in the juxtanuclear region, and the granular endoplasmic reticulae are lamellar with numerous free ribosomes. The oval or elongated mitochondria are small and dense with oblique cristae and the secretory granules are large (260 to 400 nm in diameter), electron dense, and ovoid. Further details on the fine structural cytology of the rat and human adenohypophysis are reviewed elsewhere.[16,39,40]

0-8493-8697-7/95/$0.00+$.50

Subpopulations of somatotrophs that differ in size, shape, density, and ultrastructure have, however, also been described (e.g., in mice,[41] in rats,[42] and in goats[43]) and reviewed in detail by Takahashi.[44] This heterogeneity may reflect molecular differences in the hormone synthesized or its intracellular age, differences in somatotroph receptors for hypothalamic regulatory factors, different mechanisms of intracellular signal transduction, different geographical distributions of somatotrophs within the pituitary gland, and differences due to the the maturation of pituitary cells.[44]

B. SOMATOTROPH DIFFERENTIATION

1. Pit-1

The differentiation of somatotrophs is due to a transcriptional factor, GH factor 1 (GHF-1)/pituitary-specific transcription factor (pit-1), which is also responsible for the differentiation of lactotrophs and for the maintenance of thyrotrophs.[45,46] These are the only cells in the (mouse) pituitary gland in which pit-1 is translated, although *pit-1* mRNA occurs in all pituitary cells.[47] The expression of the *pit-1* gene in normal and tumorous human pituitary glands is, however, restricted to GH-, prolactin-, and thyrotrophin-stimulating hormone (TSH)-secretory cells.[48,49]

Pit-1 is expressed in embryonic development prior to the differentiation of somatotrophs, lactotrophs, and thyrotrophs and the appearance of their hormone transcripts.[50] The recruitment of GH- and prolactin-secreting cells from acidophil stem cells and the maintenance of TSH cells is pit-1-dependent and is absent in Snell, Jackson, and Ames dwarf mice. These mice have normal somatotroph stem cells but are pit-1 deficient.[46,51-53] These deficiencies result from a point mutation in the *pit-1* gene on chromosome 11 (Snell mice) and from a gene inversion/insertion of additional DNA (Jackson mice). The *pit-1* gene is not mutated in Ames dwarfs, but its transcription is repressed by a protein encoded by a gene on chromosome 16.[51,52] Inhibition of pit-1 synthesis by *pit-1* antisense oligonucleotides also limits proliferation of somatotrophic cell lines.[54]

During organogenesis of the pituitary gland, the *pit-1* gene appears to be transiently activated, because activation of the *pit-1* gene occurs in early embryological stages of the Snell dwarf mouse but not in later embryological phases.[55] This early activation of the *pit-1* gene may be due to interactions between retinoic acid and a tissue-specific enhancer.[55] The maintenance of *pit-1* gene expression during ontogeny may, however, be dependent on pit-1 positive feedback, because a pit-1-responsive region is present in the *pit-1* gene.[56]

Pit-1 is a member of the POU-homeodomain family of transcription factors[57,58] and has binding sites on the genes encoding GH, prolactin, and TSH.[59,60] This 291-amino acid protein binds to the consensus sequence TATTCAT at two sites on the *GH* promotor and at five sites on the prolactin promoter. Pit-1 activates the *GH* promoter by displacing the nucleosome and exposing glucocorticoid and thyroid hormone response elements[45] and by interacting synergistically with other transcriptional regulators ubiquitously present in pituitary cells. In the absence of pit-1, the inactive *GH* promoter is refractory to the ubiquitous stimuli that bind to cAMP response elements or to glucocorticoid or thyroid hormone receptors. Although this transcription factor also binds to the prolactin promoter, it has lower affinity for the prolactin gene,[61,62] which may be specifically activated by a large, immunologically distinct factor, LSF-1 (lactotroph-specific factor 1).[45] The role of LSF-1 is, however, uncertain and pit-1 activation of both the *GH* and *prolactin* promoters has been demonstrated at pit-1 concentrations tenfold lower than those found in the pituitary gland.[63] Pit-1 also binds to the promoter of the *TSH* gene, but with lower affinity than to the *GH* and *prolactin* promoters, although an activation of the *TSH*-α gene is not observed in all systems.[59,63,64]

The cell-specific expression of these pituitary hormones in somatotrophs, lactotrophs, and thyrotrophs may result from the differential synthesis of different pit-1 isoforms. At least two isoforms have been identified: pit-1α and a larger protein (containing an additional 26 amino acids), pit-1β, derived from the alternate splicing of *pit-1* mRNA.[65] The smaller isoform *trans*-activates the *GH* and *prolactin* gene promoters and the *pit-1* gene, but the larger isoforms can activate only the *GH* promoter.[65,66] The differential expression of these isoforms during pituitary development has not, however, been demonstrated.[65] Thus, although pit-1 induces activation of the *GH, prolactin,* and *TSH* genes, other factors are also probably required for the phenotypic induction of somatotroph, lactotroph, and thyrotroph morphology and function. This requirement is indicated by the absence of GH, prolactin and TSH secretory cells in *rdw* rats that have normal levels of pit-1 mRNA.[67] The absence of pituitary hypoplasia in GH-deficient human patients with a pit-1 deficiency[68,69] further indicates the involvement of factors other than pit-1 in the differentiation and proliferation of pituitary cells.

2. GHRH

In addition to pit-1, GH-releasing factor (GHRH) may contribute to the cell-specific differentiation of GH-secretory pituitary cells. Indeed, pit-1 may stimulate differentiation in part via GHRH, because GHRH receptors are absent in the pituitaries of pit-1-defective Snell dwarf mice.[53] GHRH induces protein kinase A (PKA) phosphorylation and exerts posttranscriptional regulation on the *pit-1* gene, because phosphorylation decreases pit-1 affinity for the GH-1 site of the *GH* promoter but enhances dimerization at the GH-2 site.[56,70] Stimulation by GHRH also induces cAMP production, which increases pit-1 transcription by association with multiple cAMP response elements (CREs) in the *pit-1* gene.[56] Somatotroph hyperplasia therefore occurs in the pituitaries of mice transgenically expressing the *GHRH* gene.[71-74] This suggests that stimulation of *GH* gene expression also stimulates the cell division of GH cells.[44] The first appearance of GH-secreting cells in the rat pituitary gland is, however, not accelerated by GHRH,[75] indicating that it does not stimulate the differentiation of a hypothetical stem cell to a somatotroph phenotype. Indeed, the ontogenic appearance of hypothalamic GHRH[76] occurs after the first appearance of pituitary somatotrophs. The presence of somatotrophs in encephalectomized fetal rats[77] and encephalic human fetuses[78] also suggests that GHRH is not required for the development of GH-secreting cells during embryogenesis.

C. SOMATOTROPH PROLIFERATION

The pituitary gland is an "expanding organ" and every pituitary cell probably has the capacity to divide during its life span.[44] Indeed, differentiated GH cells are mitotic and self-duplicate into identical differentiated cells. Undifferentiated or stem cells may also divide and produce daughter cells, half of which may subsequently undergo terminal differentiation and proliferation. Somatotroph populations are, therefore, labile and responsive to physiological stimuli.

The number and size of somatotrophs and their secretory granules vary during ontogeny and with reproductive state[79] and are regulated by other endocrine[43] and environmental factors[80] (Figure 1). For instance, the number of GH cells and GH secretory granules in the adenohypophysis increases during fetal and postnatal development, and precedes the appearance of GH in circulation and maturational changes in growth, GH synthesis, and GH release (e.g., in human,[81] rat,[44,82,83] pig,[84] chicken,[85,86] and fish[87]). An absence or deficiency of morphologically identifiable somatotrophs is therefore causally associated with hereditary dwarfism (e.g., in human dwarfs,[88] dwarf rats,[89] Snell mice,[90] Ames mice,[91] little mice,[92,93] and Jackson mice[46]).

During normal growth, the number of somatotrophs and their volume density, surface density, cell area, and perimeter are reduced in aging rats and humans.[79,81] Although these changes do not affect trough levels of circulating GH, they are accompanied by reduced frequencies and amplitudes of episodic GH release. These morphometric changes occur

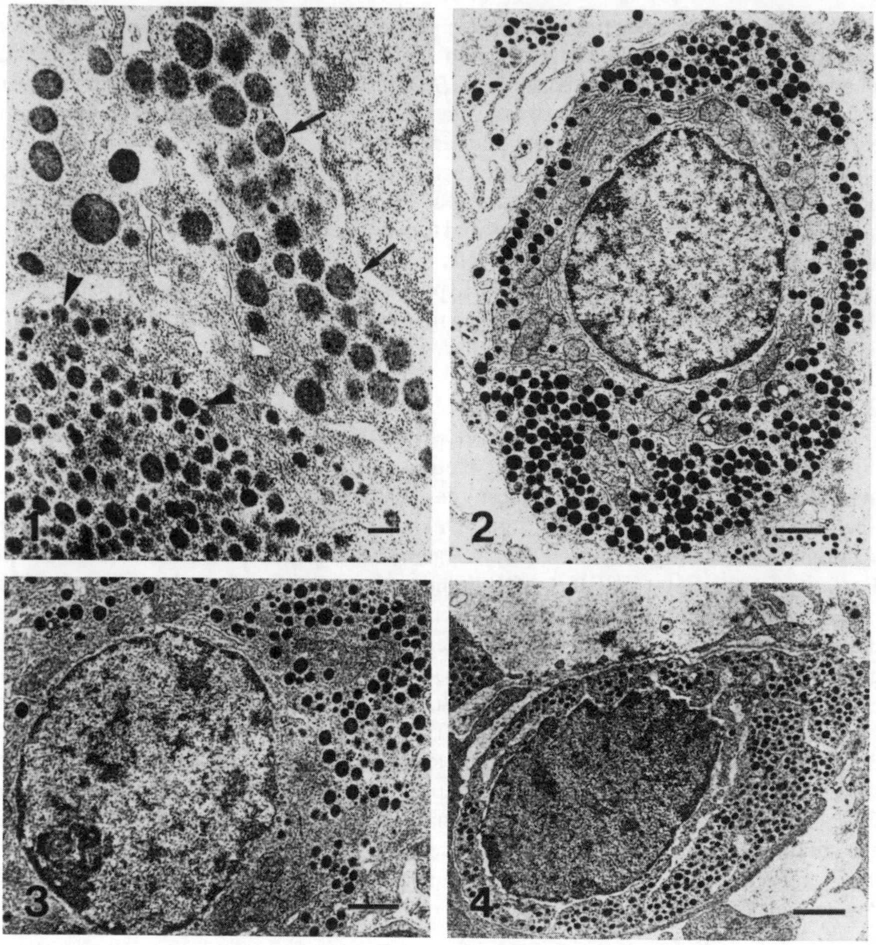

Figure 1 (1) Secretory granules in pituitary cells immunocytochemically stained with rat GH antiserum. Large secretory granules with a diameter of 250 to 350 nm (arrows) and small secretory granules with a diameter of 100 to 150 nm (arrowheads) are seen. Bar: 200 nm. (2) Type I GH cell in a 6-month-old female rat. Large secretory granules are seen throughout the cytoplasm. Bar: 1.0 μm. (3) Type II GH cell in a 6-month-old female rat. Large and small secretory granules are seen throughout the cytoplasm. Bar: 1.0 μm. (4) Type III GH cell in a 6-month-old female rat. Small secretory granules are seen throughout the cytoplasm. Bar: 1.0 μm. (From Takahashi, S., *Zool. Sci.*, 9, 901, 1992. Reproduced with permission.)

independently of changes in pituitary size and the proportions of other pituitary cells, but may reflect the progressive hypothyroidism associated with aging and senescence and are mimicked by experimental hypothyroidism.[39]

The ontogenic changes in somatotroph abundance and morphology are also sexually dimorphic in some species and thought to be induced by sex hormones and hypothalamic hormones during neonatal and postnatal life. Males have greater numbers of GH cells than do females.[41,44,94,95] The anterior pituitary gland differentiates into a male or female type in response to the absence or presence of a testis (or testosterone) during the neonatal period and to the presence or absence of an ovary (or estrogen and/or progesterone) after puberty.[41] Castration therefore reduces the size and number of GH cells in the pituitaries of adult male rats, but not in the presence of exogenous testosterone,[96] which increases somatotroph numbers in mouse pituitary gland.[41] Androgens are also known to increase the relative abundance of a somatotroph subtype (type 1)[44] (Figure 2). Estrogens also have a hyperplastic effect on rat somatotrophs but preferentially increase lactotroph populations and reduce the relative abundance of GH-secreting cells,[97] particularly type I cells[44] (Figure 2). The effects of estrogens are, however, biphasic and GH cell proliferation may be suppressed at high dose levels[98] and during estrus in the estrous cycle.[94] Indeed, ovariectomy has been shown to increase the size, cytoplasmic area, and GH content of pituitary somatotrophs,[99] even though stimulatory effects of estrogen on GH secretion are well established.[100] It is therefore possible that the actions of gonadal steroids are mediated by differential actions at pituitary and hypothalamic sites.[100]

The mitogenic activity of fully differentiated rat somatotrophs is also increased by GHRH,[71,101,102] by a cAMP-dependent action that stimulates expression of the *c-fos* protooncogene.[103] This stimulation of cell proliferation is specific to somatotrophs,[71] because GHRH inhibits the proliferation of rat mammotrophs and has no effect on the number of pituitary fibroblasts.[101] Somatotroph hyperplasia similarly occurs in patients with GHRH-secreting tumors[104] and in transgenic mice expressing the gene.[73,105-107] Chronic stimulation by GHRH induces, within days, hypertrophy and proliferation of rat somatotrophs with ultrastructural features of highly stimulated, sparsely granulated cells.[73,74,105] The size of somatotrophs during fetal development is also increased by GHRH,[75] whereas somatotroph numbers are reduced by GHRH deficiency or GHRH resistance.[108-110] The role of cAMP in inducing this proliferation is indicated by the somatotroph hyperplasia in pituitary tumors with chronically active stimulatory G pro-

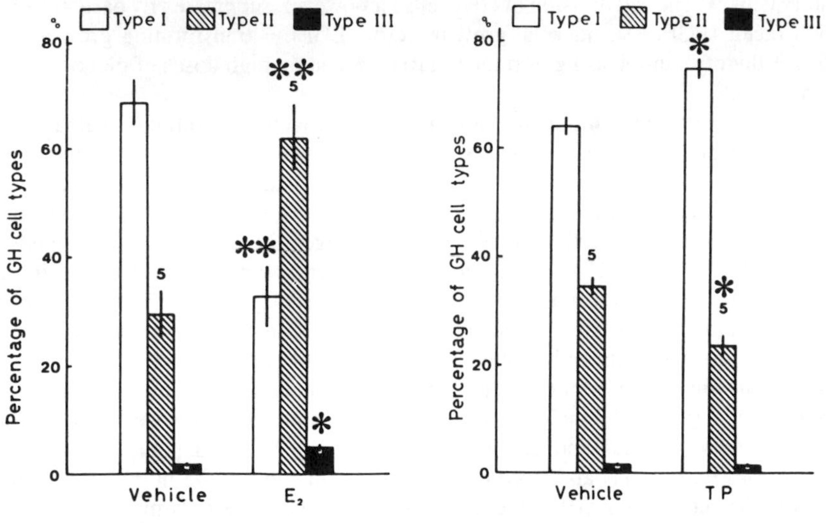

Figure 2 Plasticity of pituitary somatotroph morphology in rat pituitary glands. Male rats were injected daily for 5 days with 50 μg of 17β-estradiol (E_2), 100 μg of testosterone propionate (TP), or vehicle. Estradiol decreased the percentage of type I GH cells and increased the number of type II and type III cells, whereas testosterone increased the number of type I cells and decreased the percentage of type II cells. (Adapted from Takahashi.[44])

teins (G_s)[111] and in the pituitaries of mice in which cAMP activity is chronically elevated by transfection of cholera toxin.[112] The somatotrophic action of GHRH is also blocked when cAMP binding to CRE-binding sites in the *GH* gene promoter is impaired by CRE-binding protein (CREB) mutations.[110] Somatotroph hyperplasia induced by GHRH is also antagonized by somatostatin (SRIF), which inhibits cAMP accumulation[102] and by activin A,[113] which has no effect on c-*fos* expression.[113] The somatotroph size, nuclear volume, and GH content is increased in rats in which the hypothalamic release of SRIF is blocked by the intraventricular administration of colchicine.[114]

Somatotroph number and morphology are also regulated by GH. The protracted overproduction of GH in mice transgenically expressing the human or bovine *GH* gene results in somatotrophs of decreased size and abundance. These somatotrophs show ultrastructural features of suppression (reduction in cytoplasm size, and poor development of rough endoplasmic reticulum [ER] and Golgi).[106,115,116]

Somatotroph differentiation and proliferation in the fetus also appear to be stimulated by glucocorticoids and thyroid hormones but suppressed by pancreatic hormones (insulin and glucagon).[75,117] Somatotroph numbers are suppressed postnatally in hypothyroid mammals, in which treatment with triiodothyronine (T_3) increases somatotroph recruitment from progenitor cells. Tumorous somatotroph cell growth is also stimulated by T_3, at concentrations less than those required to induce GH production[118-121] and the formation of immature somatotrophs.[44] These actions of T_3 are dependent on the presence of glucocorticoids, which accelerate the appearance of GH cells in fetal rat pituitary glands[117] and increase somatotroph abundance.[122] These glucocorticoid actions are, however, dependent on adequate expression of the *pit-1* gene. The actions of thyroid hormones may also result from the induction of *GH* gene transcription, which may itself induce somatotroph proliferation.[44] Cytokines, for example interleukin 2 (IL-2) and IL-6, also promote the proliferation of GH_3 cells,[123] although they may inhibit the proliferation of normal pituitary cells.[123] The cell division of GH_3 cells is, however, suppressed by basic fibroblast growth factor (FGF),[124] epidermal growth factor (EGF),[125] transforming growth factor (TGF),[126] thyrotropin-releasing hormone (TRH),[125] and by high doses of glucocorticoids and estrogen.[127,128]

The influence of these mitogenic or differentiation factors on pituitary somatotrophs in lower vertebrates is largely unknown, although species differences in somatotroph regulation are likely. For instance, in contrast with mammalian species the relative abundance and volume of GH-secreting cells and their secretory granules decline during sexual maturation in domestic fowl.[25,26] These changes are accompanied by postnatal reductions in pituitary GH concentrations,[129,130] whereas pituitary GH concentrations increase during growth in mammals.[131,132] These changes may partly reflect the inhibitory (rather than stimulatory) effect of gonadal steroids on somatotroph function in birds.[133] Differences between birds and mammals in response to gonadal steroids are also indicated by the increased (rather than decreased) somatotroph numbers induced by gonadectomy in male and female birds and by the accompanying increases in the volumes of the secretory granules and endoplasmic reticulum.[134] Gonadal maturation in fish (*Pleuronectes platessa*) is also accompanied by a reduction in the number of pituitary somatotrophs and in their GH content,[135] although gonadal steroids may stimulate GH secretion in teleosts.[100] The morphology of GH cells in *Poecilia latipinna* and *Poecilia reticulata* are also related to ovarian activity. These cells are inactive in sexually quiescent fish but become progressively activated during vitellogenesis and mating, quiescent during pregnancy, and active just before, during, and after parturition.[30,136,137] The number of GH-secreting cells in the hypophyses of white-tailed deer (*Odocoileus virginianus borealis*) similarly varies on a seasonal basis.[138] The abundance and morphology of GH-secreting cells may therefore be modulated by environmental stimuli. Indeed, the stress of protein-caloric malnutrition has been shown to reduce the number of somatotrophs and their cytoplasmic and nuclear volumes.[80]

D. SOMATOTROPH DISTRIBUTION

The distribution of GH-secreting cells within the adenohypophysis differs among different animal species.[37] Whereas GH secretory cells may be widely distributed throughout the adenohypophysis of mammals (e.g., in musk shrews),[18] in many lower vertebrate species they concentrate within specific regions of the gland. For instance, in birds >90% of the somatotrophs are confined to the caudal lobe and the rest are restricted to a ventral region of the cephalic lobe.[25,27,139] A regional distribution of somatotrophs in the anterior lobe has also been demonstrated in reptiles and amphibia.[140] Most of the GH cells in the newt pars distalis are, for instance, confined to dorsal regions and central regions below the intermediate lobe.[141] In fish,[142] most of the GH cells are restricted to the proximal pars distalis and often in close association with gonadotropin-secreting cells (e.g., in barbel, *Barbus barus,*[143] salmon,[144] *Tilapia,*[145] *Aequidens pulcher* and *Poecilia latipinna,*[146] seabass, *Dicentrachus labrax,*[87] and plaice, *Pleuronectes platessa*[135]). This spatial distribution may reflect the neuropeptidergic innervation of this region and paracrine actions of GHRH and other peptides on somatotroph proliferation.[147] Somatotrophs are also confined to the proximal pars distalis in the lamprey, *Petromyzon marinus*.[148]

Somatotrophs are less compartmentalized in the pituitaries of mammals, but specific spatial distributions of GH-secreting cells have been described.[149-151] Thus in the rat pituitary, whereas somatotrophs are evenly distributed sagittally and rostrocaudally they are unevenly distributed dorsoventrally. These cells are not found near the intermediate lobe nor in the anteroventral portion of the gland.[42] These cells are usually situated along sinusoids,[44] but in the mouse, somatotrophs predominate in regions that are poorly vascularized (in anterolateral wings) and are least abundant in rostral and caudal areas, which are richly vascularized.[151]

The geographic distribution of somatotrophs within the adenohypophysis may therefore be due to regional patterns of blood supply within the gland. Moreover, because portal vessels may arise in different areas of the median eminence, in which hypophysiotropic factors may be differentially released,[152] the geographic distribution of GH-secreting cells may reflect an uneven distribution of factors regulating somatotroph function and of functionally different GH cell types.[42] Indeed, the basal and GHRH-induced release of GH from the rat pituitary gland is related to the geographical location of the somatotrophs.[42] This spatial distribution of somatotrophs may also be related to the distribution of other adenohypophysial cell types and paracrine regulators of somatotroph function. Somatotrophs in the rat and mouse pituitary, for instance, are found in close association with adrenocorticotropic hormone (ACTH) cells,[37,153] which have paracrine actions on GH release.[153] However, while corticotrophs and gonadotrophs are distributed throughout the pars distalis and the pars tuberalis of the ferret pituitary gland, somatotrophs are confined to the pars distalis.[21]

E. SOMATOTROPH HETEROGENEITY

The morphological and functional heterogeneity of pituitary somatotrophs is well established and occurs early in ontogeny.[84] Somatotrophs may be heterogeneous with respect to overall size, density, and ultrastructure and with respect to their degree of granulation and rate of degranulation.[21,84,154-156] Some somatotrophs also appear to be binucleated.[21] The somatotrophs in adult rats consist of type I cells, containing large, evenly distributed secretory granules (about 350 nm in diameter); type II cells, containing smaller secretory granules (100 to 300 nm in diameter); and type III somatotrophs with secretory granules <150 nm in diameter (Figure 1).[15,16] These cell types account for approximately 68, 22, and 10% of all the GH-secreting cells in males and 44, 47, and 9%, respectively, in females. The occurrence of type III cells is, however, more frequent (15 to 25% of cells) in neonatal rats. These cell types may also be induced by endocrine factors: androgens increase and estrogens decrease the proportion of type I somatotrophs, whereas thyroid hormones induce the recruitment of type III somatotrophs.[44] These changes may result from the specific proliferation of a somatotroph cell type and the cell death of other cell types or from the interconversion of cell types.

Similar somatotroph subtypes are also present in other species[13,21,27,142] and these structural variants may represent maturational phases of a single cell type.[15,40] The type I cells are round, polygonal, triangular, or oval in shape and contain round or oval nuclei at central or slightly eccentric positions. The swollen mitochondria in these cells are large and scattered among the ER cisternae, which are moderately dilated. Golgi occur near the nuclei of these cells and sometimes contain parallel lamella (Figure 1). In contrast, type II cells are characterized by large and small secretory granules, which are present in equal proportions in individual cells. These cells may therefore represent an intermediate somatotroph type. These cells differ in their mitochondria and rough ER and have perinuclear cytoplasm devoid of secretory granules. These mitochondria are not as swollen as in type I cells and the cisternae of their ER are mostly collapsed and arranged parallel to each other (Figure 1). Type III somatotrophs differ from type II cells in having a small cell volume and containing only small secretory granules. These cells have vesiculated ER with round cisternae and a perinuclear clear zone devoid of granules that contains moderately developed Golgi (Figure 1). The number of granules in these actively secreting cells is, however, variable and is greater in adults than in infants.[16,114]

Somatotrophs may also be functionally heterogeneous. They may differ in precursor uptake[157] and have differential GH responses to GHRH, somatostatin (SRIF), and insulin-like growth factor I (IGF-I).[42,158-163] The basal level of GH release from somatotrophs is also heterogeneous[159] and in unstimulated glands only a fraction of the somatotrophs may release GH, although the number of "quiescent somatotrophs" is reduced following GHRH or TRH stimulation.[164] This heterogeneity may be due to differences in the intracellular metabolism of somatotrophs. Differences in the intracellular content of GH preferentially released under basal or stimulated conditions may contribute to the heterogeneity of pituitary somatotrophs, because cell types that autonomously release GH differ in the type and amount of GH released. For instance, one population is known to preferentially release small amounts of newly synthesized GH, whereas another preferentially releases large amounts of stored GH, even without extrinsic stimulation.[165] Different cell types may also secrete different molecular variants of the GH molecule.[44]

Somatotroph heterogeneity may also reflect differences in their intracellular signal transduction mechanisms. For instance, only 9% of rat pituitary somatotrophs contain protein kinase C (PKC), which is known to be involved in cell proliferation in some types of cells. Somatotrophs lacking PKC may, therefore, not have the capacity to proliferate.[44]

Somatotroph heterogeneity may also reflect paracrine influences from other endocrine cells. Cell-to-cell communication plays an important role in pituitary function and adenohypophysial cells are intimately associated with each other and form specialized intercellular junctions *in situ*.[166-168] Adenohypophysial cells probably communicate with each other via gap junctions and through the release of paracrine factors.[153] For instance, the *in vitro* GH response of somatotroph aggregates to epinephrine and vasoactive intestinal peptide (VIP) (but not to GHRH) is greatly increased when the cells are reaggregated with other adenohypophysial cells, as a result of facilitatory factors from lactotrophs, corticotrophs, thyrotrophs, gonadotrophs, or folliculostellate cells.[169-171]

F. MAMMOSOMATOTROPHS

In addition to these GH-secreting cells, plurihormonal cells secreting GH and prolactin have been identified[172] within human adenomas[173] and rat pituitary tumor lines[174,175] and in the normal pituitary glands of humans,[173,183] rats,[176,177] musk shrews,[178] bats,[22] mice,[179,180] sheep,[181] cattle,[10,182] and frogs.[28] Mammosomatotrophs possess receptors, second messenger systems and secretory characteristics of both somatotrophs and lactotrophs, although where differences exist there is more resemblance to somatotrophs.[177] This heterogeneity in signal transduction systems is likely to account for the heterogeneous GH responses of somatotroph subpopulations to GH-releasing and GH-inhibiting factors.[160] Mammosomatotrophs are also heterogeneous in their degree of granulation, degranulation, and granular content.[155] For instance, GH and prolactin may be colocalized within these cells or localized in separate secretory granules. When intermixed in the same granules (which may be larger than those in lactotroph or somatotrophs[181]), the two hormones appear to be distributed in different parts (the periphery or center) of the granules. Mammosomatotrophs also tend to cluster in the central region of pituitary lobules, whereas lactotrophs and somatotrophs predominate in peripheral areas.[184]

1. Abundance

Mammosomatotrophs may constitute a substantial proportion of pituitary cells, as in the human (25 to 50% of all pituitary cells[184]), bovine (9 to 26% of cells[10,185]), and rat (8 to 15% of cells[172,176]) adenohypophysis, or be rare occurrences, as in sheep (>0.2% of all cells[155,181]), ferrets,[21] and some mice (1% of cells[41,151]). The number of these cells may, however, vary according to age, gender, and physiological state.

An age-related increase in the relative proportion of mammosomatotrophs has, for instance, been observed during the development of the bovine pituitary gland.[186] Sexual dimorphism in mammosomatotroph abundance is also observed in bovines, because these cells account for 9% of pituitary acidophils in bulls but they represent most, if not all, of the acidophils in lactating cow pituitary glands.[186-188] The abundance of these cells in cycling females is also greater than that in bulls and is selectively increased during the early luteal phase of the estrous cycle[187] and by estrogens.[188] Estrogens similarly increase the induction of mammosomatotrophs in tumorous pituitary cells.[160] The abundance and secretory activity of these cells also varies seasonally. In Japanese house bats[22,178] these cells hypertrophy during pregnancy and account for most of the prolactin-secreting cells during hibernation and reproductive quiescence. Pregnant women also have greater numbers of mammosomatotrophs than nonpregnant women.[189] The abundance of these cells is, however, reduced in rats during pregnancy and late lactation and is greater in females than in males.[190] The number of sheep mammosomatotrophs has also been shown to increase dramatically from <0.2% of all pituitary cells to about 3% following the enzymatic dispersion of pituitary cells and their *in vitro* culture.[155] The abundance of mammosomatrophs would thus appear to be dynamically regulated.

2. Proliferation

The abundance of mammosomatotrophs in most species may be regulated by gonadal steroids (Figure 3). Indeed, the proportion of these cells in the pituitaries of steers is greatly increased following castration,[182] suggesting an inhibitory action of testosterone and/or other testicular factors on mammosomatotroph proliferation. Estrogen, conversely, appears to induce mammosomatotroph proliferation, because it increases the relative proportion of these cells in the pituitaries of male rats and pituitary cell cultures.[191] The number of mammosomatotrophs may also be regulated by hypothalamic peptides. Luteinizing hormone-releasing hormone (LHRH), for example, increases the proportion of prolactin-secreting cells in rat pituitary cells.[161,192] The massive hyperplasia of mammosomatotrophs in the pituitaries of mice transgenically expressing the GHRH gene[71,72,74] also indicates a

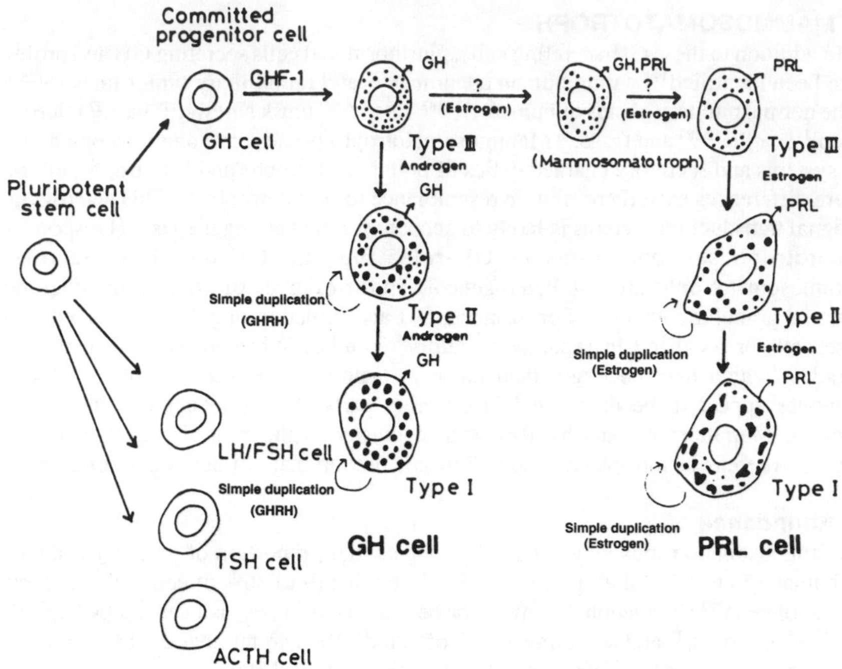

Figure 3 Possible relationships among growth hormone (GH), prolactin (PRL), and mammosomatotroph cells. Pituitary glands consist of GH cells, PRL cells, gonadotrophs (luteinizing hormone [LH]/follicle-stimulating hormone [FSH] cells), thyrotrophs (TSH cells), corticotrophs (ACTH cells), and folliculostellate cells (not shown). Pituitary-specific transcription factor, GHF-1 (pit-1), is known to be involved in the development of GH and PRL cells and possibly TSH cells. The conversion of GH cells is stimulated by androgens and that of PRL cells by estrogens. This conversion may be bidirectional. Pituitary cells proliferate by the mode of self-duplication. GH-releasing hormone (GHRH) stimulates the mitosis of GH cells, and estrogen stimulates that of PRL cells. Mammosomatotrophs may be transitional cells between GH cells and PRL cells. The transition of GH stem cells into PRL stem cells may be stimulated by estrogens. (From Takahashi, S., *Zool. Sci.*, 9, 901, 1992. Reproduced with permission.)

role for GHRH. Moreover, because pure somatotrophs are rare in these transgenic animals, the induction of mammosomatotrophs may derive from the transformation of GH-secreting cells. The progressive increase in the abundance of sheep mammosomatotrophs during *in vitro* culture may also indicate the presence of hypothalamic factors that normally inhibit this conversion.[155]

3. Differentiation

Mammosomatotrophs may be distinct cell types or result from the fusion of prolactin and GH cells,[181] especially as they tend to be multinucleated.[184] The plasticity of these multipotential cells also suggests they may serve as transitional cell types in the interconversion of GH to prolactin cell types and vice versa[174] or represent a "stem" cell capable of transformation into either cell type[172,176,184,193] (Figure 3).

Mammosomatotrophs may differentiate from somatotrophs that acquire the ability to synthesize and secrete prolactin. Indeed, full-length transcripts of the prolactin gene are normally present in somatotrophs, although these transcripts are specifically degraded in the nucleus and are not normally processed and exported into the cytoplasm.[194] The possibility that mammosomatotrophs differentiate from somatotrophs is also supported by the earlier ontogenic appearance of GH cells in the pituitary gland.[47,186,195-197] Indeed, the ontogenic appearance of prolactin cells coincides with the appearance of cells that initially cosecrete GH.[196,198] Most prolactin cells are also derived from GH-producing precursors by extinction of *GH* gene expression. Complete ablation of somatotrophs by expression of *GH*-diphtheria toxin and *GH*-thymidine fusion genes inserted into the germ line of transgenic mice results in an 80 to 90% loss of the lactotroph cells. Conversely, expression of a prolactin thymidine kinase fusion gene does not result in somatotroph ablation, suggesting that the majority of prolactin-expressing cells are postmitotic somatotrophs.[179,199] The repression of *GH* gene expression is normally mediated by the binding of two nuclear proteins (PREB and ssPREB) to a proximal repressor element (PRE) of the *GH* promoter[200] or by specific DNA methylation.[201] The importance of the latter mechanism is demonstrated by the increased proportion of GH-secreting cells in a clonal strain of GH- and prolactin-secreting cells (GH_4C_1 cells) by the pharmacological induction of genomic hypomethylation.[202] Mammosomatotrophs may, therefore, function as an intermediate in the conversion of GH-secreting cells into prolactin cells, which subsequently differentiate into mammosomatotrophs that secrete only prolactin.[184] In rats the conversion of somatotrophs to mammosomatotrophs is thought to occur in a somatotroph subpopulation and to be triggered neonatally by direct pituitary actions of a small milk peptide that promotes the translation of prolactin gene transcripts.[184,203-205] Hormones may also induce the differentiation of somatotrophs into mammosomatotrophs, because estrogen increases the relative proportion of these cells in rats.[206]

Although somatotrophs appear to be converted into mammosomatotrophs that subsequently differentiate into mammotroph cell types, both GH and prolactin cells may not be terminally differentiated and they may retain the capacity to "transdifferentiate" and be functionally interconvertible.[184] This pituitary plasticity is supported by the reciprocal changes in the proportions of GH and prolactin cells in different physiological states. For instance, during pregnancy and early lactation the relative proportion of prolactin-secreting cells is increased, although these cells apparently revert to traditional GH cells after weaning.[182,189,190,205] Reciprocal changes in the numbers of GH- and prolactin-secreting cells also occur in cultured pituitary cells treated with estrogen or TRH, both of which increase the proportion of prolactin-secreting cells and decrease the proportions of GH-secreting cells.[160,174,207,208] This recruitment of prolactin-secreting cells is abolished in the presence of cycloheximide, indicating that its induction is regulated posttranscriptionally.[207] The relative number of these GH-secreting cells is, however, increased by cortisol and progesterone, which reduce the proportion of prolactin cells.[174,188,208] Increased GH-secreting cells are also induced in MET/W_{15} tumor cells cultured with estrogen, which induces a reciprocal shift in the proportion of prolactin cells.[209] This plasticity therefore permits adaptive changes in pituitary function to occur in the absence of dramatic changes in the division or number of pituitary cells. Acidophil cells may therefore undergo cyclical changes in secretory function, as indicated in Figure 3.

G. THYROSOMATOTROPHS

In addition to mammosomatotrophs plurihormonal cells secreting GH, prolactin, and thyrotropin (thyroid-stimulating hormone [TSH]) or GH and glycoprotein subunits,[210] or GH and TSH, have been described in pituitary adenomas.[211,212] Degranulating GH cells in the pituitaries of hypothyroid rats are also thought to develop into TSH-secreting cells.[39] Like mammosomatotrophs, these thyrosomatotrophs may also be intermediate cell types, capable of transdifferentiation into pure somatotroph or thyrotroph populations, because these cells may also interconvert.[39] The expression of the *GH* gene in thyrotrophs is, however, normally minimal.[213]

REFERENCES

1. **Martin, C. R.,** *Endocrine Physiology,* Oxford University Press, New York, 1985.
2. **Hojvat, S., Baker, G., Kirsteins, L., et al.,** *Brain Res.,* 239, 543, 1986.
3. **Ogilvie, S., Buhl, W. C., Olson, J. A., et al.,** *Endocrinology (Baltimore),* 126, 3271, 1990.
4. **Pilavdzic, D., Chiu, B., Kovacs, K., et al.,** *Endocr. Pathol.,* 4, 48, 1993.
5. **Horvath, S., Palkovits, M., Gorcs, T., et al.,** *Brain Res.,* 481, 8, 1989.
6. **Pelletier, G., Robert, F., and Hardy, J.,** *J. Clin. Endocrinol. Metab.,* 46, 534, 1978.
7. **Zimmerman, E. A., Defendini, R., and Frantz, A. G.,** *J. Clin. Endocrinol. Metab.,* 38, 577, 1974.
8. **Girod, C. and Dubois, M. P.,** *Cell Tiss. Res.,* 172, 145, 1976.
9. **Dacheux, F.,** *Cell Tiss.Res.,* 207, 277, 1980.
10. **Fumagalli, G. and Zanini, A.,** *J. Cell Biol.,* 100, 2019, 1985.
11. **Dacheux, F. and Dubois, M. P.,** *Cell Tiss. Res.,* 174, 245, 1976.
12. **Parry, D. M., McMillen, I. C., and Willcox, D. L.,** *Cell Tiss. Res.,* 194, 327, 1978.
13. **Shirasawa, N., Kihara, H., and Yoshimura, F.,** *Cell Tiss. Res.,* 240, 315, 1985.
14. **Dada, M. O., Campbell, G. T., and Blake, C. A.,** *J. Endocrinol.,* 101, 87, 1984.
15. **Kurosumi, K., Koyama, T., and Tasaka, H.,** *Arch. Histol. Jpn.,* 49, 227, 1986.
16. **Kurosumi, K.,** *J. Electron Microsc. Tech.,* 19, 42, 1991.
17. **Roux, M., Bartke, A., Dumont, F., et al.,** *Cell Tiss. Res.,* 223, 415, 1982.
18. **Naik, D. R. and Dominic, C. J.,** *Am. J. Anat.,* 134, 145, 1972.
19. **Beauvillain, J. C., Mazzuca, M., and Dubois, P. M.,** *Cell Tiss. Res.,* 184, 343, 1977.
20. **Girod, C., Lheritier, M., Trouillas, J., et al.,** *Acta Anat.,* 114, 248, 1982.
21. **Mohanty, B., Takahara, H., Tachibana, T., et al.,** *Cell Tiss. Res.,* 273, 427, 1993.
22. **Ishibashi, T. and Shiino, M.,** *Endocrinology (Baltimore),* 124, 1056, 1989.
23. **Nakamura, F., Suzuki, Y., and Yoshimura, F.,** *Cell Tiss. Res.,* 244, 627, 1986.
24. **Harrisson, F.,** *Anat. Embyrol.,* 154, 185, 1978.
25. **Malamed, S., Gibney, J. A., Loesser, K. L., et al.,** *Cell Tiss. Res.,* 239, 87, 1985.
26. **Malamed, S., Gibney, J. A., and Scanes, C. G.,** *Cell Tiss. Res.,* 251, 581, 1988.
27. **Tai, S. W. and Chadwick, A.,** *Proc. Leeds Philos. Lit. Soc.,* 10, 209, 1977.
28. **Castano, J. P., Ruiz-Navarro, A., Torronteras, R., et al.,** *Tissue Cell,* 25, 165, 1993.
29. **Malagon, M. M., Castano, J. P., Dobado-Berrios, P. M., et al.,** *Gen. Comp. Endocrinol.,* 84, 461, 1991.
30. **Young, G. and Ball, J. N.,** *Gen. Comp. Endocrinol.,* 52, 86, 1983.
31. **Batten, T. F. C. and Wigham, T.,** *Cell Tiss. Res.,* 237, 585, 1984.
32. **Leatherland, J. F.,** *Z. Zellforsch. Mikrosh. Anat.,* 104, 318, 1970.
33. **Leatherland, J. F.,** *Can. J. Zool.,* 50, 835, 1972.
34. **Chang, J. P., Cook, H., Freedman, G. L., et al.,** *Gen. Comp. Endocrinol.,* 77, 256, 1990.
35. **Leatherland, J. F., Ball, J. N., and Hyder, M.,** *Cell Tiss. Res.,* 149, 245, 1974.
36. **Chambolle, P., Kat, O., Olivereau, M., et al.,** *Gen. Comp. Endocrinol.,* 43, 85, 1981.
37. **Watanabe, Y. G.,** *Anat. Embyrol.,* 172, 277, 1985.
38. **Moriatry, G. C.,** *J. Histochem. Cytochem.,* 21, 855, 1973.
39. **Horvath, E. and Kovacs, K.,** *J. Electron Microsc. Tech.,* 8, 401, 1988.
40. **Takahashi, S.,** *Cell Tiss. Res.,* 266, 275, 1991.
41. **Yamaji, A., Sasaki, F., Iwama, Y., et al.,** *Anat. Rec.,* 233, 103, 1992.
42. **Perez, F. M. and Hymer, W. C.,** *Endocrinology (Baltimore),* 127, 1877, 1990.
43. **Sanchez, J., Navarro, J. A., Bernabe, A., et al.,** *Histol. Histopathol.,* 8, 83, 1993.
44. **Takahashi, S.,** *Zool. Sci.,* 9, 901, 1992.
45. **Karin, M., Castrillo, J., and Theill, L. E.,** *Trends Genet.,* 6, 92, 1990.
46. **Li, S., Crenshaw, E. B., Rawson, E. J., et al.,** *Nature (Lond.),* 347, 528, 1990.
47. **Simmonds, D. M., Voss, J. W., Ingraham, H. A., et al.,** *Genes Dev.,* 4, 695, 1990.
48. **Asa, S. L., Puy, L. A., Lew, A. M., et al.,** *J. Clin. Endocrinol. Metab.,* 77, 1275, 1993.
49. **Friend, K. E., Chiou, Y., Laws, E. R., et al.,** *J. Clin. Endocrinol. Metab.,* 77, 1281, 1993.
50. **Dolle, P., Castrillo, J. L., Theill, L. E., et al.,** *Cell,* 60, 809, 1990.
51. **Camper, S. A., Saunders, T. L., Katz, R. W., et al.,** *Genomics,* 8, 586, 1990.
52. **Buckwalter, M. S., Katz, R. W., and Camper, S. A.,** *Genomics,* 10, 515, 1991.
53. **Lin, C. J., Lin, S. C., Chang, C. P., et al.,** *Nature (Lond.),* 360, 765, 1992.
54. **Castrillo, J., Theill, L. E., and Karin, M.,** *Science,* 253, 197, 1991.
55. **Rhodes, S. J., Chen, R. P., Dimattia, G. E., et al.,** *Genes Dev.,* 7, 913, 1993.

56. **McCormick, A., Brady, H., Theil, L. E., et al.,** *Nature (Lond.),* 345, 829, 1990.
57. **Castrillo, B. L., Bodner, M., and Karin, M.,** *Science,* 243, 814, 1989.
58. **Ingraham, H. A., Chen, R., Mangalam, H. J., et al.,** *Cell,* 55, 519, 1988.
59. **Gordon, D. F., Haugen, B. R., Sarapura, V. D., et al.,** *Mol. Cell Endocrinol.,* 96, 75, 1993.
60. **Nelson, C., Albert, V. R., Elsholtz, H. P., et al.,** *Science,* 239, 1400, 1988.
61. **Fox, S. R., Jong, M. T. C., Casanova, J., et al.,** *Mol. Endocrinol.,* 4, 1069, 1990.
62. **Larkin, S., Tait, S., Treacy, M., et al.,** *Eur. J. Biochem.,* 191, 605, 1990.
63. **Mangalam, H. J., Albert, V. R., Ingraham, H. A., et al.,** *Genes Dev.,* 3, 946, 1989.
64. **Mason, M. E., Friend, K. E., Copper, J., et al.,** *Biochemistry,* 32, 8932, 1993.
65. **Theill, L. E., Hattori, K., Lazzaro, D., et al.,** *EMBO J.,* 11, 2261, 1992.
66. **Konzak, K. E. and Moore, D. D.,** *Mol. Endocrinol.,* 6, 241, 1992.
67. **Shibayama, K., Ohyama, Y., Ono, M., et al.,** *J. Endocrinol.,* 138, 307, 1993.
68. **Pfaffle, R. W., Dimattia, G. E., Parks, J. S., et al.,** *Science,* 257, 1118, 1992.
69. **Radovick, S., Nations, M., Du, Y. F., et al.,** *Science,* 257, 1115, 1992.
70. **Kapiloff, M. S., Farkash, Y., Wegner, M., et al.,** *Science,* 253, 786, 1991.
71. **Mayo, K. E., Hammer, R. E., Swanson, L. W., et al.,** *Mol. Endocrinol.,* 2, 606, 1988.
72. **Stefaneanu, L., Kovacs, K., Horvath, E., et al.,** *Endocrinology (Baltimore),* 125, 2710, 1989.
73. **Asa, S. L., Kovacs, K., Stefaneanu, L., et al.,** *Endocrinology (Baltimore),* 131, 2083, 1992.
74. **Asa, S. L., Kovacs, K., Stefaneanu, L., et al.,** *Proc. Soc. Exp. Biol. Med.,* 193, 232, 1990.
75. **Hemming, F. J., Begeot, M., Dubois, M. P., et al.,** *Endocrinology (Baltimore),* 114, 2107, 1984.
76. **Rodier, P. M., Kates, B., White, W. A., et al.,** *J. Comp. Neurol.,* 291, 363, 1990.
77. **Rieutort, M. and Jost, A.,** *Endocrinology (Baltimore),* 98, 1123, 1976.
78. **Begeot, M., Dubois, M. P., and Dubois, P. M.,** *Neuroendocrinology,* 24, 208, 1977.
79. **Console, G. M., Dumm, C. L. A. G., and Goya, R. G.,** *Mech. Ageing Dev.,* 70, 45, 1993.
80. **Herbert, D. C., Yashiro, T., Muraki, T., et al.,** *Anat. Rec.,* 235, 121, 1993.
81. **Sun, Y. X., Fenoglio, C. M., Pushparaj, N., et al.,** *Hum. Pathol.,* 15, 169, 1984.
82. **Smets, G., Velkeniers, B., Finne, E., et al.,** *J. Histochem. Cytochem.,* 35, 335, 1987.
83. **Chatelain, A., Dupouy, J. P., and Dubois, M. P.,** *Cell Tiss. Res.,* 196, 409, 1976.
84. **Torronteras, R., Castano, J. P., Ruiz-Navarro, A., et al.,** *J. Neuroendocrinol.,* 5, 257, 1993.
85. **Malamed, S., Gibney, J. A., Cain, L. D., et al.,** *Cell Tiss. Res.,* 272, 369, 1993.
86. **McCann-Levorse, L. M., Radecki, S. V., Donoghue, D. J., et al.,** *Proc. Soc. Exp. Biol. Med.,* 202, 109, 1993.
87. **Cambre, M., Mareels, G., Corneille, S., et al.,** *Gen. Comp. Endocrinol.,* 77, 408, 1990.
88. **Asa, S. L., Kovacs, K., Halasz, A., et al.,** *Endocr. Pathol.,* 3, 93, 1992.
89. **Kineman, R. D., Chen, T. C., and Frawley, L. S.,** *Endocrinology (Baltimore),* 125, 2305, 1989.
90. **Snell, G. D.,** *Proc. Natl. Acad. Sci. U.S.A.,* 15, 733, 1929.
91. **Eicher, E. M. and Beamer, W. G.,** *J. Hered.,* 67, 87, 1976.
92. **Wilson, D. B. and Wyatt, D. P.,** *Histol. Histopathol.,* 7, 451, 1992.
93. **Cheng, T. C., Beamer, W. G., Phillips, J. A., et al.,** *Endocrinology (Baltimore),* 113, 1669, 1983.
94. **Oishi, Y., Okuda, M., Takahashi, H., et al.,** *Anat. Rec.,* 235, 111, 1993.
95. **Baker, B. L. and Gross, D. S.,** *Am. J. Anat.,* 153, 193, 1978.
96. **Gross, D. S.,** *Am. J. Anat.,* 158, 507, 1980.
97. **Ho, K. Y., Thorner, M. O., Krieg, R. J., Jr., et al.,** *Endocrinology (Baltimore),* 123, 1405, 1988.
98. **Amara, J. F. and Dannies, P. S.,** *Endocrinology (Baltimore),* 112, 1141, 1983.
99. **Carretero, J., Sanchez, D., Sanchez, F., et al.,** *Histochem. J.,* 22, 683, 1990.
100. **Harvey, S.,** *The Growth, Development and Metabolism of Vertebrates* (Eds. Schreibman, M. P., Scanes, C. G., and Pang, P. K. T.). Academic Press, New York, 1993, p. 151.
101. **Shinkai, T., Ooka, H., and Noumura, T.,** *Neurosci. Lett.,* 123, 13, 1991.
102. **Billestrup, N., Swanson, L. W., and Vale, W.,** *Proc. Natl. Acad. Sci. U.S.A.,* 83, 6854, 1986.
103. **Billestrup, N., Mitchell, R. L., Vale, W., et al.,** *Mol. Endocrinol.,* 1, 300, 1987.
104. **Asa, S. L., Sheithaur, B. W., Bilbao, S. M., et al.,** *J. Clin. Endocrinol. Metab.,* 588, 796, 1984.
105. **Stefaneanu, L., Kovacs, K., Horvath, E., et al.,** *Endocr. Pathol.,* 4, 131, 1993.
106. **Stefaneanu, L., Kovacs, K., Horvath, E., et al.,** *Endocrinology (Baltimore),* 126, 608, 1990.
107. **Kovacs, K., Stefaneanu, L., Asa, S., et al.,** *Molecular and Clinical Advances in Pituitary Disorders* (Ed. Melmed, S.). Endocrine Research and Education, Los Angeles, p. 41, 1993.
108. **Downs, T. R. and Frohman, L. A.,** *Endocrinology (Baltimore),* 129, 59, 1991.
109. **Maiter, D., Underwood, L. E., Martin, J. B., et al.,** *Endocrinology (Baltimore),* 128, 1100, 1991.

110. Struthers, R. S., Vale, W. W., Arias, C., et al., *Nature (Lond.)*, 350, 622, 1991.
111. Landis, C. A., Masters, S. B., Spada, A., et al., *Nature (Lond.)*, 340, 692, 1989.
112. Burton, F. H., Hasel, K. W., Bloom, F. E., et al., *Nature (Lond.)*, 350, 74, 1991.
113. Billestrup, N., Gonzalez-Manchon, C., Potter, E., et al., *Mol. Endocrinol.*, 4, 356, 1990.
114. Carretero, J., Sanchez, F., Rubio, M., et al., *Histol. Histopathol.*, 7, 673, 1992.
115. Stefaneanu, L., Kovacs, K., Bartke, A., et al., *Lab. Invest.*, 68, 584, 1993.
116. Stefaneanu, L., Kovacs, K., and Bartke, A., *Endocr. Pathol.*, 4, 73, 1993.
117. Hemming, F. J., Aubert, M. L., and Dubois, P. M., *Endocrinology (Baltimore)*, 123, 1230, 1988.
118. Halperin, Y., Surks, M. I., and Shapiro, L. E., *Endocrinology (Baltimore)*, 126, 2321, 1990.
119. Gourdji, D., Tougard, C. and Tixier-Vidal, A., *Front. Neuroendocrinol.*, 7, 317, 1982.
120. DeFesi, C. H. and Surks, M. I., *Endocrinology (Baltimore)*, 108, 259, 1981.
121. Kitagawa, S., Obata, T., Willingham, M. C., et al., *Endocrinology (Baltimore)*, 120, 2591, 1987.
122. Nogami, H. and Tachibana, T., *Endocrinology (Baltimore)*, 132, 517, 1993.
123. Arzt, E., Buric, R., Stelzer, G., et al., *Endocrinology (Baltimore)*, 132, 459, 1993.
124. Black, E. G., Logan, A., Davis, J. R. E., et al., *J. Endocrinol.*, 127, 39, 1990.
125. Schonbrunn, A., Krasnoff, M., Westendorf, J. M., et al., *J. Cell Biol.*, 85, 786, 1980.
126. Ramsdell, J. S., *Endocrinology (Baltimore)*, 128, 1981, 1991.
127. Daniels, M., White, M. C., and Kendall-Taylor, P., *J. Endocrinol.*, 114, 503, 1987.
128. Gourdji, D., Laverriere, J. N., Passegue, E., et al., *Cell Biol. Toxicol.*, 8, 29, 1992.
129. Hoshino, S., Suzuki, M., and Yamamoto, K., *Gunma Symp. Endocrinol.*, 17, 15, 1980.
130. Foltzer, C., Harvey, S., and Mialhe, P., *J. Endocrinol.*, 113, 57, 1987.
131. Carmignac, D. F., Robinson, I. C. A. F., Enberg, B., et al., *J. Endocrinol.*, 138, 267, 1993.
132. Walker, P., Dissault, J. H., Alvarado-Urbina, G., et al., *Endocrinology (Baltimore)*, 101, 782, 1977.
133. Harvey, S., Fraser, R. A., and Lea, R. W., *Crit. Rev. Poultry Biol.*, 3, 239, 1991.
134. Milicevic, Z., Micic, M., Isakovic, K., et al., *Poultry Sci.*, 66, 741, 1987.
135. Power, D. M., *Gen. Comp. Endocrinol.*, 85, 358, 1992.
136. Sokol, H. W., *J. Morphol.*, 109, 219, 1961.
138. Schulte, B. A., Seal, U. S., Plotka, E. D., et al., *Am. J. Anat.*, 159, 369, 1980.
139. Mikami, S. and Takahashi, H., *Jpn. J. Vet. Sci.*, 49, 601, 1987.
140. Doerr-Schott, J., *Gen. Comp. Endocrinol.*, 28, 487, 1976.
141. Capantico, E. and Guastalla, A., *Gen. Comp. Endocrinol.*, 86, 197, 1992.
142. Schreibman, M. P., *Vertebrate Endocrinology: Fundamentals and Biomedical Implications* (Eds. Pang, P. K. T. and Schreibman, M. P.). Academic Press, New York, 1986, p. 11.
143. Toubeau, G., Poilve, Baras, E., et al., *Gen. Comp. Endocrinol.*, 83, 35, 1991.
144. Komourdjian, M. P. and Idler, D. R., *Gen. Comp. Endocrinol.*, 37, 343, 1979.
145. Magahama, Y., Olivereau, M., Farmer, S. W., et al., *Gen. Comp. Endocrinol.*, 40, 389, 1981.
146. Batten, T. F. C., *Gen. Comp. Endocrinol.*, 63, 139, 1986.
147. Moons, L., Cambre, M., Ollevier, F., et al., *Gen. Comp. Endocrinol.*, 73, 270, 1989.
148. Wright, G., *Cell Tiss. Res.*, 246, 23, 1986.
149. Halmi, N. S., Parson, J. A., Erlandsen, S. L., et al., *Cell Tiss. Res.*, 158, 497, 1975.
150. Herbert, D. C. and Silverman, A. Y., *Cell Tiss. Res.*, 230, 233, 1983.
151. Sasaki, F. and Iwama, Y., *Endocrinology (Baltimore)*, 122, 1622, 1988.
152. Reymond, M. J., Speciale, S. G., and Porter, J. C., *Endocrinology (Baltimore)*, 112, 1958, 1983.
153. Denef, C., Baes, M., and Schramme, C., *Front. Neuroendocrinol.*, 9, 115, 1986.
154. Synder, G., Hymer, W. C., and Synder, J., *Endocrinology (Baltimore)*, 101, 788, 1977.
155. Thorpe, J. R. and Wallis, M., *J. Endocrinol.*, 129, 417, 1991.
156. Ohlsson, L., Lindstrom, P., and Norlund, R., *Mol. Cell. Endocrinol.*, 59, 47, 1988.
157. Hopkins, C. R. and Farquhar, M. G., *J. Cell Biol.*, 59, 276, 1973.
158. Chen, C., Israel, J. M., and Vincent, J. D., *Neuroendocrinology*, 50, 679, 1989.
159. Dobado-Berrios, P. M., Ruiz-Navarro, A., Torronteras, R., et al., *J. Histochem. Cytochem.*, 40, 1715, 1992.
160. Mogg, R. J. and Boockfor, F. R., *Mol. Cell Endocrinol.*, 87, 1, 1992.
161. Hoeffler, J. P. and Frawley, L. S., *Endocrinology (Baltimore)*, 120, 791, 1987.
162. Hoeffler, J. P., Hicks, S. A., and Frawley, L. S., *Endocrinology (Baltimore)*, 120, 1936, 1987.
163. Frawley, L. S. and Neill, J. D., *Neuroendocrinology*, 39, 484, 1984.

164. **Kineman, R. D., Faught, W. J., and Frawley, L. S.,** *Endocrinology (Baltimore),* 127, 2229, 1990.
165. **Chen, T. T., Kineman, R. D., Betts, J. G., et al.,** *Endocrinology (Baltimore),* 408, 493, 1989.
166. **Fletcher, W. H., Anderson, N. C., and Everett, J. W.,** *J. Cell Biol.,* 67, 469, 1975.
167. **Herbert, D. C.,** *Anat. Rec.,* 114, 2107, 1979.
168. **Saunders, S. L., Reifel, C. W., and Shin, S. H.,** *Acta Anat.,* 114, 74, 1982.
169. **Allaerts, W., Engelborghs, Y., Van Oostveldt, P., et al.,** *Endocrinology (Baltimore),* 127, 1517, 1990.
170. **Baes, M., Allaerts, W., and Denef, C.,** *Endocrinology (Baltimore),* 120, 685, 1987.
171. **Baes, M. and Denef, C.,** *Endocrinology (Baltimore),* 120, 280, 1987.
172. **Frawley, L. S.,** *Trends Endocrinol. Metab.,* 1, 31, 1989.
173. **Lloyd, R., Anagnostou, D., Cano, M., et al.,** *J. Clin. Endocrinol. Metab.,* 66, 1103, 1988.
174. **Boockfor, F. R. and Schwartz, L. K.,** *Endocrinology (Baltimore),* 122, 762, 1988.
175. **Tashjian, A. H., Jr., Bancroft, F. C. and Levine, L.,** *J. Cell Biol.,* 47, 61, 1970.
176. **Frawley, L. S., Boockfor, F. R., and Hoeffler, J. P.,** *Endocrinology (Baltimore),* 116, 734, 1985.
177. **Kashio, Y., Chomczynski, P., Downs, T. R., et al.,** *Endocrinology (Baltimore),* 127, 1129, 1990.
178. **Ishibashi, T. and Shirno, M.,** *Anat. Rec.,* 223, 185, 1989.
179. **Behringer, R. R., Mathews, L., Palmiter, R. D., et al.,** *Genes Dev.,* 2, 453, 1988.
180. **Sasaki, F. and Iwama, Y.,** *Cell Tiss. Res.,* 256, 645, 1989.
181. **Thorpe, J. R., Ray, K. P., and Wallis, M.,** *J. Endocrinol.,* 124, 67, 1990.
182. **Kineman, R. D., Faught, W. J., and Frawley, L. S.,** *Endocrinology (Baltimore),* 128, 2229, 1991.
183. **Losinki, N. E., Horvath, E., Kovaks, K., et al.,** *Anat. Ariz.,* 172, 11, 1991.
184. **Frawley, L. S. and Boockfor, F. R.,** *Endocr. Rev.,* 12, 337, 1991.
185. **Hashimoto, S., Fumagalli, G., Zanini, A., et al.,** *J. Cell Biol.,* 105, 1579, 1987.
186. **Kineman, R. D., Faught, W. J., and Frawley, L. S.,** *J. Endocrinol.,* 134, 91, 1992.
187. **Kineman, R. D., Henricks, D. M., and Faught, W. J.,** *Endocrinology (Baltimore),* 129, 1221, 1991.
188. **Kineman, R. D., Faught, W. J., and Frawley, L. S.,** *Endocrinology (Baltimore),* 130, 3289, 1992.
189. **Stefaneanu, L., Kovacs, K., Lloyd, R. V., et al.,** *Virchows Arch. B.,* 62, 291, 1992.
190. **Porter, T. E., Hill, J. B., Wiles, C. D., et al.,** *Endocrinology (Baltimore),* 127, 2789, 1990.
191. **Boockfor, F. R., Hoeffler, J. P., and Frawley, L. S.,** *Am. J. Physiol.,* 250, E103, 1986.
192. **Frawley, L. S. and Hoeffler, J. P.,** *Peptides,* 9, 825, 1988.
193. **Nikitovitch-Winer, M. B., Atkin, J., and Maley, B. E.,** *Endocrinology (Baltimore),* 121, 625, 1987.
194. **Farrow, S. M.,** *J. Endocrinol.,* 138, 363, 1993.
195. **Hoeffler, J. P., Bookfor, F. R., and Frawley, L. S.,** *Endocrinology (Baltimore),* 117, 187, 1985.
196. **Mulcahey, J. J. and Jaffe, R. B.,** *J. Clin. Endocrinol. Metab.,* 66, 24, 1987.
197. **Voss, J. W. and Rosenfeld, M. G.,** *Cell,* 70, 527, 1992.
198. **Asa, S. L., Kovacs, K., Horvath, E., et al.,** *Neuroendocrinology,* 48, 423, 1988.
199. **Borrelli, E., Heyman, R. A., Arias, C., et al.,** *Nature (Lond.),* 339, 538, 1989.
200. **Pan, W. T., Liu, Q., and Bancroft, C.,** *J. Biol. Chem.,* 265, 7022, 1990.
201. **Gaido, M. C. and Strobl, J. S.,** *Biochim Biophys. Acta,* 1008, 234, 1989.
202. **Cherington, P. V. and Tashjian, A. H.,** *Endocrinology (Baltimore),* 13, 418, 1983.
203. **Porter, T. E., Chapman, L. E., Van Dolah, J. M., et al.,** *Endocrinology (Baltimore),* 128, 792, 1991.
204. **Porter, T. E.,** *Endocrinology (Baltimore),* 129, 2707, 1991.
205. **Porter, T. E. and Wiles, C. D.,** *Endocrinology (Baltimore),* 129, 1215, 1991.
206. **Stratmann, I. E., Ezrin, C., and Sellers, E. A.,** *Cell Tiss. Res.,* 152, 229, 1974.
207. **Porter, T. E., Ellerkmann, E., and Frawley, L. S.,** *Mol. Cell Endocrinol.,* 84, 23, 1992.
208. **Boockfor, F. R., Hoeffler, J. P., and Frawley, L. S.,** *Endocrinology (Baltimore),* 117, 418, 1985.
209. **Lloyd, R. V., Coleman, K., Fields, K., et al.,** *Cancer Res.,* 47, 1087, 1987.
210. **Osamura, R. Y.,** *J. Electron Microsc. Tech.,* 19, 57, 1991.
211. **Horvath, E., Kovacs, K., Scheithauer, B. W., et al.,** *Ultrastruct. Pathol.,* 5, 171, 1983.
212. **Thapar, K., Stefaneanu, L., Kovacs, K., et al.,** *Endocr. Pathol.,* 4, 1, 1993.
213. **Lira, S. A., Crenshaw, E. B., Glass, C. K., et al.,** *Proc. Natl. Acad. Sci. U.S.A.,* 85, 4755, 1988.

Chapter 4

Growth Hormone Synthesis

Stephen Harvey

CONTENTS

I. INTRODUCTION

In this chapter the biosynthesis of growth hormone (GH) within somatotroph cells is discussed. The synthesis (and release) of GH in extrapituitary (reproductive, immune, and neural) tissues is reviewed in Chapters 21, 22, and 26, respectively. The biosynthesis of GH is regulated by numerous transcriptional and translational factors, some of which have been detailed elsewhere.[1–5]

II. THE GH GENE

Pituitary GH is expressed, in humans, by the *hGH-N* gene, a member of the *GH*-chorionic somatomammotropin (*hCS*) gene family.[6] This family consists of five similar genes (*hGH-N*, *hCS-L*, *hCS- A*, *hGH-V*, and *hCS-B*) aligned on the long arm of chromosome 17, of which only the *hGH-N* gene is expressed in somatotroph cells (Figure 1). The *hGH-N* gene, which is contained within a 2.6-kb DNA fragment,[7] is composed of five exons and four introns and 3′ and 5′ untranslated regions (including a 500-bp promoter region). This gene is similar in structure to other mammalian *GH* genes[8] but is smaller than the chicken *GH* gene (3.5 kb), which contains expanded intron sequences.[9] The

0-8493-8697-7/95/$0.00+$.50

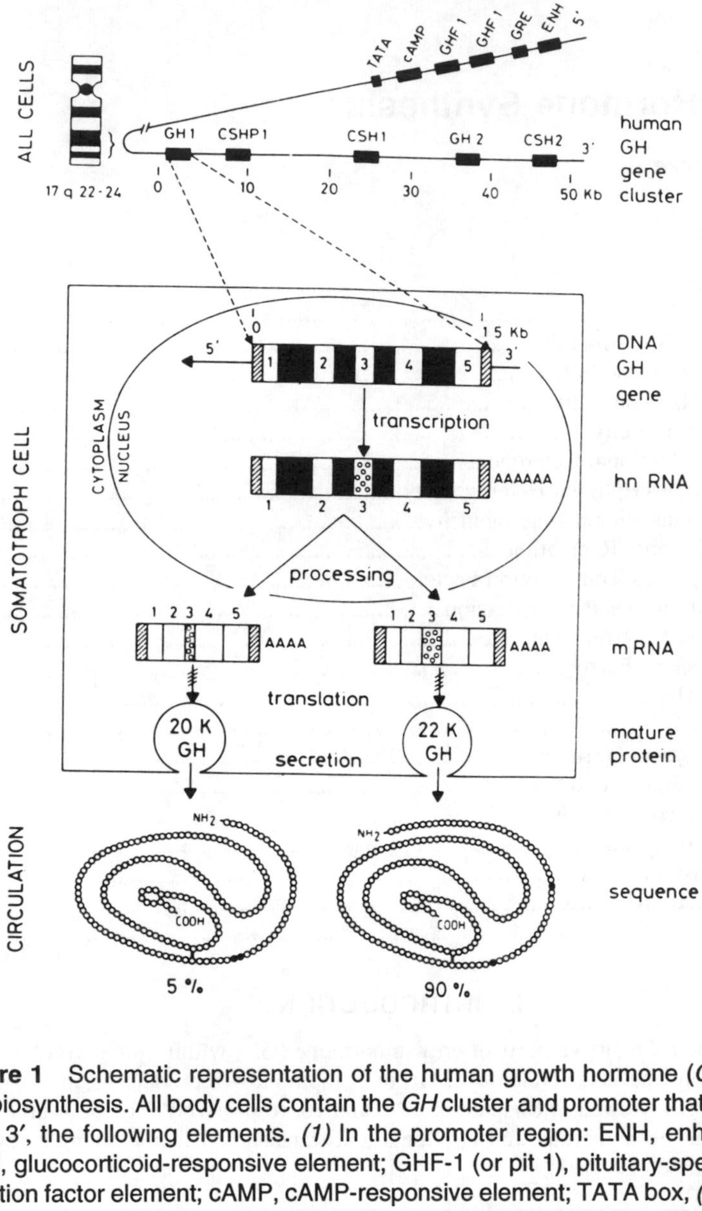

Figure 1 Schematic representation of the human growth hormone (*GH*) gene and GH biosynthesis. All body cells contain the *GH* cluster and promoter that contain, from 5′ to 3′, the following elements. *(1)* In the promoter region: ENH, enhancer region; GRE, glucocorticoid-responsive element; GHF-1 (or pit 1), pituitary-specific GH transcription factor element; cAMP, cAMP-responsive element; TATA box, *(2)* In the gene cluster region: *GH1*, gene encoding 22K or normal GH; *CSHP1*, *CSH1*, and *CSH2* encode placental lactogen forms and are expressed only in the placenta. *GH-2* encodes the GH variant expressed solely in the placenta. The *GH-1* gene is expressed only in the somatotroph cells of the pituitary. It contains five exons (white blocks numbered) and four introns. The RNA transcript is processed in two alternative forms by an alternative splicing at the level of exon 3. This process leads the somatotroph to deliver two proteins to the circulation, that is, the 22K or normal GH and the 20K variant, which lacks 15 amino acids (residues 31 to 47). (From Casanueva, F. F., Physiology of growth hormone secretion and action, *Endocrinol. Metab. Clin. N. Am.*, 21(3), 483–518, 1992. Copyright by W. B. Saunders. Reproduced with permission.)

hGH-V gene encodes a protein with structural homologies to pituitary GH but is expressed only in the placenta. The *hCS* genes express placental lactogens, which have close homology (85%) to hGH (Figure 1, Chapter 1). Placental lactogens in subprimate mammals more closely resemble prolactin, having evolved by duplication of the prolactin gene rather than of the *GH* gene.[10] Consequently only a single *GH* gene is present on chromosome 17 in most vertebrates, although the possibility that viviparous fish or reptiles may possess other genes belonging to the *hCS* family has yet to be determined. An additional, nonallelic *GH* gene is, however, present in salmonid pituitary glands[11–16] and both *GH* genes are equally expressed in Atlantic salmon.[17] A third *GH* gene, a pseudogene, is also present in Chinook salmon, but it resides on the Y chromosome and is found only in males.[18] Another gene is also present in the pituitary of teleost fish (salmon and flounder) that expresses a protein, somatolactin, that is distinct from but related to GH and prolactin (Figure 2, Chapter 1).[19] Somatolactin is synthesized and secreted from the pars intermedia,[20] unlike the synthesis of GH in the proximal pars distalis and the synthesis of prolactin in the rostral lobe. Antisera raised against this peptide have no cross-reactivity with proteins present in frog, pigeon, or rat pituitary glands, suggesting it is not synthesized in higher vertebrates.[21–23] The chromosomal location of the *GH* gene(s) may also vary in different vertebrate groups, because the *GH* gene[24] is on the long arm of chromosome 1 in chickens.

III. *GH* GENE ABNORMALITIES

The structure of the *GH* gene is defective in some individuals because of gene deletions. These gene abnormalities result in GH deficiency or the synthesis of defective GH. For instance, dwarfism associated with isolated GH deficiency type IA (IGHD IA) is characterized by a *GH* gene deficiency (reviewed by Abdul-Latif *et al.*[25]). These deficiencies include gene deletions of 6.7, 7.0, and 7.6 kb, resulting in the loss of the entire *GH* gene and of most of the 5′ or 3′ flanking regions.[26–29] Larger deletions obliterating multiple *GH* and/or *CS* genes have also been identified[30–32] (Figure 2). The genetic dysfunction responsible for these deletions is largely unknown, but the 6.7-kb deletion may be due to uneven crossing over.[26] The 6.7-kb deletion appears to be the most prevalent in patients with IGHD IA,[33] <10% of which have deletions of 7.0 and 7.6 kb.

Mutations or deletions of a single base pair can also eliminate *GH* expression. In humans, a G → A transversion creates a stop codon in the signal peptide, thus encoding a severely truncated, biologically inactive GH molecule.[34] Conversely, a G → C transition at the intron 4 splice site results in the retention of intron 4, usage of an alternate translational stop codon, and thus a truncated GH molecule lacking exon 5.[34] The spontaneous dwarf rat is also characterized by a point mutation at a splice site resulting in premature termination of translation.[35]

IV. *GH* GENE TRANSCRIPTION

The transcription of the *GH* gene is by RNA polymerase II. Transcription is initiated from a 5′ start codon downstream from a TATA box and an adenosine transcription start point. Polyadenylation signal sequences (AATAAA) at the 3′ end terminate transcription of the gene. The coding regions of the human[36,37] rat,[38,39] porcine,[40] bovine,[41] and ovine[42] genes are highly conserved but differ from those in birds[9] and some fish.[16,43]

Transcription of the *GH* genes is regulated by enhancer or silencer *cis*-acting DNA sequence elements that associate with *trans*-acting transcriptional regulators (Figure 3). The promoter regions of the rat and human *GH* genes extend 500 bp 5′ to the start of gene transcription and show extensive sequence similarities, although transcriptional regula-

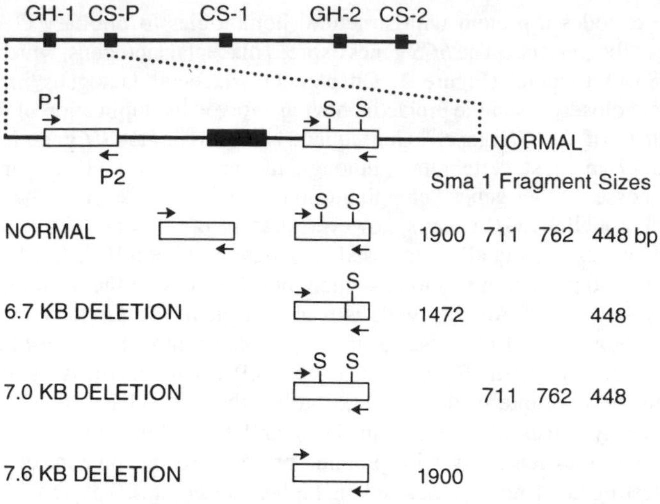

Figure 2 Growth hormone (*GH*) and chorionic somatomammotropin (*CS*) gene deletions. The normal distribution of the different genes and gene deletions in the *GH-CS* cluster. The schematic shows the various types of deletions that involve *GH-1* gene alone, the larger deletion that encompasses the *CS-1*, *CS-2*, and *GH-2* genes, the combined deletion that involves all the genes of the cluster, sparing only the *CS-P* gene, and the 45-kb deletion that spares only the CS-2 gene. (From Abdul-Latif, H., Brown, M. R., and Parks, J. S., Characteristics of GH-1 deletions by PCR, *Endocrinologist*, 3(6), 409–414, 1993. Reprinted with permission.)

tion is mainly regulated by sequences downstream of –320. The proximal region of the promoter encompasses a TATA box and upstream binding sites for ubiquitous and pituitary-specific transcription factors.[44,45] The promoter region of the chicken *GH* gene has no overall homology with those in mammals but contains a short (24-bp) sequence that is highly homologous to proximal binding sites in rat GH for the GHF-1/pit-1 transcription factor.[9] Short nucleotide sequences in the promoter region of teleost GH genes are also similar to consensus core DNA-binding sites for GHF-1/pit-1.[43] The promoter regions of fish *GH* genes lack overall homology with mammalian *GH* genes, but have short consensus sequences in proximal regions flanking their TATA boxes, suggesting a similarity of regulation.[43] The control of *GH* gene transcription in lower vertebrates may, however, differ significantly from that in mammals, in which the genetics of *GH* gene expression has been reviewed (see Parks[46]).

The primary transcription product of the mammalian *GH* gene is a premessenger RNA that is approximately 1300 nucleotides long and contains five exons and four intron sequences. The nucleotide sequences of fish genes are also larger than in mammals (1.7 kb in *Tilapia nilotica*, 1.4 kb in *Salmo salar*. 2.4 to 2.5 kb in carp, and 2.6 kb in other salmonids).[15,16,43,47] The primary transcripts of the *GH* genes in *Tilapia* and *Salmo* species also differ from those in mammals in having six exons and five introns.[43] During transcription of the *GH* gene the primary transcript is polyadenylated at the 3′ end and capped at the 5′ end by the addition of 7-methylguanine triphosphate and by methylation (Figure 3). The intron sequences are removed minutes later, beginning with intron 1,[48] and the exons are spliced to produce mature messenger RNA (mRNA), which has a half-life of 40 to 60 min.[49,50] The RNA precursors are thought to remain in the nucleus, in which splicing occurs,[51,52] although intron lariats have been located in the cytoplasm, where they may be degraded.[53] The release of mature mRNA into cytoplasm is a two-step process that

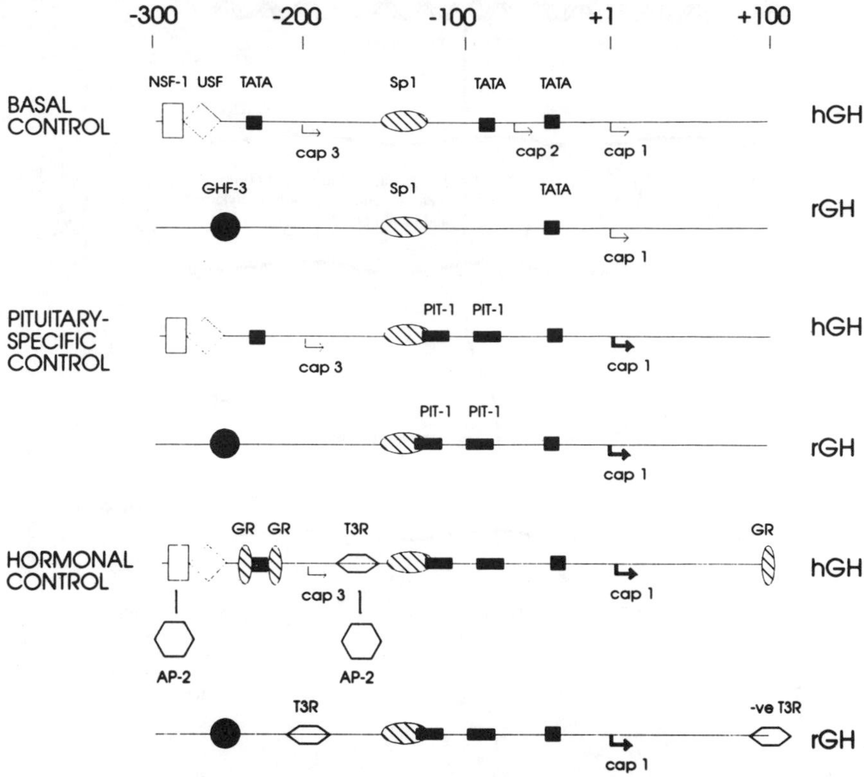

Figure 3 Model for the control of human (h) and rat (r) growth hormone (GH) gene transcription by trans-acting factors. The scale refers to base pairs relative to the transcription initiation site determined *in vivo* (cap 1). This site is indicated by a wider arrow when it is more active. GR, Glucocorticoid receptor; T3R, thyroid hormone receptor; NSF-1, nuclear specific factor 1; USF, upstream stimulatory factor; SP1, GH box-binding protein; GHF-3, GH factor 3; PIT-1, pituitary-specific transcription factor 1; AP-2, activator peptide 2. (Based on Rousseau[3] and Lemaigre et al.[87])

involves release from the nuclear matrix by an ATP-dependent mechanism and translocation through nuclear pore complexes, by a mechanism requiring ATP hydrolysis.[51,52]

V. *GH* mRNA TRANSLATION

The subsequent intracellular sequence of protein biosynthesis, processing, and trafficking (Figure 4) may take several hours to complete,[54–56] although newly synthesized hormone may be secreted minutes after the onset of translation.[57–59] Translation is initiated in cytoplasm by the interaction of 40S ribosomal complexes with the 5′ end of GH mRNA, followed by the binding of 60S ribosomal units.[4] This association with free ribosomes stimulates GH synthesis by directing the assembly of amino acids, according to the GH mRNA nucleotide sequence. A preprotein "prepro GH" of approximately 225 amino acids is first produced and subsequently processed to yield the secreted protein.[60] The removal of the signal peptide does not, however, ensure secretion of the processed hormone, because truncated GH moieties lacking the signal sequence remain in the pituitary gland.[61] The initial signal peptide or presequence of the preprotein causes the

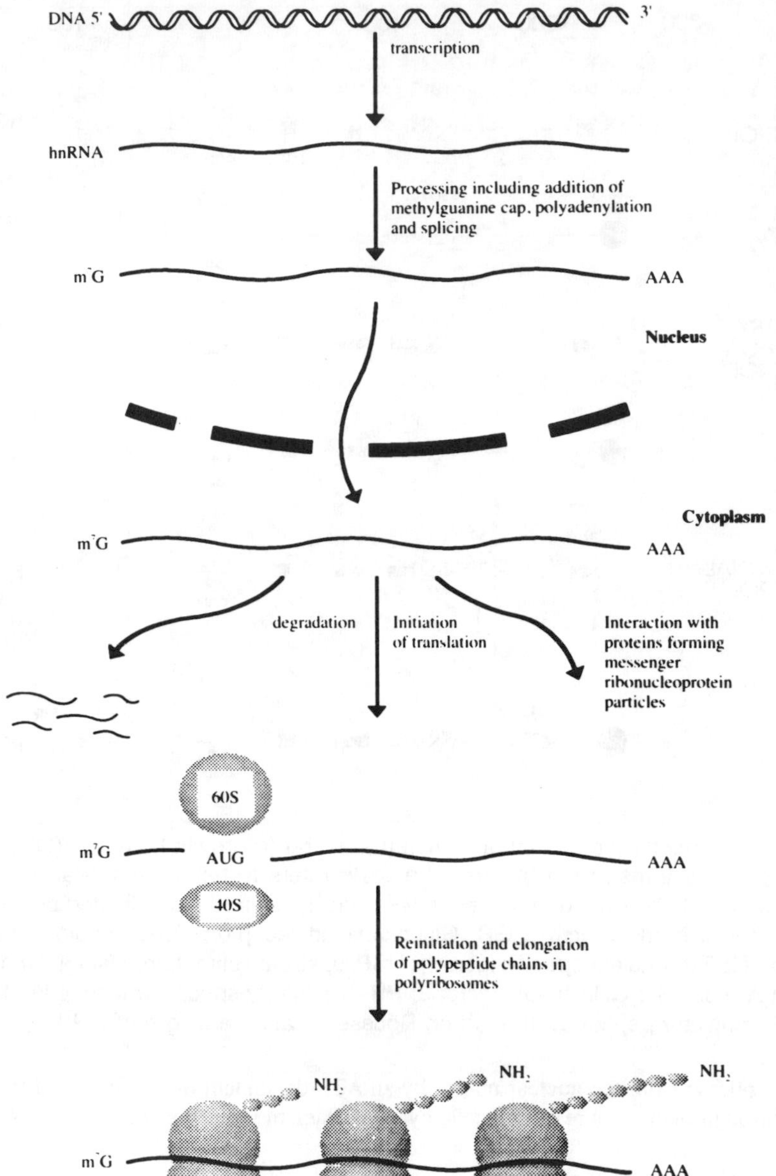

Figure 4 The protein synthetic pathway from gene transcription to polypeptide chain production on ribosomes. Following gene transcription in the nucleus, primary transcripts (heterogeneous RNA, hnRNA) undergo processing during which a 7-methylguanine cap (m^7G cap) is added to the 5′ end of RNA and a poly(A) tract is added to the 3′ region. Introns are removed by splicing and mature RNA transcripts move through nuclear pores into the cytoplasm, where they may be translated immediately, degraded, or may interact with proteins to form messenger ribonucleoprotein particles. Translation of mRNA is initiated by interaction of 40S and 60S ribosomal complexes with the initiator codon (AGG) followed by reinitiation with additional ribosomes interacting with mRNA to form polyribosomes containing a number of elongating polypeptide chains. (From Farrow, S. M., *J. Endocrinol.,* 138, 363, 1993. Reproduced by permission of the Journal of Endocrinology, Ltd.)

mRNA-ribosome complex to attach to the endoplasmic reticulum (ER) and facilitates translocation into the intracisternal space of the ER for processing and storage.[62] The cisternae therefore become more prominent in actively synthesizing somatotrophs and the number of free and membrane-bound polysomes increases.[63] The 20- to 30-amino acid signal peptide is cleaved by proteolytic processing in the ER, in which posttranslational modification of the GH sequence (including disulfide bond formation) may occur. The processing of the hormone may also continue in the Golgi, into which it is translocated by an energy-dependent process. The stimulation of Golgi function is accompanied by its enlargement and occupation of a large somatotroph area and the production of numerous multivesicular bodies that aggregate in the Golgi region.[63] In addition to processing, sorting of the GH moieties and their packaging into secretory granules also occur in the Golgi.[64] The transport of GH into these vesicles may be facilitated by "binding proteins" located in the Golgi. Further processing of the hormone may also occur within the granular storage compartments of somatotrophs. Although newly synthesized granules or cytoplasmic hormone may be released without storage,[59,65–67] most of the GH synthesized is stored within the secretory granules of GH-secreting cells. These vesicles are stored in the cytoplasm or juxtaposed to the cell membrane prior to release or crinophagic destruction. The amount of GH stored within somatotrophs appears to participate in the regulation of *GH* gene expression, which is inversely related to the degree of somatotroph granulation.[68] Moreover, the depletion of pituitary GH stores is a signal for GH gene expression when GH release is stimulated by GH-releasing hormone (GHRH), although GHRH is also able to increase gene transcription independently of GH release.[69] The accumulation of multivesicle organelles in the Golgi and their transformation into lamellar bodies or lytic vesicles containing Golgi membranes also ensures that overproduction of GH by excess Golgi activity does not occur following secretagogue stimulation.[63] The amount of GH synthesized in somatotrophs accounts for approximately 10 to 20% of all cell protein under basal conditions, although this proportion is increased by GH secretagogues.[58] In contrast, GH accounts for <1.0% of the protein synthesised in rat pituitary tumor cells.[70–73]

VI. GH HETEROGENEITY

Although pituitary GH is transcribed from a single gene, a family of structurally heterogeneous GH moieties exists in all vertebrate classes.[70,74,75] This heterogeneity is due to transcriptional, posttranscriptional, and posttranslational events.

A. TRANSCRIPTIONAL HETEROGENEITY

Two alternate promoters in the *hGH* gene have been identified: -54 (cap 2) and -197 (cap 3) from the transcriptional initiation site (cap 1).[76] Putative TATA boxes have also been located upstream of these alternate sites. The initiation of transcription from these sites would result in transcripts with novel exon 1 sequences.

B. POSTTRANSCRIPTIONAL HETEROGENEITY

Posttranscriptional modifications of the primary *GH* gene transcript can result in GH heterogeneity. These modifications include differences in mRNA splicing, because multiple splice sites have been described for the *hGH* and bovine growth hormone (*bGH*) genes, the use of which may be regulated by mRNA secondary structures. Alternate splicing of *hGH* mRNA results in the production of a 20-kDa GH, missing amino acids 32–46 (Figure 1), although this is produced much less (ninefold) than full-length hGH.[77] Differential splicing of rat *GH* mRNA can also result in the deletion of exon 3, or the use of an alternative splice within exon 3, resulting in the synthesis of GH molecules lacking amino acids 32 to 71 and 42 to 46, respectively.[6,78] Heterogeneity may also result from the incomplete cleavage of intron sequences from the primary gene transcript. Intron D, for instance, is sometimes retained in the bovine *GH* transcript as a result of a 35-kDa *trans*-acting factor that binds to *cis* sequences located in exon 5.[79] Intron D retention is also observed in a proportion of *hGH-V* and *hCS-A* transcripts.[80] Heterogeneity may also result from differences in polyadenylation of the primary *GH* gene transcript. Multiple 3′ cleavage/polyadenylation sites exist in the bovine and salmon *GH* genes,[43,81] which consequently may modify mRNA stability and the efficiency of translation, because polyadenylation is required for efficient termination of transcription and the initiation of translation.

C. POSTTRANSLATIONAL HETEROGENEITY

Posttranslational modifications of GH include deamination and proteolytic processing, acetylation, glycosylation, phosphorylation, aggregation, protein binding, and the formation of multiple variants differing in mass and charge. These modifications are detailed in Chapter 1. These heterogeneous forms have qualitative and quantitative differences in biological activity[6,70,75,82] and can be differentially expressed in a sexually dimorphic way,[61,83] during growth[61] and pregnancy,[84] and may be differentially synthesized[58] and released[85,86] by provocative stimuli.

VII. TRANSCRIPTIONAL REGULATION

The transcription of the *GH* gene is regulated by many pituitary-specific enhancers and repressors, by ubiquitous transcription factors, and by endocrines (Figure 3) and has been reviewed by Theill and Karin.[5]

A. UBIQUITOUS TRANSCRIPTION FACTORS

Transcription factors present in both nonpituitary (HeLa) and pituitary (GC) cells mediate basal levels of gene transcription. DNA footprinting experiments with extracts of HeLa or pituitary cells have demonstrated the presence of two protected regions located at –116/–140 (bound by GHF-2) and –241/–290 (bound by GHF-3)[87] of the *hGH* gene. Similar regions at –129/–147 and –219/–239 have been localized in the *rGH* gene.[88] GHF-2 is identical to the Sp-1 transcription factor (also known as the GC box-binding protein) in both rats and humans, whereas GHF-3 corresponds to two proteins: upstream stimulatory factor 1 (USF-1) (binding –253/–266) and nuclear factor 1 (NF-1) (binding –268/–290) in humans.[44,45,88–92] The roles of these factors are unclear.[5] The factor(s) corresponding to the GHF-3 footprint in rats has not been conclusively identified,[5] but appears to correspond to a 43-kDa protein acting as a docking site between nucleotides –239 and –219 for at least five other transcriptional proteins.[91] The plasticity of *GH* gene transcription in the rat may therefore be enhanced if each of these transcriptional factors exerts different transcriptional effects. The proximal region of this GHF-3 binding site (–263/–290) can also be bound by another ubiquitous factor, activation protein 2 (AP-2).[92] AP-2 can also bind to the *hGH* promoter within the GHF-3 binding site, between nucleotides –142 and –168.[93] This indicates that NF-1 and AP-2 compete for overlapping sites in the *GH* promoter. AP-2 mediates cAMP and phorbol ester responses of some gene promoters but its role in *GH* transcription is unknown.[5] However, as oligo AP-2 is unable to reduce the size of the GHF-3 footprint, AP-2 may exert only minimal transcriptional control on the *hGH* gene.[89,90]

Although GHF-2 and GHF-3 can drive low levels of *hGH* gene transcription in nonpituitary cells, their role in pituitary GH synthesis is unclear. Fusion of pituitary GC cells and fibroblast cells extinguishes *hGH* gene expression and also results in a loss of GHF-3-binding activity, suggesting a close relationship between GHF-3 and GH synthesis.[94] Moreover, mutation of the GHF-3-binding region of the *rGH* gene reduces transcription by 30%.[91] However, Lemaigre et al.[90] observed GHF-3 binding in non-*GH*-expressing cell types and adequate *hGH* gene transcription in the absence of this factor. Conversely, the GHF-2/Sp-1-binding site must be intact for maximal *rGH* expression in GC cells.[91] This dependence is curious, as the Sp-1-binding site overlaps the distal pit-1 response element and Sp-1 binding and transcriptional stimulation are observed only in the presence of very low levels of pit-1,[90,91] and GHF-1 and Sp-1 binding may be mutually exclusive.[5] A stepwise process of promoter activation involving both factors has thus been suggested.[91]

The binding of these ubiquitous factors to the *GH* promoter in pituitary cells is thought to result from an "opening" of the chromatin structure, mediated by other transcriptional regulators.[5] In nonpituitary cells the inability of these factors to activate the *GH* promoter is thought to result from a chromatin barrier that excludes access to their DNA-binding sites (Figure 5). The transcriptional regulators opening the *GH* chromatin structure may include the hormone-activated nuclear receptors for thyroid hormones, retinoic acid, and glucocorticoid hormones (Figure 5). The binding of these activated nuclear receptors may cause rearrangements of the nucleosomes over the *GH* promoter, thereby allowing access for pit-1 and other regulatory factors. The specificity of *GH* gene transcription may thus depend on the composition of the nuclear receptor population and the availability of their ligands and the presence of pituitary-specific transcriptional regulators.

An additional novel transcription factor may also be involved in regulation of the *rGH* promoter. Lipkin et al.[95] systematically examined evolutionary well-conserved regions in the proximal *GH* promoter and tested the effects of their disruption on transcription in a somatotroph (GC) cell line. The functionally most important sequence (–95/–110), designated as the GHZ box, is unusually well conserved across species, and a mutation in this site decreased transgene expression more than 100-fold. The same group of investigators identified a factor (Zn-15) binding to this site, which has an unusual DNA-binding domain consisting of three $CysX_{2-4}CysX_{11-16}HisX_{3-6}His$ zinc fingers. Schaufele et al. suggest Zn-15 is thus likely to correspond to a binding activity referred to as GHF-5[88,91] and its synergy with the pituitary specific transcription factor pit-1/GHF-1 may be required for effective activation of the *GH* gene.[95] These ubiquitous *trans*-acting factors are responsible for the basal activity of the *GH* gene and can stimulate transcription from the promoter *in vitro* when derived from nonpituitary cell extracts, indicating that these factors are not restricted to the pituitary gland.

B. PITUITARY-SPECIFIC EXPRESSION

Pituitary expression of *GH* is largely controlled by pit-1/GHF-1, which has been located in normal and tumorous rat and human pituitaries and normal trout pituitaries,[96–99] but not in fat, brain, fibroblasts, hepatocytes, placenta, or HeLa cells.[98,100,101] This factor is able to activate the *GH* gene promoter in HeLa (nonpituitary) cells, which do not normally produce pit-1 or GH.[102] Its deficiency or dysfunction in subsets of human and mouse (Snell) dwarfs is also associated with an absence or deficiency of GH biosynthesis,[100,103,104] whereas its ontogenic appearance in the pituitary gland correlates with the activation of the *GH* gene.[2]

Pit-1 is a 33-kDa polypeptide that recognizes the nucleotide sequence TTATG/CCAT and may exist in two variant forms. A cDNA encoding a variant isoform of pit-1/GHF-1 has been isolated and designated pit-1β. This naturally occurring isoform contains an insertion of 26 amino acids in the *trans*-activational domain, a consequence of utilization of an alternative 3′ splice acceptor site at the end of the first intron.[105] Pit-1β activates only a subset of the known pit-1/GHF-1 target genes;[106] however, this isoform acts as a more potent inducer of *GH* than the pit-1α form originally isolated.[105] This may be accounted

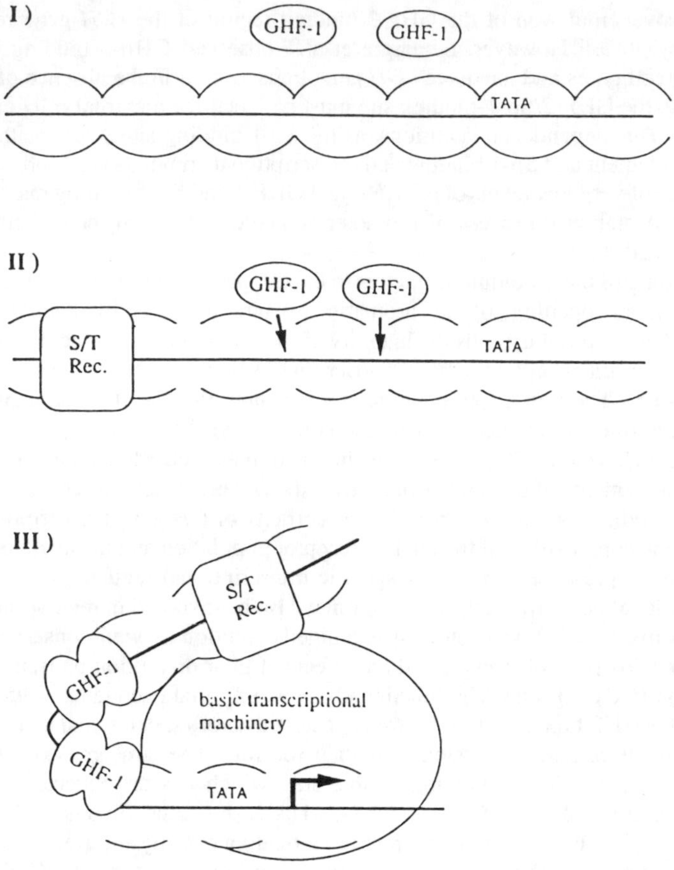

Figure 5 Model for transcriptional control of *GH* gene expression *in vivo.* (I) Closed arches illustrate a closed chromatin structure over the *GH* promoter in all cell types except somatotrophs and mammosomatotrophs. GHF-1 is excluded from its binding site. (II) A somatotroph-specific steroid and/or thyroid hormone receptor activity (S/T Rec.) binds to the *GH* promoter and causes nucleosome rearrangement, thereby providing GHF-1 access to GHF-1-binding sites. (III) Binding and multimerization of GHF-1 result in bending of DNA, which facilitates the relative positioning of upstream binding factors and the assembly of an active transcriptional complex, resulting in transcription of the gene. (From Theill L. E. and Karin, M., Transcriptional control of growth hormone expression and anterior pituitary development, *Endocr. Rev.*, 14(6), 670–689, 1993. Copyright by the Endocrine Society. Reprinted with permission.)

for by the different ways in which the two isoforms act with other transcription factors. Generation of variant transcription factors by differential splicing of a primary transcript could provide an additional mechanism for differential gene regulation.

Pit-1 activation of the *GH* promoter in the *rGH* and *hGH* genes is mediated by DNA elements at two adjacent sites approximately located at positions –105/–132 and –94/–66 from the transcription start site. Both sites are necessary for effective promoter function in tumorous rat somatotrophs (GH_3 and GC cells) and in mice transfected with the *rGH* promoter.[107] In contrast, only a single binding site has been identified in the trout *GH* gene, at position –160/–169.[97] These sites include zinc finger protein-binding elements that interact with pit-1 in stimulating *GH* gene expression.[108] The binding of pit-1 to these sites may directly stimulate DNA bending and promoter activity,[5] which may also result from the displacement of the nucleosome and exposure of the gene to other stimulatory factors.[109,110] Moreover, because the proximal pit-1-binding site overlaps the cap 2 TATA box, the binding of pit-1 limits the initiation of transcription to the cap 1 site.[109] Pit-1 is also capable of binding to the promoters of the *hCS-A*, *hCS-B*, and *hGH-V* genes[87,89,111,112] and of inducing, through the proximal binding site, *hCS-A* gene expression in GC transfected cells.[113] In contrast to *hGH*, the cap 2 site may be active in *hCS* and/or *hGH-2*, as *hCS* transcripts initiated at the cap 2 homolog occur in the human placenta.[114] Alternate transcription sites regulated by pit-1 may therefore provide a mechanism for tissue-specific and/or temporally specific production of transcription products and GH isoforms. Pit-1 also enhances the promoter activity of the somatolactin gene in teleost pituitaries and may regulate the development of somatolactin-producing cells.[115]

Pit-1 expression alone, however, is insufficient to induce GH synthesis, because immortalized somatotroph precursors contain adequate levels of pit-1 but inactive *GH* promoters.[116] Pit-1 modification or enhancement or additional factors may be cooperatively required for *GH* gene transcription. Moreover, truncated *rGH* promoters that contain only the two pit-1-binding sites are only minimally active in transgenic mice.[107] Additional factors may, therefore, be required for *GH* gene transcription, and DNA sequences 1.7 kb upstream of the pit-1 response elements are required for optimal *GH* expression.[107] One of these factors may be Sp-1, which has overlapping binding sites within the *GH* promoter.[91] Pit-1 also appears to act as a cofactor in thyroid hormone-induced *GH* gene transcription.[117] The demonstration of GH biosynthesis in the placenta, brain, and pineal gland also indicates that factors other than pit-1 are capable of inducing *GH* gene transcription.

C. TISSUE-SPECIFIC REPRESSION

Although pit-1 stimulates *GH* transcription in pituitary somatotrophs, the *pit-1* gene may also be negatively regulated in nonpituitary tissue, because the extinction of *GH* expression in GH_3 cells caused by fusion with mouse fibroblasts is accompanied by a loss of *pit-1* expression and DNA binding.[118,119] Pit-1 is, therefore, absent from most extrapituitary tissues, although it is present in hemopoietic and lymphoid tissues,[98] which also express the *GH* gene.[120–126] The pituitary-specific repression of the *hCS-A*, *hCS-B*, and *hGH-V* genes is thought to result from the binding of another regulator (pituitary-specific factor, PSF-1) to a site 2 kb upstream of the transcriptional start site. This binding site is lacking in the *hGH-N* gene, allowing its specific expression in the pituitary gland.[127] Specific expression of the *hGH-N* and chorionic somatomammotropin genes in the placenta reflects its deficiency of PSF-1, which is otherwise ubiquitously present in extraplacental tissues.

Repression of *rGH* gene expression in nonpituitary cells is mediated by site-specific and gene-specific DNA methylation of cytosine bases[128–130] and hypomethylation is correlated with increased *GH* expression in pregnancy.[131] A tissue-specific correlation between hypomethylation and gene activity has also been reported for *hGH*.[132] *rGH* gene expression was similarly correlated with an unmethylated *Tha* I site (CGCG), located 144 bp of the *GH* mRNA transcription site. This site was entirely methylated in DNA prepared from non-GH-producing tissues (spleen, kidney, liver, and brain).[129] In tissues not expressing the *GH* gene, such as the liver, a higher level of methylation has been demonstrated in comparison to that seen in the pituitary.[133] In addition to this tissue-specific effect, blocking the action of methyltransferase can increase GH mRNA levels three- to eightfold in GH_3 cells.[134] This effect likely occurs within the 5′-flanking DNA of the *GH* promotor,[128] as demonstrated for many other genes.[135]

Repression of the *GH* gene may also occur from the binding of nuclear factor 1 or activator peptide (AP-2) to silencer sequences –309 to –266 from the transcription start site.[92,94,136–138] An additional repressive factor (PREB), found only in nonpituitary cells, can also bind to the *GH* promoter between nucleotides –169 and –152.[137] Another repressor factor of unknown identity also binds to the rGH promoter upstream of the nucleotide at position –500.[136] Activin may also be a silencer, because it inhibits pit-1 binding to the promoter and consequently inhibits *GH* mRNA accumulation and somatotroph proliferation.[139] However, most of the factors inhibiting *GH* gene transcription do not hinder access of stimulatory factors to their binding sites, because the silencer elements of the gene do not overlap with positive elements. The actions of the inhibitory transcription factors must therefore be neutralized by stimulatory factors for activation of *GH* gene transcription.

D. ENDOCRINE FACTORS

The biosynthesis of GH is not autonomous and is regulated by many factors that exert stimulatory or inhibitory control (Table 1), although GH responsiveness may be modified by age and sex and by reproductive status and other physiological states.[140,141] These factors include ions (e.g., calcium[69] and sodium[142]), metabolites (e.g., free fatty acids[143]), neurotransmitters (e.g., catecholamines[144] and indolamines[145]), second messengers (e.g., cAMP,[146] inositol phosphates,[147] prostaglandins,[148] calcium[149]), neuropeptides (e.g., GHRH,[66] SRIF,[58] TRH,[146] and VIP[150]), growth factors (e.g., EGF, FGF,[151] and IGF-1[152]), pituitary hormones (e.g., prolactin[153]), thyroid hormones (e.g., triiodothyronine [T_3][154]), adrenal hormones (e.g., cortisol[155]), pancreatic hormones (e.g., insulin[152]) and gonadal hormones (e.g., estrogens,[84,156] androgens,[157] and activin[158]). These factors may interact synergistically (e.g., glucocorticoid and thyroid hormones, with or without GHRH) or antagonistically (e.g., GHRH and IGF-I),[159] and/or act transcriptionally (e.g., GHRH, thyroid hormones, glucocorticoids, and insulin) or translationally (e.g., cortisol), although for many the mechanism and locus of action are not known.

1. GHRH

GHRH may not be obligatory for *GH* gene transcription, because low levels of GH mRNA occur in the pituitary glands of GHRH-resistant *CrL* mice and *dw* rats.[160] *GH* gene transcription is, however, increased by GHRH. GHRH increases *GH* pre-RNA accumu-

Table 1 **Modulation of *GH* Gene Expression**

Regulator/Condition	Species	GH mRNA	Ref.
Growth hormone (rat)	Mice	↓	251
Growth hormone (human)	Mice	↓	246,247
Growth hormone (bovine)	Mice	↓	252
	Human (tumor)	←	253
GH-releasing hormone	Human	↑	131,254
	Rat	↑	53,170,254
	Cow	↑	216
	Chicken	↑←	255
Somatostatin	Rat	←	53,169
	Rat	↓	166,183
Triiodothyronine	Human	↑	195
	Rat	↑	201
	Cow	↓	216
	Fish	↑	198
Thyroxine	Rat	↑	256
Glucocorticoid	Human	↑	196

Table 1 (continued) **Modulation of *GH* Gene Expression**

Regulator/Condition	Species	GH mRNA	Ref.
	Rat	↑	228
	Cow	↓	216
Insulin-like growth factor I	Human	↓	238,254
	Rat	↓	254
Insulin	Human	↓	196
	Rat	↓	238,254,257
Activin	Rat	↓	250
Estradiol	Rat	↓	258
	Rat	↑	259
	Bovine	←	216
Morphine	Rat	↓	260
Tumor necrosis factor	Sheep	↑	261
Epidermal growth factor	Rat	←	254
Fibroblast growth factor	Rat	←	254
Retinoic acid	Rat	↑	223,224
Ethanol	Rat	↓	262
	Rat	←	263
Dopamine	Human	←	253
	Rat	←	183
Phorbol ester	Cow	↑	147
Cyclic AMP	Human	↑	253
	Rat	↑	175,182,264
Calcium	Rat	↑	265
Aging	Mouse	↓	266
	Rat	↓	267–269
(fetal)	Human	↑	141
	Rat	↑	270
	Sheep	←	271
(pubertal)	Rat	↑	272
(embryonic)	Chicken	↑	273
(neonatal)	Chicken	↓	273
(larval)	Bullfrog	↑	274
(neonatal)	Bullfrog	↓	274
(larval)	Sea bream	↑	275
Pregnancy	Human	↓	276
	Rat	↓	131
	Hamster	←	60
	Mouse	←↑	277
Lactation	Rat	↓	131
Estrous cycle	Hamster	←	60
Sexual dimorphism	Rat	↑♂	278,279
Food restriction	Sheep	↑	280
High-protein diet	Sheep	←	281
Obesity	Rat	↓	282
Seawater adaptation	Salmon	↑	283
Adrenalectomy	Rat	↓	256
Thyroidectomy	Rat	↓	256

Note: ↑ Increased; ↓, decreased, ←, unchanged.

lation in the nucleus, *GH* mRNA accumulation in the cytoplasm, and GH accumulation in somatotroph cells *in vitro*[53,58,69,147,161,162] and *in vivo*.[163–165] *GH* gene transcription is, conversely, suppressed by GHRH deficiency.[160,163,166–168] Stimulation of the transcription rate occurs within 10 min of GHRH exposure, although cytoplasmic *GH* mRNA levels may not be increased for several hours and may not be maximally increased until after 2 or 3 days of continuous stimulation *in vitro*.[169,170] The synthesis of GH is not induced, however, for at least 4 h after GHRH stimulation,[58,66] reflecting the temporal accumulation of cytoplasmic *GH* mRNA. Although GHRH induces a rapid rise in intracellular calcium concentrations, *GH* gene transcription occurs independently of this rise.[69]

The transcriptional action of GHRH is primarily mediated by cAMP (Figure 6), which increases GH synthesis *in vivo* and *in vitro*.[58,66,69,146,161,164,165,171,172] Activation of the adenylate cyclase-protein kinase A system stimulates *GH* gene expression[69,147,173] through a cAMP response element (CRE) located between nucleotides –82 and –9 of the *hGH* gene,[69,147,174,175] between nucleotides –104 and –51 of the *rGH* gene,[104] and between nucleotides –183 and 0 of the *bGH* gene.[176] A CRE at nucleotide –1256 has also been identified by Thomas et al.[177] in the *rGH* gene. Consensus sequences within these CREs have yet to be identified.[174,175]

The transcriptional activation induced by cAMP may be directly mediated through the CREs on *GH* gene promoters or by the activation of other transcriptional factors, such as pit-1. Indeed, CRE responsiveness of the *GH* gene is partially dependent on the pit-1 site, although independent of thyroid hormone or glucocorticoid activation.[174,176]

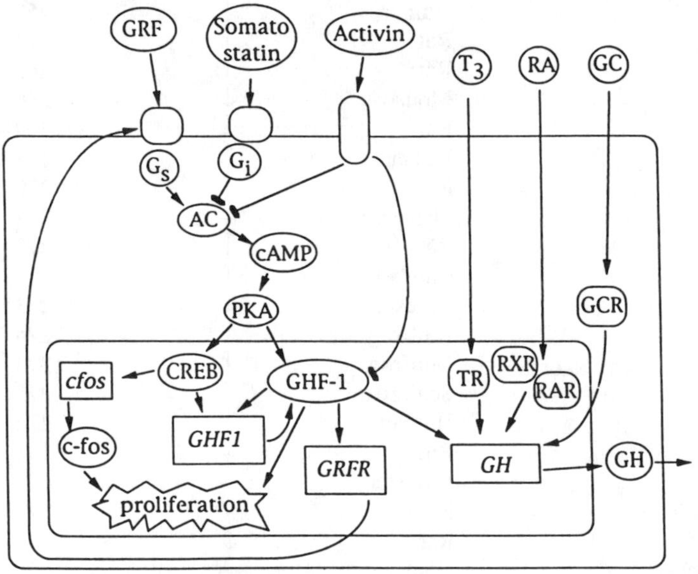

Figure 6 Schematic diagram of signal transduction and somatotroph-specific gene expression. Arrow, activation; bar, inhibition; box and italics, gene; AC, adenylate cyclase; CREB, cAMP response element-binding protein; G_s and G_i, G proteins; GC, glucocorticoid; GCR, GC receptor; GHF-1, GH factor 1; GHRHR, GHRH receptor; RA, retinoic acid; RAR, RA receptor, RXR, retinoic X receptor; TH, thyroid hormone; TR, thyroid hormone receptor. (From Theill, L. E. and Karin, M., Transcriptional control of growth hormone expression and anterior pituitary development, *Endocr. Rev.*, 14(6), 670–689, 1993. Copyright by the Endocrine Society. Reprinted with permission.)

The CRE of *rGH* is also colocalized with the binding site for pit-1.[104,178] The *pit-1* gene itself also contains two CREs that may bind a CRE-binding protein (CREB) or activation peptide 1 (AP-1).[179] AP-2 may also have a role in *GH* gene transcription, because it is activated by cAMP[180] and AP-2 response elements overlapping those of nuclear factor 1 (NF-1) (–142 to –168) and a thyroid hormone responsive element (TRE) (–262 to –290) are present in the *rGH* and *hGH* promoters.[180] This role may, however, be minimal, because the levels of AP-2 in pituitary cells are low[174,180] and phorbol esters, which stimulate AP-2 activity, are unable to induce *rGH* expression.[180] A duality of second-messenger systems may, however, be involved in integrating the response of somatotrophs to GHRH, because phorbol ester-induced activation of protein kinase C increases *GH* mRNA levels in bovine pituitary cells.[147]

2. SRIF

The accumulation of cAMP in GHRH-stimulated pituitary somatotrophs is antagonized by SRIF, by antagonism of adenylate cyclase activity.[147,181,182] GHRH-induced *GH* transcription is, therefore, reported to be antagonized by SRIF[147,183] (Figure 6), although SRIF does not antagonize cAMP-induced GH expression. This inhibitory action of SRIF is accompanied by a reduction in cytoplasmic GH mRNA, although nuclear GH mRNA is increased in the absence of increased pre-RNA levels.[184] SRIF may therefore inhibit the GHRH-stimulated release of mRNA into the cytoplasm. The participation of SRIF in gene transcription is, however, controversial and not confirmed by others.[166,185–187] SRIF is also unable to reduce GH synthesis or pituitary GH content in nonstimulated mammalian[58,183] or teleostean[188] pituitary glands, possibly because it concomitantly reduces GH release. This is, however, also controversial, because SRIF has been shown to increase basal GH mRNA levels in tumorous human pituitary cells.[189]

3. Thyroid Hormones

Thyroid hormones are essential for GH synthesis in many mammalian species[65,190] and *GH* mRNA, GH synthesis, and/or plasma GH levels are dramatically suppressed in hypothyroid rats[191] and human subjects.[192,193] Moreover, exogenous thyroid hormones stimulate *GH* gene transcription in thyroidectomized rats,[194] rat mammosomatotroph cell lines,[162] cultured human pituitary adenoma cells,[195] and transfected GC cells[196] and also reverse the inhibitory effects of hypothyroidism in humans.[192,197] Thyroid hormones also increase GH mRNA levels in the pituitaries of lower vertebrates (e.g., fish[198]).

Tissue-specific and T_3 response elements in the *rGH* gene are tightly linked[199] and transcriptional activation of the rat *GH* promoter fully accounts for the increase in GH mRNA synthesis induced by T_3.[50,200] The induction of cytoplasmic GH mRNA by T_3 is specific and unrelated to the general increase in ribosomal RNA and increase in protein synthesis induced by thyroid hormones.[201,202] *GH* gene transcription in rat somatotrophs is increased within minutes of T_3 stimulation and is maximal within 1 h[49,202] and occurs in parallel with receptor occupancy.[49] Increased cytoplasmic GH mRNA levels occur several hours after the increase in gene transcription and are maximal 48 to 72 h after hormonal stimulation. In the absence of thyroid hormones, GH mRNA and GH synthesis are suppressed, although the size of the GH mRNA transcribed is paradoxically increased by 100 to 150 nucleotides, as a result of an extension of the poly(A) tail.[191,203] The lengthening of the poly(A) segment may stabilize the GH mRNA transcript[4,204] and increase its translational efficiency, to augment the suppressed rate of GH synthesis. The half-life of GH mRNA in hypothyroid rats is 10 to 12 h, whereas it is 5 h in euthyroid rats.[203] Thyroid hormones in the presence of dexamethasone also increase the stability of GH mRNA in pituitary tumor cells,[205] although in the absence of glucocorticoids thyroid hormones increase the rate of degradation. Thyroid hormones may also affect the processing of the primary transcript, as T_3 increases the relative number of transcripts containing the third intron without altering the rate of processing.[48]

Thyroid hormone response elements (TREs) that bind thyroid hormone receptor (TR) complexes are widely located on the *GH* gene promoter.[206,207] T_3 receptors specifically interact with segments in the *hGH* promoter upstream of the transcription start site, between nucleotides –124 and –87[208] and between –290 and –129.[209] TR binding to the region between –109 to –172 of the *rGH* promoter has also been observed,[207,208,210–212] although the nucleotide regions between –254 and –241 and between –209 and –166 are necessary for T_3 induction.[49,199] Within these regions, the sequences between nucleotides –180 and –173 of the *hGH* gene and between –206 and –191 of the *rGH* gene are conserved.[49,199] Within these regions of the *GH* promoter, each TRE may consist of three domains (A, B, and C). These domains reflect direct repeats (A and B) and an overlapping palindromic domain (C) of half-site consensus sequences, as identified between nucleotides –190 and –167 in the *rGH* gene.[176,206,213] The palindromic sequence appears to enhance the activity of the direct repeated sequence. The stimulatory activity of this TRE on gene transcription is independent of its position within the *rGH* promoter but its degree of activity is position dependent[206] and probably dependent on other transcriptional factors. Indeed, Ye et al.[199] identified a TRE between nucleotides –206 and –198 that required the presence of an upstream pit-1-binding site, although pit-1-independent TREs have also been identified.[214] Synergism between T_3 and cAMP, which activates pit-1 (Figure 6), has also been demonstrated.[175] The synergism between thyroid hormones and pit-1 in the presence of TRs is also increased by stimulation of both protein kinase A and protein kinase C.[5] The interaction of TREs with pit-1 may reflect TR-induced DNA bending and the resulting alteration in the distance between the pit-1 site and the TRE within the promoter region.[110] Specific interactions between pit-1 and the TRE may also account for this synergistic response.

In addition to TREs that stimulate *GH* gene transcription, some of the multiple TREs identified confer a negative T_3 response.[215] For instance, a reduction in *GH* mRNA content occurs in primary monolayer cultures of bovine pituitaries treated with T_3.[216] Inhibitory regulation by thyroid hormones is also well established in primates, in which T_3 downregulates *GH* expression from the *hGH* promoter[154,196,217] or is ineffective.[176,196] Because TREs in the *hGH* gene are closer to the region (nucleotide –106 to –148) that binds GHF-1 or Sp-1, an overlap of binding sites or interference with bound GHF-1 or Sp-1 is thought to account for the inhibitory effect of T_3 on transcription. However, despite this inhibitory effect on the *hGH* gene promoter, thyroid hormone can also exert a stimulatory effect on *GH* gene expression in primates[49,190] and human pituitary adenoma cells.[195] These paradoxical actions may result from transcriptional actions outside the promoter region[154] and the balance between these opposing actions may normally determine GH expression.

In addition to TREs within the promoter region of the *GH* gene, TREs are also associated with the structural gene. 3′-Flanking sequences of the *hGH* gene have, for instance, been shown to mediate a stimulatory effect of T_3 on *GH* expression, in contrast to the inhibitory TRE of the *hGH* gene promoter.[154] Another TRE has also been located in the 3′-untranslated/3′-flanking DNA regions and its binding of TR inhibits promoter activity without affecting transcription termination sites or the stability of the gene transcript.[218] This TRE is, however, a stimulatory regulator when placed 5′ to the transcription site, indicating the importance of the positional location of these response elements. The third intron of the rat *GH* gene also has a TRE, which may account for the stimulated transcriptional effects of T_3 observed when exposed to the entire *GH* gene rather than to just the promoter region.[219] Other elements contained within an exon of the *hGH* gene or in 3′-flanking domains also exist[220] and these may be capable of interacting in a promoter-specific manner with elements contained in the 5′ flanking DNA to downregulate the response to thyroid hormones.

In addition to the activation of TREs within the *GH* gene, TRs by themselves may also regulate *GH* transcription by protein-protein interactions. The stimulation of *GH* expression from *hGH* genes with mutated TREs may therefore result from the binding of TR to the pit-1 protein.[91] However, in the absence of the TRE, TR-induced transcription is minimal.

4. Retinoic Acid

Retinoic acid (vitamin A) receptors belong to the same gene family as thyroid hormone receptors and can dimerize with TRs.[221] It is, therefore, not surprising that retinoic acid stimulates *GH* gene transcription in normal and tumorous rat pituitary somatotrophs, alone and in synergy with thyroid hormones[222–224] and glucocorticoids.[223] This synergy with T_3 and the requirement for the portion of the promoter containing a TRE (nucleotide –209) suggest retinoic acid may act through the TRE, to which it is capable of binding and activating.[225] Retinoic acid may, however, activate *GH* gene expression in pituitary cells via retinoic acid receptors (RARs), which can bind to a TRE, identified in the rat *GH* promoter.[5] Activation of the *GH* promoter may also result from ligand binding to RAR and TR heterodimers with the retinoic X receptor (RXR). Conversely, the activity of retinoic acid may be suppressed by the COUP protein, another member of the nuclear receptor family, because this represses RAR activity.[5,226]

5. Glucocorticoids

Glucocorticoids are reported to stimulate *GH* gene expression and to interact synergistically with thyroid hormones and retinoic acid in the production of *GH* mRNA and stimulation of GH synthesis in human and rat pituitary cells.[49,196,202,223] This synergy partly reflects the requirement of glucocorticoid response elements (GREs) proximal to the promoter for T_3-induced transcription.[227] The stimulation of *GH* transcription by glucocorticoids may, therefore, be mediated by GREs within the *GH* gene promoter binding glucocorticoid-receptor complexes.[228] Although consensus sequences have not been described,[229] partial homology exists between GREs located at nucleotides –97 to –111 from the transcription start site and between positions –250 to –264.[227] A region between –240 and –210 bp 5′ from the *GH* mRNA cap site is known to influence the dexamethasone responsiveness of the rat GH promoter, but the nature of the effects of glucocorticoids on the promoter are unknown.[207] Glucocorticoid response elements also occur in the *hGH* gene within 290 bp upstream and 251 bp downstream of the promoter.[230] A glucocorticoid response element is also present on the first intron of the *GH* gene (between positions +86 and +115),[231,232] through which glucocorticoids may primarily activate gene transcription,[232] especially as the *hGH* and *rGH* promoters may be unresponsive to glucocorticoids.[176] Glucocorticoid-induced GH expression does, however, result from an increase in transcription and occurs in the presence of a protein synthesis inhibitor, but not when RNA polymerase II is inactivated.[228] An induced protein is, however, essential for the formation or accumulation of mature cytoplasmic *GH* mRNA, because glucocorticoids are unable to increase cytoplasmic *GH* mRNA levels when protein synthesis is blocked.[229] This transcription-independent mechanism, which is independent of DNA binding,[233] may also be suppressed by calcium ions[185] and may result in *GH* mRNA stabilization by the construction of a longer poly(A) tail[234,235] and an increase in translational efficiency.[201,202] The stability of *GH* mRNA is therefore a locus for factors affecting *GH* gene expression and the inhibitory effect of TRH on GH synthesis and *GH* mRNA accumulation in cytoplasm results from a decrease in mRNA stability by reducing its half-life from 24 to 15 h.[236] Cytoplasmic levels of *GH* mRNA are also increased by a number of other factors, including activators of the inositol phosphate-protein kinase C second-messenger system,[147] and suppressed by IGF-I,[152,237,238] although their mechanisms or sites of action are still uncertain.

Table 2 **Sequence Homologies in Insulin–Responsive Genes**

Gene	5′ Position	Sequence	3′ Position	Match	Strand
GH	–287	ATGGCCTGCGG	–277		Coding
c-*fos*	–296	ATGTCCTAATA	–306	6/11	Noncoding
PEPCK	–202	GAGGCCTCAGG	–88	7/11	Noncoding
PEPCK	–78	GAGGCCTCAGG	–202	7/11	Noncoding
Amy-2.2	–223	ATGGCCTCAGA	–223	8/11	Coding
Amy-2.2	–185	ATGGCCTCAGA	–175	8/11	Coding

Source: Prager, D., Gebremedhin, S., and Melmed, S., An insulin-induced DNA-binding protein for the human growth hormone gene, *J. Clin. Invest.*, 85, 1680–1685, 1990. Copyright by the American Society for Clinical Investigation. Reprinted with permission.

6. IGF-I

The secretion of GH is suppressed by IGF-I, by actions at the level of *GH* gene transcription. Recombinant human IGF-I has been shown to suppress basal and stimulated *GH* mRNA levels in human pituitary adenoma cells,[239] and this suppressive effect is removed on addition of an IGF-I receptor antibody.[152] This effect does not appear to be mediated by the second messengers cAMP or PKC and may be exerted directly.[240] Moreover, as GH exerts most of its peripheral actions via IGFs, IGF-I-induced *GH* mRNA suppression may represent an autoregulatory component in GH regulation. The induction of IGF-I within the pituitary gland[152,237,238,241–245] may therefore account for the inhibitory action of GH on *GH* mRNA levels.[239,246,247]

7. Insulin

Although insulin can directly regulate the expression of several genes at the transcriptional level, an insulin-responsive DNA sequence has yet to be identified. Transfection studies using a *hGH* promoter-chloramphenicol acetyltransferase (CAT) reporter construct were used to localize insulin-responsive elements within the 500 bp of the *GH* promoter.[248] In this study insulin was shown to suppress directly the basal and stimulated expression of the *hGH* gene as has also been reported for the *rGH* gene.[152,243] Alignment of the promoter region of three other insulin-responsive genes with the *hGH* promoter reveals some homologous regions that may mediate these actions on gene transcription (Table 2).

Using a mobility shift assay, Prager et al.[248] identified an insulin-induced DNA-protein complex binding to the –252/–290 region of the *GH* promoter, which corresponds to the USF/NF-1-binding region. This insulin-induced factor acts as a repressor of transcription as insulin suppresses the *GH* promoter. Insulin also selectively attenuates the stimulatory effects of T_3 on GH mRNA levels, an effect mediated at both the transcriptional and posttranscriptional level by attenuating the transcription start site and increasing mRNA degradation.[249]

8. Activin

Transcription of the *GH* gene is also modulated by activin and inhibin through a mechanism that involves modulation of GHF-1 expression and/or activity.[5] Activin, a member of the transforming growth factor β family of growth factors, inhibits basal and GHRH-induced GH biosynthesis,[250] lowers *GH* mRNA levels, and decreases the transcriptional activity induced by the *GH* promoter. This inhibition of transcription probably results from reduced GHF-1 synthesis, and reduced GHF-1 binding to distal GHF-1 recognition elements in the *GH* promoter[5,139] (Figure 6).

REFERENCES

1. **Casanueva, F. F.,** *Endocrinol. Metab. Clin. N. Am.*, 21, 483, 1992.
2. **Dolle, P., Castrillo, J. L., Theill, L. E., et al.,** *Cell*, 60, 809, 1990.
3. **Rousseau, G. G.,** *Horm. Res.*, 37, 88, 1992.
4. **Farrow, S. M.,** *J. Endocrinol.*, 138, 363, 1993.
5. **Theill, L. E. and Karin, M.,** *Endocr. Rev.*, 14, 670, 1993.
6. **Parks, J. S.,** *Acta Pediatr. Scand.*, 349, 127, 1989.
7. **Fiddes, J. C., Seeburg, P. H., DeNoto, F. M., et al.,** *Proc. Natl. Acad. Sci. USA*, 76, 4294, 1979.
8. **Miller, W. L. and Eberhardt, N. L.,** *Endocr. Res.*, 4, 97, 1983.
9. **Tanaka, M., Hosokawa, Y., Watahiki, M., et al.,** *Gene*, 112, 235, 1992.
10. **Karin, M., Castrillo, J., and Theill, L. E.,** *Trends Genet.*, 6, 92, 1990.
11. **Johansen, B., Johnsen, O. C., and Valla, S.,** *Gene*, 77, 317, 1989.
12. **Lorens, J. B., Nerland, A. H., Lossius, I., et al.,** *Nucleic Acids Res.*, 17, 2352, 1989.
13. **Sekine, S., Mizukami, T., Saito, A., et al.,** *Biochim. Biophys. Acta*, 1009, 117, 1989.
14. **Male, R., Nerland, A. H., Lorens, J. B., et al.,** *Biochim. Biophys. Acta*, 1130, 345, 1992.
15. **Agellon, L. B., Davies, S. L., Lin, C., et al.,** *Mol. Reprod. Dev.*, 1, 11, 1988.
16. **Male, R., Nerland, A. H., Lorens, J. B., et al.,** *Biochim. Biophys. Acta*, 1130, 345, 1992.
17. **Lorens, J. B., Nerland, A. H., Aasland, R., et al.,** *J. Endocrinol.*, 11, 167, 1993.
18. **Du, S. J., Devlin, R. H., and Hew, C. L.,** *DNA Cell. Biol.*, 12, 7739, 1993.
19. **Ono, M., Takayama, Y., Rand-Weaver, M., et al.,** *Proc. Natl. Acad. Sci. U.S.A.*, 87, 4330, 1990.
20. **Kakizawa, S., Kaneko, T., Hasegawa, S., et al.,** *Gen. Comp. Endocrinol.*, 91, 298, 1993.
21. **Rand-Weaver, M., Swanson, P., Kawauchi, H., et al.,** *J. Endocrinol.*, 133, 393, 1992.
22. **Takayama, Y., Ono, M., Rand-Weaver, M., et al.,** *Gen. Comp. Endocrinol.*, 83, 366, 1991.
23. **Kaneko, T., Kakizawa, S., Yada, T., et al.,** *Cell Tiss. Res.*, 272, 11, 1993.
24. **Shaw, E. M., Schoffner, R. N., Foster, D. N., et al.,** *Poultry Sci.*, 69(Suppl. 1), 122, 1991.
25. **Abdul-Latif, H., Brown, M. R., and Parks, J. S.,** *Endocrinologist*, 3, 409, 1993.
26. **Vnencak-Jones, C. L. and Phillips, J. A.,** *Science*, 250, 1745, 1990.
27. **Vnencak-Jones, C. L., Phillips, J. A., and De Fen, W.,** *J. Clin. Endocrinol. Metab.*, 70, 1550, 1990.
28. **Mullis, P. E., Akinci, A., Kanaka, C., et al.,** *Pediatr. Res.*, 31, 532, 1992.
29. **Mullis, P. E. and Brickell, P. M.,** *Clin. Endocrinol.*, 37, 89, 1992.
30. **Wurzel, J., Parks, J. S., Herd, J. E., et al.,** *DNA*, 1, 251, 1982.
31. **Akinci, A., Kanaka, C., Eble, A., et al.,** *J. Clin. Endocrinol. Metab.*, 75, 437, 1992.
32. **Goosens, M., Brauner, R., and Czernichow, P.,** *J. Clin. Endocrinol. Metab.*, 62, 712, 1986.
33. **Kamijo, T. and Phillips, J. A.,** *J. Clin. Endocrinol. Metab.*, 74, 786, 1992.
34. **Cogan, J. D., Phillips, J. A., Sakati, N., et al.,** *J. Clin. Endocrinol. Metab.*, 76, 1224, 1993.
35. **Takeuchi, T., Suzuki, H., Sakurai, S., et al.,** *Endocrinology (Baltimore)*, 126, 31, 1990.
36. **DeNoto, J. M., Moore, D. D., and Goodman, H. M.,** *Nucleic Acids Res.*, 9, 3719, 1981.
37. **Seeberg, P. H.,** *DNA*, 239, 1982.
38. **Barta, A., Richards, R. I., Baxter, J. D., et al.,** *Proc. Natl. Acad. Sci. U.S.A.*, 78, 4867, 1981.
39. **Page, G. S., Smith, S., and Goodman, H. M.,** *Nucleic Acids Res.*, 9, 2087, 1981.
40. **Vize, P. D. and Wells, J. R. E.,** *Gene*, 55, 339, 1987.
41. **Woychick, R. P., Camper, S. A., Lyons, R. H., et al.,** *Nucleic Acids Res.*, 10, 7197, 1982.
42. **Orian, J. M., O'Mahoney, J. V. O., and Brandon, M. R.,** *Nucleic Acids Res.*, 16, 9046, 1988.
43. **Ber, R. and Daniel, V.,** *Gene*, 113, 245, 1992.
44. **Bodner, M. and Karin, M.,** *Cell*, 50, 267, 1989.
45. **Lefevre, C.,** *EMBO J.*, 6, 971, 1987.
46. **Parks, J. S., Abdul-Latif, H., Kinoshita, E., et al.,** *Horm. Res.*, 40, 54, 1993.
47. **Hong, Y. and Schartl, M.,** *Biochim. Biophys. Acta*, 1174, 2885, 1993.
48. **Gardner, D. G., Cathala, G., Lan, N., et al.,** *DNA*, 7, 537, 1988.
49. **Samuels, H. H., Aranda, A., Casanova, J., et al.,** *Recent Prog. Horm. Res.*, 44, 53, 1988.
50. **Yaffe, B. M. and Samuels, H. H.,** *J. Biol. Chem.*, 259, 6284, 1984.
51. **Schroder, H. C., Trolltsch, D., Friese, V., et al.,** *J. Biol. Chem.*, 262, 8917, 1987.
52. **Schroder, H. C., Bachman, M., Diehl-Seitert, B., et al.,** *Prog. Nucleic Acid Res.*, 34, 89, 1987.
53. **Morel, G., Dihl, F., and Gossard, F.,** *Mol. Cell Endocrinol.*, 65, 81, 1989.
54. **Burgess, T. L. and Kelly, R. B.,** *Annu. Rev. Cell Biol.*, 3, 243, 1987.
55. **Chen, E. Y., Laio, Y. C., Smith, D. H., et al.,** *Genomics*, 4, 479, 1989.

56. **Khanna, A. S. and Waisman, D. M.,** *Hormones and Their Actions* (Eds. Cooke, B. A., King, R. J. B., and Van der Molen, H. J.). Elsevier Science Publishers BV, Amsterdam, 1988, p. 117.
57. **Aizawa, T. and Hinkle, P. M.,** *Endocrinology (Baltimore)*, 116, 73, 1985.
58. **Fukata, J., Diamond, D. J., and Martin, J. B.,** *Endocrinology (Baltimore)*, 117, 457, 1985.
59. **Stachura, M. E.,** *Endocrinology (Baltimore)*, 98, 580, 1976.
60. **Southard, J., Sanchez-Jiminez, F., Campbell, G., et al.,** *Endocrinology (Baltimore)*, 129, 2965, 1991.
61. **McAndrew, S. J., Chen, N., Wiehl, P., et al.,** *J.Biol.Chem.*, 266, 20965, 1991.
62. **Kurosumi, K.,** *J. Electron Microsc. Tech.*, 19, 42, 1991.
63. **DeVirgilus, G., Meldolesi, J., and Clementi, F.,** *Endocrinology (Baltimore)*, 83, 1278, 1988.
64. **Shields, D.,** *Peptide Biosynthesis and Processing* (Ed. Fricker, D.) CRC Press, Boca Raton, 1991, p. 37.
65. **Stachura, M. E., Tyler, J. M., and Kent, P. G.,** *Endocrinology (Baltimore)*, 119, 1245, 1986.
66. **Stachura, M. E., Tyler, J. M., and Farmer, P. K.,** *Endocrinology (Baltimore)*, 116, 698, 1985.
67. **Stachura, M. E.,** *Endocr. Res.*, 12, 69, 1986.
68. **Synder, G., Hymer, W. C., and Synder, J.,** *Endocrinology (Baltimore)*, 101, 788, 1977.
69. **Barinaga, M., Bilezikjian, L. M., Vale, W. W., et al.,** *Nature (Lond.)*, 314, 279, 1985.
70. **Baumann, G.,** *Basic and Clinical Aspects of Growth Hormone* (Ed. Bercu, B. B.) Plenum Press, New York, 1988, p. 13.
71. **Gautvik, K. M. and Kriz, M.,** *Endocrinology (Baltimore)*, 98, 344, 1976.
72. **Gourdji, D., Tougard, C., and Tixier-Vidal, A.,** *Front. Neuroendocrinol.*, 7, 317, 1982.
73. **Hinkle, P. M.,** *Secretory Tumors of the Pituitary Gland* (Eds. Black, P., Zervas, H., Ridgeway, E. C., and Martin, J. B.). Raven Press, New York, 1984, p. 25.
74. **Aramburo, C., Donoghue, D., Montiel, J. L., et al.,** *Life Sci.*, 47, 945, 1990.
75. **Lewis, U. J., Singh, R. N. P., Lewis, L. J., and Abadi, N.,** *Basic and Clinical Aspects of Growth Hormone* (Ed. Bercu, B. B.) Plenum Press, New York, 1988, p. 43.
76. **Courtois, S. J., Lafontaine, D. A., and Rousseau, G. G.,** *J. Biol. Chem.*, 267, 19736, 1992.
77. **Estes, P. A., Cooke, N. E., and Liebhaber, S. A.,** *J. Biol. Chem.*, 267, 14902, 1992.
78. **Howland, D., Farrington, M., Taylor, W., et al.,** *Biochem. Biophys. Res. Commun.*, 147, 650, 1987.
79. **Sun, Q., Hampson, R. K., and Rottman, F. M.,** *J. Biol. Chem.*, 268, 15659, 1993.
80. **Macleod, J. N., Lee, A. K., Liebhaber, S. A., et al.,** *J. Biol. Chem.*, 267, 14219, 1992.
81. **Goodwin, E. C. and Rottman, F. M.,** *J. Biol. Chem.*, 267, 16330, 1992.
82. **Brostedt, P., Luthman, M., Wide, L., et al.,** *Endocrinology (Baltimore)*, 122, 241, 1990.
83. **McAndrews, J., Stroud, C., Deaver, D., et al.,** *72nd Annu. Meet. Endocr. Soc.,* Georgia, 255, 1990.
84. **Sinha, Y. N., Klemcke, H. G., Maurer, R. R., et al.,** *Endocrinology (Baltimore)*, 127, 410, 1990.
85. **Wehrenberg, W. B., Chatelain, P., and Baird, A.,** *Proc. Soc. Exp. Biol. Med.*, 182, 107, 1986.
86. **Yokoyama, S. and Friesen, H. G.,** *Endocrinology (Baltimore)*, 119, 2097, 1986.
87. **Lemaigre, F. P., Coutois, S. J., Durviaux, S. M., et al.,** *J. Steroid Biochem.*, 34, 79, 1989.
88. **Schaufele, F., Cassill, J. A., West, B. L., et al.,** *J. Biol. Chem.*, 265, 14592, 1990.
89. **Lemaigre, F. P., Courtois, S. J., Lafontaine, D. A., et al.,** *Eur. J. Biochem.*, 181, 555, 1989.
90. **Lemaigre, F. P., Lafontaine, D. A., Courtois, S. J., et al.,** *Mol. Cell Biol.*, 10, 1811, 1990.
91. **Schaufele, F., West, B. L., and Teudelhuber, T. L.,** *J. Biol. Chem.*, 265, 17189, 1990.
92. **Courtois, S. J., Lafontaine, D. A., Lemaigre, F. P., et al.,** *Nucleic Acids Res.*, 18, 57, 1990.
93. **Imagawa, M., Chiu, R., and Karin, M.,** *Cell*, 51, 251, 1987.
94. **Triputti, P., Guerin, S. L., and Moore, D. D.,** *Science*, 241, 1205, 1988.
95. **Lipkin, S. M., Naar, A. M., Kalla, K. A., et al.,** *Genes. Dev.*, 7, 1674, 1993.
96. **Yamada, S., Hata, J., and Yamashita, S.,** *J. Biol. Chem.*, 268, 24361, 1993.
97. **Argenton, F., Vianello, S., Bernardini, S., et al.,** *Biochem. Biophys. Res. Commun.*, 1360, 1366, 1923.
98. **Delhase, M., Vergani, P., Malur, A., et al.,** *Eur. J. Immunol.*, 23, 951, 1993.
99. **Lloyd, R. V., Jin, L., Kulig, E., et al.,** *Endocr. Pathol.*, 4, 146, 1993.
100. **Pfaffle, R. W., Dimattia, G. E., Parks, J. S., et al.,** *Science*, 257, 1118, 1992.
101. **West, B. L., Datanzaro, D. F., Mellon, S. H., et al.,** *Mol. Cell Biol.*, 7, 1193, 1987.
102. **Castrillo, J., Theill, L. E., and Karin, M.,** *Science*, 253, 197, 1991.
103. **Voss, J. W. and Rosenfeld, M. G.,** *Cell*, 70, 527, 1992.
104. **Li, S., Crenshaw, E. B., Rawson, E. J., et al.,** *Nature (Lond.)*, 347, 528, 1991.
105. **Konzak, K. E. and Moore, D. D.,** *Mol. Endocrinol.*, 6, 241, 1992.

106. **Theill, L. E., Hattori, K., Lazzaro, D., et al.,** *EMBO J.*, 11, 2261, 1992.
107. **Lira, S. A., Kalla, K. A., Glass, C. K., et al.,** *Mol. Endocrinol.*, 7, 694, 1993.
108. **Lipkin, S., Naar, A. M., Kalla, K. A., et al.,** *Genes. Dev.*, 7, 1674, 1993.
109. **Verrijzer, C. P., van Oosterhout, A., van Weperen, W. W., et al.,** *EMBO J.*, 10, 3007, 1991.
110. **Tansey, W. P., Schaufele, F., Heslewood, M., et al.,** *J. Biol. Chem.*, 268, 14906, 1993.
111. **Nickel, B. E., Nachtigal, M. W., Bock, M. E., et al.,** *Mol. Cell. Biochem.*, 106, 101, 1991.
112. **Nickel, B. E., Bock, M. E., Nachtigal, M. W., et al.,** *Mol. Cell Endocrinol.*, 91, 159, 1993.
113. **Nachtigal, M. W., Nickel, B. E., Klassen, M. E., et al.,** *Nucleic Acids Res.*, 17, 4327, 1989.
114. **Selvanayagam, C. S., Tsai, S. Y., Tsai, M. J., et al.,** *J. Biol. Chem.*, 259, 14642, 1984.
115. **Ono, M., Harigai, T., Kaneko, T., et al.,** *Mol. Endocrinol.*, 8, 109, 1994.
116. **Lew, D., Brady, H., Klausing, K., et al.,** *Genes. Dev.*, 7, 683, 1993.
117. **Schaufele, F., West, B. L., and Baxter, J. D.,** *Mol. Endocrinol.*, 6, 656, 1992.
118. **McCormick, A., Wu, D., Castrillo, J., et al.,** *Cell*, 55, 379, 1988.
119. **Supowit, S. C., Ramsey, T., and Thompson, E. B.,** *Mol. Endocrinol.*, 6, 786, 1992.
120. **Weigent, D. A., Baxter, J. B., and Blalock, J. E.,** *Brain Behav. Immun.*, 6, 365, 1992.
121. **Binder, G., Revskoy, S., and Gupta, D.,** *J. Endocrinol.*, 140, 137, 1994.
122. **Stephanou, A., Knight, R. A., and Lightman, S. L.,** *Neuroendocrinology*, 53, 628, 1991.
123. **Hattori, N., Shimatsu, A., Sugita, M., et al.,** *Biochem. Biophys. Res. Commun.*, 168, 396, 1990.
124. **Weigent, D. A., Baxter, J. B., Wear, W. E., et al.,** *FASEB J.*, 2, 2812, 1988.
125. **Weigent, D. A. and Blalock, J. E.,** *Cell. Immunol.*, 135, 55, 1991.
126. **Weigent, D. A., Riley, J. E., Galin, F. S., et al.,** *Proc. Soc. Exp. Biol. Med.*, 198, 643, 1991.
127. **Nachtigal, M. W., Nickel, B. E., and Cattini, P. A.,** *J. Biol. Chem.*, 268, 8473, 1993.
128. **Gaido, M. C. and Strobl, J. S.,** *Biochim. Biophys. Acta*, 1008, 234, 1989.
129. **Strobl, J. S., Dannies, P. S., and Thompson, E. B.,** *Biochemistry*, 25, 3640, 1986.
130. **Estienne, M. J., Schillo, K., Hileman, S., et al.,** *Endocrinology (Baltimore)*, 126, 1934, 1990.
131. **Kumar, V. and Biswas, D. K.,** *J. Biol. Chem.*, 263, 12645, 1988.
132. **Hjelle, B. L., Phillips, J. A., and Seeburg, P. H.,** *Nucleic Acids Res.*, 10, 3459, 1982.
133. **Moore, D. D., Walker, M. D., Diamond, D. J., et al.,** *Recent Prog. Horm. Res.*, 38, 197, 1982.
134. **Lan, N. C.,** *J.Biol.Chem.*, 259, 11601, 1984.
135. **Busslinger, M., Hurst, J., and Flavell, R. A.,** *Cell*, 34, 197, 1983.
136. **Guerin, S. L., Anzivino, M. J., Roy, R. J., et al.,** *Eur. J. Biochem.*, 213, 399, 1993.
137. **Pan, W. T., Liu, Q., and Bancroft, C.,** *J. Biol. Chem.*, 265, 7022, 1990.
138. **Roy, R. J., Gosselin, P., Anzivino, M. J., et al.,** *Nucleic Acids Res.*, 20, 401, 1992.
139. **Struthers, R. S., Gaddykurten, D., and Vale, W. W.,** *Proc. Natl. Acad. Sci. U.S.A.*, 89, 11451, 1992.
140. **Yanai, R. and Nagasawa, H.,** *Endocrinol. Jpn.*, 19, 185, 1972.
141. **Suganuma, N., Seo, H., Yamamoto, N., et al.,** *Am. J. Obstet. Gynecol.*, 160, 729, 1989.
142. **Kato, M., Hattori, M., and Suzuki, M.,** *Am. J. Physiol.*, 254, E476, 1988.
143. **Yen, S. and Tashjian, A. H., Jr.,** *Endocrinology (Baltimore)*, 109, 17, 1981.
144. **Cheung, C. Y., Melmed, S., and Braunstein, G. D.,** *Brain Res.*, 270, 165, 1983.
145. **Griffiths, D., Bjoro, T., Gautvik, K., et al.,** *Acta Physiol. Scand.*, 131, 43, 1987.
146. **Dannies, P. S. and Tashjian, A. H.,** *Endocrinology (Baltimore)*, 106, 1532, 1980.
147. **Tanner, J. W., Davis, S. K., McArthur, D. H., et al.,** *J. Endocrinol.*, 126, 109, 1990.
148. **Gautvik, K. M. and Kriz, M.,** *Endocrinology (Baltimore)*, 98, 352, 1976.
149. **Lussier, B. T., French, M. B., Moor, B. C., et al.,** *Endocrinology (Baltimore)*, 128, 570, 1991.
150. **Aizawa, T. and Hinkle, P. M.,** *Endocrinology (Baltimore)*, 116, 909, 1985.
151. **Schonbrunn, A., Krasnoff, M., Westendorf, J. M., et al.,** *J. Cell Biol.*, 85, 786, 1980.
152. **Yamashita, S., Weiss, M., and Melmed, S.,** *J. Clin. Endocrinol. Metab.*, 63, 730, 1986.
153. **MacLeod, R. M. and Lehmeyer, J. E.,** *Proc. Soc. Exp. Biol. Med.*, 145, 1128, 1974.
154. **Morin, A., Louette, J., Voz, M. L. J., et al.,** *Mol. Cell Endocrinol.*, 71, 261, 1990.
155. **Miller, W. L. and Leisti, S.,** *Endocrinology (Baltimore)*, 115, 249, 1984.
156. **Simard, J., Hubert, J. F., Hosseinzadeh, T., et al.,** *Endocrinology (Baltimore)*, 119, 2004, 1986.
157. **Loche, S., Corda, R., Lampis, A., et al.,** *Clin. Endocrinol.*, 25, 195, 1986.
158. **Billestrup, N., Gonzalez-Manchon, C., Potter, E., et al.,** *Mol. Endocrinol.*, 4, 356, 1990.
159. **Lamberts, S. W. J. and Oosterom, R.,** *Horm. Res.*, 14, 137, 1985.
160. **Downs, T. R. and Frohman, L. A.,** *Endocrinology (Baltimore)*, 129, 59, 1991.
161. **Billestrup, N., Swanson, L. W., and Vale, W.,** *Proc. Natl. Acad. Sci. U.S.A.*, 83, 6854, 1986.
162. **Chomczynski, P., Brar, A., and Frohman, L. A.,** *Endocrinology (Baltimore)*, 123, 2276, 1988.
163. **Hu, Z., Friberg, R., and Barkan, A. L.,** *Am. J. Physiol.*, 265, E236, 1993.

164. **Cella, S. G., Locatelli, V., De Gennaro, V., et al.,** *Endocrinology (Baltimore)*, 116, 574, 1985.
165. **Cella, S. G., Locatelli, V., De Gennaro, V., et al.,** *Endocrinology (Baltimore)*, 119, 1164, 1986.
166. **Sugihara, H., Minami, S., Okada, K., et al.,** *Endocrinology (Baltimore)*, 132, 1225, 1993.
167. **Cella, S. G., Locatelli, V., Mennini, T., et al.,** *Endocrinology (Baltimore)*, 127, 1625, 1990.
168. **Maiter, D., Underwood, L. E., Martin, J. B., et al.,** *Endocrinology (Baltimore)*, 28, 1100, 1991.
169. **Barinaga, M., Yamonoto, G., Rivier, C., et al.,** *Nature (Lond.)*, 314, 279, 1983.
170. **Gick, G. G., Zeytin, F. N., Brazeau, P., et al.,** *Proc. Natl. Acad. Sci. U.S.A.*, 81, 1553, 1984.
171. **Burton, F. H., Hasel, K. W., Bloom, F. E., et al.,** *Nature (Lond.)*, 350, 74, 1991.
172. **Landis, C. A., Masters, S. B., Spada, A., et al.,** *Nature (Lond.)*, 340, 692, 1989.
173. **Canny, B. J., Rawlings, S. R., and Leong, D. A.,** *Endocrinology (Baltimore)*, 130, 211, 1992.
174. **Dana, S. and Karin, M.,** *Mol. Endocrinol.*, 3, 815, 1989.
175. **Copp, R. P. and Samuels, H. H.,** *Mol. Endocrinol.*, 3, 790, 1989.
176. **Brent, G. A., Harney, J. W., Moore, D. D., et al.,** *Mol. Endocrinol.*, 2, 792, 1988.
177. **Thomas, M. J., Freeland, T. M., and Strobl, J. S.,** *Mol. Cell Biol.*, 10, 5378, 1990.
178. **Keech, C. A., Jackson, S. M., Siddiqui, S. K., et al.,** *Mol. Endocrinol.*, 6, 2059, 1992.
179. **McCormick, A., Brady, H., Theil, L. E., et al.,** *Nature (Lond.)*, 345, 829, 1990.
180. **Imagawa, M., Chiu, R., and Karin, M.,** *Cell*, 51, 251, 1987.
181. **Bilezikjian, L. M. and Vale, W. W.,** *Endocrinology (Baltimore)*, 113, 1726, 1983.
182. **Simard, J., Labrie, F., and Gossard, F.,** *DNA*, 5, 263, 1986.
183. **Wood, D. F., Docherty, K., Ramsden, D. B., et al.,** *Mol. Cell Endocrinol.*, 52, 257, 1987.
184. **Morel, G., Chavassieux, P., Barenton, B., et al.,** *Cell Tiss. Res.*, 273, 279, 1993.
185. **Gick, G. G. and Bancroft, C.,** *Endocrinology (Baltimore)*, 120, 1986, 1987.
186. **Herman, V., Weiss, M., Becker, D., et al.,** *Endocr. Pathol.*, 1, 236, 1990.
187. **Levy, A. and Lightman, S. L.,** *J. Mol. Endocrinol.*, 1, 19, 1988.
188. **Yada, T. and Hirano, T.,** *J. Comp. Physiol.B.*, 162, 575, 1992.
189. **Hofland, L. J., Velkeniers, B., Vanderlely, A. J., et al.,** *Clin. Endocrinol.*, 37, 240, 1992.
190. **Root, A. W., Shulman, D., Root, J., et al.,** *Acta Endocrinol. (Copenh.)*, 113, 367, 1986.
191. **Jones, P. M., Burrin, J. M., Ghatel, M. A., et al.,** *Endocrinology (Baltimore)*, 126, 1374, 1990.
192. **Valcavi, R., Dieguez, C., Preece, M., et al.,** *Clin. Endocrinol.*, 27, 85, 1987.
193. **Chernausek, S. D. and Turner, R.,** *J. Physiol.*, 114, 968, 1989.
194. **Mulloy, A. L., Smith, T. J., and Stachura, M. E.,** *Horm. Metab. Res.*, 24, 466, 1992.
195. **Chomczynski, P., Soszynski, P., and Frohman, L. A.,** *J. Clin. Endocrinol. Metab.*, 77, 281, 1993.
196. **Issacs, R. E., Findell, P. R., Mellon, P., et al.,** *Mol. Endocrinol.*, 1, 569, 1987.
197. **Valcavi, R., Dieguez, C., Azzarito, C., et al.,** *Clin. Endocrinol.*, 26, 453, 1987.
198. **Moav, B. and Mckeown, B. A.,** *Horm. Metab. Res.*, 24, 10, 1992.
199. **Ye, Z., Forman, B. M., Aranda, A., et al.,** *J. Biol. Chem.*, 263, 7821, 1988.
200. **Wright, P. A., Crew, D. D., and Spindler, S. R.,** *Mol. Endocrinol.*, 2, 536, 1988.
201. **Wegnez, M., Schacter, B. S., Baxter, J. D., et al.,** *DNA*, 1, 145, 1982.
202. **Evans, R. M., Birnberg, N. C., and Rosenfeld, M. G.,** *Proc. Natl. Acad. Sci. U.S.A.*, 79, 7659, 1982.
203. **Murphy, D., Pardy, K., Seah, V., et al.,** *Mol. Cell Biol.*, 12, 2624, 1992.
204. **Narayan, P. and Towle, H. C.,** *Mol. Cell Biol.*, 5, 2642, 1985.
205. **Diamond, D. J. and Goodman, H. M.,** *J. Mol. Biol.*, 181, 41, 1985.
206. **Brent, G. A., Williams, G. R., Harney, J. W., et al.,** *Mol. Endocrinol.*, 5, 542, 1991.
207. **Flug, F., Copp, P. R., Casanova, J., et al.,** *J. Biol. Chem.*, 263, 6373, 1987.
208. **Leidig, F., Shepard, A. R., Zhang, W., et al.,** *J. Biol. Chem.*, 267, 913, 1992.
209. **Barlow, J. W., Voz, M. L. J., Eliard, P. H., et al.,** *Proc. Natl. Acad. Sci. U.S.A.*, 83, 9021, 1986.
210. **Larsen, J. L.,** *J. Biol. Chem.*, 267, 10583, 1992.
211. **Glass, C. K., Holloway, J. M., Devary, O. V., et al.,** *Cell*, 54, 313, 1988.
212. **Wight, P. A., Crew, M. D., and Spindler, S. R.,** *J. Biol. Chem.*, 262, 5659, 1987.
213. **Brent, G. A., Harney, J. W., Chen, Y., et al.,** *Mol. Endocrinol.*, 3, 1996, 1989.
214. **Suen, C. S. and Chin, W. W.,** *Mol. Cell Biol.*, 13, 1719, 1993.
215. **Crone, D. E., Kim, H., and Spindler, S. R.,** *J. Biol. Chem.*, 265, 10851, 1990.
216. **Silverman, B., Kaplan, S., Grumbach, M., et al.,** *Endocrinology (Baltimore)*, 122, 1236, 1988.
217. **Cattini, P. A. and Eberhardt, N. L.,** *Nucleic Acids Res.*, 15, 1297, 1987.
218. **Zhang, W. G., Brooks, R. L., Silversides, D. W., et al.,** *J. Biol. Chem.*, 267, 15056, 1992.
219. **Sap, J., de Magistris, L., Stunnenberg, H., et al.,** *EMBO J.*, 9, 887, 1990.

220. **Brooks, R. L., Silverside, D. W. and Eberhardt, N. L.,** *74th Annu. Meet. Endocr. Soc.*, San Antonio, TX, 372, 1990.
221. **Lazar, M. A.,** *Endocr. Rev.*, 14, 184, 1993.
222. **Garcia-Vallalba, P., Au-Fliegner, M., Samuels, H., et al.,** *Biochem. Biophy. Res. Commun.*, 191, 580, 1993.
223. **Bedo, G., Santisteban, P., and Aranda, A.,** *Nature (Lond.)*, 339, 231, 1989.
224. **Morita, S., Matsuo, K., Tsuruta, M., et al.,** *J. Endocrinol.*, 125, 251, 1990.
225. **Umesono, K., Giguiere, V., Glass, C. K., et al.,** *Nature (Lond.)*, 336, 262, 1988.
226. **Tran, P., Zhang, X. K., Salbert, G., et al.,** *Mol. Cell Biol.*, 12, 4666, 1992.
227. **Treacy, M. N. and Rosenfeld, M. G.,** *Annu. Rev. Neurosci.*, 15, 139, 1992.
228. **Spindler, S. R., Mellon, S. H., and Baxter, J. D.,** *J. Biol. Chem.*, 257, 11627, 1982.
229. **Strobl, J. S., Van Eys, G. J. J. M., and Thompson, E. B.,** *Mol. Cell. Endocrinol.*, 66, 71, 1989.
230. **Eliard, P. H., Marchand, M. J., Rousseau, G. G., et al.,** *DNA*, 4, 409, 1985.
231. **Nyborg, J. K., Nguyen, A. P., and Spindler, S. R.,** *J. Biol. Chem.*, 259, 12377, 1984.
232. **Slater, E. P., Rabenau, O., Karin, M., et al.,** *Mol. Cell Biol.*, 5, 2984, 1985.
233. **Rousseau, G. G., Eliard, P. H., Barlow, J. W., et al.,** *J. Steroid Biochem.*, 27, 149, 1987.
234. **Gertz, B. J., Gardner, D. G., and Baxter, J. D.,** *Mol. Endocrinol.*, 1, 933, 1987.
235. **Paek, I. and Axel, R.,** *Mol. Cell Biol.*, 7, 1496, 1987.
236. **Laverriere, J. N., Morin, A., Tixier-Vidal, A., et al.,** *EMBO J.*, 2, 1493, 1983.
237. **Namba, H., Morita, S., and Melmed, S.,** *Endocrinology (Baltimore)*, 124, 1794, 1989.
238. **Yamashita, S. and Melmed, S.,** *Endocrinology (Baltimore)*, 118, 176, 1986.
239. **Yamashita, S., Slanina, S., Kado, H., et al.,** *Endocrinology (Baltimore)*, 118, 915, 1986.
240. **Morita, S., Yamashita, S., and Melmed, S.,** *Endocrinology (Baltimore)*, 121, 2000, 1987.
241. **Fagin, J. A., Brown, A., and Melmed, S.,** *Endocrinology (Baltimore)*, 122, 2204, 1988.
242. **Melmed, S. and Yamashita, S.,** *Endocrinology (Baltimore)*, 118, 1483, 1986.
243. **Yamashita, S. and Melmed, S.,** *J. Clin. Invest.*, 79, 449, 1987.
244. **Ezzat, S. and Melmed, S.,** *J. Endocrinol. Invest.*, 13, 691, 1990.
245. **Fagur, J. A., Fernandez-Megia, C., and Melmed, S.,** *Endocrinology (Baltimore)*, 125, 2385, 1989.
246. **Stefaneanu, L., Kovacs, K., Horvath, E., et al.,** *Endocrinology (Baltimore)*, 126, 608, 1990.
247. **Stefaneanu, L., Kovacs, K., and Bartke, A.,** *Endocr. Pathol.*, 4, 73, 1993.
248. **Prager, D., Gebremedhin, S., and Melmed, S.,** *J. Clin. Invest.*, 85, 1680, 1990.
249. **Prager, D., Weber, M. M., Gebremedhin, S., et al.,** *J. Endocrinol.*, 137, 107, 1993.
250. **Billestrup, N., Gonzalez-Manchon, C., Potter, E., et al.,** *Mol. Endocrinol.*, 4, 356, 1990.
251. **Mathews, L. S., Hammer, R. E., Brinster, R. L., et al.,** *Endocrinology (Baltimore)*, 123, 433, 1988.
252. **Stefaneanu, L., Kovacs, K., Bartke, A., et al.,** *Lab. Invest.*, 68, 584, 1993.
253. **Davis, J. R. E., Wilson, E. M., Vidal, M. E., et al.,** *J. Clin. Endocrinol. Metab.*, 69, 270, 1989.
254. **Melmed, S.,** *The Brain as an Endocrine Organ* (Eds. Cohen, M. P. and Foa, P. P.). Springer-Verlag, Berlin, 1992, p. 193.
255. **Vasilatos-Younken, R., Tsao, P. H., Foster, D. N., et al.,** *J. Endocrinol.*, 135, 371, 1992.
256. **Martinoli, M. G. and Pelletier, G.,** *Endocrinology (Baltimore)*, 125, 1246, 1989.
257. **Prager, D., Yamasaki, H., Weber, M. M., et al.,** *J. Clin. Invest.*, 90, 2117, 1992.
258. **Song, J., Jin, L., and Lloyd, R. V.,** *Cancer Res.*, 49, 1247, 1989.
259. **Jin, L., Song, J., and Lloyd, R. V.,** *Proc. Soc. Exp. Biol. Med.*, 192, 225, 1989.
260. **Dobado-Berrios, P., Li, S., Garciade Yebenes, E., et al.,** *J. Neuroendocrinol.*, 5, 553, 1993.
261. **Nash, A. D., Brandon, M. R., and Bello, P. A.,** *Mol. Cell Endocrinol.*, 84, R31, 1992.
262. **Emanuele, M. A., Tentler, J. J., Kirsteins, L., et al.,** *Endocrinology (Baltimore)*, 131, 2077, 1992.
263. **Soszynski, P. A. and Frohman, L. A.,** *Endocrinology (Baltimore)*, 131, 2603, 1992.
264. **Clayton, R. N., Bailey, L. C., Abbot, S. D., et al.,** *J. Endocrinol.*, 110, 51, 1986.
265. **White, B. A., Baauerle, L. R., and Bancroft, F. C.,** *J. Biol. Chem.*, 256, 5942, 1981.
266. **Crew, M. D., Spindler, S. R., Walford, R. L., et al.,** *Endocrinology (Baltimore)*, 121, 1251, 1987.
267. **Ahmad, I., Steggles, A. W., Carrillo, A. J., et al.,** *J. Cell. Biochem.*, 43, 59, 1990.
268. **Takahashi, S.,** *Zool. Sci.*, 9, 901, 1992.
269. **Martinoli, M. G., Oullet, J., Rheaume, E., et al.,** *Neuroendocrinology*, 54, 607, 1991.
270. **Nogami, H. and Tachibana, T.,** *Endocrinology (Baltimore)*, 132, 517, 1993.
271. **Merei, J. J., Rao, A., Clarke, I. J., and McMillen, I. C.,** *Acta Endocrinol.*, 129, 263, 1993.

272. **Hu, Z. Y., Friberg, R. D., and Barkan, A. L.,** *Am. J. Physiol.*, 265, E236, 1993.
273. **McCann-Levorse, L., Radecki, S., Donoghue, D. J., et al.,** *Proc. Soc. Exp. Biol. Med.*, 202, 109, 1993.
274. **Takahashi, N., Kikuyama, S., Gen, K., et al.,** *J. Mol. Endocrinol.*, 9, 283, 1992.
275. **Funkenstein, B., Tandler, A., and Cavari, B.,** *Mol. Cell Endocrinol.*, 87, R7, 1992.
276. **Stefaneanu, L., Kovacs, K., Lloyd, R. V., et al.,** *Virchows Arch.(B)*, 62, 291, 1992.
277. **Linzer, D. I. H. and Talamantes, F.,** *J. Biol. Chem.*, 260, 9574, 1985.
278. **Ahmad, I., Steggles, A. W., Carrillo, A. J., et al.,** *Mol. Cell Endocrinol.*, 65, 103, 1989.
279. **Wehrenberg, W. B. and Giustina, A.,** *Endocr. Rev.*, 13, 299, 1992.
280. **Landefeld, T. D., Ebling, F. J. P., Suttie, J. M., et al.,** *Endocrinology (Baltimore)*, 125, 351, 1989.
281. **Clarke, I. J., Fletcher, T. P., Pomares, C. C., et al.,** *J. Endocrinol.*, 138, 421, 1993.
282. **Ahmad, I., Finkelstein, J. A., Downs, T. R., et al.,** *Neuroendocrinology*, 58, 332, 1993.
283. **Yada, T., Kobayashi, T., Urano, A., et al.,** *J. Exp. Zool.*, 262, 420, 1992.

Chapter 5

Growth Hormone Storage

Stephen Harvey

CONTENTS

I. INTRODUCTION

Prior to release, growth hormone (GH) may be stored in secretory granules in heterogeneous compartments within GH-secreting cells. Quantitative and qualitative aspects of GH storage are briefly reviewed in this chapter.

II. GH STORAGE

After biosynthesis in the endoplasmic reticulum (ER), processing of GH in the Golgi cisternae results in its packaging into small immature granules, which become distributed along Golgi stacks[1] (Figure 1). In addition, small dense secretory granules may arise in the rigid lamella ("GERL") associated with the Golgi[2] (Figure 2). Newly synthesized GH that is not immediately required is stored in these cytoplasmic secretory granules for mobilization at a later time in response to an appropriate stimulus. Approximately 40% of the GH produced is stored in normal somatotrophs, although only minimal stores exist in tumorous somatotrophs.[3] Storage of GH occurs in granules that are typically spherical (Figure 3), although rod-shaped granules have also been described[4,5] (Figure 3). Both types of granule are extruded from the cell by exocytosis and are therefore secretory granules. Both types of granule may also fuse with homogeneous granules within the cytoplasm, resulting in large, spherical or branchlike structures (Figure 3). In musk shrews the fusion of these granules is most common in adults, particularly lactating females. The spherical granules typically have a diameter of 300 to 400 nm, although granules 700 to 1200 nm in diameter have been observed in human, goat, and cow pituitary glands[6-8] and are found only in GH-secreting pituitary cells. The size of the granules and their GH content vary with respect to age and physiological state.[5,7] Somatotrophs and secretory granules increase with age in rats[2,9] and pigs[10] and are larger in goats during lactation than in anestrus and pregnancy,[7] indicative of GH hyperactivity.[8] In mammosomatotrophs these granules may also contain prolactin, although separate prolactin- and GH-containing granules may be present within individual cells.[1] The secretory granules in mammosomatotrophs are, however, considerably smaller (50 to 100 nm) than those found in traditional mammotrophs (400 to 600 nm) or somatotrophs (200 to 400 nm).[11-14]

A. INTRACELLULAR STORAGE POOLS

Newly synthesized GH in secretory vesicles can be directed by the Golgi apparatus[15] to a microtubular transport system[16] toward the cell membrane for immediate release

0-8493-8697-7/95/$0.00+$.50

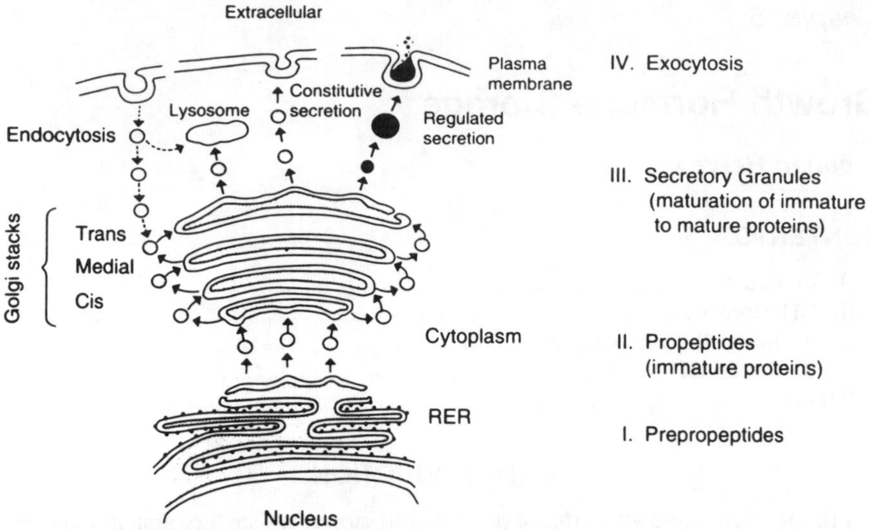

Figure 1 Schematic diagram depicting the subcellular transport and the exocytic and endocytic pathways in protein-secreting eukaryotic cells. RER, Rough endoplasmic reticulum. Proteins synthesized on polysomes attach to the endoplasmic reticulum as prepropeptides (I) and are discharged as propeptides into the cisternal space (II). Proteins are transported from the endoplasmic reticulum by a process involving the formation and shuttling of vesicles. Transportation is first to the "cis" stacks of the Golgi apparatus and then (by cycles of vesicle budding and fusion) to the "medial" and "trans" stacks, where the secretory granules are formed (III). Secretory protein can be discharged by either a regulated or nonregulated (constitutive) pathway (IV). Various endocytic exocytic vesicles are indicated (e.g., trafficking to and from the plasma membrane, to and from Golgi stacks, and from the lysosome). (From Saez, J. M. and Andre, J., *Pediatric Endocrinology*, [Ed. Bertrand, J., Rappaport, P., and Sizonenko, P. C.], Williams & Wilkins, Baltimore, 1993, p. 46. Reprinted with permission.)

(Figure 2). However, unless released, the granular storage of GH may exist for only a finite time, to control the volume of hormone secreted.[2] The number of secretory granules may thus be subject to autophagy (crinophagy) by lysosomes (Figure 2), the activity of which increases after thyroidectomy.[17] Crinophagy may also be induced within secretory granules without lysosomal hydrolysis. The contents of GH secretory granules in musk shrews, for instance, are progressively lost following fusion, leaving "empty" granules within the cytoplasm.[4,13] There are thus functionally separate pools of GH storage in granular compartments and these differ in their age and responsiveness to releasing stimuli, and in terms of the molecular size of the hormone stored. Although these pools may partly reflect differential storage and different GH moieties in specialized somatotroph subtypes,[18,19] they also occur within individual somatotrophs.[20-23]

The immediate release pool (IRP) has complex structure[20,22-26] (Figure 4) and consists of secretory granules juxtaposed to the cell membrane, that are (B, Figure 4) or are not (A, Figure 4) attached to the microtubular transport system. The IRP includes attached granules (C, Figure 4) translocating to the cell surface after release from the Golgi or after mobilization from the intracellular storage compartment (D, Figure 4).[20] The first or labile pool is available for immediate release and is exhaustible, whereas the second is more stable and available for prolonged release at an increased rate under constant stimulation.[27] Secretagogues that stimulate both compartments (e.g., GHRH, TRH, and adenylate

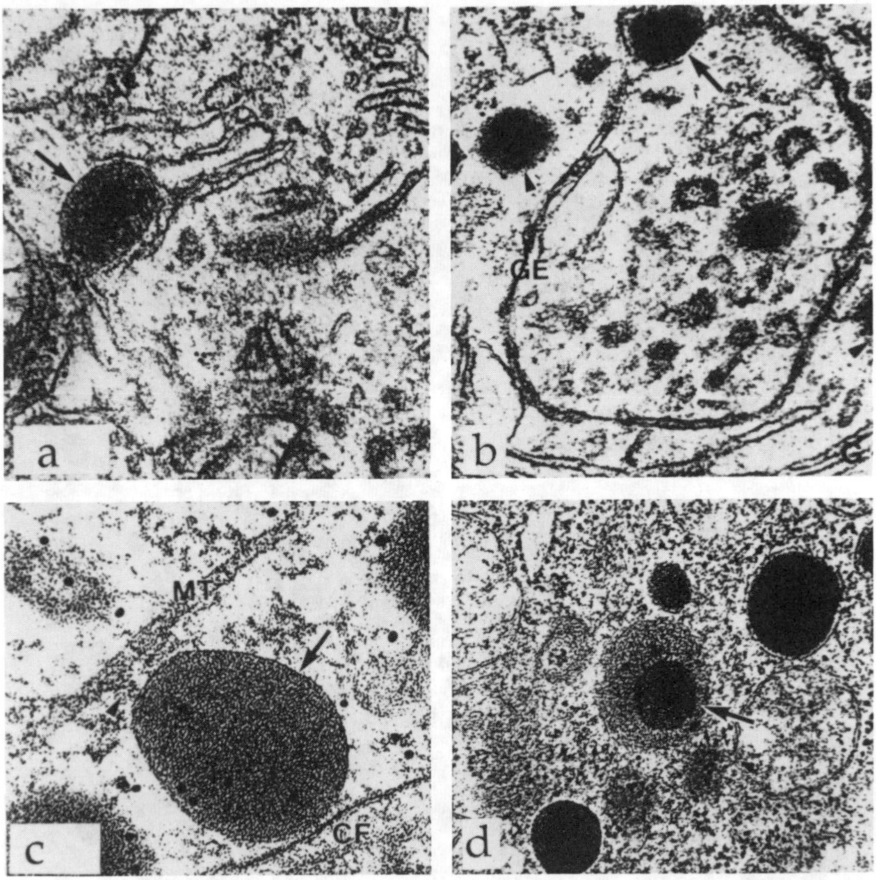

Figure 2 Secretory granule formation and secretion in rat pituitary somatotrophs. (a) Secretory granule formation (arrow) within a saccule of the Golgi stack. Growth hormone (GH) immunoreactivity is shown by gold particles. Magnification: ×80,000. (b) A secretory granule forming in the cavity of GERL (arrow) and free, matured granules (arrowheads) are labeled with gold particles showing the immunoreactivity to GH. GERL is shaped like a circle (GE), but the Golgi stack (G) is arranged in a parallel manner. ×80,000. (c) A secretory granule (arrow) immunoreactive to GH moving to the cell surface (CF) is connected to the microtubule (MT) with some amorphous substance (arrowhead). Magnification: ×16,000. (d) A part of a normal GH cell showing crinophagy. A lysosome (arrow) contains a round dark granule with positive immunoreaction to GH. Magnification: ×54,000. (From Kurosumi, K., Ultrastructural immunocytochemistry of the adenohypophysis in the rat: a review, *J. Electron Microsc. Tech.*, 19, 42, 1991. Copyright 1991 by Wiley-Liss, Inc. Reprinted by permission of John Wiley & Sons, Inc.)

cyclase activators[28-31]) therefore induce temporally biphasic patterns of GH release, in which the first phase exceeds the second in the amount of GH released. The replenishment of the immediate release pool is partly by an energy-dependent mobilization of the intracellular storage pool, and by the preferential entry of newly synthesized hormone.[21,32,33] The amount of newly synthesized hormone in the immediate release pool is, however, only 10 to 15% of that in the intracellular storage compartment,[33] from which GH is

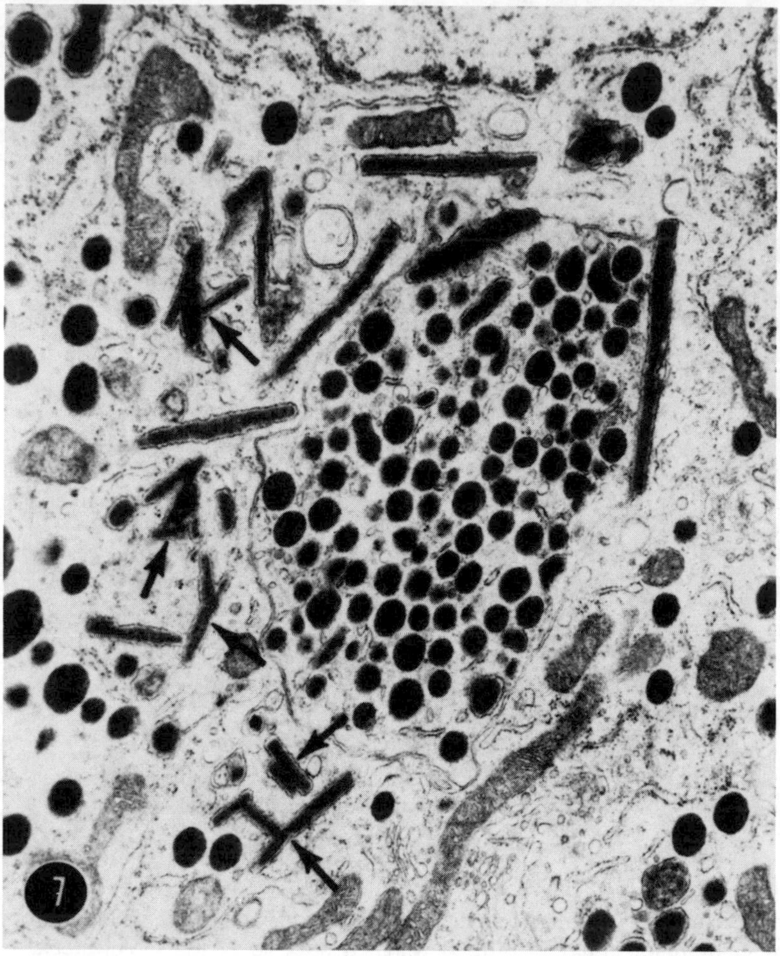

Figure 3 Secretory granules in musk shrew pituitary glands. Note both spherical and rod-shaped secretory granules and the fusing of rod-shaped granules to form spreading branchlike structures (arrows). Magnification: ×15,000. (From Ishibashi, T. and Shiino, M., Unique features of secretory granules observed in the pituitary growth hormone-secreting (GH) cells of the musk shrew (*Suncus murinus* L.), *Cell Tiss. Res.*, 251, 111–116 (Fig. 7), 1988. Copyright by Springer-Verlag. Reprinted with permission.)

preferentially released by GH-releasing stimuli.[18,24-27,34] Thus, whereas GHRH doubles the amount of newly synthesized hormone released (from a basal release rate of 37% of that stored in the IRP to a stimulatory rate of 77%), it triples the amount released from the larger stores of intracellular GH (from a basal release rate of 15% to a stimulated release rate of 45%[34]).

Under basal conditions the amount of stored GH remains relatively constant[34] and the amount of hormone released (approximately 0.09% of stores per minute[35]) is replenished by synthesis, reflecting a close relationship between these parameters. Under basal conditions newly synthesized hormone is preferentially released, although GH release is randomly derived from both pituitary storage pools and is not by a first-in/first-out

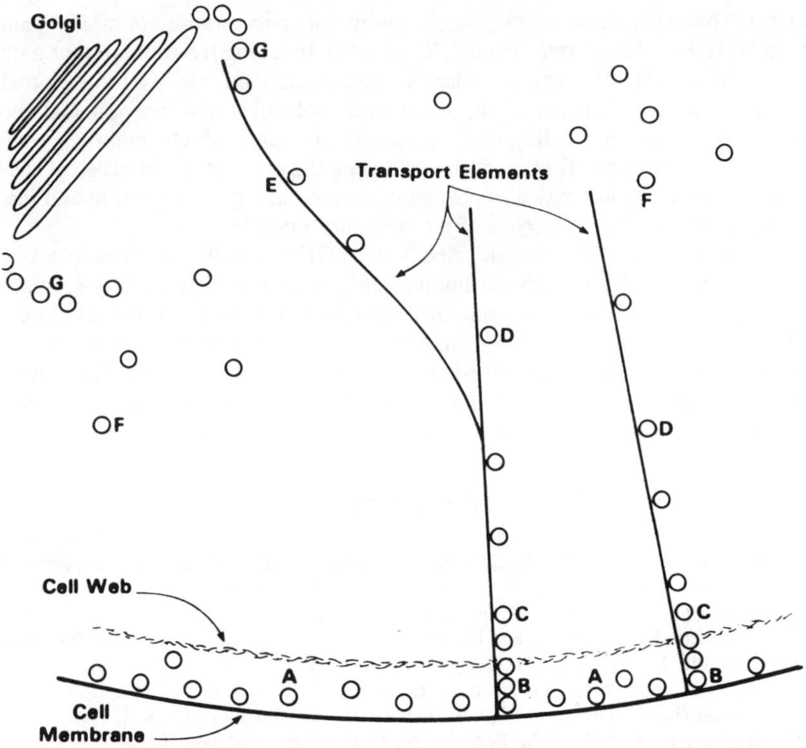

Figure 4 Hypothetical model of functionally defined storage granule positions in the somatotroph. (A) Immediate release pool (IRP) granule at cell surface but no longer attached to transport system; (B) IRP granule at cell surface and still attached to transport system; (C) IRP granule attached to transport system and near cell surface; (D) granule from intracellular storage compartment attached to transport system; (E) granule attached to transport system immediately after release from Golgi system; (F) Intracellular storage compartment granules not attached to transport system; and (G) granules recently released from Golgi system and available to become F or E granules. Position of cell web is speculative. (From Stachura, M. E. and Tyler, J. M., *Endocr. Res.*, 16(1), 1–15, 1990. Reprinted by courtesy of Marcel Dekker, Inc.)

mechanism. The mean turnover rate for GH molecules in the rat pituitary gland is 12.8 h.[35] Only a fraction (approximately 50%) of the cells that release GH are dependent, however, on newly synthesized hormone for basal secretion; the rest lack an IRP and are capable of mobilizing intracellular stores of GH in the absence of stimulation.[18,19] The heterogeneity of GH storage is thus also due to the cellular heterogeneity of pituitary somatotrophs.

B. STORED GH MOIETIES

Most of the GH released under basal conditions is monomeric,[36-38] although much (perhaps 70%) of the GH stored is dimerized or high molecular weight heteropolymers[39] or protein-bound[36,40] GH moieties. This aggregation may be due to the presence of zinc in the secretory granules, which stabilizes the hormone, possibly by chelation to intermolecular disulfide bonds.[39] Indeed, zinc and human GH (hGH) are present in the secretory granules of somatotrophs in approximately equimolar concentrations (4 mM[41,42]) and zinc binds to hGH with a moderate affinity (K_a ~1 μM), inducing the formation of a dimeric complex ($(Zn^{2+}$-hGH$)_2$.[43,44] This complex is more stable than free hGH to denaturants.[41] The cryptic binding domains of the dimerized molecules also prevent GH receptor activation in tissues and cells proximal to the pituitary, which might otherwise be stimulated by the sudden efflux of hormone during the exocytotic discharge of secretory granules.[44,45] Aggregation may also change intragranular osmotic pressure and facilitate the intracellular movement or lysis of the secretory granule.

Aggregated (large) and nonaggregated (small) GH molecules are both released from the pituitary gland,[46-48] although the amount released can be differentially regulated.[35,49] The secreted form of GH is less stable than the pituitary form and is thought to undergo modification during release.[50] These changes may account for the increased bioactivity of secreted GH[49] and may be partly mediated by cytosolic thiols (e.g., glutathione:disulfide oxidoreductase) that increase GH release from secretory granules and reduce oligomer disulfides and disulfides of secretory granule membrane proteins.[51]

REFERENCES

1. **Saez, J. M. and Andre, J.,** *Pediatric Endocrinology* [Ed. Bertrand, J., Rappaport, P., and Sizonenko, P. C.], Williams & Wilkins, Baltimore, 1983, p. 46.
2. **Kurosumi, K.,** *J.Electron Microsc. Tech.*, 19, 42, 1991.
3. **Chomczynski, P., Brar, A., and Frohman, L. A.,** *Endocrinology (Baltimore)*, 123, 2276, 1988.
4. **Ishibashi, T. and Shiino, M.,** *Cell Tiss. Res.*, 251, 111, 1988.
5. **Malamed, S., Gibney, J. A., Cain, L. D., et al.,** *Cell Tiss. Res.*, 272, 369, 1993.
6. **Yamaguchi, S., Shirasawa, N., and Nogami, H.,** *Jpn. Biomed. Res.*, 9, 1, 1988.
7. **Sanchez, J., Navarro, J. A., Bernabe, A., et al.,** *Histol. Histopathol.*, 8, 83, 1993.
8. **Horvath, E. and Kovacs, K.,** *J. Electron Microsc. Tech.*, 8, 401, 1988.
9. **Carretero, J., Sanchez, F., Rubio, M., et al.,** *Histol. Histopathol.*, 7, 673, 1992.
10. **Dada, M. O., Campbell, G. T., and Blake, C. A.,** *J. Endocrinol.*, 101, 87, 1984.
11. **Frawley, L. S. and Boockfor, F. R.,** *Endocr. Rev.*, 12, 337, 1991.
12. **Yamaji, A., Sasaki, F., Iwama, Y., et al.,** *Anat. Rec.*, 233, 103, 1992.
13. **Ishibashi, T. and Shiino, M.,** *Anat. Rec.*, 223, 185, 1989.
14. **Nikitovitch-Winer, M. B., Atkin, J., and Maley, B. E.,** *Endocrinology (Baltimore)*, 121, 625, 1987.
15. **Farmer, P. K., Tyler, J. M., and Stachura, M. E.,** *Mol. Cell. Endocrinol.*, 62, 253, 1989.
16. **Zor, U.,** *Endocr. Rev.*, 4, 1, 1983.
17. **Yang, H. J., Ozawa, H., and Kurosumi, K.,** *J. Clin. Electron Microsc.*, 22, 269, 1989.
18. **Chao, C. G., Hoeffler, J. P., and Frawley, L. S.,** *Life Sci.*, 42, 701, 1987.
19. **Chen, T. T., Kineman, R. D., Betts, J. G., et al.,** *Endocrinology (Baltimore)*, 408, 493, 1989.
20. **Stachura, M. E. and Tyler, J. M.,** *Endocr. Res.*, 16, 1, 1990.
21. **Stachura, M. E., Tyler, J. M., and Kent, P. G.,** *Endocrinology (Baltimore)*, 125, 444, 1989.
22. **Stachura, M. E., Tyler, J. M., and Kent, P. G.,** *Endocrinology (Baltimore)*, 120, 1719, 1987.
23. **Stachura, M. E., Costoff, A., and Tyler, T. M.,** *Neuroendocrinology*, 42, 383, 1986.
24. **Stachura, M. E., Tyler, J. M., Kent, P. G., et al.,** *Endocr. Res.*, 12, 171, 1986.
25. **Stachura, M. E.,** *Endocr. Res.*, 12, 69, 1986.
26. **Stachura, M. E.,** *Mol. Cell. Endocrinol.*, 44, 37, 1986.
27. **Stachura, M. E., Tyler, J. M., and Kent, P. G.,** *Endocrinology (Baltimore)*, 119, 1245, 1986.
28. **Cronin, M. J. and Canonico, P. L.,** *Biochem. Biophys. Res. Commun.*, 129, 404, 1985.
29. **Cronin, M. J., MacLeod, R. M., and Canonico, P. L.,** *Neuroendocrinology*, 40, 332, 1985.
30. **Aizawa, T. and Hinkle, P. M.,** *Endocrinology (Baltimore)*, 116, 909, 1985.
31. **Badger, T. M., Millard, W. J., McCormick, G. F., et al.,** *Endocrinology (Baltimore)*, 115, 1432, 1984.
32. **Stachura, M. E.,** *Endocrinology (Baltimore)*, 111, 1769, 1982.
33. **Sheppard, M. S. and Bala, R. M.,** *Can. J. Physiol. Pharmacol.*, 65, 515, 1987.
34. **Fukata, J., Diamond, D. J., and Martin, J. B.,** *Endocrinology (Baltimore)*, 117, 457, 1985.
35. **Stachura, M. E.,** *Endocrinology (Baltimore)*, 98, 580, 1976.

36. **Stachura, M. E. and Frohman, L. A.,** *Endocrinology (Baltimore)*, 92, 1708, 1973.
37. **Frohman, L. A., Burek, L., and Stachura, M. E.,** *Endocrinology (Baltimore)*, 91, 262, 1972.
38. **Wehrenberg, W. B., Chatelain, P., and Baird, A.,** *Proc. Soc. Exp. Biol. Med.*, 182, 107, 1986.
39. **Farrington, M. and Hymer, W. C.,** *Endocrinology (Baltimore)*, 126, 1630, 1990.
40. **Ymer, S. I. and Herington, A. C.,** *Mol. Cell. Endocrinol.*, 41, 153, 1985.
41. **Cunningham, B. C., Ultsch, M., de Vos, A., et al.,** *Science*, 254, 821, 1991.
42. **Thorlacius-Ussing, O.,** *Neuroendocrinology*, 45, 233, 1987.
43. **Dattani, M. T., Hindmarsh, P. C., Brook, C. G. D., et al.,** *Endocrinology (Baltimore)*, 133, 2803, 1993.
44. **Wells, J. A. and DeVos, A. M.,** *Annu. Rev. Biophys. Biomol. Struct.*, 351, 22329, 1993.
45. **Wells, J. A., Cunningham, B. C., Fuh, G., et al.,** *Recent Prog. Horm. Res.*, 48, 253, 1993.
46. **Yokoyama, S. and Friesen, H. G.,** *Endocrinology (Baltimore)*, 119, 2097, 1986.
47. **Stolar, M. W., Amburn, K., and Baumann, G.,** *J. Clin. Endocrinol. Metab.*, 59, 212, 1982.
48. **Stolar, M. W., Baumann, G., Vance, M. L., et al.,** *J. Clin. Endocrinol. Metab.*, 59, 235, 1984.
49. **Russell, S. M.,** *Neuroendocrinology*, 33, 67, 1981.
50. **Vodian, M. A. and Nicoll, C. S.,** *J. Endocrinol.*, 80, 69, 1979.
51. **Lorenson, M. Y. and Jacobs, L. S.,** *Endocrinology*, 110, 1164, 1982.

Chapter 6

Growth Hormone Release: Mechanisms

Stephen Harvey

CONTENTS

I. INTRODUCTION

The process of growth hormone (GH) release may result from either constitutive or regulated secretion of newly synthesized or older stored hormone in secretory granules.[1] Constitutive release involves the continuous fusion of secretory vesicles with the plasma membrane in a calcium-independent manner. Regulated secretion, in contrast, results from the intermittent mobilization of secretory granules and their calcium-dependent exocytotic discharge. These pathways may release GH moieties differentially sorted into separate vesicles containing different GH moieties.[2]

The release of GH from pituitary somatotrophs also involves the release of nongranular or cytoplasmic hormone, which may partly result from the lysis of secretory granules within the cytoplasm.[3–5] Most studies on GH release, however, consider only the release of GH from granular storage pools. The release of granular and nongranular GH may, nevertheless, be induced by similar mechanisms and releasing factors, because GH release from tumorous somatotrophs deficient in secretory granules (e.g., GH_3 cells) appears to be similar to that induced in normal somatotrophs.[6,7]

II. GH RELEASE

A. EXOCYTOSIS

The release of granular GH occurs via exocytosis, which is accomplished in several steps.[8] Secretory granules are first translocated toward the plasma membrane along microtubules, by a process facilitated by the association or contraction of microfilaments[9,10] that link the secretory granules to the microtubules and with the plasma membrane (Figure 2, Chapter 5).[11] Movement of the granules along the microtubular guides may also be mediated by an amorphous kinesin-like substance that also links microtubules with secretory granules (Figure 2, Chapter 5).

In areas in which inward dimpling of the plasma membrane occurs, the secretory granules focally fuse with the plasma membrane (Figure 1), leading to their fission and formation of single exocytotic pores (Figure 2). This pore is subsequently enlarged into a full opening or "Omega figure" and the eventual flattening of the secretory vesicle membrane into the plane of the plasma membrane results in the discharge of the granule contents into extracellular space[12] (Figure 2). The remnants of the secretory granule membrane may then be internalized and either degraded or recycled (Figure 1, Chapter 5). This membrane retrieval process ensures the cell surface area remains relatively constant.

0-8493-8697-7/95/$0.00+$.50

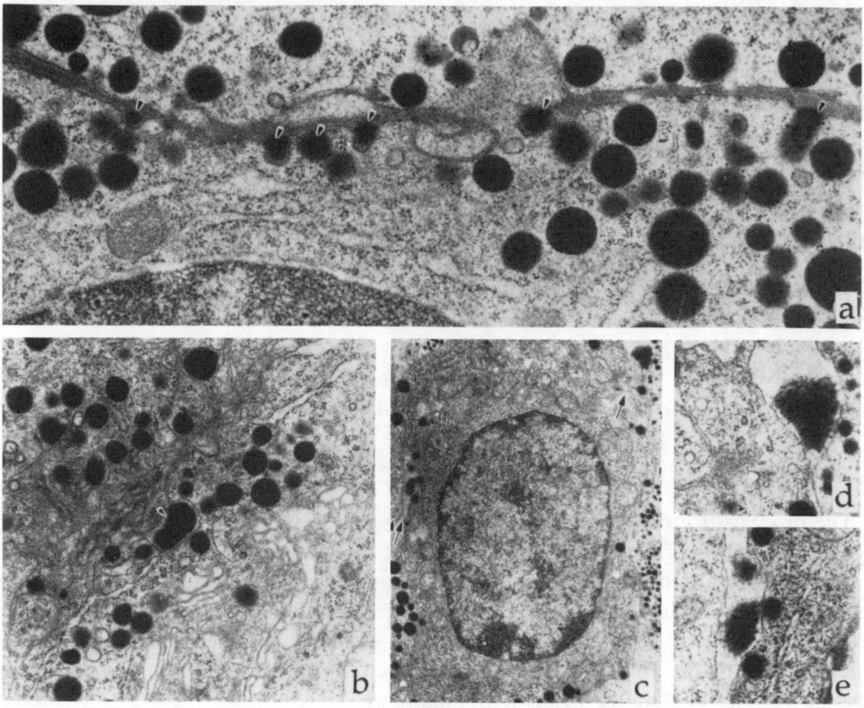

Figure 1 Exocytosis of growth hormone (GH) secretory granules from rat somatotrophs. (a) Exocytotic secretory granules (arrowheads) discharged into the perivascular space. (b) Ultrastructure of secretion from somatotrophs into blood vessels. Exocytotic secretory granules (arrowhead) sometimes aggregate with each other. Magnification: ×18,000. (c) A somatotroph 2 min after GH-releasing factor (GRF) stimulation. Magnification: ×7,000. (d and e) Higher magnifications of (c) (shown as single and double arrows, respectively). Note the GH-immunopositive electron-dense masses with irregular shapes in the intracellular space. Magnification: ×32,000. (From Shimada, O. and Tosaka-Shimada, H., Morphological analysis of growth hormone release from rat somatotrophs into blood vessels by immunogold electron microscopy. *Endocrinology*, 125(5), 2677–2682, 1989. Copyright by the Endocrine Society. Reprinted with permission.)

The exocytotic discharge of GH secretory granules from rat somatotrophs appears to lack polarity and occurs widely over the plasma membrane and in regions apposed to folliculostellate and endocrine cells and in regions in apposition with blood vessels and in portions with or without apposition to basement membranes[13] (Figure 1). Exocytosis of GH from frog pituitary glands similarly occurs widely over the plasma membrane.[14] This differs from the polar release of prolactin, which occurs toward the basal lamina.[14,15] Furthermore, whereas single secretory granules are discharged from prolactin cells, GH secretory granules typically undergo simultaneous multigranular release, which may involve the fusion of some secretory granules prior to discharge[13,14] (Figure 2, Chapter 5).

After extrusion from the somatotroph the contents of the secretory granules gradually dissolve,[16] traverse the fenestrated capillary vessels, and enter the bloodstream. It has also been suggested that the contents of the secretory granules may be transported *en masse* by

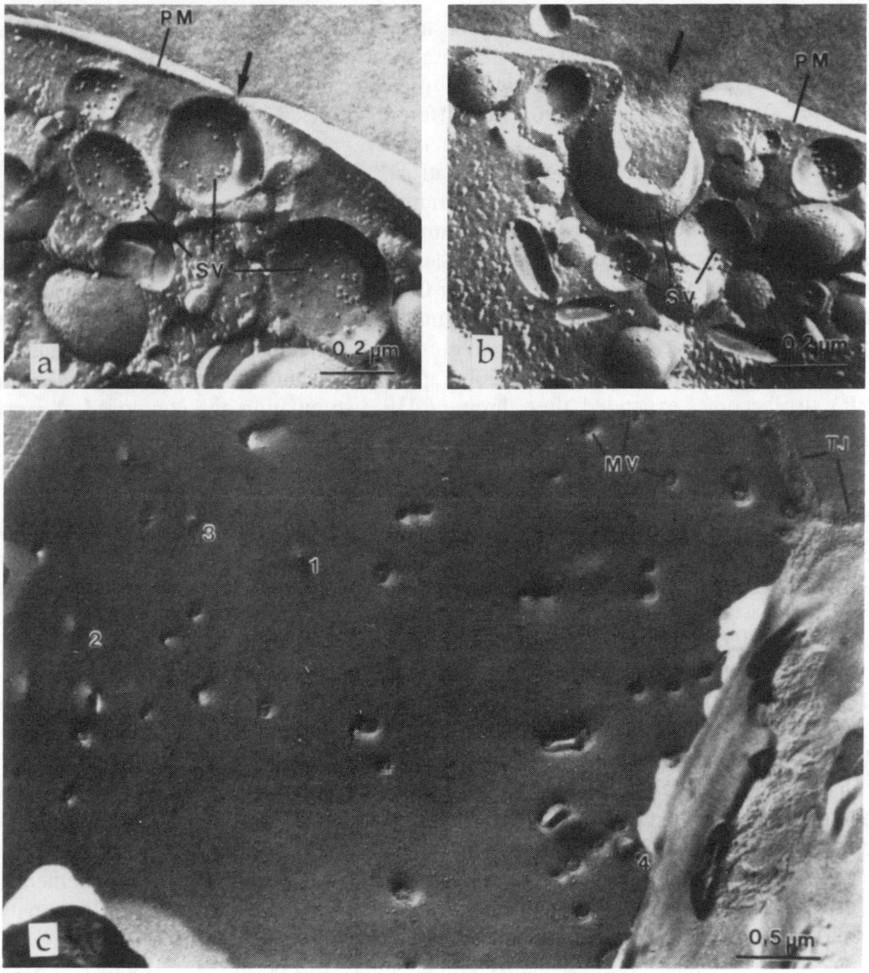

Figure 2 Exocytosis of growth hormone (GH) secretory granules from rat somatotrophs. (a) Secretory vesicle (SV) captured at an early stage of fusion with the plasma membrane (PM) and displaying a single narrow exocytotic pore (arrow). (b) Secretory vesicle at an advanced stage of exocytosis (arrow). (c) The extracellular face of the plasma membrane of the pituitary cell frozen at high pressure. The exocytotic sites appear as rounded stalks or small, volcano-like projections. The possible sequence of events associated with the formation and enlargement of exocytotic pores is indicated by numbers 1 to 4. Also labeled are microvilli (MV) and a tight junction (TJ). (From Draznin, B., Dahl, R., Sherman, N., et al., *J. Clin. Invest.* 81, 1042, 1988. Reproduced from the *Journal of Clinical Investigation*, 81, 1042, 1988. By copyright permission of the American Society for Clinical Investigation.)

interstitial fluid flow through the interstitium into blood vessels.[13] The contents of GH secretory granules sometimes form aggregates of irregular shape when released by exocytosis[13,14] (Figure 1). These electron-dense GH-immunopositive masses have no membrane-like structure but are transported through the extracellular space, in which these masses may coalesce. After passing the extracellular connective tissue involving the basement membrane, the GH mass passes through the fenestrae of the capillaries and the spaces between epithelial cells of the blood vessels and subsequently enters the bloodstream.

The morphological and physiological mechanisms involved in GH release are, however, still unclear and may differ in different species. For instance, the number and volume density of large cytoplasmic secretory granules in the frog pituitary are diminished following secretagogue stimulation.[14,17] However, when rat GH release is induced by GHRH, the cell volume may transiently decrease by 5 to 6%,[18] despite the accumulation of multivesicular bodies, immature granules, and coated vesicles in the Golgi apparatus, within 2 to 30 min.[13,16] The number of cytoplasmic secretory granules is also increased between 2 and 10 min after GH-releasing hormone (GHRH) stimulation and these accumulate under the plasma membrane, which becomes undulated with exocytotic and endocytotic figures or cavities.[13,16] The number of microtubular elements is also increased, especially near the Golgi apparatus. The rough endoplasmic reticulum (ER) also becomes expanded after chronic GHRH stimulation and composed of arrays of dilated cisternae around the nucleus or between secretory granules, which are recruited and condensed in the Golgi area.[17] These morphometric changes are indicative of increased GH release and later of increased GH synthesis and increased GH degradation.[10]

In contrast with GHRH, SRIF has no effect on somatotroph size, the number of secretory vessels, their GH content, or the number or distribution of microtubules, although it does block GHRH-induced GH release.[19-23] This antagonism is thought to result from a loss of submembrane secretory granules and the appearance of parallel bundles of microfilaments in the space between the plasma membrane and the secretory granules. The loss of submembrane secretory granules may result from an accumulation of lysosomes and increased degradation of stored hormone.[19] Somatostatin thus appears to inhibit GH release mainly at the level of the plasma membrane, probably through changes in the distribution of microfilaments.

The effects of these peptides on GH release are likely to be mediated by second messengers, stimulated or inhibited after the binding[24,25] of GHRH or SRIF to the plasma membrane. This possibility is supported by the finding that calcium-dependent exocytosis may be mediated directly by cAMP, which can provide a synergistic enhancement of GHRH-induced GH release.[26] This action of cAMP most likely occurs close to the granule-plasma membrane fusion site.

More direct actions on the secretory granules may also occur.[27] For instance, a Mg^{2+}-ATPase has been identified on GH secretory granules although it does not appear to act as a protein pump and causes an increase in intragranular osmotic pressure and lysis by water uptake, as in other tissues. Membrane receptors for GHRH coupled to an adenylate cyclase-protein kinase also are present on secretory granule membranes[28,29] and GHRH is internalized in somatotrophs and translocated to the secretory granules.[30] The activation of these receptors may result in the phosphorylation of proteins associated with the secretory granules by the cAMP-dependent protein kinase thought to participate in the process leading to exocytosis.[28,29] Small GTP-binding proteins of the Rab family are also reported to be associated with secretory vesicles[31] and are also involved in regulated exocytosis at sites distal to second-messenger generation.[32] These binding proteins are thought to be responsible for the targeting or fusion of secretory vesicles with the plasma membrane, from which they dissociate during exocytosis. Indeed, the introduction of a *rab-3a* antisense oligonucleotide into the cytosol of anterior pituitary cells suppresses calcium-dependent exocytosis.[26] The release of GH from secretory granules is also stimulated by purine nucleotides,[27,33] although their role is uncertain. Thus, although intracellular calcium is clearly an important signal for GH secretion, other small molecules in the cell cytosol may effectively act to override or to enhance the action of calcium.[31]

B. REGULATION OF GH RELEASE

It is now well established that the release of GH by exocytosis is calcium (Ca^{2+}) dependent and an increase in cytosolic Ca^{2+} is a prerequisite for GH secretion.[26] Somatotrophs are electrically active and display action potentials (spontaneous and triggered) that are mainly Ca^{2+} dependent, although sodium (Na^+)-dependent action potentials characterize 10% of GH-secreting cells.[34] Each somatotroph has spontaneous oscillations of intracellular Ca^{2+} and GH release, with distinct pulse frequencies (every 5 to 30 s) and GH amplitude (between 50 and 450 nm).[35,36] These pulses are generated asynchronously from cell to cell and the amount released correlates directly with both the frequency and amplitude of changes in the intracellular Ca^{2+} concentration. This spontaneous release of GH is partially dependent on extracellular calcium (as is stimulated GH release), which may induce intracellular signaling by information encoded by the frequency and amplitude of the Ca^{2+} oscillations. Voltage-sensitive calcium channels (VSCCs) mediate extracellular Ca^{2+} entry into these cells and are responsible for the depolarizing currents induced by GHRH and other secretagogues.[37–39] Calcium ionophores, calcium channel agonists, media of high extracellular potassium (K^+) concentration, and other Ca^{2+}-mobilizing secretagogues (e.g., cAMP analogs, prostaglandins, phorbol esters, and diacylglycerol analogs) therefore stimulate GH secretion.[35,36,40–42] This stimulation is due to Ca^{2+} oscillations of increased amplitude and frequency[35] resulting from the rapid influx (and efflux) of extracellular Ca^{2+} through L-type VSCCs and is independent of Ca^{2+} sequestration from intracellular stores. Consequently, extracellular Ca^{2+} depletion, Ca^{2+} current blockers, or Ca^{2+} channel blockade can inhibit basal and stimulated GH release, by mechanisms independent of cAMP formation.[39] Inactivation of the membrane-bound enzymes (serine proteases) that promote Ca^{2+} channel stability similarly inhibits GH release.[43–46]

The action of GHRH in inducing extracellular Ca^{2+} entry into somatotrophs may be mediated via a number of mechanisms[40–42] (Figure 3). After binding to its membrane receptor, the activation of a stimulatory G protein (G_s) may directly interact with VSCCs to change voltage sensitivity. The G_s protein may also rapidly activate adenylate cyclase activity, cAMP accumulation, and protein kinase A (PKA) activity. Cyclic AMP and activators of the PKA system are potent GH secretagogues and they phosphorylate voltage-insensitive Na^+ channels, leading to depolarization and the opening of the VSCCs and the influx of extracellular calcium. The entry of Ca^{2+} through the VSCCs is linked to increased Na^+ conductance[37,47] and GH release is conversely suppressed in Na^+ free media,[47] although this is not a consistent finding.[39]

The entry of extracellular Ca^{2+} increases cytosolic Ca^{2+} concentrations, which inhibits adenylate cyclase activity via an inhibitory G protein (G_i)[48,49] and inhibits cAMP-dependent activation of the Na^+ and Ca^{2+} channels.[50] These events lead to repolarization of the plasma membrane, the closing of the VSCC's and the activation of protein kinase C (PKC) and PKC-dependent Ca^{2+}-ATPase in plasma and ER membranes to restore the cytosolic Ca^{2+} concentration. These events also involve the participation of voltage- and Ca^{2+}-gated K^+ channels, the closure of which enhances Ca^{2+} influx and stimulates GH release.[51,52] The efflux of K^+ through these channels is delayed by GHRH, promoting Ca^{2+} entry into somatotrophs and depolarization.[53] The increased intracellular Ca^{2+} concentration then activates the Ca^{2+}-sensitive K^+ channels[40,41] and the efflux of K^+ is increased, causing membrane depolarization. GHRH also stimulates K^+ outflow, even in the absence of extracellular Ca^{2+}, and this occurs subsequent to the brief phasic release of GH.[53] The brevity of this GH response is therefore similar to that induced by K^+ depolarization and reflects the biphasic or rhythmical effect of GHRH on Ca^{2+} fluxes and the rapid induction of Ca^{2+} efflux.[54]

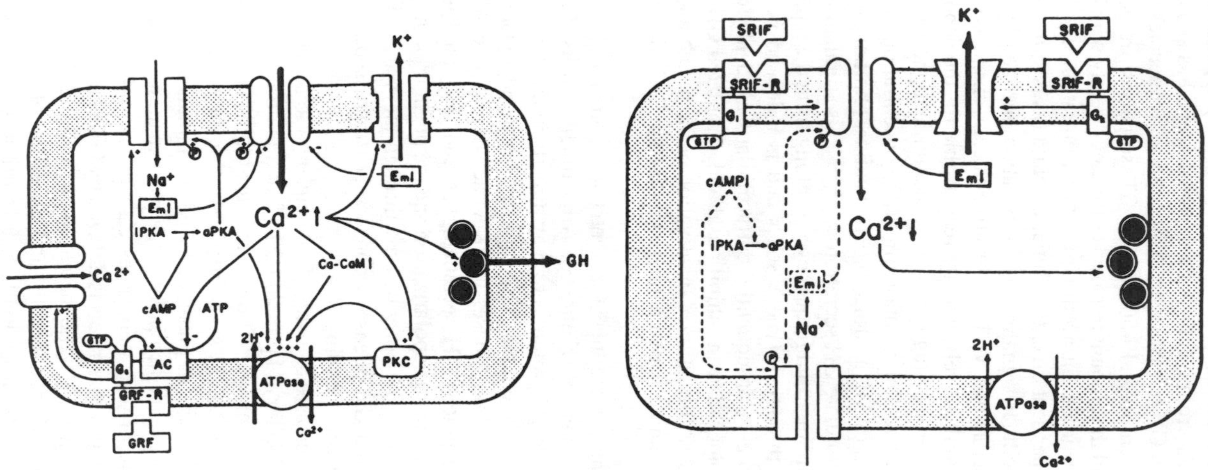

Figure 3 Models depicting (a) the pathways responsible for the growth hormone (GH) releasing hormone (GHRH or GRF, GH-releasing factor)-induced increase in $[Ca^{2+}]_i$ and GH release in rat somatotrophs; and (b) pathways responsible for the somatostatin (SRIF)-induced decrease in $[Ca^{2+}]_i$ and inhibition of GH release in rat somatotrophs. AC, adenylate cyclase; E_M, membrane potential (upward arrow, less negative; downward arrow, more negative); G_G, GTP-binding protein G_S; G_i, GTP-binding protein G_i; G_K, GTP-binding protein G_K; GRF-R, GRF receptor; aPKA, active cAMP-dependent protein kinase; iPKA, inactive cAMP-dependent protein kinase; +, stimulation; -, inhibition; P, channel phosphorylation. Dashed arrows denote decreased stimulatory effects due to the moderate reduction in cAMP accumulation seen in GHRH-challenged somatotrophs in the presence of SRIF. (From Lussier, B. T., French, M. B., Moor, B. C., and Kraicer, J., Free intracellular Ca^{2+} concentration and growth hormone (GH) release from purified rat somatotrophs. III. Mechanism of action of GH-releasing factor and somatostatin, *Endocrinology*, 128(1), 592–603, 1991. Copyright by the Endocrine Society. Reprinted with permission.)

The release of GH in response to GHRH and other GH secretagogues is tightly coupled to the efflux of Ca^{2+}, which has been suggested to play a role in exocytosis because GHRH-induced Ca^{2+} efflux occurs from cells cultured in the absence of extracellular Ca^{2+}.[55] Calcium efflux through Ca^{2+}-ATPase pumps is also stimulated by PKC activators, which are potent stimulators of GH secretion. The mobilization of intracellular Ca^{2+} therefore also appears to participate in the GH secretory process and potentiates GH release induced by many secretagogues.[55,56] The mobilization of intracellular Ca^{2+} may be induced by cAMP accumulation and especially by phosphatidylinositol metabolism, which results in the formation of diacylglycerol and inositol triphosphate, both of which sequester Ca^{2+} from intracellular pools and stimulate GH release.[52] Consequently, when phospholipase activity is inhibited, basal and GHRH-induced GH release is suppressed,[57] although French et al.[58] suggest phosphoinositide hydrolysis is not essential for GHRH action. Phosphatidylinositol hydrolysis results in the formation of arachidonic acid (a precursor for prostaglandins and leukotrienes) and activation of the inositol phosphate-PKC pathway results in the stimulation of the adenylate cyclase-PKA system,[59,60] because prostaglandins increase cAMP production and GH release in rats.[57,61] The formation of arachidonic acid may in addition facilitate growth hormone release by promoting the fusion of secretory granules to the plasma membrane and by weakening portions of the granule membrane.[8]

The stimulation of GH release by GHRH, PKA, and PKC activators and by other secretagogues is blocked by SRIF,[40,42,60] by mechanisms that antagonize intracellular Ca^{2+} accumulation, and the amplitude of Ca^{2+} fluxes[35] (Figure 3). Somatostatin decreases Ca^{2+} mobilization from intracellular organelles[62,63] and blocks the influx of extracellular Ca^{2+} in a manner mimicked by Ca^{2+} inhibitors, with which it interacts in a nonadditive manner.[34] SRIF reduces the activation rate of Ca^{2+} currents, decreases Ca^{2+} conductance, and increases voltage-dependent K^+ conductance and thereby hyperpolarizes somatotrophs.[39,40,42,64] Somatostatin is also able to hyperpolarize the somatotroph directly, even when the K^+ current is blocked.[65] Moreover, although SRIF does not inhibit adenylate cyclase activity in somatotrophs it promotes phosphodiesterase activity[49] and antagonizes GHRH-stimulated cAMP accumulation,[66,67] reducing the phosphorylation of Ca^{2+} and Na^+ channels and Ca^{2+} and Na^+ conductance.[48] These inhibitory effects of SRIF on electrical conductance appear to occur in all somatotrophs, even in heterogeneous populations.[38,39] The electrophysiological effects of GHRH, in contrast, are observed only in a subset (perhaps 30 to 40%) of the GH-secreting cells in the rat pituitary gland.

Although SRIF blocks the Ca^{2+} influx induced by most secretagogues it does not block K^+-induced Ca^{2+} entry into somatotrophs, despite blocking K^+-induced GH release. Somatostatin is thus also able to suppress GH release through calcium-independent mechanisms. Increased intracellular Ca^{2+} concentrations are therefore not necessarily stimulatory for GH release.[48] Indeed, basal and secretagogue-induced GH secretion is markedly suppressed in media of supraphysiological Ca^{2+} concentration, by mechanisms that are additive with and in parallel with those of SRIF.[63] Excessive Ca^{2+} is thought to inhibit GH release by preventing microtubule-associated intracellular hormone transport and by inhibition of enzymes responsible for intracellular transport processes.

The influx of extracellular Ca^{2+} and the mobilization of intracellular Ca^{2+} stimulate GH release in a temporally biphasic manner. The first phase of release occurs with no detectable lag period and is complete within minutes.[12,13,68] A second phase with a lower amplitude of release is then induced and persists until the stimulatory signal is inactivated or until GH stores deplete. The release of GH from the first phase is from the immediate release pool (IRP) of secretory granules located just below the plasma membrane.[63,69] This is closely correlated with membrane depolarization and is K^+ inducible and Ca^{2+} dependent and can occur in the absence of adenylate cyclase activation. This phase does, however, have a cAMP-dependent component linked to the rapid mobilization of secretory granules from the Golgi region and the preferential replenishment of the IRP with newly synthesized GH.[10] This store therefore largely accounts for basal GH release, because newly synthesized hormone is preferentially released from unstimulated somatotrophs.[67,70] This store is thus dependent on thyroid hormones for continued GH synthesis[20] and GH release from this store is preferentially suppressed when protein synthesis is blocked by cyclohexamide.[70,71] The size of the immediate release pool is therefore increased during SRIF inhibition, because SRIF inhibits GH release and not GH synthesis. The expanded immediate release pool is thus responsible for the exaggerated "rebound" GH secretion that occurs immediately after SRIF withdrawal.[21,72] The inhibitory effect of insulin-like growth factor I (IGF-1) on GH release is also likely to be targeted at this store, as it only inhibits the synthesis of newly synthesized GH and has lower potency *in vitro* than *in vivo*.[73]

The second phase of GH release is primarily mediated from deep intracellular storage compartments and is cAMP and protein kinase C sensitive,[69,74,75] but unlike the first it is not accompanied by an increase in Ca^{2+} efflux.[55] Most of the hormone in this store is not newly synthesized and GHRH and other secretagogues are therefore able to stimulate GH release when intracellular protein synthesis is blocked.[70,71] The ability of GH secretagogues to stimulate GH release is, however, blocked when the process that translocates secretory granules to the plasma membrane is inhibited at the level of the Golgi[10,32] or when movement along microtubules is suppressed by excess Ca^{2+}.[63]

Cellular proteins (e.g., receptors, channels, enzymes) are therefore of great importance in the processes that lead to GH release. Steroid hormones and thyroid hormones, which have genomic sites of action, are therefore likely to participate in these processes. Steroid hormones may also alter membrane fluidity and the efficiency of ion channels and receptor-mediated processes. These hormones may also have direct electrophysiological effects, because membrane receptors for them have been described.[76] Indeed, triiodothyronine (T_3) may inhibit GH release from somatotroph cells through a membrane receptor that blocks the entry of extracellular Ca^{2+} and blocks the Ca^{2+}-sensitive K^+ current.[77]

REFERENCES

1. **Shields, D.,** *Peptide Biosynthesis and Processing* (Ed. Fricker, D.). CRC Press, Boca Raton, Florida, 1991 p. 37.
2. **Green, R. and Shields, D.,** *J. Cell Biol.*, 99, 97, 1984.
3. **Young, G. and Ball, J. N.,** *Gen. Comp. Endocrinol.*, 52, 86, 1983.
4. **Leatherland, J. F.,** *Z. Zellforsch. Mikrosh. Anat.*, 104, 318, 1970.
5. **Leatherland, J. F.,** *Can. J. Zool.*, 50, 835, 1972.
6. **Hinkle, P. M.,** *Secretory Tumors of the Pituitary Gland* (Eds. Black, P., Zervas, H., Ridgeway, E. C. and Martin, J. B.). Raven Press, New York, 1984, p. 25.
7. **Gourdji, D., Tougard, C., and Tixier-Vidal, A.,** *Front. Neuroendocrinol.*, 7, 317, 1982.
8. **Creutz, M. J.,** *Recent Prog. Horm. Res.*, 37, 325, 1981.
9. **Zor, U.,** *Endocr. Rev.*, 4, 1, 1983.
10. **Farmer, P. K., Tyler, J. M., and Stachura, M. E.,** *Mol. Cell Endocrinol.*, 62, 253, 1989.
11. **Kurosumi, K.,** *J. Electron Microsc. Tech.*, 19, 42, 1991.
12. **Draznin, B., Dahl, R., Sherman, N., et al.,** *J. Clin. Invest.*, 81, 1042, 1988.
13. **Shimada, O. and Tosaka-Shimada, H.,** *Endocrinology (Baltimore)*, 125, 2677, 1989.
14. **Castano, J. P., Ruiz-Navarro, A., Torronteras, R., et al.,** *Tissue Cell*, 25, 165, 1993.
15. **Kelly, R. B.,** *Science*, 230, 25, 1985.
16. **Harrisson, F.,** *Anat. Embryol.*, 154, 185, 1978.
17. **Malagon, M. M., Castano, J. P., Dobado-Berrios, P. M., et al.,** *Gen. Comp. Endocrinol.*, 84, 461, 1991.
18. **Engstrom, K. G. and Ohlsson, L.,** *Biochem. Biophys. Acta*, 1135, 318, 1992.
19. **Asa, S. L., Felix, I., Kovacs, K., et al.,** *Endocr. Pathol.*, 1, 228, 1990.
20. **Stachura, M. E.,** *Mol. Cell. Endocrinol.*, 44, 37, 1986.

21. **Stachura, M. E., Tyler, J. M., and Farmer, P. K.,** *Endocrinology (Baltimore)*, 123, 1476, 1988.
22. **Stachura, M. E.,** *Endocr. Res.*, 12, 69, 1986.
23. **Stachura, M. E., Costoff, A., and Tyler, T. M.,** *Neuroendocrinology*, 42, 383, 1986.
24. **Abribat, T., Boulanger, L., and Gaudreau, P.,** *Brain Res.*, 528, 291, 1990.
25. **Katakami, H., Berelowitz, M., Marbach, M., et al.,** *Endocrinology (Baltimore)*, 117, 557, 1985.
26. **Mason, W. T., Dickson, S. L., and Leng, G.,** *Acta Paediatr. Scand.*, 82, 84, 1993.
27. **Lorenson, M. Y. and Jacobs, L. S.,** *Endocrinology (Baltimore)*, 114, 717, 1984.
28. **Lewin, M. J. M., Reyl-Desmars, F., and Ling, N.,** *C.R. Acad. Sci. Paris*, 297, 123, 1983.
29. **Lewin, M. J. M., Reyl-Desmars, F., and Ling, N.,** *Proc. Natl. Acad. Sci. U.S.A.*, 80, 6538, 1983.
30. **Morel, G., Mesguich, P., and Dubois, P. M.,** *Neuroendocrinology*, 38, 123, 1984.
31. **Burgoyne, R. D.,** *Annu. Rev. Physiol.*, 52, 647, 1990.
32. **Davidson, J. S., Eales, A., Roeske, R. W., et al.,** *FEBS Lett.*, 326, 219, 1993.
33. **Lemay, A., Labrie, F., and Drouin, D.,** *Can. J. Biochem.*, 52, 327, 1974.
34. **Israel, J. M. and Vincent, J. D.,** *Front. Neuroendocrinol.*, 11, 339, 1990.
35. **Cuttler, L., Glaum, S. R., Collins, B. A., et al.,** *Endocrinology (Baltimore)*, 130, 945, 1992.
36. **Holl, R. W., Thorner, M. O., and Leong, D. A.,** *Endocrinology (Baltimore)*, 122, 2927, 1988.
37. **Kato, M. and Suzuki, M.,** *Brain Res.*, 476, 145, 1989.
38. **Chen, C., Israel, J. M., and Vincent, J. D.,** *J. Physiol.*, 408, 493, 1989.
39. **Chen, C., Israel, J. M., and Vincent, J. D.,** *Neuroendocrinology*, 50, 679, 1989.
40. **Lussier, B. T., Wood, D. A., French, M. B., et al.,** *Endocrinology (Baltimore)*, 128, 583, 1991.
41. **Lussier, B. T., French, M. B., Moor, B. C., et al.,** *Endocrinology (Baltimore)*, 128, 570, 1991.
42. **Lussier, B. T., French, M. B., Moor, B. C., et al.,** *Endocrinology (Baltimore)*, 128, 592, 1991.
43. **Makara, G. G., Szentendrei, T., and Rappay, G.,** *Mol. Cell. Endocrinol.*, 52, 63, 1987.
44. **Nagy, I., Makara, G. B., Horvath, G., et al.,** *Endocrinology (Baltimore)*, 116, 1426, 1985.
45. **Nagy, I., Makara, G. B., Garamvolygi, V., et al.,** *Biochem. Pharmacol.*, 40, 887, 1990.
46. **Rappay, G. Y., Nagy, I., Makara, G. B., et al.,** *Life Sci.*, 34, 337, 1983.
47. **Kato, M., Hattori, M., and Suzuki, M.,** *Am. J. Physiol.*, 254, E476, 1988.
48. **Boyajian, C. L. and Cooper, D. M. F.,** *Cell Calc.*, 11, 299, 1990.
49. **Giannattasio, G., Bianchi, R., Spada, A., et al.,** *Endocrinology (Baltimore)*, 120, 2622, 1987.
50. **Dufy, B., Jaken, S., and Barker, J. L.,** *Endocrinology (Baltimore)*, 121, 793, 1987.
51. **Chen, C., Heyward, Plk Zhang, J., et al.,** *Neuroendocrinology*, 59, 1, 1994.
52. **Rawlings, S. R. and Mason, W. T.,** *Biotechnology in Growth Regulation* (Eds. Heap, R. B., Prosser, C. J., and Lamming, G. E.) Butterworths, London, 1988, p. 35.
53. **Ohlsson, L. and Lindstrom, P.,** *Acta Physiol. Scand.*, 137, 237, 1989.
54. **Nussinovitch, I.,** *J. Physiol.*, 395, 303, 1988.
55. **Ohlsson, L. and Lindstrom, P.,** *Endocrinology (Baltimore)*, 126, 488, 1990.
56. **Judd, A. M., Koike, K., Yasumoto, T., et al.,** *Neuroendocrinology*, 42, 197, 1986.
57. **Cronin, M. J., MacLeod, R. M., and Canonico, P. L.,** *Neuroendocrinology*, 40, 332, 1985.
58. **French, M. B., Lussier, B. T., Moor, B. C., et al.,** *Mol. Cell. Endocrinol.*, 72, 221, 1990.
59. **Summers, S. T. and Cronin, M. J.,** *Arch. Biochem. Biophys.*, 262, 12, 1988.
60. **Summers, S. T., Canonico, P. L., MacLeod, R. M., et al.,** *Eur. J. Pharmacol.*, 111, 371, 1985.
61. **Cronin, M. J. and Canonico, P. L.,** *Biochem. Biophys. Res. Commun.*, 129, 404, 1985.
62. **Kracier, J. and Chow, A. E. H.,** *Endocrinology*, 111, 1173, 1982.
63. **Stachura, M. E., Tyler, J. M., Kent, P. G., et al.,** *Endocr. Res.*, 12, 171, 1986.
64. **Yamashita, S., Shibuya, N., and Ogata, E.,** *Proc. Natl. Acad. Sci. U.S.A.*, 83, 6198, 1986.
65. **Nussinovitch, I.,** *Brain Res.*, 504, 136, 1989.
66. **Narayanan, N., Lussier, B., French, M., et al.,** *Endocrinology (Baltimore)*, 124, 484, 1989.
67. **Sheppard, M. S., Moor, B. C., and Kracier, J.,** *Endocrinology (Baltimore)*, 117, 2364, 1985.
68. **DeVirgilus, G., Meldolesi, J., and Clementi, F.,** *Endocrinology (Baltimore)*, 83, 1278, 1988.
69. **Stachura, M. E. and Tyler, J. M.,** *Endocr. Res.*, 16, 1, 1990.
70. **Chen, T. T., Kineman, R. D., Betts, J. G., et al.,** *Endocrinology (Baltimore)*, 408, 493, 1989.
71. **Sheppard, M. S. and Bala, R. M.,** *Can. J. Physiol. Pharmacol.*, 65, 515, 1987.
72. **Kraicer, J., Cowan, J. S., Sheppard, M. S., et al.,** *Endocrinology (Baltimore)*, 119, 2047, 1986.
73. **Sheppard, M. S., Eatock, B. A., and Bala, R. M.,** *Can. J. Physiol. Pharmacol.*, 65, 2302, 1987.
74. **Aizawa, T. and Hinkle, P. M.,** *Endocrinology (Baltimore)*, 116, 73, 1985.
75. **Stachura, M. E., Tyler, J. M., and Kent, P. G.,** *Endocrinology (Baltimore)*, 125, 444, 1989.
76. **McEwen, B. S.,** *Trends Endocrinol. Metab.*, 2, 62, 1991.
77. **du Pont, J. S.,** *Acta Endocrinol. (Copenh.)*, 123, 51, 1990.

Chapter 7

Growth Hormone Release: Stimulatory Hypothalamic Control

CONTENTS

I. INTRODUCTION

The regulation of growth hormone (GH) release is primarily by stimulatory and inhibitory hypothalamic drive, mediated through the pituitary actions of GH-releasing hormone (GHRH) and somatostatin (SRIF). These hypophysiotrophic factors are released in response to information about the internal and external environment, which is conveyed to this final common axis through a variety of neuropeptides and neurotransmitters. These neuroactive agents may also be released into hypophysial portal plasma and directly regulate somatotroph function or modulate pituitary responsiveness to GHRH or SRIF stimulation. The hypothalamic control of GH release thus results from complex interactions at hypothalamic and pituitary sites by numerous stimulatory and inhibitory factors. The participation of GHRH in GH secretion is reviewed in this chapter.

II. GROWTH HORMONE-RELEASING HORMONE

Growth hormone-releasing hormone is synthesized in neurons located in the arcuate nucleus and to a lesser extent in the ventromedial hypothalamic nuclei,[1-5] and is released into pituitary portal circulation from nerve terminals in the median eminence.[6] Many of the secretory granules in nerve terminals contain not only GHRH but also dopamine, γ-aminobutyric acid (GABA), neurotensin (NT), galanin, acetylcholine, α-melanotropin, melanin-concentrating hormone, and neuropeptide Y (NPY),[7-14] which are likely modulators of GHRH release or GHRH action.[15] Some of these factors may also regulate GHRH synthesis or release within the hypothalamus, because the perikarya and axon terminals of GHRH neurons are synaptically associated with neuronal fibers containing acetylcholine,[12] dopamine,[10,11,14,16] somatostatin, substance P and enkephalin 8,[17] neuropeptide Y,[18,19] thyrotropin-releasing hormone (TRH),[20] melanin-concentrating hormone and α-melanotropin,[12,13] or galanin.[21]

A. GHRH RELEASE

Electrical stimulation of the arcuate or ventromedial nuclei evokes GHRH and GH release,[22-25] whereas electrical stimulation of the median eminence is comparatively ineffective.[23,26] The preferential release of GH after stimulation of the arcuate and ven-

0-8493-8697-7/95/$0.00+$.50

tromedial nuclei probably reflects the SRIF content in these areas and stimulated SRIF release, because the SRIF content in these nuclei is much lower than in the median eminence. The increased GH secretion that occurs after the electrical stimulation of perikarya in the arcuate nucleus is independent of the electrical frequency but is related to the duration of stimulation.[23]

The release of GHRH from the hypothalamus is also induced by K^+ depolarization[27-30] and occurs via Ca^{2+}, phosphatidylinositol, protein kinase C, and adenylate cyclase-protein kinase A pathways.[28,31-33] Release is also enhanced after hypophysectomy,[34] thyroidectomy,[35] α_1- and α_2-adrenergic stimulation[36,37] or β-adrenergic blockade,[38] and by serotoninergic[39] and dopaminergic [30] stimulation, glucose deficiency,[40] and by opiates,[41] prostaglandins,[43,44] and GABA,[45] but is reduced by insulin-like growth factor I (IGF-I),[46] glucose,[47] and SRIF.[48-50] The release of GHRH may also be directly suppressed by inhibitory GHRH collateral neurons, because intracerebroventricular (i.c.v.) injections of GHRH suppress GH release,[51] and the iontophoretic application of GHRH to hypothalamic neurons inhibits their firing rate. The release of GHRH into pituitary portal circulation is also largely concurrent with GH pulses in peripheral plasma[52-55] and simultaneously increased during stress[56] and reduced in GH-deficient streptozocin-diabetic rats.[57]

The pivotal role of GHRH in triggering GH release is also indicated by the inhibition of basal and pulsatile GH release in GHRH-immunized rats,[58,59] cows,[60,61] sheep,[62] and pigs.[63,64] GHRH immunoneutralization similarly blocks the pubertal surge in GH secretion[65] and the GH responses induced by interleukin 1,[66] enkephalins,[67,68] opiates,[41,61] GABA,[67] α-adrenergic stimulation,[62,69,70] serotoninergic stimulation,[71] arginine,[61] dietary deprivation,[60,72,73] SRIF withdrawal,[74] and SRIF immunoneutralization.[75] The blockade of the GH responses to pentagastrin following GHRH pretreatment is also thought to indicate GHRH mediation, because GHRH pretreatment similarly blocks subsequent GH responses to GHRH stimulation.[76] The increased GH responses of rats with anterolateral hypothalamic deafferentations to i.c.v. injections of acetylcholine, substance P, dopamine (D), and norepinephrine (NE) are also thought to result from direct effects on GHRH cells or on neural circuits impinging on them, because GHRH is primarily responsible for GH secretion in these SRIF-depleted animals.[77] The increased frequency of pulsatile GH release in SRIF-depleted rats with periventricular preoptic anterior hypothalamic lesions[78] is similarly thought to be due to GHRH, and the suppression of GH secretion in these animals suggests a reduction in GHRH rather than an increase in SRIF occurs during stress.[78] The suppression of GH secretion in rats after the i.c.v. administration of GH, neurotensin, glucagon, or endotoxin is also likely to be due to suppression of GHRH release rather than to increased SRIF tone, because these stimuli do not enhance median eminence SRIF concentrations.[79] The impaired pulsatile GH secretion in rats with ventromedial hypothalamus-arcuate nucleus lesions is, similarly, restored by exogenous GHRH.[80] The impaired GH response of monosodium glutamate-treated rats to morphine and to pentobarbital anesthesia[81] also suggests these stimuli are mediated through GHRH neurons. The release of GHRH from the hypothalamus is, reciprocally, reduced by exogenous GH, in a feedback manner.[27] The reduction in GHRH release from GH-treated hypothalamic fragments is greater than the suppression of GH content.[27,82,83]

Lesioning of the arcuate nucleus similarly reduces the pituitary GH content[84] and suppresses pulsatile GH secretion,[80,85] as does monosodium glutamate-induced GHRH depletion,[81,86-88] which also blocks NMA-induced GH release.[89] Basal, sleep-related, and pulsatile GH secretion is similarly suppressed by specific GHRH receptor antagonists.[90,91]

B. GHRH SYNTHESIS

Expression of the *GHRH* gene in hypothalamic tissue results in the synthesis of a 107/108-amino acid precursor (prepro-GHRH), which undergoes posttranslational processing to yield two peptides [GHRH(1–44)-NH_2 and GHRH(1–40)-OH[92-94]]. Both of these peptides are abundantly present in the hypothalamus,[95] both induce GH release with comparable potency,[96,97] and both are probably involved in the physiological regulation of GH secretion. In the absence of hypothalamic GHRH, the synthesis and secretion of GH is suppressed and growth is severely retarded.[98,99]

Expression of the *GHRH* gene occurs at embryonic day 17 in the rat[100,101] and is thought to induce the proliferation of somatotrophs[102,103] and the synthesis and release of pituitary GH in fetal animals[104,105] (see Chapters 3 and 4). In the absence of GHRH, serum GH levels are reduced from the 19th day of gestation,[101,106] indicating roles for GHRH in fetal somatotroph function. Neurons immunoreactive for GHRH are first detected in the arcuate and ventromedial nuclei of the human fetus in the 29th week of gestation and immunoreactive fibers are detectable in the median eminence from the 31st week.[107] Fetal GHRH is thus thought to play a major role in the control of GH secretion during the perinatal period but the differentiation and development of pituitary somatotrophs occur prior to its ontogenic appearance in the hypothalamus.[107]

The synthesis, content (and release) of GHRH increases postnatally and with sexual maturation[108] and subsequently declines with aging,[109-111] possibly because of a reduction in dopaminergic tone.[112] This age-related reduction in GHRH content parallels a reduction in *GHRH* mRNA, indicating an impairment of GHRH neurons in aged rats.[109]

The hypothalamic content of GHRH or *GHRH* mRNA is also sexually dimorphic in rats; it is greatest in males[113,114] and is increased by testosterone.[114,115] Castration, therefore, suppresses *GHRH* mRNA levels in both the arcuate nucleus and the ventromedial hypothalamic nucleus and is restored by testosterone and nonaromatizable androgens but not by estradiol.[114] Estradiol may, nevertheless, modulate GHRH release, because 20 to 30% of GHRH-containing neurons in the rat arcuate nucleus have nuclear receptors for [^{3}H]estradiol.[116] Indeed, estrogen is reported to partially restore immunoreactive GHRH levels in the hypothalami of castrated males to the levels seen in gonad-intact females.[117] Gonadal steroid-induced differences in *GHRH* gene transcription are therefore likely to account partly for the sexually dimorphic patterns in pulsatile GH release.[101,108,110,117-119] The influence of gonadal steroids on *GHRH* gene expression has been reviewed by Argente and Chowen.[120]

The hypothalamic content of *GHRH* mRNA during pulsatile release is highest during trough periods of GH release and lowest during episodic bursts of GH secretion,[121] indicating an inverse feedback relationship with GH secretion. The expression of the *GHRH* gene is similarly increased in GH-deficient mice[113,122] and hypophysectomized rats[34,123-125] and reduced by GH treatment.[34,126,127] The ability of GH to inhibit GHRH synthesis may, however, be age related and greater in adolescents than in adults[128] and neonates,[126,129] in which GH secretion is less (see Chapter 12).

The feedback effect of GH on GHRH also appears to be sexually dimorphic in rats, because GHRH is able to elicit uniform GH responses during hGH infusions, whereas human GH (hGH) infusions in male rats disturb their intermittent pattern of responsiveness to GHRH and prolong the period of somatotroph refractoriness.[118] Median eminence concentrations of GHRH are reduced whereas SRIF concentrations are elevated by GH in both male and female rats, but these alterations are more pronounced in males.[130] This feedback effect of GH on hypothalamic GHRH may be mediated through cholinergic, opioid, and SRIF-producing neurons.[83,131] The content of hypothalamic GHRH is also reduced in Ames dwarf mice[132] and GHRH and *GHRH* mRNA levels are reduced in rats during *N*-methyl-D-aspartate (NMDA) receptor blockade, in concert with GH suppression.[133] The content of hypothalamic GHRH is also decreased after hypophysectomy,[29,34] despite the increased levels of *GHRH* mRNA, indicating that the rates of *GHRH* gene transcription and translation or the rates of GHRH synthesis and release may be differentially regulated. A similar difference may also account for the reduction in *GHRH* mRNA levels in the hypothalami of streptozocin-diabetic rats, which occurs independently of tissue GHRH levels.[57]

The hypothalamic content of GHRH is similarly reduced by thyroidectomy,[134] whereas GHRH mRNA levels are increased by thyroid hormone deficiency.[35] These changes with thyroid status are likely also to reflect changes in GH secretion, because *GHRH* mRNA levels are restored in the hypothalami of thyroid-deficient rats following exogenous GH administration.[35,127] The transcription of the *GHRH* gene may also be regulated independently of GH secretion, because both are diminished in rats that are genetically obese,[22,94] diabetic,[135] or dietarily restricted.[136] In contrast, the content of hypothalamic GHRH is reduced while GH secretion is increased after central α_2-adrenergic stimulation.[37]

C. GHRH RECEPTORS

The actions of GHRH on somatotroph function are mediated through high-affinity, low-capacity membrane receptors[137-140] and possibly also by low-affinity, high-capacity sites.[141,142] GHRH-binding sites are abundantly present in pituitary tissue and only partial occupancy is required to elicit maximal GH responses.[143] Although these sites are likely to be associated with somatotroph membranes, the slight stimulation of prolactin,[144-146] gonadotropin subunits,[147] gonadotropins,[148] and thyrotropin[149] release by GHRH suggests they may be additionally present on other pituitary cell types.

These GHRH receptors are essential for the induction of GH responses to GHRH stimulation, because GHRH (unlike cAMP) is unable to induce GH release in the GH-deficient *lit* dwarf mouse, which produces GHRH but has defective GHRH receptors.[150-152] The receptor-associated resistance to GHRH in the *lit* mouse results from a mutation in one of the G protein subunits that prevents GHRH receptor interaction. A similar defect also occurs in dwarf *dw* rats, although GHRH is able to induce cAMP and GH release to a limited extent in these mutants.[153,154]

The number of high-affinity GHRH receptors is reduced during aging in rats, prior to the age-related reduction in pituitary sensitivity to GHRH stimulation and reduced GH secretion.[139] The number of GHRH receptors is also reduced in the obese Zucker rat, in which GH secretion is impaired.[155] The number, but not affinity, of pituitary GHRH receptors is also related to adrenal function, being reduced two- to fourfold by adrenalectomy and increased by glucocorticoid hormones.[156,157] The reduced somatotroph response to GHRH stimulation in glucocorticoid deficiency[158,159] and the exaggerated GH responsiveness following glucocorticoid therapy[160,161] is therefore likely to reflect down- and upregulation, respectively, of pituitary GHRH receptors. Receptor downregulation may also be induced by GHRH itself.

Continuous or repetitive stimulation of the pituitary by GHRH may attenuate episodic GH release[162] and cause GH response attenuation.[163-166] The desensitization of somatotroph responsiveness may facilitate the induction of pulsatile episodes of GH release (see Chapter 12). An absence of this desensitization mechanism may also be causally related to persistent elevations of GH release and the pathogenesis of hypersomatotropism.[167] Desensitization occurs contemporaneously with, but independently of, GH depletion[168-170] and may be unrelated to the GHRH dose.[171] This attenuation of GH responsiveness does not result from GH autoregulation at pituitary sites.[172,173] Autoregulation at hypothalamic sites by GH, GHRH, or insulin-like growth factor I (IGF-I), resulting in increased SRIF secretion, may, however, partly account for GH response attenuation to GHRH *in vivo*.[174,175] This possibility is, however, unlikely because the release of pituitary thyrotropin-stimulating hormone (TSH) is not impaired during constant GHRH stimulation.[176,177] The persistence of pulsatile GH secretion in normal men during a 24-h GHRH infusion is similarly likely to reflect constitutive episodes of SRIF withdrawal rather than the intermittent induction of SRIF secretion.[178,179]

GH response attenuation *in vitro* and possibly *in vivo* is, however, induced by a downregulation of GHRH receptors[171,180,181] and an uncoupling of the regulatory protein (G_S) linked to adenylate cyclase,[182,183] which impairs GH release and *GH* gene expression.[184] Although only 10 to 20% of the GHRH receptors need to be occupied for maximal GH release,[180] this uncoupling of GHRH receptors accounts for the unresponsiveness of downregulated somatotrophs, in which the loss of GHRH receptors is no more than 50%. Refractoriness to GHRH does not, however, impair GH responses to other secretagogues (e.g., to stress,[185] arginine,[186] GH-releasing peptide [GHRP],[164,166] TRH,[187] or hypoglycemia[188]) although different secretagogues may release GH from signal-specific intracellular pools.[169,174] Indeed, GH responses to these heterologous stimuli may be enhanced rather than diminished.

Similarly, although SRIF inhibits GHRH action, it prevents desensitization to GHRH[189] and rapidly restores the attenuated GH response to GHRH in desensitized cells.[190,191] Moreover, by allowing an accumulation of GH stores, SRIF exposure potentiates subsequent GH responses to GHRH,[189,192,193] by a mechanism that augments the GHRH-induced increase in intracellular free Ca^{2+} concentrations.[194] Alternatively, continuous or repetitive stimulation of somatotrophs with GHRH can increase the number and amplitude of GH secretory episodes[162,195,196] and potentiate GH responses to subsequent GHRH challenge[197] or to SRIF withdrawal[198] or to insulin hypoglycemia.[199] This priming effect of GHRH not only occurs in GH-deficient subjects[98,196,200] and in poorly responsive old men,[201] but in normal individuals [202-204] and in neonates[205] and in young[69,206] and old adult rats,[201] in sheep[207,208] and cattle,[207,209,210] and in cultured rat pituitary cells.[183,197,205] These effects probably result from GH-induced synthesis and expansion of the GHRH-releasable pool of stored GH.[199]

In addition to actions on GH release mediated by specific membrane receptors, GHRH may also regulate GH release through its affinity for heterologous receptors (e.g., receptors for vasoactive intestinal peptide [VIP], secretin, peptide histidine isoleucine [PHI], peptide histidine methionine, oxytomodulin, gastric inhibitory peptide [GIP], and glucagon).[211] Indeed, in birds the stimulation of GH release by human and rat GHRH may be mediated through VIP-like receptors, because *GHRH* gene transcription may not occur in avian species.[212]

D. GHRH ACTIONS

GHRH stimulates the *in vivo* synthesis and release of GH and the proliferation and differentiation of somatotrophs, as detailed in Chapters 3 through 6. Somatotroph responses to GHRH *in vitro* primarily involve an increase in the amount of GH released rather than an increase in the number of GH-secreting cells.[213,214] These somatotroph responses are GHRH dependent and are deficient or defective in states of GHRH deficiency[215,216] or GHRH receptor resistance.[152,153,217] Persistent exposure to GHRH or its transgenic overproduction, conversely, results in increased GH synthesis and release and in somatotroph hyperplasia.[217-220]

The stimulation of somatotroph function in response to a GHRH pulse is demonstrated by a rapid, dose-dependent rise in plasma GH concentrations. GHRH stimulation has therefore been advocated as a physiological test of pituitary GH reserve and a potential tool for evaluating clinical disorders of growth.[221] The usefulness of this is, however, limited by the marked variability in the amplitude of the GH response within and between individuals.[222-224] GH responses to GHRH are also highly variable in cattle,[165] pigs,[225] rats,[226] monkeys,[227] and chickens[228] and may partly reflect individual differences in pituitary GH stores.[214] Indeed, the range of GH responses to GHRH in normal subjects is far greater than the GH response to other secretagogues. Interindividual variability in GH responsiveness to GHRH appears to be independent of the GHRH dose [229] and occurs between individuals that do not appear to differ in any clinically important features. In most studies, intraindividual variability is nearly as great (approximately 66%) as interindividiual variability, although consistent intraindividual GH responses to GHRH have been reported.[230] This variability may, however, be related to ultradian fluctuations in somatotroph responsiveness and to stimulatory and inhibitory hypothalamic tone. This possibility is supported by the relationship between the magnitude of the GH response and the basal GH concentration prior to GHRH stimulation.[231,232] The GH response appears to be greater during periods of rising GH secretion than during peak or nadir periods of GH secretion. The GH response to GHRH is thus less variable (and increased) during cholinergic stimulation and reduced SRIF secretion.[232,233] The GH response to GHRH is similarly enhanced in pigs, rats, rabbits, and chickens when SRIF tone is genetically[214] or experimentally diminished[164,234,235] and attenuated when SRIF secretion is enhanced.[59,236] SRIF may thus be responsible for inducing cyclical GH responsiveness to GHRH stimulation,[164] because increased pulsatile GH release has been observed in cattle,[165] rats,[164,234] and humans[237,238] during GHRH treatment. This pulsatility of GH secretion in response to repeated GHRH stimulation is augmented when tonic SRIF tone is withdrawn for brief periods.[198] The maintenance of GH responsiveness to repeated GHRH stimulation during SRIF withdrawal[239] also suggests the pituitary has a resilient capacity and that the phenomenon of GH refractoriness largely results from SRIF antagonism rather than GHRH receptor downregulation or pituitary GH depletion.

The GH response to GHRH varies in both its amplitude and its temporal induction. In most instances, GH concentrations in plasma are elevated within minutes of GHRH stimulation. The GH response may, however, be delayed by up to 90 min in some normal individuals[240] and in some laboratory animals.[225] These delayed responses do not appear to be normal episodic surges of GH release because they are larger than episodic GH peaks and occur in pigs lacking episodic GH release.[225]

The somatotroph response to GHRH stimulation also varies during ontogeny. During late gestation the fetal pituitary is highly responsive to GHRH and has a five- to eightfold greater response to GHRH than the pituitaries of perinatal and newborn rats.[104] The somatotrophs of fetal sheep in midgestation are similarly more responsive to GHRH stimulation than those in late gestation,[241] which in turn are more responsive than those in neonatal lambs.[105] Neonatal and juvenile individuals are similarly more responsive to GHRH than adults[221,242] and a decrease in GH responsiveness with advancing age has similarly been observed in cattle.[243] A further decline in GH responsiveness to GHRH stimulation occurs in the elderly,[111,244,245] although this may reflect an age-related reduction in GHRH release and decreased pituitary GH stores, because repeated GHRH stimulation restores GH responsiveness.[237] Indeed, although it is well established that *in vivo* GH responsiveness is reduced with aging (e.g., in rats[246,247] or in humans[112,204,244]), the *in vitro* responsiveness (in cAMP production and GH release) of somatotrophs may be only slightly diminished.[242,244,248,249] The diminished *in vivo* GH response to GHRH is thus likely to reflect increased SRIF tone[110] and the GH response of young and old rats to GHRH does not differ following SRIF immunoneutralization.[250] This age-related reduction in *in vivo* sensitivity may also reflect a reduction in dopaminergic tone, because the GH response is enhanced in young but not elderly men after domperidone pretreatment.[112,251]

The ability of GHRH to stimulate somatotroph function is cycloheximide dependent[252] and is likely to be modulated by neurocrines and neuroendocrines of hypothalamic or pituitary origin in endocrine or paracrine ways. Somatostatin, for instance, antagonizes GHRH-induced GH release, but not GHRH-induced GH synthesis (see below). Thyrotropin-releasing hormone also interacts at postreceptor pituitary sites to potentiate GH responsiveness to GHRH stimulation (e.g., in humans,[187,253] cattle,[254,255] sheep,[256] rats,[257,258] and chickens[259]), although TRH may act within the brain to inhibit sleep-, arginine-, hypoglycemia-, glucagon-, and dopamine-induced GH secretion.[260-262] Additive effects of VIP with maximal stimulatory doses of GHRH have also been observed in human somatotrophinomas, in which dopamine inhibits GHRH-induced GH release.[252] The production of GHRH by pituitary adenomas[263,264] may also modify GH responsiveness, because ectopic GHRH production abolishes the pituitary GH response to exogenous GHRH challenge.[265] The pituitary GH response to GHRH is also dependent on peripheral endocrine and metabolic factors. The *in vivo* GH response to GHRH is, for instance, greater in male rats than in females, as a result of male somatotrophs having a greater secretory capacity and greater sensitivity to GHRH simulation.[266,267] This sexually dimorphic response reflects a stimulatory action of testosterone, in males, because testicular feminized rats are more responsive to GHRH than female rats.[268] The GH responses to GHRH is, moreover, reduced after castration in male rats, because testosterone, diols, and other testicular factors promote GHRH-induced GH secretion.[269-271] This is accomplished by an increase in secretory capacity to GHRH stimulation and does not involve a change in pituitary sensitivity.[272] Testosterone and dihydrotestosterone similarly augment GHRH-induced GH release from bovine anterior pituitary cells, at doses that have no effect on basal levels of GH release.[273] 5α-androstane-3α, 17β-diol (3α-diol) or 17β-estradiol, in contrast, have no effect on the GH response of these cells to GHRH stimulation.[273] Ovariectomy similarly reduces the GH responsiveness of female rats to GHRH stimulation.[274] However, whereas testosterone increases pituitary adenylate cyclase activity and GH content, estrogen may be inhibitory *in vivo*[275-277] although stimulatory *in vitro*.[160,278-280] This may result from interactions with other modulators of GH release and differential actions at hypothalamic and pituitary sites.[281]

Although estrogens generally stimulate GH secretion in primates, the GH response of children to GHRH is not augmented by stilbestrol[282] and the GH responsiveness does not vary in women throughout the menstrual cycle.[283,284] The GH response of premenopausal women to GHRH is, nevertheless, greater than in age-matched men and within women the magnitude of the GH response correlates with circulatory estrogen levels.[245]

Thyroid hormones also regulate GHRH-induced GH secretion,[285] which is blunted by hypothyroidism and stimulated by exogenous triiodothyronine (T_3).[286-288] Adrenal steroids also potentiate pituitary GH responses to GHRH stimulation.[160,161,213] This may result from an action of glucocorticoids in potentiating GHRH-induced cAMP accumulation, at a step preceding cAMP formation,[289] and is readily demonstrated *in vitro*.[213] Glucocorticoid hormones may, however, have additional effects at hypothalamic sites that result in a reduced pituitary sensitivity to GHRH stimulation.[290,290a] In the absence of glucocorticoids, basal and GHRH-induced GH secretion is diminished.[291] Pancreatic hormones may also modulate pituitary responsiveness, because exaggerated GH responses to GHRH occur in diabetic rats.[292] This is concordant with the inhibitory effect of insulin on GH secretion.[293] In contrast, peripheral plasma SRIF, like hypothalamic SRIF, appears to suppress GH responsiveness to GHRH stimulation,[294] although Plewe et al.[295] suggest GH responsiveness is unrelated to peripheral plasma SRIF. Impaired somatotroph responsiveness to GHRH is also induced by hyperglycemia[296,297] but increased by hypoglycemia,[278] although hyperglycemia and nutrient intake paradoxically increase GH responsiveness in obese patients.[123] Elevated plasma free fatty acid concentrations following feeding also suppress GH responsiveness to GHRH stimulation,[298,299] by effects mediated directly at the pituitary.[300]

E. MECHANISM OF GHRH ACTION

GHRH receptors are probably coupled through guanine nucleotide-binding proteins[301] to at least four intracellular signal transduction systems: the adenylate cyclase-protein kinase A system,[302-304] the inositol phosphate-diacylglyerol-protein kinase C system,[305-307] the Ca^{2+}-calmodulin system,[308-310] and the arachidonic acid-eicosanoid system.[311-313] The roles of these interacting messengers in the control of GH release are detailed in Chapter 6.

The stimulation of somatotrophs with GHRH results in an accumulation of intracellular cAMP.[289,314,315] The adenylate cyclase-protein kinase A system is likely to be the principal signal transduction mechanism mediating GHRH actions on somatotroph function, because cAMP-responsive regions are located in the *GH* gene promoter and cAMP itself is able to induce *GH* gene transcription.[316-318] Constitutive overstimulation of adenylate cyclase also results in somatotroph hyperplasia and gigantism in mice,[319] whereas the overexpression of a cAMP binding protein that interferes with cAMP-binding to the *GH* promoter results in somatotroph depletion and dwarfism.[320] The involvement of the cyclic AMP (cAMP) system in mediating GHRH-induced GH release is indicated by the inability of cAMP or agents elevating cAMP levels to augment the GH response to a maximal stimulatory dose of GHRH, although they are effective in the absence of GHRH or at submaximal GHRH doses.[303,321] The increased accumulation of cAMP is also likely to promote somatotroph function by increasing intracellular Ca^{2+} concentrations.[322] Indeed, because GHRH augments cAMP production even in Ca^{2+} free solution or when extracellular Ca^{2+} entry is blocked and GH secretion suppressed,[322-324] the role of cAMP may be to increase Ca^{2+} entry into somatotrophs and facilitate Ca^{2+}-induced GH release.[309,322,325]

The activation of the inositol-phospholipid pathway in response to GHRH stimulation is thought to potentiate cAMP accumulation and mobilize intracellular calcium stores,[321,326] although these are controversial findings.[94,307,327] Activation of adenylate cyclase by protein kinase C may result from an interference with the G_i protein regulating adenylate cyclase activity.[328] Rapid increases in intracellular Ca^{2+} following GHRH stimulation of somatotrophs also result from an influx of extracellular Ca^{2+} through voltage-sensitive channels.[322,329,330] The influx of Ca^{2+} may be caused by membrane depolarization following an increase in membrane Na^+ conductance, because GHRH-induced GH release is blocked in the presence of tetrodotoxin, a voltage-sensitive Na^+ channel blocker.[325] The importance of extracellular Ca^{2+} in mediating GHRH-induced GH release is also indicated by its attenuation in Ca^{2+}-deficient media[331] and in the presence of Ca^{2+} channel blockers[32,323] and by its potentiation by Ca^{2+} ionophores.[32] Moreover, Honegger et al.[32] found that specific inhibition of the Ca^{2+}-ATPase that mobilizes intracellular Ca^{2+} had no effect on GHRH-induced GH release, suggesting extracellular Ca^{2+} is more important than intracellular Ca^{2+} in provoking GH release.

The intracellular effects of Ca^{2+} on GH secretion are probably modulated by its binding to and activation of calmodulin, which in turn regulates the activity of calmodulin-dependent protein kinases.[94] Calmodulin antagonism can suppress GHRH-induced GH release and cAMP accumulation, indicating a mutual stimulatory relationship between these systems.[32,332,333] The action of GHRH in inducing GH release is also correlated with a rapid and enhanced rate of Ca^{2+} efflux from rat somatotrophs.[334]

The activation of the inositol-phospholipid system is also likely to induce arachidonic acid formation, as a result of the hydrolysis of diacylglycerol or from the catabolism of arachidonic acid-containing phospholipids. The formation of arachidonic acid is increased by GHRH[335] and the GH response to GHRH is diminished when arachidonic acid formation is blocked.[43,336,337] The GH-releasing activity of intracellular arachidonic acid may, however, be dependent on its metabolism by cyclooxygenase, lipoxygenase, and epoxygenase pathways. Prostaglandins and prostacyclins[311,312,337,338] stimulate GH release in a manner additive with cAMP and GHRH[267,303,339] and the GH response to GHRH is reduced in the presence of cyclooxygenase inhibitors.[336,339] Lipoxygenase inhibitors also attenuate GHRH-induced GH release and epoxygenase inhibition likewise reduces GH release and pituitary cAMP production.[313,336,340] Hydroxyeicosatetraenoic acids (5-HETE and 15-HETE) and the epoxygenase metabolite 5,6-epoxyeicosatrienoic acid also stimulate GH secretion,[313,340] further indicating a role for the arachidonic acid system in mediating GHRH-induced GH release.

In addition to SRIF (see Chapter 8), the actions of GHRH on GH release are antagonized by nitric oxide (NO),[341] which is produced within the pituitary gland. The GH response to GHRH is thus potentiated by hemoglobin, which strongly binds NO but has no effect on basal GH release. Potentiated GH responses to GHRH also occur in the presence of NO synthesis inhibitors, whereas diminished responses occur in the presence of NO-releasing factors. Because NO synthesis is induced by cAMP, increased NO synthesis following GHRH stimulation may provide a mechanism to balance excessive GH release subtly.

F. GHRH METABOLISM

The actions of GHRH on somatotroph function are partially regulated by GHRH metabolism within hypothalamic pituitary tissues and extracellular fluids.

The half-life of GHRH(1–44)-NH_2 in plasma is estimated at between 7 and 50 min,[342-344] with a metabolic clearance rate of approximately 200 l/m^2/day. The inactivation of GHRH results from its metabolism by dipeptidyl-peptidase type IV (DPP-IV) and/or trypsin-like endopeptidases,[344-346] which remove the amino-terminal peptide (Tyr-Ala) and cleave GHRH at positions 11 and 12, respectively. The GHRH cleavage product GHRH(3–44)-NH_2 has a three- to fourfold longer half-life than GHRH and hence most of the immunoreactive GHRH in peripheral plasma is likely to be biologically inactive.

The inactivation of GHRH may also result from its internalization within target cells, and GHRH uptake into monkey[347] and rat[348,349] somatotrophs (and non-GH-secreting cells) has been demonstrated. This uptake is thought to result from both receptor-mediated endocytosis over coated pits and from nonspecific, non-receptor-mediated uptake. These processes appear to be stimulated by SRIF, which would provide an additional mechanism for SRIF antagonism of GH-induced GH release. The intracellular fate of internalized GHRH may include its trafficking to secretory granules or the nucleus or its incorporation into lysosome-like organelles and subsequent degradation by lysosomal enzymes. The binding of GHRH to secretory granules may protect against lysosomal degradation, whereas the binding of GHRH to nuclear membranes may indicate GHRH effects on the translocation of messenger RNA from the nucleus to the cytoplasm of somatotrophs.[348] The binding of GHRH to plasma membrane secretory granules and nuclear membranes occurs within a 2- to 20-min period whereas the cytoplasmic matrix is labeled within 5 min and GHRH uptake into lysosomes occurs after 15 to 30 min.[348]

G. EXTRAHYPOTHALAMIC GHRH

In addition to the arcuate nucleus and ventromedial nucleus of the brain, GHRH immunoreactivity has been demonstrated in the cerebral cortex and in hindbrain regions,[28,350,351] from where it is released by depolarizing stimuli.[28] GHRH may, therefore, also have physiological roles unrelated to GH secretion. GHRH is also present in the peripheral nervous system[352] and in most extraneural tissues,[353,354] including those of the gastrointestinal tract.[355-357] It is present in the pituitary,[347] adrenal, thyroid, pancreas, lung, and kidney,[353] and in the ovary,[358] testis,[358] and placenta[113,359,360] and in immune tissues.[361] It is also present in tumors[362,363] and may causally induce excessive GH secretion and acromegaly.[364-366] Circulating GHRH and GH levels are not, however, directly correlated in untreated acromegalics or in patients treated with dopamine or with radiotherapy.[367] Moreover, whereas GH and GHRH levels in peripheral plasma are correlated in normal children during sleep,[368] approximately 30% of the episodic GH and GHRH pulses are unrelated. Similarly, although the secretion of GHRH from the hypothalamus is stimulated by hypoglycemia, peripheral plasma GHRH levels are elevated shortly after glucose ingestion[369] or food intake,[370] when pituitary GH secretion is suppressed. Similarly, while insulin hypoglycemia and α-adrenergic stimulation enhance GH secretion they have no effect on GHRH levels in systemic circulation, whereas concordant increases in GH and GHRH plasma concentrations are induced by L-dopa[371-373] and discordant changes in GH and GHRH levels follow amino acid or ornithine administration.[371,373] The role of peripheral plasma GHRH in the physiological regulation of GH secretion is, therefore, uncertain, especially as its concentration in normal individuals (approximately <50 pg/ml) is less than in portal plasma (approximately 200 to 1000 pg/ml) and it is composed of multiple immunoreactive forms,[374] most of which may be biologically inactive.[344,346] The processing of the GHRH precursor is also variable and tissue specific [375] and the expression of the *GHRH* gene in hypothalamic and extrahypothalamic tissues also appears to be differentially regulated[369,374] and is probably driven by different gene promoters.[376]

REFERENCES

1. **Lin, H. D., Bollinger, J., Ling, N., et al.,** *J. Clin. Endocrinol. Metab.*, 58, 1197, 1984.
2. **Bloch, B., Brazeau, P., Ling, N., et al.,** *Nature (Lond.)*, 301, 607, 1983.
3. **Werner, H., Okon, E., Fridkin, M., et al.,** *J. Clin. Endocrinol. Metab.*, 63, 47, 1986.
4. **Ibata, Y., Okamura, H., Makino, S., et al.,** *Brain Res.*, 370, 136, 1986.
5. **Frohman, L. A., Downs, T. R., Chomczynski, P., et al.,** *Acta Pediatr. Scand.*, 367, 81, 1990.
6. **Niimi, M., Takahara, J., Sato, M., et al.,** *Peptides*, 10, 605, 1989.
7. **Muller, E. E. and Nistico, G.,** *Brain Messengers and the Pituitary*. Academic Press, New York, 1989.
8. **Hokfelt, T., Meister, B., Everitt, B., et al.,** *Integrative Neuroendocrinology: Molecular, Cellular and Clinical*, [Ed. McCann, S. M. and Weiner, R. I.], Karger, Basel, 1987, p. 1.
9. **Sawcenko, P. E., Swanson, L. W., Rivier, J., et al.,** *J. Comp. Neurol.*, 237, 100, 1985.
10. **Meister, B., Hokfelt, T., Vale, W. W., et al.,** *Acta Physiol. Scand.*, 124, 133, 1985.
11. **Okamura, H., Murakami, S., Chihara, K., et al.,** *Neuroendocrinology*, 41, 177, 1985.
12. **Risold, P. Y., Fellman, D., Lenys, D., et al.,** *Neurosci. Lett.*, 100, 23, 1989.
13. **Fellmann, D., Bugnon, C., and Risold, P. Y.,** *Neurosci. Lett.*, 74, 275, 1987.
14. **Meister, B., Hokfelt, T., Vale, W. W., et al.,** *Neuroendocrinology*, 42, 237, 1986.
15. **Meister, B. and Hulting, A. L.,** *Neuroendocrinology*, 46, 387, 1987.
16. **Niimi, M., Takahara, J., Sato, M., et al.,** *Neuroendocrinology*, 55, 92, 1992.
17. **Daikoku, S., Hisano, S., Kawano, H., et al.,** *Neuroendocrinology*, 47, 405, 1988.
18. **Ciofi, P., Croix, D., and Tramu, G.,** *Neuroendocrinology*, 45, 425, 1987.
19. **Ciofi, P., Tramu, G., and Bloch, B.,** *Neuroendocrinology*, 51, 429, 1990.
20. **Shioda, S., Kohara, H., and Nakai, Y.,** *Brain Res.*, 40, 355, 1987.
21. **Niimi, M., Takahara, J., Sato, M., et al.,** *Neuroendocrinology*, 51, 572, 1990.
22. **Martin, J. B.,** *Endocrinology (Baltimore)*, 91, 107, 1972.
23. **Dickson, S. L., Leng, G., and Robinson, I. C. A. F.,** *Brain Res.*, 623, 95, 1993.
24. **Kato, M., Suzuki, M., and Kakegawa, T.,** *Brain Res.*, 280, 69, 1983.
25. **McIntyre, H. B. and Odell, W. D.,** *Neuroendocrinology*, 16, 8, 1974.
26. **Malven, P. V.,** *Endocrinology (Baltimore)*, 97, 808, 1975.
27. **Miki, N., Ono, M., Miyoshi, H., et al.,** *Life Sci.*, 44, 469, 1985.
28. **Fernandez, G., Cacicedo, L., Lorenzo, M. J., et al.,** *Endocrinology (Baltimore)*, 125, 1991, 1989.
29. **Katakami, H., Downs, T. R., and Frohman, L. A.,** *Endocrinology (Baltimore)*, 120, 1079, 1987.
30. **Kitajima, N., Chihara, K., Abe, H., et al.,** *Endocrinology (Baltimore)*, 124, 69, 1989.
31. **Baes, M. and Vale, W. W.,** *Endocrinology (Baltimore)*, 124, 104, 1989.
32. **Honegger, J., D'Urso, R., Besser, G. M., et al.,** *Endocrinology (Baltimore)*, 129, 11, 1991.
33. **Cugini, C. D., Millard, W. J., and Leidy, J. W., Jr.,** *Endocrinology (Baltimore)*, 129, 1355, 1991.
34. **Chomczynski, P., Downs, T. R., and Frohman, L. A.,** *Mol. Endocrinol.*, 2, 236, 1988.
35. **Downs, T. R., Chomczynski, P. and Frohman, L. A.,** *Mol. Endocrinol.*, 4, 402, 1990.
36. **Tsagarakis, S., Ge, F., Rees, L. H., et al.,** *J. Neuroendocrinol.*, 1, 129, 1989.
37. **Gil-Ad, I., Laron, Z., and Koch, Y.,** *J. Endocrinol.*, 131, 381, 1991.

38. **Chihara, K., Kodama, H., Kaji, H., et al.,** *J. Clin. Endocrinol. Metab.*, 61, 229, 1985.
39. **Aulakh, C. S., Hill, J. L., and Murphy, D. L.,** *Neuroendocrinology*, 59, 35, 1994.
40. **Baes, M. and Vale, W. W.,** *Neuroendocrinology*, 51, 202, 1990.
41. **Wehrenberg, W. B., Bloch, B., and Ling, N.,** *Neuroendocrinology*, 41, 13, 1985.
43. **Kasting, N. W. and Martin, J. B.,** *Neuroendocrinology*, 39, 201, 1984.
44. **Ojeda, S. R., Negro-Vilar, A., Arimura, A., et al.,** *Neuroendocrinology*, 31, 1, 1980.
45. **Murakami, Y., Kato, Y., Koshiyama, H., et al.,** *Brain Res.*, 407, 405, 1987.
46. **Shibasaki, T., Yamauchi, N., Hotta, M., et al.,** *Regul. Peptides,* 15, 47, 1986.
47. **Lewis, B. M., Dieguez, C., Ham, J., et al.,** *J. Neuroendocrinol.*, 1, 1989.
48. **Yamaguchi, N., Shibasaki, T., Ling, N., et al.,** *Regul. Peptides*, 33, 71, 1991.
49. **Sugihara, H., Minami, S., and Wakabayashi, I.,** *J. Endocrinol.*, 122, 583, 1989.
50. **Murakami, Y., Kato, Y., Kabayama, Y., et al.,** *Endocrinology (Baltimore)*, 120, 311, 1987.
51. **Lumpkin, M. D., Samson, N. K., and McCann, S. M.,** *Endocrinology (Baltimore)*, 116, 2070, 1985.
52. **Thomas, G. B., Cummings, J. T., Francis, H., et al.,** *Endocrinology (Baltimore)*, 128, 1151, 1991.
53. **Plotsky, P. M. and Vale, W.,** *Science*, 230, 461, 1985.
54. **Frohman, L. A.,** *Acta Pediatr. Scand. (Suppl.)*, 343, 3, 1988.
55. **Frohman, L. A., Downs, T. R., Clarke, I. J., et al.,** *J. Clin. Invest.*, 86, 17, 1991.
56. **Aguila, M. C., Pickle, R. L., Yu, W. H., et al.,** *Neuroendocrinology*, 54, 515, 1991.
57. **Plotsky, P. M.,** *Neuroendocrinology*, 53, 433, 1991.
58. **Wehrenberg, W. B., Brazeau, P., Luben, R., et al.,** *Endocrinology (Baltimore)*, 111, 2147, 1982.
59. **Sato, M., Chihara, K., Kita, T., et al.,** *Neuroendocrinology*, 50, 139, 1989.
60. **Armstrong, J. D., Cohick, W. S., Harvey, R. W., et al.,** *Domest. Anim. Endocrinol.*, 10, 315, 1993.
61. **Moore, K. L., Armstrong, J. D., Harvey, R. W., et al.,** *Domest. Anim. Endocrinol.*, 9, 125, 1992.
62. **Magnan, E., Cataldi, M., Guillaume, V., et al.,** *Endocrinology (Baltimore)*, 134, 562, 1994.
63. **Armstrong, J. D., Esbenshade, K. L., Coffey, M. T., et al.,** *Domest. Anim. Endocrinol.*, 7, 191, 1990.
64. **Armstrong, J. D., Esbenshade, K. L., Johnson, J. L., et al.,** *J. Anim. Sci.*, 68, 427, 1990.
65. **Simpson, R. B., Armstrong, J. D., Harvey, R. W., et al.,** *J. Anim. Sci.*, 69, 4914, 1991.
66. **Payne, L. C., Obal, F., Opp, M. R., et al.,** *Neuroendocrinology*, 56, 118, 1992.
67. **Murakami, Y., Kato, Y., Kabayama, Y., et al.,** *Endocrinology (Baltimore)*, 117, 787, 1985.
68. **Murakami, Y., Kato, Y., Kabayama, Y., et al.,** *Proc. Soc. Exp. Biol. Med.*, 178, 151, 1985.
69. **Cella, S. G., Locatelli, V., De Gennaro, V., et al.,** *Endocrinology (Baltimore)*, 119, 1164, 1986.
70. **Miki, N., Ono, M., and Shizume, K.,** *Endocrinology (Baltimore)*, 114, 1950, 1984.
71. **Murakami, Y., Kato, Y., Kabayama, Y., et al.,** *Endocrinology (Baltimore)*, 119, 1089, 1986.
72. **Armstrong, J. D., Cohick, W. S., Harvey, R. W., et al.,** *Domest. Anim. Endocrinol.*, 10, 315, 1993.
73. **Kirby, C. J., Armstrong, J. D., Huff, B. G., et al.,** *J. Anim. Sci.*, 71, 3033, 1993.
74. **Clark, R. G., Carlsson, L. M. S., Rafferty, B., et al.,** *J. Endocrinol.*, 119, 397, 1988.
75. **Thomas, C. R., Groot, K., and Arimura, A.,** *Endocrinology (Baltimore)*, 116, 2174, 1985.
76. **Garcia-Rojas, J. F., Mangas, A., Barba, A., et al.,** *J. Endocrinol. Invest.*, 14, 241, 1991.
77. **Kakucska, I. and Makara, G. B.,** *Endocrinology (Baltimore)*, 113, 318, 1983.
78. **Willoughby, J. O., Koblar, S., Jervois, P. M., et al.,** *Neuroendocrinology*, 36, 358, 1983.
79. **Fukata, J., Kasting, N. W., and Martin, J. B.,** *Neuroendocrinology*, 40, 193, 1985.
80. **Tannenbaum, G. S., Eikelboom, R., and Ling, N.,** *Endocrinology (Baltimore)*, 113, 1173, 1983.
81. **Millard, W. J., Martin, J. B., Audet, J., et al.,** *Endocrinology (Baltimore)*, 110, 540, 1982.
82. **Ganzetti, I., De Gennaro, V., Redaelli, M., et al.,** *Peptides*, 7, 1011, 1986.
83. **Ganzetti, I., Petraglia, F., Capuano, I., et al.,** *J. Endocrinol. Invest.*, 10, 241, 1987.
84. **Wakabayashi, I., Inoue, S., Satoh, S., et al.,** *Brain Res.*, 346, 70, 1985.
85. **Sato, M., Takahara, J., Niimi, M., et al.,** *J. Neuroendocrinol.*, 2, 555, 1990.
86. **Maiter, D., Underwood, L. E., Martin, J. B., et al.,** *Endocrinology (Baltimore)*, 28, 1100, 1991.
87. **Corder, R., Saudan, P., Mazlan, M., et al.,** *Neuroendocrinology*, 51, 85, 1990.
88. **Bruhn, T. O., Tresco, P. A., and Jackson, I. M. D.,** *Peptides*, 12, 957, 1991.
89. **Acs, Z., Lonart, G., and Makara, G. B.,** *Neuroendocrinology*, 52, 156, 1990.

90. **Jaffe, C. A., Friberg, R. D., and Barkan, A. L.,** *J. Clin. Invest.*, 92, 695, 1993.
91. **Gaylinn, B. D., Harrison, J. K., Zysk, J. R., et al.,** *Mol. Endocrinol.*, 7, 77, 1993.
92. **Gubler, U., Monahan, J. J., Lomedico, P. T., et al.,** *Proc. Natl. Acad. Sci. U.S.A.*, 80, 4311, 1983.
93. **Mayo, K. E., Cerelli, G. M., Rosenfeld, M. G., et al.,** *Nature (Lond.)*, 314, 464, 1985.
94. **Frohman, L. A., Downs, T. R., and Chomczynski, P.,** *Front. Neuroendocrinol.*, 13, 344, 1992.
95. **Bloch, B., Baird, A., Ling, N., et al.,** *Endocrinology (Baltimore)*, 118, 156, 1986.
96. **Wehrenberg, W. B. and Ling, N.,** *Biochem. Biophys. Res. Commun.*, 115, 525, 1983.
97. **Spiess, J., Rivier, J., and Vale, W.,** *Nature (Lond.)*, 303, 532, 1983.
98. **Takano, K., Hizuka, N., Tanaka, I., et al.,** *Endocrinol. Jpn.*, 32, 511, 1985.
99. **Locatelli, V., Cella, S. G., Loche, S., et al.,** *Life Sci.*, 35, 1989, 1984.
100. **Rodier, P. M., Kates, B., White, W. A., et al.,** *J. Comp. Neurol.*, 291, 363, 1990.
101. **Jansson, J. O., Ishikawa, K., Katakami, H., et al.,** *Endocrinology (Baltimore)*, 120, 525, 1987.
102. **Billestrup, N., Swanson, L. W., and Vale, W.,** *Proc. Natl. Acad. Sci. U.S.A.*, 83, 6854, 1986.
103. **Shinkai, T., Ooka, H., and Noumura, T.,** *Neurosci. Lett.*, 123, 13, 1991.
104. **Baird, A., Wehrenberg, W. B., and Ling, N.,** *Reg. Pept.*, 10, 23, 1984.
105. **de Zegher, F., Styne, D. M., Daaboul, J., et al.,** *Endocrinology (Baltimore)*, 124, 124, 1989.
106. **Guillaume, V., Boudouresque, F., Grino, M., et al.,** *Peptides*, 7, 393, 1986.
107. **Bloch, B., Gaillard, R. C., Brazeau, P., et al.,** *Regul. Peptides*, 8, 21, 1984.
108. **Argente, J., Chowen, J. A., Zeitler, P., et al.,** *Endocrinology (Baltimore)*, 128, 2369, 1991.
109. **Colonna, V. D. G., Zoli, M., Cocchi, D., et al.,** *Peptides*, 10, 705, 1989.
110. **Ge, F., Tsagarakis, S., Rees, L. H., et al.,** *J. Endocrinol.*, 123, 53, 1989.
111. **Bando, H., Zhang, C., Takada, Y., et al.,** *Acta Endocrinol. (Copenh.)*, 124, 31, 1991.
112. **Giusti, M., Lomeo, A., Marini, G., et al.,** *Horm. Res.*, 27, 134, 1987.
113. **Mizobuchi, M., Frohman, M. A., Downs, T. R., et al.,** *Mol. Endocrinol.*, 5, 476, 1991.
114. **Zeitler, P., Argente, J., Chowen-Breed, J. A., et al.,** *Endocrinology (Baltimore)*, 127, 1362, 1990.
115. **Varela, C., Caciedo, L., Fernandez, G., et al.,** *Neuroendocrinology*, 54, 340, 1991.
116. **Shirasu, K., Stumpf, W. E., and Sar, M.,** *Endocrinology (Baltimore)*, 127, 344, 1990.
117. **Gabriel, S. M., Millard, W. J., Koenig, J. I., et al.,** *Neuroendocrinology*, 50, 299, 1989.
118. **Carlsson, L. M. S., Clark, R. G., and Robinson, I. C. A. F.,** *J. Endocrinol.*, 126, 27, 1990.
119. **Jansson, J. O., Eden, S., and Isaksson, O.,** *Endocr. Rev.*, 6, 128, 1985.
120. **Argente, J. and Chowen, J. A.,** *Horm. Res.*, 40, 48, 1993.
121. **Zeitler, P., Tannenbaum, G. S., Clifton, D. K., et al.,** *Proc. Natl. Acad. Sci. U.S.A.*, 88, 8920, 1991.
122. **Frohman, M. A., Downs, T. R., Chomczynski, P., et al.,** *Mol. Endocrinol.*, 3, 1529, 1989.
123. **Levy, A., Matovelle, M. C., Lightman, S. L., et al.,** *Brain Res.*, 579, 1, 1992.
124. **De Gennaro Colonna, V., Cattaneo, E., Cocchi, D., et al.,** *Peptides*, 9, 985, 1988.
125. **Mayo, K. E.,** *Advances in Growth Hormone and Growth Factor Research* (Eds. Muller, E. E., Cocchi, D. and Locatelli, V.). Pythagora Press, Rome, 1989, p. 217.
126. **Cella, S. G., De Gennaro Collona, V., Locatelli, M., et al.,** *J. Endocrinol.*, 199, 1990.
127. **Wood, T. L., Berelowitz, M., Gelato, M. C., et al.,** *Neuroendocrinology*, 53, 298, 1991.
128. **Leidy, J. W., Jr., McDermott, M. T., and Robbins, R. J.,** *Neuroendocrinology*, 51, 400, 1990.
129. **Lanes, R., Nieto, C., Bruguera, C., et al.,** *Biol. Neonate*, 56, 252, 1989.
130. **Maiter, D. M., Gabriel, S. M., Koenig, J. I., et al.,** *Neuroendocrinology*, 51, 174, 1990.
131. **Torsello, A., Panzeri, G., Cermenati, P., et al.,** *J. Endocrinol.*, 117, 273, 1988.
132. **Phelps, C. J., Dalcik, H., Endo, H., et al.,** *Endocrinology (Baltimore)*, 133, 3034, 1993.
133. **Cocilovo, L., Colonna, V. D., Zoli, M., et al.,** *Neuroendocrinology*, 55, 416, 1992.
134. **Katakami, H., Downs, T. R., and Frohman, L. A.,** *J. Clin. Invest.*, 77, 1704, 1986.
135. **Tannenbaum, G. S.,** *Endocrinology (Baltimore)*, 108, 76, 1981.
136. **Bruno, J. F., Olchovsky, D., White, J. D., et al.,** *Endocrinology (Baltimore)*, 127, 2111, 1990.
137. **Campbell, R. M., Lee, Y., Rivier, J., et al.,** *Peptides*, 12, 569, 1991.
138. **Veliicelebi, G., Patthi, S., Provow, S., et al.,** *Endocrinology (Baltimore)*, 118, 1278, 1986.
139. **Abribat, T., Deslauriers, N., Brazeau, P., et al.,** *Endocrinology (Baltimore)*, 128, 633, 1991.
140. **Gaudreau, P., Boulanger, L., and Abribat, T.,** *J. Med. Chem.*, 35, 1864, 1992.
141. **Abribat, T., Boulanger, L., and Gaudreau, P.,** *Brain Res.*, 528, 291, 1990.
142. **Abribat, T., Finkelstein, J. A., and Gaudreau, P.,** *Regul. Peptides*, 36, 263, 1991.
143. **Velicelebi, G., Santacroce, T. M., and Harpold, M. M.,** *Biochem. Biophys. Res. Commun.*, 126, 33, 1985.
144. **Goldman, J. A., Molitch, M. E., Thorner, M. O., et al.,** *J. Endocrinol. Invest.*, 10, 397, 1987.

145. **Serri, O.,** *Clin. Endocrinol. (Oxford)*, 27, 675, 1987.
146. **Law, G. J., Ray, K. P., and Wallis, M.,** *FEBS Lett.*, 166, 189, 1984.
147. **Beck-Peccoz, P., Bassetti, M., Spada, A., et al.,** *J. Clin. Endocrinol. Metab.*, 61, 541, 1985.
148. **Knepel, W., Schwaninger, M., Wesemeyer, G., et al.,** *Endocrinology (Baltimore)*, 120, 732, 1987.
149. **Arimura, A., Culler, M. D., Turkelson, C. M., et al.,** *Peptides*, 4, 107, 1983.
150. **Mayo, K. E.,** *Molecular and Clinical Advances in Pituitary Disorders* (Ed. Melmed, S.). Endocrine Research and Education, Los Angeles, 1993, p. 287.
151. **Clark, R. G. and Robinson, I. C. A. F.,** *J. Endocrinol.*, 106, 1, 1985.
152. **Jansson, J., Downs, T. R., Beamer, W. G., et al.,** *Science*, 232, 511, 1986.
153. **Downs, T. R. and Frohman, L. A.,** *Endocrinology (Baltimore)*, 129, 59, 1991.
154. **Brain, C. E., Chomczynski, P., Downs, T. R., et al.,** *Endocrinology (Baltimore)*, 129, 3410, 1991.
155. **Tannenbaum, G. S., Lapointe, M., Gurd, W., et al.,** *Endocrinology (Baltimore)*, 127, 3087, 1990.
156. **Seifert, H., Perrin, M., Rivier, J., et al.,** *Nature (Lond.)*, 313, 487, 1985.
157. **Seifert,, Perrin, M., Rivier, J., et al.,** *Endocrinology (Baltimore)*, 117, 424, 1985.
158. **Wehrenberg, W. B., Baird, A., and Ling, N.,** *Science*, 221, 556, 1983.
159. **Giustina, A., Romanelli, G., Candrina, R., et al.,** *J. Clin. Endocrinol. Metab.*, 68, 120, 1989.
160. **Webb, C. B., Szabo, M., and Frohman, L. A.,** *Endocrinology (Baltimore)*, 113, 1191, 1983.
161. **Vale, W., Vaughan, J., Yamamoto, G., et al.,** *Endocrinology (Baltimore)*, 112, 1553, 1983.
162. **Sassolas, G., Garry, J., Cohen, R., et al.,** *J. Clin. Endocrinol. Metab.*, 63, 1016, 1986.
163. **Hulse, J. A., Rosenthal, S. M., Cuttler, L., et al.,** *J. Clin. Endocrinol. Metab.*, 63, 872, 1986.
164. **Clark, R. G., Carlsson, L. M. S., Trojnar, J., et al.,** *J. Neuroendocrinol.*, 1, 249, 1989.
165. **Enright, W. J., Zinn, S. A., Chapin, L. T., et al.,** *Proc. Soc. Exp. Biol. Med.*, 184, 483, 1987.
166. **Sartor, O., Bowers, C. Y., and Chang, D.,** *Endocrinology (Baltimore)*, 116, 952, 1985.
167. **Spada, A., Elahi, F. R., Arosio, M., et al.,** *J. Clin. Endocrinol. Metab.*, 585, 1987.
168. **Rittmaster, R. S., Loriaux, D. L., and Merriam, G. R.,** *Neuroendocrinology*, 45, 118, 1987.
169. **Richardson, S. B. and Twente, S.,** *Am. J. Physiol.*, 254, E358, 1988.
170. **Ceda, G. P. and Hoffman, A. R.,** *Endocrinology (Baltimore)*, 116, 1334, 1985.
171. **Arsenijevic, Y., Riverst, R. W., Eshkol, A., et al.,** *Endocrinology (Baltimore)*, 121, 1487, 1987.
172. **Clark, R. G. and Robinson, I. C. A. F.,** *J. Endocrinol.*, 106, 281, 1985.
173. **de Zegher, F., Bettendorf, M., Grumbach, M. M., et al.,** *Neuroendocrinology*, 52, 429, 1990.
174. **Kelijman, M. and Frohman, L. A.,** *J. Clin. Endocrinol. Metab.*, 71, 157, 1990.
175. **Shibasaki, T., Hotta, M., Masuda, A., et al.,** *J. Clin. Endocrinol. Metab.*, 60, 1265, 1985.
176. **Hulse, J. A., Rosenthal, S. M., Cuttler, L., et al.,** *J. Clin. Endocrinol. Metab.*, 63, 872, 1986.
177. **Webb, C. B., Vance, M. L., Thorner, M. O., et al.,** *J. Clin. Invest.*, 74, 96, 1984.
178. **Vance, M. L., Kaiser, D. L., Evans, W. S., et al.,** *J. Clin. Invest.*, 75, 1584, 1985.
179. **Vance, M. L., Kaiser, D. L., Evans, W. S., et al.,** *J. Clin. Endocrinol. Metab.*, 60, 370, 1985.
180. **Bilezikjian, L. M., Seifert, H., and Vale, W.,** *Endocrinology (Baltimore)*, 118, 2045, 1986.
181. **Wehrenberg, W. B., Seifert, H., Bilezikjian, L. M., et al.,** *Neuroendocrinology*, 43, 266, 1986.
182. **Simard, J. and Labrie, F.,** *Mol. Cell. Endocrinol.*, 46, 79, 1986.
183. **Edwards, C. A., Dieguez, C., Ham, J., et al.,** *J. Endocrinol.*, 116, 185, 1988.
184. **Miller, H. A., III, Rogers, G., and Frawley, L. S.,** *Life Sci.*, 43, 629, 1988.
185. **Schulte, H. M., Allolio, B., Frohwein, U. S., et al.,** *Clin. Endocrinol.*, 25, 511, 1986.
186. **Page, M. D., Dieguez, C., Valcavi, R., et al.,** *Clin. Endocrinol.*, 28, 551, 1988.
187. **Sartorio, A., Spada, A., Bochicchio, D., et al.,** *Neuroendocrinology*, 44, 470, 1986.
188. **Barbetti, F., Crescenti, C., Negri, M., et al.,** *J. Clin. Endocrinol. Metab.*, 70, 1371, 1990.
189. **Simard, J., Lefevre, G., and Labrie, F.,** *Peptides*, 8, 263, 1987.
190. **Clayton, R. N. and Bailey, L. C.,** *J. Endocrinol.*, 112, 69, 1987.
191. **Soya, H. and Suzuki, M.,** *Endocrinology (Baltimore)*, 122, 2492, 1988.
192. **Kraicer, J., Cowan, J. S., Sheppard, M. S., et al.,** *Endocrinology (Baltimore)*, 119, 2047, 1986.
193. **Tannenbaum, G. S., Painson, J., Lengyel, A. M., et al.,** *Endocrinology (Baltimore)*, 124, 1380, 1989.
194. **Soya, H., Suzuki, M., and Kato, M.,** *Biochem. Biophys. Res. Commun.*, 172, 276, 1990.
195. **Hizuka, N., Takano, K., Shizume, K., et al.,** *Acta Endocrinol.*, 110, 17, 1985.
196. **Rochicciolo, P. E., Tauber, M. T., Uboldi, F., et al.,** *J. Clin. Endocrinol.*, 63, 1100, 1986.
197. **Heiman, M. L., Murphy, W. A., Wekola, M. V., et al.,** *Biochem. Biophys. Res. Commun.*, 124, 217, 1984.
198. **Sato, M., Takahara, J., Fujioka, Y., et al.,** *Endocrinology (Baltimore)*, 123, 1928, 1988.

199. Vance, M. L., Kaiser, D. L., Rivier, J., et al., *J. Clin. Endocrinol. Metab.*, 62, 591, 1986.
200. Hizura, N., Takano, K., Shizume, K., et al., *Endocrinol. Jpn.*, 31, 697, 1984.
201. Iovino, M., Monteleone, P., and Steardo, L., *J. Clin. Endocrinol. Metab.*, 69, 910, 1989.
202. Vance, M. L., Kaiser, D. L., Martha, P. M., Jr., et al., *J. Clin. Endocrinol. Metab.*, 68, 22, 1989.
203. Brain, C., Hindmarsh, P. C., Brook, C. G. D., et al., *Clin. Endocrinol. (Oxford)*, 28, 543, 1988.
204. Sassolas, G., Biot-Laporte, S., Cohen, R., et al., *Peptides*, 7(Suppl. 1) 281, 1986.
205. Cella, S. G., Locatelli, V., De Gennaro, V., et al., *Endocrinology (Baltimore)*, 116, 574, 1985.
206. Heiman, M. L., Nekola, M. V., Murphy, W. A., et al., *Endocrinology (Baltimore)*, 116, 410, 1985.
207. Hart, I. C., Chadwick, P. M. E., James, S., et al., *J. Endocrinol.*, 105, 189, 1985.
208. Barenton, B., Duclos, M., Diaz, J., et al., *Reprod. Nutr. Dev.*, 27, 491, 1987.
209. Moseley, W. M., Krabill, L. F., Friedman, A. R., et al., *J. Endocrinol.*, 104, 433, 1985.
210. Enright, W. J., Chapin, L. T., Moseley, W. M., et al., *J. Dairy Sci.*, 69, 344, 1986.
211. Rosselin, G., *Peptides*, 7, 89, 1986.
212. Harvey, S., Fraser, R. A., and Lea, R. W., *Crit. Rev. Poultry Biol.*, 3, 239, 1991.
213. Mulchahey, J. J., Di Blasio, A. M., and Jaffe, R. B., *J. Clin. Endocrinol. Metab.*, 66, 395, 1988.
214. Kineman, R. D., Chen, T. C., and Frawley, L. S., *Endocrinology (Baltimore)*, 125, 2035, 1989.
215. Schriock, E. A., Lustig, R. H., Rosenthal, S. M., et al., *J. Clin. Endocrinol. Metab.*, 58, 1043, 1984.
216. Carmignac, D. F. and Robinson, I. C. A. F., *J. Endocrinol.*, 127, 69, 1990.
217. Stefaneanu, L., Kovacs, K., Horvath, E., et al., *Endocr. Pathol.*, 4, 131, 1993.
218. Brar, A. K., Brinster, R. L., and Frohman, L. A., *Endocrinology (Baltimore)*, 125, 801, 1989.
219. Hammer, R. E., Brinster, R. L., Rosenfeld, M. G., et al., *Nature (Lond.)*, 315, 413, 1985.
220. Asa, S. L., Kovacs, K., Stefaneanu, L., et al., *Proc. Soc. Exp. Biol. Med.*, 193, 232, 1990.
221. Ross, R. J. M., Grossman, A., Besser, G. M., et al., *Acta Endocrinol. (Copenh.)*, 113 (Suppl. 279), 123, 1986.
222. Thorner, M. O., Spiess, J., Vance, M. L., et al., *Lancet*, 24, 1983.
223. Vance, M. L., Borges, J. L. C., Kaiser, D. L., et al., *J. Clin. Endocrinol. Metab.*, 58, 838, 1984.
224. Fornito, M. C., Calogero, A. E., Mongioi, A., et al., *J. Neuroendocrinol.*, 2, 87, 1990.
225. Kraft, L. A., Baker, P. K., Ricks, C. A., et al., *Domest. Anim. Endocrinol.*, 2, 133, 1985.
226. Wehrenberg, W. B., Ling, H., Bohlen, P., et al., *Biochem. Biophys. Res. Commun.*, 109, 562, 1982.
227. Almeida, O. F. X., Schulte, H. M., Rittmaster, R. S., et al., *J. Clin. Endocrinol. Metab.*, 58, 309, 1984.
228. Leung, F. C. and Taylor, J. E., *Endocrinology (Baltimore)*, 113, 1913, 1983.
229. Gelato, M. C., Rittmaster, R. S., Pescovitz, O. H., et al., *J. Clin. Endocrinol. Metab.*, 61, 223, 1985.
230. Hotta, M., Shibasaki, T., Masuda, A., et al., *Endocrinol. Jpn.*, 3248, 673, 1985.
231. Devesa, J., Lima, L., Lois, N., et al., *Clin. Endocrinol. (Oxford)*, 30, 367, 1989.
232. Suri, D., Hindmarsh, P. C., Brain, C. E., et al., *Clin. Endocrinol. (Oxford)*, 33, 399, 1990.
233. Mazza, E., Imperiale, E., Procopio, M., et al., *J. Endocrinol. Invest.*, 11(Suppl. 2), 92, 1988.
234. Mori, T., Inoue, S., Minami, S., et al., *Biomed. Res.*, 7, 371, 1986.
235. Stachura, M. E., Tyler, J. M., and Farmer, P. K., *Endocrinology (Baltimore)*, 123, 1476, 1988.
236. Scanes, C. G. and Harvey, S., *Gen. Comp. Endocrinol.*, 76, 256, 1989.
237. Corpas, E., Harman, S. M., Pineyro, M. A., et al., *J. Clin. Endocrinol. Metab.*, 75, 530, 1992.
238. Vance, M. L., Kaiser, D. L., Evans, W. S., et al., *J. Clin. Invest.*, 75, 1584, 1985.
239. Wehrenberg, W. B., Brazeau, P., Luben, R., et al., *Neuroendocrinology*, 36, 489, 1983.
240. Borges, J. L. C., Blizzard, R. M., Gelato, M. C., et al., *Lancet*, ii, 119, 1983.
241. Ohmura, E., Jansen, A., Chernick, V., et al., *Endocrinology (Baltimore)*, 114, 299, 1984.
242. Szabo, M. and Cuttler, L., *Endocrinology (Baltimore)*, 118, 69, 1986.
243. Johke, T., Hodate, K., Ohashi, S., et al., *Endocrinol. Jpn.*, 31, 55, 1984.
244. Shibasaki, T., Shizume, K., Nakahara, M., et al., *J. Clin. Endocrinol. Metab.*, 58, 212, 1984.
245. Lang, I., Schernthaner, G., Pietschmann, P., et al., *J. Clin. Endocrinol. Metab.*, 65, 535, 1987.
246. Ceda, G. P., Valenti, G., Butturini, U., et al., *Endocrinology (Baltimore)*, 118, 2109, 1986.
247. Sonntag, W. E., Hylka, V. W., and Meites, J., *Endocrinology (Baltimore)*, 113, 2305, 1983.
248. Robberecht, P., Gillard, M., Waelbroeck, M., et al., *Neuroendocrinology*, 44, 429, 1986.
249. Niimi, M., Takahara, J., Kawanishi, K., et al., *Endocrinol. Jpn.*, 32, 559, 1985.

250. **Sonntag, W. E. and Gough, M. A.,** *Neuroendocrinology*, 47, 482, 1988.
251. **Giusti, M., Mazzocchi, G., Sessarego, P., et al.,** *Clin. Endocrinol.*, 21, 339, 1984.
252. **White, M. C., Daniels, M., Kendall-Taylor, P., et al.,** *J. Endocrinol.*, 105, 269, 1985.
253. **Adams, E. F., Bhuttacharji, S. C., Halliwell, C. L. J., et al.,** *Clin. Endocrinol.*, 21, 709, 1984.
254. **Hodate, K., Johke, T., and Ohashi, S.,** *Endocrinol. Jpn.*, 32, 375, 1985.
255. **Lapierre, H., Petitclerc, D., Pelletier, G., et al.,** *Can. J. Anim. Sci.*, 70, 175, 1990.
256. **Davis, S. L., Hill, K. M., Ohlson, D. L., et al.,** *J. Anim. Sci.*, 42, 1244, 1976.
257. **Borges, J. L. C., Uskavitch, D. R., Kaiser, D. L., et al.,** *Endocrinology (Baltimore)*, 113, 1519, 1983.
258. **Szabo, M., Ruestow, P. C., and Kramer, D. E.,** *Endocrinology (Baltimore)*, 117, 330, 1985.
259. **Harvey, S. and Scanes, C. G.,** *Horm. Metab. Res.*, 17, 113, 1985.
260. **Maeda, K., Kato, Y., Chihara, K., et al.,** *J. Clin. Endocrinol. Metab.*, 41, 408, 1975.
261. **Brown, P. M., Bacchus, R., Sachs, L., et al.,** *Clin. Endocrinol. (Oxford)*, 10, 481, 1979.
262. **Chihara, K., Kato, Y., Maeda, K., et al.,** *J. Clin. Endocrinol. Metab.*, 44, 1094, 1977.
263. **Levy, A. and Lightman, S. L.,** *J. Clin. Endocrinol. Metab.*, 74, 1474, 1992.
264. **Wakabayashi, I., Inokuchi, K., Hasagawa, O., et al.,** *J. Clin. Endocrinol. Metab.*, 74, 357, 1992.
265. **Losa, M., Schopohl, J., and von Werder, K.,** *J. Endocrinol. Invest.*, 16, 69, 1993.
266. **Ho, K. Y., Leong, D. A., Sinha, Y. N., et al.,** *Am. J. Physiol.*, 250, E650, 1986.
267. **Cronin, M. J. and Rogol, A. D.,** *Brain Res.*, 31, 984, 1984.
268. **Batson, J. M., Krieg, R. J., Martha, P. M., et al.,** *Endocrinology (Baltimore)*, 124, 444, 1989.
269. **Ohlsson, L., Isaksson, O., and Jansson, J.-O.,** *J. Endocrinol.*, 113, 249, 1987.
270. **Wehrenberg, W. B., Baird, A., Ying, S., et al.,** *Biol. Reprod.*, 32, 369, 1985.
271. **Aguilar, E., Pinilla, L., and Tena-Sempere, M.,** *Neuroendocrinology*, 57, 132, 1993.
272. **Ho, K. Y., Thorner, M. O., Krieg, R. J., Jr., et al.,** *Endocrinology (Baltimore)*, 123, 1405, 1988.
273. **Hassan, H. A., Merkel, R. A., Enright, W. J., and Tucker, H. A.,** *Domest. Anim. Endocrinol.*, 9, 209, 1992.
274. **Aguilar, E. and Pinilla, L.,** *Neuroendocrinology*, 54, 286, 1991.
275. **Evans, W. S., Krieg, R. J., Limber, E. R., et al.,** *Am. J. Physiol.*, 249, E276, 1985.
276. **Akira, S., Wakabayashi, I., Hitsoshi, S., et al.,** *Neuroendocrinology*, 47, 116, 1988.
277. **Shulman, D. I., Sweetland, M., Duckett, G., et al.,** *Endocrinology (Baltimore)*, 120, 1047, 1987.
278. **Hertz, P., Silberman, M., Even, L., et al.,** *Endocrinology (Baltimore)*, 125, 581, 1989.
279. **Fukata, J. and Martin, J. B.,** *Endocrinology (Baltimore)*, 119, 2256, 1986.
280. **Simard, J., Hubert, J. F., Hosseinzadeh, T., et al.,** *Endocrinology (Baltimore)*, 119, 2004, 1986.
281. **Gabriel, S., Hunnicut, E. H., Millard, W. J., et al.,** *Neuroendocrinology*, 57, 63, 1993.
282. **Ross, R. J. M., Grossman, A., Davies, P. S. W., et al.,** *Clin. Endocrinol.*, 27, 155, 1987.
283. **Gelato, M. C., Pescovitz, O. H., Cassorla, F., et al.,** *J. Clin. Endocrinol. Metab.*, 59, 197, 1984.
284. **Evans, W. S., Borges, J. L. C., Vance, M. L., et al.,** *J. Clin. Endocrinol. Metab.*, 59, 1006, 1984.
285. **Root, A. W., Shulman, D., Root, J., et al.,** *Acta Endocrinol.(Copenh.)*, 113, 367, 1986.
286. **Wakabayashi, I., Tonegawa, Y., Ihara, T., et al.,** *Neuroendocrinology*, 41, 306, 1985.
287. **Edwards, C. A., Dieguez, C., and Scanlon, M. L.,** *J. Endocrinol.*, 121, 31, 1989.
288. **Dieguez, C., Jordan, V., Harris, P., et al.,** *J. Endocrinol.*, 109, 53, 1986.
289. **Michel, D., Lefevre, G., and Labrie, F.,** *Life Sci.*, 35, 597, 1984.
290. **Nakagawa, K., Akikawa, K., Matsubara, M., et al.,** *J. Clin. Endocrinol. Metab.*, 60, 306, 1985.
290a. **Burguera, B., Muruais, C., Penalva, A., et al.,** *Neuroendocrinology*, 51, 51, 1990.
291. **Guistina, A., Romanelli, G., Bossini, S., et al.,** *Horm. Metab. Res.*, 21, 693, 1989.
292. **Serri, O. and Brazeau, P.,** *Neuroendocrinology*, 46, 162, 1987.
293. **Melmed, S.,** *J. Clin. Invest.*, 73, 1425, 1984.
294. **Liapi, C., Evain-Brion, D., Argente, J., et al.,** *Acta Endocrinol. (Copenh.)*, 113, 1, 1985.
295. **Plewe, G., Scheider, C., Kurtz, V., et al.,** *Horm. Metab. Res.*, 481, 1985.
296. **Davies, R. R., Turner, S., and Johnson, D. G.,** *Clin. Endocrinol.*, 21, 477, 1984.
297. **Masuda, A., Shibasaki, T., Nakahara, M., et al.,** *J. Clin. Endocrinol. Metab.*, 60, 523, 1985.
298. **DeMarinis, L., Mancini, A., Zuppi, P., et al.,** *Psychoneuroendocrinology*, 16, 361, 1991.
299. **Imaki, T., Shibasaki, T., Shizume, K., et al.,** *J. Clin. Endocrinol. Metab.*, 60, 290, 1985.
300. **Casanueva, F. F., Villanueva, L., Dieguez, C., et al.,** *J. Clin. Endocrinol. Metab.*, 65, 634, 1987.

301. Struthers, R. S., Perrin, M. H., and Vale, W., *Endocrinology (Baltimore)*, 124, 24, 1989.
302. Lewin, M. J. M., Reyl-Desmars, F., and Ling, N., *Proc. Natl. Acad. Sci. U.S.A.*, 80, 6538, 1983.
303. Brazeau, P., Ling, N., Esch, F., et al., *Biochem. Biophys. Res. Commun.*, 109, 558, 1982.
304. Michel, D., Lefevre, G., and Labrie, F., *Mol. Cell Endocrinol.*, 33, 255, 1983.
305. Summers, S. T., Canonico, P. L., MacLeod, R. M., et al., *Eur. J. Pharmacol.*, 111, 371, 1985.
306. Judd, A. M., Koike, K., Yasumoto, T., et al., *Neuroendocrinology*, 42, 197, 1986.
307. French, M. B., Lussier, B. T., Moor, B. C., et al., *Mol. Cell Endocrinol.*, 72, 221, 1990.
308. Dufy-Barbe, L., Bresson, L., Sartor, P., et al., *Endocrinology (Baltimore)*, 131, 1436, 1992.
309. Puttagunta, A. L., Chik, C. L., Girard, M., et al., *J. Endocrinol.*, 135, 343, 1992.
310. Schofl, C., Sandow, J., and Knepel, W., *Am. J. Physiol.*, 253, E591, 1987.
311. Katakami, H., Kato, Y., Matsushita, N., et al., *Neuroendocrinology*, 38, 1, 1984.
312. Kato, Y., Dupre, J., and Beck, J. C., *Endocrinology (Baltimore)*, 93, 135, 1973.
313. Cashman, J. R., *Life Sci.*, 44, 1387, 1989.
314. Lewin, M. J. M., Reyl-Desmars, F., and Ling, N., *C.R. Acad. Sci. Paris*, 297, 123, 1983.
315. Cronin, M. J., Rogol, A. D., MacLeod, R. M., et al., *Am. J. Physiol.*, 244, E346, 1983.
316. Tanner, J. W., Davis, S. K., McArthur, N. H., et al., *J. Endocrinol.*, 125, 109, 1999.
317. Barinaga, M., Bilezikjian, L. M., Vale, W. W., et al., *Nature (Lond.)*, 314, 279, 1985.
318. Clayton, R. N., Bailey, L. C., Abbot, S. D., et al., *J. Endocrinol.*, 110, 51, 1986.
319. Burton, F. H., Hasel, K. W., Bloom, F. E., et al., *Nature (Lond.)*, 350, 74, 1991.
320. Struthers, R. S., Vale, W. W., Arias, C., et al., *Nature (Lond.)*, 350, 622, 1991.
321. Ray, K. P. and Wallis, M., *Mol. Cell. Endocrinol.*, 58, 243, 1988.
322. Lussier, B. T., Wood, D. A., French, M. B., et al., *Endocrinology (Baltimore)*, 128, 570, 1991.
323. Bilezikjian, L. M. and Vale, W. W., *Endocrinology (Baltimore)*, 113, 1726, 1983.
324. Spence, J. W., Sheppard, M. S., and Kraicer, J., *Endocrinology (Baltimore)*, 106, 764, 1980.
325. Kato, M., Hattori, M., and Suzuki, M., *Am. J. Physiol.*, 254, E476, 1988.
326. Summers, S. T. and Cronin, M. J., *Arch. Biochem. Biophys.*, 262, 12, 1988.
327. French, M. B., Moor, B. C., Lussier, B. T., et al., *Mol. Cell Endocrinol.*, 79, 139, 1991.
328. Gordeladze, J. O., Sletholt, K., Thorn, N. A., et al., *FEBS Lett.*, 177, 665, 1988.
329. Lussier, B. T., French, M. B., Moor, B. C., et al., *Endocrinology (Baltimore)*, 128, 592, 1991.
330. Holl, R. W., Thorner, M. O., and Leong, D. A., *Endocrinology (Baltimore)*, 122, 2927, 1988.
331. Kato, M. and Suzuki, M., *Jpn. J. Physiol.*, 36, 1225, 1986.
332. Schettini, G., Cronin, M. J., Hewlett, E. L., et al., *Endocrinology (Baltimore)*, 115, 1308, 1984.
333. Merritt, J. E., Dobson, P. R. M., Wojikiewicz, R. J. H., et al., *Biosci. Rep.*, 4, 995, 1984.
334. Login, I. S., Judd, A. M., and MacLeod, R. M., *Endocrinology (Baltimore)*, 118, 239, 1986.
335. Canonico, P. L., Speciale, C., Sortino, M. A., et al., *Life Sci.*, 38, 267, 1985.
336. Wandscheer, D. E., Bihoreau, C., Bertrand, P., et al., *J. Neuroendocrinol.*, 2, 439, 1990.
337. Cavagnini, F., Di Landro, A., Invitti, C., et al., *Metabolism*, 26, 193, 1977.
338. Szabo, M. and Frohman, L. A., *Endocrinology (Baltimore)*, 96, 955, 1975.
339. Fafeur, V., Gouin, E., and Dray, F., *Biochem. Biophys. Res. Commun.*, 126, 725, 1985.
340. Snyder, G. D., Yadagiri, P., and Falck, J. R., *Am. J. Physiol.*, 256, E221, 1989.
341. Kato, M., *Endocrinology (Baltimore)*, 131, 2133, 1992.
342. Boulanger, L., Roughly, P., and Gaudreau, P., *Peptides*, 13, 681, 1992.
343a. Su, C. M., Jensen, L. R., Heimer, E. P., et al., *Horm. Metab. Res.*, 23, 15, 1991.
343. Frohman, L. A., Thominet, J. L., Webb, C. B., et al., *J. Clin. Invest.*, 73, 1304, 1983.
344. Frohman, L. A., Downs, T. R., Williams, T. C., et al., *J. Clin. Invest.*, 78, 906, 1986.
345. Kubiak, T. M., Kelly, C. R., and Krabill, L. F., *Drug Metab. Dispos.*, 17, 393, 1989.
346. Frohman, L. A., Downs, T. R., Heimer, E. P., et al., *J. Clin. Invest.*, 83, 1533, 1989.
347. Morel, G., Mesguich, P., and Dubois, P. M., *Neuroendocrinology*, 38, 123, 1984.
348. Morel, G., *Endocrinology (Baltimore)*, 129, 1497, 1991.
349. Mentlein, R., Buchholz, C., and Krisch, B., *Cell Tiss. Res.*, 258, 309, 1989.
350. Leidy, J. W. and Robbins, R. J., *J. Clin. Endocrinol. Metab.*, 62, 372, 1986.
351. Fellman, D., Bugnon, C., and Lavry, G. N., *Neurosci. Lett.*, 58, 91, 1985.
352. Jozsa, R., Korf, H. W., and Merchenthaler, *Cell Tiss. Res.*, 247, 441, 1987.
353. Shibasaki, T., Kiyosawa, Y., Masuda, A., et al., *J. Clin. Endocrinol. Metab.*, 59, 263, 1984.
354. Losa, M., Wolfram, G., Mojto, J., et al., *J. Clin. Endocrinol. Metab.*, 70, 62, 1990.
355. Bruhn, T. O., Mason, R. T., and Vale, W. W., *Endocrinology (Baltimore)*, 117, 1710, 1985.
356. Arimura, A., Culler, M. D., Matsumoto, K., et al., *Peptides*, 5(Suppl. 1), 41, 1084.

357. **Christofides, N. D., Stephanou, A., Suzuki, H., et al.,** *J. Clin. Endocrinol. Metab.*, 59, 747, 1984.
358. **Moretti, C., Fabbri, A., Gnessi, L., et al.,** *J. Endocrinol. Invest.*, 13, 301, 1990.
359. **Baird, A., Wehrenberg, W. B., Bohlen, P., et al.,** *Endocrinology (Baltimore)*, 117, 1598, 1985.
360. **Margioris, A. N., Brockmann, G., Bohler, H., et al.,** *Endocrinology (Baltimore)*, 126, 151, 1990.
361. **Stephanou, A., Knight, R. A., and Lightman, S. L.,** *Neuroendocrinology*, 53, 628, 1991.
362. **Rivier, J., Spiess, J., Thorner, M., et al.,** *Nature (Lond.)*, 300, 276, 1982.
363. **Guillemin, R., Brazeau, P., Bohlen, P., et al.,** *Science*, 218, 585, 1982.
364. **Bostwick, D. G., Quan, R., Hoffman, A. R., et al.,** *Am. J. Pathol.*, 117, 167, 1984.
365. **Thorner, M. O., Frohman, L. A., Leong, D. A., et al.,** *J. Clin. Endocrinol. Metab.*, 59, 846, 1985.
366. **Barkan, A. L., Shenker, Y., Grekin, R. J., et al.,** *J. Clin. Endocrinol. Metab.*, 63, 1057, 1986.
367. **Penny, E. S., Penman, E., Price, J., et al.,** *Br. Med. J.*, 289, 453, 1984.
368. **Tapanainen, P., Rantala, H., Leppaluoto, et al.,** *Pediatr. Res.*, 26, 404, 1989.
369. **Kashio, Y., Chihara, K., Kita, T., et al.,** *J. Clin. Endocrinol. Metab.*, 64, 92, 1987.
370. **Sopwith, A. M., Penny, E. S., Besser, G. M., et al.,** *Clin. Endocrinol.*, 22, 337, 1985.
371. **Brion, D. E., Donnadieu, M., Liapi, C., et al.,** *Horm. Res.*, 24, 116, 1986.
372. **Tapanainen, P., Knip, M., Lautala, P., et al.,** *J. Clin. Endocrinol. Metab.*, 67, 845, 1988.
373. **Donnadieu, M., Evain-Brion, D., Tonon, M. C., et al.,** *J. Clin. Endocrinol. Metab.*, 60, 1132, 1985.
374. **Sasaki, A., Sato, S., Yumita, S., et al.,** *J. Clin. Endocrinol. Metab.*, 68, 180, 1989.
375. **Frohman, L. A., Downs, T. R., Kashio, Y., et al.,** *Endocrinology (Baltimore)*, 127, 2149, 1990.
376. **Gonzalez-Crespo, S. and Boronat, A.,** *Proc. Natl. Acad. Sci. U.S.A.*, 88, 8749, 1991.

Chapter 8

Growth Hormone Release: Inhibitory Hypothalamic Control

S. Harvey

CONTENTS

I. INTRODUCTION

Somatostatinergic neurons occur throughout the central and peripheral nervous systems, although only those with perikarya in the periventricular nucleus (PeVN) and paraventricular nucleus (PVN) of the anterior hypothalamus are thought to participate in growth hormone (GH) regulation.[1,2] These neurons have axon terminals in the median eminence, through which somatostatin (SRIF) is released into hypophysial portal circulation and transported to the adenohypophysis. SRIF is also produced by numerous normal and tumorous peripheral tissues (see Reichlin,[3] Berelowitz,[4] and Polak and Bloom[2] for reviews) and is released into peripheral plasma. The concentration of peripheral plasma SRIF is, however, less than 50% of that in pituitary portal circulation.[5,6] However, whereas basal, arginine-, and insulin-induced GH secretion in humans may be suppressed in pathophysiological somatostatinoma syndromes, peripheral plasma SRIF is thought to have a minimal role, if any, in GH regulation. Most of the SRIF-like immunoreactivity in peripheral plasma has a much larger molecular mass than the tetradecapeptide $SRIF_{14}$ and also includes a variety of other biologically inert forms.[7,8] The discordance between the levels of SRIF in systemic plasma with those in pituitary portal circulation and the discordant changes in the pattern of SRIF release into systemic and portal serum[5] also suggest a minor role for peripheral SRIF in GH regulation.

II. SRIF RELEASE

Electrical stimulation of the anterior or PeVN hypothalamus inhibits GH secretion[9,10] and concentrations of SRIF in pituitary portal plasma are increased >100-fold during electrical stimulation of the preoptic area.[11] The electrical stimulation of brain stem nuclei (locus coeruleus, nucleus fractus solitarius, lateral reticular nucleus, central gray of the pons), midbrain areas[12] (midbrain gray, raphe nucleus), limbic areas[10,13,14] (ventral premammillary nucleus, organum vasculosum of the lamina terminalis, bed nucleus of the stria terminalis, medial amygdala), and lateral hypothalamic areas[10] also inhibits GH secretion through activation of neurons in the PeVN nucleus. The suppression of GH secretion induced by the electrical stimulation of these brain areas is blocked by lesions of the PeVN nucleus, suggesting conveyance through the SRIF neurons in the hypotha-

0-8493-8697-7/95/$0.00+$.50

lamic nuclei. The ability of SRIF antibodies to elevate circulating GH levels in rats with PeVN lesions, nevertheless, suggests that somatostatinergic fibers outside this region have physiological roles in GH regulation.[15] Surgical isolation of the medial basal hypothalamus or removal of the preoptic hypothalamus also results in a disinhibition of GH secretion, increasing the frequency of pulsatile GH release and elevating trough GH levels.[16–18]

Electrical stimulation of the ventromedial (VMN) hypothalamic nucleus has no effect on the secretion rate or concentration of SRIF in portal blood, whereas electrical stimulation of the median eminence (ME) dramatically increases SRIF release.[19] Lesioning of the ME conversely blocks the inhibitory hypothalamic control of GH secretion.[20] The SRIF fibers in the ventromedial nucleus are, therefore, thought to terminate outside the ME, especially as they do not accumulate tracers of retrograde transport injected into the ME.[21] These neurons, therefore, probably have local rather than neuroendocrine actions in regulating GH release.

The release of SRIF from hypothalamic neurons in response to membrane depolarization has also been shown, *in vitro*, by K^+ stimulation (reviewed by Robbins[22]). This response is not sexually dimorphic[23] and is Na^+ dependent[24–27] and mediated through Ca^{2+}-calmodulin mechanisms[25,28–31] that stimulate adenylate cyclase[30,32] and protein kinase C.[25] SRIF release is also stimulated by dopamine[9,33,34] through D_2 but not D_1 receptors and by epoxygenase products of arachidonic acid.[35] Intracerebroventricular injections of norepinephrine (NE) also induce SRIF release *in vivo* via an α-adrenergic mechanism,[34,36,37] whereas catecholamine synthesis inhibition reduces SRIF concentrations in pituitary portal plasma.[9] These observations contrast, however, with the diminished SRIF release induced by NE[38] or α_2-adrenergic stimulation[9,37,39–41] and β_2-adrenergic antagonism.[24] These discordant findings may also reflect biphasic effects of NE, which is stimulatory at low doses and inhibitory at high doses.[41] This biphasic response may reflect SRIF autoregulation[42] or NE-induced inhibition of acetylcholine release.[41] The activation of muscarinic receptors provokes SRIF release in rats,[27,34,41,43,44] although inhibitory cholinergic effects have also been reported in primates.[45,46] The inhibitory action of acetylcholine may be mediated through γ-aminobutyric acid (GABA)[47,48] neurons.

Increased SRIF secretion is also induced by excitatory amino acids (e.g., glutamate and aspartic acid) through *N*-methyl-D-aspartate (NMDA)-type receptors,[25,26] and by melatonin[49] and by a number of neuropeptides. In addition to SRIF[5,42] these include glucagon,[50] bombesin,[51] neurotensin,[28,50,52] substance P,[52] enkephalins,[52] vasoactive intestinal peptide,[50,53] thyrotropin-releasing hormone (TRH),[54] corticotropin-releasing hormone,[55,56] luteinizing hormone-releasing hormone (LHRH),[47] and GH-releasing hormone (GHRH).[55,57,58] Increased SRIF release is also induced by thyrotropin,[59] thyroid hormones,[60] insulin-like growth factor I (IGF-I)[61] and by GH.[59,61–65] The short-loop regulation of SRIF secretion by GH is specific to hypothalamic neurons and not observed in extrahypothalamic nuclei.[64,66] Decreased SRIF release is, in contrast, induced by serotonin and GABA,[27,49,67] by adrenocorticotropin,[59] and by hypothyroidism[60] and hyperglycemia.[68,69]

Physiological factors regulating SRIF release include nutritional stimuli, because SRIF release is increased during restricted feeding,[70] glucocytopenia,[71–73] and by insulin-induced hypoglycemia.[74] The elevated plasma SRIF concentrations in food-deprived or malnourished rats is causally associated with a suppression of pulsatile GH release.[75] The release of SRIF from hypothalamic tissue is also pulsatile[76] and greatest during the night.[77] Phasic pulses of SRIF occur in portal plasma 180° out of phase with GHRH[74,78] and are superimposed on a sexually dimorphic pattern of tonic SRIF release.[79]

A physiological role for SRIF in regulating GH secretion is indicated by the increase in basal and stimulated plasma GH concentrations and augmented growth rate in growing animals[80–83] following SRIF receptor antagonism or SRIF immunoneutralization.[84–86] Growth is similarly augmented in rats following SRIF withdrawal by anterolateral hypothalamic deafferentation.[87] In rats, the ability of SRIF immunization to enhance basal GH secretion is greater in prepubertal, pubertal, and adult animals than in neonates or fetuses, indicating the ontogenic acquisition of inhibitory hypothalamic control.[88,89] SRIF immunization similarly augments the GH response of aged rats to GHRH and central nervous system (CNS)-active drugs (morphine and dopaminergic agonists) more than in young animals.[90,91] SRIF immunoneutralization elevates mean 24-h and trough GH levels,[75,89,92–94] indicating its importance in the generation of pulsatile GH release,[79] although pulsatile GH release persists in the absence of inhibitory SRIF tone as a result of GHRH stimulation.[79,95] Both $SRIF_{14}$ and $SRIF_{28}$ are thought to participate in regulating trough GH concentrations of pulsatile episodes of GH release because immunoneutralization of both peptides has comparable effects on plasma GH concentrations.[96] In addition to inhibiting basal GH secretion, SRIF release is also thought to mediate suppressed GH secretion in rats during acute immobilization[97] or electroshock stress,[98] during pentobarbital[99] or ether[100,101] anesthesia, and during starvation,[102] because these responses are augmented following SRIF immunoneutralization. The inhibitory GH responses to lateral-hypothalamic-medial forebrain stimulation,[103] and to hypo- and hyperthyroidism,[104] α-nonadrenergic blockade,[105] opiate receptor blockade,[106] and exogenous GH[86] are also reduced in SRIF-immunized rats.

Physiological roles of SRIF in GH regulation are also indicated in animals with PeVN lesions, in which alterations in the pulsatile pattern of GH secretion occur.[107] However, as the alterations in GH pulse amplitude and pulse frequency are transient, PeVN and amygdaloid SRIF may not have a significant role in regulating basal episodic GH secretion.[108–110] The blockade of bombesin's inhibition of stimulated GH release in SRIF-depleted cysteamine-treated rats[51] also suggests the action of bombesin is mediated through increased SRIF tone.

III. SRIF SYNTHESIS

Expression of the *SRIF* gene[111] occurs in the central nervous system and in numerous peripheral tissues in a tissue-specific way.[112,113] In the hypothalamus the 600-bp mRNA encodes a 116-amino acid precursor, preprosomatostatin,[114] which undergoes endopeptidase cleavage to yield a tetradecepeptide ($SRIF_{14}$) and an amino-terminal extended form ($SRIF_{28}$) that can be cleaved to $SRIF_{14}$ and $SRIF_{28[1-12]}$.[112,115] All of these peptides are secreted in the ME into portal circulation,[19,116–118] although they can be selectively stored or differentially released by different stimuli.[119] Whereas $SRIF_{28[1-12]}$ may be biologically inactive, both $SRIF_{14}$ and $SRIF_{28}$ bind to specific receptors in the brain and pituitary gland.[120] Both SRIF moieties inhibit GH secretion[121] although $SRIF_{28}$ is longer acting and more potent than $SRIF_{14}$.[122,123]

Expression of the *SRIF* gene in the frontal cortex, parietal cortex, and striatum of the rat brain declines with aging, although not in hypothalamic tissues,[124] in which the SRIF content increases,[125–128] although Martinoli et al.[129] reported an age-related decline in hypothalamic SRIF mRNA. Transcription of the *SRIF* gene and the hypothalamic content of SRIF is reduced by hypophysectomy[130–132] and induced by GH,[130,133] without increasing the number of SRIF-producing cells.[133]

The hypothalamic control of SRIF is also reduced in GH-deficient Snell dwarf mice.[134] The autoregulatory effects of GH on the content of hypothalamic SRIF appear to be greater in adolescents than in adults[125] and greater in males than in females.[135] The SRIF content of the rat hypothalamus is sexually dimorphic and greater in males,[129] as is the rate of *in vitro* release.[127,136] Factors other than GH are, however, also responsible for the regulation of SRIF biosynthesis, because GH is unable to restore SRIF concentrations completely in the hypothalamus of hypophysectomized rats.[125]

SRIF gene transcription in the rat PeVN hypothalamus is also reduced by castration and is increased by testosterone, through the activation of androgen receptors rather than through its aromatization to estradiol.[137,138] A stimulatory effect of estrogens on SRIF mRNA accumulation has, however, also been observed.[138] The processing of the *SRIF* gene transcripts is responsive to thyroid hormone regulation, which leads to the preferential synthesis of $SRIF_{28}$ production.[139]

In addition to hormones, *SRIF* gene transcription is regulated by neurotransmitters. Norepinephrine, for instance, reduces the SRIF content of the rat hypothalamus by an α_2-adrenergic pathway.[126] α_1-Adrenoreceptor blockade, conversely, stimulates SRIF synthesis and prevents the reduction in tissue SRIF content induced by superior cervical sympathectomy.[140] Dopamine D_2 receptor activation also induces *SRIF* gene expression,[138] as may other factors that elevate intracellular cAMP levels. A cAMP response element is present in the *SRIF* gene promoter[141] and cAMP induces *SRIF* gene transcription.[142,143]

IV. SRIF METABOLISM

SRIF is rapidly metabolized in plasma with a half life of <2 min in humans and a metabolic clearance rate of about 50 ml/kg/min.[144,145] The *in vivo* metabolism of $SRIF_{14}$ is even faster in rats, with a half-life of <0.5 min, although the *in vitro* disappearance of $SRIF_{14}$ in plasma is almost 1 h.[146] The *in vivo* and *in vitro* metabolism of $SRIF_{28}$ is, however, slower (by fourfold) than the degradation of $SRIF_{14}$.[146] The metabolism of SRIF may also occur intracellularly in target cells. The uptake of $SRIF_{14}$ and $SRIF_{28}$ into lysosomal and cytoplasmic compartments of pituitary cells[147–151] indicates an intracellular route of processing. The enzymatic cleavage of SRIF does not immediately result in a loss of biological activity, because the amino-terminal residue is not required for receptor recognition and $SRIF_{13}$ is rapidly generated in plasma.[7,152,153] Moreover, although $SRIF_{14}$ has a cyclical structure, caused by the disulfide bridge linking the two cysteines at positions 3 to 14, linear SRIF and analogs lacking a cyclical structure have biological activity and only the amino acids in positions 7–10 appear to be obligatory for receptor activation.[154]

V. SRIF RECEPTORS

The actions of SRIF on somatotroph function are mediated by high-affinity, low-capacity receptors in pituitary and hypothalamic sites.[155] The receptor-binding affinities and biological potencies of SRIF moieties and analogs are closely correlated.[156,157] The greater inhibitory effect of $SRIF_{28}$ on GH secretion is thus correlated with a binding affinity that is greater than that for $SRIF_{14}$.[158,159] The two moieties may, however, act at different receptor types, because the downregulation of brain receptors for $SRIF_{28}$ does not induce a downregulation of $SRIF_{14}$ -binding sites.[160] The binding of $SRIF_{14}$ and $SRIF_{28}$ to rat brain membranes is also differentially modulated by nucleotides and ions.[159,161]

The receptors for $SRIF_{14}$ may also be composed of heterogeneous populations of membrane-bound and cytosolic subtypes[162,163] that differ in molecular size,[164] function,[165] and binding characteristics.[163,166] Indeed, pituitary SRIF receptors have ligand specificities that are clearly different from those in the brain.[167,168] This tissue-specific heterogeneity may partly arise from alternate splicing of SRIF receptor gene transcripts, generating at least six receptor subtypes that differ between 45 and 60% in amino acid sequence identity.[169–171]

The numbers of SRIF receptors in target cells are homologously regulated. The abundance and affinity of SRIF-binding sites in the brain increase when endogenous SRIF is depleted, although pituitary SRIF receptors are more resistant.[172] The inhibitory effect of SRIF on hormone release from the rat pituitary is, however, diminished after prolonged exposure[173,174] and a disinhibition of GH release also occurs in response to continuous SRIF tone in humans.[175]

The desensitization of $SRIF_{14}$ action in rat pituitary glands is accompanied by a marked reduction (by two orders of magnitude) in the ability of SRIF to inhibit the GH responses to GHRH and phosphodiesterase inhibition by 3-isobutyl-1-methylxanthine.[176] Refractoriness develops after 12 h of exposure to doses of SRIF of 10 n*M* or more and becomes maximal at 48 h. Restoration of SRIF responsiveness follows a similar time-course on removal of the peptide. Refractoriness is not induced over a 36-h period if SRIF is delivered to pituitary cells in 15-min pulses every 2 h. Refractoriness to $SRIF_{28}$ also occurs but it is less marked and does not prevent responsiveness to $SRIF_{14}$ and, conversely, the pituitary response to $SRIF_{28}$ is not impaired in glands refractory to $SRIF_{14}$.[176] Both $SRIF_{14}$ and $SRIF_{28}$ also downregulate SRIF receptors in mouse pituitary tumor cells,[174] in which $SRIF_{28}$ is more effective than $SRIF_{14}$.[123,158]

The number of SRIF receptors in human adenomas is also correlated with the patient's ability to respond to SRIF *in vivo*.[177] Pituitary SRIF-binding sites are also downregulated in obese Zucker rats, in which endogenous SRIF tone may be elevated.[178] An upregulation of SRIF-binding sites by $SRIF_{14}$ has, conversely, been observed in tumorous rat somatotrophs (GH_4C_1 cells), independently of the cellular responsiveness to SRIF action.[179]

The abundance of SRIF receptors in pituitary cells is also subject to heterologous regulation. The binding of SRIF to rat pituitary membranes is, for instance, increased by GH,[180] and by thyroid hormones[181] and 17β-estradiol,[182] but downregulation occurs in response to thyroidectomy[183] and adrenalectomy.[184] The number of SRIF binding sites in the pituitaries of adrenalectomized rats is partially restored by glucocorticoid replacement,[184] although glucocorticoids downregulate SRIF-receptors in tumorous pituitary cells,[185] which are unresponsive to gonadal steroids.[185]

The abundance of pituitary SRIF receptors may also change during ontogeny, in response to changes in GH secretion (see Chapter 12). An age-related increase in SRIF binding to pituitary membranes has been observed and directly correlated with the ability of SRIF to inhibit secretion.[186,187] The diminished GH release during aging may, however, reflect alterations in postreceptor mechanisms, because some authors report that the number and affinity of pituitary and hypothalamic SRIF receptors do not increase during ontogeny, despite reduction in GH responsiveness to GHRH stimulation.[188,189]

VI. MECHANISM OF SRIF ACTION

Receptor-mediated actions of SRIF may involve a variety of G protein-coupled signal transduction systems that may be specific for different receptor subtypes or be linked to one or more effector systems.[190] This linkage with G proteins is indicated by characteristic amino acid sequences for G protein-coupled receptors within the structure of the SRIF receptor.[190] These intracellular mechanisms include inhibition of the adenylate cyclase system and inhibition of Ca^{2+} mobilization.[191,192]

It is now well established that populations of SRIF receptors are linked to an inhibitory G protein (G_i) that antagonizes the effects of the stimulatory G protein (G_s) on adenylate cyclase activity. SRIF thus activates GTP hydrolysis and overcomes forskolin-stimulated adenylate cyclase activity in Cyc-S49 myeloma cells, which possess functional SRIF receptors and G_i proteins but lack G_s subunits.[193] This coupling to G_i is also indicated by the ability of pertussis toxin, which specifically antagonizes G_i, to abolish the inhibitory effects of SRIF on GHRH-induced cAMP accumulation and GH release in rat pituitary cells.[194–197]

The actions of SRIF, mediated through one or more pertussis-sensitive G proteins, may not, however, directly involve inhibition of adenylate cyclase[194]. The ability of SRIF to activate G_i is also impaired in the presence of nonhydrolyzable GTP analogs.[188,190] The inhibition of basal GH release by SRIF is not, however, accompanied by reduced adenylate cyclase activity and SRIF fails to inhibit the stimulatory effect of forskolin on cellular cyclic AMP (cAMP) accumulation in pituitary cells.[198–202] Moreover, whereas SRIF can completely abolish GHRH-induced GH release, it may not inhibit[203] or only partially attenuate GHRH-induced cAMP accumulation.[199–202,204] In addition, although the binding of SRIF to its receptor normally suppresses adenylate cyclase activity, some SRIF receptors are coupled to mechanisms that increase intracellular cAMP accumulation.[165] Actions independent of cAMP are, therefore, also likely to participate in SRIF signal transduction. These mechanisms include an inhibition of intracellular Ca^{2+} concentrations (see Chen and Clarke[205] for review).

Basal and GHRH-stimulated intracellular Ca^{2+} concentrations in pituitary cells are rapidly reduced by SRIF,[206–212] independently of cAMP levels.[208,209] This inhibitory effect of SRIF is sensitive to pertussis toxin[210,211,213] and involves a reduced influx of extracellular Ca^{2+} through voltage-gated channels.[208,209,214,215] SRIF markedly reduces the amplitude of both T- and L-type Ca^{2+} currents. SRIF reversibly decreases the amplitude of the T current and its time-dependent inactivation and reduces its half-maximal inactivation potential.[205] The amplitude of the L-type current is reduced independently of voltage- or time-dependent activation.[205] This results from a hyperpolarization of cell membranes within 30 s of SRIF stimulation.[191] This hyperpolarization occurs even during extracellular Ca^{2+} depletion and is independent of Na^+ and Cl^- gradients. This hyperpolarization and the reduction in intracellular Ca^{2+} concentration are, however, completely blocked when the extracellular K^+ concentration is elevated from 4.6 to 5 m*M* or when K^+ channels are blocked pharmacologically.[190,191,216] The magnitude of the hyperpolarization also decreases exponentially with increasing extracellular K^+, indicating a causal relationship.[205] The activation of membrane G proteins by activated SRIF receptors may directly open K^+ channels and thereby hyperpolarize the cell membrane, closing voltage-sensitive Ca^{2+} channels and secondarily reducing intracellular Ca^{2+} concentrations. The activation of K^+ channels is blocked by okadaic acid, a selective inhibitor of the serine/threonine protein phosphatase 2A.[217] SRIF would thus appear to stimulate K^+ channels through dephosphorylation of the channel or a closely associated regulatory molecule. The ability of SRIF to stimulate the activity of tyrosine phosphatase[218,219] supports this view, as does the ability of SRIF to dephosphorylate epidermal growth factor-phosphorylated membranes in a tumorous cell line.[220] The participation of ion channels in the mechanism of SRIF action is indicated in Figure 1.

In addition to cAMP and intracellular Ca^{2+}, SRIF may regulate somatotroph function by modulating the activity of the inositol phosphate-diacylglycerol-protein kinase C signal transduction pathway. This possibility is supported by SRIF antagonism of phorbol ester induced Ca^{2+} accumulation and GH secretion,[201,207,208] although SRIF does not attenuate GHRH-induced phosphatidylinositol labeling[221,222] or diacylglycerol production.[223]

The intracellular action of SRIF may also be mediated through the arachidonic acid-eicosanoid pathway, because it blocks Ca^{2+} and GH responses induced by arachidonic acid derivatives[221,224] and diminishes arachidonic acid production.[225] Lipoxygenase metabolites have also been shown to be intracellular messengers linking SRIF to K^+ channel modulation in GH_4C_1 pituitary cells.[226]

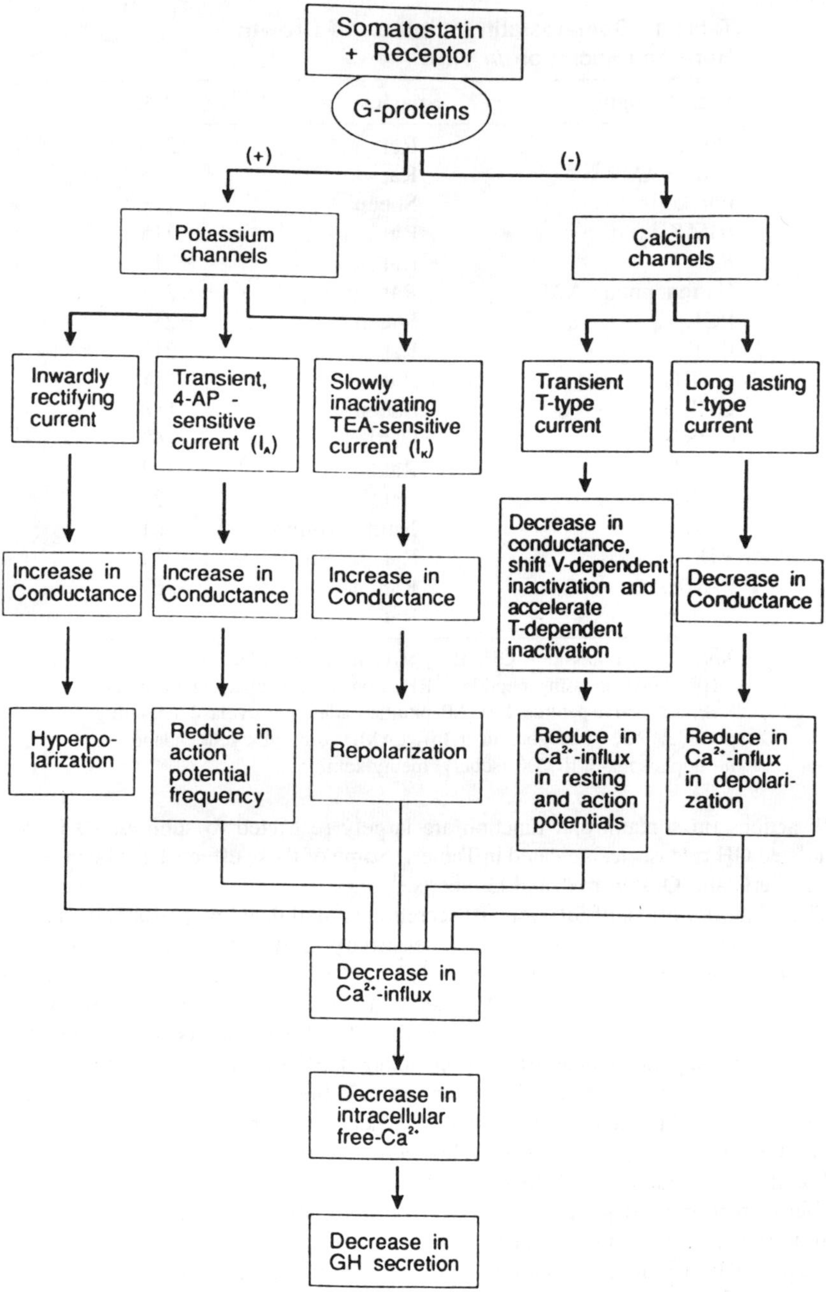

Figure 1 A hypothetical model of intracellular signaling pathways and ion channels in somatotrophs for somatostatin. (From Chen, C. and Clarke, I. J., *Growth Regul.*, 2, 167, 1992. Reprinted with permission.)

VII. SRIF ACTIONS

Although SRIF indirectly inhibits *GH* gene transcription,[227] it has little, if any, direct effect on GH synthesis[228,229] (see Chapter 4) or somatotroph proliferation (see Chapter 3).

Table 1 **Somatostatin Inhibition of Growth Hormone Secretion *In Vitro***

Secretagogue	Species	Ref.
None	Rat	236
$(Bu)_2$ cAMP	Rat	215
Forskolin	Sheep	198
IBMX	Rat	215
K^+	Rat	215
Ca ionophone A23187	Rat	215
PGE_1	Sheep	259
PGE_2	Rat	215
GHRH	Rat	251
GHRH	Sheep	188
GHRH	Cattle	260
GHRP	Rat	123
$GHRP_{28}$	Rat	123
TRH	Human (tumor)	261
VIP	Rat	202
PACAP	Rat	262
Bombesin	Rat	263

Note: K^+, Potassium; GHRH, growth hormone-releasing hormone; GHRP, GH-releasing peptide; TRH, thyrotropin-releasing hormone; VIP, vasoactive peptide; PACAP, pituitary adenylate cyclase-activating peptide; PG, prostaglandin; $(Bu)_2$ cAMP, dibutyl cyclic adenosine monophosphate; IBMX, isobutyl methylxanthine.

SRIF actions on somatotroph function are largely restricted to suppressing basal and stimulated GH release, as indicated in Table 1. Some of these effects have been reviewed by Lamberts and Oosterom[230] and Lamberts.[231]

The inhibitory effect of SRIF on GH secretion is modulated by peripheral endocrines, which increase or decrease somatotroph sensitivity. Glucocorticoid hormones, for instance, decrease the sensitivity of somatotrophs to SRIF[232–235] and prevent the inhibitory effects of IGF-I on GH secretion.[236] In the absence of glucocorticoids, inhibitory actions of IGF-I and SRIF on GH secretion are additive,[236] although mediated through different systems[229] and separate somatotroph subpopulations.[237] Although SRIF appears to suppress the GH response to most GH secretagogues (Tables 1 and 2), SRIF-responsive cells may not include all pituitary somatotrophs.[237] When somatotrophs are stimulated by SRIF the number of cells responsive to GHRH is reduced and the amount of GH released from each cell is also reduced.[234,238] SRIF is not, however, a major regulator of somatotroph proliferation in normal pituitary glands,[239] but does inhibit pituitary tumor cell (GH_3) proliferation *in vitro*[240] and has been used clinically to reduce tumor size.[231] SRIF also attenuates GHRH-induced somatotroph proliferation in normal pituitary tissue[239] and the expression of the c-*fos* proto oncogene,[241] probably as a result of impaired intracellular cAMP and Ca^{2+} levels.[1]

The inhibition of GH release by SRIF results from both morphological (a distribution of microfilaments at the plasma membrane[242]) and biochemical (changes in intracellular cAMP, Ca^{2+}, etc.) alterations in pituitary somatotrophs. This inhibitory hypothalamic control of GH release is developmentally acquired in most species,[88,187,243–249] as is the sensitivity of somatotrophs to SRIF inhibition,[187] which is increased in aging animals at postreceptor sites.[189]

Table 2 **Somatostatin Inhibition of Growth Hormone Secretion *In Vivo*[a]**

Secretagogue	Species	Ref.
None	Rat	254
None	Dog	256
None	Human	264
None[b]	Human	264
Hypothalamic extract	Rat	265
GHRH	Human	266
GHRH	Rat	267
GHRP[b]	Rat	123
GHRP	Rat	123
TRH	Pig	268
TRH	Human	269
LHRH	Human	269
Glucagon	Sheep	259
Arginine	Human	270
Arginine	Sheep	271
Insulin hypoglycemia	Human	272
Glucose	Sheep	259
Propionate	Sheep	259
Antilipolytic agents	Sheep	273
Morphine	Rat	274
Exercise	Human	275
Sleep	Human	276

Note: GHRH, growth hormone-releasing hormone; GHRP, GH-releasing peptide; TRH, thyrotropin-releasing hormone; LHRH, luteinizing hormone-releasing hormone.

[a] $SRIF_{14}$, unless indicated otherwise.
[b] $SRIF_{28}$.

In addition to these inhibitory actions on somatotroph function, SRIF may also have physiological roles that stimulate GH release. The GH response may be biphasically related to the SRIF dose and at micromolar concentrations SRIF may stimulate GH release from rat pituitaries.[250] The phasic removal of tonic SRIF secretion may also liberate somatotrophs from secretory restraint and trigger "rebound" GH release.[251–256] Although GH secretion is normally thought to be under dual hypothalamic regulation, a single releasing factor (or inhibitory release factor) has the potential to regulate pituitary function, as demonstrated for the release of gonadotrophic hormones by luteinizing hormone-releasing hormone and for the release of corticotropin by corticotropin-releasing hormone. The possibility that SRIF may be the dominant, if not sole, regulator of somatotroph function has been hypothesized by Casaneuva.[257]

Casaneuva[257] concluded that, apart from GHRH, every stimulus that increases GH secretion in humans acts through an inhibition of SRIF secretion rather than increased GHRH activity. Furthermore, abnormal GH secretion in clinical pathologies (e.g., Cushing's syndrome, obesity, anorexia nervosa) is usually thought to reflect dysfunctional somatostatinergic control or a primary pituitary lesion rather than dysfunctional GHRH secretion. In the absence of data to support a leading role for GHRH in either spontaneous or stimulated GH secretion, Casaneuva[257] suggests pulsatile GH release in humans is

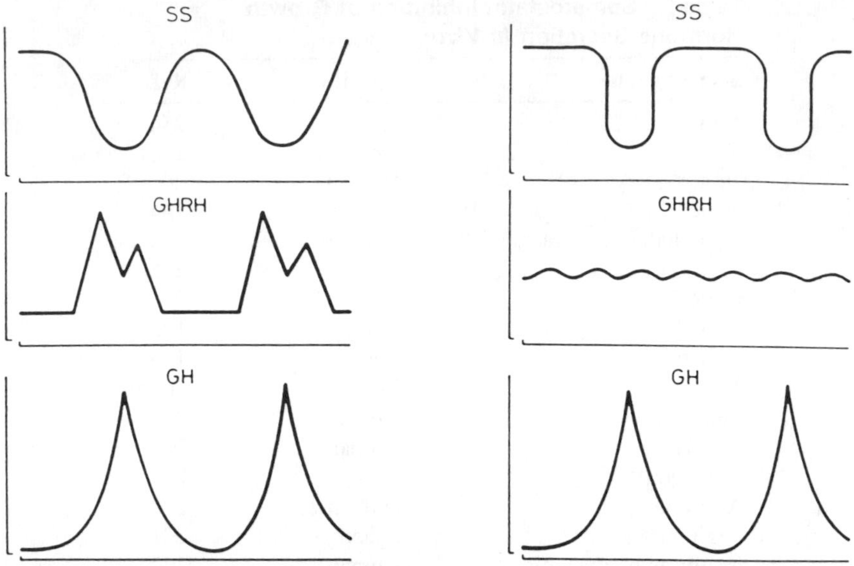

Figure 2 Postulated pattern of hypothalamic somatostatin (SS) and GHRH secretion into portal blood of male rats (left) and in humans (right). In humans the GH ultradian pulses will be regulated by the periodic drops in the tonically high somatostatin tone, while GHRH will exert a permissive role and establish the set point of the whole system. (From Casaneuva, F. F., Popovic, V., Leal-Cerro, A., et al., The physiology of growth hormone secretion, *Molecular and Clinical Advances in Pituitary Disorders* [Ed. Melmed, S.]. Endocrine Research and Education, Inc., Los Angeles, 1993, pp. 145–151. Reprinted with permission.)

regulated by SRIF (Figure 1). According to this view, a reduction in tonically high SRIF tone is responsible for generating GH discharge, in a manner comparable to the generation of prolactin pulses by dopamine withdrawal. In this model (Figure 2) GHRH only acts as a permissive factor that is necessary for somatotroph function and has a primary role in establishing the set point of the system. This model is not, however, supported by Ho et al.,[258] who reported that SRIF withdrawal alone is an ineffective generator of pulsatile GH release in humans.

REFERENCES

1. **Frohman, L. A., Downs, T. R., and Chomczynski, P.,** *Front. Neuroendocrinol.*, 13, 344, 1992.
2. **Polak, J. M. and Bloom, S. R.,** *Scand. J. Gastroenterol.*, 21(Suppl. 119), 11, 1986.
3. **Reichlin, S.,** *J. Lab. Clin. Med.*, 109, 320, 1987.
4. **Berelowitz, M.,** *Neuroendocrine Perspectives* (Eds. Muller, E. E., Macleod, R. M., and Frohman, L. A.). Elsevier Science Publishers, New York, 1985, p. 59.
5. **Plotsky, P. M.,** *Neuroendocrinology*, 53, 433, 1991.
6. **Brazeau, P., Epelbaum, J., and Benoit, R.,** *Central Nervous System Effects of Hypothalamic Hormones and Other Peptides* (Eds. Collu, R., et al.) Raven Press, New York, 1979, p. 367.
7. **Shoelson, S. E., Polonsky, K. S., Nakabayashi, T., et al.,** *Am. J. Physiol.*, 250, E428, 1986.
8. **Conlon, J. M., Bridgeman, M., and Alberti, K. G. M. M.,** *Anal. Biochem.*, 125, 243, 1982.
9. **Arimura, A., Turkelson, C., and Chihara, K.,** *Neuropeptides: Biochemical and Physiological Studies* [Ed. Millar, R. P.]. Churchill Livingstone, Edinburgh, 1981, p. 15.

10. **Kato, M., Suzuki, M., and Kakegawa, T.,** *Endocrinology (Baltimore)*, 116, 382, 1985.
11. **Chihara, K., Arimura, A., Kubli-Garfias, C., et al.,** *Endocrinology (Baltimore)*, 105, 1416, 1979.
12. **Koibuchi, N., Kato, M., Kakegawa, T., et al.,** *Endocrinology (Baltimore)*, 122, 659, 1988.
13. **Martin, J. B., Tannenbaum, G., Willoughby, J. O., Renaud, L. P., and Brazeau, P.,** *Hypothalamic Hormones* (Eds. Motta, M., Crosignani, P. G., and Martini, L.). Academic Press, New York, 1975, p. 217.
14. **Chihara, K., Arimura, A., and Schally, A. V.,** *Endocrinology (Baltimore)*, 104, 1434, 1979.
15. **Kaler, L. W., Vale, W., and Critchlow, V.,** *Brain Res.*, 447, 384, 1988.
16. **Rice, R. W. and Critchlow, V.,** *Endocrinology (Baltimore)*, 99, 970, 1976.
17. **Willoughby, J. O., Terry, L. C., Brazeau, P., et al.,** *Brain Res.*, 127, 137, 1977.
18. **Chihara, K., Kato, Y., Ohgo, S., et al.,** *Neuroendocrinology*, 18, 192, 1975.
19. **Millar, R. P., Sheward, W. J., Wegener, I., et al.,** *Brain Res.*, 260, 334, 1983.
20. **Critchlow, V., Rice, R. W., Abe, K., et al.,** *Endocrinology (Baltimore)*, 103, 817, 1978.
21. **Ishikawa, K., Taniguchi, Y., Kurosumi, K., et al.,** *Endocrinology (Baltimore)*, 121, 94, 1987.
22. **Robbins, R. J.,** *Acromegaly* (Eds. Robbins, R. J. and Melmed, S.). Plenum, New York, 1987, p. 75.
23. **Critchlow, V., Dyke, A., and Kaler, L. W.,** *Brain Res.*, 398, 347, 1986.
24. **Richardson, S. B. and Twente, S.,** *Endocrinology (Baltimore)*, 126, 1043, 1990.
25. **Reichlin, S.,** *Scand. J. Gastroenterol.*, 21(Suppl. 119), 1, 1986.
26. **Tapia-Arancibia, L. and Astier, H.,** *Endocrinology (Baltimore)*, 123, 2360, 1988.
27. **Peterfreund, R. A. and Vale, W. W.,** *Endocrinology (Baltimore)*, 112, 526, 1983.
28. **Maeda, K. and Frohman, L. A.,** *Endocrinology (Baltimore)*, 106, 1837, 1980.
29. **Clarke, M. J. O. and Gillies, G. E.,** *J. Endocrinol.*, 116, 349, 1988.
30. **Tapia-Arancibia, L., Pares-Herbute, N., and Astier, H.,** *Neuroendocrinology*, 49, 555, 1989.
31. **Honegger, J., D'Urso, R., Besser, G. M., et al.,** *Endocrinology (Baltimore)*, 129, 11, 1991.
32. **Cugini, C. D., Jr., Millard, W. J., and Leidy, J. W., Jr.,** *Endocrinology (Baltimore)*, 129, 1355, 1991.
33. **Kitajima, N., Chihara, K., Abe, H., et al.,** *Endocrinology (Baltimore)*, 124, 69, 1989.
34. **Chihara, K., Arimura, A., and Schally, A. V.,** *Endocrinology (Baltimore)*, 104, 1656, 1979.
35. **Junier, M.-P., Dray, F., Blair, I., et al.,** *Endocrinology (Baltimore)*, 126, 1534, 1990.
36. **Devesa, J., Diaz, M. J., Tresguerres, J. A. F., et al.,** *J. Clin. Endocrinol. Metab.*, 73, 251, 1991.
37. **Epelbaum, J., Tapia-Arancibia, L., and Kordon, C.,** *Brain Res.*, 215, 393, 1981.
38. **Bennett, G., Edwardson, J., Marcano De Cotte, D., et al.,** *J. Neurochem.*, 32, 1127, 1979.
39. **Lima, L., Arce, V., Tresguerres, J. A. F., et al.,** *Neuroendocrinology*, 557, 1155, 1993.
40. **Minamitani, M., Chihara, K., Kaji, H., et al.,** *Endocrinology (Baltimore)*, 125, 2839, 1989.
41. **Devesa, J., Arce, V., Tresguerres, J. A. F., et al.,** *J. Clin. Endocrinol. Metab.*, 71, 1581, 1990.
42. **Richardson, S. B. and Twente, S.,** *Endocrinology (Baltimore)*, 117, 2076, 1986.
43. **Torsello, A., Panzeri, G., Cermenati, P., et al.,** *J. Endocrinol.*, 117, 273, 1988.
44. **Locatelli, V., Torsello, A., Redaelli, M., et al.,** *J. Endocrinol.*, 111, 271, 1986.
45. **Casanueva, F. F., Villanueva, L., Dieguez, C., et al.,** *J. Clin. Endocrinol. Metab.*, 62, 186, 1986.
46. **Ross, R. J. M., Ttsagarakis, S., Grossman, A., et al.,** *Clin. Endocrinol.*, 27, 727, 1988.
47. **Muller, E. E. and Nistico, G.,** *Brain Messengers and the Pituitary*, Academic Press, New York, 1989.
48. **Gamse, R., Vaccro, D., Gamse, G., et al.,** *Proc. Natl. Acad. Sci. U.S.A.*, 77, 5552, 1980.
49. **Richardson, S. B., Hollander, C. S., Prasad, J. A., et al.,** *Endocrinology (Baltimore)*, 109, 602, 1981.
50. **Shimatsu, A., Kato, Y., Matsushita, N., et al.,** *63th Annu. Meet. Endocr. Soc.*, Cincinatti, OH, 1981, 198.
51. **Wakabayashi, I., Tonegawa, Y., Shibasaki, T., et al.,** *Life Sci.*, 36, 1437, 1985.
52. **Sheppard, M. C., Kronheim, S., and Pimstone, B. L.,** *J. Neurochem.*, 32, 647, 1979.
53. **Tapia-Arancibia, L. and Reichlin, S.,** *Brain Res.*, 336, 67, 1986.
54. **Katakami, H., Arimura, A., and Frohman, L. A.,** *Endocrinology (Baltimore)*, 117, 1139, 1985.
55. **Mitsugi, N., Arita, J., and Kimura, F.,** *Neuroendocrinology*, 51, 93, 1990.
56. **Katakami, H., Arimura, A., and Frohman, L. A.,** *Neuroendocrinology*, 41, 390, 1985.
57. **Katakami, H., Arimura, A., and Frohman, L. A.,** *Endocrinology (Baltimore)*, 118, 1872, 1986.

58. **Lumpkin, M. D., Samson, N. K., and McCann, S. M.,** *Endocrinology (Baltimore)*, 116, 2070, 1985.
59. **Robbins, R. J., Leidy, J. W., Jr., and Landon, R. M.,** *Endocrinology (Baltimore)*, 117, 538, 1985.
60. **Berelowitz, M., Maeda, K., Harris, S., et al.,** *Endocrinology (Baltimore)*, 107, 24, 1980.
61. **Berelowitz, M., Szabo, M., Frohman, M., et al.,** *Science*, 212, 1279, 1981.
62. **Chihara, K., Minamitani, N., Kaji, G., et al.,** *Endocrinology (Baltimore)*, 109, 2279, 1981.
63. **Patel, Y. C.,** *Life Sci.*, 24, 1589, 1979.
64. **Berelowitz, M., Firestone, S. I., and Frohman, L. A.,** *Endocrinology (Baltimore)*, 109, 714, 1981.
65. **Sheppard, M. C., Kronheim, S., and Pimstone, B. L.,** *Clin. Endocrinol.*, 9, 583, 1978.
66. **Molitch, M. E. and Hlivyak, L. E.,** *Horm. Metab. Res.*, 12, 519, 1980.
67. **Arancibia, S. and Briozzo, P.,** *Neurosci. Lett.*, 111, 211, 1990.
68. **Lengyel, A. M. J., Nieuwenhuyzen-Kruseman, A. C., Grossman, A., et al.,** *Life Sci.*, 35, 713, 1984.
69. **Lewis, B. M., Dieguez, C., Ham, J., et al.,** *J. Neuroendocrinol.*, 1, 1989.
70. **Thomas, G. B., Cummings, J. T., Francis, H., et al.,** *Endocrinology (Baltimore)*, 128, 1151, 1991.
71. **Lengyel, A. ., Grossman, A., Nieuwenhuyzen-Kruseman, A. C., et al.,** *Neuroendocrinology*, 39, 31, 1984.
72. **Berelowitz, M., Dudlak, D., and Frohman, L. A.,** *J. Clin. Invest.*, 69, 1293, 1982.
73. **Baes, M. and Vale, W. W.,** *Neuroendocrinology*, 51, 202, 1990.
74. **Frohman, L. A., Downs, T. R., Clark, I. J., et al.,** *J. Clin. Invest.*, 36, 17, 1990.
75. **Janowski, B. A. and Wehrenberg, W. B.,** *Life Sci.*, 50, 951, 1992.
76. **Fukata, J., Kasting, N. W., and Martin, J. B.,** *Neuroendocrinology*, 40, 193, 1985.
77. **Berelowitz, M., Dudlak, D., and Frohman, L. A.,** *Endocrinology (Baltimore)*, 110, 2195, 1982.
78. **Plotsky, P. M. and Vale, W.,** *Science*, 230, 461, 1985.
79. **Tannenbaum, G. S.,** *Horm. Res.*, 29, 70, 1988.
80. **Spencer, G. S. G.,** *Reprod. Nutr. Develop.*, 27, 581, 1987.
81. **Vicini, J. L., Clark, J. H., Hurley, W. L., et al.,** *Domest. Anim. Endocrinol.*, 5, 35, 1988.
82. **Spencer, G. S. G.,** *J. R. Soc. Med.*, 77, 496, 1984.
83. **Chaplin, R. K., Kerr, D. E., and Laarveld, B.,** *Can. J. Anim. Sci.*, 64(Suppl.), 312, 1984.
84. **Wehrenberg, W. B., Ling, N., Bohlen, P., et al.,** *Biochem. Biophys. Res. Commun.*, 109, 562, 1982.
85. **Ferland, L., Labrie, F., Jobin, M., et al.,** *Biochem. Biophys. Res. Commun.*, 68, 149, 1976.
86. **Sato, M., Chihara, K., Kita, T., et al.,** *Neuroendocrinology*, 50, 139, 1989.
87. **Mitchell, J. A., Smyrl, R., Hutchins, M., et al.,** *Neuroendocrinology*, 10, 31, 1972.
88. **Khorram, O., De Palatis, L. R., and McCann, S. M.,** *Endocrinology (Baltimore)*, 113, 720, 1983.
89. **Oliver, C., Giraud, P., Lissitzky, J. C., et al.,** *Endocrinology (Baltimore)*, 110, 1018, 1982.
90. **Sonntag, W. E., Forman, L. J., Miki, N., et al.,** *Neuroendocrinology*, 33, 73, 1981.
91. **Locatelli, V., Arimura, A., Torsello, A., et al.,** *Neuroendocrinol. Lett.*, 6, 261, 1984.
92. **Schusdziarra, V., Rouiller, D., Arimura, A., et al.,** *Endocrinology (Baltimore)*, 103, 1956, 1978.
93. **Varner, M. A., Davis, S. L., and Reeves, J. J.,** *Endocrinology (Baltimore)*, 106, 1027, 1980.
94. **Tannenbaum, G. S. and Ling, N.,** *Endocrinology (Baltimore)*, 115, 1952, 1984.
95. **Steiner, R. A., Stewart, J. R., Barber, J., et al.,** *Endocrinology*, 102, 1587, 1978.
96. **Jacovidou, N. and Patel, Y. C.,** *Endocrinology (Baltimore)*, 121, 782, 1987.
97. **Benyassi, A., Gavalda, A., Armario, A., et al.,** *Life Sci.*, 52, 361, 1992.
98. **Arimura, A., Smith, W. D., and Schally, A. V.,** *Endocrinology (Baltimore)*, 98, 540, 1976.
99. **Saito, H., Ogawa, T., Ishimaru, K., et al.,** *Horm. Metab. Res.*, 11, 550, 1979.
100. **Aguila, M. C., Pickle, R. L., Yu, W. H., et al.,** *Neuroendocrinology*, 54, 515, 1991.
101. **Strbak, V., Jurcovicova, J., and Vigas, M.,** *Neuroendocrinology*, 40, 377, 1985.
102. **Chihara, K., Kato, Y., Maeda, K., et al.,** *J. Clin. Endocrinol. Metab.*, 44, 1094, 1977.
103. **Terry, L. C. and Martin, J. B.,** *Endocrinology (Baltimore)*, 109, 622, 1981.
104. **Wakabayashi, I., Tonegawa, Y., Ihara, T., et al.,** *Neuroendocrinology*, 41, 306, 1985.
105. **Chihara, K., Minamitani, N., Kaji, H., et al.,** *Endocrinology (Baltimore)*, 114, 1402, 1984.
106. **Borer, K. T., Nicoski, D. R., and Owens, V.,** *Endocrinology (Baltimore)*, 118, 844, 1986.
107. **Urman, S., Kaler, L., and Critchlow, V.,** *Neuroendocrinology*, 41, 357, 1985.
108. **Terry, L. C., Crowley, W. R. and Johnson, M. D.,** *J. Clin. Invest.*, 69, 104, 1982.

109. **Martin, J. B.,** *Proc. NIH Symp. Growth Hormone* (Ed. Raiti, S.). NIH, Bethesda, MD, 1971, p. 223.
110. **Martin, J. B., Kontor, J., and Mead, P.,** *Endocrinology (Baltimore)*, 92, 1354, 1973.
111. **Montminy, M. R., Goodman, R. H., Horovitch, S. J., et al.,** *Proc. Natl. Acad. Sci. U.S.A.*, 81, 3337, 1984.
112. **Morel, A., Nicolas, P., and Cohen, P.,** *J. Biol. Chem.*, 258, 8273, 1983.
113. **Kiyama, H. and Emson, P. C.,** *Neuroscience*, 38, 223, 1990.
114. **Andrews, P. C. and Dixon, J. E.,** *Scand. J. Gastroenterol.*, 21(Suppl. 119), 22, 1986.
115. **Van Itallie, C. M. and Fernstrom, J. D.,** *Endocrinology (Baltimore)*, 113, 1210, 1983.
116. **Charpenet, G. and Patel, Y. C.,** *Endocrinology (Baltimore)*, 116, 1863, 1985.
117. **Pierotti, A. R., Harmar, A. J., Tannahill, L. A., et al.,** *Neurosci.Lett.*, 57, 215, 1985.
118. **Sheward, W. J., Benoit, R., and Fink, G.,** *Neuroendocrinology*, 38, 88, 1984.
119. **Bakhit, C., Koda, L., Benoit, R., et al.,** *J. Neurosci.*, 4, 411, 1984.
120. **Moyse, E., Benoit, R., Enjalbert, A., et al.,** *Regul. Peptides*, 9, 129, 1984.
121. **Jacovidou, N. and Patel, Y. C.,** *Endocrinology (Baltimore)*, 121, 782, 1987.
122. **Tannenbaum, G. S., Ling, N., and Brazeau, P.,** *Endocrinology (Baltimore)*, 111, 101, 1982.
123. **Bowers, C. Y., Momany, F. A., Reynolds, G. A., et al.,** *Endocrinology (Baltimore)*, 114, 1537, 1984.
124. **Florio, T., Ventra, C., Postiglione, A., et al.,** *Brain Res.*, 557, 64, 1991.
125. **Leidy, J. W., Jr., McDermott, M. T., and Robbins, R. J.,** *Neuroendocrinology*, 51, 400, 1990.
126. **Gil-Ad, I., Laron, Z., and Koch, Y.,** *J. Endocrinol.*, 131, 381, 1991.
127. **Argente, J., Chowen, J. A., Zeitler, P., et al.,** *Endocrinology (Baltimore)*, 128, 2369, 1991.
128. **Walker, P., Dissault, J. H., Alvarado-Urbina, G., et al.,** *Endocrinology (Baltimore)*, 101, 782, 1977.
129. **Martinoli, M. G., Oullet, J., Rheaume, E., et al.,** *Neuroendocrinology*, 54, 607, 1991.
130. **Rogers, K. V., Vician, L., Steiner, R. A., et al.,** *Endocrinology (Baltimore)*, 122, 586, 1988.
131. **Rodgers, K. V., Vician, L., Steiner, R. A., et al.,** *Endocrinology (Baltimore)*, 121, 90, 1987.
132. **Terry, L. C. and Crowley, W. R.,** *Endocrinology (Baltimore)*, 107, 1771, 1980.
133. **Bertherat, J., Timsit, J., Bluetpajot, M. T., et al.,** *J. Clin. Invest.*, 91, 1783, 1993.
134. **Phelps, C. J. and Hoffman, G. E.,** *Peptides*, 8, 1127, 1987.
135. **Maiter, D. M., Gabriel, S. M., Koenig, J. I., et al.,** *Neuroendocrinology*, 51, 174, 1990.
136. **Ge, F., Tsagarakis, S., Rees, L. H., et al.,** *J. Endocrinol.*, 123, 53, 1989.
137. **Argente, J., Chowen-Breed, J. A., Steiner, R. A., et al.,** *Neuroendocrinology*, 52, 342, 1990.
138. **Zorrilla, R., Simard, J., Rheaume, E., et al.,** *Neuroendocrinology*, 52, 527, 1990.
139. **de los Frailes, M. T., Sanchez Franco, F., Lorenzo, M. J., et al.,** *Neuropeptides*, 15, 25, 1990.
140. **Cardinali, D. P., Esquifino, A. I., Arce, A., et al.,** *Neuroendocrinology*, 59, 42, 1994.
141. **Montminy, M. R., Sevarino, K. A., Wagner, J. A., et al.,** *Proc. Natl. Acad. Sci. U.S.A.*, 83, 6682, 1986.
142. **Montminy, M. R., Low, M. J., Tapia-Arancibia, L., et al.,** *J. Neurosci.*, 6, 1171, 1986.
143. **Gonzalez, G. A. and Montminy, M. R.,** *Cell*, 59, 675, 1989.
144. **Ho, L. T., Lam, H. C., Wang, J. T., et al.,** *Chinese J. Physiol.*, 29, 91, 1986.
145. **Sheppard, M., Shapiro, B., Pimstone, B., et al.,** *J. Clin. Endocrinol. Metab.*, 48, 50, 1979.
146. **Patel, Y. C. and Wheatley, T.,** *Endocrinology (Baltimore)*, 112, 220, 1983.
147. **Mesguich, P., Benoit, R., Dubois, P. M., et al.,** *Cell Tiss. Res.*, 252, 419, 1988.
148. **Morel, G., Mesguich, P., Dubois, M. P., et al.,** *Neuroendocrinology*, 36, 291, 1983.
149. **Quere, M., El May, A., Bouzakoura, C., et al.,** *Gen. Comp. Endocrinol.*, 60, 187, 1985.
150. **Morel, G., Pelletier, G., and Heisler, S.,** *Endocrinology (Baltimore)*, 119, 1972, 1986.
151. **Mentlein, R., Buchholz, C., and Krisch, B.,** *Cell Tiss. Res.*, 258, 309, 1989.
152. **McMartin, C. and Purdon, G. E.,** *J. Endocrinol.*, 77, 67, 1978.
153. **Marki, F., Schenkel, L., Petrack, B., et al.,** *FEBS Lett.*, 127, 22, 1981.
154. **Wass, J. A. H.,** *Horm. Res.*, 29, 86, 1988.
155. **Epelbaum, J., Agid, F., Agid, Y., et al.,** *Horm. Res.*, 31, 45, 1989.
156. **Schonbrunn, A., Rorstad, O. P., Westendorf, J. M., et al.,** *Endocrinology (Baltimore)*, 113, 1559, 1983.
157. **Srikant, C. B. and Patel, Y. C.,** *Endocrinology (Baltimore)*, 110, 2138, 1982.
158. **Srikant, C. B. and Heisler, S.,** *Endocrinology (Baltimore)*, 117, 271, 1985.
159. **Srikant, C. B. and Patel, Y. C.,** *Nature (London)*, 294, 259, 1981.
160. **Wang, H., Dichter, M., and Reisine, T.,** *Mol. Pharmacol.*, 38, 357, 1990.
161. **Srikant, C. B., Dahan, A., and Craig, C.,** *Regul. Peptides*, 27, 181, 1990.
162. **Lewin, M. J. M.,** *Scand. J. Gastroenterol.*, 21(Suppl. 119), 42, 1986.

163. **Srikant, C. B. and Patel, Y. C.,** *Somatostatin* (Ed. Patel, Y. C.). Plenum Press, New York, 1985, p. 291.
164. **Bruno, J. F. and Berelowitz, M.,** *Endocrinology (Baltimore)*, 124, 831, 1989.
165. **Markstein, R., Stockli, K. A., and Reubi, J. C.,** *Neurosci. Lett.*, 104, 13, 1989.
166. **Srikant, C. B. and Patel, Y. C.,** *Somatostatin* (Ed. Patel, Y. C.). Plenum Press, New York, 1985, p. 89.
167. **Heiman, M. L., Murphy, W. A., and Coy, D. H.,** *Neuroendocrinology*, 429, 1987.
168. **Srikant, C. B. and Patel, Y. C.,** *Endocrinology (Baltimore)*, 116, 1717, 1985.
169. **Bruno, J. F., Xu, Y., Song, J., et al.,** *Endocrinology (Baltimore)*, 133, 2561, 1993.
170. **Reisine, T., Kong, H., Raynor, K., et al.,** *Molecular and Clinical Advances in Pituitary Disorders* (Ed. Melmed, S.). Endocrine Research and Education, Inc., Los Angeles, CA, 1993, p. 293.
171. **Bruns, C., Raulf, F., Rohrer, L., et al,** *Molecular and Clinical Advances in Pituitary Disorders*, (Ed. Melmed, S.). Los Angeles, Endocrine Research and Education, Inc., 1993, 299.
172. **Srikant, C. and Patel, C.,** *Endocrinology (Baltimore)*, 115, 990, 1984.
173. **Smith, M., Yamamoto, G., and Vale, W.,** *Mol. Cell. Endocrinol.*, 37, 311, 1984.
174. **Heisler, S. and Srikant, C. B.,** *Endocrinology (Baltimore)*, 117, 217, 1985.
175. **Kelijman, M. and Frohman, L. A.,** *J. Clin. Endocrinol. Metab.*, 71, 157, 1990.
176. **Smith, M. A., Yamamoto, G., and Vale, W. W.,** *Mol. Cell. Endocrinol.*, 37, 311, 1984.
177. **Ikuyama, S., Nawata, H., Kato, K., et al.,** *J. Clin. Endocrinol. Metab.*, 62, 729, 1986.
178. **Abribat, T., Finkelstein, J. A., and Gaudreau, P.,** *Regul. Peptides*, 36, 263, 1991.
179. **Presky, D. H. and Schonbrunn, A.,** *J. Biol. Chem.*, 263, 714, 1988.
180. **Katakami, H., Berelowitz, M., Marbach, M., et al.,** *Endocrinology (Baltimore)*, 117, 557, 1985.
181. **Hinkle, P., Perrone, M., and Schonbrunn, A.,** *Endocrinology (Baltimore)*, 108, 199, 1981.
182. **Kimura, N., Hayafuju, C., Konagaya, H., et al.,** *Endocrinology (Baltimore)*, 119, 1028, 1986.
183. **Martin, D., Epelbaum, J., Bluet-Pajot, M. T., et al.,** *Neuroendocrinology*, 41, 476, 1985.
184. **Rodriguez, M. N., Gomez-Pan, A., and Arilla, E.,** *Endocrinology (Baltimore)*, 123, 1147, 1988.
185. **Schonbrunn, A.,** *Endocrinology (Baltimore)*, 110, 1147, 1982.
186. **Enjalbert, A., Tapia-Arancibia, L., Rieutort, M., et al.,** *Endocrinology (Baltimore)*, 110, 1634, 1982.
187. **Cuttler, L., Welsh, J. B., and Szabo, M.,** *Endocrinology (Baltimore)*, 119, 152, 1986.
188. **Sirvio, J., Jolkkonen, J., Pitkanen, A., et al.,** *Comp. Biochem. Physiol.*, 87A, 355, 1987.
189. **Spik, K. and Sonntag, W. E.,** *Neuroendocrinology*, 50, 489, 1989.
190. **Schonbrunn, A.,** *Molecular and Clinical Advances in Pituitary Disorders* (Ed. Melmed, S.). Endocrine Research and Education, Inc., Los Angeles, CA, 1993, p. 305.
191. **Schonbrunn, A. and Koch, B. D.,** *Somatostatin* (Ed. Reichlin, S.). Plenum Press, New York, 1987, p. 121.
192. **Schonbrunn, A.,** *Metabolism*, 39(Suppl. 2), 96, 1990.
193. **Jakobs, K. H., Aktories, K., and Schultz, G.,** *Nature (Lond.)*, 303, 177, 1983.
194. **Boyd, R. S., Ray, K. P., and Wallis, M.,** *J. Endocrinol.*, 1, 179, 1988.
195. **Epelbaum, J., Enjalbert, A., Krantic, S., et al.,** *Endocrinology (Baltimore)*, 121, 2177, 1987.
196. **Cronin, M. J., Hewlett, E. L., Evans, W. S., et al.,** *Endocrinology (Baltimore)*, 114, 904, 1984.
197. **Cronin, M. J., Rogol, A. D., Myers, G. A., et al.,** *Endocrinology (Baltimore)*, 113, 209, 1983.
198. **Ray, K. P., Gomm, J. J., Law, G. J., et al.,** *Mol. Cell. Endocrinol.*, 45, 175, 1986.
199. **Bilezikjian, L. M. and Vale, W. W.,** *Endocrinology (Baltimore)*, 113, 1726, 1983.
200. **Law, G. J., Ray, K. P., and Wallis, M.,** *FEBS Lett.*, 179, 12, 1985.
201. **Michel, D., Lefevre, G., and Labrie, F.,** *Mol. Cell. Endocrinol.*, 33, 255, 1983.
202. **Koch, B. D. and Schonbrunn, A.,** *Endocrinology (Baltimore)*, 114, 1784, 1984.
203. **Law, G. J., Ray, K. P., and Wallis, M.,** *FEBS Lett.*, 166, 189, 1984.
204. **Bicknell, R. J. and Schofield, J. G.,** *Mol. Cell. Endocrinol.*, 22, 85, 1981.
205. **Chen, C. and Clarke, I. J.,** *Growth Regul.*, 2, 167, 1992.
206. **Holl, R. W., Thorner, M. O., and Leong, D. A.,** *Endocrinology (Baltimore)*, 122, 2927, 1988.
207. **Holl, R. W., Thorner, M. O., and Leong, D. A.,** *Am. J. Physiol.*, 256, E375, 1989.
208. **Lussier, B. T., French, M. B., Moor, B. C., et al.,** *Endocrinology (Baltimore)*, 128, 592, 1991.
209. **Lussier, B. T., Wood, D. A., French, M. B., et al.,** *Endocrinology (Baltimore)*, 128, 583, 1991.
210. **Simard, J., Lefevre, G., and Labrie, F.,** *Peptides*, 8, 263, 1987.
211. **Koch, B. D., Dorflinger, L. J., and Schonbrunn, A.,** *J. Biol. Chem.*, 260, 13138, 1985.

212. **Rawlings, S. R., Hoyland, J., and Mason, W. T.,** *Cell Calc.*, 12, 403, 1991.
213. **Mollard, P., Vacher, P., Dufy, B., et al.,** *Endocrinology (Baltimore)*, 123, 721, 1988.
214. **Casenueva, F. F., Villaneuva, J. L., Cabranes, J. A., et al.,** *J. Clin. Endocrinol. Metab.*, 59, 526, 1984.
215. **Kraicer, J. and Chow, A. E. H.,** *Endocrinology (Baltimore)*, 111, 1173, 1982.
216. **Deweille, J. R., Fosset, M., Epelbaum, J., et al.,** *Biochem. Biophys. Res. Commun.*, 187, 1007, 1992.
217. **White, R. E., Schonbrunn, A., and Armstrong, D. L.,** *Nature (Lond.)*, 351, 570, 1991.
218. **Pan, M. G., Florio, T., and Stork, P. J. S.,** *Science*, 256, 1215, 1992.
219. **Tahiri-Jouti, N., Cambillau, C., Viguerie, N., et al.,** *Am. J. Physiol.*, 262, G1007, 1992.
220. **Hierowski, M., Liebow, C., du Sapin, K., et al.,** *FEBS Lett.*, 179, 252, 1985.
221. **Fafeur, V., Gouin, E., and Dray, F.,** *Biochem. Biophys. Res. Commun.*, 126, 725, 1985.
222. **Canonico, P. L., Cronin, M. J., Thorner, M. O., et al.,** *Am. J. Physiol.*, 245, E587, 1983.
223. **Yajima, A. Y. and Saito, T.,** *J. Biol. Chem.*, 261, 2684, 1986.
224. **Snyder, G. D., Yadagiri, P., and Falck, J. R.,** *Am. J. Physiol.*, 256, E221, 1989.
225. **Canonico, P. L., Speciale, C., Sortino, M. A., et al.,** *Life Sci.*, 38, 267, 1985.
226. **Armstrong, D. L. and White, R. E.,** *Trends Neurosci.*, 15, 403, 1992.
227. **Sugihara, H., Minami, S., Okada, K., et al.,** *Endocrinology (Baltimore)*, 132, 1225, 1993.
228. **Barinaga, M., Bilezikjian, L. M., Vale, W. W., et al.,** *Nature (Lond.)*, 314, 279, 1985.
229. **Namba, H., Morita, S., and Melmed, S.,** *Endocrinology (Baltimore)*, 124, 1794, 1989.
230. **Lamberts, S. W. J. and Oosterom, R.,** *Horm. Res.*, 14, 137, 1985.
231. **Lamberts, S. W. J.,** *Endocr. Rev.*, 9, 417, 1989.
232. **Vale, W., Vaughan, J., Yamamoto, G., et al.,** *Endocrinology (Baltimore)*, 112, 1553, 1983.
233. **Fukata, J. and Martin, J. B.,** *Endocrinology (Baltimore)*, 119, 2256, 1986.
234. **Mulchahey, J. J., Di Blasio, A. M., and Jaffe, R. B.,** *J. Clin. Endocrinol. Metab.*, 66, 395, 1988.
235. **Schonbrunn, A.,** *Endocrinology (Baltimore)*, 110, 1147, 1982.
236. **Lamberts, S. W. J., Den Holder, F., and Hofland, L. J.,** *Endocrinology (Baltimore)*, 124, 905, 1989.
237. **Hoeffler, J. P., Hicks, S. A., and Frawley, L. S.,** *Endocrinology (Baltimore)*, 120, 1936, 1987.
238. **Hoeffler, J. P. and Frawley, L. S.,** *Endocrinology*, 120, 791, 1987.
239. **Billestrup, N., Swanson, L. W. and Vale, W.,** *Proc. Natl. Acad. Sci. U.S.A.*, 83, 6854, 1986.
240. **Pelicci, G., Pagliacci, M. C., Lanfrancone, L., et al.,** *J. Endocrinol. Invest.*, 13, 657, 1990.
241. **Billestrup, N., Mitchell, R. L., Vale, W., et al.,** *Mol. Endocrinol.*, 1, 300, 1987.
242. **Asa, S. L., Felix, I., Kovacs, K., et al.,** *Endocr. Pathol.*, 1, 228, 1990.
243. **Brazeau, P., Ling, N., Bohlen, P., et al.,** *Proc. Natl. Acad. Sci. U.S.A.*, 79, 7909, 1982.
244. **Silverman, B. L., Bettendorf, M., Kaplan, S. L., et al.,** *Endocrinology (Baltimore)*, 124, 84, 1989.
245. **Blanchard, M. M., Goodyer, C. G., Charrier, J., et al.,** *Endocrinology (Baltimore)*, 122, 2114, 1988.
246. **Goodyer, C. G., Sellen, J. M., Fuks, M., et al.,** *Reprod. Nutr. Dev.*, 27(2B), 461, 1987.
247. **Blanchard, M. M., Goodyer, C. G., Charrier, J., et al.,** *Reprod. Nutr. Dev.*, 27(2B), 471, 1987.
248. **Morimoto, M., Kawakami, F., Makino, S., et al.,** *Neuroendocrinology*, 47, 459, 1988.
249. **Rieutort, M.,** *J. Endocrinol.*, 89, 355, 1981.
250. **Sawano, S., Kokubu, T., and Ohashi, S.,** *Endocrinol. Jpn.*, 23, 541, 1976.
251. **Richardson, S. B. and Twente, S.,** *J. Endocrinol.*, 138, 369, 1993.
252. **Login, I. S. and Judd, A. M.,** *Endocrinology (Baltimore)*, 119, 1703, 1986.
253. **Cowan, J. S., Moor, B., Chow, A., et al.,** *Endocrinology (Baltimore)*, 113, 1056, 1983.
254. **Sugihara, H., Minami, S., and Waskabayashi, I.,** *J. Endocrinol.*, 122, 583, 1989.
255. **Weiss, J., Cronin, M. J., and Thorner, M. O.,** *Am. J. Physiol.*, 253, E508, 1987.
256. **Cowan, J. S., Gaul, P., Moor, B. C., et al.,** *Can. J. Physiol. Pharmacol.*, 62, 199, 1984.
257. **Casaneuva, F. F., Popovic, V., Leal-Cerro, A., et al.,** *Molecular and Clinical Advances in Pituitary Disorders* [Ed. Melmed, S.], Endocrine Research and Education, Inc., Los Angeles, 1993, p. 145.
258. **Ho, P. J., Kletter, G. B., Hopwood, N. J., et al.,** *Acta Endocrinol. (Copenh.)*, 129, 414, 1993.
259. **Bryce, D., Yeh, M., Funderburk, C., et al.,** *Diabetes*, 24, 842, 1975.
260. **Padmanabhan, V., Enright, W. J., Zinn, S. A., et al.,** *Domest. Anim. Endocrinol.*, 4, 243, 1987.
261. **Le Dafniet, M., Garnier, P., Brandi, A. M., et al.,** *Horm. Res.*, 21, 235, 1985.
262. **Goth, M. I., Lyons, C. E., Canny, B. J., et al.,** *Endocrinology (Baltimore)*, 130, 939, 1992.
263. **Rivier, C., Rivier, J., and Vale, W.,** *Endocrinology (Baltimore)*, 102, 519, 1978.

264. **Goodyer, C. G., Branchaud, C. L., and Lefebvre, Y.,** *J. Clin. Endocrinol. Metab.*, 76, 1259, 1993.
265. **Szabo, M. and Frohman, L. A.,** *Endocrinology (Baltimore)*, 96, 955, 1975.
266. **Davies, R. R., Turner, S. J., Orskov, H., et al.,** *Clin. Endocrinol. (Oxford)*, 23, 271, 1985.
267. **Wehrenberg, W. B., Ling, N., Bohlen, P., et al.,** *Biochem. Biophys. Res. Commun.*, 109, 562, 1982.
268. **Anderson, L. L., Ford, J. J., Klindt, J., et al.,** *Proc. Soc. Exp. Biol. Med.*, 196, 194, 1991.
269. **Merola, B., Colao, A., Cataldi, M., et al.,** *Horm. Res.*, 37, 18, 1992.
270. **Siler, T. M., Vanderberg, G., Yen, S. S. C., et al.,** *J. Clin. Endocrinol. Metab.*, 37, 632, 1973.
271. **Davis, S. L.,** *J. Anim. Sci.*, 40, 911, 1975.
272. **Hall, R., Besser, G. M., Schally, A. V., et al.,** *Lancet*, 2, 581, 1973.
273. **Redekopp, C., Livesey, J., and Donald, R. A.,** *J. Endocrinol. Invest.*, 7, 277, 1984.
274. **Martin, J. B., Audet, J., and Saunders, A.,** *Endocrinology (Baltimore)*, 96, 839, 1974.
275. **Prange-Hansen, A., Orskow, H., Seyer-Hansen, K., et al.,** *Br. Med. J.*, 3, 522, 1973.
276. **Parker, D. C., Rossman, L. G., and Siler, T. M.,** *J. Clin. Endocrinol. Metab.*, 38, 496, 1974.

Chapter 9

Growth Hormone Release: Integrative Hypothalamic Control

S. Harvey

CONTENTS

0-8493-8697-7/95/$0.00+$.50

I. INTRODUCTION

Growth hormone (GH) release from the pituitary gland is generally not autonomous. Instead, its secretion appears to be regulated by a myriad of interacting, multihierarchical factors that dynamically modify somatotroph function at neural and humeral interfaces in neurocrine, neuroendocrine, endocrine, and paracrine ways and the influence of these factors may differ during ontogeny, sexual maturation, and senescence according to physiological state and extrinsic and intrinsic stimuli.

II. HYPOTHALAMIC CONTROL

The control of GH release is primarily mediated through the hypothalamus, which acts as a final common pathway for afferent information and exerts positive or negative tone through the differential release of GH-releasing and GH-inhibiting factors (see Muller and Nistico[1] for review). These aminergic and peptidergic factors are synthesized within hypothalamic neurons and are released in the median eminence into pituitary portal circulation and carried by long portal vessels to the adenohypophysis, where they interact with specific somatotroph receptors. The hypothalamus can also influence somatotroph function through neuroendocrine factors that are released in the neurohypophysis and transported to the adenohypophysis through short portal vessels.[1] The hypothalamus may additionally communicate with anterior pituitary somatotrophs by the release of neuroendocrines into cerebrospinal fluid (CSF) in the third ventricle, because the pituitary gland is bathed in CSF.[2,3] Hypothalamic (and extrahypothalamic) nuclei may also regulate somatotroph function through small tracts of fine peptidergic and aminergic fibers (including substance P, galanin, cholecystokinin [CCK], somatostatin [SRIF], calcitonin gene-related peptide [CGRP], serotonin [5-HT] and norepinephrine [NE] neurons) that directly penetrate the adenohypophysis (see Ju et al.[4] for review) and terminate on GH-secreting cells.[5–7] The activity of these hypothalamic hypophysiotropic neurons is regulated by central nervous system (CNS)-acting stimuli (e.g., stress, photoperiod, and behavior), neural oscillators, and by humeral feedback from peripheral factors (e.g., metabolites, cytokines and growth factors) and pituitary hormones, although pituitary responsiveness may be age and sex related and dependent on nutritional state and physiological condition.

Stimulatory and inhibitory neural information is transmitted to these hypophysiotropic neurons by a variety of amines (e.g., epinephrine [E], NE, dopamine [DA], 5-HT, histamine [H], γ-aminobutyric acid [GABA], and acetylcholine [ACh]), through a number of receptor types (e.g., α_1-, α_2-, and β-adrenergic receptors, D_1 and D_2 DA receptors, muscarinic and nicotinic acetylcholine receptors, 5-HT_1 and 5-HT_2 serotonin receptors, and H_1 and H_2 histaminergic receptors[1]), and by neuroexcitatory amino acids (e.g., glutamate and aspartate) through *N*-methyl-D-aspartate (NMDA) receptor systems and by a variety of neuropeptides (e.g., GH-releasing hormone [GHRH], SRIF, neurotensin [NT], substance P [SP], dynorphin, galanin, CCK, CGRP, calcitonin, endorphins, enkephalins, arginine vasopressin [AVP], oxytocin, corticotropin-releasing hormone [CRH], thyrotropin-releasing hormone [TRH], neuropeptide Y [NPY], gastrin, gastrin-releasing peptide [GRP], vasoactive intestinal peptide [VIP], and activin) and hormones (e.g., GH, glucocorticoids, estrogens, androgens, progestogens, thyroid hormones, and insulin[1]). Neural information may be transmitted directly to hypothalamic perikarya or to receptors located on nerve fibers or nerve terminals. Information transfer may also be indirect and mediated by one or more interneuron relays and may originate from outside the hypothalamus.[8,9] The induction of positive or negative hypothalamic drive, therefore, results from a summation of numerous agonistic and antagonistic messengers and provides a mechanism for subtle fine-tuning of somatotroph function. The influence of these neural signals on hypothalamic function may be dependent on multisynaptic or postsynaptic interactions and may be modulated by changes in neuronal sensitivity.

In addition to actions within the hypothalamus, many of these neural modulators of GHRH and SRIF secretion are also released into hypophysial portal circulation and directly regulate somatotroph function or moderate pituitary responsiveness to GHRH or SRIF stimulation. The hypothalamic control of GH release thus results from complex interactions at hypothalamic and pituitary sites by numerous stimulatory and inhibitory factors. In this chapter, the integration of these signals is reviewed, with particular emphasis on the control of GH secretion in humans and laboratory animals. Additional recent reviews on the neuroendocrine regulation of GH secretion include those by Mason et al.,[10] Devesa et al.,[11] Frohman et al.,[12] Casaneuva et al.,[13] and Muller.[14] The comparative regulation of GH has been reviewed by Harvey[18] and Buonomo and Baile[19] and is not considered in this chapter.

III. SRIF-GHRH INTERACTIONS

Although SRIF and GHRH inhibit or stimulate somatotrophs directly, SRIF also modulates the pituitary GH response to GHRH while GHRH modulates the GH response to SRIF. Both factors may also interact at hypothalamic sites to regulate heterologous release or gene expression.

Within the pituitary gland SRIF and GHRH interact to regulate GH release (see Lamberts and Oosterom[20] for review), principally through the dual antagonistic regulation of adenylate cyclase[21–23] and Ca^{2+} influx.[24] The balance between SRIF and GHRH is usually inhibitory (apart from during fetal and neonatal development) because SRIF (on a molar basis) has a tenfold greater potency than GHRH.[17,24] Antagonistic interactions with SRIF are, however, essential in promoting somatotroph responsiveness to GHRH[25–27] and vice versa. The GH response to GHRH can, therefore, be "paradoxically" enhanced in the face of high SRIF tone.[28] The GH response to GHRH is, accordingly, attenuated when endogenous SRIF tone is impaired by hypothalamic deafferentation, lesioning of the periventricular nucleus, or after passive SRIF immunoneutralization.[29] SRIF is thus thought to prevent somatotrophs from becoming refractory to GHRH stimulation.

The refractoriness of somatotrophs to GHRH probably results from an attenuation of cAMP production.[27,30,31] It is, however, generally considered that desensitization results from the uncoupling of the stimulatory G protein (G_s)[32] or from events distal to cAMP production[23] but is independent of Ca^{2+} uptake.[31] Clayton and Bailey, nevertheless, suggest desensitization may result from an impairment at steps beyond cAMP generation and extracellular Ca^{2+} influx.[26] GHRH-induced desensitization can thus be overcome by directly stimulating adenylate cyclase with forskolin or by exogenous cAMP mimetics.[31,33] The activation of the inhibitory G protein (G_i) by SRIF may, therefore, permit recoupling of the G_s protein with the GHRH receptor and restore somatotroph responsiveness to GHRH stimulation.[27] However, Simard et al.[27] suggest the ability of SRIF to prevent GHRH desensitization occurs independently of cAMP accumulation, because desensitized adenylate cyclase activity occurs in response to GHRH (or prostaglandins) in the presence of SRIF. The protective action of SRIF is, nevertheless, blocked by pertussis toxin, suggesting a role for a GTP-binding protein,[35] and may be linked to augmented influx of extracellular Ca^{2+}.[35] This resensitization action of SRIF appears to be directed exclusively at the release of GH and does not restore attenuated GH synthesis in GHRH-refractory somatotrophs.[25]

In the absence of GHRH the pituitary becomes desensitized to the inhibitory actions of SRIF.[36,37] This leads to a supersensitivity of adenylate cyclase to stimulatory agents, through a process requiring protein synthesis.[38] Because of these phenomena "rebound"

GH secretion occurs *in vivo* and *in vitro* after the withdrawal of SRIF inhibition[39,40] and the magnitude of this GH surge is related to the amount of prior or concurrent GHRH stimulation.[41–43] Consequently, in the absence of endogenous GHRH, rebound GH secretion does not occur in hypophysial stalk-transected animals[41,44] and is mostly diminished in GHRH-immunized rats.[45,46] The possibility that rebound secretion may be independent of GHRH was, however, proposed by Janowski and Wehrenberg,[47] because these authors observed rebound GH secretion following SRIF immunoneutralization in fasted rats immunized against GHRH and the magnitude of this response was comparable with that in controls not receiving GHRH antibodies.

In most studies, the magnitude of "rebound" GH secretion is also related to the duration and magnitude of prior GH suppression[48,49] and reflects the potency of SRIF analogs[50] and SRIF actions at hypothalamic and pituitary sites.[45] The *in vivo* withdrawal of SRIF during high GHRH tone therefore results in rebound secretion of greater magnitude than that induced solely by GHRH stimulation,[39,51,52] indicating an induction of endogenous GHRH release. The secretory bursts induced during SRIF withdrawal are, however, suppressible by exogenous SRIF.[39,52]

At the pituitary level, exposure of somatotrophs for 30 to 60 min to physiological, subnanomolar concentrations of SRIF increases subsequent basal and GHRH-induced GH secretion.[49,53–55] This postsomatostatin hypersecretion of GH is dependent on both Ca^{24} influx[56] and Ca^{2+} efflux.[55] The magnitude of the *in vitro* rebound of GH release is not, however, enhanced by prolonged SRIF exposure.[57] After SRIF withdrawal, the magnitude of the rebound GH release is also less than the magnitude of GH suppression during stimulation. It has, therefore, been suggested that rebound secretion cannot be explained solely by storage of intracellular GH during SRIF infusion and suggests the involvement of a process of GH degradation and/or an inhibition of GH synthesis.[57] The GH response of somatotrophs to GHRH is, therefore, dependent on their previous, concurrent, or future exposure to SRIF.[58] The temporal patterning of hypothalamic GHRH/SRIF signals to pituitary somatotrophs may thus be a major determinant for dictating the frequency and amplitude of pulsatile GH release.[28]

The acute withdrawal of endogenous SRIF is also likely to induce pulsatile GH release and to augment GH secretion by triggering the phasic discharge of GHRH by the hypothalamus. This possibility is indicated by the acute suppression of the phasic GH release induced by SRIF immunoneutralization in GHRH-immunoneutralized rats.[59] An interaction between SRIF and GHRH in central sites is also indicated by the sustained increase of immunoreactive GHRH secretion into hypophysial portal blood following intracerebroventricular (i.c.v.) administration of SRIF antibodies.[60] The release of GHRH into hypophysial portal plasma is also reduced in sheep by long-acting SRIF analogs,[61] which similarly lower peripheral plasma GHRH levels in humans.[62] SRIF has also been found to reduce directly the release of GHRH from cultured rat hypothalamic fragments.[63] These analogs have no effect, however, on the metabolism of GHRH in plasma.[42] It is, therefore, likely that SRIF exerts tonic inhibitory control over GHRH secretion, and this possibility is supported by the stimulation of GHRH release *in vivo* and *in vitro* following cysteamine depletion of hypothalamic SRIF.[64] Cysteamine similarly increases the number (by 126%) and intensity of GHRH-immunoreactive neurons in the arcuate nucleus, particularly within the caudal tier, and in the median eminence.[65] Cysteamine depletion also increases transcription of the *GHRH* gene in the arcuate nucleus.[66] These GHRH responses may, however, indicate suppressed GHRH release, because cysteamine treatment lowers circulating GH levels and inhibits *GHRH* gene transcription in the arcuate nucleus,[65] suggesting a stimulatory effect of SRIF on GHRH neurons.

The stimulatory effect of SRIF immunoneutralization on GH secretion[67,69] and the increased GH secretion in rats following i.c.v. SRIF administration[69–71] have also been attributed to increased GHRH secretion, being suppressed by anti-GHRH serum. The participation of GHRH in the GH response to SRIF is also indicated by its absence in rats in which the ventromedial arcuate nucleus is specifically destroyed by monosodium glutamate.[71] This central action of SRIF on GHRH release is mediated by α-adrenergic and GABAergic pathways but not by opiate mechanisms.[72] These central actions of SRIF may, however, reflect ultrashort-loop inhibitory feedback effects of SRIF on the regulation of SRIF neurons.[73–76] Indeed, whereas injections of SRIF around GHRH neurons in the medial basal hypothalamus have no significant effect on GH secretion, SRIF injections among SRIF neurons in the preoptic-anterior hypothalamus stimulate GH release.[77] Increased GHRH release induced by SRIF is, therefore, likely to reflect an inhibition of endogenous SRIF and the distribution of GHRH neurons.[59] A tonic inhibitory effect of SRIF on GHRH neurons is supported by the increased *in vitro* release of GHRH from the SRIF-depleted hypothalami of rats with anterolateral hypothalamic deafferentations or electrical lesions of the medial preoptic area.[78] The elevated plasma GH concentrations in these animals are, concordantly, suppressed by GHRH immunization.[78]

The possible interaction of SRIF with GHRH neurons is supported by the demonstration of high-affinity SRIF receptors on GHRH perikarya in the arcuate nucleus.[71,79,80] The number of these receptors is increased following cysteamine administration, indicating an upregulation in the absence of SRIF.[66] These receptors are likely to be activated by neurocrine or paracrine mechanisms, because SRIF fibers (possibly from local SRIF neurons) innervate GHRH neurons in the arcuate nucleus.[81–83] The nerve terminals of these SRIF neurons synapse with both the cell bodies and dendrites of GHRH neurons.[84] The communication between SRIF and GHRH neurons also appears to be bidirectional, because GHRH fibers closely approach and synapse with SRIF neurons in the periventricular preoptic hypothalamus.[81,84] These synapses are thus likely to mediate the stimulatory actions of GHRH on *SRIF* gene transcription,[85] content,[86,87] release,[73,88,89] and pituitary portal blood concentrations.[90] This interaction between GHRH and SRIF neurons is also indicated by the reduced SRIF immunoreactivity in the median eminence of rats with arcuate nucleus lesions.[87] These effects are thus likely to mediate the inhibitory effects of i.c.v. GHRH on GH secretion.[91] These effects are also likely to reflect the tonic actions of locally produced GHRH on SRIF neurons, because GHRH antibodies block SRIF synthesis and release in fetal hypothalamic cultures.[86] These actions may also be mediated by β-endorphin interneurons,[92] because GHRH and β-endorphin antibodies suppress basal GRIF release by pathways that appear to require calmodulin or arachidonic acid activation and the mobilization of intracellular Ca^{2+} stores.[73,89,92] The action of GHRH on SRIF release is not, however, mediated by dopaminergic, adrenergic, or cholinergic pathways.[88,92]

IV. NONSPECIFIC GH-RELEASING FACTORS

In addition to GHRH and SRIF, the release of pituitary GH is also induced or suppressed by a variety of other hypothalamic peptides or amines with GH-releasing or GH-inhibiting activity. These factors may be relatively "nonspecific" in that they probably affect the secretion of most if not all adenohypophysial hormones at pituitary or hypothalamic sites. These factors may also nonspecifically regulate GH release by vasoactive actions on pituitary portal vasculature,[93] and thereby regulate transportation of hypothalamic signals in portal plasma. These factors may, however, have physiological relevance, because GH secretion is incompletely suppressed in GHRH- and SRIF-immunized rats.[67] It is also well established that GHRH may account for only 60 to 70% or less of the pulsatile surges in GH release.[94,95] Indeed, in adolescents with short stature the nocturnal peaks of GH secretion are unrelated to plasma GHRH levels, whereas increased plasma GHRH precedes stress-induced GH release.[96] The participation of non-GHRH-dependent mechanisms in the stimulation of GH secretion is also indicated by the GH responses to α-adrenergic stimulation (propranolol, epinephrine, or clonidine infusions) during periods of somatotroph refractoriness to GHRH stimulation.[97] The separate and additive effects of GHRH and insulin hypoglycemia on GH release also suggest the participation of different pathways in GH stimulation.[98,99]

In addition to GHRH, neuropeptides that stimulate GH release directly include TRH, VIP, bombesin, CCK, peptide histidine isoleucine (PHI), substance P (SP), angiotensin II (AII), galanin, motilin, glucagon, gonadotropin-releasing hormone (GnRH), delta sleep-inducing peptide (DSIP), NPY, atrial natriuretic factor (ANF), α-melanocyte-stimulating hormone (α–MSH), adrenocorticotropic hormone (ACTH), AVP, and NT, whereas GH-inhibitory factors include secretin, GRP, gastrin-inhibiting peptide (GIP), pancreatic polypeptide, calcitonin, CRH, interleukin 1(IL-1), insulin, insulin-like growth factor I(IGF-I), IGF-II, galanin, NT, and NPY.

Many of these hypothalamic neuropeptides also interact centrally to regulate GH secretion and they induce actions that are complementary or antagonistic to their pituitary actions. For instance, NPY, galanin, TRH, ANF, calcitonin, GRP, substance P, and opioid peptides modulate the activity of SRIF neurons, whereas GRP, CRH, NPY, galanin, substance P, NT, and glucagon modulate GHRH neurons.[100,101] Some of these peptides (e.g., galanin) may be colocalized with GRF or SRIF in secretory granules[102] and exert pre- or postsynaptic effects on the activity of GHRH and SRIF neurons,[103,104] whereas others (e.g., TRH[105] and substance P[106]) may regulate GHRH or SRIF neuronal activity by axosomatic or axodendritic contacts. Some of these GH-releasing factors also have pituitary sites of action, such as bombesin, opioids, motilin, CCK, VIP, PHI, IL-1, and AII,[101] as do IGF-I and activin A,[107,108] which are GH-inhibitory release factors. Some of these actions result from interactions in the pituitary gland with GHRH and SRIF, as a result of heterologous receptor regulation or cross-talk between the intracellular messengers involved. For instance, TRH modulates SRIF receptors[109] and TRH,[110] AVP, and PHI[111] interact synergistically with GHRH at postreceptor sites. The inhibition of GH secretion induced by SRIF is similarly antagonized by VIP, TRH, and secretin.[112] These interactions reflect the pleuri-regulation of the adenylate cyclase-protein kinase A system (e.g., by GHRH, AVP, and PHI)[110] and the additive, synergistic, or antagonistic effects of dissimilar mediators (e.g., calcium and the protein kinase C system).

Thus, under physiological conditions the differential regulation of pituitary function is probably related to differences in the abundance or affinity of neuropeptides and neuropeptide receptors or receptor subtypes, the geographical location of specific pituitary cell types, paracrine relationships between cells, regionalized hypophysial blood flow, and differential modulation by neurotransmitters and feedback stimuli. The actions of some of these neurotransmitters and neuropeptides in this complex control system are reviewed below, although a detailed consideration of their actions lies outside the scope of this chapter. The actions of some of these factors have also been reviewed elsewhere.[15–17,101,113,114]

A. GHRH-LIKE PEPTIDES

Neuropeptides that structurally resemble GHRH are likely to have direct GH-releasing activity, which may be mediated through specific pituitary receptors or through interactions with the GHRH receptor,[115] despite their lower affinity.[116] These peptides include members of the VIP family (e.g., VIP, peptide histidine methionine [PHM], PHI, secretin, CRH, gastrin-inhibiting peptide, GIP, and glucagon) (see Said[117] for review) and PACAP (pituitary adenylate cyclase-activating polypeptide), which has 68% homology with VIP in its 28 amino-terminal amino acids.

1. VIP/PHI/PHM

The VIP transcript encodes two peptides, VIP and peptide histidine isoleucine (PHI), in rats and peptide histidine methionine (PHM) in humans that are present in neurons terminating in the median eminence.[118,119] Both peptides are present in pituitary portal plasma[118] and have adenylate cyclase-linked pituitary receptors[115,117] in rats[120,121] and in acromegalic patients.[122–124] VIP stimulates GH release *in vivo* in acromegalics but not in normal individuals.[122] The *in vitro* release of GH from human pituitary tumors,[125,126] goat pituitaries,[127] and bovine pituitaries[128] and from rat somatomammotrophs[129] and normal[120,121,130] and glucocorticoid-sensitized somatotrophs[131] is also increased by VIP. This stimulation of GH release is dose related and accompanies increased cAMP accumulation but is less than that induced by GHRH. VIP may also stimulate GH release by actions within the hypothalamus[132] possibly mediated by a reduction in SRIF release.[133] PHI and PHM are also able to provoke GH release *in vivo*[130] and *in vitro*,[124,134] although they are less effective than VIP.[134]

2. Secretin

Although secretin has sequence homology with the amino acid sequence of GHRH, it has little (if any) stimulating effect on GH release from rat pituitary glands.[130] It may, however, act as a GHRH receptor antagonist or act within the CNS to inhibit basal and GHRH-induced GH release *in vivo*.[135] Secretin does, however, antagonize the inhibition of GH secretion induced by SRIF.[112]

3. Glucagon

Although glucagon induces hyperglycemia, it stimulates GH release in humans,[136] prior to the subsequent decline in blood glucose concentrations. In contrast, glucagon inhibits basal GH secretion in rats,[135] by a central mechanism that stimulates SRIF release.[137,138]

4. Corticotropin-Releasing Hormone

In rats, CRH inhibits GH secretion at central[139] sites by increasing SRIF release,[140] although CRH does not directly inhibit GH release from the rat pituitary.[141] In contrast, CRH enhances GH secretion in acromegalic patients[142] and GH release from adenomas,[126,143] although CRH- antagonism of GHRH-induced GH release has also been observed in normal subjects and patients with acromegaly.[144–146] Increased GH secretion during stress[147] is thus unlikely to be mediated by CRH.

5. Pituitary Adenylate Cyclase-Activating Polypeptide

PACAP peptides of 27 or 38 amino acids are synthesized in hypothalamic neurons that terminate in the median eminence[148,149] and interact with hypothalamic[150] and pituitary[151] binding sites, for which $PACAP_{38}$ has greater affinity than $PACAP_{27}$. These peptides directly stimulate GH release from rat pituitary glands,[152] although they are less effective than GHRH and also stimulate the release of other pituitary hormones. The stimulatory effect of PACAP includes an increase in the number of GH-secreting cells and the amount of GH secreted per cell.[153]

6. Gastric Inhibitory Peptide

Gastric inhibitory polypeptide has yet to be localized in the nervous system, but it is similar in structure to VIP and PHI and modulates GH secretion. Although GIP does not directly affect GH release from pituitary cells, it acts centrally to stimulate GH secretion,[185] despite inhibiting basal and GHRH-induced GH release following systemic administration.[135]

B. GHRH-UNRELATED PEPTIDES

1. Thyrotropin-Releasing Hormone

Thyrotropin-releasing hormone not only stimulates pituitary TSH secretion but the release of GH in all vertebrate groups (see Harvey[154] for detailed review). The GH-releasing activity of TRH in humans is generally considered to be a paradoxical response associated with pathological conditions of GH dysfunction. In these conditions, TRH receptors may be abnormally expressed in somatotrophs and may reflect the induction of somatolactotroph cells. An impairment of the inhibitory hypothalamic control of GH secretion has also been suggested to be at least partially responsible for anomalous GH responses to TRH. Pituitary somatotrophs are responsive to TRH stimulation *in vitro* and although the somatotrophs of in situ pituitary glands may be TRH resistant, ectopically transplanted pituitary glands are readily responsive. The *in vivo* GH response to TRH stimulation is similarly potentiated by SRIF immunoneutralization or electrical lesioning of SRIF-containing hypothalamic nuclei and is most pronounced in fetal pituitaries that are insensitive to SRIF inhibition. "Paradoxical" GH responses to TRH may also occur in GHRH-sensitized somatotrophs, in which heterologous receptor or postreceptor modifications potentiate GH responsiveness to TRH and promote synergistic interactions with GHRH signal transduction pathways. GH responsiveness to TRH may also occur as a result of a reduction in thyroid hormone secretion, which permits an upregulation of TRH receptors. In addition to stimulating GH release directly, TRH may also regulate GH release by enhancing serum GHRH levels.[157] However, whereas TRH stimulates the GH-releasing activity of the avian hypothalamus it appears to stimulate inhibitory hypothalamic control in mammalian species and i.c.v. injections of TRH inhibit GH release.[158]

Although the somatotroph response has been considered to be paradoxical, TRH is a physiological regulator of GH secretion in birds[154] and may also be a physiological regulator of GH release in mammals. This possibility is supported by the suppression of GH secretion in TRH-immunized rats.[155] Moreover, in neonatal rats the increased GH secretion induced by milk and nursing is thought to be due to a milk-borne GH-releasing factor (GRF) indistinguishable from TRH, indicating that TRH is a novel physiological GRF.[156]

2. Luteinizing Hormone-Releasing Hormone

Luteinizing hormone-releasing hormone (LHRH) is a potent GHRH in teleosts, in which it may be physiologically involved in controlling GH release (see Harvey[159] for review). Somatotroph secretory responses to LHRH have also been observed in rats[160] and in humans, particularly in depressed patients[161] and acromegalics.[142,162,163] Inhibitory effects of LHRH on GH secretion have, however, also been reported. Robberecht et al.[164] observed that LHRH displayed a transient inhibitory effect on GH release from aggregated pituitary cells, which was followed by a rebound of GH secretion after withdrawal of the peptide.

3. Angiotensin II

Renin-angiotensin systems exist in the brain and pituitary gland,[165] in which AII receptors have also been found. Although LHRH stimulates the release of AII from gonadotrophs, the stimulatory effects of LHRH on GH release persist in the presence of AII receptor antagonism.[164] Pituitary somatotrophs do, however, directly release GH in response to AII stimulation.[164] Plasma GH concentrations in humans are also elevated by systemic AII injections.[166] Central effects of AII have also been observed, through which AII inhibits plasma GH concentrations in rats, even when GH secretion is already suppressed by dopaminergic antagonism.[167]

4. Bombesin

Bombesin, a tetradecapeptide, is widely distributed throughout the CNS and is released from hypothalamic nerve endings.[168] A hypophysiotropic role for bombesin is indicated by the stimulation of GH release in steroid-primed rats following its systemic injection,[169] although it appears to be without effect in humans.[170,171] Bombesin also stimulates GH release from bovine pituitary cells in a manner that is additive with acetylcholine[172] and synergistic with GHRH.[110] Bombesin-induced GH release may be mediated through a pathway activated by TRH and, like TRH, this appears to involve activation of phospholipase C.[173] Bombesin also stimulates GH release from tumorous rat somatomammotrophs.[174] In addition to bombesin, its analog (ranatensin) and related peptides (e.g., neuromedin C and neuromedin B30) also stimulate GH release from rat pituitary cell aggregates, by mechanisms that are potentiated by estradiol, 5α-dihydrotestosterone, and glucocorticoids.[175] In contrast to these stimulatory actions, bombesin inhibits basal GH secretion in rats when injected intracerebroventricularly[176–178] and reduces GH responses induced by opiate stimulation, monoamine antagonism, or prostaglandin E_2.[178] This has been shown to result from an action within the hypothalamus that stimulates SRIF release into pituitary portal circulation.[179]

5. Gastrin-Releasing Peptide

Gastrin-releasing peptide has been located in the rat hypothalamus and its C-terminal portion has sequence homology with bombesin. Moreover, like bombesin, i.c.v. GRP inhibits basal, pulsatile, and GHRH-induced GH secretion,[180–182] by a mechanism that is blocked by SRIF depletion or SRIF immunoneutralization[180,181] and is primarily due to activities of the C-terminal fragment.[183] This mechanism involves the stimulation of SRIF release[180] and is unrelated to GHRH actions at the level of the pituitary.[180]

6. Gastrin

Gastrin-like immunoreactivity has been located throughout the CNS-hypothalamus-pituitary axis and gastrin has been shown to modulate pituitary function. Systemic injections of gastrin are, however, ineffective in regulating GH release in rats. Pentagastrin is also ineffective in women, although it elevates GH secretion in men.[184]

7. Cholecystokinin

Cholecystokinin is present in high concentrations in the pituitary, the median eminence, and hypothalamus[1] and CCK receptors are present in brain[186] and pituitary membranes.[186,187] The possibility that CCK may be a regulator of GH secretion is indicated by the dose-related rise in serum GH levels following its i.c.v. administration in ovariectomized rats[188] and by its stimulation of GH release from pituitary glands *in vitro*.[189,190] However, Malarkey et al.[191] found no effect of CCK on GH release from rat and human pituitary cells and Karashima et al.[178] found i.c.v. CCK completely suppressed GH response to chlorpromazine and prostaglandin E_2. The influence of CCK on GH secretion is therefore unclear.

8. Neurotensin

Neurotensin has been located in the para- and periventricular nuclei of the hypothalamus and in the median eminence and pituitary gland.[14] Although systemic NT injections do not elevate GH secretion in gonadectomized rats,[192,193] GH secretion is enhanced in normal estrogen-progesterone-primed rats.[194–196] This GH response is not modified by SRIF or GHRH immunoneutralization,[197] indicating that it occurs independently of SRIF or GHRH release and is not directly mediated at pituitary sites.[198] The blockade of NT-induced GH secretion by diphenhydramine, a histamine antagonist, suggests its effect is mediated through a histaminergic mechanism.[194]

9. Substance P

Substance P-containing neurons and cells are located in the hypothalamus and pituitary gland,[14] as are SP-binding sites. Roles for SP in GH regulation are indicated by stimulatory and inhibitory effects on GH release. Although at physiological concentrations SP and other tachykinins (neurokinin A and neurokinin B) do not appear to stimulate GH release directly from rat pituitary cells,[199] nonspecific SP receptor antagonists inhibit GH release at molar concentrations and stimulate release at higher concentrations.[200] This effect may, however, result from agonism or antagonism of the receptors for related peptides (e.g., bombesin), because more specific SP receptor antagonists are without effect.[200] In marked contrast, central or systemic injections of SP lower circulating GH concentrations,[199] whereas central actions of SP antagonists elevate GH release within 5 min of injection.[199] Substance P may, therefore, have a physiological role as an inhibitory regulator of GH release in the rat, especially as SP immunoneutralization promptly elevates plasma GH concentrations.[199]

In humans, the systemic infusion of SP has been observed to stimulate GH secretion by a mechanism that greatly augments maximal GH responses to GHRH stimulation.[201] Stimulatory, unexplained effects of SP on *in vivo* GH secretion have, however, also been reported in rats.[194,202]

10. Motilin

A hypophysiotropic role for motilin is indicated by its high concentration in the ventromedial hypothalamus, arcuate nucleus, and median eminence.[14] This 22-amino acid peptide stimulates the *in vitro* release of GH from rat pituitary cells at nanomolar concentrations.[203,204] Plasma GH concentrations are also elevated by systemic injections of motilin, at high (100 μg/rat) dose levels. In contrast, infusions of motilin into the third ventricle of the brain depress GH secretion, possibly in an ultrashort-loop feedback way or via increased SRIF release.[204] The possibility that motilin may be a physiological GH-releasing factor is supported by the depressed plasma GH concentrations following motilin immunoneutralization.[204]

11. Galanin

Galanin-like immunoreactivity is richly present in cell bodies in the arcuate, paraventricular, and supraoptic nuclei of the hypothalamus, from which immunoreactive fibers that contain GHRH[205] terminate in the median eminence. The hypothalamic concentration of galanin and galanin mRNA is sexually dimorphic and greatest in females.[206] The release of galanin into portal circulation is indicated by its elevated concentrations.[207–209] Galanin is also found in anterior pituitary cells within somatotroph secretory granules[210] and is released in response to TRH, but not GHRH.[210,211] Galanin-binding sites are also present in the rat hypothalamus and median eminence,[212] suggesting roles at hypothalamic and pituitary sites. The actions of galanin on GH secretion are, however, complex.

At the pituitary level, galanin has been reported to have no effect on GH release from rat pituitary cells[213–215] and to lack pituitary-binding sites,[214] although stimulatory effects have been observed by others.[216–218] Lindstrom and Savendahl[219] found that galanin rapidly stimulates GH release from rat somatotrophs, by a cAMP, K^+-independent mechanism linked to increased Ca^{2+} efflux.[218] This mechanism is thus coupled to intracellular Ca^{2+} mobilization and not membrane depolarization and is dissimilar to GHRH-induced GH release, which is reduced in rats in the presence of galanin,[213,217,219] but not in humans.[214]

Circulating GH concentrations are elevated after systemic galanin administration, although it is much less potent than GHRH, with which it synergistically interacts.[214,220–223] Galanin does not, however, interact synergistically with GHRH in hyperparathyroid patients, in which somatostatin tone may be enhanced.[224] The GH response to galanin is also reduced in glucocorticoid-treated patients, because glucocorticoids increase SRIF tone and decrease GHRH release.[234] Galanin is also unable to enhance GH responsiveness to GHRH in acromegalic patients, possibly because of anomalous receptors in tumorous somatotrophs.[225,226] GH responsiveness to galanin is also greater in females than in males and is estrogen dependent and declines with aging, unlike the GH response in males.[235]

Although intravenous injections of galanin elevate circulating GH levels, the stimulation of GH release is far less than that induced by i.c.v. injections,[227] suggesting hypothalamic loci are the principal target sites of galanin action. Because galanin was found not to modify SRIF release from rat hypothalamic fragments, Hulting et al.[214] suggest this stimulation reflects an action mediated via a stimulatory effect on GHRH neurons, as demonstrated by Kitajima et al.[228] The i.c.v. administration of GHRH antiserum is thus able to block galanin-induced GH secretion.[229,230] This induction of GHRH release by galanin is thought to be mediated by catecholaminergic interneurons,[231] because inhibition of epinephrine or norepinephrine synthesis completely blocks the GH response.[231] However, Delemarre-van de Waal et al.[232] found galanin mRNA in GHRH neurons of the rat hypothalamus, suggesting direct autocrine or intracrine actions on GHRH release. The possibility that galanin is a physiological regulator of GHRH-induced GH release in the rat is indicated by the alterations in pulsatile GH secretion (reduced pulse height and increased pulse frequency) following galanin immunoneutralization within the brain.[231,233]

Galanin may additionally be a regulator of hypothalamic SRIF, because the GH response of rats to i.c.v. galanin is blunted rather than enhanced in rats pretreated with neostigmine, cysteamine, or SRIF antibodies to deplete SRIF tone.[236] The GH response to galanin thus differs from GHRH-induced GH secretion, which is potentiated by SRIF depletion.[236] Tanoh et al.[236] therefore suggest that the stimulation of GHRH release induced by galanin is not direct but mediated through an inhibition of SRIF release and a disinhibition of GHRH neurons. This mechanism is also thought to be responsible for the stimulatory effect of insulin-induced hypoglycemia on GH secretion, because this is mediated through a mechanism partly independent of GHRH.[98] The blunted GH response to galanin after arginine administration[234] is similarly thought to indicate the participation of SRIF disinhibition in its mechanism of action, because arginine is an inhibitor of SRIF release.[235] The ability of galanin to amplify GH responses in humans to maximal doses of GHRH,[205,237] to reinstate the GH response to repeated GHRH administration,[222] and to abolish the inhibitory effect of cholinergic blockade on GHRH-induced GH secretion[238] is further evidence that SRIF inhibition is involved in galanin-induced GH secretion. This possibility is also supported by the potentiation of the GH response to GHRH and galanin administration in the presence of cholinergic agonism, which inhibits SRIF secretion.[239]

12. Neuropeptide Y

A hypophysiotropic role for NPY in GH regulation is indicated by its wide distribution in the CNS and concentration in the hypothalamus.[14,240] It is present in neurons of the arcuate nucleus in fibers that project to the median eminence, where NPY terminals are in close juxtaposition with hypophyseal portal capillaries, into which it is released.[240–242] NPY is also present in fibers that project to the preoptic hypothalamus.[243] Although NPY directly inhibits basal and GHRH-induced GH secretion by human pituitary somatotropic tumors,[244] it induces a dose-related increase in GH release from rat[245] and goldfish[246] pituitary cells. This action contrasts, however, with the inhibitory effect of intravenous NPY injections on plasma GH concentrations,[245] suggesting actions at hypothalamic sites opposite to those at the pituitary. This possibility is supported by the inhibition of circulating GH concentrations in intact and ovariectomized female rats following i.c.v. NPY administration.[245,247] Because NPY is localized in hypothalamic nuclei regulating GHRH and SRIF secretion, this inhibitory effect may result from a reduction in GHRH release or increase in SRIF secretion, or both. Evidence for SRIF mediation is demonstrated by the elevation in circulating GH levels after the injection of NPY antibodies into the third ventricle of intact male and ovariectomized female rats.[248] NPY is, moreover, able to stimulate SRIF release directly from median eminence fragments, although the dose-response curve is biphasic and reduced at high concentrations.[248] The inhibitory central effects of avian and bovine pancreatic polypeptides (APP and BPP) on GH release[249] are thus likely to reflect the actions of endogenous NPY, which closely resembles APP and BPP. The SRIF-releasing action of NPY is blocked by α_1- and β-adrenergic receptor blockade but is unaffected by α_2-adrenergic receptor blockade.[248] The possibility that NPY also acts by directly increasing hypothalamic GH-releasing activity is indicated by its stimulation through Y_2 receptors of gonadotropin-releasing hormone release in goldfish.[246]

13. Vasopressin and Oxytocin

Although Brattleboro rats with a genetic AVP deficiency have normal plasma GH levels and are normally responsive to hypothalamic stimulation,[14] AVP and oxytocin (OT) are present in hypothalamic nuclei and in the median eminence and in long and short hypophysial portal vessels and may participate in GH regulation.[250] However, whereas stimulatory effects of AVP on GH release have been observed in primates,[251,252] injections of AVP antibodies into the third ventricle of rats have no effect on basal GH secretion.[250] A physiological role for OT as an inhibitory GH-releasing factor is, however, suggested by the sustained increase in circulating GH levels following i.c.v. administration of OT antibodies.[250]

14. Adrenocorticotropin and Melanocyte-Stimulating Hormone

Expression of the proopiomelanocorticotropin (*POMC*) gene generates a number of peptides, including ACTH, α-MSH, and β-MSH. Because this gene is expressed throughout the CNS and in pituitary cells, POMC peptides might have roles in GH regulation, because both α-MSH[253–255] and ACTH[253,254] stimulate GH release in humans. This effect of ACTH results from a reduction in SRIF tone,[256] although direct effects of α-MSH on GH release from rat pituitary cells have been observed.[254]

15. Endogenous Opioid Peptides

Numerous peptides (e.g., α- and β-endorphins, derived from POMC; Leu- and Met-enkephalin, and dynorphin, derived from preenkephalin A; and dynorphin and α-neoendorphin, derived from preenkephalin B) that act through opiate receptors are widely distributed in the hypothalamus, where they probably act as interneuron messengers and hypophysiotropic regulators.[257,258] The actions of these peptides are mediated through at least six opioid receptor types (μ_1, μ_2, δ, ε, κ, and σ)[257] at hypothalamic and pituitary sites.

The participation of opiates in GH regulation is indicated by the SRIF-sensitive increase in plasma GH levels in rats following intravenous injections of morphine,[259–262] β-endorphin,[263] κ agonists,[264] and enkephalin analogs,[14,265] by effects attenuated by non-selective opiate receptor blockade.[259,263–265] This stimulatory action appears to be specifically mediated through receptors.[194,266,267] Enkephalins also enhance basal GH secretion in humans, through their higher affinity for receptors, although morphine and β-endorphin (both of which preferentially bind μ receptors) appear to be ineffective.[14,268,269] At high dose levels inhibitory effects of κ agonists on GH secretion have, however, been observed.[266,270] These effects have been thought to be largely mediated by actions at hypothalamic sites, because numerous studies have demonstrated that opiates have no direct effects on pituitary GH release.[194,263]

Hypothalamic actions of opiates are indicated by the stimulation of GH secretion following hypothalamic, intracisternal, and intracerebroventricular injection[263,265,271–274] and the antagonism of these actions by the blockade of central opiate receptors.[267,274] Hypothalamic mediation of these actions is also indicated by the inability of opiates to provoke GH release in rats with mechanical deafferentation of the medial basal hypothalamus.[262,275] Extensive pharmacological investigations have subsequently provided evidence for NMDA, cholinergic, histaminergic H_1, and nonadrenergic (via α_2-adrenoreceptors) pathways in mediating opiate-induced GH release.[14,265,272,276] These pathways may induce EOP activity, which, in turn, is likely to antagonize SRIF release or action[277] or to stimulate the release of GHRH. Although the GH response to met-enkephalin is unaffected by SRIF depletion, it is blunted in GHRH-immunized rats.[261] The GH response to morphine and β-endorphin is similarly reduced by GHRH immunoneutralization.[278] Opioids may, in contrast, stimulate GH release in humans independently of GHRH, because a met-enkephalin analog has additive or synergistic effects with a maximally stimulatory dose of GHRH on GH release,[279] suggesting an inhibition of SRIF release may be a causal mechanism in humans.

Endogenous opiates may not normally have physiological importance in GH regulation in humans, because opiate antagonism does not alter basal GH secretion[280] or alter nocturnal, sleep-associated GH release or GH responses to insulin-induced hypoglycemia[14,281] or exercise.[282] However, as opiate receptor antagonism inhibits pain-induced GH secretion[283] the release of opiates during stress may have physiological relevance. Opiates may also have a physiological role in mediating the feedback effects of GH on GHRH neurons.[284]

Roles for opiates in mediating GH responses to stress have also been reported in goats[285,286] and rats.[260,287,288] Opiate antagonism or enkephalin antibodies may not lower mean basal plasma GH levels in rats,[289,290] although Simpkins et al.[260] reported that naloxone lowered basal GH concentrations in female rats by 64%, without affecting GH pulse amplitude or pulse frequency. The frequency and amplitude of episodic GH release in hamsters is, however, reduced during opiate receptor blockade.[277] Opiates may also be physiologically involved in suckling-induced GH release, because this is blocked in postpartum rats by opiate receptor antagonism.[291]

16. Activin

Activin is a member of the transforming growth factor family of peptides and activin gene expression occurs in the pituitary gland and brain. It is, therefore, possible that activin may participate in GH regulation *in vivo*. This possibility is supported by the ability of activin A to inhibit *GH* gene expression and the proliferation of rat somatotrophs.[107] In contrast, activin A stimulates basal GH secretion from the cells of some, but not all, human somatotrophinomas *in vitro*. Pretreatment with the peptide may also partially block GHRH-stimulated GH release from GHRH-responsive somatotrophinoma cells.[292]

17. Phe-Met-Arg-Phe-amide (FMRF Amide)

The neuropeptide FMRF amide has been immunologically detected in the mammalian hypothalamus and has excitatory and inhibitory effects on neurons. The possibility that this peptide may participate in GH regulation is indicated by the increased plasma GH levels in rats following its injection into the third ventricle.[293] This central effect may reflect increased GHRH stimulation on SRIF inhibition but is unlikely to result from a direct pituitary action, because FMRF amide is ineffective when injected intravenously.

18. Delta Sleep-Inducing Peptide

Delta sleep-inducing peptide immunoreactivity is present in the rat brain[294] and this nonapeptide may regulate GH secretion. Increased GH concentrations occur in the plasma of rats between 30 and 120 min after i.c.v. DSIP administration.[295] This effect is thought to be mediated through the hypothalamus via a dopaminergic mechanism, because it is blocked by dopamine receptor antagonism. This mechanism results in a dose-related suppression of hypothalamic SRIF release.[296] DSIP may additionally stimulate GH secretion directly, because it induces GH release from incubated rat pituitary cells. This stimulation of *in vitro* GH release is, however, observed with low picomolar doses and is not observed at micromolar concentrations.[295]

19. Calcitonin

Calcitonin-like immunoreactivity has been detected in the brain with high concentrations in the posterior hypothalamus, median eminence, and pituitary gland.[297] Specific calcitonin-binding sites also occur in the brain and pituitary gland.[297] A hypophysiotropic role of calcitonin in human GH secretion is indicated by its suppression of arginine-,[298] hypoglycemia-,[299,300] protein-,[301] and GHRH-induced[299,302] GH secretion after systemic administration, which has also been reported to reduce basal plasma GH concentrations[303] and suppress pulsatile GH release.[299] Calcitonin similarly suppresses spontaneous and stimulated GH release in rats.[304,305] These inhibitory actions are likely to be mediated through the hypothalamus, because calcitonin does not directly inhibit pituitary GH release[302] although a prolonged loss of responsiveness of somatotrophs to GHRH occurs after its i.c.v. injection. This loss of somatotroph responsiveness is independent of calcitonin effects on blood calcium levels.[305] The mechanism may not be mediated by increased SRIF release, because the actions of calcitonin are not restored by SRIF immunoneutralization[305] and calcitonin does not directly induce SRIF secretion from hypothalamic fragments.[305]

20. Calcitonin Gene-Related Peptide

CGRP results from alternate processing of the calcitonin gene transcript in neural tissues,[306] in which CGRP immunoreactivity and binding sites occur in hypothalamic and extrahypothalamic areas of the brain. These binding sites differ from those of calcitonin. A role for CGRP in GH regulation is indicated by its inhibition of spontaneous GH secretory episodes following its i.c.v. administration and by the elevation in plasma GH concentrations after the i.c.v. administration of CGRP antibodies.[307]

The intracerebroventricular administration of CGRP has also been found to suppress β-endorphin, morphine, clonidine, and GHRH-induced GH release [308]. Although CGRP increases SRIF release,[309] the possibility that these actions are mediated by SRIF is uncertain, because SRIF depletion does not modify the inhibitory effect of CGRP on the GH response to GHRH.[308] This inhibitory action of CGRP is, however, unlikely to be mediated by direct effects on GH cells.[307,308] A physiological role of CGRP in GH regulation is indicated by the transient elevation in plasma GH following the injection of CGRP antibodies into the third ventricle.[307]

C. GH-RELEASING NEUROTRANSMITTERS

Somatic, dendritic, and axon terminal synapses with aminergic nerve fibers regulate the activity of GH-releasing neurons in the hypothalamus (Figure 1). The hypothalamus is a center of neural integration and receives ascending fibers from the spinal cord and the rest of the brain. Numerous stimulatory and inhibitory interneuron relays and neurotransmitter signals may modulate the activity of the GHRH and SRIF neurons. Most of these neural factors may also be released in the median eminence into hypophysial circulation

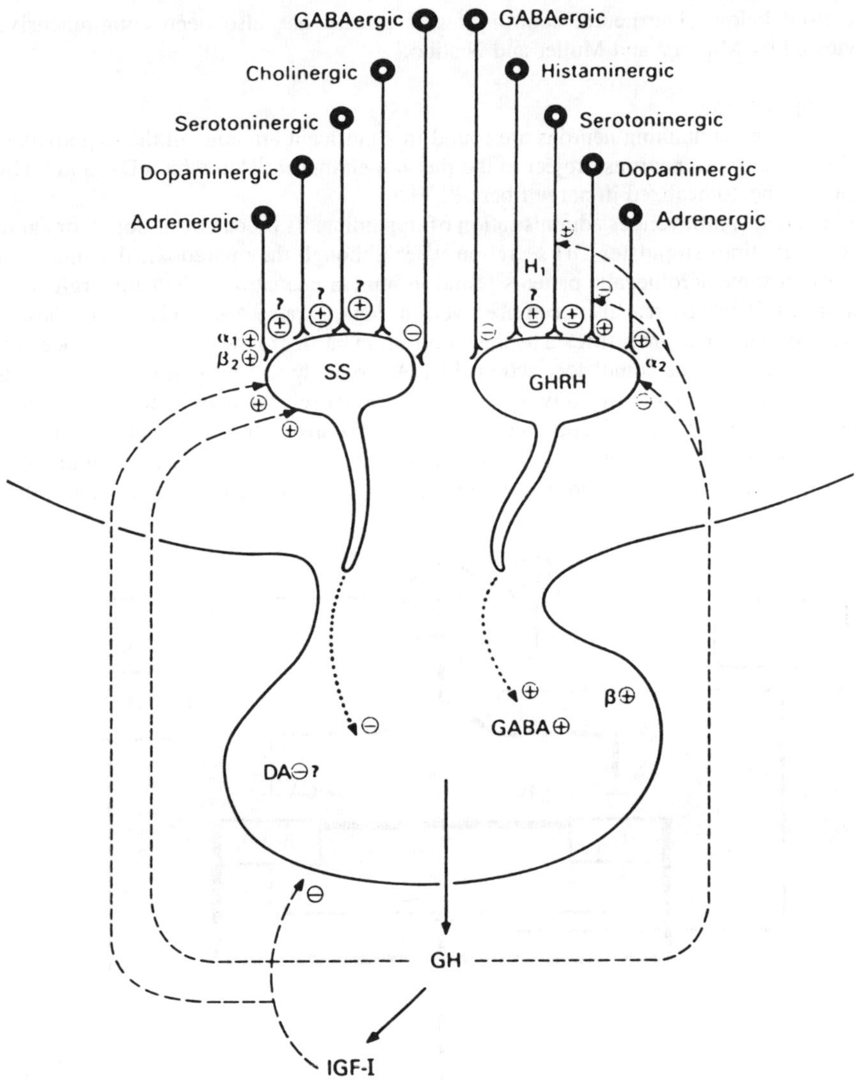

Figure 1 Main neurotransmitter influences in the control of GH secretion, as derived from animal and human studies. Key: +, stimulation; – inhibition; ?, action questionable; SS, somatostatin; DA, dopamine; GABA, γ-aminobutyric acid; IGF-I, insulin-like growth factor I; GHRH, growth hormone-releasing hormone. (From De Gennaro Colonna, V., Cella, S. G., Locatelli, V., et al., Neuroendocrine control of growth hormone secretion, *Acta Pediatr. Scand.*, 349(Suppl.), 87–92, 1989. Reprinted with permission.)

and directly stimulate or inhibit somatotroph function. The actions of neurocrines at pituitary sites may be complementary with their hypothalamic actions, although in some cases they are opposite. The GH-releasing actions of neurotransmitters may also be species dependent. The roles of many neurocrines in GH regulation are, therefore, still unclear, especially as the pharmacological stimulation or inhibition of neurotransmitter function may be relatively nonspecific. Neural influences on GH secretion are, however, reviewed below. The neural control of GH release has also been comprehensively reviewed by Muller[14] and Muller and Nistico.[1]

1. Dopamine

Dopamine-containing neurons are found in significant amounts in the hypothalamus and some of these neurons project to the median eminence,[310] in which DA and GHRH appear to be colocalized in nerve fibers.[311,312]

In humans, intravenous administration of dopamine, its precursor (L-dopa), or agonist (bromocryptine) stimulates GH secretion,[313,314] although they paradoxically inhibit GH release in some acromegalic patients[315] and in human neonates.[316] Dopaminergic stimulation of GH release has also been observed in dogs,[317] rats,[318] and mice.[319] Because the actions of dopamine are restricted by the blood-brain barrier to the median eminence or the pituitary gland,[320] the stimulatory actions of DA are likely to be restricted to these sites.

The influence of systemically injected DA on GH release may, therefore, not accurately reflect the actions of central dopaminergic systems on the hypothalamic control of GH secretion. Central actions of DA on GH release are indicated by the stimulation of GH release by L-dopa and bromocryptine (a DA agent that can cross the blood-brain

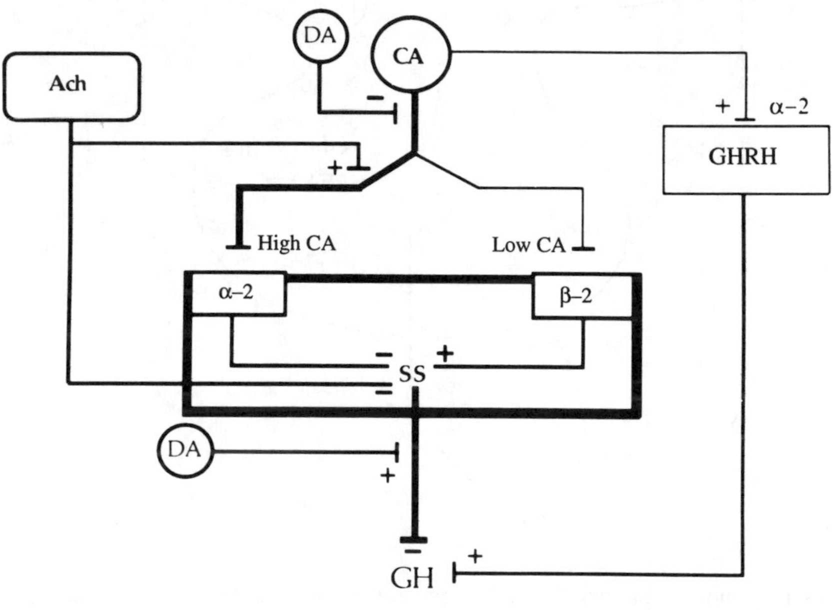

Figure 2 Central GH control by neurotransmitters. The main role is played by adrenergic input (CA) to somatostatin (SS) neurons. Depending on the amount of CA released into the synaptic cleft, either inhibitory a_2- (α2) or stimulatory β-adrenoreceptors (β-2) are activated. That, in turn, would be positively modulated by cholinergic input (Ach), but negatively by dopamine (DA). However, a direct inhibitory effect of Ach on SS neurons, or a stimulatory effect of DA on SS release at the median eminence, cannot be excluded. On the other hand, the stimulatory effect of CA on GHRH release seems to be weak and secondary. Therefore, SS appears to be the main determinant of the growth hormone (GH) pattern of secretion. Stimulation (+) and inhibition (–). (From Devesa, J., Lima, L., and Tresguerres, J. A. F., Neuroendocrine control of growth hormone secretion in humans, *Trends Endocrinol. Metab.*, 3, 175–183, 1992. Copyright 1992 by Elsevier Science, Inc. Reprinted by permission of the publishers.)

barrier). The i.c.v. injection of DA has also been shown to stimulate GH release in rats,[321] although microinjections of DA into the medial basal hypothalamus suppress GH secretion in baboons.[322] A central site of DA action is also indicated in dogs, in which nomifensine (a stimulator of DA transmission)-induced GH release is suppressed by DA receptor antagonists that can cross the blood-brain barrier but not by those that are unable to cross.[317] Dopamine may, therefore, regulate GH release by regulating the synthesis and/or release of GHRH and/or SRIF and by modulating somatotroph responses to SRIF and GHRH stimulation.

The possibility that DA-induced GH secretion is mediated by SRIF disinhibition is indicated by the potentiation of GH responses to GHRH, by effects antagonized by DA receptor blockade[310,323,324] (Figure 2). The ability of dopaminergic agents to induce GH release in response to an otherwise maximally stimulatory dose of GHRH[324–326] also suggests the action of DA is mediated through SRIF withdrawal. It is, however, also possible that DA may stimulate GHRH release, because L-dopa increases plasma GH and GHRH levels and these increases are not seen in patients with hypothalamic lesions.[323,327,328] Changes in plasma GHRH levels may not, however, reflect changes in the activity of GHRH neurons. The mechanism mediating DA-induced GH release may also involve the activation of α-adrenergic receptors, although DA also has somatotroph actions independent of α-adrenergic mechanisms.[14] CV 205–502, a specific and potent D_2 dopamine agonist, crosses the blood-brain barrier and stimulates GH release,[320] although it has no α_1- or α_2-adrenoreceptor-binding activity.

Actions of DA within the hypothalamus or median eminence, therefore, generally stimulate GH release. This stimulatory action is thus likely normally to overcome the inhibitory effect of DA on the release of GH from normal and adenomatous pituitary glands.[329–332] Inhibitory effects of DA on somatotroph cells are well established and also occur in sheep[333] and rat[334,335] pituitary glands. These D_2 receptor-mediated actions[335] are exerted at a step beyond the generation of cAMP[333] and result in an inhibition of GH synthesis[335] and blockade of basal and GHRH-induced[334] GH release. Stimulatory effects of DA on GH release, mediated through D_1 receptors, have however been reported.[336,337]

The balance between inhibitory and stimulatory action is thus likely to account for the inability of DA to modify GH secretion in some human patients.[338,339] This balance is also likely to account for the inhibitory effect of DA in human newborns and infants,[316,340] in which inhibitory SRIF tone is immature and deficient. The inhibitory effects of DA on hypoglycemia-[313] and arginine-induced[341] GH secretion may also reflect an abnormal balance between stimulatory and inhibitory hypothalamic control, because SRIF may be suppressed by these stimuli.[94,342]

2. Catecholamines

The importance of catecholamines (CAs) in GH regulation is indicated by the sustained suppression of GH secretion in reserpine-treated catecholamine-depleted animals[343–345] and pituitary tissue.[346] Plasma GH levels are also reduced in rats injected i.c.v. with 6-hydroxydopamine (6-OHDA, a CA neurotoxin),[1] although they are "paradoxically" increased after 6-OHDA lesioning of the locus coeruleus,[347] which innervates SRIF neurons in the periventricular hypothalamus. Plasma GH levels are also reduced by drugs that block the synthesis of norepinephrine or epinephrine.[348–350] This inhibition of GH secretion is thought to primarily result from increased SRIF tone and secondly from a reduction in GHRH release.[351] Antagonism of α-adrenergic receptors similarly suppresses spontaneous and secretogogue-induced GH release.[348,352,353]

These results indicate that α-adrenergic stimulation increases GH release. Inhibitory effects of α_1-agonism on GH secretion have, however, been reported in infant rats[310] and in dogs,[344,354] probably as a result of increased SRIF release. The activation of α_1-adrenoreceptors may thus have dual effects on GH secretion, because α_1 agonists reduce

GHRH secretion *in vitro*, even though they may enhance SRIF tone *in vivo*.[355] This dual effect at the hypothalamic level may be dependent on endogenous GH secretion.[356] Increased SRIF tone through α_1 activation is probably dominant when GH secretion is enhanced, although when GH release is induced by a non-GHRH-dependent mechanism the stimulatory effect of α_1 receptors on SRIF secretion may be overcome.

In contrast with α_1 agonism, the stimulation of α_2-adrenergic receptors consistently provokes GH release and pituitary GH synthesis in humans[357,358] and other mammalian species.[344,352,359–361] This stimulatory adrenergic system is defective in neonatal rats and develops during ontogeny in a sexually dimorphic way[362–364] and may be testosterone dependent.[365] This α_2-adrenergic effect (Figure 2) is thought to result from SRIF withdrawal in rodents and primates[366] and mediates the stimulatory effect of stress on GH secretion.[367] α_2-Adrenergic stimulation also stimulates GH secretion in sheep, in which pituitary-portal SRIF levels are unchanged by clonidine and in which the GH response to clonidine is unaffected by SRIF antibodies.[368] Increased GHRH release is thus thought to mediate the stimulatory effect of α_2-adrenergic stimulation in sheep, because GHRH levels in portal blood are increased by clonidine and as the GH response is blocked in GHRH-immunized sheep.[368]

α_2-Adrenergic stimulation is also thought to mediate the stimulatory effects of opioids and cholinergic agents on GH release.[265,369] The release of catecholamines during the stress of insulin hypoglycemia is similarly thought to induce GH release via α_2-adrenergic stimulation,[370] although Chapman et al.[370] found no effect of α or β blockade on normal pulsatile GH release.

The stimulatory adrenergic effects on GH secretion are unlikely to be mediated by direct actions on pituitary somatotrophs[371,372] but are thought to occur through GHRH-mediated mechanisms in rats.[372–374] α_2 agonism increases hypothalamic GHRH content[363] and GHRH release from hypothalamic tissue *in vitro*[375,376] and α_2 agonists are ineffective in GHRH-immunized rats[377] and rats with hypothalamic destruction.[321,371] Moreover, although the induction of GH release subsequently increases SRIF synthesis and inhibitory hypothalamic tone to restore plasma GH concentrations,[363] the acute GH response to α_2-receptor agonism in rats is not modified by SRIF immunoneutralization, indicating it occurs independently of SRIF withdrawal.[373] Nevertheless, whereas clonidine is ineffective in inducing GH release in rats when injected into the ventromedial hypothalamic (GHRH-secreting) nuclei, it is effective when injected into the SRIF-rich preoptic hypothalamus.[360,361,378] The application of NE in the medial basal hypothalamus of rats is also ineffective in inducing GH release, indicating α-adrenergic mechanisms triggering GH release may be mediated on SRIF perikarya rather than nerve terminals.[379] An α-adrenergic mechanism mediated by SRIF withdrawal is also implicated in the stimulatory effects of clonidine observed in clinical studies.[380]

Although α-adrenergic receptors mediate stimulatory tone, the activation of β-adrenergic receptors inhibits basal and stimulated GH release.[381–383] This is likely to be induced by an increase in the release of SRIF from the hypothalamus (Figure 2), because the effect is blocked in SRIF-immunized rats[384] and β-adrenergic antagonists have been shown to inhibit SRIF release *in vitro*.[385] The inhibitory actions of β-adrenergic agonism are also unlikely to be mediated at the pituitary, since β-adrenergic agonists directly stimulate GH release *in vitro*.[386,387] This stimulatory action may, however, account for the increased plasma GH levels induced by exogenous catecholamines in hypophysectomized rats bearing ectopic pituitary transplants, in which plasma GH levels are suppressed by CA synthesis depletion.[388]

3. Indoleamines

Both stimulatory and inhibitory[1,389,390] roles for serotonin in GH regulation have been proposed in humans and other species, largely as a result of pharmacological studies with drugs of poor specificity (see Muller[14] for review). The roles of serotonin are thus unclear, although the consensus view supports a stimulatory action. This possibility is supported by the inhibition of GH release in rats in which 5-HT synthesis[391] or 5-HT receptors[392,393] are blocked. Serotoninergic receptor blockade similarly inhibits basal and sleep-related GH secretion in humans.[394,395] Increased GH secretion is, conversely, correlated with increased brain 5-HT turnover[396] and is induced in humans, monkeys, dogs, and rats by peripheral injections of a 5-HT precursor,[393,396–401] i.c.v. injections of 5-HT or its agonist,[402] or after the administration of serotonin-releasing agents or uptake inhibitors.[403] These serotoninergic systems are functional in fetal lambs[404] and newborn rats[392] and appear to be dependent on the presence of a functional muscarinic cholinergic system.[364]

The stimulatory action of 5-HT is also dependent on α_2-adrenergic receptors on 5-HT nerve terminals, the activation of which may mediate the stimulatory effect of clonidine.[405] The activation of the CA system by 5-HT may also enhance GH secretion in a reciprocal fashion,[401,406] especially as the GH responses to the immediate precursor of serotonin (5-hydroxytryptophan) is not counteracted by 5-HT antagonists but by blockade of dopamine and α-adrenergic receptors or by central sympathectomy by 6-OHDA.[407,408] This possibility is also supported by the blunted GH response to serotoninergic stimulation in GHRH-immunized rats.[409] The stimulatory action of 5-HT may also result from SRIF withdrawal, because the inhibitory actions of 5-HT receptor antagonists are mediated by increased SRIF tone[410] and as 5-HT or its agonists directly suppress SRIF release from hypothalamic tissue *in vitro*.[411,412] The activation of the 5-HT system is thus likely to stimulate GHRH release, especially as the stimulatory effect of 5-HT on GH release is blocked in pituitary stalk-lesioned animals.[413]

The action of 5-HT on GH release could also involve direct actions on pituitary cells, because serotonin fibers penetrate the pituitary gland[414] and 5-HT is accumulated by pituitary cells.[415] A stimulatory effect of 5-HT on somatotroph function is also indicated by the ability of a 5-HT antagonist to inhibit GH release from adenomatous pituitary cells.[416] Melatonin, a 5-HT analog, also reduces the production and secretion of GH from rat pituitary cells in culture,[417] although contradictory results with rat[418] and mink[390] pituitary glands have been reported. Melatonin has similarly been found to inhibit basal, 5-HT, and GHRH-induced *in vivo* GH release in rats,[408,419] although it stimulates GH secretion in humans[418] by pathways independent of GHRH.[420]

4. Acetylcholine

The cholinergic system is now known to be a major neural pathway regulating GH release (see Casaneuva,[94] Muller,[14] and Devesa et al.[11] for reviews).

Acetylcholine and muscarinic-ACh agonists increase basal and stimulated plasma GH concentrations (e.g., in rats,[421–423] dogs,[424] humans,[425,426] and cows[427]). Muscarinic blockade, conversely, inhibits pulsatile slow-wave sleep-related and L-dopa-, apomorphine-, GHRP-, arginine-, exercise-, glucagon-, clonidine-, opioid-, and GHRH-induced GH release in humans,[326,428–438] indicating extensive cholinergic participation in GH secretion. Nicotinic cholinergic receptors are much less effective in mediating stimulatory hypothalamic control.[424]

Although cholinergic receptors are present in the median eminence and pituitary gland,[439] a peripheral site of ACh action is unlikely. Acetylcholine stimulates GH release from bovine pituitary cells by an SRIF-sensitive mechanism, but it is effective only at pharmacological (micromolar) concentrations[427,440] and millimolar concentrations of ACh receptor agonists are unable to stimulate GH release from rat pituitary cells.[449] It is, therefore, likely that the actions of ACh are largely mediated at sites within the blood-brain barrier, because muscarinic antagonists are generally ineffective at peripheral sites.[424] It is therefore possible that ACh is effective in stimulating GHRH release and/or inhibiting SRIF release. Evidence for an ACh pathway involved in GHRH release is indicated by the GH release induced by i.c.v. ACh in anterolateral deafferented (SRIF-depleted) rats.[321] However, as cholinergic agonists and antagonists modulate GHRH-induced GH release in clinical disorders,[16] the stimulatory effects of ACh are also likely to be mediated by modulation of SRIF release, because the inhibitory effect of atropine is abolished by anti-somatostatin antibodies.[449] Muscarinic antagonists, like SRIF, also restore to normal the blunted somatotroph responsiveness to repetitive bolus injections of GHRH.[441] The effects of muscarinic agonists and antagonists on GHRH-induced release are also blocked in rats with anterolateral deafferentation of the medial basal hypothalamus or treated with cysteamine to deplete SRIF stores.[449]

Acetylcholine induction of GH release thus appears to resemble α_2-adrenoreceptor mechanisms (Figure 2), because α_2 agonists also induce GH release by inhibiting SRIF release. Thus, although ACh pathways may directly innervate SRIF neurons, they may modulate SRIF activity by the activation of α_2-adrenergic neurons. The functional interconnection of these systems is indicated by the ability of α_2-adrenergic activation to overcome the inhibitory effect of muscarinic cholinergic blockade on GHRH-induced GH secretion, whereas α_2-adrenergic blockade counteracts the stimulatory action of muscarinic stimulation.[11] Moreover, because SRIF withdrawal is also thought to mediate the stimulatory effects of insulin-dependent hypoglycemia, arginine and galanin, which act through α_2 adrenoreceptors, the efficacy of these stimuli is summated or potentiated by muscarinic agonism and partly suppressed by muscarinic antagonism.[442,443] However, as hypoglycemia appears to inhibit SRIF release by actions downstream of cholinergic regulation and may lead to a complete inhibition of SRIF discharge, no further effects of ACh agonists may be demonstrable in this condition.[94]

5. Histamine

The highest concentration of histamine in the CNS is in the hypothalamus and its turnover is increased by stress and opiate stimulation (see Knigge and Warberg[444] for review). The concentrations of hypothalamic histamine are elevated by GHRH and reduced by SRIF,[445] indicating a close interaction with GH-releasing pathways. There is also evidence that histamine regulates noradrenergic neurons and α_2 receptors are present on histaminergic neurons, through which clonidine inhibits histamine synthesis and release.[446] Dense histaminergic nerve fibers are located in hypophysiotropic regions of the hypothalamus, particularly in the preoptic, suprachiasmatic, periventricular, supraoptic, paraventricular, ventromedial and arcuate nuclei, with nerve fibers that project to the median eminence and neurohypophysis. Histamine may, therefore, modulate the synthesis of GH-releasing factors, although as the histaminergic fibers in the median eminence are in the inner ependymal zone and not near the portal capillaries in the external zone, they are unlikely to modulate the release of these factors into pituitary portal circulation. It is also unlikely that histamine is released into portal circulation to act at pituitary sites, especially as histamine and H_2 receptor antagonists are without effect on GH release from rat pituitaries *in vitro*.[447,448]

A role for histamine in GH regulation is, nevertheless, suggested by the inhibition of pulsatile and morphine- and clonidine-induced GH secretion in rats following an i.c.v. injection of histamine or H_1 receptor agonists, by actions blocked by H_1 receptor antagonists.[446,450,451] This inhibitory effect of histamine is likely to be mediated by GHRH suppression, because it is equally effective in cysteamine-treated or SRIF-immunized rats.[446,451] The participation of H_2 is also indicated by H_2-induced suppression of GH secretion, although H_2 antagonists have identical inhibitory effects.[451] Intraventricular injections of histamine also inhibit GH release in dogs,[452] although H_2 antagonists are also inhibitory rather than stimulatory.[453–456] The systemic administration of histamine also increases basal and GHRH-stimulated GH release in rats and dogs.[194] These effects may be mediated through both H_1 and H_2 receptors although H_1 receptor blockade enhances plasma GH levels.[457–459] Histamine also appears to stimulate GH release in humans. Although histamine or histaminergic agents do not modify basal GH concentrations,[14] H_1 antagonists reduce the GH response to sleep, insulin, arginine, L-dopa, and opiate stimulation.[14,460,461] Moreover, whereas TRH does not normally influence GH secretion in humans, TRH-induced GH secretion is observed after histamine treatment through a H_1-sensitive pathway.[462]

The effect of histamine thus appears to be mostly stimulatory at peripheral sites and inhibitory at central sites. Histamine may, therefore, be indirectly involved in the control of GH secretion.

6. γ-Aminobutyric Acid

GABA is widely distributed throughout the CNS and is found in high concentrations in the hypothalamus, particularly the external zone of the median eminence. It is, therefore, likely that GABA acts at hypothalamic and pituitary sites and participates in GH regulation, especially as GABA receptors are located in these tissues and as it modulates the activity of dopaminergic, catecholaminergic, and serotonergic neurons (see Elias et al.[463] and Racagni et al.[464] for reviews). The actions of GABA on GH secretion are, however, complex.

Both increased (in humans[464–466] and rats[467,468]) and decreased (in rats[469–471] and humans[472–474]) basal and stimulated plasma GH levels are reported after the systemic administration of GABA and GABA analogs. These disparate actions may reflect biphasic responses to chronic or acute injections of GABA or biphasic responses to high and low GABA doses. They may also reflect biphasic effects on DA neurons,[473] as low doses given acutely are stimulatory through DA stimulation, whereas high doses or chronic GABA stimulation are inhibitory, through DA inhibition.[463] The stimulatory effect of centrally injected GABA on GH secretion, mediated through specific GABA receptors,[475] was, however, thought to be independent of DA transmission,[476–478] even though central GABA increases DA and NE release.[479] The GH-releasing effect of GABA in humans is, nevertheless, inhibited in the presence of DA receptor blockade.[479] An inhibitory effect of central GABA administration on GH secretion has, nevertheless, been reported.[480]

The increased GH release induced by GABA treatment may result from GHRH-independent pathways, because Acs et al.[468] found it is not suppressed following GHRH immunoneutralization.[468] These results, nevertheless, contrast with those of Murakami et al.,[481] who found the stimulatory GH response to GABA was completely abolished by GHRH antibodies. Stimulation with GABA also increases the GH-releasing activity of the rat hypothalamus, because GABAergic inhibition is followed by a 60% reduction in the GH-releasing activity of the rat hypothalamus.[482] This may result from a reduction in SRIF release.[465,480] The inhibitory effect of GABA may, however, be independent of hypothalamic SRIF, because it persists in rats in which SRIF release is blocked by anterolateral deafferentation of the medial hypothalamus.[471]

The stimulatory actions of GABA on GH secretion may also be directly mediated at pituitary sites. Whereas McCann et al.[470] found no effect of GABA on pituitary GH release *in vitro*, a bicuculline (GABA antagonist)-sensitive stimulatory effect of muscimol (GABA agonist) on the secretion of GH has been observed.[483,484] The GH response to prolonged muscimol stimulation downregulates but does not prevent GHRH-induced GH release.[482] The GH response is potentiated by benzodiazepine and barbiturate mechanisms and is calcium dependent.[485] GABA similarly increases GH release from the pituitaries of neonatal rats,[482] although pituitary responsiveness declines during the second and third postnatal weeks and is absent in adults.[468,485]

7. Excitatory Amino Acids

N-Methyl-D-aspartate (NMDA) receptor systems that mediate the actions of excitatory amino acids (e.g., glutamate and aspartate) are ubiquitously present within the CNS. Activation of these NMDA systems has been shown to influence the release of most neuropeptides affecting GH release.

A role for NMDA receptors is indicated by the increased GH secretion in monkeys,[486,487] rats,[468,488] sheep,[489] cattle,[490] and pigs[491] following the *in vivo* administration of *N*-methyl-D$_1$ L-aspartic acid (NMA), a synthetic agonist of NMDA receptors. Although NMDA antagonism *in vivo* reduces pituitary GH secretion *in vitro*,[492] the stimulatory action of NMA is likely to be mediated centrally or within the median eminence, because NMA fails to stimulate GH release directly from rat pituitary glands *in vitro*.[468] This action of NMA is thus likely to result from activation of NMA receptors on GHRH neurons in the median eminence,[487] and increased GHRH release, because the response to NMA is blocked in GHRH-depleted animals with monosodium glutamate (MSG) lesions of the arcuate nucleus or induced by GHRH antibodies.[468]

Selective depletion of the GHRH-like immunoreactivity and GHRH mRNA in the hypothalamus is, conversely, induced by a NMDA antagonist.[492] It is also possible that the stimulation of GHRH activity by NMA results from the stimulation of noradrenergic neurons, which in turn stimulate GHRH release.[490,493] Endogenous opiates may also mediate the stimulatory effect of NMA, because the GH response is blocked by opiate receptor antagonism.[276] The stimulatory effect of NMA may also be attenuated by a contemporaneous increase in SRIF release,[494–496] even though it does not increase SRIF mRNA content in the rat hypothalamus.[492]

In addition to glutamate and aspartate, GH release also appears to be regulated by quinolinic acid, a hepatic tryptophan metabolite, because it is found in the brain and has NMA- and kainate-like activity on central neurons. This possibility is supported by the immediate surge in GH secretion induced by intrahypothalamic injections of quinolinic acid.[497]

Taurine (2-aminoethane sulfonic acid) is one of the most abundant amino acids in the CNS and is highly concentrated in the neurohypophysis and adenohypophysis. The possibility that it participates in GH regulation is indicated by the biphasic stimulation of GH release in rats following intracerebroventricular infusion.[498,499] This may result from an interaction with endogenous opioid peptides, because the response to taurine is blocked by opiate receptor antagonism and potentiated in the presence of met-enkephalin.[499] Taurine also increases GH secretion in humans after systemic administration.[500] The biphasic GH response to taurine may indicate its stimulation of SRIF release[501] at high dose levels. This may also account for the inhibition of GH secretion in rats receiving higher doses of taurine intracerebroventricularly.[498]

Aromatic amino acids (e.g., L-5-hydroxytryptophan, L-dihydroxyphenylalanine, DL-*o*-tyrosine, and DL-m-tyrosine) are also likely to modulate GH release, because they act as amine precursors. These precursor amino acids can, nevertheless, directly inhibit GHRH-induced GH release from rat pituitaries *in vitro*.[334] This inhibitory action occurs independently of the formation of biogenic amines and is unaffected by inhibition of aromatic-L-amino-acid decarboxylase activity.[334]

Arginine, isoleucine, leucine and several other amino acids have also been found to stimulate GH release.[17] The stimulatory mechanism for most of these amino acids is unknown, although it is well established that arginine stimulates GH release in humans by suppressing endogenous SRIF secretion.[502] The administration of arginine is thus able to potentiate the GH response to maximal stimulatory doses of GHRH.[502]

REFERENCES

1. **Muller, E. E. and Nistico, G.,** *Brain Messengers and the Pituitary*, Academic Press, New York, 1989.
2. **McKinley, M. J., McAllen, R. M., Mendelsohn, F. A. O., et al.,** *Front. Neuroendocrinol.*, 11, 91, 1990.
3. **Mezey, E. and Palkovits, M.,** *Front. Neuroendocrinol.*, 7, 1, 1982.
4. **Ju, G., Liu, S., and Zhang, X.,** *News Physiol. Sci.*, 6, 26, 1991.
5. **Paden, C. M., Moffett, C. W., and Benowitz, L. I.,** *Endocrinology (Baltimore)*, 134, 503, 1994.
6. **Ju, G., Liu, S. J., and Zhang, X.,** *Neuroendocrinology*, 53(Suppl. 1), 41, 1991.
7. **Westlund, K. N., Chmielowiecs and Childs, G. V.,** *Peptides*, 4, 557, 1983.
8. **Koibuchi, N., Kato, M., Kategawa, T., et al.,** *J. Neuroendocrinol.*, 1, 209, 1989.
9. **Rice, R. W. and Critchlow, V.,** *Endocrinology (Baltimore)*, 99, 970, 1976.
10. **Mason, W. T., Dickson, S. L., and Leng, G.,** *Acta Pediatr. Scand.*, 82, 84, 1993.
11. **Devesa, J., Lima, L., and Tresguerres, J. A. F.,** *Trends Endocrinol. Metab.*, 3, 175, 1992.
12. **Frohman, L. A., Downs, T. R., and Chomczynski, P.,** *Front. Neuroendocrinol.*, 13, 344, 1992.
13. **Casanueva, F. F., Popovic, V., Leal-Cerro, A., et al.,** *Molecular and Clinical Advances in Pituitary Disorders* (Ed. Melmed, S.). Endocrine Research and Education, Inc., Los Angeles, CA, 1993, p.145.
14. **Muller, E. E.,** *Physiol. Rev.*, 67, 962, 1987.
15. **Page, M. D., Dieguez, C., and Scanlon, M. F.,** *Biotechnology in Growth Regulation* (Eds. Heap, R. B., Prosser, C. G., and Lamming, G. E.). Butterworths, London, 1988, p. 47.
16. **De Gennaro Colonna, V., Cella, S. G., Locatelli, V., et al.,** *Acta Pediatr. Scand.*, 349(Suppl.) 87, 1989.
17. **Root, R. W.,** *J. Endocrinol. Invest.*, 12(Suppl. 3), 3, 1989.
18. **Harvey, S.,** *The Endocrinology of Growth, Development, and Metabolism in Vertebrates*, (Eds. Schreibman, M. P., Scanes, C. G., and Pang, P. K. T.). Academic Press, Toronto, 1993, p. 151.
19. **Buonomo, F. C. and Baile, C. A.,** *Domest. Anim. Endocrinol.*, 7, 435, 1990.
20. **Lamberts, S. W. J. and Oosterom, R.,** *Horm. Res.*, 14, 137, 1985.
21. **Spada, A., Vallar, L., and Giannattasio, G.,** *Endocrinology (Baltimore)*, 115, 1203, 1984.
22. **Sheppard, M. S., Moor, B. C., and Kracier, J.,** *Endocrinology (Baltimore)*, 117, 2364, 1985.
23. **Bilezikjian, L. M. and Vale, W. W.,** *Endocrinology (Baltimore)*, 113, 1726, 1983.
24. **Holl, R. W., Thorner, M. O., and Leong, D. A.,** *Endocrinology (Baltimore)*, 122, 2927, 1988.
25. **Miller, H. A., III, Rogers, G., and Frawley, L. S.,** *Life Sci.*, 43, 629, 1988.
26. **Clayton, R. N. and Bailey, L. C.,** *J. Endocrinol.*, 112, 69, 1987.
27. **Simard, J., Lefevre, G., and Labrie, F.,** *Peptides*, 8, 263, 1987.
28. **Tannenbaum, G. S., Painson, J., Lengyel, A. M., et al.,** *Endocrinology (Baltimore)*, 124, 1380, 1989.
29. **Soya, H. and Suzuki, M.,** *Endocrinology (Baltimore)*, 126, 285, 1990.
30. **Bilezikjian, L. M. and Vale, W. W.,** *Endocrinology (Baltimore)*, 115, 2032, 1984.
31. **Simard, J. and Labrie, F.,** *Mol. Cell. Endocrinol.*, 46, 79, 1986.
32. **Edwards, C. A., Dieguez, C., Ham, J., et al.,** *J. Endocrinol.*, 116, 185, 1988.
33. **Ceda, G. P. and Hoffman, A. R.,** *Endocrinology (Baltimore)*, 116, 1334, 1985.
35. **Soya, H., Suzuki, M., and Kato, M.,** *Biochem. Biophys. Res. Commun.*, 172, 276, 1990.
36. **Smith, M. A., Yamamoto, G., and Vale, W.,** *Mol. Cell. Endocrinol.*, 37, 311, 1984.
37. **Kelijman, M. and Frohman, L. A.,** *J. Clin. Endocrinol. Metab.*, 71, 157, 1990.
38. **Reisine, T. D. and Takahashi, J. S.,** *J. Neurosci.*, 4, 812, 1984.
39. **Kraicer, J., Cowan, J. S., Sheppard, M. S., et al.,** *Endocrinology (Baltimore)*, 119, 2047, 1986.
40. **White, M. C., Daniels, M., Kendall-Taylor, P., et al.,** *J. Endocrinol.*, 105, 269, 1985.
41. **Plouzek, C. A., Molina, J. R., Hard, D. L., et al.,** *Proc. Soc. Exp. Biol. Med.*, 189, 158, 1988.
42. **Losa, M., Muller, O. A., Sobieszczyk, S., et al.,** *Clin. Endocrinol.*, 23, 715, 1985.
43. **Kraicer, J., Sheppard, M. S., Luke, J., et al.,** *Endocrinology (Baltimore)*, 122, 1810, 1988.
44. **Anderson, L. L., Ford, J. J., Klindt, J., et al.,** *Proc. Soc. Exp. Biol. Med.*, 196, 194, 1991.
45. **Sugihara, H., Minami, S., and Waskabayashi, I.,** *J. Endocrinol.*, 122, 583, 1989.
46. **Carmignac, D. F. and Robinson, I. C. A. F.,** *J. Endocrinol.*, 127, 69, 1990.
47. **Janowski, B. A. and Wehrenberg, W. B.,** *Life Sci.*, 50, 951, 1992.
48. **Cowan, J. S., Moor, B., Chow, A., et al.,** *Endocrinology (Baltimore)*, 113, 1056, 1983.
49. **Stachura, M. E., Tyler, J. M., and Farmer, P. K.,** *Endocrinology (Baltimore)*, 123, 1476, 1988.

50. **Redekopp, C., Livesey, J., and Donald, R. A.,** *J. Endocrinol. Invest.*, 7, 277, 1984.
51. **Clark, R. G., Carlsson, L. M. S., Rafferty, B., et al.,** *J. Endocrinol.*, 119, 397, 1988.
52. **Cowan, J. S., Gaul, P., Moor, B. C., et al.,** *Can. J. Physiol. Pharmacol.*, 62, 199, 1984.
53. **Richardson, S. B. and Twente, S.,** *J. Endocrinol.*, 128, 91, 1991.
54. **Weiss, J., Cronin, M. J., and Thorner, M. O.,** *Am. J. Physiol.*, 253, E508, 1987.
55. **Login, I. S. and Judd, A. M.,** *Endocrinology (Baltimore)*, 119, 1703, 1986.
56. **McFarlane, M. B., Bruhn, T. O., and Jackson, I. M. D.,** *Neuroendocrinology*, 57, 496, 1993.
57. **Rene, E., Willoughby, J., and Brazeau, P.,** *Regul. Peptides*, 4, 325, 1982.
58. **Richardson, S. B. and Twente, S.,** *J. Endocrinol.*, 138, 369, 1993.
59. **Miki, N., Ono, M., and Shizume, K.,** *J. Endocrinol.*, 117, 245, 1988.
60. **Plotsky, P. M. and Vale, W.,** *Science*, 230, 461, 1985.
61. **Magnan, E., Cataldi, M., Guillaume, V., et al.,** *Life Sci.*, 51, 831, 1992.
62. **Masuda, A., Shibashi, T., Kim, Y. S., et al.,** *J. Clin. Endocrinol. Metab.*, 69, 906, 1989.
63. **Yamauchi, N., Shibasak, T., Ling, N., et al.,** *Regul. Peptides*, 33, 71, 1991.
64. **Miki, N. and Ono, M.,** *69th Annu. Meet. Endocr. Soc.*, 1987, p. 118.
65. **Tannenbaum, G. S., McCarthy, C. F., Zeitler, P., et al.,** *Endocrinology (Baltimore)*, 127, 2551, 1990.
66. **Bertherat, J., Berod, A., Normand, E., et al.,** *J. Neuroendocrinol.*, 3, 115, 1991.
67. **Thomas, C. R., Groot, K., and Arimura, A.,** *Endocrinology (Baltimore)*, 116, 2174, 1985.
68. **Murakami, Y., Kato, Y., Kabayama, Y., et al.,** *Endocrinology (Baltimore)*, 120, 311, 1987.
69. **Lumpkin, M. D., Negro-Vilar, A., and McCann, S. M.,** *Science*, 211, 1072, 1981.
70. **Abe, H., Kato, Y., Iwasaki, Y., et al.,** *Proc. Soc. Exp. Biol. Med.*, 159, 346, 1978.
71. **Slama, A., Bluet-Pajot, M., Mounier, F., et al.,** *Neuroendocrinology*, 58, 178, 1993.
72. **Murakami, Y., Kato, Y., Koshiyama, H., et al.,** *Brain Res.*, 407, 405, 1987.
73. **Richardson, S. B., Twente, S., and Audhya, T.,** *Am. J. Physiol.*, 255, E829, 1988.
74. **Richardson, S. B. and Twente, S.,** *Endocrinology (Baltimore)*, 117, 2076, 1986.
75. **Spencer, G. S. G., Bass, J. J., Hodgkinson, S. C., et al.,** *Domest. Anim. Endocrinol.*, 8, 375, 1991.
76. **Peterfreund, R. A. and Vale, W.,** *Neuroendocrinology*, 39, 397, 1984.
77. **Willoughby, J. O., Brogan, M., and Kapoor, R.,** *Neuroendocrinology*, 50, 592, 1989.
78. **Katakami, H., Downs, T. R., and Frohman, L. A.,** *Endocrinology (Baltimore)*, 123, 1103, 1988.
79. **McCarthy, G. F., Beaudet, A., and Tannenbaum, G. S.,** *Neuroendocrinology*, 56, 18, 1992.
80. **Bertherat, J., Dournaud, P., Berod, A., et al.,** *Neuroendocrinology*, 56, 25, 1992.
81. **Willoughby, J. O., Brogan, M., and Kapoor, R.,** *Neuroendocrinology*, 50, 584, 1989.
82. **Liposits, Z. S., Merchenthaler, I., Paull, W. K., et al.,** *Histochemistry*, 89, 247, 1988.
83. **Daikoku, J., Hisano, S., Kawano, H., et al.,** *Neuroendocrinology*, 47, 405, 1988.
84. **Horvath, S., Palkovits, M., Gorcs, T., et al.,** *Brain Res.*, 481, 8, 1989.
85. **Zeytin, F. N., Rusk, S. F., and De Lellis, R.,** *Endocrinology (Baltimore)*, 114, 2054, 1988.
86. **de los Frailes, M. T., Cacicedo, L., Fernandez, G., et al.,** *Neuroendocrinology*, 55, 221, 1992.
87. **Sato, M., Takahara, J., Niimi, M., et al.,** *J. Neuroendocrinol.*, 2, 555, 1990.
88. **Aguila, M. C. and McCann, S. M.,** *Endocrinology (Baltimore)*, 117, 762, 1985.
89. **Aguila, M. C., Milenkovic, L., McCann, S. M., et al.,** *Neuroendocrinology*, 52, 238, 1990.
90. **Mitsugi, N., Arita, J., and Kimura, F.,** *Neuroendocrinology*, 51, 93, 1990.
91. **Katakami, H., Arimura, A., and Frohman, L. A.,** *Endocrinology (Baltimore)*, 118, 1872, 1986.
92. **Aguila, M. C. and McCann, S. M.,** *Endocrinology (Baltimore)*, 120, 341, 1987.
93. **Eckland, D. J. A. and Lightman, S. L.,** *J. Endocrinol.*, 113, R1, 1987.
94. **Casaneuva, F.,** *Endocrinol. Metab. Clin. North. Am.*, 21, 483, 1992.
95. **Tapanainen, P., Rantala, H., Leppaluoto, et al.,** *Pediatr. Res.*, 26, 404, 1989.
96. **Brion, D. E., Donnadieu, M., Liapi, C., et al.,** *Horm. Res.*, 24, 116, 1986.
97. **Valcavi, R., Dieguez, C., Page, M. D., et al.,** *Clin. Endocrinol. (Oxford)*, 29, 309, 1988.
98. **Barbetti, F., Crescenti, C., Negri, M., et al.,** *J. Clin. Endocrinol. Metab.*, 70, 1371, 1990.
99. **Page, M. D., Koppeschaar, H. P. F., Edwards, C. A., et al.,** *Clin. Endocrinol.*, 26, 589, 1987.
100. **Fukata, J., Kasting, N. W., and Martin, J. B.,** *Neuroendocrinology*, 40, 193, 1985.
101. **Rawlings, S. R. and Mason, W. T.,** *Biotechnology in Growth Regulation* (Eds. Heap, R. B., Prosser, C. J., and Lamming, G. E.). Butterworths, London, 1988, p. 35.
102. **Hokfelt, T., Meister, B., Everitt, B., et al.,** *Integrative Neuroendocrinology: Molecular, Cellular and Clinical Aspects* (Eds. McCann, S. M. and Weiner, R. I.). S. Karger, Basel, 1987, p. 1.
103. **Maiter, D. M., Hooi, S. C., Koenig, J. I., et al.,** *Endocrinology (Baltimore)*, 126, 1216, 1990.

104. **Davis, T. M. E., Burrin, J. M., and Bloom, S. R.,** *J. Clin. Endocrinol. Metab.*, 65, 1248, 1987.
105. **Shioda, S., Kohara, H., and Nakai, Y.,** *Brain Res.*, 40, 355, 1987.
106. **Daikoku, S., Hisano, S., Hitoshi, K., et al.,** *Neuroendocrinology*, 47, 405, 1988.
107. **Billestrup, N., Gonzalez-Manchon, C., Potter, E., et al.,** *Mol. Endocrinol.*, 4, 356, 1990.
108. **Bilezikjian, L. M., Corrigan, A. Z., and Vale, W.,** *Endocrinology (Baltimore)*, 126, 2369, 1990.
109. **Schonbrunn, A. and Tashijian, A., Jr.,** *J. Biol. Chem.*, 255, 190, 1980.
110. **Ingram, C. D. and Bicknell, R. J.,** *J. Endocrinol.*, 109, 67, 1986.
111. **Vigh, S. and Schaily, A. V.,** *Peptides*, 1, 241, 1984.
112. **Enjalbert, A., Epelbaum, J., Arancibia, S., et al.,** *Endocrinology (Baltimore)*, 11, 42, 1982.
113. **Joubert, D., Mouhieddine, B., Benlot, C., et al.,** *Horm. Res.*, 38, 100, 1992.
114. **Dieguez, C., Page, M. D., and Scanlon, M. F.,** *Clin. Endocrinol.*, 28, 109, 1988.
115. **Rosselin, G.,** *Peptides*, 7, 89, 1986.
116. **Seifert, H., Perrin, M., Rivier, J., et al.,** *Nature (Lond.)*, 313, 487, 1985.
117. **Said, S. I.,** *J. Endocrinol. Invest.*, 9, 191, 1986.
118. **Shimatsu, A., Kato, Y., Inoue, T., et al.,** *Neurosci. Lett.*, 43, 259, 1983.
119. **Nicosia, S., Oliva, D., Giannattasio, G., et al.,** *J. Endocrinol. Invest.*, 6, 235, 1983.
120. **Durand, D., Bluet-Pajot, M., Mounier, F., et al.,** *Reprod. Nutr. Dev.*, 27, 481, 1987.
121. **Bluet-Pajot, M., Mounier, F., Leonard, J., et al.,** *Peptides*, 8, 35, 1987.
122. **Chihara, K., Kaji, H., Minamitani, N., et al.,** *J. Clin. Endocrinol. Metab.*, 58, 81, 1984.
123. **Kato, Y., Shimatsu, A., Matsushita, N., et al.,** *Peptides*, 5, 389, 1984.
124. **Watanobe, H., Sasaki, S., Sone, K., et al.,** *J. Clin. Endocrinol. Metab.*, 72, 982, 1991.
125. **Matsushita, N., Katakami, H., Shimatus, A., et al.,** *J. Clin. Endocrinol. Metab.*, 53, 1297, 1981.
126. **Ishibashi, M. and Yamaji, T.,** *J. Clin. Endocrinol. Metab.*, 60, 985, 1985.
127. **Ogwuegbu, S. O., Hashizume, T., and Kanematsu, S.,** *Domest. Anim. Endocrinol.*, 8, 29, 1991.
128. **Hashizume, T. and Kanematsu, S.,** *Domest. Anim. Endocrinol.*, 7, 451, 1990.
129. **Kashio, Y., Chomczynski, P., Downs, T. R., et al.,** *Endocrinology (Baltimore)*, 127, 1129, 1990.
130. **Magistretti, P. J., Schonenberg, P., Kehrer, P., et al.,** *Peptides*, 7, 175, 1986.
131. **Denef, C., Schramme, C., and Baes, M.,** *Neuroendocrinology*, 40, 88, 1985.
132. **Vijayan, E., Samson, W. K., Said, S. I., et al.,** *Endocrinology (Baltimore)*, 104, 53, 1979.
133. **Epelbaum, J., Tapia-Arancibia, L., Besson, J., et al.,** *Eur. J. Pharmacol.*, 58, 493, 1979.
134. **Watanobe, H. and Takebe, K.,** *Neuropeptides*, 23, 115, 1992.
135. **Murphy, W. A., Lance, V., Sueiras-Diaz, J., et al.,** *Biochem. Biophys. Res. Commun.*, 112, 469, 1983.
136. **Sugase, T., Nonaka, K., Yoshida, T., et al.,** *Endocrinol. Jpn.*, 23, 187, 1976.
137. **Katakami, H., Kato, Y., Matsushita, N., et al.,** *Peptides*, 4, 849, 1983.
138. **Katakami, H., Kato, Y., Mastsushita, N., et al.,** *Endocrinology (Baltimore)*, 115, 1598, 1984.
139. **Rivier, C. and Vale, W.,** *Endocrinology (Baltimore)*, 114, 2409, 1984.
140. **Katakami, H., Arimura, A., and Frohman, L. A.,** *Neuroendocrinology*, 41, 390, 1985.
141. **Ono, N., Lumpkin, M. D., Samson, W. K., et al.,** *Life Sci.*, 35, 1117, 1984.
142. **Pieters, G., Hermus, A. R. M. M., Smals, A. G. H., et al.,** *J. Clin. Endocrinol. Metab.*, 58, 560, 1984.
143. **Ishibashi, M., Hara, T., Tagusagawa, Y., et al.,** *Acta Endocrinol.*, 106, 443, 1984.
144. **Watanobe, H., Sasaki, S., and Takebe, K.,** *Acta Endocrinol.*, 125, 487, 1991.
145. **Tanaka, K., Watabe, T., Yoshida, H., et al.,** *Endocrinol. Jpn.*, 31, 355, 1984.
146. **Barbarino, A., Corsello, S. M., Casa, S. D., et al.,** *J. Clin. Endocrinol. Metab.*, 71, 1368, 1990.
147. **Hokfelt, T., Fahrenkrug, J., Tatemoto, K., et al.,** *Proc. Natl. Acad. Sci. U.S.A.*, 80, 895, 1983.
148. **Koves, K., Arimura, A., Somogyvari-Vigh, A., et al.,** *Endocrinology (Baltimore)*, 127, 264, 1990.
149. **Vigh, S., Arimura, A., Koves, K., et al.,** *Peptides*, 12, 313, 1991.
150. **Gottschall, P. E., Tatsuno, I., and Arimura, A.,** *FASEB J.*, 5, 194, 1991.
151. **Gottschall, P. E., Tatsuno, I., Miyata, A., et al.,** *Endocrinology (Baltimore)*, 127, 272, 1990.
152. **Miyata, A., Arimura, A., Dahl, R. D., et al.,** *Biochem. Biophys. Res. Commun.*, 164, 567, 1989.
153. **Goth, M. I., Lyons, C. E., Canny, B. J., et al.,** *Endocrinology (Baltimore)*, 130, 939, 1992.
154. **Harvey, S.,** *J. Endocrinol.*, 125, 345, 1990.
155. **Strbak, V., Angyal, R., Jurcovicova, J., et al.,** *Biol. Neonate*, 50, 91, 1986.
156. **Kacsoh, B., Toth, B. E., and Grosvenor, C. E.,** *J. Neuroendocrinol.*, 4, 663, 1992.

157. **Hulting, A. L., Theodorsson, E., and Werner, S.,** *J. Intern. Med.*, 232, 229, 1992.
158. **Spencer, G. S. G., Aitken, W. M., Hodgkinson, S. C., et al.,** *Domest. Anim. Endocrinol.*, 9, 115, 1992.
159. **Harvey, S.,** *The Growth, Development and Metabolism of Vertebrates* (Eds. Schreibman, M. P., Scanes, C. G., and Pang, P. K. T.). Academic Press, New York, 1993, p. 151.
160. **Torres-Aleman, I., Fernandez, M., Debeljuk, L., et al.,** *Regul. Peptides*, 18, 19, 1987.
161. **Amsterdam, J. D., Winokur, A., Luchi, I., et al.,** *Psychoneuroendocrinology*, 7, 177, 1982.
162. **Rubin, A. L., Levin, S. R., Bernstein, R. I., et al.,** *J. Clin. Endocrinol. Metab.*, 37, 160, 1973.
163. **Hanew, K., Sato, S., Sasaki, A., et al.,** *Tohuku J. Exp. Med.*, 145, 161, 1985.
164. **Robberecht, W., Andries, M., and Denef, C.,** *Neuroendocrinology*, 56, 550, 1992.
165. **Hirose, S., Yamamoto, M., Kanazawa, I., et al.,** *Biomed. Res.*, 6, 23, 1985.
166. **Degli Uberti, E. C., Trasforini, G., Margutti, A., et al.,** *Metabolism*, 39, 1063, 1990.
167. **Steele, M. K., McCann, S. M., and Negro-Vilar, A.,** *Endocrinology*, 111, 722, 1982.
168. **Moody, T. W., Thoa, N. B., O'Donohue, T. L., et al.,** *Life Sci.*, 26, 1707, 1980.
169. **Rivier, C., Rivier, J., and Vale, W.,** *Endocrinology (Baltimore)*, 102, 519, 1978.
170. **Pontiroli, A. E. and Scarpignato, C.,** *Horm. Res.*, 23, 129, 1986.
171. **Morley, J. E., Varner, A. A., Modlin, I. M., et al.,** *Clin. Endocrinol. (Oxford)*, 13, 369, 1980.
172. **Bicknell, R. J. and Chapman, C.,** *Neuroendocrinology*, 36, 33, 1983.
173. **Bjoro, T., Torjesen, P. E., Ostberg, B. C., et al.,** *Regul. Peptides*, 19, 169, 1987.
174. **Westendorf, J. M. and Schonbrunn, A.,** *Endocrinology (Baltimore)*, 110, 352, 1982.
175. **Houben, H. and Denef, C.,** *Endocrinology (Baltimore)*, 126, 2257, 1990.
176. **Wakabayashi, I., Tonegawa, Y., Shibasaki, T., et al.,** *Life Sci.*, 36, 1437, 1985.
177. **Tache, Y., Brown, M., and Collu, R.,** *Endocrinology (Baltimore)*, 105, 220, 1979.
178. **Karashima, T., Okajima, T., Kato, K., et al.,** *Endocrinol. Jpn.*, 31, 539, 1984.
179. **Abe, H., Chihara, K., Minamitani, N., et al.,** *Endocrinology (Baltimore)*, 109, 229, 1981.
180. **Kentroti, S., Aguila, M. C., and McCann, S. M.,** *Endocrinology (Baltimore)*, 122, 2407, 1988.
181. **Kabayama, Y., Kato, Y., Shimatsu, A., et al.,** *Endocrinology (Baltimore)*, 115, 649, 1984.
182. **Kentroti, S. and McCann, S. M.,** *Endocrinology (Baltimore)*, 117, 1363, 1985.
183. **Kentroti, S. and McCann, S. M.,** *Brain Res. Bull.*, 33, 99, 1994.
184. **Altomonte, L., Zoli, A., Mirone, L., et al.,** *Endocrinology (Baltimore)*, 88, 334, 1986.
185. **Ottlecz, A., Samson, W. S., and McCann, S. M.,** *Peptides*, 6, 115, 1985.
186. **Wank, S. A., Pisegna, J. R., and Weerth, A. D.,** *Proc. Natl. Acad. Sci. U.S.A.*, 89, 8691, 1992.
187. **Kuwahara, T., Takamiya, M., Nagase, H., et al.,** *Peptides*, 14, 647, 1993.
188. **Vijayan, E., Samson, W. K., and McCann, S. M.,** *Brain Res.*, 172, 295, 1979.
189. **Marsumura, M., Yamanoi, A., Yamamoto, S., et al.,** *Horm. Metab. Res.*, 25, 1201, 1979.
190. **Morley, J. E., Melmed, S., Briggs, J., et al.,** *Life Sci.*, 25, 1201, 1979.
191. **Malarkey, W. B., O'Sorisio, T. M., Kennedy, M., et al.,** *Life Sci.*, 28, 2489, 1981.
192. **McCann, S. M.,** *Neuroendocrinology*, 31, 355, 1980.
193. **Vijayan, E. and McCann, S. M.,** *Life Sci.*, 26, 321, 1980.
194. **Rivier, C., Brown, M., and Vale, W.,** *Endocrinology (Baltimore)*, 100, 751, 1977.
196. **Maeda, K. and Frohman, L. A.,** *Endocrinology (Baltimore)*, 103, 1903, 1978.
197. **Ibanez, R., Mallo, F., Benitez, L., et al.,** *Life Sci.*, 53, 227, 1993.
198. **Vijayan, E. and McCann, S. M.,** *Endocrinology (Baltimore)*, 103, 1888, 1978.
199. **Arisawa, M., Snyder, G. D., DePalatis, L., et al.,** *Proc. Natl. Acad. Sci. U.S.A.*, 86, 7290, 1989.
200. **Houben, H. and Denef, C.,** *Peptides*, 14, 109, 1993.
201. **Coiro, V., Volpi, R., Capretti, L., et al.,** *Peptides*, 13, 843, 1992.
202. **Vijayan, E. and McCann, S. M.,** *Life Sci.*, 26, 321, 1980.
203. **Samson, W. K., Lumpkin, M. D., and McCann, S. M.,** *Brain Res. Bull.*, 8, 117, 1982.
204. **Samson, W. K., Lumpkin, M. D., Nilaver, G., et al.,** *Brain Res. Bull.*, 12, 57, 1984.
205. **Meister, B., Sconlon, M. F., and Hokfelt, T.,** *Neurosci. Lett.*, 119, 136, 1990.
206. **Gabriel, S. M., Kaplan, L. M., Martin, J. B., et al.,** *Peptides*, 10, 369, 1989.
207. **Lopez, F. J., Meade, E. H., and Negro-Vilar, A.,** *Brain Res. Bull.*, 24, 395, 1990.
208. **Rokaeus, A.,** *Trends Neurosci.*, 10, 158, 1987.
209. **Skofitsch, G. and Jacobowitz, D. M.,** *Peptides*, 6, 509, 1985.
210. **Hyde, J. F., Engle, M. G., and Maley, B. E.,** *Endocrinology (Baltimore)*, 129, 270, 1991.
211. **Hyde, J. F. and Keller, B. K.,** *Endocrinology (Baltimore)*, 128, 917, 1991.
212. **Melander, T., Hokfelt, T., Nilsson, S., et al.,** *Eur. J. Pharmacol.*, 124, 381, 1986.
213. **Meister, B. and Hulting, A.,** *Neuroendocrinology*, 46, 387, 1987.
214. **Hulting, A. L., Meister, B., Carlsson, L., et al.,** *Acta Endocrinol. (Copenh.)*, 125, 518, 1991.

215. **Ottlecz, A., Samson, W. K., and McCann, S. M.,** *Peptides*, 7, 51, 1986.
216. **Gabriel, S. M., Milbury, C. M., Nathanson, J. A., et al.,** *Life Sci.*, 42, 1981, 1988.
217. **Torsello, A., Sellan, R., Cella, S. G., et al.,** *Life Sci.*, 47, 1861, 1990.
218. **Sato, M., Takahara, J., Niimi, M., et al.,** *Life Sci.*, 48, 1639, 1991.
219. **Lindstrom, P. and Savendahl, L.,** *Acta Endocrinol.*, 129, 268, 1993.
220. **Loche, S., Vista, N., Ghigo, E., et al.,** *Pediatr. Res.*, 27, 405, 1990.
221. **Murakami, Y., Ohshima, K., Mochizuki, T., et al.,** *J. Clin. Endocrinol. Metab.*, 77, 1436, 1993.
222. **Arvat, E., Ghigo, E., Nicolosi, M., et al.,** *Clin. Endocrinol.*, 36, 347, 1992.
223. **Bauer, F. E., Venetikou, M., Burrin, J. M., et al.,** *Lancet*, ii, 192, 1986.
224. **Giustina, A., Bussi, A. R., Legati, F., et al.,** *Acta Endocrinol.*, 127, 504, 1992.
225. **Giustina, A., Bodini, C., Doga, M., et al.,** *J. Clin. Endocrinol. Metab.*, 74, 1296, 1992.
226. **Giustina, A., Schettino, M., Bodini, C., et al.,** *Metab. Clin. Exp.*, 41, 1291, 1992.
227. **Ottlecz, A., Snyder, G. D., and McCann, S. M.,** *Proc. Natl. Acad. Sci. U.S.A.*, 85, 9861, 1988.
228. **Kitajima, N., Chihara, K., Abe, H., et al.,** *Life Sci.*, 47, 2371, 1990.
229. **Murakami, Y., Kato, Y., Shimatsu, A., et al.,** *Endocrinology (Baltimore)*, 124, 1224, 1989.
230. **Murakami, Y., Kato, Y., Koshiyama, H., et al.,** *Eur. J. Pharmacol.*, 136, 415, 1987.
231. **Cella, S. G., Locatelli, V., De Gennarro, V., et al.,** *Endocrinology (Baltimore)*, 122, 855, 1988.
232. **Delemarre-van de Waal, H. A., Burton, K. A., Kabigting, E. B., et al.,** *Endocrinology (Baltimore)*, 134, 665, 1994.
233. **Maiter, D. M., Hooi, S. C., Koenig, J. I., et al.,** *Endocrinology (Baltimore)*, 126, 1216, 1990.
234. **Giustina, A., Licini, M., Girelli, A., et al.,** *Neuroendocrinology*, 57, 843, 1993.
235. **Negro-Vilar, A., Lopez, F., Merchenthaler, I., Liposits, Z. and Guistina, A.,** *Molecular and Clinical Advances in Pituitary Tumors* (Ed. Melmed, S.). Endocrine Research and Education, Inc., Los Angeles, CA, 1993.
236. **Tanoh, T., Shimatsu, A., Ishikawa, Y., et al.,** *J. Neuroendocrinol.*, 5, 183, 1993.
237. **Davis, T. M. E., Burrin, J. M., and Bloom, S. R.,** *J. Clin. Endocrinol. Metab.*, 65, 1248, 1987.
238. **Chatterjee, V. K. K., Ball, J. A., Proby, C., et al.,** *J. Endocrinol.*, 116, R1, 1988.
239. **Loche, S., Vista, N., Ghigo, E., et al.,** *Pediatr. Res.*, 27, 405, 1990.
240. **Chronwell, B. M.,** *Neuropeptides* (Eds. Mutt, V., et al.). Raven, Press, New York, 1989, p. 51.
241. **McDonald, J. K., Koenig, J. E., Gibbs, D. M., et al.,** *Neuroendocrinology*, 46, 538, 1987.
242. **Moore, M. R. and Black, P. M.,** *Neurosurg. Rev.*, 14, 97, 1991.
243. **Everitt, B. J., Hokfelt, T., Terenius, L., et al.,** *Neuroscience*, 11, 443, 1984.
244. **Adams, E. F., Venetikou, M. S., Woods, C. A., et al.,** *Acta Endocrinol. (Copenh.)*, 115, 149, 1987.
245. **McDonald, J. K., Lumpkin, M. D., Samson, W. K., et al.,** *Proc. Natl. Acad. Sci. U.S.A.*, 82, 561, 1985.
246. **Peng, C., Chang, J. P., Yu, K. L., et al.,** *Endocrinology (Baltimore)*, 132, 1820, 1993.
247. **Catzeflis, C., Pierroz, D., Rohner-Jeanrenaud, F., et al.,** *Endocrinology (Baltimore)*, 132, 224, 1993.
248. **Rettori, V., Milenkovic, L., Aguila, M. C., et al.,** *Endocrinology (Baltimore)*, 126, 2296, 1990.
249. **McDonald, J. K., Lumpkin, M. D., Samson, W. K., et al.,** *Peptides*, 6, 79, 1985.
250. **Franci, C. R., Anselmo-Franci, J. A., Kozlowski, G. P., et al.,** *Neuroendocrinology*, 57, 693, 1993.
251. **Heidingsfelder, S. A. and Blackard, W. G.,** *Metab. Clin. Exp.*, 17, 1019, 1968.
252. **Meyer, V. and Knobil, E.,** *Endocrinology (Baltimore)*, 79, 1016, 1966.
253. **Zahnd, G. R., Nadeau, A., and Muhlendahl, K. E.,** *Lancet*, 2, 1278, 1969.
254. **Zahnd, G. R. and Vecsey, A.,** *Front. Horm. Res.*, 4, 188, 1977.
255. **Strauch, G., Girault, D., Rifai, M., et al.,** *J. Clin. Endocrinol. Metab.*, 37, 990, 1973.
256. **Kobberling, J., Juppner, H., and Hesch, R. D.,** *Acta Endocrinol.*, 81, 263, 1976.
257. **Herz, A.,** *Central and Peripheral Endorphins* (Eds. Muller, E. E., and Genazzani, A. R.). Raven, New York, p. 43.
258. **Hokfelt, T., Vincent, S. R., Dalsgaard, C. J., et al.,** *Central and Peripheral Endorphins* (Eds. Muller, E. E. and Genazzani, A. R.). Raven, New York, 1984, p. 1.
259. **Chapman, I. M. and Willoughby, J. O.,** *Neuroendocrinology*, 57, 912, 1993.
260. **Simpkins, J. W., Millard, W. J., and Berglund, L. A.,** *Life Sci.*, 52, 1443, 1993.
261. **Murakami, Y., Kato, Y., Kabayama, Y., et al.,** *Proc. Soc. Exp. Biol. Med.*, 178, 151, 1985.
262. **Martin, J. B., Audet, J., and Saunders, A.,** *Endocrinology (Baltimore)*, 96, 839, 1974.
263. **Rivier, C., Vale, W., Ling, N., et al.,** *Endocrinology (Baltimore)*, 100, 238, 1977.
264. **Pfeiffer, A., Braun, S., Mann, K., et al.,** *J. Clin. Endocrinol. Metab.*, 62, 181, 1986.

265. **Katakami, H., Kato, Y., Masushita, N., et al.,** *Neuroendocrinology*, 33, 129, 1981.
266. **Krulich, L., Koenig, J. I., Conway, S., et al.,** *Neuroendocrinology*, 42, 75, 1986.
267. **Koenig, J. I., Mayfield, M. A., McCann, S. M., et al.,** *Life Sci.*, 34, 1829, 1984.
268. **Reid, R. L., Hoff, J. D., Yen, S. S. C., et al.,** *J. Clin. Endocrinol. Metab.*, 52, 1179, 1981.
269. **Tolis, G., Hickey, J., and Guyda, H.,** *J. Clin. Endocrinol. Metab.*, 27, 797, 1975.
270. **Krulich, L., Koenig, J. I., Conway, S., et al.,** *Neuroendocrinology*, 42, 82, 1986.
271. **Beltchetz, P. E.,** *Life Sci.*, 28, 2961, 1980.
272. **Chapman, I. M., Kapoor, R., and Willoughby, J. O.,** *Neuroendocrinology*, 57, 921, 1993.
273. **Cusan, L., Dupont, A., Kledzik, G. S., et al.,** *Nature (Lond.)*, 268, 544, 1977.
274. **Molnar, J., Marton, J., and Halasz, B.,** *J. Neuroendocrinol.*, 2, 477, 1990.
275. **Casanueva, F. F., Spampinato, S., Locatelli, V., et al.,** *Endocrinology, Neuroendocrinology, Neuropeptides II* (Eds. Stark, E., Makars, G. B., and Rappay, G. Y.). Akad. Kiado, Budapest, 1980, p. 303.
276. **Chang, W. J., Barb, C. R., Kraeling, R. R., et al.,** *Domest. Anim. Endocrinol.*, 10, 305, 1993.
277. **Borer, K. T., Nicoski, D. R., and Owens, V.,** *Endocrinology (Baltimore)*, 118, 844, 1986.
278. **Wehrenberg, W. B., Bloch, B., and Ling, N.,** *Neuroendocrinology*, 41, 13, 1985.
279. **Delitala, G., Tomasi, P. A., Palermo, M., et al.,** *J. Clin. Endocrinol. Metab.*, 69, 356, 1989.
280. **Millard, W. J. and Martin, J. B.,** *Opioid Modulation of Endocrine Function* (Eds. Delitala, G., Motta, M., and Serio, M.). Raven, New York, 1984, p. 111.
281. **Spiler, I. J. and Molitch, M. E.,** *J. Clin. Endocrinol. Metab.*, 50, 516, 1980.
282. **Thompson, D. L., Weltman, J. Y., Rogol, A. D., et al.,** *J. Appl. Physiol.*, 75, 870, 1993.
283. **Pullan, P. T., Finch, P. M., Yuen, R. W. M., et al.,** *Life Sci.*, 32, 1705, 1983.
284. **Ganzetti, I., Petraglia, F., Capuano, I., et al.,** *J. Endocrinol. Invest.*, 10, 241, 1987.
285. **Hart, I. C. and Cowie, A. T.,** *J. Endocrinol.*, 77, 16P, 1978.
286. **Tindal, J. S., Knaggs, G. S., Hart, I. C., et al.,** *Endocrinology (Baltimore)*, 76, 333, 1978.
287. **Briski, K. P., Quigley, K., and Meites, J.,** *Proc. Soc. Exp. Biol. Med.*, 177, 137, 1984.
288. **Armario, A., Garcia-Marquez, C., and Jolin, T.,** *Brain Res.*, 401, 200, 1987.
289. **Martin, J. B., Tolis, G., Wood, J., et al.,** *Brain Res.*, 168, 210, 1979.
290. **Tannenbaum, G. S., Panerai, A. E., and Friesen, H. G.,** *Life Sci.*, 25, 1983, 1979.
291. **Miki, N., Sontag, W. B., Forman, L. J., et al.,** *Proc. Soc. Exp. Biol. Med.*, 168, 334, 1981.
292. **Daniels, M., Harris, P. E., James, R. A., et al.,** *J. Endocrinol.*, 137, 329, 1993.
293. **Ottlecz, A. and Telegdy, G.,** *Neuropeptides*, 9, 161, 1987.
294. **Charnay, Y., Bouras, C., Vallet, P. G., et al.,** *Neuroendocrinology*, 49, 169, 1989.
295. **Iyer, K. S. and McCann, S. M.,** *Peptides*, 8, 45, 1987.
296. **Iyer, K. S. and McCann, S. M.,** *Neuroendocrinology*, 46, 93, 1987.
297. **Fischer, J. A., Tobler, P. H., Kaufmann, M., et al.,** *Proc. Natl. Acad. Sci. U.S.A.*, 78, 7801, 1981.
298. **Cantalamessa, L., Catania, A., Reschini, E., et al.,** *Metabolism*, 27, 987, 1978.
299. **Looij, B. J., Jr., Roelfsema, F., van der Heide, D., et al.,** *Clin. Endocrinol.*, 29, 517, 1988.
300. **Petralito, A., Lunetta, M., Liuzzi A., et al.,** *Horm. Metab. Res.*, 11, 641, 1979.
301. **Lunetta, M., Infantone, E., Spanti, D., et al.,** *J. Endocrinol. Invest.*, 4, 185, 1981.
302. **Ceda, G. P., Denti, L., Ceresini, G., et al.,** *Acta Endocrinol.(Copenh.)*, 120, 416, 1989.
303. **Zofkova, I., Hampl, P., and Nedvidkova, J.,** *Horm. Metab. Res.*, 16, 499, 1984.
304. **Minamitani, N., Chihara, K., Kaji, H., et al.,** *Endocrinology (Baltimore)*, 117, 347, 1985.
305. **Lengyel, A. M. J. and Tannenbaum, G. S.,** *Endocrinology (Baltimore)*, 120, 1377, 1987.
306. **Rosenfeld, M., Amara, S., and Evans, R.,** *Science*, 225, 1315, 1984.
307. **Fahim, A., Rettori, V., and McCann, S. M.,** *Neuroendocrinology*, 51, 688, 1990.
308. **Netti, C., Guidobono, F., Sibilia, V., et al.,** *Neuroendocrinology*, 49, 242, 1989.
309. **Brubaker, P. L. and Greenberg, G. R.,** *Endocrinology (Baltimore)*, 133, 2833, 1993.
310. **Jonsson, G., Fuxe, K., and Hokfelt, T.,** *Brain Res.*, 40, 271, 1972.
311. **Meister, B., Hokfelt, T., Vale, W. W., et al.,** *Neuroendocrinology*, 42, 237, 1986.
312. **Niimi, M., Takahara, J., Sato, M., et al.,** *Neuroendocrinology*, 55, 92, 1992.
313. **Leebaw, W. F., Lee, L. A., and Woolf, P. D.,** *J. Clin. Endocrinol. Metab.*, 47, 480, 1978.
314. **Burrow, G. N., May, P. B., Spaulding, S. W., et al.,** *J. Clin. Endocrinol. Metab.*, 45, 65, 1977.
315. **Liuzzi, A., Chiodini, P. G., Botalla, L., et al.,** *J. Clin. Endocrinol. Metab.*, 39, 871, 1974.
316. **de Zegher, F., Van den Berghe, G., Devlieger, H., et al.,** *Pediatr. Res.*, 34, 642, 1993.
317. **Casaneuva, F., Betti, R., Cocchi, D., et al.,** *Endocrinology (Baltimore)*, 108, 1469, 1981.
318. **Vijayan, E., Krulich, L., and McCann, S. M.,** *Neuroendocrinology*, 26, 174, 1978.
319. **Sinha, Y. N., Salocks, C. B., and Vanderlaan, W. P.,** *Hormone Metab. Res.*, 8, 332, 1976.
320. **Miell, J. P., Pralong, F. P., Corder, R., et al.,** *J. Clin. Endocrinol. Metab.*, 71, 1519, 1990.

321. **Kakucska, I. and Makara, G. B.,** *Endocrinology (Baltimore)*, 113, 318, 1983.
322. **Toivola, P. T. K. and Gale, C. C.,** *Neuroendocrinology*, 6, 210, 1970.
323. **Bansal, S., Lee, L. A., and Woolf, P. D.,** *J. Clin. Endocrinol. Metab.*, 53, 1273, 1981.
324. **Vance, M. L., Kaiser, D. L., Frohman, L. A., et al.,** *J. Clin. Endocrinol. Metab.*, 64, 1136, 1987.
325. **Delatalia, G. and Palermo, M.,** *Neuroendocrinology*, 1994.
326. **Delitala, G., Palermo, M., Ross, R., et al.,** *Neuroendocrinology*, 45, 243, 1987.
327. **Donnadieu, M., Evain-Brion, D., Tonan, M. C., et al.,** *J. Clin. Endocrinol. Metab.*, 60, 1132, 1985.
328. **Chihara, K., Kashio, Y., Kita, T., et al.,** *J. Clin. Endocrinol. Metab.*, 62, 466, 1986.
329. **Cronin, M. J., Thorner, M. O., Hellmann, P., et al.,** *Proc. Soc. Exp. Biol. Med.*, 175, 191, 1984.
330. **Marcovitz, S., Goodyer, C. G., Gudya, H., et al.,** *J. Clin. Endocrinol. Metab.*, 54, 6, 1982.
331. **Bression, D., Brandi, A. M., Nousbaum, A., et al.,** *J. Clin. Endocrinol. Metab.*, 5, 589, 1982.
332. **Lawton, N. F., Evans, A. J., and Weller, R. O.,** *J. Neuro. Sci.*, 49, 229, 1987.
333. **Ray, K. P., Gomm, J. J., Law, G. J., et al.,** *Mol. Cell. Endocrinol.*, 45, 175, 1986.
334. **Lindstrom, P. and Ohlsson, L.,** *Endocrinology (Baltimore)*, 120, 780, 1987.
335. **Cheung, C. Y., Melmed, S., and Braunstein, G. D.,** *Brain Res.*, 270, 165, 1983.
336. **Bluet-Pajot, M. T., Mounier, F., Durand, D., et al.,** *J. Endocrinol.*, 127, 191, 1990.
337. **Wong, A. O.-L., Chang, J. P., and Peter, R. E.,** *Am. J. Physiol.*, 264, E925, 1993.
338. **Camanni, F., Massara, F., Belforte, L., et al.,** *J. Clin. Endocrinol. Metab.*, 44, 465, 1977.
339. **Leblanc, H., Lachelin, C. G. L., Abu Fadil, S., et al.,** *J. Clin. Endocrinol. Metab.*, 43, 668, 1976.
340. **Bazan, M. C., Barontini, M., Domene, H., et al.,** *J. Clin. Endocrinol. Metab.*, 52, 314, 1981.
341. **Bansal, S. A., Lee, L. A., and Woolf, P. D.,** *J. Clin. Endocrinol. Metab.*, 53, 1273, 1981.
342. **Bellastella, A., Parlato, F., and Sinisi, A. A.,** *J. Clin. Endocrinol. Metab.*, 70, 856, 1990.
343. **Eden, S., Bolle, P., and Modigh, K.,** *Endocrinology (Baltimore)*, 105, 523, 1979.
344. **Cella, S. G., Picotti, G. B., Morgese, M., et al.,** *Life Sci.*, 34, 447, 1984.
345. **Eden, S. and Modigh, K.,** *Brain Res.*, 129, 379, 1977.
346. **Login, I. S., Judd, A. M., Thorner, M. O., et al.,** *Proc. Soc. Exp. Biol. Med.*, 171, 247, 1982.
347. **Bluet-Pajot, M. T., Mounier, F., Durand, D., et al.,** *J. Neuroendocrinol.*, 4, 9, 1992.
348. **Krulich, L., Mayfield, M. A., Steele, M. K., et al.,** *Endocrinology (Baltimore)*, 110, 196, 1982.
349. **Crowley, W. R., Terry, L. C., and Johnson, M. D.,** *Endocrinology (Baltimore)*, 110, 1102, 1982.
350. **Durand, D., Martin, S. B., and Brazeau, P.,** *Endocrinology (Baltimore)*, 100, 722, 1977.
351. **Malozowski, S., Hao, E.-H., Guang Ren, S., et al.,** *Neuroendocrinology*, 51, 455, 1990.
352. **McWilliam, J. R. and Meldrum, B. S.,** *Endocrinology (Baltimore)*, 112, 254, 1983.
353. **Chihara, K., Minamitani, N., Kaji, H., et al.,** *Endocrinology (Baltimore)*, 114, 1402, 1984.
354. **Cella, S. G., Picotti, G. B., and Muller, E. E.,** *Life Sci.*, 2785, 1983.
355. **Cella, S. G., Locatelli, V., DeGennaro, V., et al.,** *Endocrinology (Baltimore)*, 120, 1639, 1987.
356. **Muruais, J., Penalva, A., Dieguez, C., et al.,** *J. Endocrinol.*, 138, 211, 1993.
357. **Lal, S., Tolis, G., Martin, J. B., et al.,** *J. Clin. Endocrinol. Metab.*, 41, 703, 1975.
358. **Lancranjan, I. and Marbach, P.,** *Metabolism*, 26, 1225, 1977.
359. **McWilliam, J. R. and Meldrum, B. S.,** *Horm. Metab. Res.*, 17, 443, 1985.
360. **Ishikawa, K., Suzuki, M., and Kakegawa, T.,** *Endocrinol. Jpn.*, 30, 397, 1983.
361. **Jaffer, A., Daniels, W. M. U., Russell, V. A., et al.,** *Neurochem. Res.*, 17, 1255, 1992.
362. **Kacsoh, B., Opp, J. S., Crowley, W. R., et al.,** *Acta Endocrinol. (Copenh.)*, 128, 184, 1993.
363. **Gil-Ad, I., Laron, Z., and Koch, Y.,** *J. Endocrinol.*, 131, 381, 1991.
364. **Kacsoh, B. and Grosvenor, C. E.,** *J. Neuroendocrinol.*, 3, 529, 1991.
365. **Kiem, D. T., Stark, E., and Fekete, M. I. K.,** *Neuroendocrinology*, 51, 226, 1990.
366. **Devesa, J., Lois, N., Diaz, M. J., et al.,** *J. Steroid Mol. Biol.*, 40, 165, 1991.
367. **Vigas, M., Malatinsky, J., Nemeth, S., et al.,** *Metabolism*, 26, 399, 1977.
368. **Magnan, E., Cataldi, M., Guillaume, V., et al.,** *Endocrinology (Baltimore)*, 134, 562, 1994.
369. **Devesa, J., Diaz, M. J., Tresguerres, J. A. F., et al.,** *J. Clin. Endocrinol. Metab.*, 73, 251, 1991.
370. **Chapman, I. M., Kapoor, R., and Willoughby, J. O.,** *J. Neuroendocrinol.*, 5, 145, 1993.
371. **Kato, Y., Katakami, H., and Imura, H.,** *Growth and Growth Factors* (Eds. Shizume, K. and Takano, K.). Univ. of Tokyo Press, Tokyo, 1980, p. 159.
372. **Cella, S. G., Locatelli, V., De Gennaro, V., et al.,** *Endocrinology (Baltimore)*, 119, 1164, 1986.
373. **Eden, S., Eriksson, E., Martin, J. B., et al.,** *Neuroendocrinology*, 33, 24, 1981.
374. **Miki, N., Ono, M., and Shizume, K.,** *Endocrinology (Baltimore)*, 114, 1950, 1984.

375. **Kabayama, Y., Kato, Y., Murakami, Y., et al.,** *Endocrinology (Baltimore)*, 119, 432, 1986.
376. **Tsagarakis, S., Ge, F., Rees, L. H., et al.,** *J. Neuroendocrinol.*, 1, 129, 1989.
377. **Miki, N., Ono, M., and Shizume, K.,** *Endocrinology (Baltimore)*, 114, 1950, 1984.
378. **Willoughby, J. O., Chapman, I. M., and Kapoor, R.,** *Neuroendocrinology*, 57, 687, 1993.
379. **Day, T. A., Jervois, P. M., Menadue, et al.,** *Brain Res.*, 253, 213, 1982.
380. **Lima, L., Arce, V., Diaz, J., et al.,** *Clin. Endocrinol.*, 35, 129, 1991.
381. **Massara, F. and Strumia, E.,** *J. Endocrinol.*, 47, 95, 1970.
382. **Chihara, K., Kodama, H., Kaji, H., et al.,** *J. Clin. Endocrinol. Metab.*, 61, 229, 1985.
383. **Mauras, N., Blizzard, R. M., Thorner, M. O., et al.,** *Metabolism*, 36, 369, 1987.
384. **Krieg, R. J., Perkins, S. N., Johnson, J. H., et al.,** *Endocrinology (Baltimore)*, 122, 231, 1988.
385. **Richardson, S. B. and Twente, S.,** *Endocrinology (Baltimore)*, 126, 1043, 1990.
386. **Perkins, S. N., Evans, W. S., Thorner, M. O., et al.,** *Endocrinology (Baltimore)*, 117, 1818, 1985.
387. **Perkins, S. N., Evans, W. S., Thorner, M. O., et al.,** *Neuroendocrinology*, 37, 473, 1983.
388. **Esquifino, A. I., Agrasal, C., Steger, R., et al.,** *Life Sci.*, 41, 1043, 1987.
389. **Spencer, G. S. G., Dobbie, P., and Bass, J. J.,** *Domest. Anim. Endocrinol.*, 8, 383, 1991.
390. **Meunier, M., Brebion, P., Chene, N., et al.,** *J. Endocrinol.*, 119, 287, 1988.
391. **Martin, J. B.,** *Frontiers in Neuroendocrinology* (Eds. Martini, L. and Ganong, W. F.). Raven, New York, 1976, p. 128.
392. **Stuart, M., Lazarus, L., Smythe, G. A., et al.,** *Neuroendocrinology*, 22, 337, 1976.
393. **Arnold, M. A. and Fernstrom, J. D.,** *Endocrinology (Baltimore)*, 108, 331, 1981.
394. **Smythe, G. A., Compton, P. J., and Lazarus, L.,** *Growth Hormone and Related Proteins* (Eds. Muller, E. E. and Panerai, A.). Exerpta Medica, Amsterdam, 1978, p. 222.
395. **Dammacco, F., Puca, F. M., Rigillo, N., et al.,** *Horm. Metab. Res.*, 9, 244, 1977.
396. **Smythe, G. A., Duncan, M. W., Bradshaw, J. E., et al.,** *Endocrinology (Baltimore)*, 110, 376, 1982.
397. **Koulu, M.,** *J. Neural Transmission*, 55, 269, 1982.
398. **Smythe, G. A., Brandstater, J. F., and Lazarus, L.,** *Neuroendocrinology*, 17, 245, 1975.
399. **Chambers, J. W. and Brown, G. M.,** *Endocrinology (Baltimore)*, 98, 420, 1976.
400. **Dorsa, D. M. and Connors, M. H.,** *Life Sci.*, 22, 1391, 1978.
401. **Dorsa, D. M. and Connors, M. H.,** *Endocrinology (Baltimore)*, 104, 101, 1979.
402. **Willoughby, J. O., Menadue, M. F., and Liebelt, H.,** *Brain Res.*, 404, 319, 1987.
403. **Lewis, D. A. and Sherman, B. M.,** *Acta Endocrinol. (Copenh.)*, 110, 158, 1985.
404. **Marti-Henneberg, C., Gluckman, P., Kaplan, S., et al.,** *Endocrinology (Baltimore)*, 107, 1273, 1980.
405. **Aulakh, C. S., Hill, J. L., and Murphy, D. L.,** *Neuroendocrinology*, 59, 35, 1994.
406. **Nakai, Y. and Imura, H.,** *Endocrinol. Jpn.*, 21, 493, 1974.
407. **Cocchi, D., Gil-Ad, I., Panerai, A. E., et al.,** *Neuroendocrinology*, 24, 1, 1977.
408. **Attanasio, A., Bombelli, M., Kuzmanovic, D., et al.,** *Neuroendocrinol. Lett.*, 8, 275, 1986.
409. **Murakami, Y., Kato, Y., Kabayama, Y., et al.,** *Endocrinology (Baltimore)*, 119, 1089, 1986.
410. **Arnold, M. A. and Fernstrom, J. D.,** *Neuroendocrinology*, 31, 194, 1980.
411. **Peterfreund, R. A. and Vale, W. W.,** *Endocrinology (Baltimore)*, 112, 526, 1983.
412. **Richardson, S. B., Hollander, C. S., Prasad, J. A., et al.,** *Endocrinology (Baltimore)*, 109, 602, 1981.
413. **Wehrenberg, W. B., McNicol, D., Frantz, A. G., et al.,** *Endocrinology (Baltimore)*, 107, 1747, 1980.
414. **Westlund, K. N. and Childs, G. V.,** *Endocrinology (Baltimore)*, 111, 1761, 1982.
415. **Johns, M. A., Azmitia, E. C., and Kriegeer, D. T.,** *Endocrinology (Baltimore)*, 110, 754, 1982.
416. **Ishibashi, M., Fukushima, T., and Yamaji, T.,** *Acta Endocrinol.*, 109, 474, 1985.
417. **Griffiths, D., Bjoro, T., Gautvik, K., et al.,** *Acta Physiol. Scand.*, 131, 43, 1987.
418. **Valcavi, R., Dieguez, C., Azzarito, C., et al.,** *Clin. Endocrinol. (Oxford)*, 26, 453, 1987.
419. **Smythe, G. A. and Lazarus, L.,** *Horm. Metab. Res.*, 5, 227, 1973.
420. **Valcavi, R., Zini, M., Maestroni, G. J., et al.,** *Clin. Endocrinol. (Oxford)*, 39, 193, 1993.
421. **Locatelli, V., Tosello, A., Redaelli, M., et al.,** *J. Endocrinol.*, 111, 271, 1986.
422. **Bruni, J. F. and Meites, J.,** *Life Sci.*, 23, 1351, 1978.
423. **Mukherjee, A., Snyder, A. G., and McCann, S. M.,** *Life Sci.*, 27, 475, 1980.
424. **Casanueva, F. F., Betti, R., Cella, S. G., et al.,** *Acta Endocrinol.*, 103, 15, 1983.
425. **Ross, R. J. M., Ttsagarakis, S., Grossman, A., et al.,** *Clin. Endocrinol.*, 27, 727, 1988.
426. **Penalva, A., Muruais, C., Casanueva, F. F., et al.,** *J. Clin. Endocrinol. Metab.*, 70, 324, 1990.
427. **Bicknell, R. J., Young, P. W., and Schofield, J. G.,** *Mol. Cell. Endocrinol.*, 13, 167, 1979.

428. **Massara, F., Ghigo, E., Goffi, S., et al.,** *J. Clin. Endocrinol. Metab.*, 59, 1025, 1984.
429. **Jordan, V., Dieguez, C., Lafaffian, I., et al.,** *Clin. Endocrinol. (Oxford)*, 24, 291, 1986.
430. **Taylor, B. J., Smith, P. J., and Brook, C. G. D.,** *Clin. Endocrinol. (Oxford)*, 22, 497, 1985.
431. **Casanueva, F. F., Villanueva, L., Dieguez, C., et al.,** *J. Clin. Endocrinol. Metab.*, 62, 186, 1986.
432. **Degli Uberti, E. C., Trasforini, G., Salvadori, S., et al.,** *Horm. Res.*, 24, 251, 1986.
433. **Delitala, G., Grossman, A., and Besser, G. M.,** *Clin. Endocrinol. (Oxford)*, 18, 401, 1983.
434. **Peters, J. R., Evans, P. J., Page, M. D., et al.,** *Clin. Endocrinol.*, 25, 213, 1986.
435. **Arvat, E., Cappa, M., Casanueva, F. F., et al.,** *J. Clin. Endocrinol. Metab.*, 76, 374, 1993.
436. **Penalva, A., Carballo, A., Pombo, M., et al.,** *J. Clin. Endocrinol. Metab.*, 76, 168, 1993.
437. **Delitala, G., Frulio, T., Pacifico, A., et al.,** *J. Clin. Endocrinol. Metab.*, 55, 1231, 1982.
438. **Delitala, G., Maioli, M., Pacifico, A., et al.,** *J. Clin. Endocrinol. Metab.*, 5754, 1145, 1983.
439. **Burt, D. R. and Taylor, L.,** *Neuroendocrinology*, 30, 344, 1980.
440. **Young, P. W., Bicknell, R. J., and Schofield, J. G.,** *J. Endocrinol.*, 80, 203, 1979.
441. **Massara, F., Ghigo, E., Molinatti, P., et al.,** *Acta Endocrinol.*, 113, 12, 1986.
442. **Ghigo, E., Bellone, J., Imperiale, E., et al.,** *Neuroendocrinology*, 52, 42, 1990.
443. **Evans, P. J., Dieguez, C., Foord, S., et al.,** *Clin. Endocrinol.*, 22, 733, 1985.
444. **Knigge, U. and Warberg, J.,** *Acta Endocrinol. (Copenh.)*, 124, 609, 1991.
445. **Cacabelos, R., Niigawa, H., Yamatodani, A., et al.,** *Endocrinology (Baltimore)*, 122, 1269, 1988.
446. **Netti, C., Sibilia, V., Guidobono, F., et al.,** *Neuroendocrinology*, 57, 1066, 1993.
447. **Gonzales-Villapando, C., Szabo, M., and Frohman, L. A.,** *J. Clin. Endocrinol. Metab.*, 51, 1417, 1980.
448. **Yeo, T., Delitala, G., Besser, G. M., et al.,** *Br. J. Clin. Pharmacol.*, 10, 171, 1980.
449. **Locatelli, V., Torsello, A., Redaelli, M., et al.,** *J. Endocrinol.*, 111, 271, 1986.
450. **Netti, C., Guidobono, G. B., Olgiati, V. R., et al.,** *Neuroendocrinology*, 35, 43, 1982.
451. **Netti, C., Guidobono, F., Oligiati, V. R., et al.,** *Horm. Res.*, 14, 180, 1981.
452. **Rudolph, C., Richards, G. E., Kaplan, S., et al.,** *Neuroendocrinology*, 29, 169, 1979.
453. **Locatelli, V., Panerai, A. E., Cocchi, D., et al.,** *Neuroendocrinology*, 25, 84, 1978.
454. **Betti, R., Casanueva, F., Cella, S. G., et al.,** *Acta Endocrinol.*, 108, 36, 1985.
455. **Casanueva, F. F., Betti, R., Frigerio, C., et al.,** *Endocrinology (Baltimore)*, 106, 1239, 1980.
456. **Casanueva, F., Betti, R., Cocchi, D., et al.,** *Endocrinology (Baltimore)*, 108, 157, 1981.
457. **Collu, R., Du Ruisseau, P., and Tache, Y.,** *Neuroendocrinology*, 28, 178, 1979.
458. **Maeda, K. and Frohman, L. A.,** *Endocrinology (Baltimore)*, 103, 1903, 1978.
459. **Tanaka, J., Yamatodani, A., Niigawa, H., et al.,** *Endocrinol. Jpn.*, 34(Suppl.), 151, 1987.
460. **Pontiroli, A. E., Pellicciotta, G., Alberetto, M., et al.,** *Horm. Metab. Res.*, 12, 172, 1980.
461. **Pontiroli, A. E., Viberti, G., Vicari, A., et al.,** *J. Clin. Endocrinol. Metab.*, 43, 582, 1976.
462. **Knigge, U., Thuesen, B., Wollesen, F., et al.,** *J. Clin. Endocrinol. Metab.*, 58, 692, 1984.
463. **Elias, A. N., Valenta, L. J., Szekeres, A. V., et al.,** *Psychoneuroendocrinology*, 7, 15, 1982.
464. **Racagni, G., Apud, J. A., Cocchi, D., et al.,** *Life Sci.*, 31, 823, 1982.
465. **Tamminga, C., Neophytides, A., Chase, T. N., et al.,** *J. Clin. Endocrinol. Metab.*, 47, 1348, 1978.
466. **Koulu, M., Lammintausta, R., and Dahlstrom, S.,** *J. Clin. Endocrinol. Metab.*, 48, 1038, 1979.
467. **Cocchi, D., Casanueva, F., Locatelli, V., et al.,** *GABA and Benzodiazepine Receptors* (Eds. Costa, E., Dichiari, G., and Gessa, G. L.), Raven Press, New York, 1981, p. 247.
468. **Acs, Z., Lonart, G., and Makara, G. B.,** *Neuroendocrinology*, 52, 156, 1990.
469. **Martin, J. B., Durand, B., Gurd, W., et al.,** *Endocrinology (Baltimore)*, 102, 106, 1978.
470. **McCann, S. M., Vijayan, E., Negro-Vilar, A., et al.,** *Psychoneuroendocrinology*, 9, 97, 1984.
471. **Fiok, J., Acs, G., Makara, G., et al.,** *Neuroendocrinology*, 39, 510, 1984.
472. **Cavagnini, F., Invitti, C., Pinto, M., et al.,** *Acta Endocrinol.*, 91, 149, 1980.
473. **Cavagnini, F., Invitti, C., Di Landro, A., et al.,** *J. Clin. Endocrinol. Metab.*, 45, 579, 1977.
474. **Koulu, M., Lammintausta, R., and Dahlstrom, S.,** *J. Clin. Endocrinol. Metab.*, 51, 124, 1980.
475. **Vijayan, E. and McCann, S. M.,** *Endocrinology (Baltimore)*, 103, 1888, 1978.
476. **McCann, S. M., Vijayan, E., and Negro-Vilar, A.,** *Adv. Biochem. Psychopharmacol.*, 20, 237, 1980.
477. **Vijayan, E. and McCann, S. M.,** *Brain Res.*, 163, 69, 1979.
478. **Bruni, J. F., Miodusewski, R. M., Grandison, L. J., et al.,** *Fed. Proc.*, 36, 323, 1977.
479. **Cavagnini, F., Benetti, G., Invitti, C., et al.,** *J. Clin. Endocrinol. Metab.*, 51, 789, 1980.
480. **Takahara, J., Yunoki, S., Hisogi, H., et al.,** *Endocrinology (Baltimore)*, 106, 343, 1980.
481. **Murakami, Y., Kato, Y., Kabayama, Y., et al.,** *Endocrinology (Baltimore)*, 117, 787, 1985.

482. **Acs, Z., Szabo, B., Kapocs, G., et al.,** *Endocrinology (Baltimore)*, 120, 1790, 1987.
483. **Acs, Z., Makara, G. B., and Stark, E.,** *Life Sci.*, 34, 1505, 1984.
484. **Anderson, R. A. and Mitchell, R.,** *J. Endocrinol.*, 108, 1, 1986.
485. **Acs, Z., Zsom, L., and Makara, G. B.,** *Life Sci.*, 50, 273, 1991.
486. **Plant, T. M., Gay, V. L., Marshall, G. R., et al.,** *Proc. Natl. Acad. Sci. U.S.A.*, 86, 2506, 1989.
487. **Medhamurthy, R., Gay, V. L., and Plant, T. M.,** *Neuroendocrinology*, 55, 660, 1992.
488. **Mason, G. A., Bissette, G., and Nemeroff, C. B.,** *Brain Res.*, 289, 366, 1983.
489. **Estienne, M. J., Schillo, K. K., Green, M. A., et al.,** *Life Sci.*, 44, 1527, 1989.
490. **Shahab, M., Nusser, K. D., Griel, L. C., et al.,** *J. Neuroendocrinol.*, 5, 469, 1993.
491. **Barb, C. R., Derochers, G. M., Johnson, B., et al.,** *Domest. Anim. Endocrinol.*, 9, 225, 1992.
492. **Cocilovo, L., Colonna, V. D. V., Zoli, M., et al.,** *Neuroendocrinology*, 55, 416, 1992.
493. **MacDonald, M. C. and Wilkinson, M.,** *J. Neuroendocrinol.*, 4, 223, 1992.
494. **Benyassi, A., Tapia-Arancibia, L., and Arancibia, S.,** *J. Neuroendocrinol.*, 3, 429, 1991.
495. **Tapia-Arancibia, L. and Astier, H.,** *Endocrinology (Baltimore)*, 123, 2360, 1988.
496. **Tapia-Arancibia, L., Rage, F., Recasens, M., et al.,** *Eur. J. Pharmacol. Mol. Pharmacol.*, 225, 253, 1992.
497. **Nemeroff, C. B., Mason, G. A., Bissette, G., et al.,** *Neuroendocrinology*, 41, 332, 1985.
498. **Collu, R., Charpenet, G., and Clemont, M. J.,** *Can. J. Neurol. Sci.*, 5, 139, 1978.
499. **Ikuyama, S., Okajima, T., Kato, K., et al.,** *Life Sci.*, 43, 807, 1988.
500. **Mantovani, J. and Sevivo, D. C.,** *Arch. Neurol.*, 36, 672, 1979.
501. **Aguila, M. C. and McCann, S. M.,** *Endocrinology (Baltimore)*, 116, 1158, 1986.
502. **Alba-Roth, J., Muller, O. A., Schopohl, J., et al.,** *J. Clin. Endocrinol. Metab.*, 67, 1186, 1988.

Chapter 10

Growth Hormone Release: Feedback Regulation

S. Harvey

CONTENTS

I. INTRODUCTION

Unlike other pituitary hormones, growth hormone (GH) lacks specific target organs and has widespread effects on the growth and metabolism of most tissues. These actions of GH at central and peripheral sites provide short- and long-loop inhibitory feedback pathways in the regulation of GH secretion (see Muller[1] for review). These actions are subject to constraints of time, space and GH concentration and are dependent on the actions and interactions of inducible factors at pituitary and hypothalamic sites (Figure 1). The feedback regulation of GH secretion is therefore complex and involves the integration of multiple metabolic and hormonal signals.

II. ULTRASHORT-LOOP FEEDBACK

An ultrashort-loop feedback mechanism in the control of GH release is suggested by the presence of GH receptor mRNA, GH receptor immunoreactivity, and GH-binding sites in human, mouse, rat, rabbit, bovine, and chicken pituitary glands[2–8] (see Chapter 27). Moreover, exogenous GH inhibits (by 40%) basal, but not GH-releasing hormone-induced, GH release from bovine pituitary glands.[6,9] Exogenous GH treatment of rats *in vivo* has also been shown to block the *in vitro* GH response of *in situ* pituitary glands to GH-releasing hormone (GHRH). The implantation of tumorous GH-secreting cells in the rat pituitary[10] also results in impaired *GH* gene transcription, suggesting an ultrashort-

0-8493-8697-7/95/$0.00+$.50

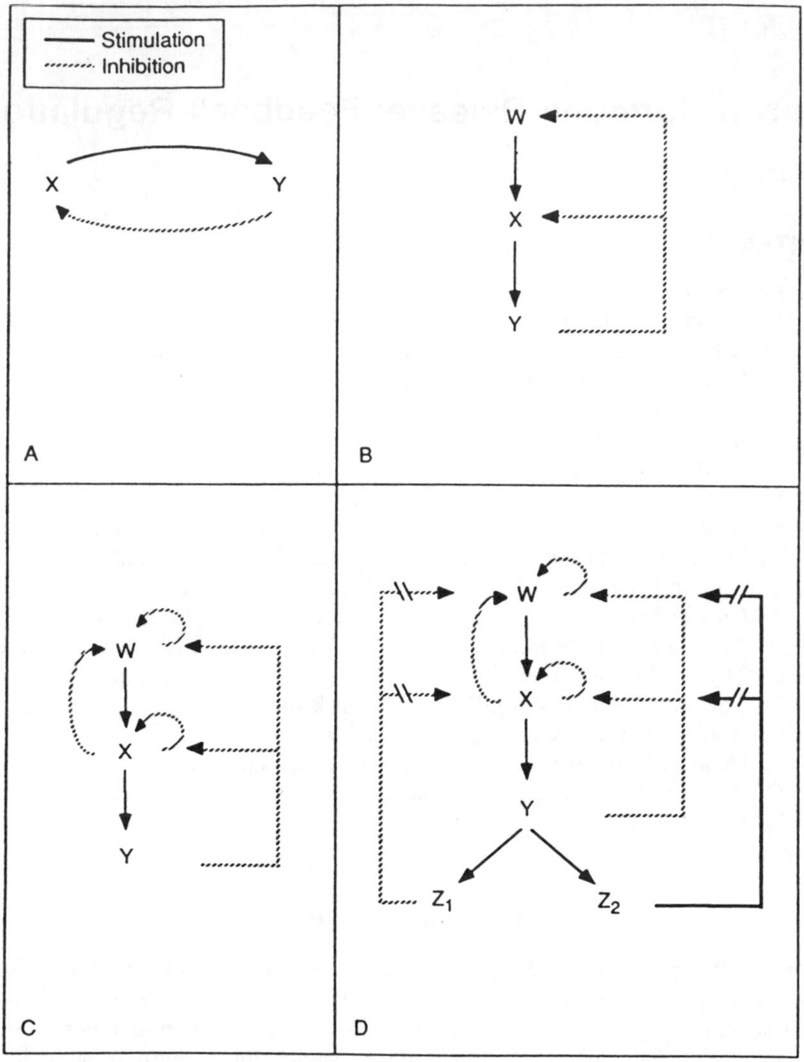

Figure 1 Models of feedback pathways in endocrine systems. (A) Simple negative feedback between the end-product hormone, Y, and the secretagogue hormone, X. In (A) through (D), solid lines represent stimulation and dashed lines represent inhibition. (B) Feedback at multiple sites. W represents the hormone product of a third site, which stimulates X and is inhibited by Y. The potency and time course of Y feedback may differ at the two sites (X and W). (C) In this scheme, the individual hormones may also exert feedback effects on themselves (ultrashort loop feedback). (D) Z_1 and Z_2 are metabolic products formed by the actions of Y. Their appearance in blood is delayed relative to X and Y. The metabolites may be either stimulatory or inhibitory at the higher sites. The break in the line is intended to suggest that the effects of Z may be mediated indirectly at other sites (e.g., at other brain nuclei). (From Widmaier, E. P., *Horm. Metab. Res.*, 24, 147, 1992. Reprinted with permission.)

loop feedback pathway in GH regulation. The blunting of the endogenous GH response to GHRH by exogenous GH in human subjects in which lipolysis and somatostatin (SRIF) secretion are blocked[11] similarly suggests autoinhibition of GH secretion at the pituitary level. Blunted pulsatile GH release[12–14] and GH responses to pituitary-acting stimuli (e.g., GHRH and thyrotropin-releasing hormone [TRH])[15–25] following exogenous GH therapy or after elevations in endogenous GH secretion[15,22,26–29] might also indicate autoregulatory actions of GH at pituitary (or hypothalamic) sites. This inhibition of GH secretion may be due to GH actions on somatotroph cells, because GH reduces the release of GH from tumorous GH-secreting (GH_3) cells.[30,31] The autoregulation of GH secretion at pituitary cells may, however, also be mediated by GH actions on extrasomatotroph cells.

Although GH receptors occur on somatotroph membranes, they are also present in lactotroph, thyrotroph, and corticotroph populations[5] and autoregulation could be mediated through paracrine factors produced in these cell types. These factors may include insulin-like growth factors (IGFs), because IGF-I and IGF-II are synthesized in the pituitary in response to GH[32–38] and IGFs directly inhibit basal pituitary GH release[37,39–45] and release induced by triiodothyroxine (T_3)[46,47] and by GHRH.[48–51] IGFs also inhibit basal and stimulated transcription of the *GH* gene.[43,45–47,52–54] Autocrine or paracrine actions of IGFs on somatotroph function may, therefore, be induced by GH, because the distribution of IGF-I receptors in pituitary glands parallels the distribution of somatotrophs.[55–58] Autocrines or paracrines distinct from IGFs may also mediate ultrashort-loop feedback regulation of GH secretion, because Stachura et al.[31,59] reported that GH and IGF-I were not the autoregulatory factors produced in GH_3 cells. The existence of an ultrashort-loop feedback pathway is also controversial, because it has not been observed in studies with rat, ovine, or chicken pituitary glands.[28,41,60–65]

III. SHORT-LOOP FEEDBACK

A short-loop autoregulatory pathway in the control of GH release is well established (Figure 2). The implantation of GH or GH-secreting cells into the median eminence or the intracerebroventricular injection of GH reduces pituitary GH concentration,[66–68] and suppresses basal, sleep-related, episodic, and pharmacologically induced GH secretion.[69–72] Basal GH secretion[20] and the pituitary GH response to centrally acting stimuli (e.g., exercise, hypoglycemia, arginine, pyridostigmine, and clonidine) are similarly reduced by systemic GH administration.[17,19,20,69,73–76] This inhibition of GH secretion may require

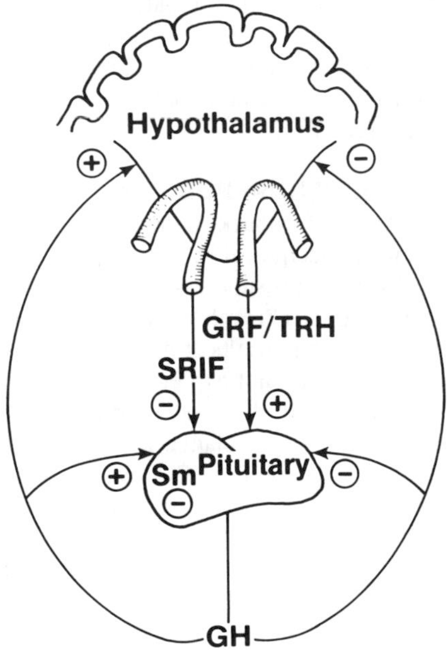

Figure 2 Autoregulation of growth hormone (GH) secretion. At hypothalamic sites, GH stimulates the release of somatostatin (SRIF) and inhibits the release of GH-releasing factor (GRF). At pituitary sites GH may inhibit GH synthesis and release directly or indirectly through increased somatomedin (Sm) activity.

several hours for its induction[77–80] and occurs independently of changes in the peripheral concentrations of other hormones (e.g., IGF-I) or inhibitory metabolites (e.g., glucose and free fatty acids [FFAs]).[19,81–85] This inhibition of GH secretion could also result from downregulated GHRH or TRH receptors on somatotroph membranes or impaired signal transduction mechanisms[8,86] or from upregulated SRIF receptors or signal transduction.[87,88] Changes in stimulatory and inhibitory hypothalamic tone are, however, likely to mediate these inhibitory actions of GH on its own secretion. This possibility is supported by the suppression of clonidine (α_2-adrenergic stimulation)-induced GH release in a pituitary-hypothalamus coincubation system following GH pretreatment of the hypothalamic tissue.[65] This suppression of GH release could, therefore, result from either reduced GHRH secretion or increased SRIF release or from a combination of both.

A. HYPOTHALAMIC ACTIONS

Although SRIF and GHRH neurons in the hypothalamus may be inaccessible to GH in peripheral plasma because of blood-brain barrier impedence, the median eminence lies outside this barrier. The retrograde flow of GH along hypophysial portal vessels and the elevated GH concentrations of portal blood[89–91] would thus facilitate modulation of SRIF- and GHRH-containing nerve terminals in the median eminence. Moreover, retrograde ependymal tanycyte transport from the hypothalamus into the third ventricle may permit the entry of pituitary GH into cerebrospinal fluid (CSF) and ultimately the access of GH to neural perikarya.[92] The entry of pituitary GH into CSF could also occur from the passage of GH through the vascular endothelium of blood vessels in the anterior pituitary gland or the pituitary stalk. The presence of GH receptors or binding proteins in the choroid plexus[93] and median eminence[94] also suggests a receptor-mediated mechanism for the entry of GH into the brain and induction of short-loop autoregulation.

Although GH is synthesized within the brain (see Chapter 28) its synthesis does not correlate with changes in pituitary GH release and hence is unlikely to mediate autoregulatory effects of GH on its neuroendocrine regulation. The receptors for GH on SRIF and GHRH neurons are, therefore, likely to be targeted by GH released from the adenohypophysis. The GH receptor gene is expressed in hypophysiotropic regions of the hypothalamus and located on SRIF cells in the arcuate and periventricular nuclei and on GHRH cells in the arcuate nucleus, supraoptic nucleus, periventricular nucleus, and ventromedial nucleus.[95,96] The responsiveness of these neurons to peripheral GH is indicated by the specific and rapid (within 40 min) induction of c-*fos* expression in the periventricular and arcuate nuclei of hypophysectomized rats following systemic GH injections.[97]

B. GHRH SUPPRESSION

An autofeedback reduction in stimulating hypothalamic tone might be expected in view of a reciprocal relationship between hypothalamic GHRH and pituitary GH. When GH secretion is suppressed by hypophysial stalk section[98] or hypophysectomy,[95,98–102] the content of *GHRH* mRNA in hypothalamic tissue is increased, particularly in the arcuate nucleus. This results from the recruitment of ventromedial hypothalamic neurons that previously did not express the *GHRH* gene[100] and by increasing the amount of GHRH mRNA produced per cell. Conversely, when GH secretion is elevated or exogenous GH is injected, hypothalamic levels of GHRH mRNA are suppressed,[99,100,103–106] in concert with reduced pituitary GH levels and impaired pituitary[103] and plasma[21,107] GH responses to provocative stimuli. Indeed, the chronic hypersecretion of GH in rats bearing GH-secreting tumors results in an 84% reduction in hypothalamic GHRH content and a 52% reduction in GHRH mRNA in the neurons of the arcuate nucleus.[105] Resection of the tumor restores normal hypothalamic GHRH mRNA levels.[105] This inhibitory effect of GH on GHRH content decreases during aging,[103] possibly as a result of reduced neurotransmission.[108] The inhibitory effect of GH on GHRH content may also be sexually dimorphic and is less in female rats than in males,[109] contributing to the sexually dimorphic pattern of pulsatile GH release (see Chapter 12). The content of hypothalamic GHRH and GHRH mRNA is also reversibly increased in hypothyroid, GH-deficient rats[82] and GHRH mRNA levels are also increased in the GH-deficient *lit/lit* mouse.[110] In the absence of GH, an increase in GHRH release or changes in the posttranscriptional regulation of GHRH synthesis or in intracellular GHRH trafficking may, however, occur and the content of GHRH in hypothalamic tissue may be unchanged[111] or reduced,[99,110,112,113] rather than enhanced. The importance of reduced GHRH in GH autoregulation is indicated by the persistence of impaired GH responses to clonidine in SRIF-immunoneutralized, GH-treated rats[69] and by the inability of intracerebroventricular (i.c.v.) GH to markedly inhibit pituitary GH responses to GHRH and TRH.[69] These observations were not, however, confirmed by Lanzi and Tannenbaum,[24] who reported that SRIF immunoneutralization reversed GH-induced blunting of endogenous GH responses to GHRH.

C. INCREASED SRIF TONE

Increased SRIF inhibition is also likely to participate in short-loop regulation of GH secretion,[114] because SRIF is also correlated with changes in GH secretion. This relationship is indicated by the reversible reduction in hypothalamic SRIF content after hypophysectomy[115–118] or pituitary stalk section[98,111] or after the administration of GH antiserum.[119] Hypophysectomy similarly reduces the expression of preprosomatostatin in the rat periventricular nucleus, whereas *SRIF* gene transcription is increased in hypophysectomized and intact animals in response to exogenous GH.[95,105,117,118] The release of SRIF into hypophysial blood *in vivo* is also increased by i.c.v. injections of GH[120] and GH induces SRIF synthesis and release in hypothalamic tissue *in vivo* and *in vitro*.[103,119,121–124] The overproduction or underproduction of GH in transgenic mice is similarly accompanied by increased and decreased levels of SRIF mRNA, respectively.[125] The chronic hypersecretion of GH from implanted GH-secreting tumors similarly increases SRIF concentrations and *SRIF* gene transcription in the rat hypothalamus.[109] The actions of GH or SRIF synthesis are, however, confined to neurons in the periventricular nuclei and are not observed in other brain nuclei[102,126] and are not observed in aged rats,[103] in which *SRIF* gene expression is reduced. The tonically suppressed GH secretion in aged rats results from reduced GHRH tone and reduced pituitary responsiveness to GH-releasing stimuli.[103]

D. GHRH-SRIF INTERACTIONS

Synaptic interactions between SRIF and GHRH neurons have been described.[127,128] SRIF receptors are located on GHRH neurons[105,129–131] and both SRIF and GHRH neurons possess GH receptors.[96,132] Cystamine-induced SRIF depletion increases hypothalamic GHRH activity,[133,134] indicating an inhibitory effect of SRIF on *GHRH* gene regulation. Indeed, the withdrawal of SRIF increases GH levels[29] and induces GHRH secretion in rats,[135] whereas increased SRIF release induced by GH feedback is likely to suppress GHRH activity,[136] because it directly inhibits GHRH release from hypothalamic tissue *in vitro*[137] and into sheep portal plasma *in vivo*.[138] The reciprocal relationship between SRIF and GHRH neurons is further indicated by the induction of SRIF secretion[139] and gene expression[140] during concomitantly elevated GHRH neuronal activity. These interactions are thus likely to be causally involved in the induction of pulsatile episodes of GH release, which are followed by periods of somatotroph quiescence (see Chapter 12).

E. INSULIN-LIKE GROWTH FACTORS

Inhibitory effects of IGFs on the hypothalamic regulation of GH secretion are well established (see Melmed[37] for review) and likely to be responsible for the suppression of pulsatile GH secretion in rats[70,72,141] and fetal pigs[142,159] induced by i.c.v. injections of IGF-I (but not IGF-II.[143,144] Although central effects of IGF-I on pulsatile GH release have not been observed in other studies in rats[144] and sheep[143,145] the action of IGF-I appears to be dependent on synergistic interactions with IGF-II).[144] Central IGFs may reflect component pathways in the short-loop regulation of GH release. This possibility is supported by the reduced levels of IGF-I mRNA in the hypothalami of hypophysectomized rats and their restoration following i.c.v. injections of human GH.[102,146] Hypophysectomy or pituitary stalk section similarly reduces IGF-II mRNA levels in the rat hypothalamus, in parallel with reduced SRIF mRNA concentrations.[98] Feedback effects of GH within the hypothalamus may thus result in IGF-I and IGF-II synthesis and promote their inhibition of GHRH synthesis[106] and release[147] and stimulation of SRIF synthesis and release.[39,72,106,141] The release of IGF-I from the median eminence into hypophysial circulation may also contribute to GH autoregulation, because IGFs may act as hypophysiotropic factors with direct inhibitory effects on GH synthesis and release.[37] These actions of IGFs within the hypothalamus-somatotroph axis are independent of peripheral IGF-I status, because hypothalamic GHRH mRNA levels are resistant to systemic infusions of IGF-I but responsive to i.c.v. IGF-I injections.[106] These inhibitory actions of IGFs on hypothalamus-somatotroph function may, however, be independent of short-loop GH autoregulation.

Although hypophysectomy reduces hypothalamic IGF-I mRNA levels and increases GHRH mRNA concentrations, the restoration of IGF-I mRNA levels by exogenous GH therapy may occur without normalization of GHRH mRNA.[102] Moreover, whereas Sato and Frohman[106] found reciprocal changes in IGF-I and GHRH mRNA levels in GH-treated GH-deficient dwarf rats, i.c.v. injections of IGF-I and GH produced discordant changes in the hypothalamic content of SRIF mRNA, suggesting the effect of GH on GHRH mRNA is independent of central IGF-I. Aguila and McCann[121] also suggest that the mechanism through which GH influences SRIF neurons differs from the one induced by IGF-I, because the induction of SRIF mRNA by IGF-I (but not by GH) appears to be mediated by increased GHRH activity.

F. NEURAL PATHWAYS

In addition to direct actions on GHRH or SRIF neurons, short-loop actions of GH on hypothalamic function may be mediated through a variety of interneuronal pathways. The suppression of GHRH activity by exogenous GH in humans may, for instance, be mediated through opioidergic pathways. The modulation of neurotransmitter content and turnover in GH-treated rats also supports this possibility (see Chapter 28). The modulation of hypothalamic SRIF in humans and rodents by GH has, moreover, been shown to be under cholinergic control[22,26,148] and blocked by pyridostigmine, an acetylcholinesterase inhibitor that increases cholinergic tone. Kelijman and Frohman,[17] conversely, suggest the effects of exogenous GH and pyridostigmine on GH secretion are mediated through independent mechanisms, because pyridostigmine overcomes the inhibitory effects of exogenous GH on basal and GHRH-induced GH release but not the suppressed GH responses to arginine, TRH, or insulin-induced hypoglycemia. These authors also suggest β-adrenergic tone may also overcome the GH-induced suppression of endogenous GH responses to GHRH and insulin[84] by inhibiting SRIF release, although β-adrenergic blockade fails to overcome exogenous GH-induced inhibition of GH responsiveness to arginine. It is, therefore, possible that GH-induced changes in neurotransmitter turnover in hypothalamic and extrahypothalamic regions of the brain are coincidental and unrelated to GH autoregulation.[149,150]

G. PULSATILE GH SECRETION

Pulsatile GH release in rats is suppressed by central[69,70,72] or peripheral[13,16,72,151] GH injections. For instance, a bolus subcutaneous GH injection that elevates GH concentrations to levels comparable to those during spontaneous GH release results in a fivefold reduction in the amplitude of GH release and may totally suppress GH pulsatility.[12,70] This suppression of GH release occurs within 1 to 2 h of GH injection and persists for approximately 3 to 4 h,[12,152] comparable to the interval between spontaneous pulses of GH release. This suppression of GH secretion is IGF-I independent but SRIF dependent. A reduction in hypothalamic GHRH may also be induced, because the passive immunoneutralization of SRIF in GH-treated rats restores spontaneous GH pulse amplitude but not the normal pattern of GH secretion. The absence of rebound GH secretion at the end of GH-induced suppression also indicates a reduction in GHRH stimulation during the period of GH suppression.[29] The 3-h blunting of the GH response to exogenous GHRH in rats receiving a single bolus GH injection also occurs with a time course comparable to the interval between spontaneous GH pulses.[24] This blunting of GHRH-induced GH release is partly induced by SRIF although SRIF does not inhibit GH synthesis (Chapter 4). At the end of the period of GH suppression, the GH response to GHRH is, therefore, augmented as a result of SRIF withdrawal and the stimulation of the accumulated stores of newly synthesized GH. Increased hypothalamic SRIF release may, therefore, occur in response to pulsatile surges of GH,[91] thereby terminating duration of the pulsatile episode. Enhanced SRIF secretion during the subsequent GH trough period[91] could thus permit an accumulation of GH stores, allowing somatotrophs to respond optimally to subsequent pulses of hypothalamic GHRH. Pulsatile GH secretion may thus result from temporal interactions between GHRH, GH, and SRIF, although the intrinsic generator of the ultradian rhythm of GH secretion may be independent of GH feedback.[153]

III. LONG-LOOP FEEDBACK

In addition to pituitary and central mechanisms participating in GH autoregulation, slower onset long-loop feedback pathways are also well established. These pathways are mediated by peripheral endocrines (Figure 3) and metabolites (Figure 4) responsive to GH action.

A. GROWTH FACTORS

IGFs are synthesized in the liver and other peripheral tissues in response to GH stimulation[154] and act at hypothalamic and pituitary receptors[57,155,156] to inhibit GH secretion. IGF-I stimulates SRIF synthesis and release,[39,72,106,141] while inhibiting GHRH synthesis and release from the hypothalamus,[106,147] and directly inhibits pituitary *GH* gene transcription[43,45,47,52–54] and inhibits basal-, cAMP-, protein kinase C-, GHRH-, and TRH-induced GH release in mammals.[48,50,51,72,157–159] IGF-II also inhibits ovine GH responses to GHRH stimulation, but is less potent than IGF-I.[50] IGF-I is, however, far less effective in inhibiting GH release than SRIF[160] and may only inhibit GH release from a subpopulation of GH cells that are not responsive to SRIF inhibition.[160] IGF-I also inhibits GH release from avian pituitary cells *in vitro*[161,162] and exogenous IGF-I inhibits GH release *in vivo* in domestic fowl.[162] The release of GH from teleost pituitary glands is similarly suppressed by IGF-I, by mechanisms that are additive with SRIF.[163] IGF-I also increases the binding capacity of SRIF receptors and increases somatotroph sensitivity to SRIF inhibition[87] by mechanisms that are thyroid hormone dependent.[37,47,164] The receptor mediation of these inhibitory effects of IGF-I on GH secretion is demonstrated by the lack of GH suppression in pituitaries with defective IGF-I receptors.[165]

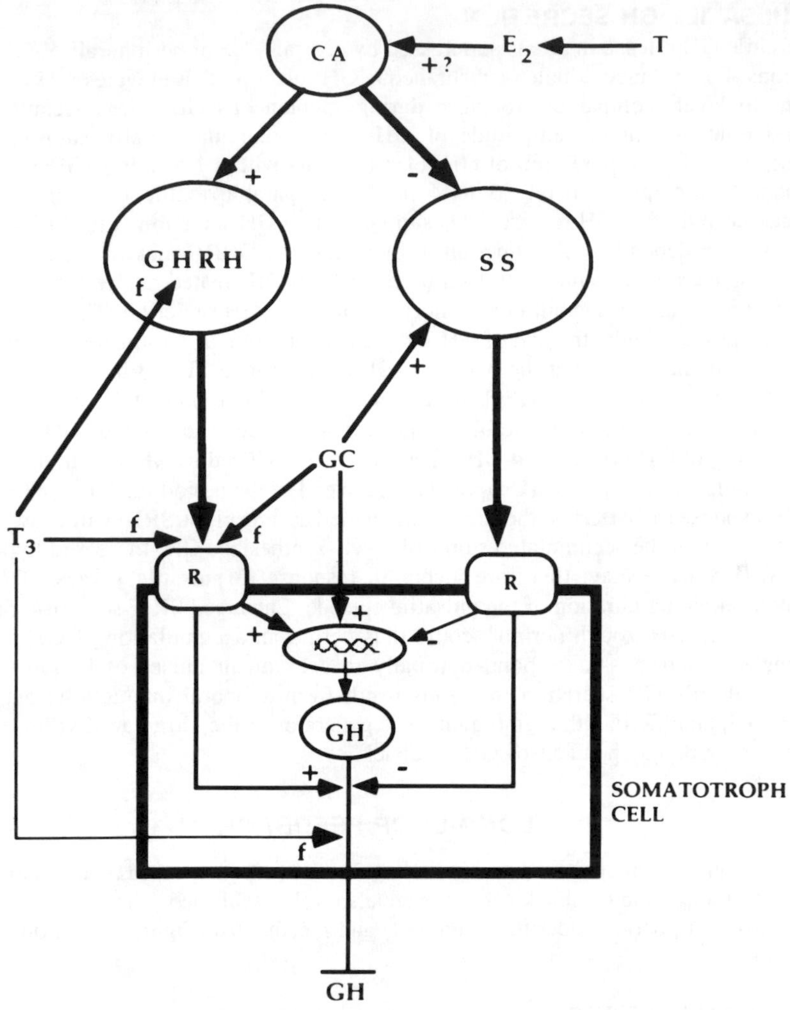

Figure 3 Growth hormone (GH) control by other hormones in humans. GHRH, after binding to its receptor (R), induces *GH* gene transcription and GH release. Both mechanisms are antagonized by somatostatin (SS). T_3 plays a facilatory role (f) in GHRH synthesis and in GHRH receptor capacity, and also facilitates the secretory process at the somatotrophs. Acutely, glucocorticoids (GCs) enhance *GH* gene transcription, and increase GHRH receptor capacity (f); however, as a more delayed effect these steroids increase SS synthesis and/or release. Sexual hormones mainly act on SS neurons, by modifying catecholamine (CA) availability, an effect most likely due to free estradiol (E_2). Therefore, the role of testosterone (T) seems to be secondary to its hypothalamic aromatization to E_2. +, Stimulation; –, inhibition. (From Devesa, J., Lima, L., and Tresguerres, J. A. F., Neuroendocrine control of growth hormone secretion in humans, *Trends Endocrinol. Metab.*, 3, 175–183, 1992. Copyright 1992 by Elsevier Science, Inc. Reprinted with permission.)

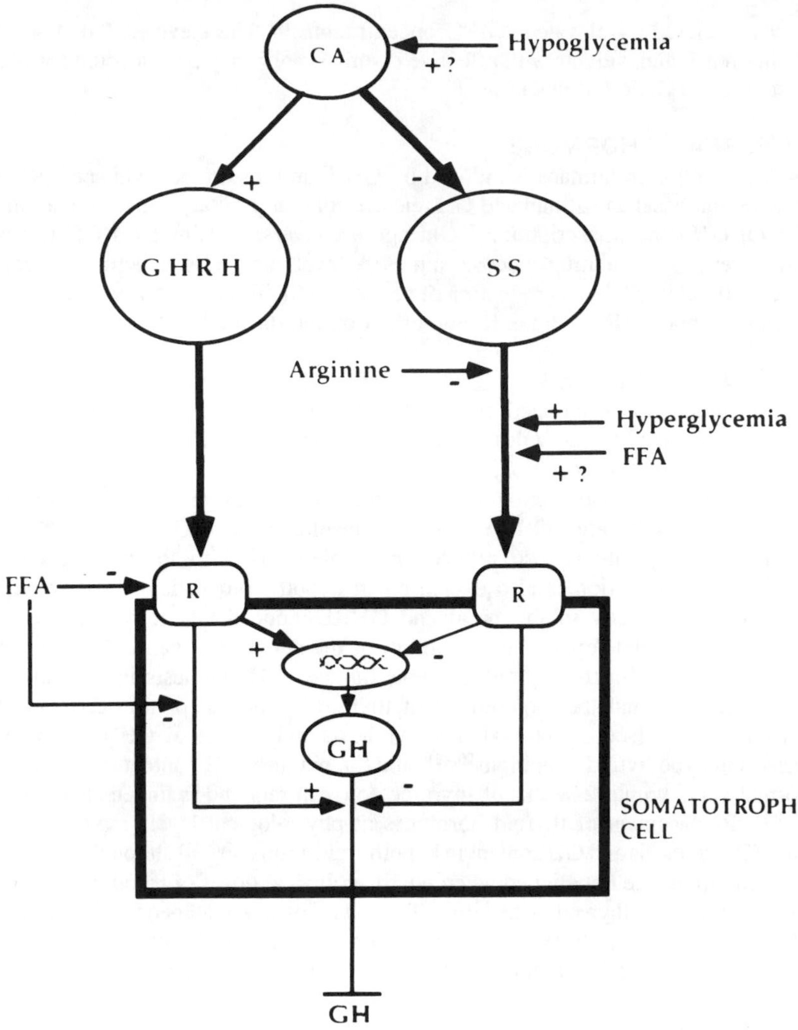

Figure 4 Growth hormone (GH) control by metabolic intermediates. Although arginine inhibits somatostatin (SS) release, hyperglycemia stimulates it. Conversely, the lack of glucose activates adrenergic transmission (CA), therefore decreasing SS release. The inhibitory effect of free fatty acids (FFAs) on GH secretion appears to be mediated by the alteration of the membrane layer of the somatotrophs, affecting both the GHRH receptor (R) and the secretory mechanisms of GH release. FFAs may also stimulate hypothalamic SS release. +, Stimulation; –, inhibition. (From Devesa, J., Lima, L., and Tresguerres, J. A. F., Neuroendocrine control of growth hormone secretion in humans, *Trends Endocrinol. Metab.*, 3, 175–183, 1992. Copyright 1992 by Elsevier Science, Inc. Reprinted with permission.)

The involvement of IGFs in long-loop GH autoregulation is also indicated by the elevated circulating GH concentrations in Laron dwarfs[166–169] and patients with anorexia nervosa[170] or liver cirrhosis,[171] in which GH is unable to induce IGF-I synthesis. Circulating IGF-I levels are also low and GH levels elevated in postmenopausal women treated with estrogen[172] and in depressed patients. In contrast, GH secretion is impaired in disease states characterized by elevated IGF-I concentrations.[114] The elevated IGF-1 levels in obese children[107] and patients with affective disorders[173,174] may thus account for blunted GH responses to GHRH stimulation.[107]

B. PANCREATIC HORMONES

Insulin secretion in humans is induced by GH[175] and insulin acts via specific receptors[176] to inhibit basal and stimulated GH release from rat pituitary cells[46,177] and directly inhibits rat *GH* gene transcription.[178] Glucagon, conversely, stimulates GH release,[179] despite increasing circulating glucose and FFA levels and despite acting centrally to increase SRIF release.[180] The production of pancreatic SRIF is not, however, regulated by GH[124] and plasma SRIF has little, if any, effect on pituitary GH release.[50,181,182]

C. THYROID HORMONES

Thyroid function is stimulated by GH[183–192] (see Chapter 26) and thyroid hormones may provide stimulatory (see Valcavi et al.[187] for review) or inhibitory (see Harvey[193] for review) feedback on GH secretion.

In euthyroid mammals, exogenous thyroid hormones may inhibit stimulated GH release *in vivo*[194] (e.g., GH responses to insulin[184,195]) to GHRH.[187,194,196,197] The 24-h GH secretory rate is also reduced in adolescents with thyrotoxicosis[198] and sleep-related GH secretion is also decreased in hypothyroid patients.[199] Thyroid hormones may also directly inhibit basal and GHRH-induced GH release from human pituitary cells[200,201] and suppress transcription of the human *GH* gene.[202] These actions contrast, however, with the thyroid hormone-induced GH synthesis in rat and bovine pituitary cells[203–207] and the requirement of thyroid hormones for transcription of the human *GH* gene[204] (see Chapter 4). The synthesis and release of GH is dramatically suppressed in hypothyroid conditions[207–210] and the pituitary GH content drops to 30% of the normal value within 2 weeks of thyroidectomy in rats and is undetectable after 24 days.[209,210] Replacement of thyroid hormones at physiological levels rapidly increases pituitary GH synthesis and GH content in hypothyroid animals[209–211] although supranormal GH concentrations are not elicited when pharmacological doses of triiodothyronine (T_3) are administered to euthyroid rats.[207,212,213] The reduction in spontaneous GH release[214,215] and blunted GH responses of hypothyroid patients to insulin, arginine, and GHRH[187,194,196,216,217] and of animals[196,218–222] are similarly reversed by thyroid hormone replacement.

The inhibitory effect of hyperthyroidism on GH secretion may partly result from the induction of IGF-I synthesis in peripheral, pituitary, or CNS tissues.[187,223–226] Thyroid hormones also modulate pituitary responsiveness to IGF-I stimulation and the refractoriness of the hypothyroid pituitary gland to IGF-I inhibition may be due to a downregulation of IGF-I receptors.[227] Thyroid hormones also inhibit GH secretion by increasing inhibitory hypothalamic tone. Although thyroxine (T_4) (or thyroidectomy) does not modify SRIF levels in hypophysial portal circulation or alter transcription of the *SRIF* gene,[222,228] T_3 is reported to stimulate SRIF release from the perfused rat hypothalamus[229] and reduces the SRIF content of cultured rat neurons.[230] Thyroid hormones also decrease the content of GHRH and GHRH mRNA in the rat hypothalamus,[222] whereas GHRH mRNA levels are elevated in the hypothalami of the thyroidectomized rats[222] and reduced by thyroxine replacement.[222] These actions on *GHRH* gene transcription may, however, be indirectly induced by the changes in GH secretion that accompany hyper- and hypothyroidism.[231] An inhibitory effect of thyroid hormones on GH secretion is also indicated by the induction or augmentation of GH release in response to TRH stimulation. Although TRH normally has minimal effects on GH secretion in humans, it is a potent GH secretagogue in pathophysiological conditions (e.g., anorexia nervosa, diabetes mellitus, and renal failure) with "low T_3" syndromes.[232,233] This stimulation of GH release may be mediated through TRH receptors on somatotroph cells that are normally downregulated by thyroid hormones.

D. THYROIDAL INHIBITION OF GH SECRETION IN BIRDS

Thyroidal activity in birds is enhanced by GH stimulation[234] and thyroid hormones provide feedback at hypothalamic and pituitary sites to regulate GH secretion (see Harvey[193] for review). Thyroid hormones thus directly inhibit basal and secretagogue-induced synthesis and release of GH in chicken pituitary glands.[235–238] Exogenous thyroid hormones also suppress basal and TRH- and GHRH-induced GH release *in vivo*,[239–244] whereas chemical or surgical thyroidectomy elevates basal and stimulated GH levels.[193,240,245] This results from increased inhibitory hypothalamic tone and reduced pituitary responsiveness to GH-releasing stimuli (GHRH or TRH), as a result of downregulated membrane receptors or impaired signal transduction pathways. Thyroid inhibition of GH secretion is also IGF-I mediated.

E. ADRENAL HORMONES

Adrenocortical function is induced by GH stimulation[246] (see Chapter 26) and hence cortisol or corticosterone might provide humeral signals in the long-loop feedback regulation of GH secretion (see Mehls et al.[247] for review). Moreover, because 70% of GH-secreting cells contain glucocorticoid receptors,[248] glucocorticoids are likely to have significant roles in the regulation of GH secretion. This possibility is supported by the glucocorticoid induction of *GH* gene transcription (see Chapter 4) and the reversible reduction in pituitary GH mRNA levels following adrenalectomy.[249] The effects of glucocorticoids on GH secretion are, however, biphasic both with respect to time and dose,[250–252] because acute glucocorticoid stimulation at low dose levels generally stimulates basal and stimulated GH secretion *in vivo*.[250,252–254] These acute actions are not immediate but induced after several hours.[255,256] This stimulation is thought to be primarily mediated at hypothalamic sites through a GHRH-dependent mechanism and a reduction in SRIF tone.[253,256,257] In contrast, high doses of glucocorticoids or long-term glucocorticoid therapy usually inhibit spontaneous and secretagogue-induced GH release, after a lag period of at least 12 h.[255,258,259] However, acute glucocorticoid treatment *in vitro* inhibits GH secretion while longer glucocorticoid treatment increases the sensitivity of somatotrophs to GH-releasing stimuli[248] and upregulate the number of GHRH receptors on pituitary membranes.[260] The actions of glucocorticoids also tend to be predominantly inhibitory at central sites through SRIF induction,[261] although Lam et al.[262] recently found a reduction in SRIF mRNA in the hypothalami of dexamethasone-treated rats. Glucocorticoid actions at the pituitary level, in contrast, tend to be stimulatory rather than inhibitory[263,264] (see Giustina and Wehrenberg[265] for review). The actions of glucocorticoids on GH secretion also differ between species and differ during ontogeny and in different physiological states.[255,256,258]

F. GONADAL HORMONES

Roles for GH in puberty,[266] ovulation,[267] and gonadal steroidogenesis[268] have been demonstrated and cyclical patterns of GH synthesis and secretion occur in parallel with cyclical reproductive activity[269] (see Chapter 23) Gonadal steroids, in turn, have feedback roles in the regulation of GH secretion[270] and steroid receptors are located in a large proportion of pituitary and hypothalamic cells.[271] These roles are therefore mediated at pituitary and hypothalamic sites and are sexually dimorphic and account for the sexually dimorphic patterns of pulsatile GH release.[272]

In rats, the pituitary GH content in males exceeds that in females[273] and males have greater GH secretory bursts.[272] Moreover, whereas *GH* mRNA is directly and indirectly elevated by androgens in male rats, estrogens have no effect on *GH* gene transcription in the bovine pituitary.[274] Androgens, but not estrogens, are able to increase the pituitary GH content in ovariectomized female rats[275] and to induce GHRH release from bovine pituitary glands.[276] Androgens are also responsible for the increased *in vitro* and *in vivo* responsiveness of male pituitaries to GHRH stimulation.[276–283] Basal and GHRH-stimulated GH release is reduced by prepubertal orchidectomy and restored by testosterone.[284,285] Ovariectomy, in contrast, does not inhibit GH secretion, although exogenous estrogen lowers basal and stimulated GH secretion *in vivo*[285] and estrogen is unable to restore the masculine pattern of GH release in castrated males.[279] The increased pulsatility of GH release in males is due to testosterone, and exogenous androgen can increase the amplitude of episodic GH release.[286] The sexually dimorphic pattern of GH secretion may also result from the greater sensitivity of female pituitaries to SRIF inhibition, although contrary findings have also been reported.[285,287]

The increased responsiveness of male rat somatotrophs to provocative stimuli may reflect differences between males and females in the hypothalamic control of GH release. Estradiol-binding sites have been located on GHRH and SRIF neurons in the rat hypothalamus[271] and expression of the *SRIF* and *GHRH* genes is stimulated by androgens[288–290] and estrogens.[291] Male and female rats have comparable concentrations of hypothalamic GHRH,[292,293] and comparable basal levels of *in vitro* GHRH release but males have greater levels of stimulated GHRH release and greater concentrations of GHRH mRNA.[99,294,295] Baseline GHRH secretion is, however, slightly elevated in females although the amplitude of GHRH release is less.[270] These differences are probably due to differences in the gonadal steroid regulation of *GHRH* gene transcription. Although minimal effects of gonadal steroids on GHRH mRNA concentrations have been reported,[296,297] Zeitler et al.[289] and Hasegawa et al.[298] suggest GHRH mRNA content is reduced by castration and restored by testosterone. Progestins also have direct stimulatory effects on pituitary GH release[299] and indirect effects mediated through GHRH neurons.[271,300]

Androgens masculinize the GH secretory pattern in ovariectomized rats, as a result of an augmentation of hypothalamic GHRH (and SRIF) mRNA levels.[298] Androgens, but not estrogens, also directly stimulate GHRH release from cultured hypothalamic neurons.[301] The content and release of hypothalamic SRIF may also be comparable in male and female rats,[109,295,302–305] although the pattern of SRIF release in the male is more pulsatile than in females.[303,306] Male rats also appear to have a greater content of hypothalamic SRIF mRNA than do female rats.[293] In both sexes, gonadectomy reduces SRIF mRNA levels,[304,305] but androgens are more effective than estrogens in the restoration of SRIF mRNA levels in castrated rats.[307] The sexually dimorphic pattern of GH secretion may thus result from differential effects of gonadal steroid feedback within the hypothalamus-somatotroph axis. Gonadal steroids also modulate the short-loop GH autoregulation.

Feedback inhibition of GHRH mRNA by GH has been shown to be sex specific in rats. The hypersecretion of GH by an implantable pituitary tumor induces a sevenfold reduction in GHRH mRNA levels in the hypothalami of male rats, but only a twofold suppression in females.[296] The inhibitory effect of GH on GHRH release is also greater in males than in females.[308] The greater feedback suppression of GH secretion in male rats[16,109,296,308] may also be SRIF mediated, because SRIF regulates *GHRH* gene transcription.

Gonadal steroids also modulate hypothalamus-somatotroph function in primates, in which resting and stimulated GH release is greater in females than males (see Kerrigan and Rogol[309] for review). This is likely to result from estrogen actions at hypothalamic sites, because estrogen has little effect on GH release from somatotrophs *in vitro*.[310] The pituitary reserve of GH, in response to GHRH stimulation, is also similar in noncycling patients with anorexia nervosa to that in normal menstruating women[311] and remains constant throughout the menstrual cycle.[312] Estrogen receptor blockade also has no effect on the pituitary GH response to GHRH in normal women.[311] Acute or chronic hypoestrogenism is therefore unlikely to be a significant factor in the regulation of pituitary somatotrophs in adult women. However, as estrogen receptor blockade increases the interval between episodic GH pulses or increases the frequency of somatotroph quiescence,[311] estrogens may normally reduce SRIF tone in women. The hypothalamus-somatotroph axis of prepubertal girls is, however, exquisitely sensitive to extremely small amounts of estrogen, which increase secretory burst frequency and augment GH pulse amplitude.[313] In contrast, although testosterone can augment circulating GH levels by increasing GH pulse amplitude and can enhance GH responses to provocative stimuli,[314,315] it is less effective than estrogen. Moreover, this effect is not observed with nonaromatizable androgens[316] or in response to acute, physiological elevations in the blood testosterone level.[317] Testosterone also appears to stimulate SRIF release through an α_2-adrenergic pathway, even though this action of testosterone is dependent on its aromatization to 17β-estradiol. The sexual dimorphism in GH secretion in primates may thus reflect different circulating levels of estrogen and tissue responsiveness at hypothalamic and pituitary sites.

G. IMMUNE FACTORS

Numerous stimulatory effects of GH on immune function have been documented (see Kelley[318] and Blalock[319] and Chapter 24 for reviews). Thymic factors and cytokines (reviewed by Blalock,[320] Spangelo and MacLeod,[321] Smith,[322] and Scarborough[323]) reciprocally modulate GH secretion at hypothalamic and pituitary sites in a feedback manner. Moreover, whereas these immune factors might contribute to the long-loop feedback regulation of GH release, their local synthesis within the CNS and pituitary gland[324] might also contribute to short- and ultrashort-loop autoregulation.

Thymic activity appears to be GH dependent (reviewed by Millington and Buckingham[325]) and to promote GH secretion. The removal of the thymus reversibly degranulates pituitary acidophils and reduces plasma GH levels.[326,327] Thymic factors, including TNRF (thymic neuroendocrine-releasing factor[328,329]) and MB-35,[330] are released by thymic epithelial cells and directly stimulate GH release from pituitaries *in vitro*, without stimulating cAMP. Thymopeptin, another peptide secreted by the thymus, has also been found to markedly enhance basal plasma GH levels in children.[331] Thymic hormones also elevate circulating GH levels in young and old rats.[332]

Interleukins (ILs), produced by macrophages, monocytes, and lymphocytes, also modulate hypothalamus-somatotroph function and cytokine receptors occur widely within CNS and pituitary tissue.[333–336] Although ILs are produced in most tissues, they also have systemic effects mediated through humeral pathways.[322] Stimulatory effects of interleukin 1 (IL-1) and IL-6 on GH release from rat and human anterior pituitary cultures have been reported.[328,337–341] The stimulation of GH release by these cytokines differs from that induced by hypothalamic releasing factors in that it is not mediated by cAMP, inositol phosphate, or calcium signal transduction pathways.[342] This stimulation also occurs despite the stimulatory effects of interleukins on IGF-I synthesis and action in other tissues.[343,344] *In vivo* GH secretion in humans is also reported to be elevated by interferon α(IFN-α) and IFN-β[331,345] and IL-6 stimulates GH secretion in chickens by a SRIF-sensitive mechanism.[346] In contrast, IL-2, at femtomolar concentrations, directly suppresses GH,[333,347] whereas IFN inhibits stimulated GH release *in vitro*[333,348,349] and *in vivo*.[350] Tumor necrosis factor (TNF) also inhibits GH release from pituitary cells[351] and lowers circulating GH concentrations in cattle.[350,352] Although IL-2 inhibits GH release *in vitro*, it stimulates GH release in cancer patients.[353] This may reflect interactions of cytokines within the immune system and/or extrapituitary sites of action. Similarly, whereas IL-1 increases GH release directly, IL-1 and IL-1β suppress plasma GH levels in rats after systemic injections[354] or after i.c.v. or arcuate nucleus injections.[355] This suppression of GH release partly reflects a reduction in the number of pituitary GHRH receptors.[354] The suppression of GH release following central injections of IL-1β and IL-1 also appears to be mediated by increased SRIF synthesis and/or release.[356–359] Tumor necrosis factor α, which synergizes with IL-1 *in vivo* and *in vitro*, also stimulates SRIF synthesis and release,[360] as does IL-6.[323] The tonic inhibition of pulsatile GH release induced by i.c.v. IL-1 has also been suggested to be mediated by corticotropin-releasing hormone through a mechanism that involves suppression of GHRH rather than stimulation of SRIF secretion.[361] In contrast, the ability of IL-2 to inhibit GH secretion has been shown to be induced by a dopaminergic pathway that increases SRIF release independently of GHRH secretion.[362] The stimulated, but not basal, release of GHRH is, however, also sensitive to IL-2 inhibition.[362] The temporal pattern of these effects on hypothalamus function, nevertheless, suggests that they are likely to be induced by chronic cytokine stimulation, but they are unlikely to be induced by acute changes in cytokine activity. Chronic changes in GH secretion in hyper- or hyposomatotropic states could thus be modulated by cytokines in feedback ways.

H. METABOLIC FACTORS

Carbohydrate and fat metabolism are regulated by GH (as reviewed in Chapters 20 and 21) and glucose and free fatty acids (FFAs) provide stimuli that contribute to the long-loop inhibition of GH secretion. The influence of nutritional factors on GH secretion has been reviewed by Ross and Buchanan[363] and is detailed in Chapter 12.

A reciprocal relationship between circulating glucose and GH concentrations is well established and has been utilized for chemical tests of hyper- and hyposomatotropic activity. Gluconeogenesis and glucose utilization are regulated by GH and hyperglycemia accompanies increased blood GH concentrations.[364–367] However, in normal subjects, elevated blood glucose concentrations normally suppress spontaneous and stimulated GH secretion.[368,369] Because this suppressive response is blocked by enhanced cholinergic tone (by pyridostigmine), it is likely to be mediated through increased SRIF secretion.[370] Reduced pituitary responsiveness to GH-releasing stimuli[371] and impaired GHRH release [372,373] may also contribute to the suppression of GH secretion during hyperglycemia. Hypoglycemia, conversely, stimulates GH secretion in humans and is likely to result from α_2-adrenergic inhibition of SRIF release.[374–376] Hypoglycemia may also stimulate GH secretion through activation of the hypothalamus-pituitary-adrenal axis, because stress provokes GH release in humans (see Chapter 12). Growth hormone is a potent lipolytic agent and elevates circulating levels of FFAs. Plasma GH and FFA levels are inversely related in humans, monkeys, sheep, and rats.[78,377–380] Feedback inhibition of GH secretion by FFAs is mediated at both hypothalamic and pituitary sites. The feedback regulation of GH secretion involves SRIF-dependent pathways,[381] although the mechanisms triggering SRIF release are unclear.[379] FFAs could disrupt normal membrane fluidity and thereby change calcium conductance.[382] However, the observation that FFAs have little or no effect on the secretion of other pituitary hormones[377,378] suggests that the feedback actions on GH are not the result of nonspecific perturbation in the plasma membrane. It is, therefore, possible that intracellular metabolites of fatty acids regulate neuronal and adenohypophysial cell activity. However, as brain cells do not utilize FFAs as metabolic fuels, this possibility is unlikely.[379]

REFERENCES

1. **Muller, E. E.,** *Horm. Res.,* 33(Suppl. 2), 90, 1990.
2. **Fraser, R. A., Siminoski, K., and Harvey, S.,** *J. Endocrinol.*, 128, R9, 1991.
3. **Fraser, R. A. and Harvey, S.,** *Endocrinology (Baltimore)*, 130, 3593, 1992.
4. **Hull, K. L., Fraser, R. A., and Harvey, S.,** *J. Endocrinol.*, 135, 459, 1992.

5. **Harvey, S., Baumbach, W. R., Sadeghi, H., et al.,** *Endocrinology (Baltimore)*, 133, 1125, 1993.
6. **Rosenthal, S. M., Silverman, B. L., and Wehrenberg, W. B.,** *Neuroendocrinology*, 53, 597, 1991.
7. **Hauser, S. D., McGrath, M. F., Collier, R. J., et al.,** *Mol. Cell Endocrinol.*, 72, 187, 1990.
8. **Sotelo, A. I., Bartke, A., and Turyn, D.,** *Acta Endocrinol. (Copenh.)*, 129, 446, 1993.
9. **Glenn, K. C.,** *Endocrinology (Baltimore)*, 118, 2450, 1986.
10. **Fagin, J. A., Brown, A., and Melmed, S.,** *Endocrinology (Baltimore)*, 122, 2204, 1988.
11. **Pontiroli, A. E., Lanzi, R., Monti, L. D., et al.,** *J. Clin. Endocrinol. Metab.*, 72, 492, 1991.
12. **Lanzi, R. and Tannenbaum, G. S.,** *Endocrinology (Baltimore)*, 130, 780, 1992.
13. **Clark, R. G., Carlson, L. M. S., and Robinson, I. C. A. F.,** *J. Endocrinol.*, 119, 201, 1988.
14. **Lea, R. W. and Harvey, S.,** *J. Endocrinol.*, 125, 409, 1990.
15. **Voderholzer, U., Laakmann, G., Hinz, A., et al.,** *Psychoneuroendocrinology*, 18, 365, 1993.
16. **Carlsson, L. M. S., Clark, R. G., and Robinson, I. C. A. F.,** *J. Endocrinol.*, 126, 27, 1990.
17. **Kelijman, M. and Frohman, L. A.,** *J. Clin. Endocrinol. Metab.*, 72, 1081, 1991.
18. **Rosenthal, S. M., Hulse, J. A., Kaplan, S. L., et al.,** *J. Clin. Invest.*, 77, 176, 1986.
19. **Nakamoto, J. M., Gestner, J. M., Press, C. M., et al.,** *J. Clin. Endocrinol. Metab.*, 62, 822, 1986.
20. **Sakuma, M. and Knobil, E.,** *Endocrinology (Baltimore)*, 86, 890, 1970.
21. **Gil-Ad, I., Klinger, B., Pertzelan, A., et al.,** *J. Endocrinol. Invest.*, 15, 735, 1992.
22. **Giustina, A., Bossoni, S., Bodini, C., et al.,** *Acta Endocrinol. (Copenh.)*, 125, 510, 1991.
23. **Varma, S., Sabharwal, P., Sheridan, J. F., et al.,** *J. Clin. Endocrinol. Metab.*, 76, 49, 1993.
24. **Lanzi, R. and Tannenbaum, G. S.,** *Endocrinology (Baltimore)*, 130, 1822, 1992.
25. **Lea, R. W., Ahene, C., Marsh, J. A., et al.,** *J. Endocrinol.*, 126, 237, 1990.
26. **Ross, R. J. M., Ttsagarakis, S., Grossman, A., et al.,** *Clin. Endocrinol.*, 27, 727, 1988.
27. **Ghigo, E., Goffi, S., Mazza, E., et al.,** *Acta Endocrinol. (Copenh.)*, 120, 598, 1989.
28. **Harvey, S. and Baidwan, J. S.,** *J. Mol. Endocrinol.*, 4, 123, 1990.
29. **Clark, R. G., Carlsson, L. M. S., Rafferty, B., et al.,** *J. Endocrinol.*, 119, 397, 1988.
30. **Harvey, S., Scanes, C. G., and Marsh, J. A.,** *Gen. Comp. Endocrinol.*, 55, 493, 1984.
31. **Lapp, C. A., Tyler, J. M., Lee, Y. S., et al.,** *In vitro Cell. Dev. Biol.*, 25, 528, 1989.
32. **Fagin, J. A., Pixley, S., Slanina, S., et al.,** *Endocrinology (Baltimore)*, 120, 2037, 1987.
33. **Binoux, M., Hossenlopp, P., Lasarre, C., et al.,** *FEBS Lett.*, 124, 178, 1981.
34. **Matthews, L. S., Norstedt, G., and Palmiter, R. D.,** *Proc. Natl. Acad. Sci. U.S.A.*, 84, 9343, 1986.
35. **Yamashita, S., Slanina, S., Kado, H., et al.,** *Endocrinology (Baltimore)*, 118, 915, 1986.
36. **Ezzat, S. and Melmed, S.,** *J. Endocrinol. Invest.*, 13, 691, 1990.
37. **Melmed, S.,** *The Brain as an Endocrine Organ* (Eds. Cohen, M. P. and Foa, P. P.). Springer-Verlag, Berlin, 1992, p. 193.
38. **Guilhaume, A., Benoit, O., Gourmelen, M., et al.,** *Pediatr. Res.*, 16, 299, 1982.
39. **Berelowitz, M., Szabo, M., Frohman, M., et al.,** *Science*, 212, 1279, 1981.
40. **Buyse, J., Decuypere, E., and Simon, J.,** *Reprod. Nutr. Dev.*, 30, 683, 1990.
41. **Goodyer, C. G., De Stephano, L., Guyda, H. J., et al.,** *Endocrinology (Baltimore)*, 115, 1568, 1984.
42. **Ceda, G. P., Hoffman, A. R., Silver, G. D., et al.,** *J. Clin. Endocrinol. Metab.*, 600, 1204, 1985.
43. **Yamashita, S. and Melmed, S.,** *Endocrinology (Baltimore)*, 118, 176, 1986.
44. **Sheppard, M. S. and Bala, R. M.,** *Can. J. Physiol. Pharmacol.*, 65, 515, 1987.
45. **Namba, H., Morita, S., and Melmed, S.,** *Endocrinology (Baltimore)*, 124, 1794, 1989.
46. **Melmed, S. and Slanina, S. M.,** *Endocrinology (Baltimore)*, 117, 532, 1985.
47. **Melmed, S. and Yamashita, S.,** *Endocrinology (Baltimore)*, 118, 1483, 1986.
48. **Blanchard, M. M., Goodyer, C. G., Charrier, J., et al.,** *Endocrinology (Baltimore)*, 122, 2114, 1988.
49. **Brazeau, P., Guillemin, R., Ling, N., et al.,** *C.R. Acad. Sci. Paris*, 295, 651, 1982.
50. **Ceda, G. P., Davis, R. G., Rosenfeld, R. G., et al.,** *Endocrinology (Baltimore)*, 120, 1658, 1987.
51. **Blanchard, M. M., Goodyer, C. G., Charrier, J., et al.,** *Reprod. Nutr. Dev.*, 27(2B), 471, 1987.
52. **Yamashita, S., Weiss, M., and Melmed, S.,** *J. Clin. Endocrinol. Metab.*, 63, 730, 1986.
53. **Yamashita, S. and Melmed, S.,** *J. Clin. Invest.*, 79, 449, 1987.
54. **Morita, S., Yamashita, S., and Melmed, S.,** *Endocrinology (Baltimore)*, 121, 2000, 1987.
55. **Walker, D. A., Hogg, A., Haynes, K., et al.,** *J. Neuroendocrinol.*, 2, 305, 1990.
56. **Rosenfeld, R. G., Pham, H., Oh, Y., et al.,** *Endocrinology (Baltimore)*, 124, 2867, 1989.

57. **Goodyer, C. G., De Stephano, L., Lai, W. H., et al.,** *Endocrinology (Baltimore),* 114, 1187, 1984.
58. **Werther, G. A., Hogg, A., Oldfield, B. J., et al.,** *Endocrinology (Baltimore),* 121, 1562, 1987.
59. **Stachura, M. E., Lapp, C. A., Tyler, J. M., et al.,** *In vitro Cell. Dev. Biol.*, 25, 482, 1990.
60. **Richman, R. A., Weiss, J. P., Hochberg, Z., et al.,** *Endocrinology (Baltimore),* 108, 2287, 1981.
61. **Lamberts, S. W. J. and Oosterom, R.,** *Horm. Res.*, 14, 137, 1985.
62. **Kraicer, J., Lussier, B., Moor, B. C., et al.,** *Endocrinology (Baltimore),* 122, 1511, 1988.
63. **Cella, S. G., De Gennaro Collona, V., Locatelli, M., et al.,** *J. Endocrinol.*, 199, 1990.
64. **de Zegher, F., Bettendorf, M., Grumbach, M. M., et al.,** *Neuroendocrinology,* 52, 429, 1990.
65. **Becker, K. and Conway, S.,** *Brain Res.*, 578, 107, 1992.
66. **Katz, S. H., Molitch, M., and McCann, S. M.,** *Endocrinology (Baltimore),* 85, 725, 1969.
67. **Krulich, L. and McCann, S. M.,** *Proc. Exp. Biol. Med.*, 121, 1114, 1966.
68. **MacLeod, R. M., De Witt, G. W., and Smith, M. C.,** *Endocrinology (Baltimore),* 82, 889, 1968.
69. **Conway, S., McCann, S. M., and Kulich, L.,** *Endocrinology (Baltimore),* 117, 2284, 1985.
70. **Tannenbaum, G. S.,** *Endocrinology (Baltimore),* 107, 2117, 1980.
71. **Advis, J. P., White, S. S., and Ojeda, S. R.,** *Endocrinology (Baltimore),* 108, 1343, 1981.
72. **Abe, H., Molitch, M. E., Van Wyk, J., et al.,** *Endocrinology (Baltimore),* 113, 1319, 1983.
73. **Page, M. D., Dieguez, C., Valcavi, R., et al.,** *Clin. Endocrinol.*, 28, 551, 1988.
74. **Mendelson, W. B., Jacobs, L. S., and Gillin, J. C.,** *J. Clin. Endocrinol. Metab.*, 56, 486, 1983.
75. **Hagen, T. C., Lawrence, A. M., and Kirsteins, L.,** *Metabolism,* 21, 603, 1972.
76. **Abrams, R. L., Grumbach, M. M., and Kaplan, S. L.,** *J. Clin. Invest.*, 50, 940, 1971.
77. **Rosenbaum, M., Leibel, R. L., and Gertner, J. M.,** *Metab. Clin. Exp.*, 38, 590, 1989.
78. **Rosenbaum, M. J., Fong, Y., Hesse, D. G., et al.,** *J. Clin. Endocrinol. Metab.*, 69, 310, 1989.
79. **Rosenthal, S. M., Kaplan, S. L., and Grumbach, M. M.,** *J. Clin. Endocrinol. Metab.*, 68, 1101, 1989.
80. **Molitch, M. E., King, L. W., Moses, A. C., et al.,** *Neuroendocrinology,* 43, 651, 1986.
81. **Pontiroli, A. E., Lanzi, R., and Pozza, G.,** *J. Clin. Endocrinol. Metab.*, 68, 956, 1989.
82. **Colonna, V. D. G., Bertola, G., Coco, C. B., et al.,** *Proc. Soc. Exp. Biol. Med.*, 196, 432, 1991.
83. **Ross, R. J. M., Borges, F., Grossman, A., et al.,** *Clin. Endocrinol. (Oxford),* 26, 117, 1987.
84. **Kelijman, M. and Frohman, L. A.,** *J. Clin. Endocrin. Metab.*, 69, 1187, 1978.
85. **Lanzi, R., Pontiroli, A. E., Monti, L. D., et al.,** *Metab. Clin. Exp.*, 39, 812, 1990.
86. **Harvey, S. and Baidwan, J. S.,** *J. Mol. Endocrinol.*, 4, 13, 1990.
87. **Katakami, H., Berelowitz, M., Marbach, M., et al.,** *Endocrinology (Baltimore),* 117, 557, 1985.
88. **Harvey, S., Baidwan, J. S., and Attardo, D.,** *J. Endocrinol.*, 27, 417, 1990.
89. **Paradisi, R., Frank, G., Magrini, O., et al.,** *J. Clin. Endocrinol. Metab.*, 77, 523, 1993.
90. **Oliver, C., Mical, R. S., and Porter, J. C.,** *Endocrinology (Baltimore),* 101, 598, 1977.
91. **Sato, M., Chihara, K., Kita, T., et al.,** *Neuroendocrinology,* 50, 139, 1989.
92. **Bergland, R. M. and Page, R. B.,** *Science,* 204, 18, 1979.
93. **Lai, Z., Emnter, M. R. P., and Nyberg, F.,** *Brain Res.*, 546, 222, 1991.
94. **Van Houten, H., Posner, B. I., and Walsh, R. J.,** *Exp. Brain. Res.*, 38, 455, 1980.
95. **Minami, S., Kamegari, J., Hasegawa, O., et al.,** *J. Neuroendocrinol.*, 5, 691, 1993.
96. **Burton, K. A., Kabigting, E. B., Clifton, D. K., et al.,** *Endocrinology (Baltimore),* 130, 958, 1992.
97. **Minami, S., Kamegai, J., Sugihara, H., et al.,** *Endocrinology (Baltimore),* 131, 247, 1992.
98. **Levy, A., Matovelle, M. C., Lightman, S. L., et al.,** *Brain Res.*, 579, 1, 1992.
99. **Chomczynski, P., Downs, T. R., and Frohman, L. A.,** *Mol. Endocrinol.*, 2, 236, 1988.
100. **Eccleston, L. M., Powell, J. F., and Clayton, R. N.,** *J. Neuroendocrinol.*, 3, 661, 1991.
101. **Mayo, K. E.,** *63rd Annu. Meet. Endocr. Soc.,* Cincinatti, OH, 1986, p. 30,
102. **Wood, T. L., Berelowitz, M., Gelato, M. C., et al.,** *Neuroendocrinology,* 53, 298, 1991.
103. **De Gennaro Collona, V., Fidone, F., Cocchi, D., et al.,** *Neurobiol. Aging,* 14, 503, 1993.
104. **De Gennaro Colonna, V., Cattaneo, E., Cocchi, D., et al.,** *Peptides,* 9, 985, 1988.
105. **Bertherat, J., Timsit, J., Bluet-Pajot, M. T., et al.,** *J. Clin. Invest.*, 91, 1783, 1993.
106. **Sato, M. and Frohman, L. A.,** *Endocrinology (Baltimore),* 133, 793, 1993.
107. **Frohman, L. A.,** *Molecular and Clinical Advances in Pituitary Disorders* (Ed. Melmed, S.), Endocrine Research and Education, Inc., Los Angeles, 1993, p. 35.
108. **Schreibman, P. M. and Scanes, C. G., Eds.,** *Development, Maturation, and Senescence of Neuroendocrine Systems*, Academic Press, New York, 1989.

109. **Maiter, D. M., Gabriel, S. M., Koenig, J. I., et al.,** *Neuroendocrinology*, 51, 174, 1990.
110. **Frohman, M. A., Downs, T. R., Chomczynski, P., et al.,** *Mol. Endocrinol.*, 3, 1529, 1989.
111. **Kamegai, J., Wakabayashi, I., Sugihara, H., et al.,** *Acta Endocrinol. (Copenh.)*, 124, 700, 1991.
112. **Merchenthaler, I. and Arimura, A.,** *Peptides*, 6, 865, 1985.
113. **Katakami, H., Downs, T. R., and Frohman, L. A.,** *Endocrinology (Baltimore)*, 120, 1079, 1987.
114. **Muller, E. E.,** *Horm. Res.*, 33(Suppl. 2), 90, 1990.
115. **Kanatsuka, A., Makino, H., Matsushima, Y., et al.,** *Neuroendocrinology*, 29, 186, 1979.
116. **Patel, Y. C.,** *Life Sci.*, 24, 1589, 1979.
117. **Fernandez-Durango, R., Arimura, A., Fishback, J., et al.,** *Proc. Soc. Exp. Biol. Med.*, 157, 235, 1978.
118. **Rogers, K. V., Vician, L., Steiner, R. A., et al.,** *Endocrinology (Baltimore)*, 122, 586, 1988.
119. **Berelowitz, M., Firestone, S. I., and Frohman, L. A.,** *Endocrinology (Baltimore)*, 109, 714, 1981.
120. **Chihara, K., Minamitani, N., Kaji, G., et al.,** *Endocrinology (Baltimore)*, 109, 2279, 1981.
121. **Aguila, M. C. and McCann, S. M.,** *Brain Res.*, 623, 89, 1993.
122. **Sheppard, M. C., Kronheim, S., and Pimstone, B. L.,** *Clin. Endocrinol.*, 9, 583, 1978.
123. **Hoffman, D. L. and Baker, B. L.,** *Proc. Soc. Exp. Biol. Med.*, 156, 265, 1977.
124. **Molitch, M. E. and Hlivyak, L. E.,** *Horm. Metab. Res.*, 12, 519, 1980.
125. **Hurley, D. L. and Phelps, C. J.,** *Endocrinology (Baltimore)*, 130, 1809, 1992.
126. **Terry, L. C. and Crowley, W. R.,** *Endocrinology (Baltimore)*, 107, 1771, 1980.
127. **Horvath, S., Palkovits, M., Gorcs, T., et al.,** *Brain Res.*, 481, 8, 1989.
128. **Liposits, Z., Merchenthaler, I., Paull, W., et al.,** *Histochemistry*, 89, 247, 1988.
129. **Bertherat, J., Dournaud, P., Berod, A., et al.,** *Neuroendocrinology*, 56, 25, 1992.
130. **Mccarthy, G. F., Beaudet, A., and Tannenbaum, G. S.,** *Neuroendocrinology*, 56, 18, 1992.
131. **Epelbaum, J., Moyse, E., Tannenbaum, G. S., et al.,** *J. Neuroendocrinol.*, 1, 109, 1989.
132. **Burton, K. A., Kabigting, E. B., Clifton, D. K., et al.,** *Progr. Abstr. 73rd Annu. Meet. Endocr. Soc.*,Washington, D.C., 1991, p. 419.
133. **Bertherat, J., Berod, A., Normand, E., et al.,** *J. Neuroendocrinol.*, 3, 115, 1991.
134. **Tannenbaum, G. S., McCarthy, C. F., Zeitler, P., et al.,** *Endocrinology (Baltimore)*, 127, 2551, 1990.
135. **Miki, N., Ono, M., and Shizume, K.,** *J. Endocrinol.*, 117, 245, 1988.
136. **Miki, N., Ono, M., Miyoshi, H., et al.,** *Life Sci.*, 44, 469, 1985.
137. **Yamaushi, N. T., Shibasaki, N., Ling, N., et al.,** *Regul. Peptides*, 3, 71, 1991.
138. **Magnan, E. V., Guillaume, B., Conte-Devolx, M., et al.,** *Ann. Endocrinol.*, 52, 167, 1991.
139. **Katakami, H., Arimura, A., and Frohman, L. A.,** *Endocrinology (Baltimore)*, 118, 1872, 1986.
140. **Zeytin, F. N., Rusk, S. F., and De Lellis, R.,** *Endocrinology (Baltimore)*, 114, 2054, 1988.
141. **Tannenbaum, G. S., Huyda, H. J., and Posner, B. I.,** *Science*, 77, 1983.
142. **Spencer, G., Macdonald, A. A., Buttle, H. L., et al.,** *Acta Endocrinol. (Copenh.)*, 124, 563, 1991.
143. **Spencer, G. S. G., Berry, C., Hodgkinson, S. C., et al.,** *Mol. Cell. Neurosci.*, 4, 538, 1993.
144. **Harel, Z. and Tannenbaum, G. S.,** *Endocrinology (Baltimore)*, 131, 758, 1992.
145. **Spencer, G. S. G., Bass, J. J., Hodgkinson, S. G., et al.,** *Domest. Anim. Endocrinol.*, 8, 155, 1991.
146. **Hynes, M. A., Van Wyk, J. J., Brooks, P. J., et al.,** *Mol. Endocrinol.*, 1, 233, 1987.
147. **Shibasaki, T., Yamauchi, N., Hotta, M., et al.,** *Regul. Peptides*, 15, 47, 1986.
148. **Torsello, A., Panzeri, G., Cermenati, P., et al.,** *J. Endocrinol.*, 117, 273, 1988.
149. **Harvey, S.,** *Avian Endocrinology* (Ed. Sharp, P.). *J. Endocrinol.*, 1993, p. 11.
150. **Lea, R. W. and Harvey, S.,** *J. Endocrinol.*, 136, 245, 1993.
151. **Willoughby, J. O., Menadue, M., Zeegers, P., et al.,** *J. Endocrinol.*, 86, 165, 1980.
152. **Carlsson, L. and Jansssen, J. O.,** *Endocrinology (Baltimore)*, 126, 6, 1990.
153. **Willoughby, J. O. and Kapoor, R.,** *J. Neuroendocrinol.*, 2, 351, 1990.
154. **Holly, J. M. P. and Wass, J. A. H.,** *J. Endocrinol.*, 122, 611, 1989.
155. **Matsuo, K., Niwa, M., Kurihara, M., et al.,** *Cell. Mol. Neurobiol.*, 9, 357, 1989.
156. **Yamasaki, H., Prager, D., Gebremedhin, S., et al.,** *Endocrinology (Baltimore)*, 128, 857, 1991.
157. **Sheppard, M. S., Eatock, B. A., and Bala, R. M.,** *Can. J. Physiol. Pharmacol.*, 65, 2302, 1987.
158. **Brazeau, P., Guillemin, R., Ling, N., et al.,** *C.R. Acad. Sci. Paris*, 295, 651, 1982.

159. **Spencer, G. S. G., Macdonald, A. A., Carlyle, S. S., et al.,** *Reprod. Nutr. Develop.*, 31, 585, 1991.
160. **Hoeffler, J. P., Hicks, S. A., and Frawley, L. S.,** *Endocrinology (Baltimore)*, 120, 1936, 1987.
161. **Perez, F. M., Malamed, S., and Scanes, C. G.,** *IRCS Med. Sci.*, 13, 871, 1985.
162. **Buonomo, F. C., Lauterio, T. J., Baile, C. A., et al.,** *Gen. Comp. Endocrinol.*, 66, 274, 1987.
163. **Perez-Sanchez, J., Weil, C., and Lebail, P. Y.,** *J. Exp. Zool.*, 262, 287, 1992.
164. **Fagin, J. A., Fernandez-Megia, C., and Melmed, S.,** *Endocrinology (Baltimore)*, 125, 2385, 1989.
165. **Prager, D., Yamasaki, H., Weber, M. M., et al.,** *J. Clin. Invest.*, 90, 2117, 1992.
166. **Laron, Z., Klinger, B., Jensen, L. T., et al.,** *Clin. Endocrinol.*, 355, 145, 1991.
167. **Laron, Z., Blum, W., Chatelain, P., et al.,** *J. Pediatr.*, 122, 241, 1993.
168. **Laron, Z.,** *Arch. Dis. Child.*, 68, 345, 1993.
169. **Rosenbloom, A. L., Savage, M. O., Blum, W. F., et al.,** *Acta Pediatr. Scand.*, 81, 121, 1992.
170. **Frankel, R. J. and Jenkins, J. S.,** *Acta Endocrinol. (Copenh.)*, 78, 209, 1975.
171. **Salerno, F., Locatelli, V., and Muller, E. E.,** *Clin. Endocrinol.*, 27, 183, 1987.
172. **Kelly, J. J., Rajkovic, I. A., O'Sullivan, A. J., et al.,** *Clin. Endocrinol.*, 39, 561, 1993.
173. **Brown, G. M., Seggie, J. A., Chambers, J. W., et al.,** *Psychoneuroendocrinology*, 3, 131, 1978.
174. **Nemeroff, C. B. and Krishnan, K. R. R.,** *Neuroendocrinology* (Ed. Nemeroff, C. B.). CRC Press, Boca Raton, 1991, p. 413.
175. **Blackard, W. G. and Andrews, S. S.,** *Current Topics in Experimental Endocrinology* (Eds. James, V. H. T. and Martini, L.). Academic Press, New York, 1974, p. 129.
176. **Turyn, D. and Dellacha, J. M.,** *IRCS Med. Sci.*, 8, 580, 1980.
177. **Melmed, S.,** *J. Clin. Invest.*, 73, 1425, 1984.
178. **Ivarie, R., Baxter, J., and Morris, J.,** *J. Biol. Chem.*, 256, 4520, 1981.
179. **Sugase, T., Nonaka, K., Yoshida, T., et al.,** *Endocrinol. Jpn.*, 23, 187, 1976.
180. **Katakami, H., Kato, Y., Matsushita, N., et al.,** *Endocrinology (Baltimore)*, 115, 1598, 1984.
181. **Liapi, C., Evain-Brion, D., Argente, J., et al.,** *Acta Endocrinol. (Copenh.)*, 113, 1, 1985.
182. **Plewe, G., Scheider, C., Kurtz, V., et al.,** *Horm. Metab. Res.*, 481, 1985.
183. **Jorgensen, J. O. L., Pedersen, S. A., Laurberg, P., et al.,** *J. Clin. Endocrinol. Metab.*, 69, 1127, 1989.
184. **Grunfeld, C., Sherman, B. M., and Cavalieri, R. R.,** *J. Clin. Endocrinol. Metab.*, 67, 1111, 1988.
185. **Kuhn, E. R., Van Osselaer, P., Siau, O., et al.,** *J. Endocrinol.*, 109, 215, 1986.
186. **Rumsey, T. S., Kahl, S., Norton, S. A., et al.,** *Domest. Anim. Endocrinol.*, 7, 125, 1990.
187. **Valcavi, R., Zini, M., and Portioli, I.,** *J. Endocrinol. Invest.*, 15, 313, 1992.
188. **Hochberg, Z., Maor, G., Lewinson, D. et al.,** *Biotechnology of Growth*, Lamming, Ed., Butterworths, London, 1992, p. 123.
189. **Massa, G., de Zegher, F., and Vanderschueren-Lodeweyckx, M.,** *Clin. Endocrinol.*, 34, 205, 1991.
190. **Lippe, B. M., Van Herle, A. G., La Franjie, S. H., et al.,** *J. Clin. Endocrinol. Metab.*, 46, 12, 1975.
191. **Jorgensen, J. O. L., Moller, J., Skakkeboek, N. E., et al.,** *Horm. Res.*, 38, 63, 1992.
192. **McClean, C. and Laarveld, B.,** *Can. J. Anim. Sci.*, 71, 1053, 1991.
193. **Harvey, S.,** *Endocrinology of Birds: Molecular to Behavioural* (Eds. Wada, M., Ishi, S., and Scanes, C. G.). Japanese Scientific Press, Tokyo, 1990, p. 111.
194. **Dieguez, C., Jordan, V., Harris, P., et al.,** *J. Endocrinol.*, 109, 53, 1986.
195. **Brauman, H., Smets, P., and Corvilain, J.,** *J. Clin. Endocrinol. Metab.*, 36, 1162, 1973.
196. **Root, A. W., Shulman, D., Root, J., et al.,** *Acta Endocrinol. (Copenh.)*, 113, 367, 1986.
197. **Wakabayashi, I., Tonegawa, Y., Ihara, T., et al.,** *Neuroendocrinology*, 41, 306, 1985.
198. **Finkelstein, J. W., Boyar, R. M., and Hellman, L.,** *J. Clin. Endocrinol. Metab.*, 38, 634, 1974.
199. **Sasaki, N., Tsuyusaki, T., Nakamura, H., et al.,** *Endocrinol. Jpn.*, 32, 39, 1985.
200. **Adams, E. F., Bray Kovich, I. E., and Mashiter, K.,** *J. Clin. Endocrinol. Metab.*, 53, 381, 1981.
201. **Mulchahey, J. J., Di Blasio, A. M., and Jaffe, R. B.,** *J. Clin. Endocrinol. Metab.*, 66, 395, 1988.
202. **Cattini, P. A., Anderson, T. R., Baxter, J. D., et al.,** *J. Biol. Chem.*, 261, 13367, 1986.
203. **Brent, G. A., Harney, J. W., Moore, D. D., et al.,** *Mol. Endocrinol.*, 2, 792, 1988.
204. **Samuels, H. H., Forman, B. M., Horowitz, Z. D., et al.,** *Annu. Rev. Physiol.*, 51, 623, 1989.
205. **Ezzart, S., Laks, D., Oster, J., et al.,** *Endocrinology (Baltimore)*, 128, 937, 1991.
206. **Ye, Z., Forman, B. M., Aranda, A., et al.,** *J. Biol. Chem.*, 263, 7821, 1988.
207. **Coulombe, P., Ruel, J., and Dussault, J. H.,** *Endocrinology (Baltimore)*, 107, 2027, 1980.
208. **Peake, G. T., Birge, C. A., and Daughaday, W. H.,** *Endocrinology (Baltimore)*, 92, 487, 1973.

209. **Coiro, V., Braverman, L. E., Christianson, D., et al.,** *Endocrinology (Baltimore)*, 195, 641, 1979.
210. **Hervas, F., Morreale, G., and Escobar Del Rey, F.,** *Endocrinology (Baltimore)*, 97, 91, 1975.
211. **Mulloy, A. L., Smith, T. J., and Stachura, M. E.,** *Horm. Metab. Res.*, 24, 466, 1992.
212. **Coulombe, P., Schwartz, H. L., and Oppenheimer, J. H.,** *J. Clin. Invest.*, 62, 1020, 1978.
213. **Hinkle, P. M. and Goh, K. B. C.,** *Endocrinology (Baltimore)*, 110, 1725, 1982.
214. **Chernausek, S. D. and Turner, R.,** *J. Pediatr.*, 114, 968, 1989.
215. **Buchanan, C. R., Stanhope, R., Adlard, P., et al.,** *Clin. Endocrinol.*, 29, 427, 1988.
216. **Valcavi, R., Jordan, V., Dieguez, C., et al.,** *Clin. Endocrinol.*, 24, 693, 1986.
217. **Williams, T., Maxon, H., Thorner, M. O., et al.,** *J. Clin. Endocrinol. Metab.*, 61, 454, 1985.
218. **DeGennaro-Colonna, V., Bertola, G., Coco, C. B., et al.,** *Proc. Soc. Exp. Biol. Med.*, 193, 214, 1990.
219. **Root, J. L., Duckett, G. E., Sweeland, M., et al.,** *Endocrinology (Baltimore)*, 116, 1703, 1985.
220. **Hendrich, C. E. and Porterfield, S. P.,** *Proc. Soc. Exp. Biol. Med.*, 201, 296, 1992.
221. **Bruhn, T. O., Mcfarlane, M. B., Deckey, J. E., et al.,** *Endocrinology (Baltimore)*, 131, 2615, 1992.
222. **Jones, P. M., Burrin, J. M., Ghatel, M. A., et al.,** *Endocrinology (Baltimore)*, 126, 1374, 1990.
223. **Rodriguez-Arnao, J., Miell, J. P., and Ross, R. J. M.,** *Trends Endocrinol.Metab.*, 4, 169, 1993.
224. **Miell, J. P., Taylor, A. M., Zini, M., et al.,** *J. Clin. Endocrinol. Metab.*, 76, 950, 1993.
225. **Nantosalonen, K., Muller, H. L., Hoffman, A. R., et al.,** *Endocrinology (Baltimore)*, 132, 781, 1993.
226. **Elsasser, T. H., Rumsey, T. S., and Kahl, S.,** *Domest. Anim. Endocrinol.*, 10, 71, 1993.
227. **Matsuo, K., Yamashita, S., Niwa, M., et al.,** *Endocrinology (Baltimore)*, 126, 550, 1990.
228. **Gillioz, P., Giraud, P., Conte-Devoix, P., et al.,** *Endocrinology (Baltimore)*, 104, 1407, 1979.
229. **Berelowitz, M., Maeda, K., Harris, S., et al.,** *Endocrinology (Baltimore)*, 107, 24, 1980.
230. **de los Frailes, M., Cacicedo, L., Lorenzo, M. J., et al.,** *Endocrinology (Baltimore)*, 123, 898, 1988.
231. **Downs, T. R., Chomczynski, P., and Frohman, L. A.,** *Mol. Endocrinol.*, 4, 402, 1990.
232. **Harvey, S.,** *J. Endocrinol.*, 125, 345, 1990.
233. **Wartofsky, L. and Burman, K. D.,** *Endocr. Rev.*, 3, 164, 1982.
234. **Kuhn, E. R., Decuypere, E., Huybrechts, L. M., and Darras, V. M.,** *Endocrinology of Birds* (Ed. Wada, M.). Japan Sciences Society Press, Tokyo, 1990, p. 129.
235. **Denver, R. J. and Harvey, S.,** *J. Endocrinol.*, 131, 39, 1991.
236. **Kuhn, E. R., Herremans, M., Dewil, E., et al.,** *Reprod. Nutr. Dev.*, 31, 431, 1991.
237. **Donoghue, D. J. and Scanes, C. G.,** *Poultry Sci.*, 68, 46, 1989.
238. **Donoghue, D. J., Perez, F. M., Diamante, B. S. A., et al.,** *Domest. Anim. Endocrinol.*, 7, 35, 1990.
239. **Harvey, S., Decuypere, E., Darras, V. M., et al.,** *Reprod. Nutr. Dev.*, 31, 451, 1991.
240. **Harvey, S., Klandorf, H., and Scanes, C. G.,** *J. Endocrinol.*, 124, 215, 1990.
241. **Harvey, S. and Baidwan, J. S.,** *J. Mol. Endocrinol.*, 4, 127, 1990.
242. **Harvey, S., Hoshino, S., and Suzuki, M.,** *Gen. Comp. Endocrinol.*, 65, 92, 1987.
243. **Harvey, S.,** *J. Endocrinol.*, 96, 329, 1983.
244. **Scanes, C. G. and Harvey, S.,** *Gen. Comp. Endocrinol.*, 73, 477, 1989.
245. **Harvey, S., Sterling, R. J., and Klandorf, H.,** *Gen. Comp. Endocrinol.*, 50, 275, 1983.
246. **Carsia, R. V., Weber, H., King, D. B., et al.,** *Endocrinology (Baltimore)*, 117, 928, 1985.
247. **Mehls, O., Tonshoff, B., Kovacs, G., et al.,** *Acta Pediatr. Scand.*, 82, 77, 1993.
248. **Kononen, J., Honkaniemi, J., Gustafsson, J., et al.,** *Mol. Cell. Endocrinol.*, 93, 97, 1993.
249. **Martinoli, M. G. and Pelletier, G.,** *Endocrinology (Baltimore)*, 125, 1246, 1989.
250. **Casanueva, F. F., Burguera, D., Tome, M. A., et al.,** *Neuroendocrinology*, 47, 46, 1988.
251. **Ceda, G. P., Davis, R. G., and Hoffman, A. R.,** *Acta Endocrinol. (Copenh.)*, 114, 465, 1987.
252. **Casanueva, F. F., Burguera, B., Murvais, C., et al.,** *J. Clin. Endocrinol. Metab.*, 70, 234, 1990.
253. **Martul, P., Pineda, J., Dieguez, C., et al.,** *J. Clin. Endocrinol. Metab.*, 75, 536, 1992.
254. **Burguera, B., Muruais, C., Penalva, A., et al.,** *Neuroendocrinology*, 51, 51, 1990.
255. **Muruais, C., Cordido, F., Morales, M. J., et al.,** *Clin. Endocrinol.*, 35, 485, 1991.
256. **Popovic, V., Damjanovic, S., Micic, D., et al.,** *Neuroendocrinology*, 58, 465, 1993.
257. **Miell, J., Corder, R., Miell, P. J., et al.,** *J. Endocrinol.*, 131, 75, 1991.
258. **Bueguera, B., Muruais, C., Penalva, A., et al.,** *Neuroendocrinology*, 51, 51, 1990.
259. **Bozzola, M., Locatelli, F., Gambarana, D., et al.,** *Horm. Res.*, 36, 183, 1991.
260. **Seifert, H., Perrin, M., Rivier, J., et al.,** *Endocrinology (Baltimore)*, 117, 424, 1985.
261. **Giustina, A., Bussi, A. R., Lincini, M., et al.,** *Horm. Res.*, 37, 212, 1992.

262. **Lam, K. S. L., Srivastava, G., Tam, S., et al.,** *Neuroendocrinology*, 58, 325, 1993.
263. **Wehrenberg, W. B., Baird, A., and Ling, N.,** *Science*, 221, 556, 1983.
264. **Wehrenberg, W., Janowski, B., Piering, A. W., et al.,** *Endocrinology (Baltimore)*, 126, 3200, 1990.
265. **Giustina, A. and Wehrenberg, W. B.,** *Trends Endocrinol. Metab.*, 3, 306, 1992.
266. **Arsenijevic, Y., Wehrenberg, W. B., Conz, A., et al.,** *Endocrinology (Baltimore)*, 124, 3050, 1989.
267. **Davis, S. R., Smith, J. F., and Gluckman, P. D.,** *Reprod. Fertil. Dev.*, 2, 173, 1990.
268. **Schemm, S. R., Deaver, D. R., Griel, L. C., Jr., et al.,** *Biol. Reprod.*, 42, 815, 1990.
269. **Landerfeld, T. D. and Suttie, J. M.,** *Endocrinology (Baltimore)*, 125, 1474, 1989.
270. **Wehrenberg, W. B. and Giustina, A.,** *Endocr. Rev.*, 13, 299, 1992.
271. **Shirasu, K., Stumpf, W. E., and Sar, M.,** *Endocrinology (Baltimore)*, 127, 344, 1990.
272. **Jansson, J. O., Eden, S., and Isaksson, O.,** *Endocr. Rev.*, 6, 128, 1985.
273. **Birge, C. A., Peake, G. T., Mariz, I. K., et al.,** *Endocrinology (Baltimore)*, 81, 195, 1967.
274. **Silverman, B., Kaplan, S. L., Grumbach, M. M., et al.,** *Endocrinology (Baltimore)*, 122, 1236, 1988.
275. **DePippo, V. A. and Powers, C. A.,** *Endocrinology (Baltimore)*, 129, 1696, 1991.
276. **Hassan, H. A., Merkel, R. A., Enright, W. J., et al.,** *Domest. Anim. Endocrinol.*, 9, 209, 1992.
277. **Cronin, M. J. and Rogol, A. D.,** *Biol. Reprod.*, 31, 984, 1984.
278. **Wehrenberg, W. B., Baird, A., Ying, S. Y., et al.,** *Biol. Reprod.*, 32, 369, 1985.
279. **Evans, W. S., Krieg, R. J., Limber, E. R., et al.,** *Am. J. Physiol.*, 249, E276, 1985.
280. **Batson, J. M., Krieg, R. J., Martha, P. M., et al.,** *Endocrinology (Baltimore)*, 124, 444, 1989.
281. **Arsenijevic, Y., Riverst, R. W., Eshkol, A., et al.,** *Endocrinology (Baltimore)*, 121, 1487, 1987.
282. **Hoeffler, J. P. and Frawley, L. S.,** *Endocrinology (Baltimore)*, 119, 1037, 1986.
283. **Ho, K. Y., Leong, D. A., Sinha, Y. N., et al.,** *Am. J. Physiol.*, 250, E650, 1986.
284. **Ohlsson, L., Isaksson, O., and Jansson, J.,** *J. Endocrinol.*, 113, 249, 1987.
285. **Hertz, P., Silbermann, M., Even, L., et al.,** *Endocrinology (Baltimore)*, 125, 581, 1989.
286. **Liu, L., Merrian, G. R., and Sherins, R. J.,** *J. Clin. Endocrinol. Metab.*, 64, 651, 1987.
287. **Schettini, G. F. T., Meucci, O., Landolfi, E., et al.,** *Brain Res.*, 439, 322, 1988.
288. **Zeitler, P., Chowen-Breed, J. A., Argente, J., et al.,** *72nd Ann. Meet. Endocr. Soc.*, Atlanta, GA, 1990, p. 53.
289. **Zeitler, P., Argente, J., Chowen-Breed, J. A., et al.,** *Endocrinology (Baltimore)*, 127, 1362, 1990.
290. **Chowen-Breed, J. A., Steiner, R. A., and Clifton, D. K.,** *Endocrinology (Baltimore)*, 125, 357, 1989.
291. **Zorilla, R. and Rheaume, E.,** *72nd Ann. Meet. Endocr. Soc.*, Atlanta, GA, 1990, p. 50.
292. **Jansson, J. O., Ishikawa, K., Katakami, H., et al.,** *Endocrinology (Baltimore)*, 120, 525, 1987.
293. **Argente, J., Chowen, J. A., Zeitler, P., et al.,** *Endocrinology (Baltimore)*, 128, 2369, 1991.
294. **Gabriel, S. M., Millard, W. J., and Koenig, J. I.,** *Neuroendocrinol.*, 50, 299, 1989.
295. **Ge, F., Tsagarakis, S., Rees, L. H., et al.,** *J. Endocrinol.*, 123, 53, 1989.
296. **Maiter, D., Koenig, J. I., and Kaplan, L. M.,** *Endocrinology (Baltimore)*, 128, 1709, 1991.
297. **Colonna, V. D. G., Zoli, M., Cocchi, D., et al.,** *Peptides*, 10, 705, 1989.
298. **Hasegawa, O., Sugihara, H., Minami, S., et al.,** *Peptides*, 13, 475, 1992.
299. **Fukata, J. and Martin, J. B.,** *Endocrinology (Baltimore)*, 119, 2256, 1986.
300. **Sar, M.,** *Endocrinology (Baltimore)*, 123, 1110, 1988.
301. **Fernandez, G., Sanchez-Franco, F., de los Frailes, M. T., et al.,** *Regul. Peptides*, 42, 135, 1992.
302. **Gross, D. S.,** *Am. J. Anat.*, 158, 507, 1980.
303. **Clark, R. G. and Robinson, I. C. A. F.,** *J. Endocrinol.*, 106, 281, 1985.
304. **Werner, H., Koch, Y., Baldino, F., et al.,** *J. Biol. Chem.*, 263, 7666, 1988.
305. **Zorrilla, R., Simard, J., Rheaume, E., et al.,** *Neuroendocrinology*, 52, 527, 1990.
306. **Wehrenberg, W. B., Ling, N., Bohlen, P., et al.,** *Biochem. Biophys. Res. Commun.*, 109, 562, 1982.
307. **Argente, J., Chowen-Breed, J. A., Steiner, R. A., et al.,** *Neuroendocrinology*, 52, 342, 1990.
308. **Conway, S., Moherek, R., Mauceri, H., et al.,** *Endocrinology (Baltimore)*, 125, 2475, 1989.
309. **Kerrigan, J. R. and Rogol, A. D.,** *Endocr. Res.*, 13, 281, 1992.
310. **Bethea, C. L. and Freesh, F.,** *Endocrinology (Baltimore)*, 129, 2110, 1991.
311. **Devesa, J., Lois, N., Arce, V., et al.,** *J. Steroid Biochem. Mol. Biol.*, 40, 165, 1991.
312. **Benito, P., Avila, L., Corpas, M. S., et al.,** *J. Endocrinol. Invest.*, 14, 265, 1991.
313. **Mauras, N., Rogol, A. D., and Veldhuis, J. D.,** *Pediatr. Res.*, 28, 626, 1990.

314. **Merimee, T. J. and Fineberg, S. E.,** *J. Clin. Endocrinol. Metab.*, 33, 896, 1971.
315. **Martin, L. G., Grossman, M. S., Connor, T. B., et al.,** *Acta Endocrinol.*, 91, 201, 1979.
316. **Link, R., Blizzard, R. M., Evans, W. S., et al.,** *J. Clin. Endocrinol. Metab.*, 62, 159, 1986.
317. **Foster, C. M., Hopwood, N. J., Hassing, J. M., et al.,** *Pediatr. Res.*, 1989, 320, 1989.
318. **Kelley, K. W.,** *Neuropeptides and Immunopeptides: Messengers in a Neuroimmune Axis* (Eds. O'Dorsio, M. S. and Panerai, A.). New York Academy of Sciences, New York, 1990, p. 95.
319. **Blalock, J. E. (Ed.),** *Neuroimmunoendocrinology*, S. Karger, Basel, 1992, p. 1.
320. **Blalock, J. E.,** *Physiol. Rev.*, 69, 1, 1988.
321. **Spangelo, B. L. and MacLeod, R. M.,** *Trends Endocrinol. Metab.*, 2, 408, 1990.
322. **Smith, E. M.,** *Neuroimmunoendocrinology* (Ed. Blalock, J. E.). S. Karger, Basel, 1992, p. 154.
323. **Scarborough, D. E.,** *Neuropepetides and Immunopeptides: Messengers in a Neuroimmune Axis* (Eds. O'Dorsio, S. M. and Panerai, A.). New York Academy of Sciences, New York, 1990, p. 169.
324. **Benveniste, E. N.,** *Neuroimmunoendocrinology* (Ed. Blalock, J. E.). S. Karger, Basel, 1992, p. 107.
325. **Millington, G. and Buckingham, J. C.,** *Endocrinology (Baltimore)*, 133, 163, 1992.
326. **Bianchi, E., Pierpaoli, W., and Sorkin, E.,** *J. Endocrinol.*, 51, 1, 1971.
327. **Michael, S. D., Taguchi, O., and Nishizuka, Y.,** *Biol. Reprod.*, 22, 343, 1980.
328. **Spangelo, B. L., Judd, A. M., Ross, P. C., et al.,** *Endocrinology (Baltimore)*, 121, 2035, 1987.
329. **Spangelo, B. L., Ross, P. C., Judd, A. M., et al.,** *J. Neuroimmunol.*, 25, 37, 1989.
330. **Badamchian, M., Wang, S., Spangelo, B. L., et al.,** *Prog. Neuroendoimmunol.*, 3, 258, 1990.
331. **Angioni, S., Iori, G., Cellini, M., et al.,** *Acta Endocrinol. (Copenh.)*, 127, 237, 1992.
332. **Goya, R. G., Quigley, K. L., Takahashi, R., et al.,** *Mech. Ageing Dev.*, 49, 119, 1989.
333. **Farrar, W., Hill, J., Hael-Bellan, A., et al.,** *Immunol. Rev.*, 100, 361, 1987.
334. **Marguette, C., Ban, E., Fillon, G., et al.,** *Neuroendocrinology*, 52, 49, 1990.
335. **Farrar, W. L., Kilian, P. L., Ruff, M. R., et al.,** *J. Immunol.*, 139, 459, 1987.
336. **Takao, T., Tracey, D. E., Mitchell, W. M., et al.,** *Endocrinology (Baltimore)*, 127, 3070, 1990.
337. **Spangelo, B. L., Judd, A. M., Isakson, P., et al.,** *Endocrinology (Baltimore)*, 125, 575, 1989.
338. **Spangelo, B. L. and MacLeod, R. M.,** *Prog. Neurol. Endocrinol. Immunol.*, 3, 167, 1990.
339. **Malarkey, W. B. and Zvara, B. J.,** *J. Clin. Endocrinol. Metab.*, 69, 196, 1989.
340. **Bernton, E. W., Beach, J. E., Holaday, J. W., et al.,** *Science*, 238, 519, 1987.
341. **Berczi, I.,** *Endocr. Pathol.*, 1, 197, 1990.
342. **Grimaldi, M., Meucci, O., Scorziello, A., et al.,** *Life Sci.*, 51, 1243, 1992.
343. **Yateman, M., Claffey, D., Cwyfan Hughes, S. C., et al.,** *Endocrinology (Baltimore)*, 137, 151, 1993.
344. **Linkhart, T. A. and MacCharles, D. C.,** *Endocrinology (Baltimore)*, 131, 2297, 1992.
345. **D'Urso, R., Falaschi, P., Canfalone, G., et al.,** *Neurol. Endocrinol. Immunol.*, 4, 20, 1991.
346. **Buonomo, F. C.,** *74th Annu. Meet. Endocr. Soc.,* San Antonio, TX, 1992.
347. **Karanth, S. and McCann, S. M.,** *Proc. Natl. Acad. Sci. U.S.A.*, 88, 2961, 1991.
348. **Vankelecom, H., Carmeliet, P., Heremans, H., et al.,** *Endocrinology (Baltimore)*, 126, 2919, 1990.
349. **Vankelecom, H., Andries, M., Billiau, A., et al.,** *Endocrinology (Baltimore)*, 130, 3537, 1992.
350. **Gonzalez, M. C., Riedel, M., Rettori, V., et al.,** *Prog. Neuroendocrinol. Immunol.*, 3, 49, 1990.
351. **Walton, P. E. and Cronin, M. J.,** *Endocrinology (Baltimore)*, 125, 925, 1989.
352. **Elsasser, T., Caperna, T., Kenison, D. C., et al.,** *71st Annu. Meet. Endocr. Soc.,* Seattle, WA, 1989, p. 788.
353. **Atkins, M. B., Gould, M., Allegretta, J. J. L., et al.,** *J. Clin. Oncol.*, 4, 1380, 1986.
354. **Campbell, R. M., McIntyre, K., Lee, Y., et al.,** *73rd Annu. Meet. Endocr. Soc.* Washington, D.C., 1991, p. 1279.
355. **Lumpkin, M. D. and Hartmann, D. P.,** *71st Annu. Meet. Endocr. Soc.,* Seattle, WA, 1989, p. 789.
356. **Peisen, J. N., McDonnell, K. J., Bordeaux, J. W., et al.,** *75th Annu. Meet. Endocr. Soc.*, Las Vegas, Nevada, 1993, p. 415.
357. **Scarborough, D. E., Lee, S. L., and Reichlin, S.,** *Endocrinology (Baltimore)*, 124, 549, 1989.
358. **Scarborough, D. E.,** *2nd Int. Pituitary Congress*, 1989, p. 24.
359. **Rettori, V., Jurcovicova, J., and McCann, S. M.,** *Neurosci. Res.*, 18, 179, 1987.
360. **Scarborough, D. E. and Dinarello, C. A.,** *71st Annu. Meet. Endocr. Soc.,* Seattle, WA, 323, 1989.
361. **Lumpkin, M., Koenig, J., Tracey, D. E., et al.,** *71st Annu. Meet. Endocr. Soc.,* Seattle, WA, 1989, p. 1277.

362. **Karanth, S., Aguila, M. C., and McCann, S. M.,** *Neuroendocrinology*, 58, 185, 1993.
363. **Ross, R. J. M. and Buchanan, C. R.,** *Nutr. Res. Rev.*, 3, 143, 1990.
364. **Wurzburger, M. I., Prelevic, G. M., Sonksen, P. H., et al.,** *J. Clin. Endocrinol. Metab.*, 77, 267, 1993.
365. **Salomon, F., Cuneo, R. C., Hesp, R., et al.,** *N. Engl. J. Med.*, 321, 1797, 1989.
366. **Marcus, R., Butterfield, G., Holloway, L., et al.,** *J. Clin. Endocrinol. Metab.*, 70, 519, 1990.
367. **Ponting, G. A., Halliday, D., Teale, J. D., et al.,** *Lancet*, 1, 438, 1988.
368. **Davies, R. R., Turner, S., and Johnson, D. G.,** *Clin. Endocrinol.*, 21, 477, 1984.
369. **Quabbe, H. J., Burgess, S., Walz, T., et al.,** *J. Clin. Endocrinol.*, 70, 908, 1990.
370. **Balzano, S., Loche, S., Murtas, M. L., et al.,** *Horm. Metab. Res.*, 21, 52, 1989.
371. **Renier, G. and Serri, O.,** *Neuroendocrinology*, 54, 521, 1991.
372. **Berelowitz, M., Dudlak, D., and Frohman, L. A.,** *J. Clin. Invest.*, 69, 1293, 1982.
373. **Lengyel, A., Grossman, A., Nieuwenhuyzen-Kruseman, A., et al.,** *Neuroendocrinology*, 39, 31, 1984.
374. **Devesa, J., Lois, N., Diaz, M. J., et al.,** *J. Steroid Mol. Biol.*, 40, 165, 1991.
375. **Devesa, J., Diaz, M. J., Tresguerres, J. A. F., et al.,** *J. Clin. Endocrinol. Metab.*, 73, 251, 1991.
376. **Devesa, J., Arce, V., Lois, N., et al.,** *J. Clin. Endocrinol. Metab.*, 71, 1581, 1990.
377. **Casanueva, F. F., Villanueva, L., Dieguez, C., et al.,** *J. Clin. Endocrinol. Metab.*, 65, 634, 1987.
378. **Estienne, M. J., Schillo, K. K., Hileman, S. M., et al.,** *Endocrinology (Baltimore)*, 126, 1934, 1990.
379. **Widmaier, E. P.,** *Horm. Metab. Res.*, 24, 147, 1992.
380. **Alvarez, C. V., Mallo, F., Burguera, B., et al.,** *Neuroendocrinology*, 53, 185, 1991.
381. **Imaki, T., Shibasaki, T., Shizume, K., et al.,** *J. Clin. Endocrinol. Metab.*, 60, 290, 1985.
382. **Love, J. A., Saum, W. R., and McGee, R., Jr.,** *Cell. Mol. Neurobiol.*, 5, 333, 1985.
383. **Devesa, J., Lima, L. and Tresguerres, J. A. F.,** *Trends Endocrinol. Metab.*, 3, 175, 1992.

Chapter 11

Growth Hormone Release: Paracrine Control

S. Harvey

CONTENTS

I. INTRODUCTION

Morphological and biochemical features of anterior pituitary cells suggest paracrine interactions occur between somatotroph and nonsomatotroph cells which participate in regulating growth hormone (GH) synthesis and/or release. Paracrine interactions in pituitary function have previously been reviewed by Denef et al.,[1–3] O'Halloran et al.,[4] Schwartz and Cherny,[5] Jones et al.[6] and Funder.[7]

II. SOMATOTROPH TOPOGRAPHY

The topographical distribution of pituitary cell types is not random but differs in different species (see Chapter 3). This discrete, spatial distribution of cell types is thought to reflect the anatomical distribution of blood vessels within the pituitary gland and differences in the type and abundance of the hypophysiotropic releasing factors in different blood vessels.[8] The hypophysial portal vessels arise in different regions of the median eminence and hence they may receive neurosecretions from discrete hypothalamic nuclei, and these neurosecretions may be directed to specific pituitary regions by point-to-point vasculature.[9] This spatial distribution may also reflect the actions of local factors on the proliferation and differentiation of neighboring cells. The spatial relationships between cells may also be modulated by peripheral endocrines (e.g., by sex steroids, thyroid hormones, glucocorticoids, and pancreatic hormones) and by physiological state (e.g., nutrient status, age, and pregnancy).

In the rat, somatotrophs are scattered or grouped in clusters and evenly dispersed in most of the adenohypophysis but are unevenly distributed dorsoventrally and in areas adjacent to the intermediate lobe of the gland.[2] They also appear in palisades along capillary vessels. Somatotrophs are frequently in juxtaposition with corticotroph cells, which have long cytoplasmic processes that may engulf or completely encircle these GH-secreting cells.[10] Somatotrophs are also in close association with thyrotrophs[11,12] and surrounded by the long cytoplasmic extensions of folliculostellate cells.[1] Corticotrophs, thyrotrophs, and folliculostellate cells may thus communicate with somatotrophs through gap junctions[13–15] or through paracrine regulators.

0-8493-8697-7/95/$0.00+$.50

The influence of cell-to-cell communication on somatotroph function is likely to depend on somatotroph heterogeneity[12] and the heterogeneity of the juxtaposed cell types and subtypes.[6] Cell-to-cell communication is also likely to reflect the density and proximity of adjacent cells, as GH responsiveness (but not sensitivity) to GHRH is enhanced at high densities and basal GH release is greatest at low cell densities.[16,17] This basal release of GH from tissue slices is similarly less than that from dispersed pituitary cells.[12,17] Cell-to-cell contact may, therefore, affect the cellular integrity of somatotrophs, because GH synthesis or secretory granule storage may be better maintained in high-density cell concentrations than in low-density concentrations.[16]

III. PARACRINE REGULATORS

Intercellular communication between neighboring pituitary cells may be mediated through neurotransmitters, neuropeptides, hormones, cytokines, or growth factors (Tables 1 and 2). Although the pituitary is the source of peptide hormones, it also has the capacity to synthesize, secrete, and degrade some biogenic amines.[18] Somatotrophs are also electrically active and possess neurotransmitter receptors and directly respond to neurotransmitter signals.[18,19] In addition to secreting GH, prolactin, luteinizing hormone (LH), follicle-stimulating hormone (FSH), thyrotropin (TSH), and adrenocorticotropic hormone (ACTH), pituitary cells also contain at least 50 biologically active peptides, most of which appear to be synthesized within the pituitary gland (Table 2). Most of these peptides also appear to be released within the pituitary gland and to be subject to sexually dimorphic regulation and regulated by factors that modify pituitary hormone secretion (see Denef et al.,[2] Houben et al.[1,20] O'Halloran et al.,[4] Jones et al.,[6] and Spangelo and MacLeod[21] for reviews). Local roles of these peptides are indicated by their low intracellular concentrations and the presence of their receptors on pituitary cell membranes (Table 2). Putative paracrine actions of some of these factors on GH release are summarized in Table 3.

A. INTERACTIONS WITH FOLLICULOSTELLATE CELLS

Folliculostellate cells comprise 5 to 15% of adenohypophysial cells. They have a distinctive stellate shape with long extensions between other pituitary cell types and often associate to form follicles.[3] These cells typically surround follicular cavities and intercellular lacunae, through which molecules delivered by portal blood are thought to diffuse into secretory adenohypophysial cells. These cells are also characterized by intercellular (tight) junctions between adjacent folliculostellate cells and between folliculostellate

Table 1 **Neurocrines in Anterior Pituitary Cells**

Neurotransmitter	Presence	Biosynthetic Enzymes	Degradative Enzymes	Receptors
Epinephrine	–	–		+(α_1, $\alpha_2\beta$)
Norepinephrine	–	–	+	+(α_1, $\alpha_2\beta$)
Dopamine	+	+/–	+	+(D_1, D_2)
Serotonin	+	+/–		
Histamine	+	–		
Acetylcholine		+	+	+
GABA	+	–		+

Note: +, Presence; –, absence; GABA, γ-aminobutyric acid.

Source: Based on Muller et al.[18]

Table 2 **Neuropeptides, Cytokines, and Growth Factors in Anterior Pituitary Cells**

Peptide	Peptide Immunoreactivity	Peptide mRNA	Peptide Receptor
Activin		+	
Angiotensin II	+	+	+
Adipocyte growth factor	+		
Atrial natriuretic factor	+	+	+
Lys-bradykinin	+		
Met-bradykinin	+		
Calcitonin	+		+
Calcitonin gene-related peptide	+		+
Cholecystokinin	+		
Chondrocyte growth factor	+		
Corticotropin-releasing factor	+		
Cyclo(His-Pro)	+		
Dynorphin	+	+	+[EOP]
β-Endorphin	+		+[EOP]
γ-Endorphin	+		+[EOP]
α-Endorphin	+		+[EOP]
Met-enkephalin	+		+[EOP]
Leu-enkephalin	+		+[EOP]
Endothelin	+		
Epidermal growth factor	+		
Fibroblast growth factor	+	+	
Glial growth factor	+		
Galanin	+	+	
Gastrin-releasing peptide	+		
Gastrin	+		
Growth hormone-releasing hormone	+		
Insulin-like growth factor I	+	+	+
Insulin-like growth factor II		+	+
Inhibin	+	+	+
Interferon λ	+		
Interleukin 6	+	+	
Lipotropin	+		
Luteinizing hormone releasing hormone	+		
Motilin	+		
Neurotensin	+	+	+
Neurokinin A	+		+
Neuromedin B	+		
Neuromedin U	+		
Neuropeptide Y		+	
Neoendorphin	+		
Ovarian growth factor	+		
Oxytocin	+	+	+
Pancreastatin	+		

Table 2 (continued) **Neuropeptides, Cytokines, and Growth Factors in Anterior Pituitary Cells**

Peptide	Peptide Immunoreactivity	Peptide mRNA	Peptide Receptor
Somatostatin	+		
Substance P	+		+
Secretin	+		
Thyrotropin-releasing hormone	+		
Transforming growth factor α	+	+	
Tachykinin Y	+	+	
Tachykinin B	+	+	
Vasoactive intestinal peptide	+	+	+
Vasopressin	+	+	+
Vascular endothelial growth factor	+		

Note: +, Presence of endogenous opioid receptor (EOP).

Source: Based on Denef et al.,[2] O'Halloran et. al.,[4] and Houben et. al.[1]

cells and other cell types. These junctional barriers do not, however, impede molecular diffusion into follicles, cavities, or intercellular lacunae.

Although they lack secretory granules, folliculostellate cells produce bioactive molecules that are not secreted into blood but instead act locally to alter the ionic composition of interstitial fluid.[22] These cells are involved in the transport of Na^+, K^+, and Cl^- ions and may regulate the Ca^{2+} concentration in extracellular space. They may thus play a pivotal role in the secretory process of endocrine cells (see Chapter 6). Folliculostellate cells also buffer the secretory activity of pituitary endocrine cells by paracrine regulation.[23–25] In particular, these cells appear to inhibit by 50% the response of somatotrophs to GH-releasing hormone (GHRH), isoproterenol, and somatostatin (SRIF).[23] The factors mediating this inhibitory paracrine regulation are, however, still unknown.

Folliculostellate cells are cytokine producing cells and synthesize and secrete interleukin 6 (IL-6) in response to interferon γ or endotoxin.[21,26] Moreover, whereas interferon modulates GH release, it is only able to do so in the presence of folliculostellate cells.[25] Folliculostellate cells may thus modify somatotroph function through paracrine actions of IL-6, although stimulatory effects on basal GH release[27] and inhibitory effects on GHRH and epinephrine-induced GH release have been reported.[23] Growth factors (e.g., basic fibroblast growth factor and epidermal growth factor) may also mediate folliculostellate actions on somatotroph cells.[3]

B. INTERACTIONS WITH CORTICOTROPHS

The paracrine regulation of somatotrophs by corticotrophic factors is well established.[2] Although somatotrophs release GH in response to epinephrine and β-adrenergic agonists,[28] the GH response is greatly augmented by the coculture of somatotroph-enriched pituitary cells with corticotroph, especially in the presence of glucocorticoids.[29] This effect is specific and not induced when somatotrophs are cultured with gonadotrophs, thyrotrophs, or folliculostellate cells. The intracellular signal produced by corticotrophs in the presence of glucocorticoids is unknown. Corticotrophs are also thought to potentiate vasoactive intestinal peptide (VIP)-induced GH release.[30] Whereas VIP has negligible effects on GH release from purified somatotrophs, the coculture of somatotrophs with corticotrophs greatly augments VIP-induced GH release, by permissive actions of glucocorticoids. Similarly, whereas thyrotropin-releasing hormone (TRH) weakly stimulates GH release from purified somatotrophs, it antagonizes the stimulatory actions of

Table 3 **Putative Paracrine Actions on Somatotroph Function**

Regulator	Action	Ref.
Acetylcholine (muscarinic tone)	↑	47
	↓ tumor cells	48
	↓ stimulated release	28
Activin	↓	36
Angiotensin	↑ or ↓ (old rats)	39
Bradykinin	↑	49
Cholecystokinin 8	↑	50
Endothelin 3	↑	51
Epidermal growth factor	↑	52
Epidermal growth factor	↓	53
Epinephrine (α-adrenergic tone)	↑	28
Epinephrine (β-adrenergic tone)	↑	29
Fibroblast growth factor	↓ tumor cells	53
Galanin	↓ stimulated release	43
Galanin	↑	41
Gastrin-releasing peptide	↓	45
Growth hormone-releasing hormone	↑	54
Insulin-like growth factor I	↓	55
Interleukin 1	↓	56
	↑	57
Interleukin 6	↑	27
Interferon	↓	25
Interferon-γ	↓	56
Luteinizing hormone-releasing hormone	↑	40
Motilin	↑	51
Neuromedin B	↑	45
Neuromedin B30	↑	45
Neuromedin B32	↑	45
Neuromedin C	↑	45
Peptide histidine isoleucine	↑	31
Thyrotropin-releasing hormone	↑	2, 31
Vasoactive intestinal hormone	↑	31

Note: ↑, increase; ↓, decrease.

GHRH, peptide histidine isoleucine (PHI), and VIP when cocultured with corticotrophs and glucocorticoid hormones.[31]

Corticotrophs may therefore release factors that either stimulate or inhibit GH release. One of these inhibitory factors appears to be a cholinomimetic substance, which is able to activate muscarinic receptors and inhibit GH release.[32–34] The suppression of acetylcholine production by interleukin 1 might account for the stimulatory effect of this cytokine on GH release.[5]

C. INTERACTIONS WITH LACTOTROPHS

Lactotrophs also participate in the paracrine regulation of GH release. Lactotrophs synthesize and release VIP, which has been shown to exert paracrine actions on neighboring prolactin cells.[35] This peptide also increases GH secretion in reaggregates of dissociated anterior pituitary cells.[31] Lactotrophs may also mediate stimulatory effects of epinephrine on GH release, because they are responsive to epinephrine and epinephrine-

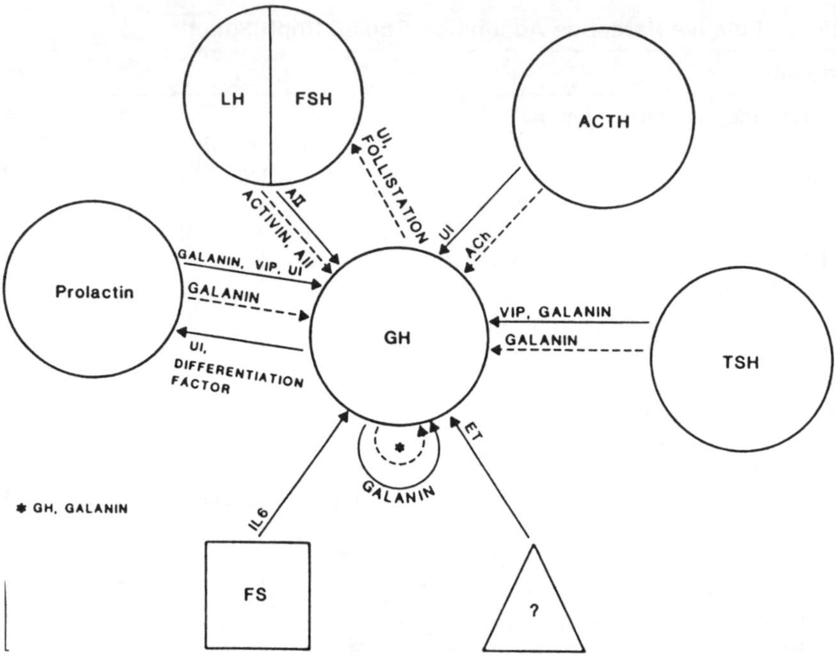

Figure 1 Possible intrapituitary interactions modulating growth hormone (GH) release from somatotrophs by factors released from lactotroph, gonadotroph, corticotroph, and thyrotroph cells. Solid lines denote stimulatory regulation whereas dashed lines indicate inhibitory regulation. FS, Folliculostellate cells; LH, luteinizing hormone; FSH, follicle-stimulating hormone; ACTH, adrenocorticotropin; TSH, thyrotropin; VIP, vasoactive intestinal peptide; UI, unidentified factor; AII, angiotensin II; Ach, acetylcholine; ET, endothelin; IL-6, interleukin 6. (From Schwartz, J. and Cherny, R., Intracellular communication within the anterior pituitary influencing the secretion of hypophysial hormones, *Endocr. Rev.,* 13(3), 453–475, 1992. Copyright 1992 by the Endocrine Society. Reprinted with permission.)

induced GH secretion is potentiated in the presence of lactotrophs.[29] This paracrine factor is not, however, likely to be prolactin.

D. INTERACTIONS WITH GONADOTROPHS

Like corticotrophs, gonadotrophs may produce paracrine factors that inhibit or stimulate GH secretion. Activin, for instance, is secreted by gonadotrophs and decreases basal and GHRH-induced GH release and inhibits *GH* gene transcription.[36–38] Gonadotrophs also produce angiotensin II (AII), which has been found to decrease the GH response of adult pituitary glands to GHRH stimulation but to potentiate basal GH release from the pituitary glands of young rats.[39] These actions of angiotensin are, however, dependent on the presence of other cell types, possibly lactotrophs, thyrotrophs, or corticotrophs, because these possess AII receptors.[5] Moreover, whereas gonadotropin-releasing hormone (GnRH) has no direct effects on pituitary somatotrophs, it stimulates GH release when endogenous opioid activity is blocked, but not when AII activity is antagonized.[39,40] The endogenous opioids synthesized in gonadotrophs are thus thought to modulate GH responses to GnRH. Endogenous opioid activity, conversely, inhibits GHRH-induced GH secretion from pituitary cell aggregates.[39]

E. INTERACTIONS WITH THYROTROPHS

Thyrotrophs may also modulate GH release by paracrine interactions with GH-secreting cells. In addition to VIP, thyrotrophs synthesize and secrete galanin (in common with lactotrophs and somatotrophs), which is reported to augment or inhibit the GH response to GHRH.[41–43]

IV. STEROID MILIEU

Topographical relationships and paracrine interactions between pituitary cells are therefore important factors in GH regulation, although these relationships are dependent on the hormonal milieu surrounding the somatotrophs. Thus the GH responses to β-agonists,[29] AII,[39] carbachol,[44] and VIP[29] are critically dependent on the presence of glucocorticoids. The ability of bombesin-like (gastrin releasing peptide and neuromedin C) and ranatensin-like (neuromedin B, neuromedin B_{30}, and neuromedin B_{32}) peptides to stimulate GH release from rat pituitary cell aggregates is similarly increased by 5-dihydrotestosterone, dexamethasone, and estradiol, even though estradiol inhibits GHRH-induced GH release.[45] Subtle control over the regulation of GH secretion therefore occurs within the pituitary gland and complements the complex integration that occurs within the CNS to provide positive or negative drive to the hypothalamus-somatotroph axis.

REFERENCES

1. **Houben, H., Tilemans, D., and Denef, C.,** *J. Endocrinol. Invest.*, 13, 855, 1990.
2. **Denef, C., Baes, M., and Schramme, C.,** *Front. Neuroendocrinol.*, 9, 115, 1986.
3. **Allaerts, W., Carmeliet, P., and Denef, C.,** *Mol. Cell Endocrinol.*, 71, 73, 1990.
4. **O'Halloran, D. J., Jones, P. M., and Bloom, S. R.,** *Mol. Cell Endocrinol.*, 75, C7, 1991.
5. **Schwartz, J. and Cherny, R.,** *Endocr. Rev.*, 13, 453, 1992.
6. **Jones, T. H., Brown, B. L., and Dobson, P. R. M.,** *J. Endocrinol.*, 127, 5, 1990.
7. **Funder, J. W.,** *Mol. Cell Endocrinol.*, 70, C21, 1990.
8. **Sasaki, F. and Iwama, Y.,** *Endocrinology*, 122, 1622, 1988.
9. **Schreibman, M. P.,** *Vertebrate Endocrinology: Fundamentals and Biomedical Implications* (Eds. Pang, P. K. T. and Schreibman, M. P.). Academic Press, New York, 1986, p. 11.
10. **Nagata, M., Mizunaga, A., Ema, S., et al.,** *Endocrinol. Jpn.*, 27, 13, 1980.
11. **Yoshimura, F. and Nogami, H.,** *Endocrinol. Jpn.*, 27, 43, 1980.
12. **Perez, F. M. and Hymer, W. C.,** *Endocrinology (Baltimore)*, 127, 1877, 1990.
13. **Fletcher, W. H., Anderson, N. C. and Everett, J. W.,** *J. Cell Biol.*, 67, 469, 1975.
14. **Herbert, D. C.,** *Anat. Rec.*, 114, 2107, 1979.
15. **Scanes, C. G. and Harvey, S.,** *Gen. Comp. Endocrinol.*, 73, 477, 1989.
16. **Sugimoto, H., Suzuki, M., Takeuchi, T., et al.,** *J. Endocrinol.*, 131, 237, 1991.
17. **Perez, F. M.,** *Endocrine J.*, 1, 49, 1993.
18. **Muller, E. E. and Nistico, G.,** *Brain Messengers and the Pituitary*, Academic Press, New York, 1989.
19. **Israel, J. M., Denef, C., and Vincent, J. D.,** *Neuroendocrinology*, 37, 193, 1983.
20. **Houben, H. and Denef, C.,** *Trends Endocrinol. Metab.*, 2, 398, 1990.
21. **Spangelo, B. L. and MacLeod, R. M.,** *Trends Endocrinol. Metab.*, 2, 408, 1990.
22. **Ferrara, N., Fuji, D. K., Goldsmith, P. C., et al.,** *Am. J. Physiol.*, 252, 304, 1987.
23. **Baes, M., Allaerts, W., and Denef, C.,** *Endocrinology (Baltimore)*, 120, 685, 1987.
24. **Allaerts, W., Engelborghs, Y., Van Oostveldt, P., et al.,** *Endocrinology (Baltimore)*, 127, 1517, 1990.
25. **Vankelecom, H., Andries, M., Billiau, A., et al.,** *Endocrinology (Baltimore)*, 130, 3537, 1992.
26. **Spangelo, B. L., MacLeod, R. M., and Isakson, P. C.,** *Endocrinology (Baltimore)*, 126, 582, 1990.
27. **Spangelo, B. L., Judd, A. M., Isakson, P., et al.,** *Endocrinology (Baltimore)*, 125, 575, 1989.
28. **Maertens, P. and Denef, C.,** *J. Endocrinol. Invest.*, 10, 13, 1987.

29. **Baes, M. and Denef, C.,** *Endocrinology (Baltimore)*, 120, 280, 1987.
30. **Shibasaki, T., Yamauchi, N., Hotta, M., et al.,** *Regul. Peptides,* 15, 47, 1986.
31. **Denef, C., Schramme, C., and Baes, M.,** *Neuroendocrinology*, 40, 88, 1985.
32. **Carmeliet, P. and Denef, C.,** *Endocrinology (Baltimore)*, 123, 1128, 1988.
33. **Carmeliet, P. and Denef, C.,** *J. Endocrinol. Invest.*, 10(Suppl. 3), 76, 1987.
34. **Carmeliet, P. and Denef, C.,** *Endocrinology (Baltimore)*, 124, 2218, 1989.
35. **Nagy, G., Mulchahey, J. J., and Neil, J. D.,** *Endocrinology (Baltimore)*, 122, 364, 1988.
36. **Billestrup, N., Gonzalez-Manchon, C., Potter, E., et al.,** *Mol. Endocrinol.*, 4, 356, 1990.
37. **Bilezikjian, L. M., Corrigan, A. Z., and Vale, W.,** *Endocrinology (Baltimore)*, 126, 2369, 1990.
38. **Kitaoka, M., Kojima, I., and Ogata, E.,** *Biochem. Biophys. Res. Commun.*, 157, 48, 1988.
39. **Robberecht, W. and Denef, C.,** *Endocrinology (Baltimore)*, 122, 1496, 1988.
40. **Robberecht, W., Andries, M., and Denef, C.,** *Neuroendocrinology*, 56, 550, 1992.
41. **Gabriel, S. M., Milbury, C. M., Nathanson, J. A., et al.,** *Life Sci.*, 42, 1981, 1988.
42. **Hulting, A. L., Meister, B., Carlsson, L., et al.,** *Acta Endocrinol. (Copenh.)*, 125, 518, 1991.
43. **Meister, B. and Hulting, A. L.,** *Neuroendocrinology*, 46, 387, 1987.
44. **Carmeliet, P., Baes, M., and Denef, C.,** *Endocrinology (Baltimore)*, 124, 2625, 1989.
45. **Houben, H. and Denef, C.,** *Endocrinology (Baltimore)*, 126, 2257, 1990.
46. **Ye, Z., Forman, B. M., Aranda, A., et al.,** *J. Biol. Chem.*, 263, 7821, 1988.
47. **Young, W., Bicknell, R. J., and Schofield, P. C.,** *J. Endocrinol.*, 80, 203, 1979.
48. **Wojikiewicz, R. J. H., Dobson, P., and Brown, B. L.,** *Biochem. Biophys. Acta*, 805, 25, 1984.
49. **Drouhault, R., Abrous, N., David, J., et al.,** *Neuroendocrinology*, 46, 360, 1987.
50. **Morley, J. E., Melmed, S. H., Briggs, J., et al.,** *Life Sci.*, 25, 1201, 1979.
51. **Samson, W. K., Skala, K. D., Alexander, B., et al.,** *Endocrinology (Baltimore)*, 128, 1465, 1991.
52. **Ikeda, H., Mitsuhashi, T., Kubota, K., et al.,** *Endocrinology (Baltimore)*, 115, 556, 1984.
53. **Schonbrunn, A., Krasnoff, M., Westendorf, J. M., et al.,** *J. Cell Biol.*, 85, 786, 1980.
54. **Robberecht, W. and Denef, C.,** *J. Endocrinol. Invest.*, 10, 69, 1987.
55. **Goodyer, C. G., De Stephano, L., Guyda, H. J., et al.,** *Endocrinology (Baltimore)*, 115, 1568, 1984.
56. **Vankelecom, H., Carmeliet, P., Heremans, H., et al.,** *Endocrinology (Baltimore)*, 126, 2919, 1990.
57. **Bernton, E. W., Beach, J. E., Holaday, J. W., et al.,** *Science*, 238, 519, 1987.

Chapter 12

Growth Hormone Release: Profiles

S. Harvey and W. H. Daughaday

CONTENTS

I. INTRODUCTION

The release of growth hormone (GH) is not static and occurs episodically throughout the day. The circulating GH concentration, therefore, reflects the amplitude, frequency and duration of episodic GH release and the half-life of GH in plasma. The amplitude and frequency of these secretory episodes are also dynamically regulated and vary during growth and sexual maturation and in most species are sexually dimorphic and modulated by nutrition, environmental and genetic factors, and by disease. These patterns of GH release are reviewed in this chapter. Some of these have also been recently reviewed by Hartman et al.[1]

II. BASAL GH RELEASE

In the absence of hypothalamic stimulation, ultrastructural and physiological evidence suggests that the synthesis and release of GH are usually negligible in the rat[2] and are accompanied by a reduction in pituitary size[3] and GH content.[4] However, transection of the hypophysial stalk fails to abolish GH secretion in rats,[3–5] in which plasma GH levels are low, but higher than the trough levels in pituitary-intact animals and similar to those in somatostatin (SRIF)-depleted rats given antiserum to GH-releasing hormone (GHRH).[6] Pituitaries ectopically transplanted under the kidney capsule also remain viable in the absence of hypothalamic regulation and continue to synthesize and release GH,[7–10] as do

0-8493-8697-7/95/$0.00+$.50

chicken pituitary glands transplanted over the pigeon crop sac.[11] Pituitary stalk section also fails to eliminate GH secretion in cattle[12] and fails to reduce serum GH concentrations in rhesus monkeys[13] and pigs,[14] in which pituitary somatotrophs are normal, even though the size and GH content of the pig pituitary gland are drastically reduced.[15] Hypothalamic deafferentation similarly reduces but does not block GH secretion in pigs[16] and rats,[4] in which pituitary GH stores are maintained at a diminished level. Stalk transection does, however, abolish episodic GH release in all of these animals.[3,5,12,14,17] The maintenance of somatotroph function in stalk-sectioned or hypothalamic deafferented animals may, therefore, reflect stimulatory actions of peripheral hormones[9] or intrinsic paracrine or autocrine regulators (see Chapters 10 and 11).

An autonomous component stimulating basal GH release is also suggested by the ability of mammalian,[18] avian,[19] and reptilian[20–22] somatotrophs to remain viable in cell cultures lasting several days or weeks[23,24] and as long as 12 months.[25–27] There is also some evidence that in the absence of hypothalamic stimulation the pituitary is able to release GH episodically in secretory bursts of low amplitude.[28,29] Indeed, such autonomous pulsatility may account for up to 35% of the GH pulses in sheep plasma.[30,31] Pulsatile GH secretion in SRIF-treated acromegalic patients unresponsive to GHRH has also been thought to be indicative of a pituitary-derived mechanism for pulse generation.[32]

III. EPISODIC GH RELEASE

The secretion of GH is characterized by periodic pulses that coincide with or precede elevations in plasma GH concentrations (Figure 1). This pulsatile pattern optimally activates signal transduction cascades in target cells and produces intermittent biological responses that retard the development of tissue resistance.[33] Pulsatile release thus reduces the net amount of extracellular GH required to elicit biological responses in target cells[10,33–35] and, by modulating the amplitude and/or frequency of GH release, provides a mechanism to dynamically attenuate or enhance the activity of specific signal transduction pathways in target cells. The biological actions of GH differ within or between cells and may reflect thresholds of GH stimulation and different requirements of GH pusatility.[33,36] This modulation of secretory events may involve both the frequency and amplitude of GH release, but because frequency and amplitude may be independently regulated,[37] modulation of both parameters may not occur contemporaneously.

The episodic release of GH results from low-amplitude, high-frequency pulses that occur within clusters or volleys as well as from irregular, infrequent GH pulses of lower amplitude.[38] The timing of these secretory events differs between individuals[30,39–42] although pulses of GH release are more synchronous in male rats and entrained to the light-dark cycle.[43] Pulsatile GH release has also been reported to be synchronous in sheep, and entrained to meal feeding.[44,45] The timing of GH pulses is thus not random, especially as long and short interpulse intervals appear to alternate in humans[46] and to occur more frequently during hours of sleep and to be susceptible to extrinsic regulation.[38]

The pulses of GH release are separated by intervening trough periods, lasting approximately 2 h in humans and 3 h in rats (Tables 1 and 2). During these troughs GH concentrations may be undetectable in plasma, indicating an absence of tonic stimulation in the control of GH release. In humans, for instance, 95% of the daily GH production occurs during 8.8 h (37%) of the day, during which most (95%) of the GH released in 10 to 12 secretory episodes is released within the first 44 min of each secretory burst.[46] The majority (96%) of GH secretion is also secreted in volleys composed of multiple (approximately four) discrete secretory bursts (Figure 2). These volleys of GH secretion are separated by 171 min, whereas their constituent individual secretory events occur every

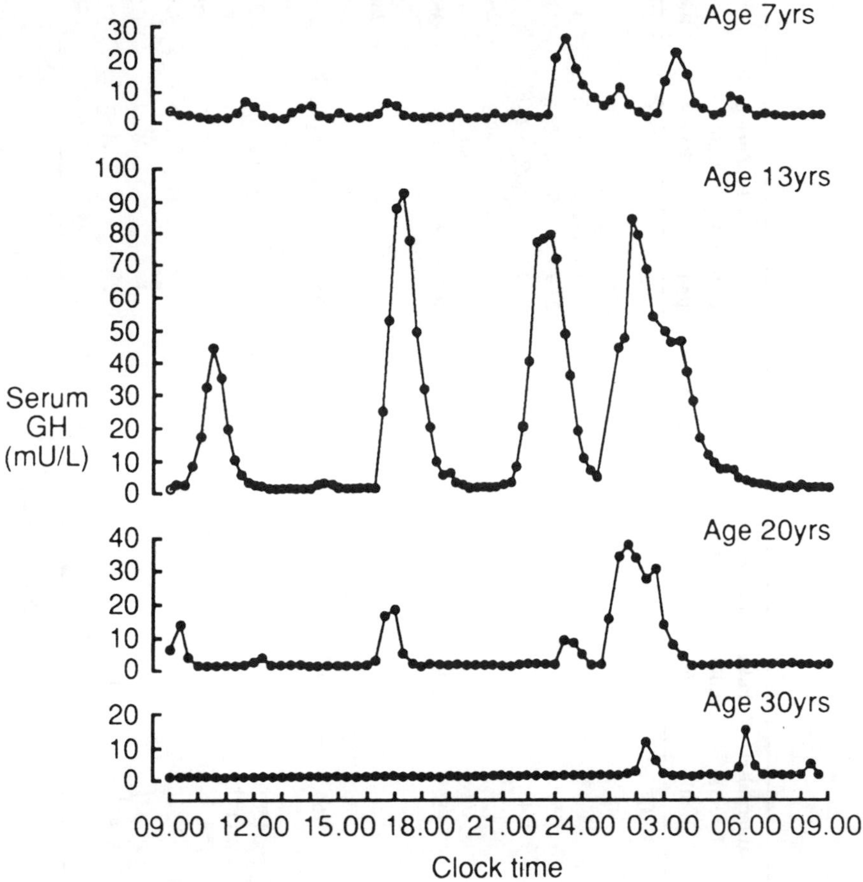

Figure 1 Changes with age in 24-h serum GH concentration profiles in males: a 7-year-old boy (top), a pubertal 13-year-old boy (upper middle) an adult aged 20 years (lower middle), and an adult aged 30 years (bottom). Note the high-amplitude GH pulses occurring during the day in the pubertal individual. (From Brook, C. G. D. and Hindmarsh, P. C., The somatotropic axis in puberty, *Endocrinol. Metab. Clin. N. Am.*, 21, 767–782, 1992. Reprinted with permission.)

36 min. Between secretory volleys GH secretion may not occur. Similar nyctohemeral rhythms of GH release with low constant GH levels between episodic pulses also occur in cattle,[47] sheep,[44,48,49] goats,[50] rabbits,[51,52] guinea pigs,[42] hamsters,[53] rats,[54,55] dogs,[56–58] monkeys,[59,60] pigs,[61,62] horses,[63,64] chickens,[65–68] and turkeys.[69–74] Low-level irregular fluctuations in the GH concentrations also occur between major secretory pulses in cattle[75] and sheep.[49] Dynamic fluctuations with amplitudes varying over three orders of magnitude have also been observed during trough periods in humans, below the baseline level of GH normally detected, indicating that GH secretion may be oscillatory rather than episodic.[76]

The neuroendocrine mechanisms that generate pulsatile GH release are still uncertain. The frequency of the secretory episodes is, however, likely to involve neural mechanisms of hypothalamic (or other central nervous system [CNS]) origin(s)[77] that activate or inhibit pulse generator systems believed to control the timing of pituitary secretory

Table 1 **Kinetics of GH Secretion in Humans**

Condition	Mean GH Conc. (μg/l)	GH Pulse Frequency (per 24 h)	GH Half-Pulse Duration (min)	Interpulse Interval (min)	Pulse Amplitude (μg/l/min)	Maximal Pulse Amplitude (μg/l)	Maximal Burst (μg/l/min)	Mass/Burst (μg/l)	Production Rate (μg/l/day)[a]	Ref.
Premature infant		31.2	45	48	2.2			106	3244	150
Term infant		30.8	25	48	1.4			38	1132	150
Preadolescent ♂	4.6	13.3	30.9	114			0.54	17.3	543 μg/24 h	473
Early adolescent ♂	4.8	10.1	33.9	159			0.64	23.9	739 μg/24 h	473
Late adolescent ♂	13.8	10.9	39.5	145			1.00	41.0	1805 μg/24 h	473
Post-pubertal ♂		8.3	31	172			0.88	28	170 μg	201
Adult ♂	1.8	12	25	87			0.3	7.1	80	46
Adult ♀		12	24	127	0.33			3.5	41	357
Adult ♀ (early follicular)		8.3	50			5.7				286
Adult ♀ (mid follicular)		7.9	58			8.9				286
Adult ♀ (mid luteal)		8.5	49				5.8			286
ISS (10-year-old ♂)	3.5	8.4	25	164	0.92			25	17.8 μg/l/kg/day↑	174
GH deficient	1.4	9.5								474
Turner's (girls)	2.2	9.4↓	23↓	138↑	0.87			19	3.9 μg/l/kg/day	299
IDDM boys	9.0	12.6	33.7	111↑			1.02	34.9	760 μg/24 h	473
IDDM adults		14.4↑	66.5↑			27.2↑	0.21↑			285
IDDM ♀		11.1↑	47.7			14.2↑	0.15↑			285
Obese ♂	0.66↓	3.2↓	30	282↑	0.016			2.1	6.7 μg/day↓	475
Fasting	6.7↑	32↑	24	45↓	0.45↑			11.0↑	371↑	404
Hyperthyroid	1.9↑	15↑	39	91↓	0.73↑		0.14↑	3.7↑	85	476

Note: ↑, increased; ↓, decreased in comparison to controls; ISS, idiopathic short stature; IDDM, insulin-dependent diabetes mellitus.

[a] Unless otherwise indicated.

Table 2 **Kinetics of GH Secretion in Animals**

Species	Basal Level (μg/l)	Pulse Frequency (per 24 h)	Pulse Duration (min)	Peak Interval (min)	Pulse Amplitude (μg/l)	Peak Amplitude (μg/l/min)	Ref.
Monkey	4.93		3.25	246	10–60		59
Horse							
Foals(♂/♀)	8–10	6–9			Higher than adults		162
Adult ♂	3–5	12.9					162
Adult ♀	3–5	3.9					162
Pig							
Adult ♂	4.0 (day); 3.8 (night)				7.6 (day); 6.7 (night)		61
Adult ♀	3.0 (day); 3.1 (night)				4.9 (day);4.9 (night)		61
Sheep							
Lambs							
fed normally	2–3.0	15.7			6.62		49
fetal	>150	17.8					477
neonatal	>10	13.2					477
Rams							
1 year	3.8	6.6			5.3		481
2 years	5.3	6.0			6.6		481
5 years	6.8	3.9			4.1		481
Bovine							
Steer	4.1	4.99			9.2		478
Early lactating cow	12.0	8.8				27.3 μg/l	479
Late lactating cow	9.0	8.0				20.2 μg/l	479
α-GHRH	3.8	0↓					108
Reindeer	2.1	4.65	6.72		6.6		308
Rabbit							
control		12.2		122		13.7	52
α-SRIF		18.8↑		75↓		38.8↑	52

Table 2 **Kinetics of GH Secretion in Animals (Continued)**

Species	Basal Level (μg/l)	Pulse Frequency (per 24 h)	Pulse Duration (min)	Peak Interval (min)	Pulse Amplitude (μg/l)	Peak Amplitude (μg/l/min)	Ref.
Hamster		17.8		75			53
Guinea Pig							
♂	0.82	6–8					42
♀	1.62	12–24					42
Steers							
feed restricted	2.8	11.3			20.4		479a
Rats	38.2	26.4	0.67	66 min	73.9		480
MSG		↓	↓	0.4↓	↓		96
Hypothyroid	9.9	62.4↑	0.28↑	24↓	9.3↓		480
i.c.v. IGF-I	35.1			186↑	150		115
α-GHRH ♂					6.4↓		114
α-GHRH ♀					12.3↓		114
Control ♂					117.9		114
Control ♀					79.9		114
α-SRIF ♂	51.5↑	15.2			137.9		114
α-SRIF ♀	78.7↑	14.8			171.3↑		114
Control ♀	56.8			77.1	150.3		270
ovx ♀	19.7			72.8	217.0		270
DHT-ovx ♀	6.7			155.0	312.2		270
Control ♂	8.3			161.3	304.5		270
α-Galanin	3.0		74↓	111↓	21–194↓		482
Chicken							
Broiler							
(17 day, ♂)	5.4	22.8			35.2		483
Broiler							
(38 day, ♂)	7.7	19.4			33.0		483

Broiler						
(17 day, ♀)	5.2	18.5			38.6	483
Broiler						
(38 day, ♀)	2.0	11.8			6.6	483
Broiler						
(28 day, ♀)	8.7		48.7	77.2	35.0	484
Broiler						
(56 day, ♀)	2.3		62.5	142.5	5.9	484
Broiler						
(84 day, ♀)	0.7		60.6	97.5	2.6	484
Turkey						
2 week, ♂	26.5	16.3	61		59.4	485
4 week, ♂	17.9	13.4	70		62.0	485
8 week, ♂	5.4	13.2	79		25.0	485
14 week, ♂	0.8	7.0	78		5.2	485
24 week, ♂	0.6	7.0	65		1.6	485
2 week, ♀	17.9		63		58.8	485
4 week, ♀	10.1		71		45.6	485
8 week, ♀	1.7		64		10.1	485
14 week, ♀	0.7		70		2.1	485
24 week, ♀	0.6		54		1.4	485

Note: ↑, Increased; ↓, decreased in comparison to controls; MSG, monosodium glutamate; α-SRIF, somatostatin antiserum; α-GHRH, growth hormone-releasing hormone antiserum; DHT, dihydrotestosterone; ovx, ovariectomized; i.c.v. IGF-I, insulin-like growth factor I injected intracerebroventricularly.

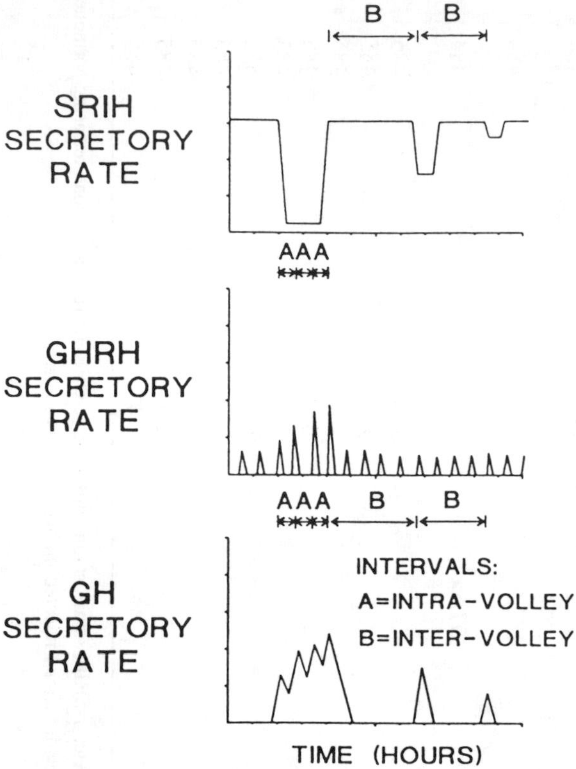

Figure 2 Hypothetical model for physiological basis of a volleyed burst-like mode of GH secretion in humans. Intravolley interburst intervals are considered to reflect frequency of bursts of GH-releasing hormone (GHRH) secretion, whereas intervolley interburst intervals represent periods of time separating nadirs of somatostatin (SRIH) secretion. Thus multiple GHRH bursts during an interval of decreased SRIH secretion may give rise to volleys of GH secretion. During periods of increased SRIH secretion, GH response to GHRH is inhibited. Frequency of GHRH release is illustrated here as constant although some physiological variability occurs, based on a mean intravolley coefficient of variation of 25 ± 1.6%. ** Combined SRIH withdrawal and GHRH secretion. (From Hartman, M. L., Faria, A. C. S., Vance, M. L., Johnson, M. L., Thorner, M. O., and Veldhuis, J. D., Temporal structure of *in vivo* growth hormone secretory events in man, *Am. J. Physiol.*, 260, E101–E110, 1991. Copyright by the American Physiological Society. Reprinted with permission.)

events.[37,78] In contrast, the amplitude of the secretory bursts is likely to reflect the responsiveness of pituitary somatotrophs to hypothalamic signals and to the duration and net magnitude of these signals.

A. HYPOTHALAMIC REGULATION

The pulsatile secretion of GH is thought to result from interactions between GHRH and SRIF at pituitary and hypothalamic sites (reviewed by Tannenbaum et al.[55,79–81]). Both peptides are essential for pulsatile release, which does not occur when either is depleted from the hypothalamus by active[59] or passive[82–85] immunoneutralization or after pharmacological GHRH blockade[86] or when SRIF tone is diminished by hypothalamic lesions[6,87] or hypophysial stalk section.[3,88] The generation of episodic GH pulses, at least in humans and rodents, probably results from phasic episodes of GHRH release, coincident with episodes of suppressed SRIF release,[89–91] both of which may be mediated through increased galanin release.[92] Reciprocal changes in hypothalamic GHRH and SRIF during pulsatile GH release have been observed[93] and GHRH and SRIF secretion may be 180° out of phase.[90,94] This temporal interaction is thought to be responsible for the volleys or clusters of pulsatile GH secretion.[38,95] It has, therefore, been suggested that the intravolley bursts of GH secretion reflect brief pulses of GHRH stimulation, during nadir periods of SRIF inhibition (Figure 2).

Interpulse GH troughs in rats and rabbits are thought to result from tonically high somatostatinergic tone and minimal GHRH release and trough GH levels can be elevated by SRIF depletion.[39,52,94,96–99] The elevated trough GH levels in SRIF-depleted rats and rabbits[52,100,101] also indicate that SRIF inhibition is primarily responsible for the low GH concentrations between GH secretory episodes, which may also reflect the inhibitory action of SRIF on the release of GHRH.[6,90,102] The increased frequency of pulsatile GH release in rabbits deficient in SRIF[52] is also consistent with an inhibitory effect of SRIF on GHRH neurons. However, as the periodic withdrawal of SRIF is insufficient for stimulating GH secretion in the absence of GHRH,[97,103,104] a close interplay between GHRH and SRIF is essential for GH pulsatility. The persistence of pulsatile GH secretion in some rats[100] and rabbits[52] injected with SRIF antibodies and in rats with hypothalamic lesions[105] suggests a pivotal role for GHRH in regulating pulsatile GH release, especially as pulsatile release is completely abolished in prepubertal heifers,[106,107] cows,[108] steers,[109] sows,[110] and gilts[111] immunized against GHRH. The close correlation of pulsatile GHRH release in hypophysial portal circulation with pulsatile GH levels in the peripheral plasma of rats and sheep[30,31,90] also suggests that the timing of episodic GH release is largely GHRH dependent. The occurrence of isolated GH secretory pulses of low amplitude between high-frequency, higher amplitude bursts of GH release is also thought to reflect the action of occasional high-amplitude GHRH pulses that trigger GH secretion during periods of tonic SRIF tone.[38] The major role of GHRH may thus be to set the amplitude of episodic GH release, although the maintenance of measurable or relatively high GH levels in some species during trough periods[112–115] also suggests GHRH involvement during trough periods of GH release. It has, therefore, been suggested[82] that GHRH alone might be capable of fully generating episodic GH bursts, especially as it can reproduce the pulsatile pattern of GH secretion *in vitro*, when intermittently delivered to rat somatotrophs.[97]

The role of somatostatin in the regulation of pulsatile GH release may be to modulate the action of GHRH at the pituitary level.[115] However, as ultradian GH rhythms are preserved in cattle,[116] humans,[117] and chickens[66] receiving continuous GHRH infusions and in men given prolonged GH-releasing peptide infusions[118] and in acromegalic patients with ectopic GHRH-secreting tumors,[1,119,119a] or GHRH-resistance,[120] rhythmical SRIF release rather than phasic GHRH stimulation has also been proposed as being primarily responsible for establishing the timing and duration of GH pulses.[79,81,121] SRIF withdrawal may, therefore, phase pulsatile release in rats[55], dogs,[57] and humans.[122] This may result from a direct pituitary disinhibition of GHRH action[123] or from a hypothalamic disinhibition of GHRH release.[124] The persistence of pulsatile release from pituitary cells continually stimulated with GHRH in the absence of SRIF,[23] nevertheless, suggests that GHRH itself can phase episodic bursts of GH release.

The pulsatility of GH release is also likely to be augmented by phasic periods of GHRH/SRIF receptor desensitization/sensitization, preventing sustained GH release[116,125–129] and potentiating intermittent responsiveness to GHRH stimulation.[66,95] The pulsatile GH response to GHRH stimulation may also be attenuated within the hypothalamus by

α_1-noradrenergic pathways in the paraventricular nucleus.[130] The pulsatility of GHRH/SRIF release may also be phased by feedback effects of GH and insulin-like growth factor I and other target gland hormones, including gonadal steroids and glucocorticoids.[131] Indeed, Harel and Tannenbaum[132] suggested that a synergistic interaction between IGF-I and IGF-II in the brain suppresses the central regulation of pulsatile GH secretion, although IGF-I or IGF-II alone does not inhibit GH peak amplitude, GH trough levels, GH interpeak interval, or mean GH concentration.[115,132]

B. PULSATILE PATTERNS

Most GH secretory episodes may be spontaneous[133] but the amplitude and frequency of GH release has been shown to vary diurnally, in concert with activity and feeding patterns, and to be modulated by sleep, stress, exercise, and postprandial glucose concentrations.[30,45,52,76,134] Pulsatile GH release also differs during growth and sexual maturation, with gender and physiological and pathological conditions (Table 1), and within individuals[135–137] but not groups,[136,138] and differs among species (Table 2). Pulsatile GH release is also modulated, both acutely and chronically, by neurotransmitters, hormones, and metabolic intermediates (amino acids, glucose, and free fatty acids), by actions mediated at hypothalamic and pituitary sites.[139] Some of these pulsatile patterns are discussed below. Pulsatile GH secretion in some pathophysiological states is summarized in Table 1, and is further discussed in Chapter 13.

1. Ontogeny

The secretory dynamics of GH release change during development, as a result of the maturation of neuroendocrine systems regulating GH release and because of changes in the abundance, differentiation, and responsiveness of pituitary somatotrophs. These ontogenic changes in GH secretion are reviewed below and have also been detailed by Ho and Hoffman[140] and by Hartman et al.[1]

a. Fetal Development

Although GH may not be required for fetal growth, GH secretion occurs in the human fetus from 70 days of gestation, the earliest age studied.[141] The pituitary GH concentration begins to rise at 10–14 weeks and is maximal between weeks 35 and 40.[142,143] The reported GH levels in fetal plasma are higher than observed in later life but as they mostly reflect measurements of cord serum GH levels at the time of caesarian section, they may not accurately reflect levels in the undisturbed fetus.[144] Fetal somatotrophs are, nevertheless, highly responsive to GH-releasing stimuli from the ninth week of gestation[145] and the increased somatotroph activity during early gestation is associated with an increased responsiveness to GHRH stimulation and minimal SRIF inhibition.[144] The plasma GH concentration is high during midgestation, reaching peaks of >100 μg/l at 20 to 24 weeks, but progressively declines during the third trimester and the neonatal period. These changes in GH secretion are not maternally mediated,[146] nor are they likely to be regulated by placental GH-releasing factors.[147] These ontogenic changes in GH secretion are, however, dependent on the fetal brain, because pituitary and serum GH levels are abnormally low in anencephalic fetuses.[143] An intact hypothalamus has similarly been demonstrated to be required for GH secretion in fetal lambs.[148,149] It is, therefore, thought that immature hypothalamic neurons are synthesizing and releasing GHRH at a high rate during most of the second trimester,[144] causing the proliferation and maturation of pituitary somatotrophs and provoking unrestrained GH release.[150]

Both GHRH and SRIF are present in the fetal hypothalamus and both increase in abundance toward term,[151–153] suggesting a maturation of the neuroendocrine control system. However, the release of GH from fetal pituitaries is not attenuated by equimolar SRIF until weeks 12 to 13 of gestation and even then the inhibition is incomplete.[144] Newborns also have blunted GH responses to SRIF and the acquisition of inhibitory hypothalamic tone in humans and rats is accomplished only neonatally.[144,154–156] The gradual reduction in the responsiveness of the pituitary to GHRH stimulation toward term is thus not due to SRIF and is unlikely to be due to pituitary GH depletion. Instead, a partial loss of GHRH receptors or an uncoupling of GHRH receptors from signal transduction systems[144,157] or a maturation of the signal transduction pathway[158,159] is likely to be causally involved in this perinatal change in GH secretion.

Although fetal serum GH concentrations *in utero* are elevated in comparison with neonatal, pubertal, and adult GH concentrations, they are similarly characterized by high-amplitude, high-frequency pulsatile GH release.[160] Spontaneous pulsatile release in premature male and female human infants between 32 and 33 weeks of gestation is similar and does not differ in GH pulse frequency (6.2/24 h), incremental pulse amplitude (22.3 μg/l), pulse duration (125 min), mean GH level (36.0 μg/l), or nadir (25.2 μg/l) GH concentrations.[161] Pulsatile GH secretion in the human fetus thus differs from that in sheep, in which the pulsatile pattern is sexually dimorphic (highest in males), and characterized by exaggerated amplitude of the pulses and elevated interpulse nadirs.[148,149] The hypersomatotropism of perinatal and neonatal human infants also contrasts with the low GH concentrations in horses at birth.[162]

The GH production rates, secretory pulse amplitudes, and the mass of GH released per pulse are higher in preterm human infants than in term neonates, although the frequency and duration of episodic GH release are identical.[150] These changes in pulsatile GH release may result from increased GHRH activity and increased somatotroph responsiveness to GHRH stimulation.[144,157,159] A decrease in negative feedback or inhibitory tone may also be causal, because serum insulin-like growth factor I (IGF-I) concentrations are low in preterm infants, reflecting their reduced numbers of hepatic GH receptors.[150,161] The ability of IGF-I to inhibit GH release from fetal pituitary cells is also less than in neonates and adults.[163] This may partly reflect a relative thyroid hormone deficiency of the fetus, because the inhibitory effects of IGF-1 are restored in the presence of thyroid hormones.[163] The somatotrophs of premature infants are also not fully responsive to the inhibitory actions of IGF-I[163,164] and SRIF[144,165–167] and are deficient in SRIF receptors.[168] Somatotroph function in the fetal pituitary also differs from that in late life in that sleep-induced surges of GH release are not established until 2 to 3 months after birth [169] and the GH responses to thyrotropin-releasing hormone (TRH),[170] glucose,[171] and L-dopa[154] are paradoxical. In fetal lambs the rapid decline in circulating GH concentrations at birth may also be independent of SRIF inhibition, because it is thought to reflect the increased concentrations of free fatty acids that result from the induction of nonshivering thermogenesis.[172] Free fatty acids directly inhibit fetal GH release in sheep and are more effective than SRIF.[167,173] Maternal nutrients are also likely to regulate the GH responsiveness of fetal sheep pituitaries to GHRH stimulation because basal and stimulated GH secretion levels in twin lambs are higher than those in nontwinned lambs.[159]

b. Neonatal and Juvenile Development

The GH concentrations in human infants are high immediately after birth and decline during the neonatal and prepubertal period.[174] Indeed, cohorts of prepubertal children may lack wakeful GH pulses and secrete GH only during sleep.[175–177] Veldhuis et al.[174] also found age to be a negative predictor of GH secretory burst mass and duration and of daily GH secretion rate in a population of prepubertal boys (also see Table 1). Basal circulating GH concentrations similarly decline neonatally in rats,[178] mice,[179] pigs,[180–182] cattle,[183] and dogs,[184,185] although GH concentrations are already low at birth in horses.[162] The decline in GH concentrations in the horse may, therefore, occur earlier than in other species (namely, before parturition) and indicates an earlier maturation of the neuroendocrine control of GH secretion.

The hypersomatotropism of the newborn reflects augmented GH synthesis and release[168,186–188] when SRIF receptors and SRIF tone are minimal.[168,189] The responsiveness of somatotrophs to GHRH gradually declines during the neonatal period,[186,187] whereas SRIF sensitivity increases[155,156,168,190] and circulating GH concentrations decline.[178,187,191–193] This decline reflects the acquisition of inhibitory hypothalamic control,[155,156,190] which is likely to reflect maturational interactions of the adrenergic, serotonergic, opioidergic, and cholinergic pathways in the central nervous system (CNS) that regulate GHRH and SRIF neurons.[192–196]

The postnatal suppression of GH secretion may also be a consequence of weaning. Prior to weaning, the increased GH secretion in newborn and suckling rat pups is partially maternally mediated. This maternal modulation of GH secretion partly results from milk-borne factors that stimulate pituitary GH release *in vivo* and *in vitro*.[191] However, suckling, even without milk ingestion, elevates serum GH levels in neonatal rats and some aspect of maternal activity during suckling, including the provision of warmth, is involved in the GH response of neonates.[197] The neonatal pattern of GH secretion may also be modified by suckling, because GH secretion in fetal lambs and weaned lambs is pulsatile, whereas it is not pulsatile in lambs during suckling.[198]

Postnatal changes in GH secretion may also reflect ontogenic changes in the number and heterogeneity of pituitary somatotrophs, because the relative number of type II and type III somatotrophs (Chapter 3) is decreased from 1 to 6 months of age in rats, when pituitary GH concentrations are maximal.[199]

c. Sexual Maturation

The transition of human juveniles into adolescence is associated with elevations of blood GH concentrations during both wakefulness and sleep periods and with a consistent increase in the 24-h GH secretion rate. This increase in GH secretion is responsible for the pubertal growth spurt.[175] The striking increase in GH secretion in pubertal children is largely induced by a two to threefold increase in GH pulse amplitude that occurs without changing pulse duration or frequency, which remains unchanged at 180 to 200 min.[174,200–204] This increase in pulse amplitude is induced by rising circulating concentrations of gonadal steroids[177] and occurs earlier in children with precocious puberty.[205] The suppression of gonadotropin secretion in such individuals is also accompanied by a suppression in GH secretion.[175] The induction of puberty by exogenous androgen or gonadotropin-releasing hormone conversely induces high amplitude GH secretion,[175,206] if androgens are elevated for a considerable portion of the 24-h period.[207] Estrogens are also effective in inducing GH secretion in pubertal boys.[200]

The secretion of GH in the male rat is also characterized by augmented release during puberty[208] and by the development of sexually dimorphic patterns of pulsatile GH release. Prior to 33 days of age, pulsatile GH secretion is similar in males and females and low nadir GH concentrations are interspaced with infrequent, low-amplitude pulses. However, in early puberty, between 33 and 40 days of age, GH pulse amplitude increases more than tenfold in both sexes.[112] The duration of GH pulses during the pubertal period is significantly greater in males than in females, a pattern that continues into adulthood. By late puberty, between 41 and 54 days of age, GH pulse amplitudes are similarly increased twofold in both sexes. Nadir GH concentrations are also increased at this time in both sexes, with females having a higher baseline compared to males. However, only in rats over 54 days of age is the typical pattern of low basal GH secretion and high-amplitude, low-frequency GH pulses observed in males.[112] These developmental changes in GH secretion are preceded by increased concentrations of testosterone and estradiol in the plasma of the male and female rats, respectively. These changes in secretion are likely to result from increased GHRH activity and increased somatotroph responsiveness to GHRH stimulation.[208,209]

d. Aging

Even though basal concentrations may be unchanged,[210,211] a reduction in pulsatile GH secretion occurs in humans during aging (see Corpas et al.[212] and Ho and Hoffman[140] for reviews). With each advancing decade between the ages of 20 and 70 years, the GH production rate decreases by 14%[213–216] and the mean 24-h GH concentrations in the elderly are only 12 to 20% of those during late puberty.[140] This reflects the reduced amplitude, frequency and duration of GH secretion in adults over 50 years of age,[176,213,214,217,218] particularly during the first 3 to 4 h of sleep. Nocturnal GH secretion is similarly reduced in aged rhesus monkeys.[60] Nocturnal GH secretion in the elderly, nevertheless, remains elevated in comparison with diurnal GH secretion.[211,219]

This impairment of GH release is probably due to a reduction in the release of GHRH, because elderly men have a decreased reserve of hypothalamic GHRH.[210,216] The age-related decline in GH secretion can also be reversed by prolonged GHRH treatment.[211,220] Impaired GH secretion in the elderly has also been suggested[221] to result from a reduction in cholinergic transmission in the brain, leading to increased hypothalamic SRIF release. This possibility is also supported by the potentiated GH response to GHRH induced by arginine,[222] which inhibits SRIF secretion.[223] An age-related decline in dopamine turnover in the elderly may also contribute to impaired GH secretion, because dopamine is a stimulator of GH release. An impairment of the neuroendocrine mechanisms controlling GH release in the elderly is also demonstrated by their "paradoxical" GH responses to TRH.[224] This age-related decline in GH secretion has also been correlated with reduced testosterone secretion[211] and the stimulatory actions of testosterone at hypothalamic and pituitary sites. In addition, the reduced somatotroph responsiveness of elderly subjects to GHRH stimulation[210,216,218,225,226] also suggests the suppression of GH secretion in the elderly partly results from a reduction in pituitary sensitivity to GH secretagogues. Indeed, the GH responses to galanin,[227] insulin-induced hypoglycemia, exercise, arginine, L-dopa, and apomorphine are also reduced in the elderly.[140,212,224] The improved responsiveness of elderly patients to GHRH after theophylline treatment[228] also suggests an intrinsic defect in signal transduction occurs in the pituitary during aging, because theophylline increases intracellular cAMP concentrations. The number and size of pituitary somatotrophs and their GH content may also be reduced in the elderly.[229]

Growth hormone secretion also declines from about 12 months of age in rats, in which GH pulse amplitude is reduced.[230–232] Aging is similarly associated with reduced *in vivo* and *in vitro* GH responses to GHRH in rats.[233–235] This may result from an uncoupling of the GHRH signal transduction pathway,[168] although the reduced number of somatotrophs and reduced GH and GH mRNA content in the pituitaries of old rats[168,232,236,237] may also be involved. Increased SRIF tone in aged rats may also be a causal factor in the reduced responsiveness of their somatotrophs.[233,238] Increased levels of SRIF synthesis and release have been observed in the hypothalamus of aged rats[239,240] and SRIF tone is greater in old rats[241] because of their increased number and affinity of pituitary SRIF receptors[168,233,242,243] and pituitary SRIF content.[232] This increase in inhibitory hypothalamic tone may also be accompanied by an age-related decline in the synthesis and content of hypothalamic GHRH.[244–247] This may result from a reduction in the content of neurotransmitters and neurotransmitter receptors involved in the control of GHRH synthesis and release.[246]

Aging similarly reduces basal and stimulated GH secretion *in vivo* and *in vitro* in cattle[183,248] and pigs[182] and *in vivo* in dogs.[184,249] This decline partly reflects a reduction in the sensitivity of pituitary somatotrophs to GHRH stimulation. The defective GH secretion in old dogs is characterized by a reduced frequency of pulsatile GH release[57,58,250] and results from both hypothalamic and pituitary components. These defects are not irreversible and can be restored by the actions of adrenergic drugs at CNS and pituitary sites.[250]

2. Gender

In many species (Table 2), the pulsatile pattern of GH release is sexually dimorphic (e.g., in humans,[251,252] rats,[54,208] mice,[179] hamsters,[53] pigs,[253,254] cattle,[255] sheep,[256] and horses[162]). This dimorphism may account for sexual differences in the rate of growth,[257,258] especially as exogenous GH facilitates growth promotion in hypophysectomized male and female rats when administered so as to mimic the masculine pattern of GH release.[113]

The GH secretion pattern in male rats displays high-amplitude GH pulses at precise 3.3-h intervals, with low or undetectable nadir GH levels. Similar patterns are also observed in the males of other mammalian species. In contrast, GH release in female rats is less pulsatile and characterized by GH pulses of lower amplitude and higher frequency and interpulse GH concentrations higher than those in males[54,114,208,259,260] (Figure 3). This sexual dimorphism in GH secretion is primarily due to gender differences in the mode of hypothalamic SRIF/GHRH signaling to pituitary somatotrophs[114,259] (see Jansson et al.[208] for review). In female rats, SRIF secretion appears to be continu-

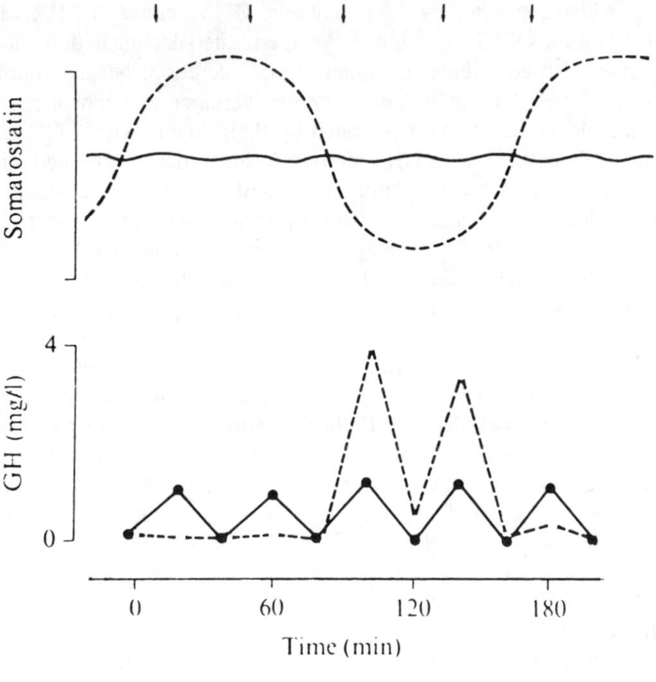

Figure 3 Sexual dimorphism in pulsatile GH secretion and GH responsiveness to GH-releasing factor (GRF) stimulation. *Bottom*: Two typical examples of the GH responses in male and female rats to pulses of human GRF (1–29)-NH2 (arrows) given every 40 min (male rat, open symbols and broken line; female rat, solid symbols and line). *Top*: Different patterns of somatostatin secretion into portal blood may explain the different endogenous GH secretory patterns and the responses to exogenous GRF in males and females. (From Clark, R. G. and Robinson, I. C. A. F., Growth hormone response to multiple injections of a fragment of human growth hormone-releasing factor in conscious male and female rats, *J. Endocrinol.*, 106, 281–289, 1985. Reproduced by permission of the Journal of Endocrinology, Ltd.)

ous whereas it is cyclical in males (Figure 3) and phase synchronized with alternating pulses of GHRH. The secretion of GHRH in males is, therefore, also more rhythmical than in females, although steady-state release of GHRH in females is higher than in males. These patterns may reflect sexual differences in the content of hypothalamic GHRH[261,262] and SRIF,[262] different levels of GHRH mRNA[263,264] and SRIF mRNA,[238,264,265] and dimorphic responses to GH feedback.[259] These dimorphic patterns of GHRH and SRIF release may also reflect sexual differences in gene responsiveness to gonadal steroids[265–267] and sexual differences in the steroidal melieu[112] and in the abundance and type of steroid hormone receptors in neural hypothalamic nuclei.[268] Androgens, for instance, masculinize the GH secretory pattern in ovariectomized female rats and augment the content of hypothalamic GHRH (and SRIF) mRNA.[269,270] Androgens are thus thought to regulate the amplitude of GH pulses, especially as neonatal or prepubertal gonadectomy of male rats causes a reduction in GH pulse height to levels observed in female rats.[271,272] Exposure to androgens during critical periods of postnatal development appears to induce the male pulsatile pattern of GH release and the GH secretory pattern in androgen resistant (testicular feminized) rats resembles that of female rats.[273] The inherent pattern of GH release is thus likely to be that observed in female rats.

The developmental changes in pulsatility are likely to result from sexually differentiated GHRH gene expression but they may not reflect the adult steroid milieu, indicating a developmental window to their actions.[112] Sexually dimorphic GH secretion is likely to be patterned during neonatal or juvenile development. Pulsatile GH secretion before 33 days of age is similar in male and female rats, although between 33 and 40 days of age the duration of GH pulse amplitude is increased in males but the male and female adult patterns of GH release are not observed until after 54 days of age.[113]

In addition to androgens, the sexual dimorphism of GH secretion in the rat may be partly due to estrogen, which feminizes the male pattern of spontaneous and GHRH-stimulated GH secretion and somatic growth.[274] Trough GH levels and pulse frequency are increased but pulse amplitude is reduced by an increase in somatostatinergic tone.[266,274] Estrogens are responsible for the maintenance of high trough GH levels, which are attenuated in females by ovariectomy and by estrogen receptor blockade.[274] Estrogen may, therefore, also act on GHRH-secreting cells in the arcuate-ventromedial and preoptic areas of the hypothalamus, in which estrogen receptors have been localized.[275]

In addition to the actions of estrogens and androgens on the neuroendocrine mechanisms regulating hypothalamic GHRH and SRIF secretion, the sexual dimorphism in GH release may also result from direct pituitary actions of estrogens and androgens[248,276–279] and dimorphic pituitary responses to GH-releasing factors.[280–282] These pituitary actions of testosterone may also result from its conversion to dihydrotestosterone (DHT), but not from its formation of diol compounds.[248] These sexual dimorphic responses are evident in adult rat pituitaries but not in prepubertal tissues and reflect the neonatal exposure of the pituitary gland to sex steroids. In response to GHRH, neonatal oophorectomy of female rats increases the subsequent amplitude and amount (area under the response curve) of GH released from adult pituitary glands *in vitro*,[279] comparable to *in vitro* changes in GH release.[271,272] This increase in GH release is also comparable to the increase observed from male rat pituitary glands. The female characteristic of low GH pulse amplitude is thus exerted only when intact neonatal ovaries imprint this pattern of GH release. The failure of prepubertal gonadectomy to modify *in vitro* pulsatile GH release[279] indicates that these gonadal steroid effects occur only during critical periods of development. Neonatal testosterone is similarly thought to imprint the pituitary gland of male rats for high-amplitude, low-frequency GH pulses, which can be induced in gonadectomized neonatal females following testosterone treatment.[279] The sexually dimorphic pattern of GH release may also reflect differences in the abundance and type of somatotrophs in male and female pituitary glands and the actions of estrogens and androgens on somatotroph proliferation and differentiation[283,284] (see Chapter 3).

Sexual dimorphism in pulsatile GH secretion is also observed in humans. Pulsatile GH release in men differs from that in rats in that it is less pronounced and premenopausal females (rather than males) have higher and more frequent spontaneous GH peaks than males during waking and during sleep.[175,177,252,285] Indeed, mean GH levels are positively correlated with free serum estradiol levels (but not with free testosterone) in both young and old women and in young and old men.[175] Significant relationships between serum estrogen levels and GH secretion have also been observed during puberty in girls.[177] Pulsatile GH secretion, particularly pulse amplitude, is also greatest in periovulatory and luteal phases of the menstrual cycle than during early follicular phases, in concert with estradiol levels.[286] Exogenous estrogen is also associated with increased circulating GH levels in prepubertal girls,[287] premenopausal women,[288] postmenopausal women,[289] and men.[290] This may partly result from an estrogen-induced increase in *GH* gene transcription[175] and increased pituitary responsiveness to releasing stimuli,[175,227,291,292] although estrogens do not appear to augment pituitary GH responses to GHRH.[293,294] The stimulatory action of estrogen may, therefore, be primarily exerted at hypothalamic sites and is likely to involve an increased frequency of SRIF withdrawal,[286] possibly resulting from a reduction in IGF-I production.[140,291]

The physiological importance of estrogen in regulating pulsatile GH secretion is demonstrated by the reduction in the number and height of nocturnal GH pulses following estrogen receptor antagonism.[295] A role for estrogen is also indicated by the patterns of GH secretion in patients with ovarian dysgenesis (Turner's syndrome). This disease is characterized by short stature, and is accompanied by reduced mean serum GH concentrations and in pubertal girls with reductions in the frequency and duration of pulsatile GH release that are partially restored by estrogen therapy.[296,298–300] Thus, while androgens are able to induce pulsatile GH release independent of their conversion to estrogens,[301] they are also likely to augment GH secretion in males and females by activation of estrogen-mediated mechanisms. Indeed, the increased GH secretion (enhanced GH mass per secretory burst, increased maximal rate of GH secretion, and increased pulse frequency) in pubertal boys following androgen receptor blockade[302] is also thought to result from the increased estrogenic activity of androgens. Androgens also increase the height of serum GH concentration peaks without altering their number.[204,303] This is achieved by an increase in the total mass of GH released per pulse and an accelerated maximal role of GH release within each secretory episode.[206] Both gonadal steroids are, therefore, likely to causally induce the increase in pulsatile release during sexual maturation in both males and females.[177,201,204,206,302–305] In both sexes, the augmented GH secretion that occurs at puberty appears to result from a steroid induced increase in pulse amplitude that occurs almost independently of GH pulse frequency.[177,204] This increase in pulse amplitude probably reflects the actions of gonadal steroids on somatotroph sensitivity to GH-releasing factors, which may also result from increased GHRH release or diminished SRIF tone.[204]

3. Seasonality

Although 24-h mean serum GH concentrations may not differ in humans in different seasons,[138] Malarkey et al.[306] noted that both daytime and nocturnal GH secretion was consistently higher in the fall than in the spring. Seasonal differences in pulsatile patterns of GH secretion occur, however, in cattle,[307] sheep,[45,159] and reindeer,[308] but not in horses.[64] These seasonal variations appear to correlate more with changes in food intake than with changes in photoperiod or other environmental parameters. Seasonal patterns of GH secretion have also been observed in feral birds (see Scanes et al.[309] for review), in which secretion is correlated with reproductive activity, migration, and environmental temperatures.[309–313]

4. Sleep

In humans, baboons, and sheep,[49] but not in rats,[80] goats,[314] guinea pigs,[42] rabbits,[52] or hamsters,[53] the largest spontaneous secretory bursts are related to the first part of sleep, although wakeful peaks also occur, particularly in younger adults.[315,316] In humans, GH secretion in the evening is no greater than that during the day[317] and nocturnal GH pulse amplitude is increased.[136] GH pulse frequency also increases to 1.2 pulses/h during the night.[318] Mean GH concentrations and secretory rates are higher during stage 3 and 4 sleep compared to stages 1 and 2 and rapid eye movement sleep.[318] GH secretory rates and peripheral GH concentrations thus correlate with sleep stage, with lags of 4.5 and 16 min, respectively, suggesting that maximal GH release occurs within minutes of the onset of stage 3 or 4 sleep and is probably mediated by GRF.[318] This contrasts with the 24-h profile of GH in wild boars and domestic pigs,[61] in which less pulsatile GH release occurs during the night. This rhythmicity correlates in adults with the onset of slow-wave sleep rather than to clocktime or light-dark cycles,[137,319–321] although in short children GH secretion appears to correlate with neurohormonal changes occurring at night time rather than with specific sleep stage or sleep stage sequence.[322] This nocturnal increase in GH secretion can be mimicked by GHRH but is not attenuated by SRIF.[323] This nocturnal increase may also be related to the short-term stress of nutritional deprivation.[49]

5. Stress

Stressors generally increase GH release in primate species[324–330] although GH levels have been reported to decrease,[328,331–333] remain unchanged,[306,334–336] or show paradoxical responses to GH secretagogues.[328,337] Stress also generally enhances GH secretion in pigs, sheep, cattle, and hamsters,[53,338–341] but reduces GH secretion in most rodents.[342] Stress may, however, also increase GH secretion in rats,[343] suppress GH release in cattle[344] and humans,[345] or fail to influence GH secretion in sheep.[346–348] Variable GH responses to stress occur in fish[349] and birds.[11,350] The variability of these observations may reflect blood sampling at different times relative to pulsatile GH release[306] or reflect the number, severity, duration, and interaction of stressful stimuli on GH secretion and individual differences in perception and susceptibility. The magnitude of increased GH responses to psychological stress is, for instance, related to individual differences in anxiety and neurosis.[333,351]

In mammals, stress-induced surges of GH secretion may result from acute increases in sympathetic tone and hypothalamus-pituitary-adrenal activity, mediated through the actions of opioids,[348,352,353] adrenocorticotropin,[354] catecholamines,[316] and glucocorticoids[355,356] (see Chapters 10 and 11). A role for glucocorticoids in acutely increasing GH secretion during stress is indicated by the action of short-term glucocorticoid excess on pulsatile GH release in normal men.[357] GH secretory burst frequency is increased by glucocorticoids (from 12 to 18 episodes every 24 h), whereas GH interburst interval is reduced (from 127 to 79 min) and the daily secretion rate is enhanced (from 41 to 101 μg/l/day) (Table 1). This stimulation of GH secretion is thought to reflect a decrease in SRIF tone, increased pituitary responsiveness to GHRH stimulation, and increased *GH* gene transcription. These stimulatory actions are, however, opposite to the inhibitory effects of long-term glucocorticoids on GH secretion, as might be expected in response to persistent stressors. The inhibitory tone of persistent stressors on GH secretion may result from neuroendocrine dysfunctions[337] at central and peripheral sites and changes in somatotroph responsiveness to releasing factors.[355,356] Increased GH secretion is, however, frequently seen in responses to physical or psychological stressors that do not result in adrenal activation and vice versa (reviewed by Brown et al.[328]). These dissociated GH responses to stress may reflect neurohypophysial[358] or opioidergic[340] mechanisms or other neuroendocrine responses impinging on the hypothalamus-somatotroph axis, resulting in alterations in GHRH and SRIF release.[359] In rats, for instance, acute and chronic stressors normally inhibit GH secretion and increased SRIF concentrations occur in the hypothalami and hypophysial portal circulation[343,360] of stressed rats.[361] This suppression may also be mediated by central effects of increased CRH (corticotropin-releasing hormone) secretion.[362,363] Impaired GH secretion in these animals may also result from stress-induced alterations in pituitary sensitivity to hypothalamic regulatory factors.[364]

6. Exercise

Acute bouts of exercise increase GH secretion in humans[365–367] and increase the proportion of the 20-kDa variant in plasma.[365,368] This increase is related to the intensity and duration of the exercise, aerobic stress, work output during exercise, muscle mass used, and training state.[369,370] The pulsatile mode of GH release is, for instance, increased approximately twofold when the intensity of exercise exceeds the lactate threshold.[371,372] This increase occurs despite simultaneous increases in both glucose and insulin and is not directly due to lactate stimulation.[369,373]

Cholinergic and opioidergic neural pathways are thought to mediate the increase in GH secretion during exercise and exercise recovery,[374–380] probably as a result of decreased SRIF release and increased GHRH secretion. The decreased pulsatile release of GH in obese individuals and elderly subjects has also been considered by Rogol et al.[370] to partially reflect their decreased activity and insufficient exercise. Nevertheless, Pyka et al.[381] found exercise failed to modulate GH secretion in healthy elderly men and women.[381] The influence of exercise on GH secretion is less well known for other species. Exercise is, however, associated with increased GH secretion in horses,[64] pigs,[382] and birds[383] and with reduced GH secretion in rats and mice.[384]

7. Nutrition

Nutritional states have profound effects on GH secretion (see Ross and Buchanan[385] for a clinical review and Chapter 13 for human pathophysiological conditions associated with malnutrition). Pulsatile GH secretion is, for instance, suppressed in rats during food deprivation[5,386] and has been considered to reflect increased SRIF secretion.[387] Normal pulsatile patterns of GH release are restored in fasted rats after the passive immunoneutralization of endogenous SRIF.[388,389] However, the maintenance of pulsatile GH secretion in fasted rats infused with GHRH indicates the persistence of low-level pulsatile SRIF release in these animals,[386] which may be induced independently of metabolite concentrations in plasma.[5] Indeed, Janowski et al.[386] suggest the suppression of pulsatile release in fasted rats is due to a reduction in GHRH and is not a consequence of the increased SRIF[386] that occurs independently of plasma metabolite concentrations.[5] Bruno et al.[297] similary concluded that food deprivation results in a reduction in hypothalamic SRIF without altering its pattern of release, suggesting that the accompanying[297] decrease in hypothalamic GHRH is responsible for the absence of spontaneous GH peaks. The elevated trough levels of GH in food-deprived rats also suggest a decrease in SRIF tone during food deprivation, although the increased trough GH concentrations after SRIF immunoneutralization imply SRIF inhibition is not absent. Caloric deprivation in dogs, in contrast, elevates GH secretion as a result of reduced hypothalamic SRIF tone,[390] resulting in an increase in GH pulse frequency and amplitude.[391] Starvation or undernutrition similarly increases GH secretion in primates,[392,393] calves,[394] heifers,[106] steers,[307,395] sheep,[30,44,396] reindeer,[308] pigs,[397] rabbits,[384] goats,[50] chickens,[398,399] turkeys,[73,400] and fish,[401,402] but not in dairy heifers,[183] mice,[403] and hamsters.[53]

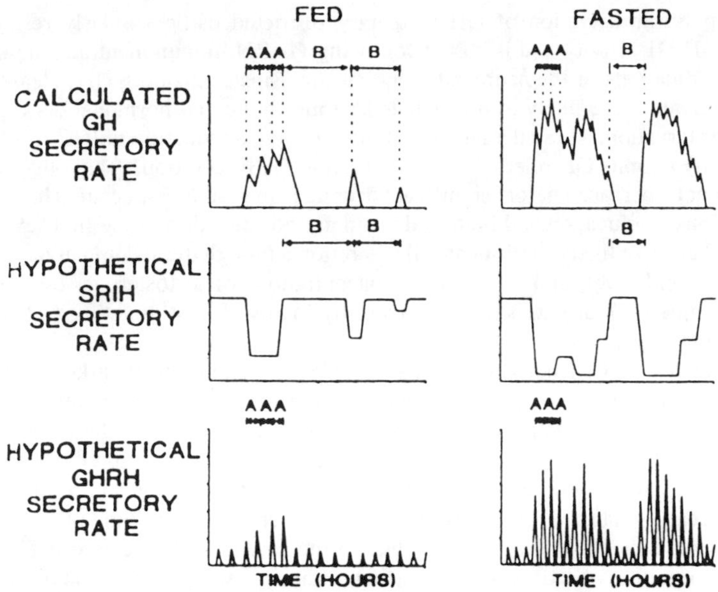

Figure 4 Hypothetical model for the physiological basis of fasting-induced increases in burstlike volleys of GH secretion in humans. *Top:* Typical patterns of GH secretory rates derived by deconvolution anaysis on a fed day (left) and on the second day of a fast (right). Intervals between GH secretory burst are defined as either intravolley (denoted by "A") or intervolley (denoted as "B") intervals. *Middle and bottom*: Hypothetical patterns of somatostatin (SRIH) and GHRH secretion, derived from analysis of the GH interburst intervals. Intravolley interburst intervals are considered to reflect the frequency of bursts of GHRH secretion, whereas intervolley interburst intervals represent periods of time separating nadirs of SRIH secretion. Thus multiple GHRH bursts during an interval of decreased SRIH secretion may give rise to volleys of GH secretion. During periods of increased SRIH secretion, GH response to GHRH is inhibited. Decreased mean intra- and intervolley interburst intervals in fasting subjects probably reflect an increased frequency of GHRH release and prolonged nadirs of SRIH secretion. The frequency of GHRH is illustrated here as constant, although some physiological variability probably occurs, based on mean intravolley interburst interval coefficients of variation of 23% (control) and 26% (fasting). (From Hartman, M. L., Veldhuis, J. D., Johnson, M. L., Lee, M. M., Alberti, K. G. M. M., Samojlike, E., and Thorner, M. O., Augmented growth hormone (GH) secretory burst frequency and amplitude enhanced GH secretion during a two-day fast in normal men, *J. Clin. Endocrinol. Metab.*, 74(4), 757–765, 1992. Copyright by the Endocrine Society. Reprinted with permission.)

In humans, fasting-induced GH secretion results from an amplification of GH pulse frequency and amplitude. Two days of fasting induces a fivefold increase in the 24-h GH production rate in men (from 7 to 37 μg/l distribution volume) and a twofold increase in the number of daily secretory bursts (from 14 to 32), and in the mass of GH secreted during each burst (from 6.3 to 11 μg/l). Fasting, therefore, reduces the periods of secretory quiescence between the complex volleys of multiple GH pulses and increases the frequency of the individual secretory bursts within these secretory volleys.[404] These changes in pulsatile GH release are thought to result from an increased frequency of GHRH release and from longer more pronounced periods of SRIF withdrawal (Figure 4). The increased pulsatile secretion of GH in dietary restricted cattle similarly results from increased GHRH activity and is absent following GHRH immunoneutralization.[106]

The modifications of the neuroendocrine mechanisms controlling GH release during dietary deprivation are likely to be mediated by metabolic and hormonal derangements induced by starvation. These derangements may involve a reduction in IGF-I,[405] because IGF-I directly inhibits GH release while stimulating SRIF secretion,[406] although pulsatile GH secretion is increased before significant decreases in total IGF-I occur. The increased concentrations of glucagon and branched chain amino acids during the first few days of fasting[407,408] are also likely to stimulate GH secretion although the accompanying increase in free fatty acid level, and the reduced concentrations of testosterone, estradiol, and triiodothyronine (T_3), are unlikely to be causally involved in stimulating GH secretion during fasting.[404]

Fasting also increases mean and nadir serum GH concentrations in turkeys,[73] although as a result of increased GH pulse frequency rather than of GH pulse amplitude. The normal frequency of pulsatile GH release is, however, restored in these birds within 30 min of refeeding. Fasting also increases the kinetics of GH secretion in sheep, but by increasing GH pulse amplitude rather than pulse frequency or pulse duration[30,45] and without increasing nadir GH concentrations.[45] The increased amplitude of pulsatile GH release in fasted sheep is not, however, due to increased GHRH release but reflects a reduction in inhibitory SRIF tone,[30] increasing pituitary responsiveness to GHRH stimulation. Feeding has, conversely, been shown to reduce the GH responsiveness of sheep to GHRH challenge.[409] Feeding also lowers mean GH concentrations and the amplitude of GH release in cattle[394,410,411] and sheep,[44,45] in which the GH concentrations are lowest after meals in the morning and increase toward the afternoon meal, probably due to an increase in pituitary sensitivity to GHRH.[412] This inhibitory response to feeding is also neurally mediated and can be induced in anticipation of meal feeding.[44,412,413] Excess caloric intake also blunts the pituitary GH response of rats to GH-releasing factors.[414] These results contrast, however, with the reduced responsiveness of human somatotrophs to GH secretagogues during short-term fasting.[393] Low food intake also elevates GH secretion in reindeer,[308] as a result of increased GH pulse duration, because pulse frequency and amplitude are unaffected.

a. Glucose

During malnutrition a reduction in blood-borne glucose is causally related to changes in GH secretion. Hypoglycemia, for instance, stimulates GH secretion in primates,[415] dogs,[416] pigs,[61,417] chickens,[418] and fish,[419] but is ineffective in cats[420] and mice[403,421] and is inhibitory in sheep[422] and rats.[423,424] Hyperglycemia, conversely, inhibits GH secretion in primates[139,385,425] and pigs[417] but not in cats,[420] cattle,[426] mice,[403] or chickens[418] and hyperglycemia may stimulate GH secretion in sheep.[427]

In humans, it is the relative fall in glucose that provokes GH release rather than the absolute concentrations, even if the fall in glucose concentration merely restores normoglycemia from a hyperglycemic state, as in diabetic patients.[425,428] Interference with the intracellular utilization of glucose, by administration of 2-deoxyglucose (2-DG), can also stimulate GH release in humans.[429] These responses are induced through hindbrain glucoreceptors that respond to glucoprivation[430] and impinge on the hypothalamic neurons secreting GHRH and SRIF. Because the hypoglycemia-induced rise in GH strongly activates adrenergic transmission, an α_2-adrenoreceptor-mediated inhibition of SRIF release is likely to be a causal factor in this response.[139,431,432] The hypothalamus can, however, also respond directly to glucoprivation. The release of both GHRH and SRIF from rat and mouse hypothalamic fragments is increased in media of low glucose concentrations and by 2-DG,[433–436] although the release of GHRH from the mouse hypothalamus is more sensitive than SRIF to glucoprivation. This stimulation of GHRH release is not simply due to membrane depolarization but is related to a reduction in glycolysis and probably to the activation of glutamate receptors, as a result of glucopenia-induced glutamate release.[435,437] Glucopenia has also been shown to stimulate hypothalamic SRIF release without modulating GHRH secretion in sheep *in vivo*.[422] The differential actions of glucopenia in different species may therefore reflect differences in the synthesis and release of GHRH and SRIF.

An extrapituitary site of glucopenic action is also indicated by the unresponsiveness of rat pituitary somatotrophs to media of high or low glucose concentration,[438,439] although the somatotrophs of fish appear to respond directly to glucoprivation.[419] The stimulated release of GH from rat somatotrophs is, however, suppressed when exposed to prolonged glucose excess,[438] and pituitary responsiveness thereby contributes to the inhibition of GH secretion induced in diabetic rats.[440,441] Hyperglycemia is also reported to acutely reduce the amplitude of GH surges in rats,[442] although King et al.[443] reported that sustained hyperglycemia did not alter pulsatile release in these animals. The reduced pituitary responsiveness of hyperglycemic rats to GHRH[439,444] is, however, correlated with increased SRIF synthesis, even though an accompanying increase in GHRH synthesis also occurs. In addition to increasing SRIF release from the hypothalamus, hypoglycemia dramatically increases *SRIF* gene transcription without altering *GHRH* gene transcription, which would result in a complete collapse of pulsatile GH secretion *in vivo* and reduce pituitary responsiveness to GHRH stimulation.[439] A reduction in the molar ratio between GHRH and SRIF has also been observed after the incubation of mouse hypothalamic fragments in media of high glucose concentration, although as a result of reduced GHRH release rather than from increased SRIF secretion,[435] as observed from incubated rat hypothalami.[433,434] A somatostatinergic pathway, activated through cholinergic neurons,[425] is also thought to be responsible for the inhibitory effect of hyperglycemia on GH secretion in humans.[385,425] In contrast, hyperglycemia has no effect on GH secretion in pigs[382] or mice.[403]

b. Free Fatty Acids

Free fatty acids (FFAs) inhibit GH secretion in humans,[445–447] whereas acute FFA suppression is followed by spontaneous GH release[445] and enhances GH responses to provocative stimuli.[447] GH secretion and plasma FFA concentrations are also inversely correlated in rats[448] and sheep[173,449,450] but inhibit GH secretion in birds only at high dose levels.[452–454] This inhibition of GH secretion induced by FFAs is thought to be mediated through increased SRIF tone,[139,444,445] although FFA suppression of GH release occurs in fetal sheep[172] that are unresponsive to SRIF. The release of SRIF from cultured rat hypothalamic neurons is, moreover, directly suppressed by FFAs, while GHRH release is increased.[455,456] The synthesis and content of SRIF in fetal hypothalamic nuclei is also reduced by FFAs *in vitro*.[456] Central pathways mediating the inhibitory effects of FFAs are therefore still uncertain. Direct pituitary inhibition is, however, thought to be involved in FFA suppression of GH.[457] This possibility is indicated by the inhibition of GH secretion by FFAs in rats depleted of hypothalamic SRIF by passive immunoneutralization or medial hypothalamic ablation.[448] Inhibitory effects of FFAs on the release of GH from ectopic pituitary glands transplanted into hypophysectomized rats have also been observed.[448] This pituitary action of FFAs may result from modifications in the lipid bilayer of pituitary membranes, impeding receptor activation or signal transduction.

c. Amino Acids

Protein deficiency impairs both spontaneous and GHRH-induced GH release in rats,[458–460] reducing GH peak amplitude and mean GH concentrations. Dietary protein restriction does not, however, lower trough GH levels or GH interpeak intervals or disturb the ultradian rhythm of GH or SRIF. This suppression of GH secretion may, therefore, partly result from a reduction in hypothalamic GHRH synthesis[461] and from an increase in inhibitory SRIF tone.[458] This suppression may also reflect the reduced size and GH content of the pituitary gland in protein-restricted rats.[458] This suppression is not, however, merely due to the caloric restriction of protein deficiency, as spontaneous GH secretory profiles in pair-fed controls are similar to those in animals.[458] A decrease in caloric intake in cows fed low protein diets is, however, thought to account for the increased plasma GH concentrations in these animals.[462] Dietary and caloric restriction similarly increase GH secretion in chickens,[399,463] in which amino acids have little, if any, effect on GH release.[454] Essential amino acids, particularly arginine, ornithine, histidine, and lysine, are well-established stimulators of GH release in men.[385,464] Branched chain amino acids also augment GH secretion in baboons,[407] in which they may phase shift the nocturnal peak of GH secretion, increasing early morning blood GH concentrations. Amino acids also increase GH secretion in ruminants,[465,466] rats,[458] rabbits,[51,384] and fish,[419] but are ineffective in cats,[420] dogs,[467] and birds.[454] In men, arginine augments the GH response to a maximal dose of GHRH[468] and increases GH release from pituitary glands refractory to GHRH stimulation.[468,469] The stimulation of GH release may, therefore, be mediated through the hypothalamus and probably involves a reduction in SRIF secretion[223,386,464,468] or an increase in dopaminergic tone,[468] since dopamine administration similarly stimulates basal and GHRH-induced GH secretion[470,471] and, like arginine, inhibits hypoglycemia-induced GH secretion.[468,472]

REFERENCES

1. **Hartman, M. L., Veldhuis, J. D., and Thorner, M. O.,** *Horm. Res.*, 40, 37, 1993.
2. **Thomas, C. R., Groot, K., and Arimura, A.,** *Endocrinology (Baltimore)*, 116, 2174, 1985.
3. **Kamegai, J., Wakabayashi, I., Sugihara, H., et al.,** *Acta Endocrinol. (Copenh.)*, 124, 700, 1991.
4. **Locatelli, V., Cella, S. G., Loche, S., et al.,** *Life Sci.*, 35, 1989, 1984.
5. **Mounier, F., Bluet-Pajot, M. T., Durand, D., et al.,** *Horm. Res.*, 31, 266, 1989.
6. **Katakami, H., Downs, T. R., and Frohman, L. A.,** *Endocrinology (Baltimore)*, 123, 1103, 1988.
7. **Jansson, J. O., Carlsson, L., and Isaksson, O. G. P.,** *Endocrinology (Baltimore)*, 116, 95, 1985.
8. **Krieg, R. J., Johnson, J. H., and Adler, R. A.,** *Endocrinology (Baltimore)*, 125, 2273, 1989.
9. **Bridges, R. S. and Millard, W. J.,** *Horm. Behav.*, 22, 194, 1988.
10. **Adler, R. A.,** *Endocr. Rev.*, 7, 302, 1986.
11. **Harvey, S.,** *Progress in Non-mammalian Brain Research* (Eds. Nistico, G. and Bolis, L.) CRC Press, Boca Raton, FL, 1983, p. 307.
12. **Plouzek, C. A., Molina, J. R., Hard, D. L., et al.,** *Proc. Soc. Exp. Biol. Med.* 189, 158, 1988.
13. **Krey, L. C., Lu, K. H., Butler, W. R., et al.,** *Endocrinology (Baltimore)*, 96, 1088, 1975.
14. **Klindt, J., Ford, J. J., Berarrdinelli, J. G., et al.,** *Proc. Soc. Exp. Biol. Med.*, 172, 503, 1983.
15. **Anderson, L. L., Ford, J. J., Klindt, J., et al.,** *Proc. Soc. Exp. Biol. Med.*, 196, 194, 1991.
16. **Molina, J. R., Klindt, J., Ford, J. J., et al.,** *Proc. Soc. Exp. Biol. Med.*, 183, 163, 1986.
17. **Anderson, L. L.,** *Am. J. Physiol.*, 232, E497, 1977.
18. **Daniels, M., White, M. C., and Kendall-Taylor, P.,** *J. Endocrinol.*, 114, 503, 1987.
19. **Perez, F. M., Malamed, S., and Scanes, C. G.,** *Gen. Comp. Endocrinol.*, 65, 408, 1987.
20. **Denver, R. J. and Licht, P.,** *J. Exp. Zool.*, 251, 306, 1989.
21. **Denver, R. J. and Licht, P.,** *J. Exp. Zool.*, 246, 293, 1988.
22. **Denver, R. J. and Licht, P.,** *J. Exp. Zool.*, 247, 146, 1988.
23. **Deslauriers, N., Gaudreau, P., and Brazeau, P.,** *Regul. Peptides,* 20, 261, 1988.
24. **Chen, C., Heyward, P., Zhang, J., et al.,** *Neuroendocrinology*, 59, 1, 1994.
25. **Chomczynski, P., Soszynski, P., and Frohman, L. A.,** *J. Clin. Endocrinol. Metab.*, 77, 281, 1993.
26. **Chomczynski, P., Brar, A., and Frohman, L. A.,** *Endocrinology (Baltimore)*, 123, 2276, 1988.

27. **Kashio, Y., Chomczynski, P., Downs, T. R., et al.,** *Endocrinology (Baltimore)*, 127, 1129, 1990.
28. **Stewart, J. K., Clifton, D. K., Koerker, D. J., et al.,** *Endocrinology (Baltimore)*, 116, 1, 1985.
29. **Shin, S. H.,** *Life Sci.*, 31, 597, 1982.
30. **Thomas, G. B., Cummings, J. T., Francis, H., et al.,** *Endocrinology (Baltimore)*, 128, 1151, 1991.
31. **Frohman, L. A., Downs, T. R., Clarke, I. J., et al.,** *J. Clin. Endocrinol. Metab.*, 39, 385, 1974.
32. **Riedel, M., Gunther, T., Muhlen, A. V., et al.,** *Clin. Endocrinol.*, 37, 233, 1992.
33. **Brabant, G., Prank, K., and Schofl, C.,** *Trends Endocrinol. Metab.*, 3, 183, 1992.
34. **Hindmarsh, P. C., Matthews, D. R., Stratton, I., et al.,** *Clin. Endocrinol.*, 36, 165, 1992.
35. **Isgaard, J., Carlsson, L., Isaksson, O. G. P., et al.,** *Endocrinology (Baltimore)*, 123, 2605, 1988.
36. **Jorgenson, J. O., Moller, N., Lauritzen, T., et al.,** *J. Clin. Endocrinol. Metab.*, 70, 1616, 1990.
37. **Veldhuis, J. D., Iranmanesh, A., Johnson, M. L., et al.,** *J. Clin. Endocrinol. Metab.*, 71, 1616, 1990.
38. **Veldhuis, J. D. and Johnson, M. L.,** *Front. Neuroendocrinol.*, 11, 363, 1990.
39. **Carlsson, L. and Janssen, J. O.,** *Endocrinology (Baltimore)*, 126, 6, 1990.
40. **Clark, R. G., Carlsson, L. M. S., and Robinson, I. C. A. F.,** *J. Endocrinol.*, 114, 399, 1987.
41. **Jansson, J. and Frohman, L. A.,** *Endocrinology (Baltimore)*, 120, 1551, 1987.
42. **Gabrielsson, B., Fairhall, K. M., and Robinson, I. C. A. F.,** *J. Endocrinol.*, 124, 371, 1990.
43. **Tannenbaum, G. S. and Martin, J. B.,** *Endocrinology (Baltimore)*, 98, 562, 1976.
44. **Driver, P. M. and Forbes, J. M.,** *J. Physiol.*, 317, 413, 1981.
45. **Bocquier, F., Kann, G., and Theireiz, M.,** *Anim. Prod.*, 51, 115, 1990.
46. **Hartman, M. L., Faria, A. C. S., Vance, M. L., et al.,** *Am. J. Physiol.*, 260, E101, 1991.
47. **Blom, A. K., Halse, K., and Hove, K.,** *Acta Endocrinol. (Copenh.)*, 82, 758, 1976.
48. **Davis, S. L., Ohlson, D. L., Klindt, J., et al.,** *Am. J. Physiol.*, 233, 519, 1977.
49. **Laurentie, M. P., Barenton, B., Charrier, J., et al.,** *Endocrinology (Baltimore)*, 125, 642, 1989.
50. **Hart, I. C., Flux, D. S., Andrews, P., et al.,** *Horm. Metab. Res.*, 7, 35, 1975.
51. **McIntyre, H. B. and Odell, W. D.,** *Neuroendocrinology*, 16, 8, 1974.
52. **Minamitani, N., Chihara, K., Kaji, H., et al.,** *J. Neuroendocrinol.*, 1, 147, 1989.
53. **Borer, K. T., Kelch, R. P., and Hayashida, T.,** *Neuroendocrinology*, 35, 349, 1982.
54. **Eden, S.,** *Acta Physiol. Scand.*, 104, 1, 1978.
55. **Tannenbaum, G. S.,** *Horm. Res.*, 29, 70, 1988.
56. **Hochberg, Z., Amit, T., and Zadik, Z.,** *J. Clin. Endocrinol. Metab.*, 72, 236, 1991.
57. **Cowan, J. S., Gaul, P., Moor, B. C., et al.,** *Can. J. Physiol. Pharmacol.*, 64, 884, 1984.
58. **Takahashi, Y., Ebihara, S., Nakamura, Y., et al.,** *Endocrinology (Baltimore)*, 109, 262, 1981.
59. **Steiner, R. A., Stewart, J. K., Barber, J., et al.,** *Endocrinology (Baltimore)*, 102, 1587, 1978.
60. **Kaler, L. W., Gliessman, P., Craven, J., et al.,** *Endocrinology (Baltimore)*, 119, 1281, 1986.
61. **Claus, R., Bingel, A., Hofacker, S., et al.,** *Live Prod. Sci.*, 25, 247, 1990.
62. **Dubreuil, P., Pelletier, G., Petitclerc, D., et al.,** *Domest. Anim. Endocrinol.*, 4, 299, 1987.
63. **Cahill, C. M. and Hayden, T. J.,** *J. Endocrinol.*, 123(Suppl.), Abst. 84, 1989.
64. **Thompson, D. L., Rahmaniam, M. S., DePew, C. L., et al.,** *J. Anim.Sci.*, 70, 1201, 1992.
65. **Vasilatos-Younken, R. and Leach, R. M., Jr.,** *Growth*, 50, 84, 1986.
66. **Vasilatos-Younken, R., Tsao, P. H., Foster, D. N., et al.,** *J. Endocrinol.*, 135, 371, 1992.
67. **Johnson, R. J., Fairclough, R. J., and Cahill, L. P.,** *Br. Poultry Sci.*, 28, 103, 1987.
68. **Scanes, C. G., Carsia, R. V., Lauterio, T. J., et al.,** *Life Sci.*, 34, 1127, 1984.
69. **Anthony, N. B., Vasilatos-Younken, R., Emmerson, D. A., et al.,** *Poultry Sci.*, 69, 2057, 1990.
70. **Vasilatos-Younken, R., Bacon, W. L., and Nestor, K. E.,** *Poultry Sci.*, 67, 826, 1988.
71. **Lilburn, M. S., Bacon, W. L., Sacco, R. E., et al.,** *Poultry Sci.*, 69, 1215, 1990.
72. **Shaw, S. N., Bacon, W. L., Vasilatos-Younken, R., et al.,** *Gen. Comp. Endocrinol.*, 68, 331, 1987.
73. **Anthony, N. B., Vasilatos-Younken, R., Bacon, W. L., et al.,** *Poultry Sci.*, 69, 801, 1990.
74. **Bacon, L., Vasilatos-Younken, R., Nestor, K. E., et al.,** *Gen. Comp. Endocrinol.*, 75, 417, 1989.
75. **Christian, L. E., Everson, D. O., and Davis, S. L.,** *J. Anim. Sci.*, 46, 699, 1978.
76. **Winer, L. M., Shaw, M. A., and Baumann, G.,** *J. Clin. Endocrinol. Metab.*, 70, 1688, 1990.
77. **Scanes, C., Lauterio, T., and Buonomo, F. C.,** *Avian Endocrinology: Environmental and Ecological Perspectives* (Eds. Mikami, S., Homma, K., and Wada, M.). Springer-Verlag, Berlin, 1983, p. 307.

78. **Veldhuis, J. D., Faria, A. C. S., Vance, M. L., et al.,** *Acta Pediatr. Scand.*, 63, 347, 1988.
79. **Tannenbaum, G. S., Painson, J. C., Lapointe, M., et al.,** 72nd Annu. Meet. Endocr. Soc., Atlanta, GA, 1990, p. 106.
80. **Tannenbaum, G. S., Martin, J. B., and Colle, E.,** *Endocrinology (Baltimore)*, 98, 562, 1976.
81. **Tannenbaum, G. S.,** *Acta Pediatr. Scand.*, 372(Suppl.), 5, 1991.
82. **Ono, M., Miki, N., and Demura, H.,** *Endocrinology (Baltimore)*, 129, 1791, 1991.
83. **Wehrenberg, W. B., Brazeau, P., Luben, R., et al.,** *Endocrinology (Baltimore)*, 111, 2147, 1982.
84. **Miki, N., Ono, M., and Shizume, K.,** *Endocrinology (Baltimore)*, 114, 1950, 1984.
85. **Frohman, L. A., Downs, T. R., Katakami, H., and Jansson, J.,** *Growth Hormone: Basic and Clinical Aspects* (Eds. Isaksson, O., Binder, C., Holl, K., and Hokfelt, B.). Elsevier, Amsterdam, 1987, p. 63.
86. **Lumpkin, M. D. and McDonald, J. K.,** *Endocrinology (Baltimore)*, 124, 1522, 1989.
87. **Karteszi, M., Fiok, J., and Makara, G. B.,** *J. Endocrinol.*, 94, 77, 1982.
88. **Levy, A., Matovelle, M. C., Lightman, S. L., et al.,** *Brain Res.*, 579, 1, 1992.
89. **Plotsky, P. M.,** *Neuroendocrinology*, 53, 433, 1991.
90. **Plotsky, P. M. and Vale, W.,** *Science*, 230, 461, 1985.
91. **Fukata, J., Kasting, N. W., and Martin, J. B.,** *Neuroendocrinology*, 40, 193, 1985.
92. **Maiter, D. M., Hooi, S. C., Koenig, J. I., et al.,** *Endocrinology (Baltimore)*, 126, 1216, 1990.
93. **Zeitler, P., Tannenbaum, G. S., Clifton, D. K., et al.,** *Proc. Natl. Acad. Sci. U.S.A.*, 88, 8920, 1991.
94. **Tannenbaum, G. S. and Ling, N.,** *Endocrinology (Baltimore)*, 115, 1952, 1984.
95. **Hindmarsh, P. C., Brain, C. E., Robinson, C. A. F., et al.,** *Clin. Endocrinol.*, 35, 353, 1991.
96. **Maiter, D., Underwood, L. E., Martin, J. B., et al.,** *Endocrinology (Baltimore)*, 28, 1100, 1991.
97. **Sato, M., Takahara, J., Fujioka, Y., et al.,** *Endocrinology (Baltimore)*, 123, 1928, 1988.
98. **Sato, M., Chihara, K., Kita, T., et al.,** *Neuroendocrinology*, 50, 139, 1989.
99. **Frohman, L. A., Downs, T. R., Clark, I. J., et al.,** *J. Clin. Invest.*, 36, 17, 1990.
100. **Terry, L. C. and Martin, J. B.,** *Endocrinology (Baltimore)*, 109, 622, 1981.
101. **Tannenbaum, G. S., Epelbaum, J., Colle, E., et al.,** *Endocrinology (Baltimore)*, 102, 1909, 1978.
102. **Miki, N., Ono, M., and Shizume, K.,** *J. Endocrinol.*, 117, 245, 1988.
103. **Ho, P. J., Kletter, G. B., Hopwood, N. J., et al.,** *Acta Endocrinol. (Copenh.)*, 129, 414, 1993.
104. **Weiss, J., Cronin, M. J., and Thorner, M. O.,** *Am. J. Physiol.*, 253, E508, 1987.
105. **Thorner, M. O. and Cronin, M. J.,** *Neuroendocrine Perspectives* (Eds. Muller, E. E., Macleod, R. M., and Frohman, L. A.). Elsevier, Amsterdam, 1985, p. 95.
106. **Armstrong, J., Cohick, W. S., Harvey, R. W., et al.,** *Domestic Anim. Endocrinol.*, 10, 315, 1993.
107. **Simpson, R. B., Armstrong, J. D., Harvey, R. W., et al.,** *J. Anim. Sci.*, 69, 4914, 1991.
108. **Moore, K. L., Armstrong, J. D., Harvey, R. W., et al.,** *Domest. Anim. Endocrinol.*, 9, 125, 1992.
109. **Trout, W. E. and Schanbacher, B. D.,** *J. Endocrinol.*, 125, 123, 1990.
110. **Armstrong, J., Esbenshade, K. L., Coffey, M. T., et al.,** *Domest. Anim. Endocrinol.*, 7, 191, 1990.
111. **Armstrong, J. D., Esbenshade, K. L., Johnson, J. L., et al.,** *J. Anim. Sci.*, 68, 427, 1990.
112. **Gabriel, S. M., Roncancio, J. R., and Ruiz, N. S.,** *Neuroendocrinology*, 56, 619, 1992.
113. **Pampori, N. A., Agrawal, A. K., and Shapiro, B. H.,** *Acta Endocrinol.(Copenh.)*, 124, 283, 1991.
114. **Painson, J.-C. and Tannenbaum, G. S.,** *Endocrinology (Baltimore)*, 128, 2858, 1991.
115. **Harel, Z. and Tannenbaum, G. S.,** *Neuroendocrinology*, 56, 161, 1992.
116. **Campbell, R. M.,** *J. Anim. Sci.*, in press, 1994.
117. **Vance, M. L., Martha, P. M., Furlanetto, R., et al.,** *J. Clin. Endocrinol. Metab.*, 68, 22, 1989.
118. **Huhn, W. C., Hartman, M. L., Pezzoli, S. S., et al.,** *J. Clin. Endocrinol. Metab.*, 76, 1202, 1993.
119. **Vance, M. L., Kaiser, D. L., Evans, W. S., et al.,** *J. Clin. Invest.*, 75, 1584, 1985.
119a. **Barkan, A. L., Stred, S. E., Reno, K., et al.,** *J. Clin. Endocrinol. Metab.*, 69, 1225, 1989.
120. **Riedel, M., Gunther, T., von zur Muhlen, A., et al.,** *Clin. Endocrinol.*, 37, 233, 1992.
121. **Kraicer, J., Sheppard, M. S., Luke, J., et al.,** *Endocrinology (Baltimore)*, 122, 1810, 1988.
122. **Gelato, M. C., Oldfield, E., Loriaux, D. L., et al.,** *J. Clin. Endocrinol. Metab.*, 71, 585, 1990.
123. **Kraicer, J., Cowan, J. S., Sheppard, M. S., et al.,** *Endocrinology (Baltimore)*, 119, 2047, 1986.

124. **Robinson, I. C. A. F., Jeffrey, S., and Clark, R. G.,** *Acta Pediatr. (Suppl.)*, 367, 87, 1990.
125. **Gelato, M. C., Rittmaster, R. S., Pescovitz, O., et al.,** *J. Clin. Endocrinol. Metab.*, 61, 223, 1985.
126. **Soya, H. and Suzuki, M.,** *Endocrinology (Baltimore)*, 122, 2492, 1988.
127. **Soya, H. and Suzuki, M.,** *Endocrinology (Baltimore)*, 126, 285, 1990.
128. **Soya, H., Suzuki, M., and Kato, M.,** *Biochem. Biophys. Res. Commun.*, 172, 276, 1990.
129. **Sato, K., Hotta, M., Kageyama, J., et al.,** *Biochem. Biophys. Res. Commun.*, 167, 360, 1990.
130. **Mounier, F., Bluet-Pajot, M., Durand, D., et al.,** *Neuroendocrinology*, 59, 29, 1994.
131. **Frohman, L. A., Downs, T. R., and Chomczynski, P.,** *Front. Neuroendocrinol.*, 13, 344, 1992.
132. **Harel, Z. and Tannenbaum, G. S.,** *Endocrinology (Baltimore)*, 131, 758, 1992.
133. **Martin, J. B.,** *Front. Neuroendocrinol.*, 4, 129, 1976.
134. **Tapanainen, P., Rantala, H., Leppaluoto, et al.,** *Pediatr. Res.*, 26, 404, 1989.
135. **Weill, J., Dherbomez, M., Fialdes, P., et al.,** *Horm. Res.*, 34, 9, 1990.
136. **Albertsson-Wikland, K. and Roseberg, S.,** *Acta Endocrinol. (Copenh.)*, 126, 109, 1992.
137. **Van Cauter, E. and Refetoff, S.,** *Clin. Endocrinol.*, 8, 381, 1985.
138. **Saini, S., Hindmarsh, P. C., Matthews, D. R., et al.,** *Clin. Endocrinol.*, 34, 455, 1991.
139. **Devesa, J., Lima, L., and Tresguerres, J. A. F.,** *Trends Endocrinol. Metab.*, 3, 175, 1992.
140. **Ho, K. K. Y. and Hoffman, D. M.,** *Horm. Res.*, 40, 80, 1993.
141. **Gluckman, P. D., Grumbach, M. M., and Kaplan, S. L.,** *Endocr. Rev.*, 2, 363, 1981.
142. **Suganuma, N., Seo, H., Yamamoto, N., et al.,** *Am. J. Obstet. Gynecol.*, 160, 729, 1989.
143. **Kaplan, S. L., Grumbach, M. M., and Shepard, T. H.,** *J. Clin. Invest.*, 51, 3080, 1972.
144. **Goodyer, C. G., Branchaud, C. L., and Lefebvre, Y.,** *J. Clin. Endocrinol. Metab.*, 76, 1259, 1993.
145. **Goodyer, C. G., Sellen, J. M., Fuks, M., et al.,** *Reprod. Nutr. Dev.*, 27(2B), 461, 1987.
146. **de Zegher, F., Vanderschueren-Lodeweyckx, M., Spitz, B., et al.,** *J. Clin. Endocrinol. Metab.*, 71, 520, 1990.
147. **Nagashima, K., Yagi, H., Suzuki, S., et al.,** *Biol. Neonate*, 49, 307, 1986.
148. **Gluckman, P. D., Mueller, P., Kaplan, S. L., et al.,** *Endocrinology (Baltimore)*, 104, 162, 1979.
149. **Bassett, N. S. and Gluckman, P. D.,** *J. Endocrinol.*, 109, 307, 1986.
150. **Wright, N. M., Northington, F. J., Miller, J. D., et al.,** *Pediatr. Res.*, 32, 286, 1992.
151. **Bresson, J. L., Clavequin, M. C., Fellman, D., et al.,** *Neuroendocrinology*, 39, 68, 1983.
152. **Chayvialle, A., Poulin, C., Dubois, P. M., et al.,** *Acta Endocrinol. (Copenh.)*, 94, 1, 1980.
153. **Ackland, J., Ratter, S., Bourne, G. L., et al.,** *Regul. Peptides*, 5, 95, 1983.
154. **Delitala, G., Meloni, T., Masala, A., et al.,** *Biomed. Res.*, 29, 13, 1978.
155. **Rieutort, M.,** *J. Endocrinol.*, 89, 355, 1981.
156. **Khorram, O., Depalatis, L. R., and McCann, S. M.,** *Endocrinology (Baltimore)*, 113, 720, 1983.
157. **Gooodyer, C. G., Branchaud, C. L., and Lefebvre, Y.,** *J. Clin. Endocrinol. Metab.*, 76, 1265, 1993.
158. **Baird, A., Wehrenberg, W. B., and Ling, N.,** *Regul. Peptides*, 10, 23, 1984.
159. **de Zegher, F., Styne, D. M., Daaboul, J., et al.,** *Endocrinology (Baltimore)*, 124, 124, 1989.
160. **de Zegher, F., Devlieger, H., and Veldhuis, J. D.,** *J. Clin. Endocrinol. Metab.*, 76, 1177, 1993.
161. **Miller, J. D., Wright, N. M., Esparza, A., et al.,** *J. Clin. Endocrinol. Metab.*, 75, 1508, 1992.
162. **Stewart, F., Goode, J. A., and Allen, W. R.,** *J. Endocrinol.*, 138, 81, 1993.
163. **Ezzat, S., Laks, D., Oster, J., et al.,** *Endocrinology (Baltimore)*, 128, 937, 1991.
164. **Ceda, G. P., Davis, R. G., Rosenfeld, R. G., et al.,** *Endocrinology*, 120, 1658, 1987.
165. **Blanchard, M., Goodyer, C. G., Charrier, J., et al.,** *Endocrinology (Baltimore)*, 122, 2114, 1988.
166. **Cuttler, L., Welsh, J., and Szabo, M.,** *Pediatr. Res.*, 19, 604, 1985.
167. **de Zegher, F., Daaboul, J., Grumbach, M., et al.,** *Endocrinology (Baltimore)*, 124, 1114, 1989.
168. **Parenti, M., Cocchi, D., Ceresoli, G., et al.,** *J. Endocrinol.*, 128, 1967, 1991.
169. **Shaywitz, B. A., Finkelstein, J., Hellman, L., et al.,** *Pediatrics*, 48, 103, 1971.
170. **Ho, K. Y., Evans, W. S., and Thorner, M. O.,** *Clin. Endocrinol. Metab.*, 14, 1, 1985.
171. **Cornblath, M., Parker, M. C., Reisner, S. H., et al.,** *J. Clin. Endocrinol. Metab.*, 25, 209, 1965.
172. **Ball, K. T., Power, G. G., Gunn, T. R., et al.,** *Endocrinology (Baltimore)*, 131, 337, 1992.
173. **Bassett, N. S. and Gluckman, P. D.,** *J. Dev. Physiol.*, 9, 301, 1987.
174. **Veldhuis, J. D., Blizzard, R. M., Rogol, A. D., et al.,** *J. Clin. Endocrinol. Metab.*, 74, 766, 1992.
175. **Brook, C. G. D. and Hindmarsh, P. C.,** *Neuroendocrinology*, 21, 767, 1992.
176. **Finkelstein, J. W., Roffwarg, H., Boyar, R. M., et al.,** *J. Clin. Endocrinol. Metab.*, 35, 665, 1972.

177. **Wennink, J. M. B., Delemarre-van de Waal, H. A., Schoemaker, R., et al.,** *Acta Endocrinol. (Copenh.)*, 124, 129, 1991.
178. **Rieutort, M.,** *J. Endocrinol.*, 60, 261, 1976.
179. **Sinha, Y. N., Selby, F. W., Lewis, U. J., et al.,** *Endocrinology (Baltimore)*, 91, 784, 1972.
180. **Dubreuil, P., Lapierre, H., Petitclerc, D., et al.,** *Domest. Anim. Endocrinol.*, 5, 157, 1988.
181. **Farmer, C., Lapierre, H., Matte, J. J., et al.,** *Domest. Anim. Endocrinol.*, 10, 249, 1993.
182. **Klindt, J. and Stone, R. T.,** *Growth*, 48, 1, 1984.
183. **Lapierre, H., Farmer, C., Girard, C., et al.,** *Domest. Anim. Endocrinol.*, 9, 199, 1992.
184. **Nap, R. C., Mol, J. A., and Hazewinkel, H. A. W.,** *Domest. Anim. Endocrinol.*, 10, 237, 1993.
185. **Cocola, F., Udeschini, G., Secchi, C., et al.,** *Endocrinology (Baltimore)*, 140, 1919.
186. **Cella, S. G., Locatelli, V., De Gennaro, V., et al.,** *Endocrinology (Baltimore)*, 116, 574, 1985.
187. **Cozzi, M. G., Zanini, A., Locatelli, V., et al.,** *Biochem. Biophys. Res. Commun.*, 138, 1223, 1986.
188. **Szabo, M. and Cuttler, L.,** *Endocrinology (Baltimore)*, 118, 69, 1986.
189. **Cuttler, L., Welsh, J. B., and Szabo, M.,** *Endocrinology (Baltimore)*, 119, 152, 1986.
190. **Oliver, C., Giraud, P., Lissitzky, J. C., et al.,** *Endocrinology (Baltimore)*, 110, 1018, 1982.
191. **Kacsoh, B., Terry, L. C., Meyers, J. S., et al.,** *Endocrinology (Baltimore)*, 125, 1326, 1989.
192. **Cella, S. G., Locatelli, V., De Gennaro, V., et al.,** *Endocrinology (Baltimore)*, 119, 1164, 1986.
193. **Cella, S. G., Locatelli, V., DeGennaro, V., et al.,** *Endocrinology (Baltimore)*, 120, 1639, 1987.
194. **Bero, L. A. and Kuhn, C. M.,** *J. Pharmacol. Exp. Ther.*, 237, 137, 1986.
195. **Kuhn, C. M. and Schanberg, S. M.,** *J. Pharmacol. Exp. Ther.*, 217, 152, 1981.
196. **Kacsoh, B., Toth, B. E., and Grosvenor, C. E.,** *Neuroendocrinology*, 57, 195, 1993.
197. **Kacsoh, B., Meyers, J. S., Crowley, W. R., et al.,** *J. Endocrinol.*, 124, 233, 1990.
198. **Barenton, B., Duclos, M., Diaz, J., et al.,** *Reprod. Nutr. Dev.*, 27, 491, 1987.
199. **Takahashi, S.,** *Cell Tiss. Res.*, 266, 275, 1991.
200. **Blizzard, R. M., Martha, P. M., Kerrigan, J. R., et al.,** *J. Endocrinol. Invest.*, 12, 65, 1989.
201. **Martha, P. M., Goorman, K. M., Blizzard, R. M., et al.,** *J. Clin. Endocrinol. Metab.*, 74, 335, 1992.
202. **Carlsson, L. M. S., Rosberg, S., Vitangcol, R. V., et al.,** *J. Clin. Endocrinol. Metab.*, 77, 356, 1993.
203. **Hindmarsh, P. C., Matthews, D. R., and Brook, C. G. D.,** *Clin. Endocrinol.*, 29, 35, 1988.
204. **Mauras, N., Blizzard, R. M., Link, K., et al.,** *J. Clin. Endocrinol. Metab.*, 64, 596, 1987.
205. **Mansfield, M. J., Rudlin, C. R., and Crigler, J. F.,** *J. Clin. Endocrinol. Metab.*, 66, 3, 1988.
206. **Ulloa-Aguirre, A., Blizzard, R., Garcia-Rubi, E., et al.,** *J. Clin. Endocrinol. Metab.*, 71, 846, 1990.
207. **Foster, C. M., Hopwood, N. J., Hassing, J. M., et al.,** *Pediatr. Res.*, 26, 320, 1989.
208. **Jansson, J. O., Eden, S., and Isaksson, O.,** *Endocr. Rev.*, 6, 128, 1985.
209. **Jansson, J., Albertson-Willand, K., Isakson, O., et al.,** *Endocrinology (Baltimore)*, 114, 1287, 1984.
210. **Bando, H., Zhang, C., Takada, Y., et al.,** *Acta Endocrinol. (Copenh.)*, 124, 31, 1991.
211. **Corpas, E., Harman, S. M., Pineyro, M. A., et al.,** *J. Clin. Endocrinol. Metab.*, 75, 530, 1992.
212. **Corpas, E., Harman, S. M., and Blackman, M. R.,** *Endocr. Rev.*, 14, 20, 1993.
213. **Carlson, H. E., Gillin, J. C., Gorden, P., et al.,** *J. Clin. Endocrinol. Metab.*, 34, 1102, 1972.
214. **Rudman, D., Vintner, M. H., Rogers, C. M., et al.,** *J. Clin. Invest.*, 67, 1361, 1981.
215. **Iranmanesh, A., Lizarralde, G., and Veldhuis, J. D.,** *J. Clin. Endocrinol. Metab.*, 73, 1081, 1991.
216. **Leppaluoto, J., Tapanainen, P., and Knip, M.,** *J. Clin. Endocrinol. Metab.*, 65, 1035, 1987.
217. **Ho, K. Y., Evans, W. S., Blizzard, R. M., et al.,** *J. Clin. Endocrinol. Metab.*, 64, 51, 1987.
218. **Shibasaki, T., Shizume, K., Nakahara, M., et al.,** *J. Clin. Endocrinol. Metab.*, 58, 212, 1984.
219. **Vermeulen, A.,** *J. Clin. Endocrinol. Metab.*, 64, 884, 1987.
220. **Iovino, M., Monteleone, P., and Steardo, L.,** *J. Clin. Endocrinol. Metab.*, 69, 910, 1989.
221. **Ceda, G. P., Ceresini, G., Denti, L., et al.,** *Acta Endocrinol. (Copenh.)*, 124, 516, 1991.
222. **Ghigo, E., Goffi, E., Nicolosi, M., et al.,** *J. Clin. Endocrinol. Metab.*, 71, 1481, 1990.
223. **Alba-Roth, J., Muller, O. A., Schopohl, J., et al.,** *J. Clin. Endocrinol. Metab.*, 67, 1186, 1988.
224. **Chiodera, P., Gnudi, A., Delsignore, R., et al.,** *Neuroendocrinol. Lett.*, 211, 1986.
225. **Shibasaki, T., Shizume, K., Nakahara, M., et al.,** *J. Clin. Endocrinol. Metab.*, 58, 212, 1984.
226. **Lang, I., Schernthaner, G., Pietschmann, P., et al.,** *J. Clin. Endocrinol. Metab.*, 65, 535, 1987.
227. **Giustina, A., Licini, M., Bussi, A. R., et al.,** *J. Clin. Endocrinol. Metab.*, 76, 1369, 1993.
228. **Coiro, V., Volpi, R., Cavazzini, U., et al.,** *J. Gerontol.*, 46, 155, 1991.
229. **Sun, Y. X., Fenoglio, C. M., Pushparaj, N., et al.,** *Hum. Pathol.*, 15, 169, 1984.

230. **Sonntag, W. E., Steger, R. W., Forman, L., et al.,** *Endocrinology (Baltimore)*, 107, 1875, 1980.
231. **Takahashi, S., Gottschall, P. E., Quigley, K. L., et al.,** *Neuroendocrinology*, 46, 137, 1987.
232. **Deslauries, N., Gaudreau, P., Abribat, T., et al.,** *Neuroendocrinology*, 53, 439, 1991.
233. **Sonntag, W. E., Hylka, V. W., and Meites, J.,** *Endocrinology (Baltimore)*, 113, 2305, 1983.
234. **Robberecht, P., Gillard, M., Waelbroeck, M., et al.,** *Neuroendocrinology*, 44, 429, 1986.
235. **Ceda, G. P., Valenti, G., Butturini, U., et al.,** *Endocrinology (Baltimore)*, 118, 2109, 1986.
236. **Takahashi, S.,** *Acta Endocrinol. (Copenh.)*, 127, 531, 1992.
237. **Takahashi, S., Kawashima, S., Seo, H., et al.,** *Endocrinol. Jpn.*, 37, 827, 1990.
238. **Martinoli, M. G., Oullet, J., Rheaume, E., et al.,** *Neuroendocrinology*, 54, 607, 1991.
239. **Ge, F., Tsagarakis, S., Rees, L. H., et al.,** *J. Endocrinol.*, 123, 53, 1989.
240. **Sonntag, W. E., Boyd, R. L., and Booze, R. M.,** *Neurobiol. Aging*, 11, 409, 1990.
241. **Sonntag, W. E., Forman, L. J., Miki, N., et al.,** *Neuroendocrinology*, 33, 73, 1981.
242. **Sonntag, W. E. and Gough, M. A.,** *Neuroendocrinology*, 47, 482, 1988.
243. **Spik, K. and Sonntag, W. E.,** *Neuroendocrinology*, 50, 489, 1989.
244. **Morimoto, N., Kawakami, F., Makino, S., et al.,** *Neuroendocrinology*, 47, 459, 1988.
245. **De Gennaro, C. V., Zoli, M., Cocchi, D., et al.,** *Peptides*, 10, 705, 1989.
246. **De Gennaro Collona, V., Fidone, F., Cocchi, D., et al.,** *Neurobiol. Aging*, 14, 503, 1993.
247. **De Gennaro Collona, V., Zoli, M., Cocchi, D., et al.,** *Peptides*, 10, 705, 1989.
248. **Hassan, H. A., Merkel, R. A. X., and Tucker, H. A.,** *Domest. Anim. Endocrinol.*, 9, 209, 1992.
249. **Cella, S. G., Moiraghi, V., Minuto, F., et al.,** *Acta Endocrinol. (Copenh.)*, 121, 177, 1989.
250. **Cella, S. G., Arce, V. M., Pieretti, F., et al.,** *Neuroendocrinology*, 57, 432, 1993.
251. **Ho, K. Y., Evans, W. S., Blizzard, R. M., et al.,** *J. Clin. Endocrinol. Metab.*, 64, 51, 1987.
252. **Quabbe, H. J., Schilling, E., and Helge, H.,** *J. Clin. Endocrinol. Metab.*, 26, 1173, 1966.
253. **Arbona, J. R., Marple, D. N., Russell, et al.,** *J. Anim. Sci.*, 66, 3068, 1988.
254. **Siers, D. G. and Swieger, L. A.,** *J. Anim. Sci.*, 32, 1229, 1973.
255. **Plouzek, C. A. and Trenkle, A.,** *Domest. Anim. Endocrinol.*, 8, 63, 1991.
256. **Davis, S. L. and Michael, B.,** *J. Anim. Sci.*, 38, 795, 1974.
257. **Jansson, J. O., Albertson-Wilkland, K., Eden, S., et al.,** *Acta Endocrinol. (Copenh.)*, 99, 24, 1982.
258. **Isaakson, O. G. P., Eden, S., and Jansson, J. O.,** *Annu. Rev. Physiol.*, 47, 483, 1985.
259. **Carlsson, L. M. S., Clark, R. G., and Robinson, I. C. A. F.,** *J. Endocrinol.*, 126, 27, 1990.
260. **Badger, T. M., Millard, W. J., Owens, S. M., et al.,** *Endocrinology (Baltimore)*, 128, 1065, 1991.
261. **Jansson, J. O., Ishikawa, K., Katakami, H., et al.,** *Endocrinology (Baltimore)*, 120, 525, 1987.
262. **Gabriel, S. M., Millard, W. J., Koenig, J. I., et al.,** *Neuroendocrinology*, 50, 299, 1989.
263. **Maiter, D., Koenig, J. I., and Kaplan, L. M.,** *Endocrinology (Baltimore)*, 128, 1709, 1991.
264. **Uchiyama, T., Abe, H., Shakutsui, S., et al.,** 72nd Annu. Meet. Endocr. Soc., Atlanta, GA, 1990, p. 50.
265. **Chowen-Breed, J., Steiner, R. A., and Clifton, D. K.,** *Endocrinology (Baltimore)*, 125, 357, 1989.
266. **Werner, H., Koch, Y., Baldino, F., Jr., et al.,** *J. Biol. Chem.*, 263, 7666, 1988.
267. **Zeitler, P., Argente, J., Chowen-Breed, J. A., et al.,** *Endocrinology (Baltimore)*, 127, 1362, 1990.
268. **Roselli, C. E., Handa, R. J., and Resko, J. A.,** *Neuroendocrinology*, 49, 449, 1989.
269. **Gabriel, S. M., Millard, W. J., Koenig, J. I., et al.,** *Neuroendocrinology*, 50, 2299, 1989.
270. **Hasegawa, O., Sugihara, H., Minami, S., et al.,** *Peptides*, 13, 475, 1992.
271. **Jansson, J. and Frohman, L. A.,** *Endocrinology (Baltimore)*, 120, 1551, 1987.
272. **Jansson, J. and Frohman, L. A.,** *Endocrinology (Baltimore)*, 121, 1417, 1987.
273. **Millard, W. J., Politch, J. A., Martin, J. B., et al.,** *Endocrinology (Baltimore)*, 119, 2655, 1986.
274. **Painson, J.-C., Thorner, M. O., Krieg, R. J., et al.,** *Endocrinology (Baltimore)*, 130, 511, 1992.
275. **Shirasu, K., Stumpf, W. E., and Sar, M.,** *Endocrinology (Baltimore)*, 127, 344, 1990.
276. **Simard, J., Hubert, J. F., Hosseinzadeh, T., et al.,** *Endocrinology (Baltimore)*, 119, 2004, 1986.
277. **Fukata, J. and Martin, J. B.,** *Endocrinology (Baltimore)*, 119, 2256, 1986.
278. **Herbert, D. C. and Sheridan, P. J.,** *Biol. Reprod.*, 28, 377, 1983.
279. **Fishman, A., Hertz, P., and Hochberg, Z.,** *Neuroendocrinology*, 57, 782, 1993.
280. **Clark, R. G. and Robinson, I. C. A. F.,** *J. Endocrinol.*, 106, 281, 1985.
281. **Schettini, G., Florio, T., Meucci, O., et al.,** *Brain Res.*, 439, 322, 1988.
282. **Cronin, M. J. and Rogol, A. D.,** *Brain Res.*, 31, 984, 1984.

283. **Ho, K. Y., Leong, D. A., Sinha, Y. N., et al.,** *Am. J. Physiol.*, 250, E650, 1986.
284. **Takahashi, S.,** *Zool. Sci.*, 9, 901, 1992.
285. **Asplin, C. M., Fariai, A. C. S., Carlsen, E. C., et al.,** *J. Clin. Endocrinol. Metab.*, 69, 239, 1989.
286. **Faria, A. C. S., Bekenstein, L. W., Booth, R. A., et al.,** *Clin. Endocrinol.*, 36, 591, 1992.
287. **Mauras, N., Rogol, A. D., and Veldhuis, J. D.,** *J. Clin. Endocrinol. Metab.*, 69, 1053, 1989.
288. **Thompson, R. G., Rodriguez, A., Kowarski, A., et al.,** *J. Clin. Invest.*, 51, 3193, 1972.
289. **Dawson-Hughes, B., Stern, D., Goldman, J., et al.,** *J. Clin. Endocrinol. Metab.*, 63, 424, 1986.
290. **Frantz, A. G. and Rabkin, M. T.,** *J. Clin. Endocrinol. Metab.*, 25, 1470, 1965.
291. **Wiedemann, E., Schwartz, E., and Frantz, A. G.,** *J. Clin. Endocrinol. Metab.*, 42, 942, 1976.
292. **Merimee, T. J. and Fineberg, S. E.,** *J. Clin. Endocrinol. Metab.*, 33, 896, 1971.
293. **Gelato, M., Malozowski, S., Caruso-Nicoleti, M., et al.,** *J. Clin. Endocrinol. Metab.*, 63, 174, 1986.
294. **Wehrenberg, W. B. and Giustina, A.,** *Endocr. Rev.*, 13, 299, 1992.
295. **Perlow, M., Sassin, J., Boyar, R., et al.,** *Metabolism*, 22, 1269, 1973.
296. **Mauras, N., Rogol, A. D., and Veldhuis, J. D.,** *Pediatr. Res.*, 28, 626, 1990.
297. **Bruno, J. F., Olchovsky, D., White, J. D., et al.,** *Endocrinology (Baltimore)*, 127, 2111, 1990.
298. **Schober, E., Frisch, H., Waldhauser, F., et al.,** *Acta Endocrinol. (Copenh.)*, 120, 442, 1989.
299. **Veldhuis, J., Sotos, J. F., and Sherman, B. M.,** *J. Clin. Endocrinol. Metab.*, 73, 71, 1991.
300. **Massarano, A. A., Brook, D. G., and Hindmarsh, P. C.,** *Arch. Dis. Child.*, 64, 587, 1989.
301. **Ulloa-Aguirre, A., Blizzard, R., Carcia-Rubi, E., et al.,** *J. Clin. Endocrinol. Metab.*, 71, 846, 1990.
302. **Metzger, D. L. and Kerrigan, J. R.,** *J. Clin. Endocrinol. Metab.*, 76, 1147, 1993.
303. **Liu, L., Merrian, G. R., and Sherins, R. J.,** *J. Clin. Endocrinol. Metab.*, 64, 651, 1987.
304. **Martha, P. M., Jr., Rogol, A. D., Veldhuis, J. D., et al.,** *J. Clin. Endocrinol. Metab.*, 69, 563, 1989.
305. **Kerrigan, J. R. and Rogol, A. D.,** *Endocr. Res.*, 13, 281, 1992.
306. **Malarkey, W. B., Hall, J. C., Pearl, D. K., et al.,** *J. Clin. Endocrinol. Metab.*, 73, 1089, 1991.
307. **Kazmer, G. W., Canfield, R. W., and Bean, B.,** *J. Anim. Sci.*, 70, 503, 1992.
308. **Suttie, J. M., White, R. G., and Littlejohn, R. P.,** *Gen. Comp. Endocrinol.*, 85, 36, 1992.
309. **Scanes, C. G., Lauterio, T., and Buonomo, F.,** *Avian Endocrinology: Environmental and Ecological Perspectives* (Eds. Mikami, S. I., Homma, K., and Wada, M.). Springer-Verlag, Berlin, 1983, p. 307.
310. **Siverin, B., Viebke, P. A., Westin, J., et al.,** *Gen. Comp. Endocrinol.*, 73, 404, 1989.
311. **Harvey, S., Scanes, C. G., and Sharp, P. J.,** *Gen. Comp. Endocrinol.*, 48, 411, 1982.
312. **John, T. M., George, J. C., and Scanes, C. G.,** *Gen. Comp. Endocrinol.*, 51, 44, 1983.
313. **Stokkan, K. A., Harvey, S., Klandorf, H., et al.,** *Gen. Comp. Endocrinol.*, 58, 76, 1985.
314. **Tindal, J. S., Knaggs, G. S., Hart, I. C., et al.,** *Endocrinology (Baltimore)*, 76, 333, 1978.
315. **Colle, M., Rosenzweig, P., Bianchetti, G., et al.,** *Horm. Res.*, 35, 30, 1991.
316. **Muller, E. E.,** *Physiol. Rev.*, 67, 962, 1987.
317. **Holl, R. W., Schwartz, U., Schauwecker, P., et al.,** *J. Clin. Endocrinol. Metab.*, 77, 216, 1993.
318. **Holl, R. W., Hartman, M. L., Veldhuis, J. D., et al.,** *J. Clin. Endocrinol. Metab.*, 72, 854, 1991.
319. **Follenius, M., Brandenberger, G., Simon, C., et al.,** *Sleep*, 11, 546, 1988.
320. **Obal, F., Payne, L., Kapas, L., et al.,** *Brain Res.*, 557, 149, 1991.
321. **Obal, F., Payne, L., Opp, M., et al.,** *Am. J. Physiol.*, 263, R1078, 1992.
322. **Buzi, F., Zanotti, P., Tiberti, A., et al.,** *J. Clin. Endocrinol. Metab.*, 77, 1495, 1993.
323. **Steiger, A., Guldner, J., Hemmeter, U., et al.,** *Neuroendocrinology*, 56, 566, 1992.
324. **Gerra, G., Volpi, R., Delsignore, R., et al.,** *Acta Endocrinol. (Copenh.)*, 126, 24, 1992.
325. **Noel, G. L., Suh, H. K., Stone, J. G., et al.,** *J. Clin. Endocrinol. Metab.*, 35, 840, 1972.
326. **Abplanalp, J. M., Livingston, L., Rose, R. M., et al.,** *Psychosom. Med.*, 39, 158, 1977.
327. **Ader, R., Felton, D., and Cohen, N.,** *Psychoneuroimmunology*, Academic Press, New York, 1991.
328. **Brown, G. M., Seggie, J. A., Chambers, J. W., et al.,** *Psychoneuroendocrinology*, 3, 131, 1978.
329. **Rose, R. M.,** *Neuroendocrinology and Psychiatric Disorder* (Eds. Brown, G. P., Koslow, S. H., and Reichlin, S.). Raven Press, New York, 1984, p. 95.
330. **Moberg, G. P.,** *Animal Stress* (Ed. Moberg, G. P.). American Physiological Society, Bethesda, MD, 1985, p. 27.
331. **Stanhope, R., Adlard, P., Hamill, G., et al.,** *Clin. Endocrinol.*, 28, 335, 1988.
332. **Powell, G. F., Brasel, J. A., Raiti, S., et al.,** *New Engl. J. Med.*, 276, 1279, 1967.
333. **Miyabo, S., Asato, T., and Mizushima, N.,** *Psychosom. Med.*, 41, 515, 1979.
334. **Okada, Y., Miyai, K., Iwatsubo, H., et al.,** *J. Clin. Endocrinol.*, 30, 393, 1970.

335. **Allen, P. I. M., Batty, K. A., and Dodd, C. A.,** *J. Endocrinol.*, 107, 163, 1985.
336. **Malarkey, W. B., Hall, J. C., Pearl, D. K., et al.,** *J. Clin. Endocrinol. Metab.*, 73, 1089, 1991.
337. **Nemeroff, C. B. and Krishnan, K. R. R.,** *Neuroendocrinology* (Ed. Nemeroff, C. B.). CRC Press, Boca Raton, FL, 1991, p. 413.
338. **Munksgaard, L. and Lovendahl, P.,** *Can. J. Anim. Sci.*, 73, 847, 1993.
339. **Eaton, L. W., Klosterman, E. W., and Johnson, R. R.,** *J. Anim. Sci.*, 27, 1785, 1968.
340. **Rushen, J., Schwarze, N., Ladewig, J., et al.,** *Physiol. Behav.*, 53, 923, 1993.
341. **Coleman, E. S., Elsasser, T. H., Kemppainen, R. J., et al.,** *Neuroendocrinology*, 58, 111, 1993.
342. **Pecile, A. and Olgiati, V. R.,** *The Endocrine Hypothalamus* (Eds. Jeffcoate, S. L. and Hutchinson, J. S. M.). Academic Press, London, 1978, p. 362.
343. **Arimura, A., Smith, W. D., and Schally, A. V.,** *Endocrinology (Baltimore)*, 98, 540, 1976.
344. **Reynaert, R., Marcus, S., De Paepe, M., et al.,** *Horm. Metab. Res.*, 8, 109, 1976.
345. **Mitchell, P., Smythe, G., Parker, G., et al.,** *Psych. Res.*, 37, 179, 1991.
346. **Cronin, M. T., Siegel, B. J., and Moberg, G. P.,** *Physiol. Behav*, 26, 887, 1981.
347. **Davis, S. L.,** *Endocrinology (Baltimore)*, 91, 549, 1972.
348. **Parrott, R. F. and Goode, J. A.,** *Domest. Anim. Endocrinol.*, 9, 141, 1992.
349. **Pickering, A. D., Pottinger, T. G., Sumpter, J. P., et al.,** *Gen. Comp. Endocrinol.*, 83, 86, 1991.
350. **Harvey, S.,** *The Endocrinology of Growth, Development and Metabolism in Vertebrates* (Ed. Schreibman, M. P., Scanes, C. G., and Pang, P. K. T.), Academic Press, New York, 1993, p. 151.
351. **Shavit, Y., Yirmiya, R. and Beilin, B.,** *The Neuroendocrine-Immune Network* (Ed. Freier, S.). CRC Press, Boca Raton, FL, 1990, p. 163.
352. **Armstrong, J. D. and Johnson, B. H.,** *J. Endocrinol.*, 121, 11, 1989.
353. **Panerai, A. E., Casaneuva, F., Martini, A., et al.,** *Endocrinology (Baltimore)*, 108, 2400, 1981.
354. **Wiedermannk, von Bardelben, Y., and Holsboer, F.,** *Neuroendocrinology*, 54, 462, 1991.
355. **Giustina, A. and Wehrenberg, W. B.,** *Trends Endocrinol. Metab.*, 3, 306, 1992.
356. **Wehrenberg, W. B., Janowski, B. A., Piering, A. W., et al.,** *Endocrinology (Baltimore)*, 126, 3200, 1990.
357. **Veldhuis, J. D., Lizarralde, G., and Iranmanesh, A.,** *J. Clin. Endocrinol. Metab.*, 74, 96, 1992.
358. **Franci, C. R., Anselmo-Franci, J., Kozlowski, G. P., et al.,** *Neuroendocrinology*, 57, 693, 1993.
359. **Farmer, C., Dubreuil, P., Couture, Y., et al.,** *Domest. Anim. Endocrinol.*, 8, 529, 1991.
360. **Hamanaka, K., Soya, H., Katayama, I., et al.,** *Neuroendocrinol. Lett.*, 133, 3, 1991.
361. **Juarez, L. M. and Meserve, L. A.,** *Growth Dev. Aging*, 52, 139, 1988.
362. **Rivier, C. and Vale, W.,** *Endocrinology (Baltimore)*, 114, 2409, 1984.
363. **Rivier, C. and Vale, W.,** *Endocrinology (Baltimore)*, 117, 2478, 1985.
364. **Armario, A., Marti, O., Gavalda, A., et al.,** *Psychoneuroendocrinology*, 18, 405, 1993.
365. **Dore, S., Brisson, G. R., Fournier, A., et al.,** *Horm. Metab. Res.*, 23, 431, 1991.
366. **Buckler, J. M. H.,** *Acta Endocrinol. (Copenh.)*, 69, 219, 1972.
367. **Kraemer, R. R., Kilgore, J. L., Kraemer, G. R., et al.,** *Med. Sci. Sports Exerc.*, 24, 1346, 1992.
368. **Dore, S., Brisson, G. R., Faurnier, A., et al.,** *Eur. J. Appl. Physiol.*, 62, 130, 1991.
369. **Luger, A., Watschinger, B., Deuster, P., et al.,** *Neuroendocrinology*, 56, 112, 1992.
370. **Rogol, A. D., Weltman, J. Y., Evans, W. S., et al.,** *Neuroendocrinology*, 21, 817, 1992.
371. **Weltman, A., Seip, R. L., Snead, D., et al.,** *Int. J. Sport Med.*, 13, 257, 1992.
372. **Felsing, N. E., Brasel, J. A., and Cooper, D. M.,** *J. Clin. Endocrinol. Metab.*, 75, 157, 1992.
373. **Sutton, J. R., Jones, N. L., and Toews, C. J.,** *Clin. Sci. Mol. Med.*, 50, 241, 1976.
374. **Bramnert, M. and Hokfelt, B.,** *Acta Endocrinol. (Copenh.)*, 115, 125, 1987.
375. **Thompson, D. L., Weltman, J. Y., Rogol, A. D., et al.,** *J. Appl. Physiol.*, 75, 870, 1993.
376. **Brillon, D. N., Nabil, D. N., and Jacobs, L. S.,** *Endocr. Res.*, 12, 137, 1986.
377. **Casenueva, F. F., Villaneuva, J. L., Cabranes, J. A., et al.,** *J. Clin. Endocr. Metab.*, 59, 526, 1984.
378. **Few, J. D. and Davies, C. T. M.,** *Eur. J. Appl. Physiol. Occup. Physiol.*, 43, 221, 1980.
379. **Casanueva, F. F., Villanueva, L., Cabranes, J. A., et al.,** *J. Clin. Endocrinol. Metab.*, 59, 526, 1984.
380. **Grossman, A., Bouloux, P., Price, P., et al.,** *Clin. Sci. Lond.*, 67, 483, 1984.
381. **Pyka, G., Wiswell, R. A., and Marcus, R.,** *J. Clin. Endocrinol. Metab.*, 75, 404, 1992.
382. **Machlin, L. J., Horino, M., Hertelendy, F., et al.,** *Endocrinology (Baltimore)*, 82, 369, 1968.
383. **John, T. M., Viswwanathan, M., George, J. C., et al.,** *Horm. Metab. Res.*, 20, 271, 1988.
384. **Garcia, J. F. and Geschwind, L. L.,** *Growth Hormone* (Eds. Pecile, A. and Muller, E. E.). Excerpta Medica Foundation, Amsterdam, 1968, p. 267.

385. **Ross, R. J. M. and Buchanan, C. R.,** *Nutr. Res. Rev.*, 3, 143, 1990.
386. **Janowski, B. A., Ling, N. C., Giustina, A., et al.,** *Life Sci.*, 52, 981, 1993.
387. **Tannenbaum, G. S., Rorstad, O., and Brazeau, P.,** *Endocrinology (Baltimore)*, 104, 1733, 1979.
388. **Tannenbaum, G., Epelbaum, J., Colle, D., et al.,** *Endocrinology (Baltimore)*, 102, 1909, 1978.
389. **Hughes, J. N., Enjalbert, A., Moyse, E., et al.,** *J. Endocrinol.*, 109, 169, 1986.
390. **Arce, V. M., Cella, S. G., Locatelli, V., et al.,** *Neuroendocrinology*, 467, 1991.
391. **Cella, S. G., Moriaghi, V., Minuto, F., et al.,** *Acta Endocrinol. (Copenh.)*, 121, 177, 1989.
392. **Ho, K. Y., Veldhius, J. S., Johnson, M. L., et al.,** *J. Clin. Invest.*, 122, 11464, 1988.
393. **Rojdmark, S.,** *Clin. Endocrinol.*, 25, 721, 1986.
394. **Mears, G. J.,** *Can. J. Anim. Sci.*, 73, 987, 1993.
395. **Brier, B. H., Bass, J. J., Butler, J. H., et al.,** *J. Endocrinol.*, 111, 209, 1986.
396. **Foster, D. L., Ebling, F. J. P., Micka, A. F., et al.,** *Endocrinology (Baltimore)*, 125, 342, 1989.
397. **Antinmo, T., Baldijao, K. A., Houpt, K. A., et al.,** *J. Anim. Sci.*, 46, 940, 1978.
398. **Scanes, C. G. and Griminger, P.,** *Exp.Zool.Suppl.*, 4, 98, 1990.
399. **Lauterio, T. J. and Scanes, C. G.,** *J. Endocrinol.*, 117, 223, 1988.
400. **Proudman, J. A. and Opel, H.,** *Poult. Sci.*, 60, 659, 1981.
401. **Sumpter, J. P., Le Bail, P. Y., Pickering, A. D., et al.,** *Gen. Comp. Endocrinol.*, 83, 94, 1991.
402. **Sumpter, J. P., Lincoln, R. F., Bye, V. J., et al.,** *Gen. Comp. Endocrinol.*, 83, 103, 1991.
403. **Schindler, W. J., Hutchins, M., and Septimus, E. J.,** *Endocrinology (Baltimore)*, 91, 483, 1972.
404. **Hartman, M. L., Veldhuis, J. D., Johnson, M. L., et al.,** *J. Clin. Endocrinol. Metab.*, 74, 757, 1992.
405. **Clemmons, D. R. and Van Wyk, J. J.,** *Clin. Endocrinol. Metab.*, 13, 113, 1984.
406. **Berelowitz, M., Szabo, M., Frohman, M., et al.,** *Science*, 212, 1279, 1981.
407. **Stewart, J. K., Koerker, D. J., and Goodner, C. J.,** *Endocrinology (Baltimore)*, 115, 1897, 1984.
408. **Aoki, T. T.,** *Joslin's Diabetes Mellitus* (Eds. Marble, A., Karel, L. P., Bradley, R. J., and Christlieb, A. R.). Lea & Febiger, Philadelphia, 1985, p. 138.
409. **Trenkle, A. and Plouzek, C.,** *67th Annu. Meet. Endocr. Soc.,* Baltimore, MD, 1985.
410. **Wheaton, J. E., Al-Raheem, S. N., Massri, Y. G., et al.,** *J. Anim. Sci.*, 62, 1267, 1986.
411. **Vasilatos, R. and Wangsness, P. J.,** *J. Nutr.*, 110, 1479, 1980.
412. **Trenkle, A.,** *J. Nutr.*, 119, 61, 1989.
413. **Balzano, S., Loche, S., Murtas, M. L., et al.,** *Horm. Metab. Res.*, 21, 52, 1989.
414. **Renier, G., Gaudreau, P., Hajjad, H., et al.,** *Neuroendocrinology*, 52, 284, 1990.
415. **Himsworth, R. L., Carmel, P. W., and Frantz, A. G.,** *Endocrinology (Baltimore)*, 91, 217, 1972.
416. **Tsushima, T., Irie, M., and Sakuma, M.,** *Endocrinology (Baltimore)*, 89, 685, 1971.
417. **Barb, C. R., Kraeling, R. R., Barrett, J. B., et al.,** *Proc. Soc. Exp. Biol. Med.*, 198, 636, 1991.
418. **Harvey, S., Scanes, C. G., Chadwick, A., et al.,** *J. Endocrinol.*, 76, 501, 1978.
419. **Rodgers, B. D., Helms, L. M., and Grau, E. G.,** *Gen. Comp. Endocrinol.*, 86, 344, 1992.
420. **Kokka, N., Garcia, J. F., Morgan, M., et al.,** *Endocrinology (Baltimore)*, 88, 359, 1971.
421. **Muller, E. E., Miedica, D., Giustina, G., et al.,** *Endocrinology (Baltimore)*, 88, 345, 1971.
422. **Frohman, L. A., Downs, T. R., Clarke, I. J., et al.,** *J. Clin. Invest.*, 86, 17, 1990.
423. **Tannenbaum, G. S., Martin, J. B., and Colle, E.,** *Endocrinology (Baltimore)*, 99, 720, 1976.
424. **Painson, J. C. and Tannenbaum, G. S.,** *Endocrinology (Baltimore)*, 117, 1132, 1985.
425. **Delitala, G., Tomasi, P. A., Palermo, M., et al.,** *J. Endocrinol. Invest.*, 13, 653, 1990.
426. **Adriaens, F. A., Miller, M. A., Hard, D. L., et al.,** *J. Dairy Sci.*, 75, 472, 1992.
427. **Wallace, A. L. C. and Bassett, J. M.,** *J. Endocrinol.*, 47, 21, 1970.
428. **Press, M., Tamborlane, W. V., Thorner, M. O., et al.,** *Diabetes*, 33, 804, 1984.
429. **Roth, J., Glick, S. M., Yalow, R. S., et al.,** *Metabolism*, 12, 577, 1963.
430. **Penicaud, L., Pajot, M. T., and Thompson, D. A.,** *Endocr. Res.*, 16, 461, 1990.
431. **Devesa, J., Diaz, M. J., Tresguerres, J. A. F., et al.,** *J. Clin. Endocrinol. Metab.*, 73, 251, 1991.
432. **Devesa, J., Arce, V., Lois, N., et al.,** *J. Clin. Endocrinol. Metab.*, 71, 1581, 1990.
433. **Berelowitz, M., Dudlak, D., and Frohman, L. A.,** *J. Clin. Invest.*, 69, 1293, 1982.
434. **Lengyel, A. M. J., Grossman, A., Nieuwenhuyzen-Kruseman, A. C., et al.,** *Neuroendocrinology*, 39, 31, 1984.
435. **Sato, M. and Frohman, L. A.,** *Neuroendocrinology*, 57, 1097, 1993.
436. **Baes, M. and Vale, W. W.,** *Neuroendocrinology*, 51, 202, 1990.

437. **Berelowitz, M., Ting, N. C., and Murray, L.,** *Endocrinology (Baltimore)*, 124, 826, 1989.
438. **Renier, G. and Serri, O.,** *Neuroendocrinology*, 54, 521, 1991.
439. **Murao, K., Sato, M., Mizobuchi, M., et al.,** *Endocrinology (Baltimore)*, 134, 418, 1994.
440. **Tannenbaum, G. S.,** *Endocrinology (Baltimore)*, 108, 76, 1981.
441. **Ndon, J. A., Giustina, A., and Wehrenberg, W. B.,** *Neuroendocrinology*, 55, 500, 1992.
442. **Tannenbaum, G. S., Martin, J. B., and Colle, E.,** *Endocrinology (Baltimore)*, 99, 720, 1976.
443. **King, R. A., Smith, R. M., and Willoughby, J. O.,** *Horm. Metab. Res.*, 18, 510, 1986.
444. **Imaki, T., Shibasaki, T., Masuda, A., et al.,** *Endocrinology (Baltimore)*, 118, 2390, 1986.
445. **Imaki, T., Shibasaki, T., Shizume, K., et al.,** *J. Clin. Endocrinol. Metab.*, 60, 290, 1985.
446. **Rosenbaum, M. J., Fong, Y., Hesse, D. G., et al.,** *J. Clin. Endocrinol. Metab.*, 69, 310, 1989.
447. **Pontiroli, A. E., Lanzi, R., Monti, L. D., et al.,** *J. Endocrinol. Invest.*, 13, 539, 1990.
448. **Alvarez, C. V., Mallo, F., Burguera, B., et al.,** *Neuroendocrinology*, 53, 185, 1991.
449. **Estienne, M. J., Schillo, K. K., Green, M. A., et al.,** *Endocrinology*, 125, 85, 1989.
450. **Hertelendy, F. and Kipnis, D. M.,** *Endocrinology (Baltimore)*, 92, 402, 1973.
451. **Irie, M., Sakuma, M., Tsushima, T., et al.,** *Proc. Soc. Exp. Biol. Med.*, 126, 708, 1987.
452. **Foltzer, C. H., Strosser, M. T., Harvey, S., et al.,** *J. Endocrinol.*, 106, 21, 1988.
453. **Harvey, S. and Scanes, C. G.,** *J. Endocrinol.*, 75, 50, 1977.
454. **Scanes, C. G. and Harvey, S.,** *Aspects of Avian Endocrinology: Practical and Theoretical Implications* (Ed. Scanes, C. G.). Texas Tech Press, Lubbock, TX, 1982, p. 173.
455. **Senaris, R. M., Lewis, M. D., Lago, F., et al.,** *Neurosci. Lett.*, 135, 80, 1992.
456. **Senaris, R. M., Lewis, M. D., Lago, F., et al.,** *J. Mol. Endocrinol.*, 10, 207, 1993.
457. **Casanueva, F. F., Villanueva, L., Dieguez, C., et al.,** *J. Clin. Endocrinol. Metab.*, 65, 634, 1987.
458. **Harel, Z. and Tannenbaum, G. S.,** *Endocrinology (Baltimore)*, 133, 1035, 1993.
459. **Cree, T. C. and Schalch, D. S.,** *Endocrinology (Baltimore)*, 117, 667, 1985.
460. **Okada, K., Suzuki, N., Sugihara, H., et al.,** *Neuroendocrinology*, 59, 380, 1994.
461. **Bruno, J., Song, J., and Berelowitz, M.,** *74th Annu. Meet. Endocr. Soc.*, San Antonio, TX, 1992, p. 393.
462. **Kung, L., Jr., Huber, J. T., Bergen, W. G., et al.,** *J. Dairy Sci.*, 67, 2519, 1984.
463. **Lauterio, T. J. and Scanes, C. G.,** *Poultry Sci.*, 67, 120, 1988.
464. **Giustina, A., Bossoni, S., Bodini, C., et al.,** *J. Clin. Endocrinol. Metab.*, 74, 1301, 1992.
465. **Kuhara, T., Ikeda, S., Ohneda, A., et al.,** *Am. J. Physiol.*, 260, E21, 1991.
466. **Davis, S. L.,** *Endocrinology (Baltimore)*, 91, 549, 1972.
467. **Hampshire, J., Altszuler, N., Steele, R., et al.,** *Endocrinology (Baltimore)*, 96, 822, 1975.
468. **Koppeschaar, H., ten Horn, C. D., Thijssen, J. H. H., et al.,** *Clin. Endocrinol.*, 36, 487, 1992.
469. **Page, M. D., Dieguez, C., Valcavi, R., et al.,** *Clin. Endocrinol.*, 28, 551, 1988.
470. **Delitala, G., Palermo, M., Ross, R., et al.,** *Neuroendocrinology*, 45, 243, 1987.
471. **Vance, M. L., Kaiser, D. L., Frohman, L. A., et al.,** *J. Clin. Endocrinol. Metab.*, 64, 1134, 1987.
472. **Woolf, P. D., Lantigua, R., and Lee, L. A.,** *J. Clin. Endocrinol. Metab.*, 49, 326, 1979.
473. **Nieves-Rivera, F., Rogol, A. D., Veldhuis, J. D., et al.,** *J. Clin. Endocrinol. Metab.*, 77, 638, 1993.
474. **Zadik, Z., Chalew, S. A., and Kowarski, A.,** *J. Clin. Endocrinol. Metab.*, 74, 801, 1992.
475. **Veldhuis, J. D., Iranmanesh, A., Ho, K. K. Y., et al.,** *J. Clin. Endocrinol. Metab.*, 72, 51, 1991.
476. **Iranmanesh, A., Lizarralde, G., Johnson, M. L., et al.,** *J. Clin. Endocrinol. Metab.*, 72, 108, 1991.
477. **Bassett, N. S. and Gluchman, P. D.,** *J. Endocrinol.*, 109, 307, 1986.
478. **Breier, B. H., Bass, J. J., Butler, J. H., et al.,** *J. Endocrinol.*, 111, 209, 1986.
479. **Vasilatos, R. and Wangsness, P. J.,** *Endocrinology (Baltimore)*, 108, 300, 1981.
479a. **Breier, B. H., Gluckman, P. D., and Bass, J. J.,** *J. Endocrinol.*, 118, 243, 1988.
480. **Bruhn, T. O., Mcfarlane, M., Deckey, J. E., et al.,** *Endocrinology (Baltimore)*, 131, 2615, 1992.
481. **Morrison, M. W., Davis, S. L., and Spicer, L. J.,** *J. Anim. Sci.*, 53, 160, 1981.
482. **Maiter, D. M., Hooi, S. C., Koenig, J. I., et al.,** *Endocrinology (Baltimore)*, 126, 1216, 1990.
483. **Johnson, R. J.,** *J. Endocrinol.*, 119, 101, 1988.
484. **Vasilatos-Younken, R. and Zarkower, P. G.,** *Growth*, 51, 171, 1987.
485. **Bacon, W. L., Vasilatos-Younken, R., Nestor, K. E., et al.,** *Gen. Comp. Endocrinol.*, 75, 417, 1989.

Chapter 13

Growth Hormone Release: Pathophysiological Dysfunction

W. H. Daughaday and S. Harvey

CONTENTS

0-8493-8697-7/95/$0.00+$.50

I. INTRODUCTION

A pathophysiological deficiency or excess of growth hormone (GH) secretion leads to the clinical condition of hypo- and hypersomatotropism, respectively. Both of these disease states may result from heterogeneous lesions in the hypothalamus-pituitary axis and both can occur before or after growth cessation. These conditions are reviewed in this chapter, which also includes a review of the dysfunctional GH secretion that occurs secondarily in central nervous system (CNS) disorders and peripheral endocrine diseases.

II. HYPOSOMATOTROPISM

A. CLINICAL MANIFESTATIONS

Growth hormone deficiency in infancy and childhood is characterized primarily by impaired skeletal growth[1] (Figure 1). In general, birth weight and birth length are near normal, but newborns with intrauterine GH or GH receptor deficiency have a slightly decreased mean birth length. In severely GH-deficient infants decreased growth velocity is well established by 3 to 6 months and subsequent growth continues at about half the normal velocity.

Body proportions with relatively short extremities remain appropriate for height, but not for age. The cranium is near normal in size, but facial features remain undeveloped, giving an appearance of excessive head size. The bridge of the nose fails to develop normally and tooth eruption is delayed. There is a retardation of bone maturation, as determined by X-rays of the hand, proportionate to the retardation of skeletal growth. This is even more pronounced if GH deficiency is combined with hypothyroidism. Total bone mineral content is reduced.

There are important changes in body composition in hypopituitarism. Body fat is increased diffusely in subcutaneous tissues. This is most prominent in early childhood and rapidly responds to GH treatment.[2] Increased subcutaneous fat and decreased muscle mass is also demonstrable in GH-deficient adults.[3] There is a decrease in subcutaneous water in adults with GH deficiency as inferred from measurements of total body electrical resistance. The degree of subcutaneous hydration is maintained in large part by subcutaneous hydrophilic proteoglycans whose synthesis is regulated by GH secretion. The converse of this situation exists with the accumulation of subcutaneous water in acromegaly responsible for acral puffiness.

Changes in carbohydrate metabolism are present in hyposomatotropism. Glucose tolerance is either normal or slightly impaired. Insulin secretion in response to oral glucose is reduced, perhaps by a direct consequence of GH deficiency on islet β cells. Because of decreased insulin secretion, there are lower liver glycogen stores, which makes young hyposomatotrophic infants prone to hypoglycemia. GH deficiency also increases peripheral sensitivity to insulin, which allows nearly normal blood sugars, despite markedly lower serum insulin concentrations.

Growth hormone deficiency also decreases protein synthesis. This can be rapidly reversed by GH administration. GH acts at many levels of protein metabolism: it promotes amino acid transport into cells, increases the expression of GH-responsive genes in the liver, and regulates the rate of mRNA translation. The net result is increased protein synthesis recognized as positive nitrogen balance, and an increase in skeletal muscle and cardiac muscle mass. Growth hormone also contributes to normal testicular development and severe intrauterine GH deficiency often leads to micropenis.

B. ETIOLOGY

GH deficiency can be the result of impaired hypothalamic synthesis of GH-releasing hormone (GHRH), or of impaired transport of GHRH to the median eminence and into

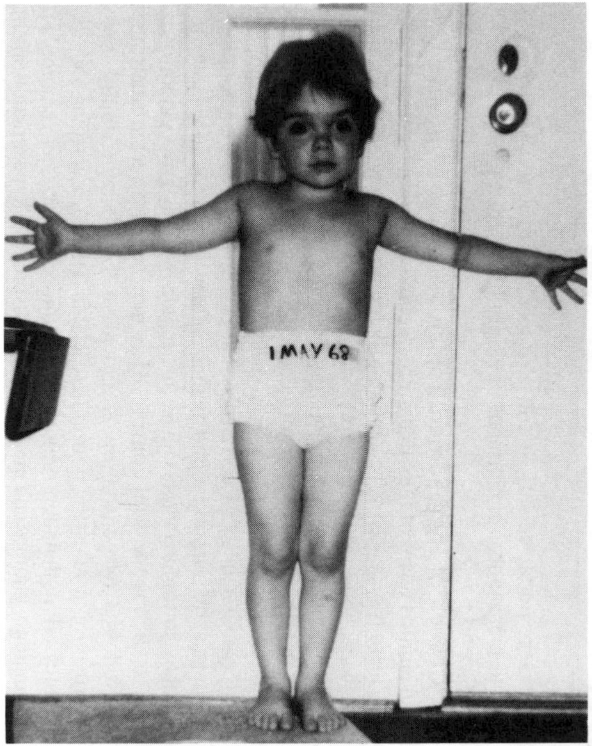

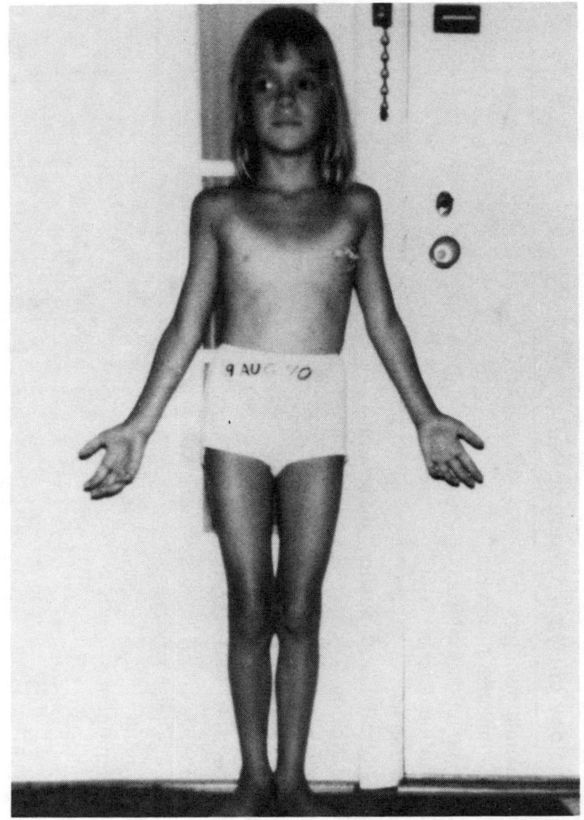

Figure 1 *Left*: A 6-year-old girl with hyposomatotropic dwarfism secondary to a craniopharyngioma and surgical treatment. Height, 94 cm. *Right*: Fifteen months later, after receiving 2.5 mg of hGH twice a week. Height, 122.6 cm. (From Daughaday, W., The anterior pituitary, in *Williams Textbook of Endocrinology* [Eds. Wilson, J. D. and Foster, D. W.], W. B. Saunders, Philadelphia, PA, 1981, p. 568. Reprinted with permission.)

Table 1 **Etiology of Hyposomatotropism**

- I. Hypothalamic disorders
  - 1. Developmental disorders
    - a. Septooptic dysplasia
    - b. Midline defects
  - 2. Birth trauma
    - a. Surgical and traumatic stalk section
    - b. Irradiation
  - 3. Granulomas
    - a. Sarcoid
    - b. Histiocytosis X
  - 4. Infections — tuberculosis, syphilis, other
  - 5. Tumors
    - a. Craniopharyngioma
    - b. Glioma
    - c. Meningioma
    - d. Metastatic
- II. Pituitary disorders
  - 1. Developmental
    - a. Pit-1 gene loss (GH, PRL, and TSH lack)
    - b. Other causes
  - 2. *GH* gene mutations and deletions
  - 3. Vascular
    - a. Postpartum necrosis
    - b. Sickle cell
    - c. Vasculitis
    - d. Diabetes
  - 4. Pituitary tumors

the hypothalamic pituitary portal vessels. It can also result from defects restricted to somatotroph cells or to more global impairment of all pituitary cells (Table 1).

1. Hypothalamic Causes of Hyposomatotropism

Both sporadic and familial forms of hypothalamic structural defects are recognized. In septooptic dysplasia, there is absence of the septum pellucidum, with underdevelopment of the corpus callosum and optic nerve, frequently with visual defects.[1] GH deficiency is usually severe and micropenis and neonatal hypoglycemia may occur.

Defective synthesis of GHRH on the basis of genetic abnormalities could lead to GH deficiency. Abnormalities of the *GHRH* gene have been looked for in a limited number of cases of suspected hypothalamic GH deficiency, but no defects have been found.

The orderly synthesis, axon transport to the median eminence, and transport down the portal vessels of hypophysiotrophic factors can be interrupted by many gross pathological conditions. In these conditions the secretion of GH is more often affected than other pituitary hormones, followed by gonadotropins. Extensive hypothalamic injury is required to produce corticotropin-releasing hormone (CRH) and thyrotropin-releasing hormone (TRH) deficiency. Hypothalamic tumors, in childhood, most often craniopharyngiomas, but less commonly gliomas and meningiomas, can produce hyposomatotropism. Metastatic tumors are a consideration in adults. Granulomatous diseases can involve both the hypothalamus and the pituitary, but the former is affected predominantly. In general, histiocytosis (Hand-Schüller-Christian disease) is the most common granuloma in children, whereas sarcoidosis is the chief consideration in adults.

Cranial irradiation is currently employed in the treatment of childhood leukemias, and often leads to hypothalamic damage resulting in hyposomatotropism. The dose employed is usually inadequate to produce significant direct adenohypophysial damage.[3]

Trauma can damage the hypothalamic GHRH transport pathways. Breech delivery or other birth traumas is a risk factor for the subsequent development of hyposomatotropism. In later life, automobile accidents and other types of blunt trauma can damage the median eminence or transect the hypophysial portal vessels. Although GH secretion is suppressed in these conditions, prolactin secretion is often elevated, reflecting the escape of lactotroph cells from chronic hypothalamic dopaminergic suppression.

A high percentage of children with pituitary dwarfism will respond to the administration of GHRH with a rise in serum GH and accelerated growth velocity. This indicates the importance of hypothalamic dysfunction in these cases, but associated pituitary pathology may also exist.

GH secretion in children may also be suppressed in adverse psychosocial environments.[4] These children have growth retardation and often exhibit bizarre feeding behaviors. Unlike children with uncomplicated protein caloric malnutrition, in psychosocial dwarfism GH levels are depressed rather than elevated. When such children are removed from their home and placed in a more supportive environment, even a metabolic research ward, normal GH secretion is restored and they start to grow.

2. Primary Pituitary Causes of Hyposomatotropism

GH deficiency can result from pathological mechanisms limited to the pituitary. In some cases, these are developmental and result in congenital pituitary aplasia.

The molecular basis of hereditary GH, prolactin, and thyrotropin (TSH) deficiency has been greatly clarified (see Phillips and Cogan[12] for review). The normal development of somatotroph, lactotroph, and thyrotroph cells is dependent on a transcription factor known as pit-1 or GHF-1.[5] Pit-1 binds to two specific regions of the promoters for *GH*, prolactin, and *TSH-β* genes (POU-S and POU-HD). This binding is essential for the normal development of the three secretory cell types as well as the expression of these genes.

Two inbred strains of dwarf mice with combined deficiencies of GH, prolactin, and TSH have been shown to have mutations in the gene for pit-1.[6] The studies of pit-1 have now been extended to patients presenting with combined GH, prolactin, and TSH deficiency. In one family, two affected siblings had the presenting clinical findings of cretinism.[8] Study of the DNA of children demonstrated homozygosity for a mutant *pit-1* gene, with C-to-T transversion at codon 172, which converted it into a stop codon. Both parents were heterozygous for this mutation. In another family with this endocrine syndrome the affected patient had one *pit-1* gene with an Arg-to-Trp substitution at codon 271 inherited from the father.[7] A normal *pit-1* gene was inherited from the mother. The product of this dominant mutant gene bound normally to POU sites, but was inactive in further transcription activity. The *pit-1* gene defect in nine Dutch families involved a substitution of alanine for proline at codon 158.[8] This mutant pit-1 also bound normally to the *GH* promoter, but had impaired transcription-promoting activity. This mutation acted like a simple recessive trait.

GH gene deletion is an important cause of severe hyposomatotropism.[9–11] The recognized cases have growth retardation before 6 months of age, which becomes progressively severe. Height is –4 to –5 SD below the mean. When treated with GH, initial response is excellent, but growth is not sustained in most cases, because of the development of high-titer GH antibodies. A common cause of this severe growth failure is deletions within the *GH* gene, which may be of variable size. The *GH* gene is one of five closely homologous genes of the *GH* family, which are linked in sequence on chromosome 17. Because of their structural similarity, unequal chromosomal cross-over of genetic material is prone to develop, resulting in deletions of the *GH* gene. In studies of 20 cases of severe GH deficiency, with height less than 4 SD below the mean, 7 were found to have *GH* gene deletions.[11] Frameshift mutation, splicing mutations, and nonsense mutations of the *GH* gene have also been identified in patients with isolated GH deficiency diseases.[12] Genetic defects of the GH receptor have also been recognized in patients with the phenotype of severe growth retardation, but with high levels of serum GH and low serum levels of insulin-like growth factor I (IGF-I) (Laron's or GH insensitivity syndrome).[13–18] These patients are markedly insensitive or refractory to administered GH. Both deletions and point mutations of GH receptor gene have been found in this condition.[15,16] A favorable growth response to administered IGF-I has been reported.[17–18]

Pituitary necrosis can lead to hyposomatotropism (Sheehan's syndrome).[19] Formerly the most common cause of this type of hypopituitarism was postpartum hemorrhage or other obstetrical crisis leading to profound hypotension. The proliferation of lactotroph cells during pregnancy within the bony confines of the sella turcica makes them vulnerable to even moderate ischemia. Cell swelling interrupts blood flow to the anterior lobe. Fortunately, as the general level of obstetrical care in the United States has improved over the past 50 years, this complication of pregnancy has become rare. Pituitary necrosis can also be caused by microangiopathy of diabetes, the inflammatory lesions of the arthritides, or the capillary occlusion of sickle cell disease.

Hypopituitarism can also result from primary tumors of the pituitary or from metastatic tumors arising elsewhere. Hypopituitarism often results from surgical or radiation therapy for pituitary tumors. The anterior pituitary can also be damaged by infections and granulomatous diseases, which affect the hypothalamus.

C. DIAGNOSIS OF GH DEFICIENCY

Diagnosis of severe GH deficiency is relatively simple when proportionate dwarfism is combined with low or unmeasurable concentrations of GH and IGF-I.[20] The diagnostic problem is much more difficult in the majority of cases, in which GH deficiency is incomplete (reviewed by Heinze et al.[21]). Pitfalls in the diagnosis of impaired GH secretion may result from the variability of GH responses to repeated tests, the existence of transient GH deficiencies, and the low GH levels found in normal short-statured children with delayed puberty[22,23]. Provocative GH testing may also not correlate with endogenous GH secretion in some children with GH deficiency or GH neurosecretory dysfunction or in children that are short-statured or normal.[24] In pediatric research centers, measurements of serum GH at 20- and 30-min intervals for 12 h at night or for the full 24 h have provided the only reliable method of determining mean serum GH concentrations and calculating GH secretion.[25–27] Even after such extraordinary efforts, the separation between GH deficiency and so-called idiopathic short stature is arbitrary and not a reliable guide to predicting response to GH treatment, especially as spontaneous GH secretion in children of short stature varies with body composition, age, and gender.[28]

Measurement of urinary GH excretion has been proposed as a more practical index of 24-h GH secretion.[27,29] While a general correlation exists between urinary GH and the known GH state of the subject, this measurement has not proved sensitive or specific. Urinary GH is dependent on variable glomerular filtration, tubular reabsorption and degradation, and considerable variation in urinary GH concentrations occurs.

Because of the impracticality of measuring the kinetics of GH secretion, most endocrinologists have resorted to provoking GH secretion by insulin-induced hypoglycemia, or by administration of arginine or clonidine. GH responsiveness to these stimuli is augmented in patients fed a hypocaloric diet, which greatly reduces the number of false-positive diagnoses. Exogenous GHRH is also frequently used for the diagnosis of hyposomatotropism and provides a simple means for differentiating between hypothalamic and pituitary GH deficiency[30–32] and repeated GHRH therapy can reverse secondary somatotroph unresponsiveness.[32] A test involving the combined administration of pyridostigmine and GHRH has also been devised, because the GH response to GHRH is remarkably potentiated by the acetylcholinesterase inhibitor, which is thought to suppress inhibitory hypothalamic tone.[33] In the absence of pyridostigmine the GH response to GHRH does not seem to be a reproducible test for the evaluation of GH release, nor is it useful to differentiate GH-deficient GHRH responders from short, normal children.[34]

On arbitrary grounds, the failure of serum GH to rise above 7 or 10 μg/l in response to provocative testing has been considered abnormal.[35] Because between 5 and 20% of normal children fail to respond to any given test, the failure of GH to respond to two provocative stimuli is usually required for diagnosis.[35] As GH treatment has been extended to a greater population of very short children, we now recognize that this form of testing is only of limited value in predicting therapeutic growth response.

D. GH TREATMENT

The treatment of GH deficiency in children was begun over 20 years ago with GH extracted from pituitaries laboriously collected at autopsy. Because of limited supplies, the use of native GH was restricted to children with the most obvious deficiency and its effectiveness was clearly established. The use of native human GH was finally abandoned in 1985 when it became evident that the prions associated with Creutzfeldt-Jakob disease escaped the purification procedures and contaminated a small number of GH lots.[29] A total of 7 cases of Creutzfeldt-Jakob disease occurred in a population of 6284 recipients of GH prepared by the U.S. National Human Pituitary Program. A similar experience has been reported from other countries. It has been recognized that subjects receiving native GH prepared in France were at greater risk. After a brief hiatus, when pituitary-derived GH was withdrawn, recombinant GH became available for treatment in 1986. At present, GHs with (somatrem) or without (somatropin) an N-terminal methionyl residue have been approved by the FDA and are equally effective.

The initial dose of recombinant human GH (rhGH) provided to GH-deficient children has generally been about 0.3 mg/kg per week in three or more divided doses given subcutaneously. The initial acceleration of growth velocity subsides after 12 to 18 months, but may be sustained with an increase in dosage. Immune resistance and complications of therapy have been extremely rare.

III. HYPERSOMATOTROPISM

A. ETIOLOGY

1. Primary Forms

As shown in Table 2, GH hypersecretion can arise primarily in the pituitary, as the result of pituitary adenoma. Despite the fact that GH secretion by adenomas often appears to respond to certain stimuli acting on the hypothalamus (glucose, arginine, etc.),[36] there is little evidence of preexisting hypothalamic dysfunction in most acromegalics, and total removal of an adenoma usually results in normal GH secretory dynamics.

When examined with suitable molecular markers, most pituitary adenomas are monoclonal.[37,38] In the few exceptions to this finding there may have been intermingling of normal and adenomatous cells. In most cases the induction of tumorigenesis is thought to result from mutations of general oncogenes (e.g., *ras*, *p53*, or *Rb*) or proteins involved in the GHRH cell signaling pathway (see Jameson,[39] Spada and Vallar,[40] Jameson and Hollenberg,[41] and Pei et al.[42] for recent reviews). In about 40% of GH-secreting adenomas a specific mutation of the Gsα protein is present.[43] Gsα is the component of a ternary complex that transduces the effect of the GHRH receptor to increase intracellular cyclic AMP. Cyclic AMP in turn mediates critical phosphorylations and eventual somatotroph

Table 2 **Etiology of Acromegaly and Gigantism**

I. Primary pituitary
  1. Adenomas, often with Gsα mutation
    a. Somatotroph cells
    b. Mixed somatotroph and lactotroph cells
    c. Somatomammotroph cell
    d. Stem cell
  2. Adenomas associated with extrapituitary disease
    a. McCune-Albright syndrome
    b. Multiple endocrine neoplasia, type 1
II. GH secretion by nonpituitary tumors
III. GHRH hypersecretion
  1. Functional (?)
  2. Eutopic — hypothalamic tumors
  3. Ectopic — bronchial adenomas, carcinoids, islet cell tumors, other
  4. Acromegaloidism (normal GH)
    a. Hyperinsulinism
    b. Tumor secretion of IGF-II
    c. Other

cell growth and GH secretion. Two point mutations of the *Gsα* gene have been recognized, which results in the constitutive activation of Gsα and an increase in cyclic AMP even in the absence of GHRH.[44]

Similar mutations in Gsα have been observed in many cells of patients with McCune-Albright syndrome.[45] This condition is characterized by polyostotic fibrous dysplasia, cafe-au-lait pigment nevi, and multiple autonomous endocrine abnormalities. When such mutations occur in the pituitary of children, gigantism occurs.

Tumor growth in some human pituitary tumors has also been thought to be mediated by the GH transcription factor, pit-1.[46] pit-1 antisense oligonucleotides have been shown to reduce GH production as well as the rate of cell proliferation in some tumor cell lines.[47] Moreover many pituitary tumors, including non-GH-secreting adenomas, are pit-1 positive and appear to contain a pit-1 mRNA variant that is not present in normal pituitary glands.[46]

Genetic mutations may also account for some adenomas in multiple endocrine neoplasia (MEN) type 1. This is a hereditary condition characterized by familial predisposition for the development of one or more pancreatic, parathyroid, or pituitary adenomas. Although most of the pituitary adenomas are prolactinomas, some secrete GH. Mutations of a tumor suppressor gene localized in chromosome 11q13 is linked to the condition.[48]

GH-secreting adenomas can arise from precursor cells at any level of differentiation.[49] If the precursor cell is already committed as a somatotroph cell, the adenoma will secrete mainly GH. If the adenoma arises from a cell that can differentiate into separate somatotrophs and lactotrophs, a mixed adenoma containing separate GH- and prolactin-staining cells results. If the terminal differentiation of tumor cells cannot occur, both hormones are within the same cell (somatomammotropin adenomas). An even more primitive precursor cell gives rise to a less differentiated cell, which on occasion contains histochemical evidence of thyrotropin and gonadotropin synthesis (stem cell adenoma). These less differentiated cells exhibit more rapid growth, local invasiveness and, on rare occasions, distant metastases. The well-differentiated pituitary adenomas often have a pattern of slow growth over decades. Pituitary adenomas rarely can arise from embryonal nests in the sphenoid sinus and pharynx.

A number of neuroendocrine tumors have also been shown to contain low levels of GH by histochemical methods, yet the secretion of GH in sufficient quantities to produce hypersomatotropism is exceedingly rare. One convincing case of acromegaly resulting from GH secretion by a neuroendocrine pancreatic tumor has been reported.[50]

Although GH-secreting adenomas appear to result predominantly from genetic mutations (Figure 2), a primary hypothalamic disturbance causing increased release of GHRH or GHRH action has also been hypothesized to lead to the adenomatous transformation of chronically stimulated somatotrophs,[51] as recently reviewed by Frohman.[52] This possibility is supported by the rare finding of normal pituitary histology in some patients with acromegaly.[51] There are also several reports of neonatal or early childhood onset gigantism associated with somatomammotroph hyperplasia, with or without adenomatous transformation.[53–55] The occurrence of "paradoxical" GH responses to centrally acting stimuli (glucose and dopaminergic agonists) comparable to those in acromegalic patients with true somatotroph hyperplasia secondary to the ectopic production of GHRH[56] is also indicative of hypothalamic dysfunction. The occasional coexistence of acromegaly, pituitary adenoma, and hypothalamic structural abnormalities (e.g., chordomas or ganglioneurocytomas) may also indicate a hypothalamic disorder leading to adenoma formation,[57] although these tumors may ectopically express the *GHRH* gene. Moreover, whereas basal GH levels return to normal after the surgical removal of GH-secreting adenomas, postoperative GH responses are often abnormal, suggesting an intrinsic hypo-

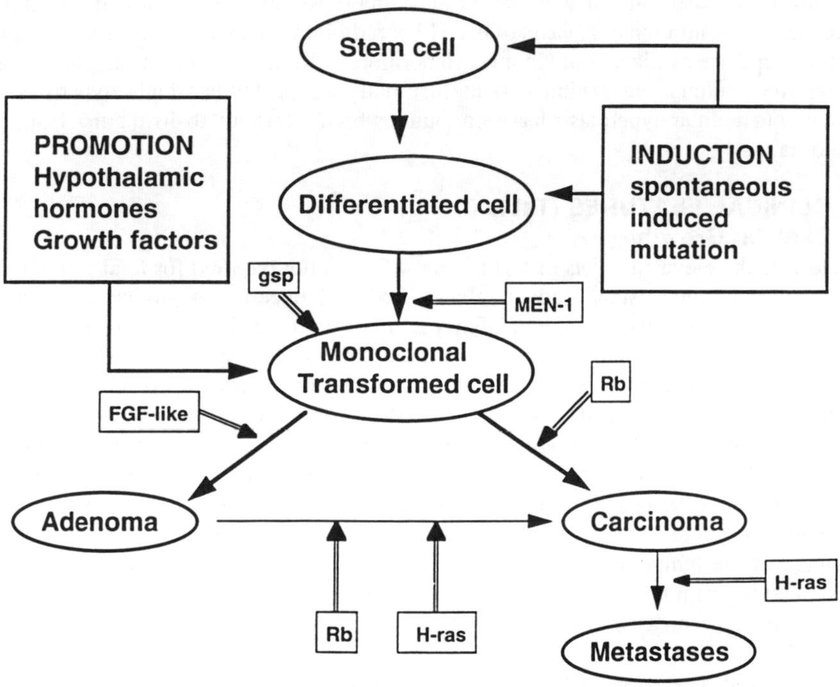

Figure 2 Multistep pituitary tumorigenesis model. Oncogene mutations depicted are based on limited information from the literature. (From Pei, L., Prager, D., Scheithauer, B., and Melmed, S., *Molecular and Clinical Advances in Pituitary Disorders* [Ed. Melmed, S.]. Endocrine Research and Education, Inc., Los Angeles, CA, 1993, p. 21. Reprinted with permission.)

thalamic disorder not corrected by tumor removal.[51] The accelerated GH pulse that is characteristic of acromegaly[58] may also remain after tumor removal.[59]

Thus, whereas pituitary neoplasia in acromegaly is likely to result primarily from spontaneous somatotroph mutations, dysfunctional hypothalamic control of somatotroph function may act as a promoter in a small number of patients.[42,52]

2. Secondary Forms

Hypersecretion of GH in this condition results from excess secretion of GHRH (see Faglia et al.[60] and Losa et al.[61] for reviews). Although theoretically this could result from a functional defect in GHRH secretion, in only one case does this seem to have led to gigantism.[62]

GHRH hypersecretion can arise from tumors of the hypothalamus and sella (eutopic). There are now seven well-documented cases of acromegaly associated with gangliocytomas or related tumors of the hypothalamus.[63] The presence of GHRH in these tumors has been established by immunocytochemistry and by immunoassay of tumor extracts.

There is immunocytological evidence of the presence of GHRH in many APUD neuroendocrine tumors, but most of these are clinically silent. More than 50 cases have been described in which ectopic sources of GHRH have resulted in acromegaly.[63,64] Most of these tumors were pulmonary carcinoids and bronchial adenomas. In addition, pancreatic islet cell adenomas, intestinal carcinoids, pheochromocytomas, and paragangliomas have been found. The diagnosis can be suspected if no pituitary tumor is found on imaging or transsphenoidal exploration. The diagnosis can be confirmed by finding increased concentrations of plasma GHRH by radioimmunoassay. Such assays are demanding and are available only in a few laboratories. In the presence of ectopic GHRH excess the pituitary can exhibit various histopathologies. Often simple hyperplasia is present, but nodular hyperplasia has been found, which is difficult to distinguish from true adenoma.

B. CLINICAL FEATURES (Table 3)

1. Skeletal Growth

Despite the elevated levels of GH in fetal life it is not required for fetal growth. GH dependence is soon established in infancy and GH excess can produce accelerated skeletal growth, which can begin in infancy and last until the closing of epiphyses late in puberty. This results in the clinical condition known as gigantism. GH excess left untreated has resulted in the remarkable giants recorded through history. The tallest medically documented giant was Robert Wadlow (the Alton giant), whose height just before his premature death at age 21 years was 272 cm (8 ft 11.09 in.) (Figure 3).[65]

When GH excess occurs after fusion of the epiphyses of long bones, acromegaly results and increase in height no longer occurs (Figure 4). Growth in bone length is restricted to several bones. Persistent mandibular growth results from proliferation of cartilage at the temporal mandibular articulation and increase in thoracic size (barrel chest) results from persisting growth at the costal cartilage junction. Thickening of the calvarium with increased prominence of the frontal ridge is often prominent and there is

Table 3 **Manifestations of Hypersomatotropism**[a]

Manifestation	Early Recognition Signs
I. Skin and connective tissue growth	
1. Soft tissue swelling	
a. Facial	+++
b. Acral	+++

2. Excess sweating	+
3. Hirsutism	
4. Laryngeal thickening (deep voice)	++
II. Skeletal and articular changes	
1. Gigantism (with childhood onset)	++
2. Costal growth (barrel chest)	+
3. Mandibular growth (prognathism)	++
4. Calvarial growth	
a. Frontal bossing	+
b. Frontal sinus overgrowth	+
5. Vertebral bony overgrowth	
6. Phalangeal bony overgrowth	
7. Articular cartilage growth	
a. Widening joint space	
b. Accelerated osteoarthritis	
III. Visceral growth	
1. Cardiac (ventricular and septal thickening)	
2. Hepatic	
3. Renal	
4. Pulmonary	
5. Thyroidal	
IV. Changes in metabolism	
1. Insulin resistance	
a. Hyperinsulinemia	
b. Impaired glucose tolerance	+
c. Diabetes mellitus	+
2. Increase in 1,25-dihydroxycholecalciferol	
3. Hypercalciuria (renal lithiasis)	
V. Associated endocrine abnormalities	
1. Hyperprolactinemia (galactorrhea)	+
2. FSH, LH	
a. Women: menstrual irregularity, amenorrhea	+
b. Men: impotence	+
3. TSH	
a. Hypersecretion (very rare)	
b. Hyposecretion (uncommon)	
4. ACTH	
a. Hypersecretion (very rare)	
b. Hyposecretion (uncommon)	
5. Vasopressin: Hyposecretion (uncommon)	
VI. Neuromuscular manifestation	
1. Headache	+++
2. Visual field loss	++
3. Sleep apnea	++
4. Peripheral nerves	
a. Entrapments (carpal tunnel)	++
b. Hypertrophic neuropathy	
5. Myopathy	

[a] These features should suggest to the physician the possibility of acromegaly.

+++, common symptoms; +, less common symptoms.

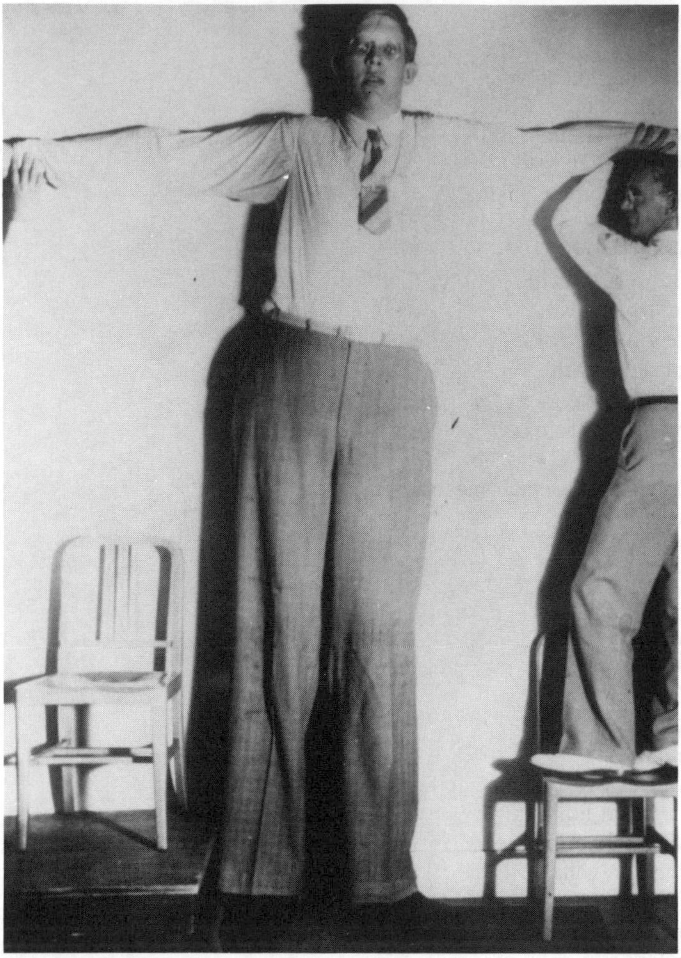

Figure 3 Extreme gigantism: Robert Wadlow, the Alton giant, being measured 3 weeks before his death at age 21 years. Height 272 cm; weight, 216 kg. (From Daughaday, W., The anterior pituitary, in *Williams Textbook of Endocrinology* [Eds. Wilson, J. D. and Foster, D. W.], W. B. Saunders, Philadelphia, PA, p. 27. Reprinted with permission.)

some increase in long bone width. Proliferation of articular cartilage can lead to alteration of the joint surface, with accelerated cartilage degeneration and periarticular bony growth, particularly in the phalanges with further impairment of joint motion.[66]

2. Soft Tissue Growth

The earliest manifestation of GH excess is an increase in synthesis of dermal hydrophilic proteoglycans. This increases the accumulation of dermal water, which is recognized most readily as puffiness of facial features and extremities (acromegaly). Because dermal proteoglycans have a half-life of about 3 weeks, regression of puffiness can occur rapidly after successful surgical correction of hypersomatotropism. If GH excess persists for a prolonged period, increased deposition of dermal collagen occurs and regresses after successful treatment is complete. The increase in extracellular matrix is not restricted to dermal tissues, but occurs also in internal organs.

Increase in the size of the viscera also occurs. The weights of heart, liver, pancreas, spleen, lungs, and kidneys are all increased. In the kidney the glomeruli are larger than normal and glomerular filtration and renal blood flow are increased.

Early in acromegaly there is hypertrophy of the cardiac muscle. Because this occurs without dilatation, it is usually unrecognized by physical examination, but is evident on

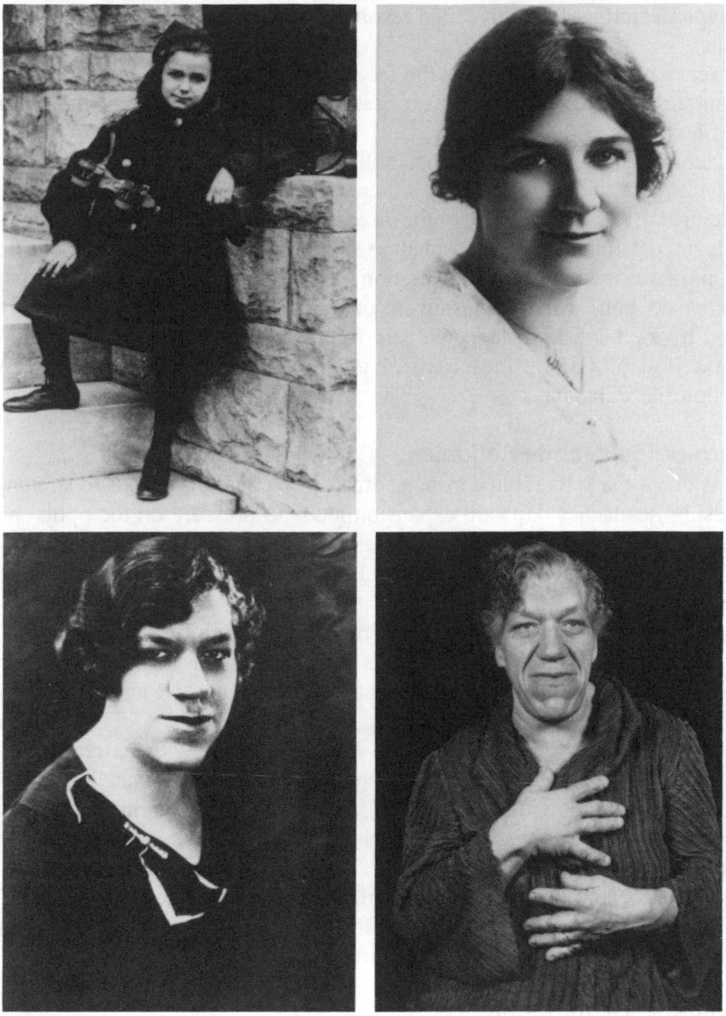

Figure 4 Progression of acromegaly. *Upper left*: At age 9 years. *Upper right*: At age 16 years with early facial coarsening. *Lower left*: At age 33 years, with established acromegaly and prognathism. *Lower right*: End-stage acromegaly with gross disfigurement at age 52 years, soon before her death from congestive heart failure. (From Daughaday, W. H., *Am. J. Med. Sci.*, 20, 133, 1956. Reprinted with permission.)

sonography as an increase in ventricular wall thickness and septal mass. These changes adversely affect cardiac function and if uncorrected contribute to the risk of cardiovascular death in acromegaly.

3. Metabolic Changes

Hypersomatotropism leads to a number of changes in metabolism (see Chapters 20 and 21). Resistance to the action of insulin in muscle, adipose tissue, and liver at first leads to a compensatory increase in the secretion of insulin required for maintenance of euglycemia. When there is any genetic or acquired limitation of the islet β cells to sustain insulin hypersecretion, hyperglycemia results and frank diabetes develops in 10 to 20% of patients.

GH excess promotes lipolysis in the fat depots and increases plasma free fatty acids (see Chapter 21). After hepatic uptake, fatty acids are returned to the circulation as increased levels of triglycerides.

GH excess also alters mineral metabolism. There is increased formation of 1,25-dihydroxycholecalciferol with resultant increase in calcium absorption from the gut and hypercalciuria. Early in the course of the disease there may be an increase in bone mineral and an elevated serum alkaline phosphatase suggestive of increased osteoblastic activity. There is an increase in the urinary excretion of hydroxyproline, which is largely derived from increased bone collagen turnover. Later in the disease, particularly if there is associated hypogonadism, osteopenia can occur. There is hyperphosphatemia resulting from the action of GH on the proximal tubule of the kidney to increase phosphate reabsorption (see Chapter 25).

4. Neurological Manifestations

Expansion of the sella turcica from a somatotroph adenoma leads to stretching of the dura, causing headaches. Upward expansion leads to pressure on the optic chiasm with bitemporal hemianopsia or other field losses. Further growth of the tumor may occlude the third ventricle, causing hydrocephalus. Lateral extension of an adenoma can lead to cranial nerve deficits.

Peripheral nerve disorders are common in acromegaly. Deposition of connective tissue matrix under the carpal ligament leads to entrapment of the median nerve with pain, paresthesia, and numbness of the thumb and first four fingers.[67] If left untreated, weakness of the intrinsic muscles of the hand occurs.

Patients with untreated gigantism, and less commonly with untreated acromegaly, develop a hypertrophic neuropathy with axonal dropout that can lead to sensory defects of a stocking-and-glove distribution.[65,68] In severe cases, neuropathic arthropathy (Charcot's) has been observed. Late muscle weakness can result from neuropathy and also from acromegalic myopathy.

Sleep apnea occurs in up to 25% of patients with established acromegaly.[69] In most cases this is obstructive, attributable to redundancy of pharyngeal tissues, soft palate and tongue. Occasionally, the apnea is of central origin.

5. Endocrine Dysfunction

Galactorrhea is observed by some women with hypersomatotropsim (see Chapter 23). This is often associated with hyperprolactinemia, which occurs in 20 to 40% of acromegalic patients. Galactorrhea is not restricted to women with hyperprolactinemia because human GH, but not nonprimate GH, has high intrinsic lactogenic activity (see Chapter 1).

Mixed cell adenomas often contain a minority of cells with immunostaining for gonadotropins, TSH, and adrenocorticotropic hormone (ACTH). It is rare for these hormones to be secreted in clinically significant amounts. The secretion of gonadotropins is frequently impaired in acromegaly. TSH and ACTH secretion is usually normal except when aggressive macroadenomas destroy normal pituitary tissue. Neurohypophysial function is almost always intact prior to ablative therapy.

C. DIAGNOSIS

In most cases the laboratory diagnosis of hypersomatotropism is straightforward. It is best to measure plasma GH 60 and 90 min after ingestion of 75 g of glucose. Glucose administration is effective in suppressing random spontaneous peaks of GH secretion in normal individuals. Serum GH concentrations below 2.5 µg/l, as measured by sensitive assays, provides strong evidence of normal GH secretion.[70] Most patients with acromegaly will have basal serum GH concentrations in excess of 5 μg/l that are nonsuppressible by glucose or that are paradoxically enhanced. In rare cases serum GH levels as high as 1000 μg/l are present in active disease and there is a loose positive correlation between tumor size and circulating GH levels. Younger patients also tend to have both larger tumors and higher concentrations of GH than are observed in older age groups.[71]

An alternative approach to diagnosis when random GH levels are borderline is to obtain serum samples for GH measurement at 20- to 30-min intervals for 12 to 24 h. Mean GH concentrations over 5 μg/l and basal levels over 2 μg/l are highly suggestive of hypersomatotropism.

Pulsatile GH secretion in acromegaly is characterized by augmented basal GH concentrations, increased pulse frequency, and an attenuation of the underlying 24-h rhythm.[58,72,73] Such patterns may be secondary to the intrinsic pathology of adenomatous somatotrophs and/or the effects of altered hypothalamic regulation.[74]

A dysfunction in GH regulation in acromegaly is well established and recognized by anomalous GH responses to clinical tests of GH function. These paradoxical responses reflect impaired neuroendocrine regulation and/or the autonomous functioning of adenomatous somatotrophs that possess receptors that are defective[75] or not normally expressed.[51] Adenomatous pituitary cells are also likely to lack the stimulatory and inhibitory paracrine relationships that normal somatotrophs have with non-GH-secreting cells and may have different electrophysiological properties.[76] The high circulating GH concentrations in these states may also contribute to paradoxical GH secretion by autoregulatory mechanisms that block somatotroph responses to provocative stimuli.

In addition to glucose, the secretion of GH in acromegalics may be abnormally responsive to TRH, with stimulation occurring in 90% of acromegalic patients.[77] Indeed, subpopulations of acromegalics that respond to TRH but not to GHRH have been identified.[78] The poor GH responsiveness of these acromegalics to GHRH may indicate the presence of an ectopic GHRH-producing tumor.[79] The blunted GH response to GHRH may reflect a desensitization of pituitary GHRH receptors,[79] although GHRH exposure sensitizes somatotrophs to TRH stimulation.[79,80] The ability of TRH to provoke GH release in acromegaly has also been suggested to be due to prolactin-sensitization and the presence of prolactin secreting cells in many tumors.[81,82] Hyperprolactinemia is present in >40% of all acromegalics[82] and the GH response to TRH can be predicted by the basal prolactin level.[83] Although paradoxical GH responses to TRH and to dopamine stimulation have been correlated[84,85] in some patients, they do not effectively predict the efficacy of bromocriptine therapy.[86]

A defect in monominergic transmission is also thought to occur in acromegaly, because clonidine lowers GH secretion in some patients.[87] Moreover, contrary to their stimulatory effects in healthy subjects, dopaminergic drugs are unable to stimulate GH release in acromegalics and paradoxically lower GH concentrations in more than 50% of patients.[88] The dopaminergic inhibition of GH secretion in acromegalics is principally in patients with mixed GH- and prolactin-secreting tumors, in which the somatotrophs are probably sensitized to dopamine inhibition by increased prolactin secretion.[81] The relatively poor GH responsiveness of acromegelics to exogenous GHRH may also be related to impaired dopaminergic regulation, because the GH response is further suppressed by exogenous dopamine[89] and is promoted in patients insensitive to bromocriptine therapy.[90] A defect in monominergic transmission may also account for the failure of centrally acting stimuli, such as glucose, to inhibit GH secretion in acromegalics. The ability of SRIF to lower (but not abolish) circulating GH concentrations in acromegalics has also been cited as evidence for a defect in the inhibitory hypothalamic control of GH release.[82]

In addition to paradoxical GH responses to clinical tests of GH secretory function, the measurement of serum IGF-I is particularly useful in the diagnosis of acromegaly (see Clemmons and Underwood[91] for review) because the IGF-I concentration is relatively

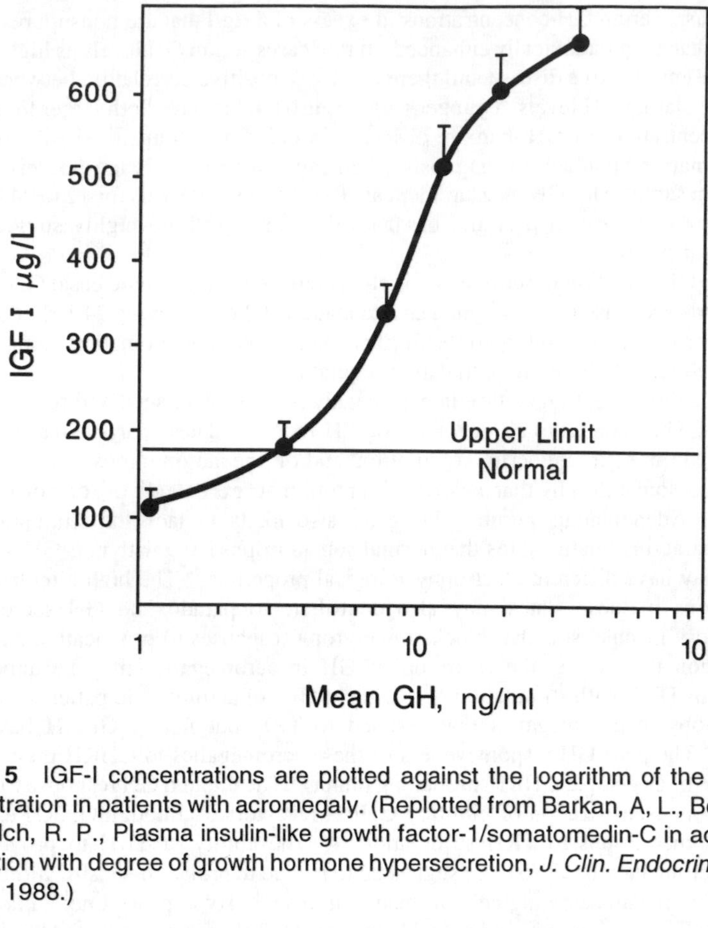

Figure 5 IGF-I concentrations are plotted against the logarithm of the mean GH concentration in patients with acromegaly. (Replotted from Barkan, A, L., Beitins, I. Z., and Kelch, R. P., Plasma insulin-like growth factor-1/somatomedin-C in acromegaly: correlation with degree of growth hormone hypersecretion, *J. Clin. Endocrinol. Metab.*, 67, 69, 1988.)

constant throughout the day. Even slight increments of mean serum GH can result in elevation of serum IGF-I.[92] As shown in Figure 5, the concentration of serum IGF-I is a sigmoidal function of the logarithm of the GH concentration. With the assays used in these studies, mean serum GH concentration in excess of 4 to 5 μg/l led to an elevation in serum IGF-I. This relationship between mean serum GH and IGF-I concentration no longer holds if resistance from nutritional deficiency or from hepatic or adrenal disease is present. Figure 5 also indicates that with mean serum GH concentrations above 40 to 60 μg/l, the rise in serum IGF-I flattens out so that a predictable correlation between serum IGF-I and serum GH is lost.

The laboratory diagnosis of GH excess in childhood presents more problems. The mean serum GH in prepubertal children is usually 4 to 5 μg/l and this may double in late puberty. Moreover, in late puberty serum IGF-I rises two- to threefold. These changes make it difficult to recognize mild GH excess in pubertal children. If the physiological concentrations of GH and IGF-I present in puberty persisted into adult life we would all be suffering from acromegaly!

In addition to hormonal tests, the diagnosis of hypersomatotropism requires proper imaging of the pituitary. In over 90% of patients magnetic resonance imaging (MRI) with contrast will reveal a pituitary tumor. When this is not present, consideration of an ectopic source of GH or GHRH is warranted.

D. PROGNOSIS

Untreated acromegaly leads to gross disfigurement that has psychological and sociological consequences. Parasellar extension can cause partial or total visual loss. Diabetes is a common complication, but its acute manifestations are controlled by insulin. There have been several studies of the life expectancy of patients with acromegaly.[93–96] Despite the fact that these series included both treated and untreated patients, the age-adjusted mortality was two to three times that of the control population. The increased mortality in these series was mostly attributable to cardiovascular and cerebrovascular causes. Often this is due to hypertension and coronary atherosclerosis, but it is likely that some cases of heart failure are attributable to cardiomyopathy. In the series reported by Wright et al.[93] from several London hospitals, it would appear that survival was improved by treatment, but this is hard to interpret because various forms of treatment were employed and effectiveness of the treatment often was not evaluated by current, more rigorous, standards. There is every reason to expect modern treatment to improve survival.

There is evidence many patients with acromegaly may be at increased risk of developing malignancies.[93] In one survey the increased risk was estimated to be significant in men, but not women,[94] but in another survey the increased risk applied to both sexes.[96] The types of tumors reported in these surveys were varied and there was no particular predominance of one type. There has been a number of retrospective studies and case reports suggesting that patients with acromegaly have increased risk for the development of colon polyps and colon carcinoma.[97] Some have found that as many as 53% of acromegalic patients have polyps[98] and 7% have colon carcinoma.[99] The risk appears to be higher in patients with a family history of colon cancer. These studies have generally come from gastrointestinal clinics and bias in case selection may exist.

E. TREATMENT OF HYPERSOMATOTROPISM

1. Surgery

There are a number of treatment options for GH-secreting tumors. Because of relative safety and minimal postoperative morbidity of transsphenoidal adenomectomy, this is the initial treatment for most patients.[100,101] The procedure is highly successful when applied to patients with microadenoma. With experienced surgeons, cure is possible in up to 80% of such patients and tumor recurrence is infrequent. Macroadenomas (>1 cm in diameter) present more of a therapeutic challenge because surgical cure rates are between 40 and 60%.

2. Radiation Treatment

Radiotherapy with carefully focused high energy γ rays or protons or neutron beams is effective, but the full response to treatment may be delayed for 3 to 10 years.[101] When the tumor is confined to the sella, damage to the hypothalamus and cranial nerves is infrequent. When the tumor extends beyond the sella, neural and vascular tissues are more vulnerable. At the present time radiotherapy is primarily used for patients who have persistent hypersomatotropism after surgery or for patients who are not considered good surgical risks or who refuse surgery.

3. Medical Treatment

Medical treatment provides an alternative for radiation therapy. The first effective medical agent was the dopaminergic drug bromocriptine.[102] Bromocriptine is remarkably effective in lowering serum prolactin to normal in most patients with prolactinoma, often with striking reduction in tumor size. The results of treatment of patients with acromegaly are much less striking. With doses up to 40 and 60 mg/day, serum GH is lowered to less than 5 μg/l in 20% of patients and another 50% have symptomatic improvement with lesser degrees of GH response. Few patients have significant shrinkage of their pituitary tumors. It would seem that patients with mild hypersomatotropism are best selected for bromocriptine treatment. Side effects are common, however, with the institution of bromocriptine treatment. At lower dosage levels, nausea, vomiting, and postural hypertension and headache are frequent at the start, but in most cases these symptoms remit with time. At higher doses, seizures, hallucinations, and other central nervous system symptoms can arise.

Pergolide is a second dopaminergic drug useful in treating acromegaly. It is available to U.S. physicians, and has been approved by the FDA for this indication. It has the advantage of a somewhat longer serum half-life, providing a more sustained blood level and somewhat fewer side effects.

Octreotide is the most effective drug in the treatment of acromegaly.[103–106] It is a synthetic analog of somatostatin (SRIF), with only 8 rather than the 14 amino acids of somatostatin. Its biological half-life has been prolonged by the introduction of two D-amino acids. When injected subcutaneously the serum half-life is 2.5 h and GH suppression is sustained for six or more hours. It also antagonizes anomalous GH responses to TRH and glucose.[107] For most patients, subcutaneous injections of 100 μg at 8-h intervals are required. Some small additional benefit can be achieved by giving the agent by constant subcutaneous injection using a pump device, or combining octreotide and bromocriptine therapy. More than 90% of patients will respond to octreotide with a substantial lowering of serum GH. Approximately 75% of treated patients with microadenoma and 50% of patients with macroadenomas will have serum GH and IGF-I concentrations considered to be in the normal range. In one clinical trial, 30% of the patients had significant shrinkage of the size of the tumors.[103] In other studies, a reduction of up to 33% in tumor size has been reported in more than half the patients.[108] Some adenomas in 20–30% of acromegalic patients are, however, poorly responsive to octreotide because of a reduced SRIF-binding capacity and reduced inhibition of adenylate cyclase activity.

In general, octreotide is well tolerated. Despite the fact that octreotide inhibits insulin secretion, abnormalities of glucose metabolism are seldom seen. If the patient is a diabetic on insulin treatment, reduction of the insulin dose is generally required. This is attributable to lowering of serum GH and glucagon. Gastrointestinal complaints, however, are common, particularly early in the course of therapy.[109] Bloating, cramps, and nausea are often experienced. Gastritis associated with *Heliobacter pylori* has been found. Stools become bulky and steatorrhea occurs. The most significant gastrointestinal side effect of octreotide therapy is the formation of new gall stones in as many as 10 to 30% of patients treated for 2 years. Octreotide both decreases biliary flow and inhibits gall bladder contraction. Although these gall stones resulting from octreotide treatment are generally asymptomatic, the concomitant use of anticholilithogenic therapy may be required.

IV. SECONDARY SOMATOTROPH DYSFUNCTION

Alterations in the synthesis and release of GH are characteristic of many disease states, in which the GH dysfunction is a consequence rather than a cause of the pathophysiology. These abberations of GH secretion do not, therefore, persist after illness recovery. The causal factors implicated in GH dysfunction differ in each pathophysiological condition, in which the severity of dysfunctional GH regulation is highly variable among individuals. Heterogeneous patterns of GH secretion may also occur in most diseases, because they do not encompass a homogeneous patient population. A heterogeneity of lesions is implicated in the etiology of most diseases and this is likely to account for the differential GH responses in "responder" — and "nonresponder" — patients to diagnostic tests of GH secretory function. In this section, GH secretion in selected disease states is considered.

A. DIAGNOSTIC TESTS OF GH SECRETION

Clinical tests of GH function facilitate the diagnosis of GH excess or deficiency syndromes and differentiate primary or secondary lesions. Abnormal GH responses to physiological stimuli that promote (sleep, exercise, fasting, and high-protein meals) or inhibit (hyperglycemia and hyperlipidemia) GH secretion may reflect pituitary and/or hypothalamic dysfunctions. Insulin-induced hypoglycemia, glucagon and clonidine (acting through α_2-adrenergic receptors), propranolol (via β-adrenergic antagonism), pyridostigmine (via cholinergic stimulation), galanin (via catecholamine pathways or SRIF suppression), L-dopa (by dopaminergic stimulation), and arginine or ornithine (by SRIF suppression) are commonly used as tests to evaluate a derangement of stimulatory hypothalamic regulation, whereas phentolamine (via β-adrenergic stimulation), chlorpromazine (via dopaminergic antagonism), cyproheptadine (via serotonergic blockade), and atropine or pirenzepine (via cholinergic antagonism) are routinely used in tests of impaired inhibitory hypothalamic control.[110–112] Although these tests primarily evaluate hypothalamic function, defective responses could, however, result from pituitary lesions. High-potency analogs of GHRH test GH secretory reserve and pituitary responsiveness, although the GHRH test lacks sensitivity and discriminating ability, because poor GH responses can be secondary to reduced hypothalamic stimulation or increased hypothalamic tone[113,114] and absent GH responses may occur in patients with ectopic GHRH-secreting tumors.[111] Similarly, while paradoxical GH responses to TRH are not only indicative of excessive GH secretion in acromegaly and GH secretagogues, and inhibiting stimuli occur in primary hypersomatotropism, they are secondary abnormalities of a variety of metabolic and psychological pathologies.[114]

B. OBESITY

Basal serum GH concentrations are low in obese men, women, and children, in which circadian, pulsatile GH release is blunted[115–117] and the number of GH secretory bursts is reduced.[117–120] Obesity is also associated with reduced GH responses to GHRH, clonidine, propranolol, L-dopa, galanin, glucagon, fasting, protein meals, methoxamine, nicotinic acid, insulin-induced hypoglycemia, exercise, and sleep[118,121–126] (see Glass et al.[127] and Vanderschueren-Lodeweyckx [128] for reviews). This reduced responsiveness is linked to a reduction in the daily GH production rate[115] and an increased rate of GH clearance, as the GH half-life is short.[118,129] This increased rate of clearance is a correlate of body mass and is not due to a deficiency in plasma GH-binding protein, which may, conversely, be elevated in obese patients.[129,130]

Although GH deficiency may lead to obesity,[118] impaired GH secretion in the obese may result from overfeeding[131] and is inversely related to the degree of obesity[127] and can be reversed in some patients by weight loss and fasting.[132–135] A primary hypothalamic dysfunction regulating GH release is therefore likely to be responsible for persistent GH dysfunction in thinned-obese patients.[134,136–138] This dysfunction is therefore likely to result from GHRH deficiency and/or SRIF excess.

An impairment of endogenous GHRH release and/or action has been linked to the hyposomatotropism of obesity[125,139] and the priming of pituitary somatotrophs with low doses of GHRH has been found to restore GH responsiveness in obese women.[139] Impaired GH secretion in genetically obese Zucker rats similarly reflects insufficient stimulation by hypothalamic GHRH.[140,141] A GHRH deficiency is, however, unlikely to account for impaired GH secretion in other obese patients,[142,143] in which an increase in SRIF tone is likely to be causal. This possibility is supported by the partial restoration of GH responsiveness to GHRH in obese patients treated with pyridostigmine[144,145] or galanin[123] to inhibit endogenous SRIF release. However, as galanin is unable to restore normal GH release completely, the suppression of GH secretion in these obese patients is likely to result from additional etiologies.

The ability of pyridostigmine and galanin to increase GH secretion in obesity suggests the GHRH/SRIF imbalance is related to dysfunctional cholinergic control.[123,146] Hypothalamic dysfunction in obesity has also been correlated with opioid dysfunction[147–149] and with a serotoninergic deficiency.[124,150] Impaired serotoninergic tone is likely to result in reduced GH secretion in humans[151,152] and brain serotonin turnover is deficient in obese animals.[153,154] A serotoninergic deficiency in obese patients is indicated by the normalization of the GH responses to arginine[150] and to insulin-induced hypoglycemia[124] after serotoninergic stimulation through specific 5-HT_2 receptors. A catecholamine pathway may be similarly linked to the central dysfunction regulating GH release, since GH responsiveness to secretagogues is increased in obese patients following propranolol treatment.[155]

The GHRH/SRIF imbalance in obesity may also result from increased feedback from elevated levels of plasma insulin and IGF-I.[156–158] Hyperinsulinemia in the obese may reduce GH synthesis and secretion at pituitary sites[159] but the restoration of normal insulin levels in obese patients is found to have no effect on GH secretion.[160] Possible roles for free fatty acids and/or glucose in mediating the impaired GH secretion in obese individuals has also been proposed, but not proved.[127,161–165] Moreover, whereas triglycerides and free fatty acids might be expected to suppress GH secretion, the magnitude of the GH response to GHRH stimulation in obese patients is unrelated[166] or directly related[125] to the plasma levels of these metabolites, indicating a lack of metabolic feedback in this disease.[125] Similarly, paradoxical GH responses to feeding occur in the obese,[165] indicating an altered sensitivity of the hypothalamus- pituitary axis to metabolic signals.

A primary somatotroph defect has also been proposed by Ghigo et al.[167] to account for the impaired GH secretion in obesity, because they found that GHRH pretreatment failed to improve somatotroph hyporesponsiveness to GHRH. The decreased GH secretion in the obese is not, however, due to a primary deficiency in pituitary GH production because impaired GH responses to secretagogues can be augmented by pharmacological doses of triiodothyronine (T_3)[168] and by cholinergic, catecholaminergic, opioidergic, and serotoninergic agents.

C. INSULIN-DEPENDENT DIABETES MELLITUS

Abnormalities in GH secretion commonly occur in patients (particularly juveniles) with type I or insulin-dependent diabetes mellitus (IDDM) (see Holly et al.,[169] Shaper[170] and Sonksen et al.[191] for review). Basal serum GH levels are elevated throughout the day[171–173] despite a slightly elevated rate of metabolic clearance in some patients,[174,175] in which the volume of GH distribution is increased. This increase results from GH secretory episodes of increased amplitude and frequency[172,175–177] leading to an increased daily GH production rate.[175,177] Inappropriate or exaggerated GH responses to sleep,[178] exercise,[179,180] dopamine,[181] arginine,[182] glucagon,[183] clonidine,[183,184] GHRH,[185–187] TRH,[173,188,189] and LHRH[190] occur in diabetic patients. GH secretion is also paradoxically increased in diabetics in response to insulin, independently of hypoglycemia.[191,192] The magnitude of these exaggerated responses appears to be correlated with the severity of diabetes,[179,192–196] although this is not a consistent finding.[187] These GH responses also occur despite hyperglycemia[174,192,197,198] and are not modified by short-term euglycemia[177,199] or metabolic control.[200–202] Elevated diurnal GH levels also persist in well-controlled diabetes.[203] This, therefore, contrasts with the impaired GH secretion in patients with non-IDDM (NIDDM)[204] but is similar to the increased GH secretion in cystic fibrosis patients with pancreatogenic diabetes mellitus (CFDM).[205]

In most patients with IDDM the metabolic clearance rate of GH is normal or reduced[206–208] whereas the GH half-life is increased.[174] The prolonged GH half-life in IDDM patients is, therefore, likely to be partly responsible for their increased plasma GH levels.

The increased frequency of pulsatile GH release in diabetics is, however, likely to be a consequence of derangements in hypothalamic regulation.[172,177]

Evidence for a reduction in SRIF tone in diabetics is based on the elevated interpulse, trough levels of plasma GH,[172] because SRIF tone results in nadir GH concentrations. A deficiency of SRIF release in diabetics is indicated by the impaired secretion of SRIF from the hypothalami of diabetic rats[209] and by the ability of an anticholinergic drug, pirenzepine, to suppress the amplitude (but not frequency) of pulsatile GH release in diabetic patients,[177,210,211] because SRIF release is enhanced by muscarinic receptor blockade.[212] An impairment of SRIF release may result from prolonged effects of glucose on hypothalamic neurons[213] or from reduced sensitivity to inhibitory feedback stimuli.[33,172,200]

The reduced somatotroph responsiveness to GH-releasing factors may also reflect an altered set point for the inhibitory effects of glucose, insulin and free fatty acids on GHRH-stimulated secretion[33,198,213] and an impairment of somatotroph responsiveness to SRIF action.[214] The hypersecretion of GH may similarly result from impaired autoregulation at pituitary sites by the low levels of insulin, and IGF-I in diabetic patients.[200] An autoregulatory effect of the increased GH levels on the hypothalamic regulation of pituitary function is also possible, although while this would inhibit GH release, it would be expected simultaneously to increase (rather than decrease) SRIF tone.[169]

D. HYPOTHYROIDISM

Hypothyroidism, which results in growth retardation, is associated with low plasma GH levels and diminished GH responses to GHRH-stimulation, both of which are normalized by thyroxine (T_4) replacement therapy.[215–217] The stimulatory effect of thyroxine results from a restoration of GH pulse amplitude and pulsatile GH release.[217] In the absence of thyroid hormone replacement, GH secretion induced by insulin hypoglycemia, arginine, clonidine, and sleep is impaired in patients with primary hypothyroidism.[217–219] The severity of this inhibition of GH secretion in hypothyroid patients is likely to be partly masked by a prolongation of the GH half-life and a reduction in the rate of GH clearance.[118,220]

The impaired GH secretion in hypothyroid patients has often been thought to reflect diminished *GH* gene transcription and depletion of pituitary GH stores, because thyroid hormones stimulate GH synthesis in rats (see Chapter 4). Thyroid hormones do not, however, increase *GH* gene transcription in humans[221] and Valcavi et al.[222] found that the somatotroph secretory capacity in patients with primary hypothyroidism is comparable to that in normal subjects. Moreover, whereas euthyroid individuals are not normally GH responsive to TRH stimulation, hypothyroid subjects are readily responsive.[223] Although this might be thought to reflect upregulation of pituitary TRH receptors in thyroid deficiency, Faggiano et al.[224] found the GH response of hypothyroid patients was not suppressed by T_4 therapy. Abnormal GH responses to TRH, gonadotropin-releasing hormone (GnRH), and CRH also occur in individuals with congenital thyrotropin deficiency.[225] It is therefore possible that blunted GH secretion in these patients results from increased SRIF tone or reduced GHRH stimulation. Pyridostigmine and arginine inhibit hypothalamic SRIF and enhance GHRH-induced GH secretion in hypothyroid patients,[222] implicating hypersomato-statinergic tone in the impairment of GH secretion.

E. HYPERTHYROIDISM

Blunted GH responses to GH secretagogues with hypothalamic (sleep, arginine, hypoglycemia, clonidine, and cholinergic agonists) and pituitary (GHRH) sites of action have also been observed in hyperthyroid patients.[226–230] These abnormal responses are reversed when euthyroidism is restored by methimazole treatment.[231] Although basal GH concentrations are not suppressed,[219,229] this may result from thyroidal actions at hypothalamic and pituitary sites to increase SRIF tone and to antagonize somatotroph responsiveness to GHRH stimulation.[232] The increased frequency and amplitude of pulsatile GH secretion in hyperthyroid patients,[228,233] and their fourfold increase in daily rate of GH production[233] in the absence of changes in GH metabolism, is, nevertheless, characteristic of reduced SRIF tone, as is the inability of cholinergic agonists to stimulate GH release.[228] The failure of clonidine to reverse the blunted GH responses in hyperthyroid patients also suggests that thyroidal excess causes an increase in SRIF release or blocks GH secretion directly.[230] However, as a cholinergic drug (pirenzepine) that is likely to increase hypothalamic somatostatinergic tone completely blunts the GH response of hyperthyroid patients to GHRH, inhibitory hypothalamic regulation is functional in these patients, even though exogenous glucose is unable to suppress GH secretion.[227,231]

F. CUSHING'S SYNDROME

An impairment of GH secretion occurs in patients with Cushing's syndrome, in which attenuated GH secretion occurs during sleep[234] and blunted GH responses occur in response to insulin-induced hypoglycemia, L-dopa, arginine, galanin, and GHRH.[235–240] This impairment is thought to be due to hypercortisolemia, because blunted GH responses to GH-releasing factors also occur in normal subjects injected with supraphysiological doses of glucocorticoids,[241–243] which can induce dwarfism in growing children[244] (see Giustina and Wehrenberg[245] for review). Normal GH responses to these factors can be induced in Cushing's patients after the normalization of plasma cortisol levels.[238,243,246–250] This inhibition of GH release probably results from increased SRIF tone,[251,252] because an acetylcholinesterase inhibitor (pyridostigmine) that is thought to inhibit SRIF secretion[253,254] is able to enhance the GH response to GHRH in some patients with Cushing's disease[255] and in subjects acutely treated with glucocorticoids.[244] Some Cushing's patients may, however, be resistant to pyridostigmine action,[239] indicating impaired cholinergic pathways following chronic glucocorticoid excess may lead to reduced GH secretion.

The defective GH secretion in Cushing's patients may also result from diminished hypothalamic GHRH, as induced in rats by glucocorticoid excess.[256] A deficiency of hypothalamic GHRH would also desensitize somatotrophs to GHRH stimulation, because GHRH priming potentiates subsequent GH responses to homologous stimulation.[239] Damage to the somatotroph cell may also contribute to the reduced GH secretion in Cushing's patients, because glucocorticoids are potent stimulators of GH gene transcription (see Chapter 4).

The suppression of spontaneous and stimulated GH secretion in Cushing's patients has also been correlated with increased plasma levels of free IGF-I, which result from reduced levels of IGF-I-binding proteins.[257] The persistence of low IGF-I-binding protein levels following the surgical treatment of Cushing's patients is also thought to be responsible for the suppression of GH, which persists for at least 12 months in some patients.[257]

G. RENAL FAILURE

Although uremia is associated with growth disturbances in children and severe wasting in adults,[258,259] basal GH levels are high or moderately elevated.[260–266] Pulsatile GH release may, conversely, be depressed in renal failure.[267] After dialysis, GH concentrations are significantly lowered.[262,263,268,269] This elevation in GH level is likely to reflect an increase in GH half-life and a reduced rate of renal clearance.[220] The abnormal prolongation of secretagogue-elevated plasma GH concentrations[270–273] may thus reflect reduced GH clearance, although Garcia et al.[270] suggest this more likely reflects a greater variability in the onset of the GH secretory response.

Paradoxical GH responses to TRH[272,274,275] or glucose loading[261,276,277] and abnormal responses to L-dopa[278] and insulin-induced hypoglycemia[279] indicate central GH secretory dysfunction in renal failure, even though normal GH responses to arginine[271] and GHRH[270,280] have been reported. Exaggerated GH responses to GHRH have also been observed,[273] but unlike in normal individuals, the GH response to GHRH in uremic patients is potentiated by erythropoietin treatment.[280] An inappropriate production of erythropoietin has, therefore, been implicated in the dysfunctional regulation of GH secretion in some uremic patients.[280] Ramirez et al.[281] similarly found elevated basal GH levels and exaggerated GH responses to GHRH in chronic hemodialysis patients, but the correction of anemia with erythropoietin blunted their GH responses to GHRH.

The increased GH secretion in uremic patients may partly result from a deficiency in feedback inhibition, because plasma IGF-I concentrations are low.[275] The removal of circulating IGF-I inhibitors[282] during dialysis is thus thought to account for the postdialysis suppression of GH secretion.

H. LIVER DISEASE

Patients with severe liver disease often have high baseline plasma GH levels.[283,284] Because the liver is a major site of GH degradation and clearance, this elevation in blood GH concentration is likely to reflect a reduced rate of metabolic clearance.[285] This may also partly result from increased pituitary responsiveness to GH-releasing factors, including GHRH[286] and TRH.[284,286–288] This hyperresponsiveness may result from the reduced IGF-I levels in patients with liver disease,[286] although in individual patients no negative relationship between plasma IGF-I concentrations and the magnitude of the GHRH-induced GH response is observed.[286] The increased GH secretion in patients with liver disease could thus be due to reduced somatostatinergic tone, although the hyperresponsiveness to GHRH occurs in patients with normal or elevated basal plasma GH levels. Salerno et al.[286] therefore conclude that decreased SRIF is not causally involved in the GH hyperresponsiveness associated with liver failure.

A central dysfunction leading to a GHRH/SRIF imbalance is, however, indicated by the abnormal GH responses to glucose[289] and arginine[290] and diminished GH responses to L-dopa[291,292] and apomorphine.[293] This dysfunction may involve cholinergic pathways, because the anomalous GH responses of cirrhotic patients to L-dopa, apomorphine, TRH, and arginine are blocked by pirenzepine, a specific muscarinic-cholinergic receptor blocker[287,294] (reviewed by Massara et al.[294a]). However, as pirenzepine crosses the blood-brain barrier very poorly, this cholinergic mechanism is likely to occur at the level of the median eminence or pituitary gland. Because pirenzepine is thought to stimulate SRIF release, cholinergic dysfunction in liver disease may inhibit SRIF release and result in a disinhibition of somatotroph function. The ability of exogenous SRIF to inhibit some of the anomalous GH responses seen in liver disease[295] supports this view.

I. PSYCHIATRIC DISORDERS

Psychiatric disease is, axiomatically, associated with neurological dysfunction and disturbances in brain neurotransmitter content which may impinge on the hypothalamus-somatotroph axis and impair GH regulation. Indeed, defective GH secretion has been observed in a variety of psychiatric disorders, as reviewed by Brown et al.[296] and by Nemeroff and Krishnan[297] and Risch.[298]

1. Affective Disorders

Patients with unipolar and bipolar depression secrete four to sixfold more GH than normal subjects during waking hours[299–301] but not during the night, because sleep and GH secretion may be disturbed in depressed patients.[302] Aberrant GH responses to TRH occur in depressed patients[303,304] but GH responses to insulin-induced hypoglycemia,[305,306] 5-hydroxytryptophan,[307] clonidine,[308,309] and GHRH[310,311] are diminished. These derangements are likely to reflect dysfunctional α_2-adrenergic and γ-aminobutyric acid pathways,[312] excessive sympathetic activity and hypercortisolemia,[308,313,314] and increased IGF-I inhibition[315] and are corrected following recovery from the depressive illness.[305]

2. Alzheimer's Disease

Patients with Alzheimer's disease have reduced CNS concentrations of SRIF[316] and, unlike patients with late onset disease,[317] the GH response to GHRH is enhanced.[318]

3. Schizophrenia

Schizophrenia is associated with hyperactivity of the mesocorticolimbic dopamine system, although the tuberoinfundibular system regulating hypothalamic function may be unimpaired.[319] The basal GH concentrations of schizophrenic patients and their GH responses to provocative stimuli (dopamine and clonidine) are reported as being elevated, reduced or normal, although GH concentrations in acute schizophrenia may differ from those with late onset disease and may be dependent on the presence or withdrawal of neuroleptic medication and on the predominance of positive or negative psychotic behavior (see Garver[320] for review). Aberrant GH secretion in schizophrenics is, however, indicated by GH responses to TRH and GnRH in most patients.[320]

4. Anorexia Nervosa

Basal plasma GH concentrations are elevated in more than half of anorectic patients.[297,321] The GH levels are inversely related to body weight and weight loss and are normalized following the resumption of normal feeding.[321,322] These changes are similar to those in malnourished individuals and are related to the weight loss and low-calorie diet of these patients (see Muller and Locatelli[323] and Newman and Halmi[321] for review) and low plasma IGF-I levels.[324,325] Although these responses are similar to those in starvation and underfeeding, significant physiological differences exist between undernourished and anorectic individuals and neurotransmitter and receptor dysfunctions within the hypothalamus-pituitary axis are likely also to be responsible for dysfunctional GH secretion.

Although basal plasma GH levels are elevated in anorexia nervosa, dysfunctional GH regulation is indicated by the diminished GH responses to sleep, insulin, L-dopa, and apomorphine. These aberrant GH responses are normalized by the refeeding of a balanced diet.[322,326–328] Dysfunctional regulation is also indicated by aberrant responses to TRH[303] and elevated GH responses to glucose,[322] although GH responses to arginine and glucagon are mostly normal.[329,330] The GH response to TRH may result from the prolonged stimulation of the pituitary gland by GHRH, which may sensitize it to TRH action.[321] Pituitary GH responsiveness to GHRH is, however, markedly enhanced in anorectic patients,[331,332] which Muller and Locatelli[323] consider to be due to reduced SRIF tone in the face of reduced plasma IGF-I levels. This possibility is partially supported by the ability of cholinergic agonists and antagonists (which inhibit and stimulate SRIF release, respectively) to potentiate and suppress the GH response to GHRH.[333,334] Cholinergic tone may, therefore, be elevated in anorexia, especially as SRIF levels are low in the cerebrospinal fluid of anorectic patients.[335] Dysfunctional GH regulation in anorexia nervosa may also result from the chronic activation of the hypothalamus-pituitary adrenal axis and reflect interactions of CRH, ACTH, and cortisol at hypothalamic and/or pituitary sites.[323]

5. Psychosocial Dwarfism

The reduced growth in psychosocial dwarfs is secondary to reduced GH secretion. In these patients GH responses to insulin-induced hypoglycemia and arginine are suppressed and, although sleep patterns are normal, sleep-associated GH secretion does not occur.[296]

These abnormalities are reversed during catch-up growth in emotionally enriched environments.[336] The primary defect appears to be related to hypothalamic dysfunction, resulting from influences from higher cortical centers.

REFERENCES

1. **Daughaday, W. H.,** in *Williams Textbook of Endocrinology*, 7th ed. (Ed. Wilson, J. D. and Foster, D. W.), W. B. Saunders, Philadelphia, 1985, p. 568.
2. **Frasier, S. D., Rudlin, C. R., Zeisell, H. J., et al.,** *Am J. Dis. Child.,* 146, 682, 1992.
3. **Constine, L. S., Woolf, P. D., Mick, G., et al.,** *N. Engl. J. Med.*, 328, 87, 1993.
4. **Powell, G. F., Brasel, J. A., and Blizzard, R. M.,** *New Engl. J. Med.*, 276, 1271, 1967.
5. **Karin, M., Castrillo, J.-L., and Theill, L. E.,** *Trends Genet.*, 6, 92, 1990.
6. **Li, S., Crenshaw, E. B., III, Rawson, E. J., et al.,** *Nature (Lond.)*, 347, 528, 1990.
7. **Radovick, S., Nations, M., Du, Y., et al.,** *Science*, 257, 1115, 1992.
8. **Pfaffle, R. W., Dimattia, G. E., Parks, J. S., et al.,** *Science*, 257, 1118, 1992.
9. **Phillips, J. A., III,** in *The Metabolic Basis of Inherited Disease,* 6th ed. (Eds. Scriver, C. R., Beaudet, A. L., and Sly, W. S.). McGraw-Hill, New York, 1989, p. 1965.
10. **Parks, J. S., Meacham, L. R., McKean, M. C., et al.,** *Pediatr. Res.*, 25, 90A, 1989.
11. **Vnencak-Jones, C. L., Phillips, J. A., III, Chen, E. Y., et al.,** *Proc. Natl. Acad. Sci. U.S.A.*, 85, 5615, 1988.
12. **Phillips, J. A., III and Cogan, J. D.,** *J. Clin. Endocrinol. Metab.*, 78, 11, 1994.
13. **Laron, Z.,** *Adv. Intern. Med. Ped.*, 51, 117, 1984.
14. **Rosenbloom, A. L., Aguirre, J. G., Rosenfeld, R. G., et al.,** *New Engl. J. Med.*, 323, 1367, 1990.
15. **Godowski, P. J., Leung, D. W., Meacham, L. R., et al.,** *Proc. Natl. Acad. Sci. U.S.A.*, 86, 8083, 1989.
16. **Amselem, S., Sobrier, M.-L., Duquesnoy, P., et al.,** *J. Clin. Endocrinol. Metab.*, 87, 1098, 1991.
17. **Laron, Z., Anin, S., Klipper-Aurbach, Y., et al.,** *Lancet*, 339, 1258, 1992.
18. **Wilton, P.,** *Acta Pediatr. (Suppl.)*, 383, 137, 1992.
19. **Daughaday, W. H.,** *Endocrine Causes of Menstruation Disorders* (Ed. Givens, J. R.). Year Book Medical Publications, Chicago, IL, 1978, p. 143.
20. **Adan, L., Souberbielle, J.-C., and Brauner, R.,** *J. Clin. Endocrinol. Metab.*, 78, 353, 1994.
21. **Heinze, H. J., Jorgensen, V., and Bercu, B. B.,** *Molecular and Clinical Advances in Pituitary Disorders* (Ed. Melmed, S.). Endocrine Research and Education, Inc., Los Angeles, CA, 1994, p. 159.
22. **Cacciari, E., Tassoni, P., Parisi, G., et al.,** *J. Clin. Endocrinol. Metab.*, 74, 1284, 1992.
23. **Saggese, G., Cesaretti, G., Giannessi, N., et al.,** *J. Endocrinol. Invest.*, 13, 475, 1990.
24. **Bercu, B. B., Shulman, D., Root, A. W., et al.,** *J. Clin. Endocrinol. Metab.*, 63, 709, 1986.
25. **Oerter, K. E., Sobel, A. M., Rose, S. R., et al.,** *J. Clin. Endocrinol. Metab.*, 75, 1413, 1992.
26. **Zadik, Z., Chalew, S. A., Gilula, Z., et al.,** *J. Clin. Endocrinol. Metab.*, 71, 1127, 1990.
27. **Albini, C. H., Quattrin, T., Vandlen, R. L., et al.,** *Pediatr. Res.*, 23, 89, 1988.
28. **Abdenur, J. E., Solans, C. V., Smith, M. M., et al.,** *J. Clin. Endocrinol. Metab.*, 78, 277, 1994.
29. **Fradkin, J. E., Schonberger, L. B., Mills, J. L., et al.,** *JAMA*, 265, 880, 1991.
30. **Bozzola, M., Tato, L., Cisternino, M., et al.,** *J. Endocrinol. Invest.*, 9, 503, 1986.
31. **Grossman, A., Savage, M. O., Blacklay, A., et al.,** *Horm. Res.*, 22, 52, 1985.
32. **Romer, T., Rymkiewicz-Kuzynska, B., Olivier, M., et al.,** *J. Clin. Endocrinol. Metab.*,72, 503, 1991.
33. **Giustina, A., Bossoni, S., Bodini, C., et al.,** *Acta Endocrinol. (Copenh.)*, 125, 510, 1991.
34. **Cavallo, L., Laforgia, N., Acquafredda, A., et al.,** *Horm. Res.*, 34, 13, 1990.
35. **Ghigo, E., Imperiale, E., Boffano, G. M., et al.,** *J. Endocrinol. Invest.*, 13, 307, 1990.
36. **Cryer, P. E., Jacobs, L. S., and Daughaday, W. H.,** *Mt. Sinai J. Med.*, 40, 402, 1973.
37. **Alexander, J. M., Biller, B. M. K., Zervas, N. T., et al.,** *J. Clin. Invest.*, 86, 336, 1990.
38. **Herman, V., Fagin, J., Gonsky, R., et al.,** *J. Clin. Endocrinol. Metab.*, 71, 1427, 1990.
39. **Jameson, J. L.,** *Molecular and Clinical Advances in Pituitary Disorders — 1993* (Ed. Melmed, S.). Endocrine Research and Education, Inc., Los Angeles, CA, 1993, p. 15.
40. **Spada, A. and Vallar, L.,** *Molecular and Clinical Advances in Pituitary Disorders* (Ed. Melmed, S.). Endocrine Research and Education, Inc., Los Angeles, CA, 1993, p. 29.

41. **Jameson, J. L. and Hollenberg, A. N.,** *Horm. Metab. Res.*, 24, 201, 1992.
42. **Pei, L., Prager, D., Scheithauer, B., and Melmed, S.** *Molecular and Clinical Advances in Pituitary Disorders* (Ed. Melmed, S.). Endocrine Research and Education, Inc., Los Angeles, CA, 1993, p. 21.
43. **Spada, A., Arosio, M., Bochicchio, D., et al.,** *J. Clin. Endocrinol. Metab.*, 71, 1421, 1990.
44. **Landis, C. A., Masters, S. B., Spada, A., et al.,** *Nature (Lond.)*, 340, 692, 1989.
45. **Weinstein, L. S., Shenker, A., Gejman, P. V., et al.,** *New Engl. J. Med.*, 325, 1688, 1991.
46. **Friend, K. E., Chiou, Y., Laws, E. R., et al.,** *J. Clin. Endocrinol. Metab.*, 77, 1281, 1993.
47. **Castrillo, J., Theill, L. E., and Karin, M.,** *Science*, 253, 197, 1991.
48. **Bystrom, C., Larsson, C., Blomberg, C., et al.,** *Proc. Natl. Acad. Sci. U.S.A.*, 87, 1968, 1990.
49. **Asa, S. L. and Kovacs, K.,** *Endocrinol. Metab. Clin. N. Am.*, 21, 553, 1992.
50. **Melmed, S., Ezrin, C., Kovacs, K., et al.,** *New Engl. J. Med.*, 312, 9, 1985.
51. **Melmed, S., Braunstein, G. D., Horvath, E., et al.,** *Endocr. Rev.*, 4, 271, 1983.
52. **Frohman, L. A.,** *Molecular and Clinical Advances in Pituitary Disorders* (Ed. Melmed, S.). Endocrine Research and Education, Inc., Los Angeles, CA, 1993, p. 35.
53. **Moran, A., Asa, S. L., Kovacs, K., et al.,** *N. Engl. J. Med.*, 323, 322, 1992.
54. **Kovacs, K., Stefaneanu, L., Asa, S. et al.,** *Molecular and Clinical Advances in Pituitary Disorders* (Ed. Melmed, S.). Endocrine Research and Education, Inc., Los Angeles, CA, 1993, p. 41.
55. **Asa, S. L., Kovacs, K., Stefaneanu, L., et al.,** *Endocrinology (Baltimore)*, 141, 895, 1992.
56. **Thorner, M. O., Perryman, R. L., Cronin, M. L., et al.,** *J. Clin. Invest.*, 70, 965, 1982.
57. **Scheithauer, B. W., Kovacs, K., Randall, R. V., et al.,** *J. Clin. Endocrinol. Metab.*, 15, 655, 1986.
58. **Barkan, A. L.,** *Trends Endocrinol. Metab.*, 3, 205, 1992.
59. **Ho, P., Chandler, W., Demott-Friberg, R., et al.,** *73rd Annu. Meet. Endocr. Soc.* Washington, D.C., 1991.
60. **Faglia, G., Arosio, M., and Bazzoni, N.,** *Endocrinol. Metab. Clin. N. Am.*, 21, 575, 1992.
61. **Losa, M., Schopohl, J., and von Werder, K.,** *J. Endocrinol. Invest.*, 16, 69, 1993.
62. **Zimmerman, D., Young, W. F., Ebersold, M. J., et al.,** *J. Clin. Endocrinol. Metab.*, 76, 216, 1993.
63. **Kovaks, K.,** *Endocr. Rev.*, 9, 357, 1987.
64. **Barkan, A. L., Shenker, Y., Grekin, R. J., et al.,** *J. Clin. Endocrinol. Metab.*, 63, 1057, 1986.
65. **Daughaday, W. H.,** *New Engl. J. Med.*, 297, 1267, 1977.
66. **Lieberman, S., Bjorkengren, A., and Hoffman, A. R.,** *Endocrinol. Metab. Clin. N. Am.*, 21, 615, 1992.
67. **O'Duffy, J. D., Randall, R. V. and MacCarty, C. S.,** *Ann. Intern. Med.*, 78, 379, 1973.
68. **Pickett, J. B. E., Layzer, R. B., Levin, S. R., et al.,** *Neurology*, 25, 638, 1975.
69. **Trotman-Dickenson, B., Weetman, A. P., and Hughes, J. M. B.,** *Quart. J. Med.*, 79(NS), 527, 1991.
70. **Imura, H., Shimatsu, A., Hattor, N., and Kato, Y.,** *Molecular and Clinical Advances in Pituitary Disorders* (Eds. Melmed, S. and Robbins, R. J.). Blackwell Publishing, Boston, MA, 1991, p. 261.
71. **Johnston, D. G., Davies, R. R., and Prescott, R. W. G.,** *J. Roy. Soc. Med.*, 78, 319, 1985.
72. **Christensen, S. E., Weeke, J., Orskov, H., et al.,** *Acta Endocrinol. (Copenh.)*, 116, 49, 1987.
73. **Barkan, A. L., Stred, S. E., Reno, K., et al.,** *J. Clin. Endocrinol. Metab.*, 69, 1225, 1989.
74. **Hartman, M. L., Veldhuis, J. D., Vance, M. L., et al.,** *J. Clin. Endocrinol. Metab.*, 70, 1375, 1990.
75. **Spada, A., Bassetti, M., Reza-Elahi, F., et al.,** *J. Clin. Endocrinol. Metab.*, 78, 411, 1994.
76. **Israel, J. M. and Vincent, J. D.,** *Front. Neuroendocrinol.*, 11, 339, 1990.
77. **Linfoot, J.,** *Endocrine Control of Growth* (Ed. Daughaday, W. H.). Elsevier, New York, 1981, p. 207.
78. **Losa, M., Sckopohl, J., Stalla, G. K., et al.,** *Clin. Endocrinol. (Oxford)*, 23, 99, 1985.
79. **Schulte, H. M., Benker, G., Windeck, R., et al.,** *J. Clin. Endocrinol. Metab.*, 61, 585, 1985.
80. **Sartorio, A., Spada, A., Bochicchio, D., et al.,** *Neuroendocrinology*, 44, 470, 1986.
81. **Lamberts, S., Klijn, J., Van Vroonkover, C., et al.,** *J. Clin. Endocrinol. Metab.*, 60, 1148, 1985.
82. **Faglia, J. A., Arosio, M., and Ambrosi, B.,** *The Pituitary Gland* (Ed. Imura, H.). Raven Press, New York, 1985, p. 363.
83. **Lamberts, S. W. J., Luizzi, A., Chiodini, P. G., et al.,** *Eur. J. Clin. Invest.*, 12, 151, 1982.
84. **Jialal, I., Naidoo, C., Nathoo, B. C., et al.,** *J. Endocrinol. Invest.*, 7, 235, 1984.

85. **Nortier, J. W. R., Croughs, R. J. M., Donker, G. H., et al.,** *Acta Endocrinol.*, 106, 175, 1984.
86. **Smals, A. E. M., Pieters, G., Smals, A. G. H., et al.,** *Acta Endocrinol. (Copenh.)*, 116, 53, 1987.
87. **Muller, E., Cavagnini, F., Martinez, A., et al.,** *Acta Endocrinol. (Copenh.)*, 107, 155, 1984.
88. **Muller, E. E.,** *Physiol. Rev.*, 67, 962, 1987.
89. **Giusti, M., Lomeo, A., Mazzocchi, G., et al.,** *J. Endocrinol. Invest.*, 8, 203, 1985.
90. **Page, M. D., Dieguez, C., and Scanlon, M. F.,** *Biotechnology in Growth Regulation* (Eds. Heap, R. B., Prosser, C. G., and Lamming, G. E.). Butterworths, London, 1988, p. 47.
91. **Clemmons, D. R. and Underwood, L. E.,** *J. Clin. Endocrinol. Metab.*, 15, 629, 1986.
92. **Oppizzi, G., Petroncini, M., Dallabonzana, D., et al.,** *J. Clin. Endocrinol. Metab.*, 63, 1348, 1986.
93. **Wright, A. D., Hill, D. M., Lowy, C., et al.,** *Quart. J. Med.*, 39, 1, 1970.
94. **Alexander, L., Appleton, D., Hall, R., et al.,** *Clin. Endocrinol. (Oxford)*, 12, 71, 1980.
95. **Nabarro, J. D. N.,** *Clin. Endocrinol. (Oxford)*, 26, 481, 1987.
96. **Bengtsson, B. A., Eden, S., Ernest, I., et al.,** *Acta. Med. Scand.*, 223, 327, 1988.
97. **Ezzat, S. and Melmed, S.,** *J. Clin. Endocrinol. Metab.*, 72, 245, 1991.
98. **Klein, I., Parveen, G., Gavaler, J. S., et al.,** *Ann. Intern. Med.*, 97, 27, 1982.
99. **Brunner, J. E., Johnson, C. C., Zafar, S., et al.,** *Clin. Endocrinol. (Oxford)*, 32, 65, 1990.
100. **Ross, D. A. and Wilson, C. B.,** *J. Neurosurg.*, 68, 854, 1988.
101. **Fahlbusch, R., Honegger, J., and Buchfelder, M.,** *Endocrinol. Metab. Clin. N. Am.*, 21, 669, 1992.
102. **Jaffe, C. A. and Barkan, A. L.,** *Endocrinol. Metab. Clin. N. Am.*, 21, 713, 1992.
103. **Ezzat, S., Snyder, P. J., Young, W. F., et al.,** *Ann. Intern. Med.*, 117, 711, 1992.
104. **Wass, J. A. H., Lytras, N., and Besser, G. M.,** *Scand. J. Gastroenterol.*, 21(Suppl. 119), 136, 1986.
105. **Lamberts, S. W. J. and Del Pozo, E.,** *Scand. J. Gastroenterol.*, 21(Suppl. 119), 141, 1986.
106. **Sadoul, J-L., Thyss, A., and Freychet, P.,** *Acta Endocrinol.*, 126, 179, 1992.
107. **Pieters, G. M., Smals, A., Smals, A. G. H., et al.,** *Acta Endocrinol. (Copenh.)*, 114, 537, 1987.
108. **Jackson, I. M. D., Barnard, L. B., and Lamberton, P.,** *Am. J. Med.*, 81, 94, 1986.
108a. **Bertherat, J., Chanson, P., Dewailly, D., et al.,** *J. Clin. Endocrinol. Metab.*, 77, 1577, 1993.
108b. **Reubi, J. C. and Landolt, A. M.,** *J. Clin. Endocrinol. Metab.*, 68, 844, 1989.
108c. **Ikuyama, S., Nawata, H., Kato, I., et al.,** *J. Clin. Endocrinol. Metab.*, 62, 729, 1986.
109. **Plockinger, U., Dienermann, D. and Quabbe, H.-J.,** *J. Clin. Endocrinol. Metab.*, 71, 1658, 1990.
110. **Dieguez, C., Page, M. D., and Scanlon, M. F.,** *Clin. Endocrinol. (Oxford)*, 28, 109, 1988.
111. **Delitala, G., Tomasi, P. A., and Virdis, R.,** *J. Endocrinol. Invest.*, 11, 441, 1988.
112. **Cocchi, D., Baldasseroni, G. C., Becherucci, P., et al.,** *J. Endocrinol. Invest.*, 12, 9, 1989.
113. **Fornito, M. C., Calogero, A. E., Mongioi, A., et al.,** *J. Neuroendocrinol.*, 2, 87, 1990.
114. **Goldman, J. A., Molitch, M. E., and Thorner, M. O.,** *J. Endocrinol. Invest.*, 10, 397, 1987.
115. **Veldhuis, J. D., Iranmanesh, A., Ho, K. K. Y., et al.,** *J. Clin. Endocrinol. Metab.*, 72, 51, 1991.
116. **Martha, P. M., Jr., Goorman, K. M., Blizzard, R. M., et al.,** *J. Clin. Endocrinol.*, 74, 335, 1992.
117. **Iranmanesh, A., Lizarralde, G., and Veldhuis, J. D.,** *J. Clin. Endocrinol. Metab.*, 73, 1081, 1991.
118. **Iranmanesh, A. and Veldhuis, J. D.,** *Neuroendocrinology*, 21, 783, 1992.
119. **Holl, R. W., Schwartz, U., Schauwecker, P., et al.,** *J. Clin. Endocrinol. Metab.*, 77, 216, 1993.
120. **Veldhuis, J. D., Iranmanesh, A., Ho, K. K. Y., et al.,** *J. Clin. Endocrinol. Metab.*, 72, 51, 1991.
121. **Davies, R. R., Turner, S. J., Cook, D., et al.,** *Clin. Endocrinol. (Oxford)*, 23, 521, 1985.
122. **Kopelman, P. G., Noonan, K., Goulton, R., et al.,** *Clin. Endocrinol. (Oxford)*, 23, 885, 1985.
123. **Loche, S., Pintus, S., Cella, S. G., et al.,** *Clin. Endocrinol. (Oxford)*, 33, 187, 1990.
124. **Bernini, G. P., Argenio, G. F., Vivaldi, M. S., et al.,** *Clin. Endocrinol. (Oxford)*, 32, 453, 1990.
125. **Csizmadi, I., Brazeau, P., and Serri, O.,** *Metabolism*, 38, 1016, 1989.
126. **Laurian, L., Oberman, Z., Ayalon, D., et al.,** *Isr. J. Med. Sci.*, 11, 482, 1975.
127. **Glass, A. R., Burman, K. D., Dahms, W. T., et al.,** *Metabolism*, 30, 89, 1981.
128. **Vanderschueren-Lodeweyckx, M.,** *Horm. Res.*, 40, 23, 1993.
129. **Veldhuis, J. D., Iranmanesh, A., Ho, K. K. Y., et al.,** *J. Clin. Endocrinol. Metab.*, 72, 51, 1991.
130. **Hochberg, Z., Hertz, P., Colin, V., et al.,** *Metabolism*, 41, 106, 1992.
131. **Sims, E. A. H., Danforth, E. Jr., and Horton, E. S.,** *Recent. Prog. Horm. Res.*, 29, 457, 1973.
132. **Kelijman, M. and Frohman, L. A.,** *J. Clin. Endocrinol. Metab.*, 66, 489, 1988.
133. **Tanaka, K., Inoue, S., Numata, K., et al.,** *Metabolism*, 39, 892, 1990.
134. **Williams, T., Berelowitz, M., and Joffe, S. N.,** *N. Engl. J. Med.*, 311, 1403, 1984.

135. **Crockford, P. M. and Salmon, P. A.,** *Can. Med. Assoc. J.*, 64, 878, 1970.
136. **Laurian, L., Oberman, Z., Hoerer, E., et al.,** *Isr. J. Med. Sci.*, 18, 625, 1982.
137. **Ball, M. F., El-Khodary, A. Z., and Canary, J. J.,** *J. Clin. Endocrinol. Metab.*, 34, 498, 1972.
138. **Kelijman, M. and Frohman, L. A.,** *J. Clin. Endocrinol. Metab.*, 66, 489, 1988.
139. **Kopelman, P. G. and Noonan, K.,** *Clin. Endocrinol. (Oxford)*, 24, 157, 1986.
140. **Tannenbaum, G. S., Lapointe, M., Gurd, W., et al.,** *Endocrinology (Baltimore)*, 127, 3087, 1990.
141. **Cocchi, D., Parenti, M., Cattaneo, L., et al.,** *Neuroendocrinology*, 57, 928, 1993.
142. **Van Vliet, G., Bosson, D., and Rummens, E.,** *Clin. Endocrinol. (Oxford)*, 27, 145, 1987.
143. **Van Vliet, G., Bosson, D., Rummens, E., et al.,** *Acta Endocrinol. (Copenh.)*, 249(Suppl.), 403, 1986.
144. **Loche, S., Pintor, C., Cappa, M., et al.,** *Acta Endocrinol.*, 120, 624, 1989.
145. **Ghigo, E., Mazza, E., Corrias, A., et al.,** *Metabolism*, 38, 631, 1989.
146. **Cordido, F., Casanueva, F. F., and Dieguez, C.,** *J. Clin. Endocrinol. Metab.*, 68, 290, 1989.
147. **Plewe, G., Schneider, U., Krause, U., et al.,** *J. Endocrinol. Invest.*, 10, 137, 1987.
148. **Finer, N., Price, P., Grossman, A., et al.,** *Horm. Metab. Res.*, 19, 68, 1987.
149. **Faccinietti, F., Bernasconi, S., Petraglia, F., et al.,** *Horm. Metab. Res.*, 20, 348, 1988.
150. **Altomonte, L., Zoli, Z., Alessi, F., et al.,** *Horm. Res.*, 27, 145, 1987.
151. **Imura, H., Nakai, Y., and Yoshimi, T.,** *J. Clin. Endocrinol. Metab.*, 36, 204, 1973.
152. **Nakai, Y., Imura, H., Sakurai, H., et al.,** *J. Clin. Endocrinol. Metab.*, 11, 489, 1974.
153. **Blundell, J. E.,** *Appetite*, 7(Suppl.), 39, 1986.
154. **Hoebel, B. G. and Leibowitz, S. F.,** *Brain Behaviour and Bodily Diseases* (Eds. Wiener, H. A., Hoffer, M. A., and Stunkard, A. J.). Raven Press, New York, 1981, p. 103.
155. **Barbarino, A., DeMarinis, L., and Troncone, L.,** *Metabolism*, 27, 275, 1978.
156. **Komorowski, J. M. and Pawlikowski, M.,** *Endocrinology (Baltimore)*, 73, 209, 1979.
157. **Loche, S., Cappa, M., Borrelli, P., et al.,** *Clin. Endocrinol. (Oxford)*, 27, 145, 1987.
158. **Metstas, M. T., Foster, G. V., Margolis, S., et al.,** *Metabolism*, 31, 1224, 1982.
159. **Isaacs, R. E., Gardner, D. G., and Baxter, J. D.,** *Endocrinology (Baltimore)*, 120, 2022, 1987.
160. **Chalew, S. A., Lozano, R. A., Armour, K. M., et al.,** *Int. J. Obes.*, 16, 459, 1992.
161. **Tanaka, K., Inoue, S., Shiraki, J., et al.,** *Metab. Clin. Exp.*, 40, 1257, 1991.
162. **Golay, A., Swislocki, A. L., Chen, Y. D., et al.,** *J. Clin. Endocrinol. Metab.*, 63, 481, 1986.
163. **Casanueva, F. F., Villanueva, L., Dieguez, C., et al.,** *J. Clin. Endocrinol. Metab.*, 65, 634, 1987.
164. **Imaki, T., Shibasaki, T., Shizume, K., et al.,** *J. Clin. Endocrinol. Metab.*, 60, 290, 1985.
165. **De Marinis, L., Mancini, A., Zuppi, P., et al.,** *Psychoneuroendocrinology*, 16, 361, 1991.
166. **Loche, S., Cappa, M., Borrelli, P., et al.,** *Clin. Endocrinol. (Oxford)*, 27, 145, 1987.
167. **Ghigo, E., Procopio, M., Maccario, M., et al.,** *Horm. Metab. Res.*, 25, 305, 1993.
168. **Londono, J. H., Gallagher, T. F. Jr., and Bray, G. A.,** *Metabolism*, 18, 986, 1969.
169. **Holly, J. M. P., Amiel, S. A., Sandhu, R. R., et al.,** *J. Endocrinol.*, 118, 353, 1988.
170. **Shaper, N. C.,** *Acta Endocrinol. (Copenh.)*, 122, 7, 1990.
171. **Hansen, A. P.,** *Dan. Med. Bull.*, 19(Suppl. 1), 1, 1972.
172. **Asplin, C. M., Fariai, A. C. S., Carlsen, E. C., et al.,** *J. Clin. Endocrinol. Metab.*, 69, 239, 1989.
173. **Vanelli, M., Bernasconi, S., Bolondi, O., et al.,** *J. Endocrinol. Invest.*, 9, 293, 1986.
174. **Mullis, P. E., Pal, B. R., Matthews, D. R., et al.,** *Clin.Endocrinol. (Oxford)*, 36, 255, 1992.
175. **Nieves-Rivera, F., Rogol, A. D., Veldhuis, J. D., et al.,** *J. Clin. Endocrinol. Metab.*, 77, 638, 1993.
176. **Hayford, J. T., Danney, M. M., Hendrix, J. A., et al.,** *Diabetes*, 29, 391, 1980.
177. **Pal, B. R., Matthews, D. R., Edge, J. A., et al.,** *Clin. Endocrinol. (Oxford)*, 38, 93, 1993.
178. **Hansen, A. P., Ledet, T., and Lundbaek, K.,** *Handbook of Diabetes Mellitus. Biochemical Pathology* (Ed. Brownlee, M.). John Wiley & Sons, Chichester, 1981, p. 231.
179. **Passa, P., Gauville, C., and Canivet, J.,** *Lancet*, ii, 72, 1974.
180. **Tamborlane, W. V., Sherwin, R. S., Koivisto, V., et al.,** *Diabetes*, 28, 785, 1979.
181. **Lorenzi, M., Karam, J., McIlroy, M., et al.,** *J. Clin. Invest.*, 65, 146, 1980.
182. **Burday, S. Z., Fine, P. H., and Schalch, D. S.,** *J. Lab. Clin. Med.*, 71, 897, 1968.
183. **Speroni, G., Ceda, G. P., Capretii, L., et al.,** *Horm. Metab. Res.*, 15, 46, 1983.
184. **Topper, E., Gertner, J., Amiel, S., et al.,** *Pediatr. Res.*, 19, 534, 1985.
185. **Pietschmann, P., Schernthaner, G., Stephenson, J., et al.,** *Horm. Metab. Res.*, 23, 379, 1991.
186. **Pietschmann, P., Schernthaner, G., Prskavec, F., et al.,** *Diabetes*, 36, 159, 1987.
187. **Krassowski, J., Felber, J. P., Rogala, H., et al.,** *Acta Endocrinol. (Copenh.)*, 117, 225, 1988.
188. **Dasmahapatra, A., Urdanivia, E., and Cohen, M. P.,** *J. Clin. Endocrinol. Metab.*, 52, 859, 1981.

189. **Ceda, G. P., Speroni, G., Dall'Aglio, E., et al.,** *J. Clin. Endocrinol. Metab.*, 55, 170, 1982.
190. **Giampietro, O., Ferdeghini, M., Miccoli, R., et al.,** *Peptide Biol. Fluid. Proc. Colloq.*, 34, 209, 1986.
191. **Sonksen, P. J., Srivastava, M. C., Tompkins, C. V., et al.,** *Lancet*, ii, 155, 1972.
192. **Sharp, P. S., Foley, F., and Kohner, E. M.,** *Diabet. Med.*, 1, 205, 1984.
193. **Kaneko, K., Komine, S., Maeda, T., et al.,** *Diabetes*, 34, 710, 1985.
194. **Sudkvist, G., Almer, L., and Pandolfi, M.,** *Acta Med. Scand.*, 215, 55, 1984.
195. **Kaneko, K., Komine, S., Maeda, T., et al.,** *Diabetes*, 34, 710, 1985.
196. **Winkler, G., Gero, L., Halmos, T., et al.,** *Acta Diabetol. Lat.*, 24, 109, 1986.
197. **Press, M., Tamborlane, W. V., Thorner, M. O., et al.,** *Diabetes*, 33, 804, 1984.
198. **Schaper, N., Tamsma, J. T., Sluiter, W., et al.,** *Acta Endocrinol. (Copenh.)*, 122(Suppl. 2), 32, 1990.
199. **Ismail, I. S., Scanlon, M. F., and Peters, J. R.,** *Clin. Endocrinol. (Oxford)*, 35, 499, 1991.
200. **Miller, J. D., Wright, N. M., Lester, S. E., et al.,** *J. Clin. Endocrinol. Metab.*, 75, 1087, 1992.
201. **Hershcopf, R., Plotnick, L. P., Kaya, K., et al.,** *J. Clin. Endocrinol. Metab.*, 54, 504, 1982.
202. **Arias, P., Kerner, W., de la Fuenta, A., et al.,** *Acta Endocrinol. (Copenh.)*, 107, 250, 1984.
203. **Johansen, K. and Hansen, A. P.,** *Diabetes*, 20, 239, 1971.
204. **Richards, N. T., Wood, S. M., Christofides, N. D., et al.,** *Diabetologia*, 27, 529, 1984.
205. **Culler, F. L. and Meacham, L. R.,** *Neuroendocrinology*, 58, 473, 1993.
206. **Boucher, B. J., Butterfield, W. J. H., and Whichelow, M. J.,** *Clin. Sci.*, 37, 721, 1969.
207. **Lipman, R. L., Taylor, A. L., Conly, P., et al.,** *Diabetes*, 21, 175, 1972.
208. **Sperling, M. A., Wollesen, F., and De Lamater, P. V.,** *Diabetologia*, 9, 380, 1973.
209. **Richardson, S. B. and Twente, S.,** *Diabetologia*, 30, 893, 1987.
210. **Pietschmann, P. and Schernthaner, G.,** *Acta Endocrinol. (Copenh.)*, 117, 315, 1988.
211. **Pietschmann, P., Schernthaner, G., and Luger, A.,** *J. Clin. Endocrinol. Metab.*, 63, 389, 1986.
212. **Richardson, S. B., Hollander, C., D'Eletto, R., et al.,** *Endocrinology (Baltimore)*, 107, 122, 1980.
213. **Lewis, B. M., Dieguez, C., Inglesias, R., et al.,** *J. Endocrinol. Invest.*, 10(Suppl. 3), 31, 1987.
214. **Cohen, R. M. and Frohman, L. A.,** *Clin.Res.*, 36, 584A, 1988.
215. **Valcavi, R., Jordan, V., Dieguez, C., et al.,** *Clin. Endocrinol.*, 24, 693, 1986.
216. **Williams, T., Maxon, H., Thorner, M. O., et al.,** *J. Clin. Endocrinol. Metab.*, 61, 454, 1985.
217. **Buchanan, C. R., Stanhope, R., Adlard, P., et al.,** *Clin. Endocrinol. (Oxford)*, 29, 427, 1988.
218. **Chernausek, S. D., Underwood, L., Utiger, R. D., et al.,** *Clin. Endocrinol. (Oxford)*, 19, 337, 1983.
219. **Velardo, A., Zizzo, G., Della Casa, L., et al.,** *Exp. Clin. Endocrinol.*, 101, 243, 1993.
220. **Owens, D., Srivastava, M. C., Tompkins, C. V., et al.,** *Eur. J. Clin. Invest.*, 3, 2814, 1973.
221. **Brent, G. A., Harney, J. W., Moore, D. D., et al.,** *Mol. Endocrinol.*, 2, 792, 1988.
222. **Valcavi, R., Valente, F., Dieguez, C., et al.,** *J. Clin. Endocrinol. Metab.*, 77, 616, 1993.
223. **Collu, R., Leboeuf, G., Letarte, J., et al.,** *J. Clin. Endocrinol. Metab.*, 44, 743, 1977.
224. **Faggiano, M., Criscuolo, T., Graziani, M., et al.,** *Clin.Endocrinol. (Oxford)*, 23, 61, 1985.
225. **Hayashizaki, Y., Miyai, K., Onishi, T., et al.,** *Horm. Metab. Res.*, 18, 849, 1986.
226. **Giustina, A., Buffoli, M. G., Ferrari, C., et al.,** *Horm. Metab. Res.*, 23, 506, 1991.
227. **Finkelstein, J. W., Boyar, R. M., and Hellman, L.,** *J. Clin. Endocrinol. Metab.*, 38, 634, 1974.
228. **Valcavi, R., Dieguez, C., Zini, M., et al.,** *Clin. Endocrinol. (Oxford)*, 35, 141, 1991.
229. **Giustina, A., Ferrari, C., Bodini, C., et al.,** *Acta Endocrinol.*, 123, 613, 1990.
230. **Guldner, J., Buffoli, M. G., Bussi, A. R., et al.,** *Horm. Res.*, 36, 192, 1991.
231. **Valcavi, R., Dieguez, C., Zini, M., et al.,** *Clin. Endocrinol. (Oxford)*, 38, 515, 1993.
232. **Casaneuva, F.,** *Acromegaly*, 21, 483, 1992.
233. **Iranmanesh, A., Lizarralde, G., Johnson, M. L., et al.,** *J. Clin. Endocrinol. Metab.*, 72, 108, 1991.
234. **Krieger, D. T. and Glick, S. M.,** *J. Clin. Endocrinol. Metab.*, 39, 986, 1974.
235. **Smals, A. E. M., Pieters, G. F. F. M., Smals, A. G. H., et al.,** *Clin. Endocrinol.*, 24, 401, 1986.
236. **Krieger, D.,** *J. Clin. Endocrinol. Metab.*, 36, 277, 1973.
237. **Hotta, M., Shibasaki, T., Masuda, A., et al.,** *Life Sci.*, 42, 979, 1988.
238. **Tyrrel, J. B., Wiener-Kronish, J., Lorenzi, M., et al.,** *J. Clin. Endocrinol. Metab.*, 44, 3193, 1977.
239. **Leal-Cerro, A., Pereira, J., Garcia-Luna, P. P., et al.,** *Clin. Endocrinol. (Oxford)*, 33, 291, 1990.
240. **Takahashi, H., Bando, H., Zhang, C., et al.,** *Acta Endocrinol. (Copenh.)*, 127, 13, 1992.
241. **Hartog, M., Gaafar, M. A., and Fraser, R.,** *Lancet*, ii, 376, 1964.

242. **Lantigua, R. A., Streck, W. F., Lockwood, D. H., et al.,** *J. Clin. Endocrinol. Metab.*, 50, 298, 1980.
243. **Guistina, A., Doga, M., Bodini, C., et al.,** *Acta Endocrinol.*, 122, 206, 1990.
244. **Blodgett, F. M., Burgin, L., Iezzoni, D., et al.,** *N. Engl. J. Med.*, 254, 636, 1956.
245. **Giustina, A. and Wehrenberg, W. B.,** *Trends Endocrinol. Metab.*, 3, 306, 1992.
246. **Giustina, A., Bossoni, S., Bodini, C., et al.,** *Horm. Res.*, 35, 99, 1991.
247. **Demura, R., Demura, H., Nunokawa, T., et al.,** *J. Clin. Endocrinol. Metab.*, 34, 852, 1972.
248. **Suda, A., Goverde, H., Kloppenborg, P., et al.,** *J. Clin. Endocrinol. Metab.*, 51, 1048, 1980.
249. **Whitehead, H. M., McKnight, J. A., Sheridan, B., et al.,** *J. Endocrinol. Invest.*, 13, 217, 1990.
250. **Leal-Cerro, A., Pumar, A., Villamil, F., et al.,** *Clin. Endocrinol. (Oxford)*, 38, 399, 1993.
251. **Wehrenberg, W., Janowski, B., Piering, A., et al.,** *Endocrinology (Baltimore)*, 126, 3200, 1990.
252. **Guistina, A., Girelli, A., Doga, M., et al.,** *J. Clin. Endocrinol. Metab.*, 71, 580, 1990.
253. **Locatelli, V., Torsello, A., Redaelli, M., et al.,** *J. Endocrinol.*, 111, 271, 1986.
254. **Giustina, A., Girelli, A., Alberti, D., et al.,** *Clin. Endocrinol. (Oxford)*, 35, 491, 1991.
255. **Del Balzo, P., Salvatori, R., and Cappa, M.,** *Clin. Endocrinol. (Oxford)*, 33, 605, 1990.
256. **Nakagawa, K., Ishizuka, T., Obara, T., et al.,** *Acta Endocrinol.*, 116, 165, 1987.
257. **Magiakou, M. A., Mastorakos, G., Gomez, M. T., et al.,** *J. Clin. Endocrinol. Metab.*, 78, 131, 1994.
258. **Broyer, M., Kleinknecht, C., Loriat, C., et al.,** *J. Pediatr.*, 84, 642, 1974.
259. **Kopple, J. B.,** *Kidney Int.*, 14, 340, 1978.
260. **Wright, A. D., Lowy, C., Fraser, T. R., et al.,** *Lancet*, 2, 798, 1968.
261. **El Bishti, M. M., Counahan, R., Bloom, S. R., et al.,** *Am. J. Clin. Nutr.*, 31, 1865, 1978.
262. **Donnay, S., Garcia-Martin, F., Martin-Escobar, E., et al.,** *Nephrol. Dial. Translant.*, 7, 246, 1992.
263. **Caufriez, A., Abramowicz, D., Vanherwegham, J. L., et al.,** *J. Endocrinol. Invest.*, 16, 691, 1993.
264. **Samaan, N. A. and Freeman, R. M.,** *Metabolism*, 257, F503, 1970.
265. **Takano, K., Hall, K., Kastrup, K. W., et al.,** *J. Clin. Endocrinol. Metab.*, 48, 371, 1979.
266. **Czernichow, P., Dauzet, C., Broyer, M., et al.,** *J. Clin. Endocrinol. Metab.*, 43, 630, 1976.
267. **Thorner, M. O. and Rogol, A. D.,** *Human Growth Hormone: Progress and Challenges* (Ed. Underwood, L. E.). Marcel Dekker, Inc., New York, 1988, p. 113.
268. **Olgaard, K., Hagen, C., and McNeilly, A. S.,** *Acta Endocrinol. (Copenh.)*, 80, 237, 1975.
269. **Stuart, M., Lazarus, L., and Hayes, J.,** *IRSC Med. Sci.*, 2, 1102, 1984.
270. **Garcia, R. V. G., Andrade, A., Perez, J., et al.,** *J. Endocrinol. Invest.*, 14, 383, 1991.
271. **Mauro, F., Sakai, T., and Sato, S.,** *Nephron*, 24, 81, 1979.
272. **Gonzalez-Barcena, D., Kastin, A., Schalch, D. S., et al.,** *J. Clin. Endocrinol. Metab.*, 36, 117, 1973.
273. **Bessarione, D., Perfumo, F., Giusti, M., et al.,** *Acta Endocrinol. (Copenh.)*, 114, 5, 1987.
274. **Gonzalez-Barcena, D., Kastin, A., Schalch, D. S., et al.,** *J. Clin. Endocrinol. Metab.*, 36, 117, 1973.
275. **Powell, D. R., Rosengeld, R. G., Baker, B. K., et al.,** *J. Clin. Endocrinol. Metab.*, 63, 1186, 1986.
276. **Wright, A. D., Lowy, C., Fraser, T. R., et al.,** *Lancet*, 2, 798, 1968.
277. **Orskov, H. and Christensen, N. J.,** *J. Clin. Invest.*, 67, 51, 1981.
278. **Ramirez, G., O'Neill, W. M., Bloomer, H. A., et al.,** *Arch. Intern. Med.*, 138, 267, 1978.
279. **Ljaiya, K.,** *Eur. J. Pediatr.*, 131, 185, 1979.
280. **Cantalamessa, L., Cremagnani, L., Orsatti, A., et al.,** *Clin. Endocrinol. (Oxford)*, 34, 85, 1991.
281. **Ramirez, G., Bittle, P. A., Sanders, H., et al.,** *J. Clin. Endocrinol. Metab.*, 78, 63, 1994.
282. **Phillips, L. S., Fusco, A. C., Unterman, T. G., et al.,** *J. Clin. Endocrinol. Metab.*, 59, 764, 1984.
283. **Hernandez, A., Zorrilla, E., and Gershberg, H.,** *J. Lab. Clin. Med.*, 73, 25, 1969.
284. **Panerai, A. E., Salerno, F., Manneschi, M., et al.,** *J. Clin. Endocrinol. Metab.*, 45, 134, 1977.
285. **Bauer, A. G. C., Lamberts, S. W. J., and Wilson, J. H. P.,** *Horm. Res.*, 16, 126, 1981.
286. **Salerno, F., Locatelli, V., and Muller, E. E.,** *Clin. Endocrinol. (Oxford)*, 27, 183, 1987.
287. **Zanoboni, A., Zecca, L., and Zanoboni-Muciaccia, V.,** *Neuroendocrinol. Lett.*, 8, 117, 1994.
288. **Salerno, F., Cocchi, D., Frigerio, C., et al.,** *J. Clin. Endocrinol. Metab.*, 51, 641, 1980.
289. **Becker, M. D., Cook, G. C., and Wright, A. D.,** *Lancet*, ii, 1035, 1969.
290. **Muggeo, M., Tiengo, A., Fedele, D., et al.,** *Acta Endocrinol. (Copenh.)*, 139, 1157, 1979.

291. **Langer, M., Masala, A., Sassu, R., et al.,** *The Endocrines and the Liver* (Eds. Langer, M., Chiandussi, L., Chopra, I. J., and Martini, L.). Academic Press, London, 1982, p. 279.
292. **Borzio, M., Caldara, R., Ferrari, C., et al.,** *Acta Endocrinol. (Copenh.)*, 447, 1981.
293. **Lal, S., Oravec, M., Aronoff, A., et al.,** *J. Neural Transmission*, 53, 7, 1982.
294. **Volpi, R., Chiodera, P., Cerri, L., et al.,** *Acta Endocrinol. (Copenh.)*, 114, 603, 1987.
294a. **Massara, F., Ghigo, E., Careddy, D., et al.,** *Endocrinology* (Ed. Molinatti, G. M. and Martini, L.), Excerpta Medica, Amsterdam, 1986, p. 55.
295. **Salerno, F., Cocchi, D., Lampertico, M., et al.,** *Horm. Metab. Res.*, 14, 482, 1982.
296. **Brown, G. M., Seggie, J. A., Chambers, J. W., et al.,** *Psychoneuroendocrinology*, 3, 131, 1978.
297. **Nemeroff, C., and Krishnan, K.,** *Neuroendocrinology* (Ed. Nemeroff, C.). CRC Press, Boca Raton, FL, 1991, p. 413.
298. **Risch, S. C.,** *Neuropeptides in Psychiatric Disorders* (Ed. Nemeroff, C.). APA Press, Washington, D.C., 1991, p. 93.
299. **Linkowski, P., Mendlewicz, J., Kerkhofs, M., et al.,** *J. Clin. Endocrinol. Metab.*, 65, 141, 1987.
300. **Schilkrut, R., Chandra, O., Osswald, M., et al.,** *Neuropsychobiology*, 1, 70, 1975.
301. **Mendlewicz, J., Linkowski, P., Kerkhofs, M., et al.,** *J. Clin. Endocrinol. Metab.*, 60, 505, 1985.
302. **Schilkrut, R., Chandra, O., Osswald, M., et al.,** *Neuropsychobiology*, 1, 70, 1975.
303. **Maeda, K., Kato, K., Ohgo, S., et al.,** *J. Clin. Endocrinol. Metab.*, 40, 501, 1975.
304. **Winokur, A., Amsterdam, J. D., Oper, J., et al.,** *Arch. Gen. Psychiatry*, 40, 525, 1983.
305. **Endo, M. and Endo, J.,** *Psychoneuroendocrinology* (Ed. Hatotani, N.). S. Karger, Basel, 1974, p. 22.
306. **Gruen, P. H., Sacher, E. J., Altman, N., et al.,** *Arch. Gen. Psychiatry*, 32, 31, 1975.
307. **Takahashi, S., Kondo, H., and Yoshimura, M.,** *Psychoneuroendocrinology* (Ed. Hatotani, N.). Basel, Karger, 1974, 32.
308. **Lesch, K. P., Laux, G., Erb, A., et al.,** *Psychiatr. Res.*, 25, 301, 1988.
309. **Siever, L. J., Uhde, T. W., Silberman, E. K., et al.,** *Psychiatr. Res.*, 6, 171, 1982.
310. **Lesch, K. P., Laux, G., Erb, A., et al.,** *Biol. Psychiatry*, 22, 1495, 1987.
311. **Lesch, K. P., Laux, G., Erb, A., et al.,** *Psychoneuroendocrinology*, 13, 255, 1988.
312. **O'Flynn, K. and Dinan, T. G.,** *Am. J. Psychiatry*, 1994.
313. **Nathan, R. S., Sachar, E. J., Asmis, G. M., et al.,** *Psychiatr.Res.*, 4, 291, 1981.
314. **Garver, D., Pandey, G., Derkmenjian, H., et al.,** *Am. J. Psychiatry*, 132, 1149, 1975.
315. **Lesch, K.-P., Laux, G., Pfuller, H., et al.,** *J. Clin. Endocrinol. Metab.*, 65, 1278, 1987.
316. **Nemeroff, C. B., Kizer, J. S., Reynolds, G. P., et al.,** *Regul. Peptides*, 25, 123, 1989.
317. **Nemeroff, C. B., Krishnan, K. R. R., Belkin, B. M., et al.,** *Neuroendocrinology*, 50, 663, 1989.
318. **Rosser, M. M., Iversen, L. L., Reynolds, G. P., et al.,** *Br. Med. J.*, 288, 961, 1984.
319. **Seeman, P., Lee, T., Chau-Wong, M., et al.,** *Nature*, 261, 717, 1976.
320. **Garver, D. L.,** *Endocrinol. Metab. Clin. N. Am.*, 17, 103, 1988.
321. **Newman, M. M. and Halmi, K. A.,** *Endocrinol. Metab. Clin. N. Am.*, 17, 195, 1988.
322. **Casper, R. and Davis, J.,** *Psychoneuroendocrinology*, 2, 105, 1977.
323. **Muller, E. E. and Locatelli, ,** *J. Endocrinol.*, 132, 327, 1992.
324. **Jarrell, J., Meltzer, S., and Tolis, G.,** *Clinical Neuroendocrinology: A Patholphysiological Approach* (Ed. Tolis, G.). Raven Press, New York, 1979, p. 355.
325. **Winterer, J., Gwirtsman, H. E., George, D. T., et al.,** *J. Clin. Endocrinol. Metab.*, 61, 693, 1985.
326. **Sherman, B., and Halmi, K.,** *Anorexia Nervosa* (Ed. Vigensky, R.). Raven Press, New York, 1977, p. 211.
327. **Kalucy, R. S., Crisp, A. H., Chard, T., et al.,** *J. Psychosomatic Res.*, 20, 595, 1976.
328. **Vigensky, R., and Loriaux, D.,** *Anorexia Nervosa* (Ed. Vigensky, R.). Raven Press, New York, 1977, p. 109.
329. **Neri, V., Ambrosi, B., Beck-Pecoz, P., et al.,** *Folia Endo.*, 25, 143, 1972.
330. **Scanlon, M. F., Peters, J. R., and Foord,** *Thyrotropin Releasing Hormone* (Eds. Griffiths, E. C. and Bennett, G. W.). Raven Press, New York, 1983, p. 303.
331. **Brambilla, F., Ferrari, E., Cavagnini, F., et al.,** *Biol. Psychiat.*, 25, 256, 1989.
332. **Rolla, M., Andreoni, A., Belliti, D., et al.,** *Biol. Psychiat.*, 29, 1079, 1991.
333. **Tamia, K., Komaki, G., Matsubayashi, S., et al.,** *J. Clin. Endocrinol. Metab.*, 70, 738, 1990.
334. **Muller, E. E., Locatelli, V., Cocchi, D., et al.,** *J. Clin. Endocrinol. Metab.*, 14(Suppl. 1), 10, 1991.
335. **Gerner, R. H. and Yamada, T.,** *Brain Res.*, 238, 298, 1982.
336. **Stanhope, R., Adlard, P., Hamill, G., et al.,** *Clin. Endocrinol. (Oxford)*, 28, 335, 1988.

Chapter 14

Growth Hormone Transport

Stephen Harvey and Kerry Hull

CONTENTS

I. INTRODUCTION

Almost all tissues, organs, and cells possess growth hormone (GH) receptors and are biologically responsive to GH stimulation. Most of these actions of GH occur at sites distant to the site of GH synthesis and are dependent on the extrapituitary distribution of GH in extracellular fluids. GH may be transported to some of its target sites in lymph and cerebrospinal fluid and possibly in milk and other secretions, but in this chapter only the transportation of GH in blood is considered.

II. GH IN BLOOD

Hormones enter (and leave) the bloodstream from (and to) perivascular spaces via small pores in attenuated capillary endothelial cells.[1] Adenohypophysial blood vessels are highly permeable to large molecules and are richly fenestrated. Micropinocytotic vesicles in endothelial cells may also transport molecules such as GH by transcytosis.[2] Entry of GH into the general circulation occurs rapidly after secretion into the venous capillaries of the pituitary gland and after transit through the lungs and heart it is distributed throughout the body by arterial vessels.

0-8493-8697-7/95/$0.00+$.50

GH is a hydrophilic protein and hence is soluble in blood plasma. A fraction of circulating GH is, nevertheless, bound to plasma protein. This fraction accounts for approximately 30 to 50% of the circulating GH at concentrations up to 10 to 15 μg/l.[3–5] At higher GH levels the proportion of protein-bound GH declines, owing to the saturation of binding sites.[3] The levels of free, bound, and total GH in plasma are therefore nonequilibrium functions of the rate of GH secretion, the volume of GH distribution, the amount, affinity, and capacity of circulating GH-binding proteins (GHBPs), and the rate of GH utilization or clearance. The circulating GH concentration is therefore dynamically regulated by a number of interactive factors at a variety of pituitary and extrapituitary sites that are susceptible to physiological regulation and pathophysiological dysfunction.

III. GH-BINDING PROTEINS

The binding of GH in plasma is to heterogeneous proteins of different molecular size, structure, and abundance, and with differential affinities for circulating GH moieties.[6–9] The structure, function, and regulation of these proteins has been reviewed by Baumann,[9–12] Kelly et al.,[6,7] Herington et al.,[13–15] Robinson et al.,[8] and Waters et al.[16–18]

A. HIGH-AFFINITY GHBP IN HUMANS

In humans, much (40 to 45%) of the 22-kDa GH monomer that is bound in plasma is bound to a 61-kDa protein that has an isoelectric point of 5.0. This single-chain glycoprotein is composed of 246 amino acids and contains 3 small disulfide loops in its amino-terminal half that bridge sequential cysteine residues.[19] This protein has at least six potential sites for N-linked glycosylation and at least half of the GH-binding protein (GHBP) mass is due to carbohydrate moieties. These moieties are not, however, required for binding, because a recombinant, nonglycosylated version of the GHBP binds GH with an affinity at least as high as the natural GHBP.[19] The binding site of the native GHBP has been mapped to the disulfide-rich region but involves a discontinuous patch composed of the extremities of four antiparallel loops.[20]

This GHBP binds 22-kDa GH with high affinity ($K_a = 10^{-9} - 10^{-8}$ mol), although with limited capacity (approximately 0.5 to 2.0 nmol/l). The formation in plasma of GH complexes of 80 to 85 kDa suggests a 1:1 binding stoichiometry for the GH:GHBP interaction,[20a] although the GH molecule has two binding sites and can simultaneously associate with two membrane receptor molecules.[21,22] This GHBP also binds the placental human GH variant (hGHV) with comparable affinity to 22-kDa hGH.[23] It also binds 20-kDa hGH, although with an affinity 20-fold less than 22-kDa hGH [24,25]. This GHBP does not, however, bind nonhuman GH moieties or human prolactin or placental lactogen.[4,10,24,26] The binding of 22-kDa hGH to this protein occurs rapidly with physiological concentrations of GH and binding is nearly complete (75 to 80% of maximum) within 5 min at 37°C.[12,24] This binding is completely reversible and dissociation occurs with a half-time of 40 min at 21°C.

The binding affinity, capacity, and specificity of this plasma protein are comparable to the binding of 22-kDa GH to membrane GH receptors, despite its smaller size.[11] This protein is, furthermore, immunologically and structurally homologous to the extracellular domain of the hepatic GH receptor.[27–29] It is therefore a truncated, soluble form of the receptor that is thought to arise from the proteolytic cleavage of the membrane receptor or from the alternate splicing of GH receptor mRNA and *de novo* synthesis.[11,12,30] However, whereas cultured human IM-9 lymphocytes "shed" the extracellular portion of their GH receptors *in vitro* following exposure to sulfhydryl inactivating reagents (which presumably activate a protease that cleaves the receptor close to the extracellular domain), these are not physiological conditions and IM-9 cells are abnormal, cancerous cells.

Moreover, whereas GH receptor gene expression is known to produce small transcripts that encode the plasma GHBP in rats and mice,[31–33] similar transcripts have not been detected in human tissues. It is therefore still uncertain how this GHBP is generated in humans. It is also unclear from which tissues this protein is derived, although the liver is likely to be a major source.[34,35]

1. Plasma Concentrations

In humans this GHBP normally circulates at a concentration of approximately 1 nmol/l and complexes about 40 to 45% of circulating 22-kDa GH under basal conditions.[14,24] The amount of GH bound to GHBP is greatest at times of peak plasma GH concentrations, following episodic bursts of GH release.[36] This may be facilitated by concomitant episodic fluctuations in serum GHBP,[37] although other studies[38] suggest the GHBP level is relatively constant throughout a 24-h period or has minimal[36] diurnal fluctuations positively[39] or negatively[40] related to the GH concentration. The GHBP level is also highly variable (>12-fold) between individuals[41–44] and largely reflects differences in binding capacity rather than binding affinity.

In healthy individuals plasma GHBP levels are maximal in the second or third decade[45–47] and at most ages the binding affinity (K_a) and capacity (B_{max}) are comparable in males and females (Table 1), although women tend to have slightly higher GHBP concentrations.[41,43,46,48–51] The binding protein is absent or undetectable in early fetal development but is measurable at low levels in cord blood collected during the last trimester of gestation. GHBP levels are low in newborns and neonates and progressively rise during infancy and the prepubertal period and during adolescence, with the largest increase occurring during the first few years of life.[41,43,44,50–53] The pubertal growth spurt is not, however, accompanied by increased serum GHBP levels.[44,47,54] Decreased GHBP levels may subsequently occur in the elderly[42] and male subjects over 40 years of age, at which time GHBP levels in females are greater than those in males.[46] This ontogenic decline is, however, controversial.[12,41]

Circulating GHBP levels may also change during pregnancy, although the overall levels in non pregnant and pregnant females are similar.[23,45,48,54] Plasma GHBP levels may, nevertheless, be increased during early pregnancy and progressively decline from the end of the first trimester to prepregnant GHBP levels.[26,45,54] This gestational decline in maternal high-affinity GH-binding capacity may reflect the rising serum concentrations of prolactin, placental lactogen, and placental GH, which may compete at high levels for the 22-kDa hGH-binding site.[23,26] This decline is, however, surprising because oral estrogen increases GHBP levels in cycling[54] and postmenopausal women.[56,57]

2. Hormonal Regulation

Gonadal steroids may also regulate plasma GHBP concentrations in males (Table 1). Testosterone, for instance, inhibits plasma GH binding by reducing the capacity (by fourfold) and affinity (by twofold) of the GHBP.[56,58] The pharmacological suppression of testosterone secretion in precocious boys[58] and the secretion of estrogen in precocious girls[59] is, conversely, associated with increased GHBP concentrations. Other hormones, including insulin[60] and thyroid hormones,[61] also increase GHBP activity (Table 1). These effects have been considered to be indirectly mediated through changes in endogenous GH secretion, although exogenous GH is reported to be ineffective,[56,62–64] or to increase[58,65,66] or decrease[66] GHBP concentrations. Diurnal and ontogenic changes in GH secretion also occur independently of serum GHBP levels.[36,37,40,44] Abnormal GHBP activity is also not characteristic of most patients with GH excess or deficiency syndromes[41,56,63,66] and the suppression of endogenous GH secretion in acromegalic patients

Table 1 **High-Affinity GHBP Concentrations in Man**

Individuals/Condition	K_a (*M*)	Capacity	GHBP Concentration	% Specific Binding of ^{125}I-Labeled GH	Ref.
Fetus				3.1	52
Fetus				3.3	189
Fetus				4.7	53,54
Fetus		176 pmol/l			36
Fetus (24–36 weeks)				0.6	42
Fetus (38–42 weeks)				2.95	42
Newborn				4.2	52
Newborn				2.58	45
Newborn				1.6	200
Neonate				1.65	48
Neonate				8.5	52
Neonate				21.4	41
Neonate				17.2	43
Neonate				7.7	45
Neonate				4.1	189
Neonate				6.4	50
Infant				16.0	50
Infant				26.3	43
Infant	1.3×10^{-9}	451 pmol/l		7.2	75
Juvenile				22.2	41
Juvenile (10–16 years)		40-504 pmol/l			44
Prepubertal ♂				28.6	54
Prepubertal ♀				28.7	54
Pubertal ♂				27.5	54
Pubertal ♀				26.1	54

Adolescent				21.3	83
Adolescent				17.9	50
Adolescent				22.2–29.3	60
Adolescent				17.1	76
Adolescent Children (2–19 years)				11.3	42
Adolescent				27.2	43
Adolescent	1.99×10^{-8}			12.7	189
Adult				19–24	38
Adult		65.8–305 pmol/l			203
Adult				14.8	63
Adult	1.17×10^{-9}	810 pmol/l		11.6	69
Adult ♂			0.63 nmol/l		199
Adult ♂	4.9×10^{8}			38.6	94
Adult ♂				26.6	41
Adult ♂				22.8	46
Adult ♂ (24–76 years)	1.08×10^{-9}			11.8	68
Adult ♂ (20–30 years)				23.2	43
Adult ♂ (20–46 years)				19.1	42
Adult ♀/♂ (38–45 years)	2.3×10^{-8}	56.6 mU/l		13–32	202
Adult ♂ (>40 years)	$3.6–7.4 \times 10^{-8}$			23.0	43
Adult ♂ (>40 years)				38.5	43
Adult ♂ (28–45 years)	1.08×10^{-9}			11.7	48
Adult ♂ (20–70 years)				19.7	51
Adult ♂				11.6	75
Adult ♀			0.99 nmol/l ↑		199
Adult ♀				23.3	41
Adult ♀ (20–80 years)				23.2	46
Adult ♀	$0.93–1.47 \times 10^{-9}$	7.56–17.24 pmol/l			171
Adult ♀				32.7	43

Table 1 **High-Affinity GHBP Concentrations in Man (Continued)**

Individuals/Condition	K_a (*M*)	Capacity	GHBP Concentration	% Specific Binding of ^{125}I-Labeled GH	Ref.
Adult ♀				10.6	48
Adult ♀	2.2×10^{-8}				172
Adult ♀ (>40 years)				22.7	46
Middle-aged ♂				19.1	46
Middle-aged ♀				22.7	46
Middle-aged ♀	1.08×10^{-9}	730 pmol/l		11.3	77
Menstrual (proliferative phase) ♀				33.4	55
Menstrual (luteal phase) ♀				35.3	55
Menstrual (menses) ♀	4.9×10^{-8}			34.6	55
Premenopausal ♀				18.9	57
Postmenopausal ♀	1.1×10^{-9}	980 pmol/l		17.1	204
Postmenopausal ♀	0.70×10^{-9}			14.7	56
Elderly				16.1	42
Elderly	2.3×10^{-8}			39.0	80
Pregnant ♀ (3rd trimester)	4.8×10^{-8} ←			29.6	55
Pregnant ♀ (2nd trimester)				35.1	55
Pregnant ♀ (1st trimester)				32.4	55
Pregnant ♀				8.8 ←	45
Pregnant ♀ (40 weeks)				20.6 ↓	41
Pregnant ♀ (20 weeks)				10.6 ←	48
Pregnant ♀				8.2 ↓	26
Pregnant ♀ (6 weeks)				13.5–22.1 ↑	26
Delayed puberty				30.6 ←	58
Precocious pubertal ♂				16.6 ↓	58
Precocious pubertal ♀				24.2 ↓	59

Estrogen-treated ♀	0.96×10^{-9}	2017 pmol/l ↑		25.6 ↑	56
Estrogen-treated ♀	0.6×10^{-9}	1384 pmol/l ←		15.6 ←	56
Estrogen-treated ♀				20–22 ↑	57
Estrogen-treated ♀				47.0 ↑	55
Estrogen-treated ♀			2.5 nmol/l ↑		199
LHRH-treated				28.1 ↑	59
LHRH-treated				21.6 ↑	58
Testosterone-treated ♂				20.9 ↓	58
Acromegalic			1.0 nmol/l		199
Acromegalic				20.8 ↓	41
Acromegalic	1.24×10^{-9}	386 pmol/l		5.4 ↓	68
Acromegalic		500 pmol/l ↓			67
Acromegalic				11.3	56
GH-deficient				10.2 ↓	58
GH-deficient				6.5 ←	66
GH-deficient (40 years)				11.4 ←	56
GH-deficient				9.7	63
GH-deficient				8.5 ↓	85
GH-deficient			0.89 nmol/l		199
GH-deficient				15.5	63
Laron dwarf				ND–45.9	95
Laron dwarf				ND	86
Laron dwarf				ND	96
Laron dwarf	5.4×10^{-8}			2–61.6 ←	94
Pygmy				6.5 ↓	90
Pygmy				8.0 ↓	45
Pygmy				<3.0 ↓	91
Pygmy (children)				14.3	209
Pygmy (adolescents)				7.7↓	209
Pygmy (adults)				6.9↓	209

Table 1 **High-Affinity GHBP Concentrations in Man (Continued)**

Individuals/Condition	*K*a (*M*)	Capacity	GHBP Concentration	% Specific Binding of ^{125}I-Labeled GH	Ref.
Short-stature children				7.38 ↓	48
Short-stature children				5.2 ↓	92
Short-stature children				↓	205
Short-stature children				11.1 ↓	65
Short-stature girl	5.1×10^{-8}	2650 μg/l ↑			201
Short-stature adult ♂	1.1×10^{-9}	750 μg/l ↑			201
GH-treated				22.5 ↑	85
GH-treated				↑	206
GH-treated				10.7 ←	56
GH-treated				←	63
GH-treated				21.1 ↑	65
GH-treated				9.5 ↑	66
GH-treated	3.7×10^{-8}			18.5 ↑	58
GH-treated				6.9 ←	64
Insulin-dependent diabetic				18.2 ↓	82
Insulin-dependent diabetic				16.8 ↓	83
Insulin-dependent diabetic				7.8 ↓	84
Insulin-dependent diabetic				22.2 ↓	60
Non-insulin-dependent diabetic				24.6 ←	82
Malnutrition				2.8 ↓	52
Fasting				42.5 ←	73
Obese				20.1 ↑	75
Obese				27.7 ↑	76
Obese	1.25×10^{-9}	1717 pmol/l		17.8 ↑	69

Anorectic ♀				↓	74
Anorectic ♀	1.39×10^{-9}	494 pmol/l		6.9 ↓	69
Refed anorectic		175 pmol/l ↑			70
Uremic				14.2 ↓	81
Uremic				19.0 ↓	41
Renal failure (nonuremic)	$7.5–9.3 \times 10^{-8}$ ↑			16.8–25.7 ↓	80
Cirrhotic				19.7 ↓	41
Cirrhotic				8.5 ↓	77
Cirrhotic				5.9 ↓	48
Cirrhotic				14.6 ↓	78
Cirrhotic	1.15×10^{-9}		529 pmol/l ↑	7.2–9.7 ↓	77
Werner's syndrome	1.8×10^{-8}		121 mU/l ↑	33.5 ↑	202
Critical illness				19.9 ↓	73
Turner's syndrome				←	210
Hypothyroid				7.3 ↓	61
Hyperthyroid				12.9 ↑	61
Clonidine-treated				↓ ←	26
SRIF- treated				16.5 ←	56
SRIF-treated				5–13 ←	68
SRIF-treated				←	67
Infected				21.1 ↓	41

Note: ↑, Increased; ↓, decreased; ←, unchanged, in comparison with corresponding controls; ND, not detected.

by octreotide has inconsistent effects on GHBP levels.[56,67,68] The influence of GH on GHBP regulation is therefore uncertain. The clinical implications of GHBP regulation have recently been reviewed by Postel-Vinay.[68a]

3. Nutritional Regulation

The GH-binding capacity of plasma is also related to nutritional status.[52,62,69,70] Short periods of fasting suppress plasma GHBP in healthy individuals,[71,72] but not in intensive care patients that already have low GHBP levels.[73] Plasma GHBP concentrations are also low in patients with anorexia nervosa but rise after refeeding an adequate diet.[52,69,70,74,75] The development of obesity is, conversely, characterized by increased levels of plasma GHBP.[50,75,76]

4. Pathophysiological Concentrations

Plasma GHBP levels are also increased in hyperthyroid subjects but reduced in hypothyroid patients.[61] Plasma GHBP levels are also reduced by liver disease,[41,75,77–79] in uremic and nonuremic patients with renal failure,[41,69,80,81] and in insulin-dependent diabetics, although not in non-insulin dependent diabetics.[60,82–84] GHBP concentrations are also low in some[67,68] but not all[24,41,56] acromegalic populations and low in some[58,65,66] but not all[56,62–64,66] GH-deficient subjects (Table 1). Similarly, whereas GHBP is deficient or absent in most populations of Laron dwarfs,[85–89] pygmies,[90,91] and pygmy-like people[88,92] and in individuals with syndromes of GH resistance and idiopathic short stature[65,93] (Table 1), some short-statured subjects have normal GHBP levels, including some Laron dwarfs.[94,95] This variability may partly reflect methodological differences in the measurement of high-affinity GHBP activity, but it is also likely to reflect heterogeneous loci of the pathological lesions and different mutations of the GH receptor gene. Indeed, in Laron dwarfism several gene deletions, base substitutions, and nonsense mutations of the receptor gene have been identified.[86,87,94,96–106] Because these patients have deficient or dysfunctional tissue GH receptors, these findings demonstrate the dependence of this GHBP on GH receptor gene expression.

B. LOW-AFFINITY GHBP IN HUMANS

A second GHBP, which is also specific for hGH, is also present in human plasma.[4,9–12] This is a single-chain protein with a molecular mass of about 100 kDa or larger and a p*I* of 7.1.[43,107] This protein is structurally and immunologically unrelated to the high-affinity binding protein or membrane GH receptor[25,28,89,107] and is present in Laron dwarfs.[86,89] It has low affinity (approximately 10^{-5} mol) but high binding capacity (approximately 15 μg/ml) for 22-kDa hGH. This protein circulates at a concentration of 700 n*M* and complexes 7 to 8% of the 22-kDa GH monomer in plasma[3] with a binding stoichiometry of 1:1.[25,107] Unlike the high-affinity GHBP and liver GH receptor,[108] this protein also binds 20-kDa hGH with equal or higher affinity than 22-kDa hGH[25] and complexes 25% of circulatory 20-kDa GH.[3] The majority (80%) of protein-bound 20-kDa GH in plasma is complexed with this protein, although it accounts for only 10 to 15% of the total GH bound in plasma.[41] This specific binding of 20-kDa GH may be to a specific binding site on the low-affinity GHBP or to a separate, specific low-affinity GHBP, because biochemical evidence suggests low-affinity GH binding may be to a heterogeneous family of plasma proteins,[10,12,25] which may include dimerized or variant binding proteins. Another low-affinity ($K_a = 3 \times 10^6$ *M*) GHBP of higher molecular mass (165 to 174 kDa) with a lower binding capacity of 2 μg/ml has also been detected in human plasma.[43]

The low-affinity GHBP (like the high-affinity GHBP) is thought to be produced in the liver, although its plasma concentration appears to be differentially regulated.[41,43,90,92] For instance, GH binding to the low-affinity protein is minimal in fetal cord blood, and is not correlated with gestational age or birth weight.[45,52] The concentration of this protein increases fourfold by 5 years of age, remains relatively constant through adolescence, and (apart from a transient decrease at puberty) it declines between 16 and 80 years of age to the level at birth.[43,45] In contrast, GH binding to the high-affinity protein is low in cord blood but increases toward term, and further increases neonatally and progressively thereafter to reach maximum levels six- to eightfold higher than those in the fetus at 23 to 25 years of age.[45,52] Moreover, whereas the high-affinity GHBP in normal boys is negatively correlated with mean 24-h GH concentration, and GH pulse amplitude, frequency, and interpulse interval, the low-affinity GHBP is unrelated to these parameters of GH release.[40] Furthermore, whereas high-affinity GH binding is low in the plasma of children with GH deficiency and increased by GH therapy, low-affinity GH binding is increased in these patients and decreased during GH treatment.[58] Similarly, whereas high-affinity GH binding is normal in the plasma of boys with pubertal delay, low-affinity GH-binding activity is increased and unlike the high affinity protein is not reduced by testosterone therapy.[58] The low-affinity GH-binding protein is also increased in boys with precocious puberty, whether treated or untreated, whereas their high-affinity GHBP levels are low before treatment and increased after testosterone suppression.[58] Similarly, whereas GH binding to the high-affinity protein in maternal plasma is relatively unaltered during pregnancy, binding to the low-affinity protein is greatly increased.[41,45] This increase in low-affinity binding may reflect the associated rise in plasma estrogen, because increased low-affinity GH binding is also seen in other conditions characterized by high estrogen levels[12,41] and is higher in women than in men.[41] Low-affinity binding is also increased in patients with liver disease, renal failure, or acromegaly, in contrast to the lower levels of the high-affinity GHBP.[41]

C. NONHUMAN GHBPS

Although guinea pigs appear to lack plasma GHBPs,[109–111] binding proteins for GH are also present in the plasma of rats, mice, rabbits, cats, dogs, pigs, goats, sheep, cattle, horses, and monkeys (Table 2). These may reflect four classes of binding proteins, on the basis of differences in binding specificity.[110] Type I GHBPs are found in rat and mouse sera. These have a relatively low affinity for hGH and lactogenic hormones (ovine prolactin and bovine prolactin) but high specificity for somatogenic GH preparations (bovine GH [bGH] and rat GH [rGH]). Type II GHBPs are found in the goat, sheep, and cow and have low affinity for hGH but an even lower affinity for bGH, rGH, or prolactin.[112] Type III GHBPs are found in the rabbit, horse, dog, cat, and pig, in which hGH binding is highest in the rabbit and least in the cat. The affinity of hGH binding by this species is approximately one order of magnitude higher than that of type I or II GHBP and is to a mixed GH/prolactin-binding site. Type IV GHBPs resemble human GHBP and are found in the monkey. This GHBP has an affinity intermediate between types II and III and specifically binds hGH to a site that does not recognize prolactin (PRL) or nonprimate GH.

In each species, the binding of hGH (but not bGH nor bPRL) to these binding proteins is increased in the presence of Mg^{2+}, at concentrations of 3.7 to 70 m*M*.[113] This cation dependence of hGH binding to plasma GHBP is similar to the binding of hGH to hepatic membrane GH receptors and cytosolic GHBP.[113] This is therefore also consistent with the sequence identity shown to exist between serum GHBP and the extracellular domain of the hepatic GH receptor. This cation dependence is also consistent with the electrostatic model proposed by Barnard and Waters[114] for the association of GH with its binding protein. The relative GH resistance associated with cation (Zn^{2+} and Mg^{2+}) deficiency may thus result from an impairment of GH binding to plasma GHBP and/or GH receptors.[113]

The binding capacities of these plasma proteins are species related and greatest in the rabbit and least in guinea pigs and type 1 animals (Table 2). The GHBP levels in these

Table 2 **GHBPs in Different Species**

Species	Ligand	K_a (*M*)	B_{max} (nmol/l)	Ref.
Human ♂	hGH	1.1×10^{-9}	0.73	110
	hGH	$2–3.0 \times 10^{-8}$	20 ng/ml	211
	hGH	0.32×10^{-9}	1.84	14
	hGH	0.34×10^{-9}		212
Monkey ♂	hGH	1.8×10^{-9}	0.17	110
Pig	hGH	6.1×10^{-9}	0.60	110
	hGH	0.6×10^{-9}	0.8	212
	hGH	2.2×10^{-9}	4.7×10^{-10} mol/mg protein	112
	pGH	2.0×10^{-7}	4.1×10^{-9} mol/mg protein	112
Cat ♂	hGH	6.0×10^{-9}	0.38	110
Dog ♂	hGH	9.2×10^{-9}	0.67	110
	hGH	0.3×10^{-8}		212
Horse	hGH	8.0×10^{-9}	0.70	110
Rabbit ♂	hGH	9.2×10^{-9}	2.50	110
	hGH	1.6×10^{-9}	3.7	213
Cow	hGH	1.8×10^{-8}	0.89	110
Sheep ♂	hGH	2.6×10^{-8}	1.72	110
	hGH	2.3×10^{-8}	6.0×10^{-10} mol/mg protein	112
Goat ♀	hGH	3.7×10^{-9}	0.85	110
Mouse ♀	hGH	1.2×10^{-8}	10.3	110
	mGH	8.3×10^{-8}	<10–>300	120
Rat ♀	hGH	3.6×10^{-8}	19.2	110
	bGH	$1.9–2.1 \times 10^{-8}$	10.0–18.0	115
	bGH	2.2×10^{-8}	64.0	123
Rat ♂	hGH	1.5×10^{-8}	2.5	110
	bGH	1.7×10^{-8}	16.0	123
Chicken	hGH	1.5×10^{-9}	1.7×10^{-10} mol/mg protein	112
	cGH	2.1×10^{-7}	1.5×10^{-10} mol/mg protein	112

species may remain relatively constant over short-term periods in individual animals although GHBP levels differ significantly between individuals and species.[10,112] Some of this variation may be related to nutritional and endocrine states, because GHBP levels in rats are increased by uremia[114a] but reduced (by 50%) by dietary deprivation and restored by refeeding.[115,116] A reduction in plasma GHBP also occurs in diabetic rats as a result of low insulin levels.[117] These changes in GHBP concentrations are entirely due to a change in the binding capacity of the binding protein. A change in capacity rather than affinity also accounts for the progressive increase in plasma GHBP levels in rats, pigs, and sheep during ontogeny.[112,115,118] Increased GHBP concentrations also progressively occur in maternal sera of rats and mice during pregnancy.[10,119–121] The amount of plasma GHBP may also be sexually dimorphic, because the GHBP level in nonpregnant females (575 ng/ml) greatly exceeds the level (300 ng/ml) in adult male rats.[113,115,122–125] Sadeghi and Wang[125a] similarly found that the amount of immunoreactive GHBP in the plasma of male rats (0.4–0.56 μg/ml) was less than that in females (1.0–1.9 μg/ml).

This sexual dimorphism may be related to the action of estrogen in females, because plasma GHBP levels are reduced in ovariectomized or tamoxifen-treated rats and restored by exogenous estrogen.[8,126] Estrogen similarly increases GHBP concentrations in normal male and female rats, but not in GH-deficient rats.[8,126] The action of estradiol is therefore GH dependent, although in normal rats the sexual dimorphism in GHBP level is not due to the equally dimorphic estrogen-dependent pattern of pulsatile GH secretion (see Chapters 9 and 12) because plasma GHBP levels in both sexes remain stable and are not affected by acute changes in endogenous or exogenous GH.[8,124] The induction of GHBP in hypophysectomized rats continuously treated with estrogen and GH also occurs in the absence of the female pulsatile pattern of GH secretion.[8] The expression of GH receptor mRNA and GHBP mRNA in the rat is, furthermore, not acutely suppressed by hypophysectomy or by exogenous GH administration.[127–129] The plasma GHBP level may, however, be GH dependent in the longer term (in the absence of estrogen),[8,124,126] and the higher GHBP levels in female rats may reflect a more chronic stimulation of secretion rather than its pulsatile GH secretory pattern.[124] The GHBP in rat plasma is GH inducible and is deficient in GH-deficient dwarf or hypophysectomized rats and increases to normal concentrations by continuous GH infusion.[124,126,128–131] Pulse-related surges of GH at 3-h intervals may also be followed 60 min later by increased GHBP levels.[129,130] The 30- to 50-fold rise in circulating GHBP in mouse serum during pregnancy also seems to be partly GH related, because it is not seen in hypophysectomized animals in which it is partly restored by GH therapy.[120,132] Plasma GHBP levels are also induced by exogenous GH in pigs, especially in prepubertal animals with constitutively low GHBP levels.[118]

In contrast with these mammalian studies, Tobar-Dupres et al.[133] have found circulating GHBP activity in chicken plasma to decline markedly between hatch and 5 weeks of age, prior to puberty. The GHBP concentration in these birds is not sexually dimorphic and is unaffected by exogenous cortisone, although it is reduced by the stress of dietary deprivation.

D. GHBP STRUCTURE

The GH-binding proteins in the plasma of these species appear to be proteins of different molecular size (e.g., 51 kDa in rabbits,[134] 42 kDa in mice,[121] 39 kDa in rats,[122] 50 to 60 kDa in pigs,[112] and 50 to 60 kDa in chickens[135,136]) that differ between species in structural and immunological relatedness.[11,27,28,134,137,138]

The binding of GH in each species may also be to heterogeneous plasma components, because proteins with differing binding affinities, molecular mass, and amount of glycosylation have been demonstrated in rat and mouse sera.[10,28,111,117,122,139,140] Most of these components do, however, appear to be related to the membrane-bound liver GH receptor.[27,141] Indeed, the GHBP in the rabbit is structurally homologous to the extracellular ligand-binding domain of the human and rabbit GH receptor and to the human GHBP.[27,28,142] The predominant plasma GHBP in the rat and mouse is also identical to the extracellular domain of the liver membrane GH receptor, although a short C-terminal hydrophilic tail of 17 and 25 amino acids (respectively) is substituted for the transmembrane and intracellular domains of the GH receptor.[31–33,143] This hydrophilic tail ensures the secretion of the GHBP in a soluble form. In these species this GHBP results from alternate splicing at the exon 7-exon 8 boundary[27,141] of GH receptor mRNA and from the translation of this truncated (1.2-kb) transcript.[122] The GHBP in the mouse, however, does differ from the membrane receptor in its affinity for GH, despite having an identical amino acid sequence.[31,32] The membrane receptor has a 20-fold greater affinity for mouse GH than the plasma binding protein.[119,120,144] Sadeghi et al.[122] have also demonstrated at least two immunoreactive GHBP moieties in rat serum that differ in the extent of glycosylation.

Whereas Frick and Goodman[145] have identified a number of short GH receptor isoforms in rat adipocyte plasma membranes, short GH receptor gene transcripts of

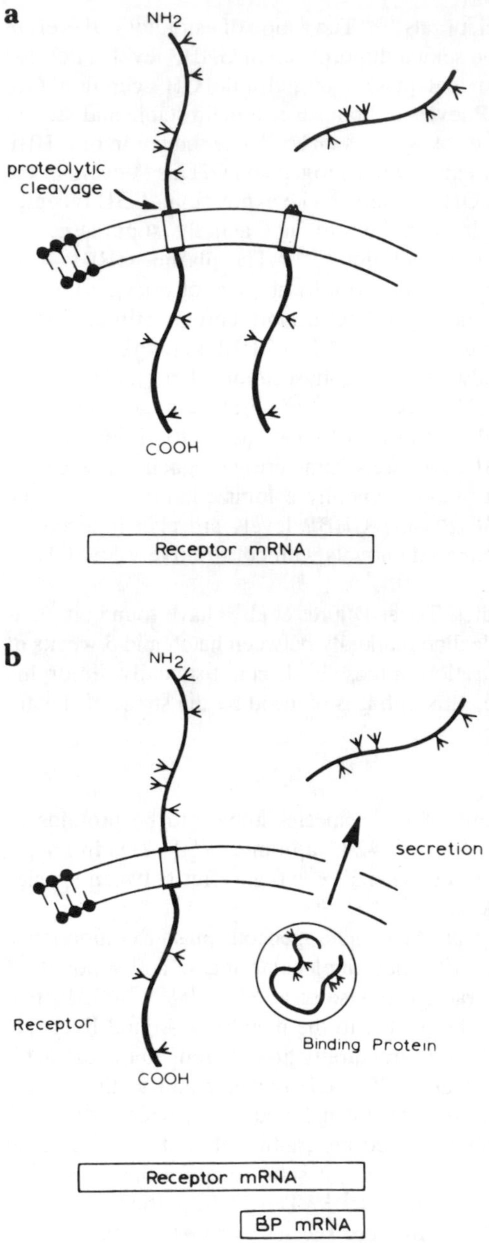

Figure 1 Two possible mechanisms for the generation of the high-affinity growth hormone binding protein (GHBP). (a) Proteolytic cleavage of the GH receptor near the transmembrane domain and shedding of the binding protein into extracellular space. (b) Separate synthesis of receptor and binding protein, with secretion of the latter. In this case, separate mRNAs for receptor and BP are produced either from two different genes (not known to exist) or through two alternative mRNA-splicing mechanisms (demonstrated in the pregnant mouse). Tree structures denote putative N-linked oligosaccharides. (From Baumann, G., Growth hormone binding proteins, *Trends Endocrinol. Metab.*, 1, 342, 1990. Copyright by Elsevier Science, Inc. Reprinted with permission.)

approximately 2.6 kb that differ from the 1.2-kb transcript in the length of the 3′ untranslated region also encode a GHBP in rat liver.[146] Short transcripts of comparable size have also been detected in other species, although in the rabbit[13,15] and chicken[147,148] these do not give rise to soluble GH-binding proteins.

E. GHBP ORIGINS

In most species it is still uncertain how GHBPs are generated (Figure 1) and it is unclear from which tissues GHBPs are derived.[6,7,11–13,15] The liver is, however, likely to be the source of GHBP production in most if not in all species.

Although the GH receptor gene, from which GHBPs are derived, is expressed in most tissues, the liver has the highest GH receptor concentration.[12] The deficiency of high-affinity GHBPs in the plasma of Laron dwarfs that lack liver GH receptors[87,90] also indicates the liver is the GHBP source. The low levels of high- and low-affinity GHBP in the plasma of patients with liver disease[41,77,78] also support this view. Changes in the production of the hepatic GH receptor and plasma GH-binding activity are also concordant during growth,[115,118] pregnancy,[120,121] nutritional status,[115,118] GH status,[129,149] and according to gender.[117,123] Gene transcripts for both the GH receptor and GHBP are present in fetal rat liver (and other tissues) and both increase in abundance pre- and postnatally and are highest in adults.[34,150–152] Both transcripts are also present in other tissues (e.g., kidney, lung, hypothalamus, and ileum) but at much lower levels and in ratios different from those in the liver.[34,151,153–155] The liver is also the source of the putative novel GHBP encoded by a 2.6-kb mRNA that differs from the 1.2-kb transcript in the length of the 3′ untranslated region. This transcript does not occur in extrahepatic tissues,[146] indicating the liver is also the source of this GHBP.

F. GHBP SYNTHESIS AND RELEASE

The GHBP in rats and mice probably arises (Figure 1) entirely from the alternate splicing of the GH receptor gene transcripts and the translation of the truncated (1.2-kb) messages.[32,33] The absence of GHBP in the media of COS-7 cells transfected with the full-length rat GH receptor cDNA[156] supports this view, especially as the mouse GHBP is abundantly present in the media of COS-7 cells transfected with the mouse GHBP cDNA, which generates only a 1.2-kb transcript.[144] The secretion of GHBP from rat liver cells or by any extrahepatic tissues has, however, yet to be demonstrated although Amit et al.[156a] recently reported the membrane shedding of GHBPs from Hep G2 human hepatoma cell lines.

In species other than rats and mice the circulating GHBP may arise from the proteolytic cleavage of the membrane receptor (Figure 1), because cultured human IM-9 lymphocytes "shed" the extracellular portion of the GH receptor *in vitro* following exposure to sulfhydryl inactivating reagents.[30] This possibility is supported by the appearance of high levels of GHBP in the media of CHO cells transfected with rabbit GH receptor cDNA, which transcribes only a 4.5-kb message for the full-length receptor.[144] This proteolytic process is likely to occur at the cell surface but may also occur proximal to the insertion of the full-length receptor in the plasma membrane. This possibility is supported by the accumulation of GHBP in the plasma of cycloheximide-treated rats deficient in membrane GH receptors, suggesting the mobilization of GHBPs from intracellular pools.[157]

G. EXTRAHEPATIC GHBPs

Although GHBPs immunologically distinct from the extracellular domain of the GH receptor are present in the liver, Lobie et al.[154,158] found GHR/GHBP immunoreactivity in the kidney and throughout the gastrointestinal tract, in skeletal and muscular systems, in skin, in the reproductive tract, in cardiovascular and respiratory systems, and in the central and peripheral nervous systems. In each tissue immunoreactivity for the binding protein was both cytoplasmic and nuclear. Large quantities of soluble GH- binding proteins have also been demonstrated in rabbit liver and kidney cytosol,[10,159–164] and in cytoplasmic and nuclear fractions of the rat,[162,165] human,[166,167] and chicken[136] liver, indicating the presence of intracellular GH-binding proteins. The source of the circulating binding protein in most species may therefore be intracellular rather than from the shedding of membrane receptors, especially as the total amount of binding protein in rabbit serum is far greater than that present in liver membranes.[168] The ubiquitous distribution of GHBPs and GHBP mRNA in rat tissues (Table 3) also suggests that most

Table 3 **Tissue Expression of Growth Hormone-Binding Proteins**

	GHBP mRNA			
Tissue[a]	**1.2 kb**	**Ref.**	**2.6 kb**	**Ref.**
Liver	+	15,150	+	146
Kidney	+	15,34,150	–	146
Skeletal muscle	+	15,34	–	146
Adrenal	+	15,215	–	146
Heart	+	15,34	–	146
Adipose	+	15,145	–	146
Mammary gland	+	15		
Ovary	+	15	–	146
Stomach	+	15		
Corpus Antrus	+	15		
Intestine	+	34,150		
Duodenum	+	15	–	146
Jejunum	+	15		
Ileum	+	15		
Colon	+	15		
Lung	+	150		
Hypothalamus	+	153		
Skin	+	34		
Tongue	+	34		
Brain	–	15		
Testis	–	15	–	146
Thymus	–	15	–	146
Spleen	–	34		

Note: +, Present; –, absent.

[a] In rats.

tissues of the body may secrete the binding proteins in circulation, including the pituitary gland (Figure 2).

The presence of GHBPs in the secretory granules of adenohypophysial cells[165,169] is indicative of the exocytotic discharge of GHBPs into the bloodstream, at least from this tissue. The colocalization of GH and GHBP in the same secretory granules of somatotroph cells also suggests that some of the high molecular weight GH moieties secreted from the pituitary gland are protein-bound variants. However, the widespread distribution of GHBP and its presence in cells lacking ready access to serum (e.g., neurons and oocytes)[154,158,170,171] suggest the release of intracellular GHBPs may be tissue specific. Indeed, the soluble GHBP in rat adipocytes is not released into incubation media and most (>90%) associates with membrane fractions of the cells.[145] It is therefore possible that GHBPs have functions at the cellular/tissue level that are independent of GH binding in plasma.

H. GHBP TURNOVER

In most individuals, plasma GHBP concentrations are relatively constant,[36,44] indicating a dynamic interplay between the rate of GHBP production and the rate of GHBP clearance. Indeed, when protein synthesis is pharmacologically blocked the concentrations of GHBP in rat plasma immediately decline at a rate consistent with first-order kinetics, in parallel with the disappearance of hepatic GH receptors.[157]

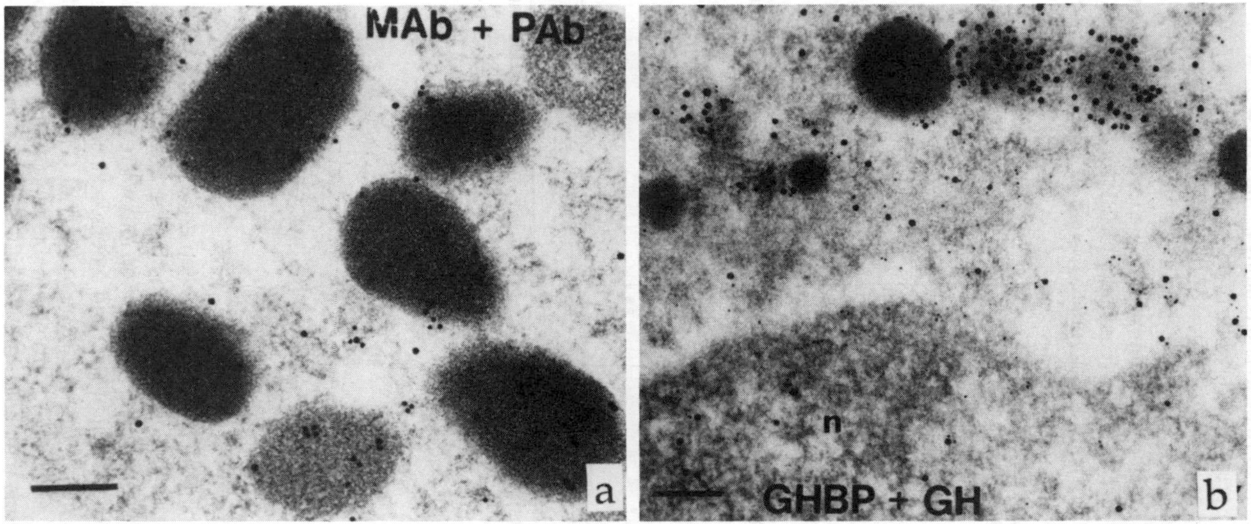

Figure 2 Localization of growth hormone (GH)-binding protein (GHBP) in secretory granules of the rat pituitary gland. (a) Colocalization of MAb 4.3 (10-nm gold particles) and polyclonal antibody to the GHBP (15-nm gold particles). Both antibodies localize to the same secretory granules. (Magnification, ×10,000; bar, = 10 μm.) (b) Colocalization of GHBP (5-nm particles) and GH (15-nm particles) antibodies in secretory granules of rat pituitary cells. There is colocalization to the secretory granules but not to the nucleus (n). (Magnification, ×9,200; bar, = 10 μm). (From Harvey, S., Baumbach, W. R., Sadeghi, H., and Sanders, C. J., Ultrastructural colocalization of growth hormone binding protein and pituitary hormones in adenohypophyseal cells of the rat, *Endocrinology (Baltimore)*, 133(3), 1125–1130, 1993. Copyright by the Endocrine Society. Reprinted with permission.)

The clearance of recombinant hGHBPs in guinea pigs occurs with a $t_{1/2}$ of 11 to 20 min,[109] although the degradative route is unknown. The molecular size of these GHBPs (22 to 28 kDa) is not large enough to inhibit renal filtration, although only <0.01% of exogenous hGHBP appears in the urine of rats in an immunoreactive form.[109] Although GHBPs with molecular size (50 to 70 kDa) and binding affinity ($K_a = 2.3 \times 10^{-8} M$) similar to those in human plasma are present in the urine of normal subjects, urinary GHBPs have considerably higher binding capacities.[172] Much of the GHBP in urine may therefore be derived from the renal tubules rather than the glomerular filtration of plasma, especially as the proportion of GHBP in urinary protein is much higher than that in plasma protein. The larger (30 to 50-kDa) size of the glycosylated GHBP molecule would, however, impair renal filtration and would likely attenuate its rate of GHBP clearance and the

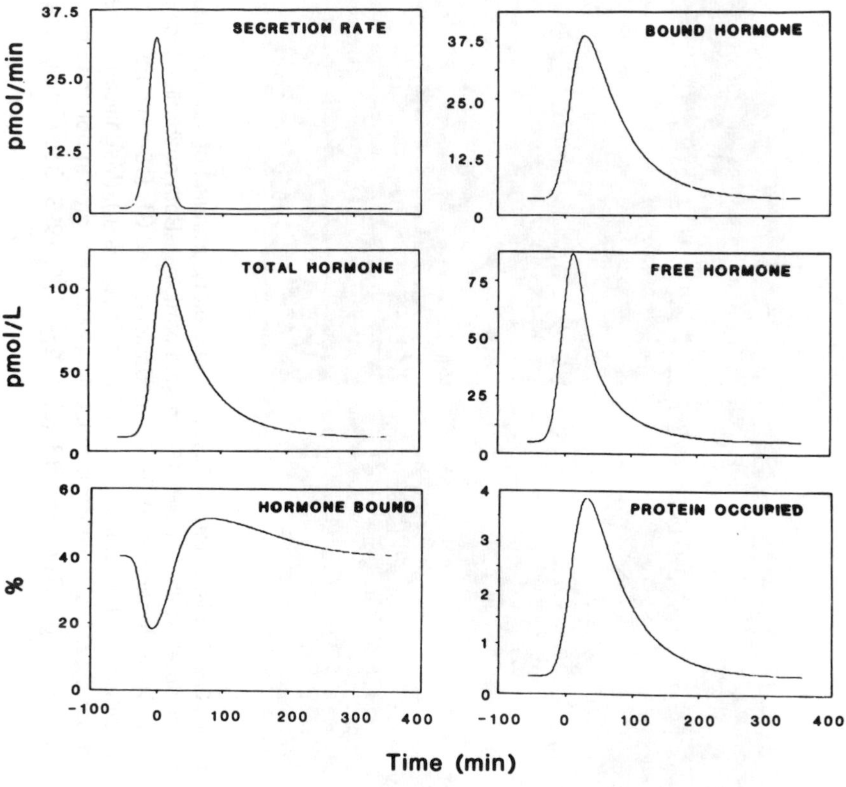

Figure 3 Model of growth hormone (GH) secretion and clearance and GH association with higher affinity GH-binding protein (BP) in plasma. Assuming a half-life of free GH of 18 min, note that a burst of GH secretion (top left) will result in a rapid increase in plasma free and total GH concentrations, and a delayed increase in protein-bound GH concentrations. The percentages of GH bound and of GHBP occupied are also shown. The presence of GH-binding protein in blood tends to prolong the GH secretory event and delay the metabolic removal of GH. The plots shown are based on an assumed equilibrium dissociation constant of 1.5 nmol/l, a binding capacity of 1.0 nmol/l, a distribution volume of 4.9l, and a mass of GH secreted of 1 nmol (22 µg). (From Iranmanesh, A. and Veldhuis, J. D., Clinical pathophysiology of the somatotropic (GH) axis in adults, *Endocrinol. Metab. Clin. N. Am.*, 21, 783–816, 1992. Copyright by W. B. Saunders. Reprinted with permission.)

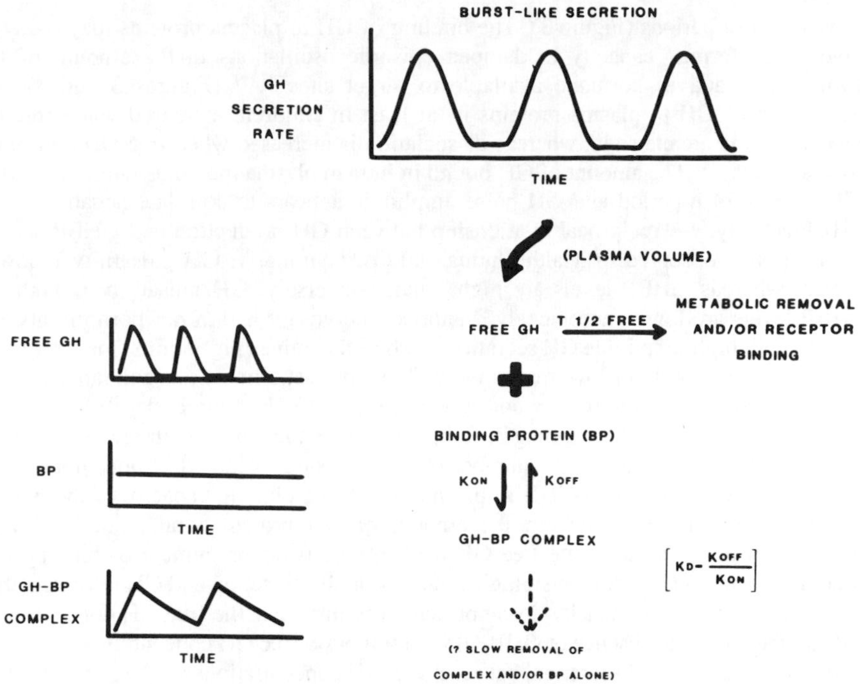

Figure 4 Schema illustrating the nonequilibrium kinetic behavior of the growth hormone (GH) axis, assuming the presence of a high-affinity GH-binding protein, GHBP, in plasma, metabolic clearance of GH from the free compartment and an episodic burstlike mode of GH secretion. (From Veldhuis, J. D., Johnson, M. L., Faunt, L. M., et al., Influence of the high-affinity growth hormone (GH)-binding protein on plasma profiles of free and bound GH and on the apparent half-life of GH, *J. Clin. Invest.*, 91, 629, 1993. Reprinted with permission.)

GHBP in rats has a $t_{1/2}$ of 83 to 120 min.[157] The complexing of GH to GHBP and the increase in molecular size (to 50 to 70 kDa) similarly prolong the $t_{1/2}$ of recombinant GHBP in guinea pigs to 75 to 94 min.[109] The complexing of GH to GHBP is also likely to limit protease access to cleavage sites within its GH-binding domain. Complexing with GH also restricts GHBP distribution to the vascular compartment,[173,174] in which it is likely to be less susceptible to protease digestion, because the turnover of full-length membrane receptors is approximately three times as fast as truncated binding proteins.[157]

I. GHBP ROLES

Binding proteins may increase ligand solubility and hence increase ligand concentrations in plasma. However, as high-affinity GHBPs have the same p*I* as GH, increasing GH solubility is unlikely to be a carrier role of plasma GHBP. Instead, as GH action is dependent on circulating GH levels, the pattern of GH secretion, the rate of GH turnover, access to extravascular space and to target cells, the number and affinity of cell surface receptors, and the binding of GH to intravascular GHBPs may modulate GH action.[15]

1. Intravascular GHBP and GH Concentrations

The release of GH is characteristically pulsatile, and GH concentrations in plasma are rapidly elevated during and shortly after each secretory episode and low or undetectable between these periods (Figure 3). The binding of GH to plasma proteins may therefore have a "buffering" capacity to dampen episodic oscillations in the amount of free (biologically active) hormone available to target sites[175–177] (Figures 3 and 4). The association of GH to plasma proteins is, at least in children, increased when free GH concentrations are elevated, whereas dissociation is increased when free GH concentrations are low.[36,37] The amount of GH bound in human plasma may thus range from 10 to 80% over a 24-h period and GH pulse amplitude appears to correlate negatively with GHBP activity.[40] A reciprocal relationship between GH production and GHBP activity has also been observed in healthy human adults.[64] Similarly, GH pulsatility is low in obesity whereas GHBP levels are high[69] and, conversely, GH pulsatility is high and GHBP levels are low in acromegaly,[68] anorexia nervosa,[69] and in newborn infants.[48] In species with high-amplitude GH secretion, such as the guinea pig[178] or the male rat,[179] the GHBP is either absent or low, respectively.[110] In contrast, female rats and rabbits have a higher GHBP concentration[110,124] and much lower GH pulse amplitude.[179,180] However, whereas the affinity of the binding protein determines the absolute and relative amount of GH bound in plasma and the number of binding sites occupied, it does not directly determine the amount of free GH in plasma.[175–177] The binding capacity of the binding protein is also a determinant of the amount of GH protein bound, but it is not a determinant of the steady state free GH concentrations or the number of binding sites occupied. However, within physiological ranges of GH secretion and GH disappearance, the binding affinity or capacity does not markedly influence the free GH concentration. Indeed, the complete absence of GHBP does not modify free GH concentrations at steady state for any rate of GH secretion, because free GH concentrations are largely controlled by the GH secretion rate, half-life, and volume of distribution.[175–177]

2. Intravascular GHBP and GH Turnover

The free GH moiety in plasma is the major, but not necessarily exclusive, moiety subject to irreversible metabolic removal. In healthy individuals the half-life ($t_{1/2}$) of free GH ranges between 2 and 12 min, with a metabolic clearance rate (MCR) of approximately 3 ml/min/kg (see Chapter 15). The complexing of GH to plasma proteins prolongs (by six- to tenfold) the half-life of the bound hormone and also slows the metabolic clearance of the binding protein.[109,173,174] The increased molecular size of the GH:GHBP complex greatly restricts its distribution volume to twice the intravascular space, whereas free GH has a much larger distribution volume and access to the entire extracellular space.[173] Protein binding therefore also diminishes the receptor-mediated clearance of GH in target cells.[181] The large size of the GH complex also prolongs the rate of GH degradation by preventing glomerular filtration within the renal tubule,[182] which is the main site of GH elimination. Protein-bound GH in the circulation may thus be a potential reservoir that may be mobilized during quiescent periods of GH secretion,[15,182] to dampen oscillations in the free GH level (Figures 3 and 4). It is, however, possible that intravascular GHBP could prevent low levels of GH passing through the capillary barrier to target tissues, thereby raising the threshold concentration of GH required for the induction of biological activity and ensuring quiescent periods of GH activity between episodic pulses of GH release.[109]

3. Intravascular GHBP and GH Action

The restriction of circulating protein-bound GH to the vascular compartment attenuates GH action by limiting GH access to receptors on target cells.[75] The high-affinity (but not low-affinity) binding protein is also capable of diminishing GH activity by competing with membrane receptors for free GH, in view of their similar structure and GH affinity.[75,163,183] Studies with recombinant-derived or purified rabbit/human GHBP show that the GHBP can inhibit GH binding to cellular receptors and also inhibits the action of GH in stimulating adipogenesis in 3T3-F 442A adipocytes[184] and insulin-like growth factor I (IGF-I) production by cultured human fibroblasts.[183] This impairment of GH action may therefore be an autoregulatory mechanism, because GH may increase GHBP concentrations.[8,124] This impairment of *in vitro* GH activity occurs, however, only at physiological doses and can be overcome by supraphysiological GH concentrations and saturation of GH binding to the binding protein. The binding of GH to GHBPs may also modulate GH action by reducing the turnover of membrane GH receptors, because the receptor has a shorter half-life when it is occupied by its ligand.[181] In recent studies Hansen et al.[185] have similarly found that hGHBP from human IM-9 lymphocytes binds both 22- and 20-kDa GH and blocks somatogenic and lactogenic actions. Similarly, Dattani et al.[185a] recently found that GHBP blocked the stimulation of Nb2 cells induced by 20 kDa hGH and 22 kDa hGH in a competitive manner.

The ontogenic increase in GHBP concentrations may therefore provide a mechanism by which to attenuate somatic growth.[150] This possibility is supported by the enhanced growth in rats immunized with GH antibodies,[186] which prevent GH from interacting with circulating GHBPs and make more GH available to tissue receptors. The attenuated pubertal growth spurt in precocious girls chronically treated with a luteinizing hormone-

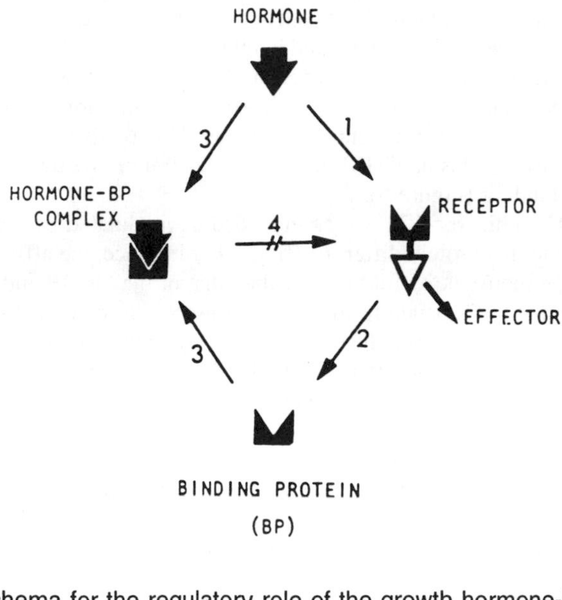

Figure 5 A schema for the regulatory role of the growth hormone-binding protein. (1) The hormone binds to its specific receptor at the cell surface, stimulating receptor upregulation and accelerating the receptor turnover. (2) Upregulation of the receptor is followed by an increase in receptor turnover rate and results in an increased secretion of binding protein, either by shedding of the extracellular domain or by corollary synthesis of the soluble fraction. (3) The secreted binding protein forms a complex with the hormone, which in itself slows hormonal clearance. (4) By forming the complex, the binding protein backregulates the receptor to dampen the hormone binding and thereby the biological effect (effector). (Reprinted from Hochberg, Z., Amit, T., and Youdin, M. B. H., The growth hormone binding protein as a paradigm of the erythropoietin superfamily of receptors, *Cell. Signal.*, 3, 85–91. Copyright 1991, with kind permission from Elsevier Science Ltd., The Boulevard, Langford Lane, Kidlington OX5 1GB U.K.)

releasing hormone analog is also thought to be due to an accompanying increase in serum GHBP levels and diminished GH binding to tissue receptors.[59] The reduced GHBP concentrations during dietary restriction[75,115] may, conversely, potentiate the lipolytic and antiinsulin-like effects of GH. Similarly, the reduced GHBP levels and elevated GH concentrations in cirrhotic patients may facilitate the binding of GH to the limited GH receptors in cirrhotic livers[78] and the extremely high levels of GHBP in obese patients may impair GH-induced lipolysis and result in excessive adiposity.[41,75]

The action of GH *in vivo* may, however, be enhanced by binding the GHBP in plasma because the rate of GH turnover is greatly reduced and GH availability is prolonged. Indeed, the growth-promoting effects of GH in normal and dwarf rats and GH-deficient children are enhanced by GHBP.[8,64,187] This prolongation of GH bioactivity may diminish the requirement for pulsatile episodes of GH release, especially as GHBP levels are positively correlated with the duration of interpulse intervals.[40] The binding of GH to plasma binding proteins not only modulates GH action but also the turnover of membrane GH receptors. Occupancy of the GH receptor is associated with a shorter half-life of the receptor[181] and the binding of GH to GHBP would reduce GH receptor turnover and GH degradation.[75] Actions of GHBP that potentiate receptor-mediated GH bioactivity are also indicated by the parallel changes in the concentrations of plasma GHBP and hepatic GH receptors,[118,123,149] particularly during ontogeny[34,115,150,151] and pregnancy,[120] especially as GHBPs are absent or nonfunctional in GH-resistant Laron dwarfs.[87,89] The positive correlations between overall body mass and height with serum GHBP levels[44,50,51,69,188,189] also support this view. It has therefore been suggested[190] that the biological role of the GHBP is to increase the availability of GH, resulting in an upregulation and increased turnover of the tissue GH receptor, and in its increased internalization and recycling. The concomitant increase in plasma GHBP impairs the action of GH by competing with the receptor for GH binding (Figure 5).

Actions of GH within specific tissues may also be modulated by binding proteins if the receptor and binding protein differ in affinity. For instance, the affinity of the hepatic GH receptor in the mouse is 20-fold greater than that of the GHBP and ligand bound to the plasma protein will dissociate from it in the presence of the receptor.[20a] Because the amount of GHBP greatly exceeds the amount of GH in the circulation of pregnant mice, the binding protein may serve to direct GH to the liver and other tissues with pregnancy-induced increases in GH receptor expression.[120,144,191]

J. EXTRAVASCULAR GHBPS

The cellular release of GHBPs or the shedding of the extracellular binding domains of membrane GH receptors (Figure 1) indicates the presence of GHBPs in extravascular spaces, in which they may also modulate GH action or metabolism. Although free GH and free GHBP must be able to traverse capillary barriers, the association of GH and GHBPs within the extravascular space is likely to prevent their diffusion into the intravascular compartment. Consequently, extravascular GHBPs would locally attract, protect, and retain high concentrations of GH around cells.[109] Thus, after a secretory pulse of pituitary GH release, GH could be trapped extravascularly until its escape in lymph, or by dilution and dissociation, or by its uptake into cells. Within the extravascular space GHBPs could also potentiate or competitively inhibit GH receptor-mediated GH actions.

In addition to plasma, high-affinity GHBPs have also been identified in milk.[192–195] These milk GHBPs differ from the GHBP in human serum in hormone specificity, molecular size, and solid-phase binding characteristics. The presence of these GHBPs in milk also suggests roles other than those of the serum GHBP. These roles may include the transport of GH through mammary epithelial cells and the modulation of GH actions within the mammary gland and/or intestine of the suckling neonates. The presence of

GHBPs in the follicular fluid of human preovulatory follicles also suggests novel GHBP roles in modulating steroidogenesis and folliculogenesis.[171]

K. INTRACELLULAR GHBPS

It is now clear that GHBPs are abundantly present in many tissues and in numerous intracellular compartments and GHBPs may have actions at the cellular or tissue level independent of GH binding in plasma. This possibility is supported by the presence of GHBP in cells without ready access to serum proteins[154,171] and by the independent regulation of hepatic and extrahepatic GHBP mRNA.[34,146,147,150]

The possible roles of GHBPs at the cellular level are, however, uncertain. The complexing of GH to GHBP within target tissues may, for instance, enhance GH stability and prolong its biological activity. Differential binding of GH to membrane GH receptors or to membrane GHBPs[195a] may also determine the intracellular compartment into which GH becomes internalized. The binding of GH to membrane proteins permits the entry of GH into at least five cellular compartments, one of which bypasses lysosomal degradation and provides a pool from which intact GH is subsequently reexternalized.[181] The binding of GH to intracellular GHBPs rather than to intracellular GH receptors[136,154,158,165,169,196] may also competitively inhibit receptor-mediated GH actions. Intracellular GHBPs may also facilitate the intracellular trafficking of GH and possibly its translocation to the nucleus and induction of genomic responses. The tight association of GHBPs within the chromatin fraction of the nucleus[154,158] supports this view, especially as there is an upstream *cis*-acting GH response element within the region of the rat liver protease inhibitor gene[197] and as a GH-responsive exon has been detected in the *IGF-I* gene.[198] It is therefore possible that the GHBP may participate in GH signal transduction, especially as GHBPs associate with membrane proteins in adipocyte cells and are not secreted.[145] It is also possible that GHBPs have roles independent of ligand binding.

REFERENCES

1. **Baulieu, E. E.,** *Hormones: From Molecules to Disease* (Eds. Baulieu, E. E. and Kelly, P. A.). Chapman & Hall, New York, 1990, p. 3.
2. **Levidiotis, M., Perry, R. A., Wintou, E. M., et al.,** *Neuroendocrinology*, 53, 222, 1991.
3. **Baumann, G., Amburn, K., and Shaw, M.,** *Endocrinology (Baltimore)*, 122, 976, 1988.
4. **Baumann, G., Vance, M. L., Shaw, M. A., et al.,** *J. Clin. Endocrinol. Metab.*, 71, 470, 1990.
5. **Barsano, C. P. and Baumann, G.,** *Endocrinology (Baltimore)*, 124, 1101, 1989.
6. **Kelly, P. A., Ali, S., Rozakis, M., et al.,** *Recent Prog. Horm. Res.*, 48, 123, 1993.
7. **Kelly, P. A., Djiane, J., Postel-Vinay, M. C., et al.,** *Endocr. Rev.*, 12, 235, 1991.
8. **Robinson, I. C. A. F., Carmignac, D. F., and Fairhall, K. M.,** *Acta Pediatr. (Suppl.)*, 391, 22, 1993.
9. **Baumann, G.,** *Trends Endocrinol. Metab.*, 1342, 1990.
10. **Baumann, G.,** *Acta Endocrinol. (Copenh.)*, 124, 21, 1991.

10a. **Baumann, G.,** *J. Endocrinol.*, 141, 1, 1994.

10b. **Baumann, G., Shaw, M. A., and Amburn, K.,** *J. Endocrinol. Invest.*, 17, 67, 1994.

11. **Baumann, G.,** *Endocr. Rev.*, 12, 424, 1991.
12. **Baumann, G.,** *Proc. Soc. Exp. Biol. Med.*, 202, 392, 1993.
13. **Herington, A. C., Tiong, T. S., and Ymer, S. I.,** *Acta Pediatr. Scand.*, 61, 1991.
14. **Herington, A. C., Ymer, S., and Stevenson, J.,** *J. Clin. Invest.*, 77, 1817, 1986.

14a. **Herington, A. C.,** *J. Pediatr. Endocrinol.*, 6, 235, 1993.

15. **Herington, A. C., Ymer, S. I., and Tiong, T. S.,** *Acta Endocrinol. (Copenh.)*, 124, 14, 1991.
16. **Waters, M. J., Barnard, R. T., Lobie, P. E., et al.,** *Acta Pediatr. Scand.*, 366(Suppl.), 60, 1990.
17. **Waters, M. J., Barnard, R., and Hamlin, G.,** *Progress in Endocrinology* (Eds. Imura, H., et al.). Elsevier Science, Amsterdam, 1988, p. 601.
18. **Waters, M. J., Spencer, S. A., Leung, D., et al.,** *Biotechnology in Growth Regulation* (Ed. Heap, B.). Butterworths, Surrey, 1989, p. 15.

19. **Fuh, G., Mulkerrin, N. G., Bass, S., et al.,** *J. Biol. Chem.*, 265, 3111, 1990.
20. **Bass, S. H., Mulkerrin, M. G., and Wells, J. A.,** *Proc. Natl. Acad. Sci. U.S.A.*, 88, 4498, 1991.
20a. **Baumann, G., Lowman, H. B., Mercado, M., et al.,** *J. Clin. Endocrinol. Metab.*, 78, 1113, 1994.
21. **Wells, J. A., Cunningham, B. C., Fuh, G., et al.,** *Recent Prog. Horm. Res.*, 48, 253, 1993.
22. **Wells, J. A. and DeVos, A. M.,** *Annu. Rev. Biophys. Biomol. Struct.*, 22329, 1993.
23. **Baumann, G., Davila, N., Shaw, M. A., et al.,** *J. Clin. Endocrinol. Metab.*, 73, 1175, 1991.
24. **Baumann, G., Stolar, M. W., Amburn, K., et al.,** *J. Clin. Endocrinol. Metab.*, 62, 134, 1986.
25. **Baumann, G. and Shaw, M. A.,** *J. Clin. Endocrinol. Metab.*, 71, 1339, 1990.
26. **Blumenfeld, Z., Barkey, R. J., Youdim, M. B. H., et al.,** *J. Clin. Endocrinol. Metab.*, 75, 1242, 1992.
27. **Leung, D. W., Spencer, S. A., Cachianes, G., et al.,** *Nature (Lond.)*, 330, 537, 1987.
28. **Baumann, G. and Shaw, M. A.,** *Biochem. Biophys. Res. Commun.*, 152, 573, 1988.
29. **Barnard, R., Quirk, P., and Waters, M. J.,** *J. Endocrinol.*, 123, 327, 1989.
30. **Trivedi, B. and Daughaday, W. H.,** *Endocrinology (Baltimore)*, 123, 2201, 1988.
31. **Smith, W. C. and Talamantes, F.,** *J. Biol. Chem.*, 262, 2213, 1987.
32. **Smith, W. C., Linzer, D. I. H., and Talamantes, F.,** *Proc. Natl. Acad. Sci.*, 85, 9576, 1988.
33. **Baumbach, W. R., Horner, D. L. and Logan, J. S.,** *Gene Dev.*, 3, 1199, 1989.
34. **Carlsson, B., Billig, H., Rymo, L., et al.,** *Mol. Cell. Endocrinol.*, 73, R1, 1990.
35. **Herington, A. C., Ymer, S., Roupas, P., et al.,** *Biochim. Biophys. Acta*, 881, 236, 1988.
36. **Carlsson, L. M. S., Rosberg, S., Vitangcol, R. V., et al.,** *J. Clin. Endocrinol. Metab.*, 77, 356, 1993.
37. **Hochberg, Z., Amit, T., and Zadik, Z.,** *J. Clin. Endocrinol. Metab.*, 72, 236, 1991.
38. **Snow, K. J., Shaw, M. A., Winer, L. M., et al.,** *J. Clin. Endocrinol. Metab.*, 70, 417, 1990.
39. **Hochberg, Z., Bick, T., and Harel, Z.,** *Endocrinology (Baltimore)*, 126, 325, 1990.
40. **Martha, P. M., Rogol, A. D., Blizzard, R. M., et al.,** *J. Clin. Endocrinol. Metab.*, 73, 175, 1991.
41. **Baumann, G., Shaw, M. A., and Amburn, K.,** *Metabolism*, 38, 683, 1989.
42. **Daughaday, W. H., Trivedi, B., and Andrews, B. A.,** *J. Clin. Endocrinol. Metab.*, 65, 1072, 1987.
43. **Tar, A., Hocquette, J. F., Souberbielle, J. C., et al.,** *J. Clin. Endocrinol. Metab.*, 71, 1202, 1990.
44. **Martha, P. M., Rogol, A. D., Carlsson, L. M. S., et al.,** *J. Clin. Endocrinol. Metab.*, 77, 452, 1993.
45. **Merimee, T. J., Russell, B., and Quinn, S.,** *J. Clin. Endocrinol. Metab.*, 75, 852, 1992.
46. **Hattori, N., Kurahachi, H., Ikikubo, K., et al.,** *Clin. Endocrinol.*, 35, 295, 1991.
47. **Merimee, T. J., Russell, B., Quinn, S., et al.,** *J. Clin. Endocrinol. Metab.*, 73, 1031, 1991.
48. **Amit, T., Barkey, R. J., Youdim, M. B. H., et al.,** *J. Clin. Endocrinol. Metab.*, 71, 474, 1990.
49. **Barnard, R., Quirk, P., and Waters, M. J.,** *J. Endocrinol.*, 123, 327, 1989.
50. **Silbergeld, A., Lazar, L., Erster, B., et al.,** *Clin. Endocrinol.*, 31, 295, 1989.
51. **Holl, R. W., Snehotta, R., Siegler, B., et al.,** *Horm. Res.*, 35, 190, 1991.
52. **Massa, G., Dezegher, F., and Vanderschueren-Lodeweyckx, M.,** *Pediatr. Res.*, 32, 69, 1992.
53. **Massa, G., Bouillon, R., and Vanderschueren-Lodeweyckx, M.,** *J. Clin. Endocrinol. Metab.*, 75, 1298, 1992.
54. **Massa, G., Bouillon, R., and Vanderschueren-Lodeweyckx, M.,** *Clin. Endocrinol.*, 37, 175, 1992.
55. **Massa, G., Igout, A., Rombauts, L., et al.,** *Clin. Endocrinol. (Oxford)*, 39, 569, 1993.
56. **Ho, K. K. Y., Valiontis, E., Waters, M. J., et al.,** *J. Clin. Endocrinol. Metab.*, 76, 302, 1993.
57. **Weissberger, A. J., Ho, K. K. Y., and Lazarus, L.,** *J. Clin. Endocrinol. Metab.*, 73, 374, 1991.
58. **Postel-Vinay, M. C., Tar, A., Hocquette, F., et al.,** *J. Clin. Endocrinol. Metab.*, 73, 197, 1991.
59. **Oliveira, S. B., Donnadieu, M., and Chaussain, J. L.,** *Horm. Res.*, 39, 42, 1993.
60. **Massa, G., Dooms, L., Bouillon, R., et al.,** *Diabetologia*, 36, 239, 1993.
61. **Amit, T., Hertz, P., Ish-Shalom, S., et al.,** *Clin. Endocrinol. (Oxford)*, 35, 159, 1991.
62. **Martha, P. M. J., Reiter, E. O., Davila, N., et al.,** *J. Clin. Endocrinol. Metab.*, 75, 1470, 1992.
63. **Ho, K. K. Y., Jorgensen, J. O. L., Valiontis, E., et al.,** *Clin. Endocrinol.*, 38, 143, 1993.
64. **Martha, P. M., Reiter, E. O., Davila, N., et al.,** *J. Clin. Endocrinol. Metab.*, 75, 1464, 1992.
65. **Fontoura, M., Mugnier, E., Brauner, R., et al.,** *Clin. Endocrinol. (Oxford)*, 37, 249, 1992.
66. **Hochberg, Z., Barkey, R., Even, L., et al.,** *Acta Endocrinol. (Copenh.)*, 125, 23, 1991.
67. **Roelen, C. A. M., Donker, G. H., Thijssen, J. H. H., et al.,** *Clin. Endocrinol. (Oxford)*, 37, 373, 1992.
68. **Amit, T., Ish-Shalom, S., Glaser, B., et al.,** *Horm. Res.*, 37, 205, 1992.
68a. **Postel-Vinay, M. C., Leger, J., Sotiropoulos, A., et al.,** *J. Pediatr. Endocrinol.*, 6, 241, 1993.
69. **Hochberg, Z., Hertz, P., Ish-Shalom, S., et al.,** *Metabolism*, 41, 106, 1992.

70. **Counts, D. R., Gwirtsman, H., Carlsson, L. M. S., et al.,** *J. Clin. Endocrinol. Metab.*, 75, 762, 1992.
71. **Baumann, G., Shaw, M. A., Merimee, T. J., et al.,** *Clin. Res.*, 36, 477A, 1988.
72. **Malumba, N., Massa, G., Ketelslegers, J. M., et al.,** *Acta Endocrinol. (Copenh.)*, 125, 409, 1991.
73. **Ross, R. J. M., Miell, J. P., Holly, J. M. P., et al.,** *Clin. Endocrinol.*, 35, 361, 1991.
74. **Murata, A., Yasuda, T., and Niimi, H.,** *Horm. Metab. Res.*, 24, 297, 1992.
75. **Amit, T., Barkey, R. J., Youdim, M. B. H., et al.,** *Metabolism*, 41, 732, 1992.
76. **Aguirre, A., Donnadieu, M., and Job, J. C.,** *Horm. Metab. Res.*, 23, 281, 1991.
77. **Baruch, Y., Amit, T., Hertz, P., et al.,** *J. Clin. Endocrinol. Metab.*, 73, 777, 1991.
78. **Hattori, N., Kurahachi, H., Ikekubo, K., et al.,** *Metabolism*, 41, 377, 1992.
79. **Baumann, G., Shaw, M. A., and Amburn, K.,** *Acta Endocrinol. (Copenh.)*, 119, 329, 1988.
80. **Maheshwari, H. G., Rifkin, I., Butler, J., et al.,** *Acta Endocrinol. (Copenh.)*, 127, 485, 1992.
81. **Postel-Vinay, M. C., Tar, A., Crosnier, H., et al.,** *Pediatr. Nephrol.*, 5, 545, 1991.
82. **Mercado, M., Molitch, M. E., and Baumann, G.,** *Diabetes*, 41, 605, 1992.
83. **Holl, R. W., Siegler, B., Scherbaum, W. A., et al.,** *J. Clin. Endocrinol. Metab.*, 76, 165, 1993.
84. **Menon, R. K., Arslanian, S., May, B., et al.,** *J. Clin. Endocrinol. Metab.*, 74, 934, 1992.
85. **Tauber, M., De Bovet, Du Portal, H., Sallerin-Caute, B., et al.,** *J. Clin. Endocrinol. Metab.*, 76, 1135, 1993.
86. **Amselem, S., Duquesnoy, P., Attree, O., et al.,** *N. Engl. J. Med.*, 321, 990, 1989.
87. **Laron, Z., Klinger, B., Erster, B., et al.,** *Acta Endocrinol. (Copenh.)*, 121, 603, 1111.
88. **Rosenbloom, A. C., Guevara Aguirre, J., Rosenfeld, R. G., et al.,** *N. Engl. J. Med.*, 323, 1367, 1990.
89. **Baumann, G., Shaw, M. A., and Winter, R. J.,** *J. Clin. Endocrinol. Metab.*, 65, 814, 1987.
90. **Baumann, G., Shaw, M. A., and Merimee, T. J.,** *N. Engl. J. Med.*, 230, 1705, 1989.
91. **Merimee, T. J., Baumann, G., and Daughaday, W. H.,** *J. Clin. Endocrinol. Metab.*, 71, 1183, 1990.
92. **Baumann, G., Shaw, M. A., Brumbaugh, R. C., et al.,** *J. Clin. Endocrinol. Metab.*, 72, 1346, 1991.
93. **Carlsson, L. M. S., Attie, K. M., Compton, P. G., et al.,** *Pediatr. Res.*, 31, 74A, 1992.
94. **Buchanan, C. R., Mahelshwari, H. G., Norman, M. R., et al.,** *Clin. Endocrinol. (Oxford)*, 35, 179, 1991.
95. **Savage, M. O., Blum, W. F., Ranke, M. B., et al.,** *J. Clin. Endocrinol. Metab.*, 77, 1465, 1993.
96. **Godowski, P. J., Leung, D. W., Meacham, L. R., et al.,** *Proc. Natl. Acad. Sci. U.S.A.*., 86, 8083, 1989.
97. **Amselem, S., Duquesnoy, P., Duriez, B., et al.,** *Hum. Mol. Genet.*, 2, 355, 1993.
98. **Amselem, S., Duquesnoy, P., and Goossens, M.,** *Trends Endocrinol. Metab.*, 2, 35, 1991.
99. **Amselem, S., Sobrier, M. L., Duquesnoy, P., et al.,** *J. Clin. Invest.*, 87, 1987, 1991.
100. **Amselem, S., Sobrier, M. L., Duquesnoy, P., et al.,** *Ann. d'Endocrinol.*, 52, 474, 1991.
101. **Amselem, S., Duriez, B., Sobrier, M. L., et al.,** *Acta Pediatr. Scand.*, 81, 132, 1992.
102. **Amselem, S., Duquesnoy, P., Sobrier, M. L., et al.,** *Acta Pediatr. Scand.*, 377, 81, 1991.
103. **Aguirre, A., Donnadieu, M., and Job, J. C.,** *Horm. Res.*, 34, 4, 1990.
104. **Duquesnoy, P., Sobrier, M. L., Amselem, S., et al.,** *Proc. Natl. Acad. Sci. U.S.A.*, 88, 10272, 1991.
105. **Fielder, P., Guevera-Aguirre, J., Rosenbloom, A., et al.,** *J. Clin. Endocrinol. Metab.*,74,743, 1992.
106. **Daughaday, W. H. and Trivedi, B.,** *Proc. Natl. Acad. Sci. U.S.A.*, 84, 4636, 1987.
107. **Baumann, G. and Shaw, M. A.,** *J. Clin. Endocrinol. Metab.*, 70, 680, 1990.
108. **McCarter, J., Shaw, M. A., Winer, L. A., et al.,** *Mol. Cell Endocrinol.*, 1990, 11, 7392.
109. **Fairhall, K., Carmignac, D., and Robinson, I.,** *Endocrinology (Baltimore)*, 131, 1963, 1992.
110. **Amit, T., Hochberg, Z., Waters, M. J., et al.,** *Endocrinology (Baltimore)*, 131, 1793, 1992.
111. **Shaw, M. A. and Baumann, G.,** *70th Annu. Meet. Endocr. Soc.*, New Orleans, LA, 1988, p. 88. (Abstract)
112. **Davis, S. L., Graf, M., Morrison, C. A., et al.,** *J. Anim. Sci.*, 70, 773, 1992.
113. **Amit, T., Hochberg, Z., and Barkey, J.,** *Biochem. J.*, 293, 345, 1993.
114. **Barnard, R. and Waters, M. J.,** *Biochem. J.*, 250, 533, 1988.
114a. **Tonshoff, B., Eden, S., Weiser, E., et al.,** *Kid. Intl.*, 45, 1085, 1994.
115. **Mulumba, N., Massa, G., Ketelslegers, J. M., et al.,** *Acta Endocrinol. (Copenh.)*, 125, 409, 1991.
116. **Straus, D. S. and Takemoto, D. C.,** *Mol. Endocrinol.*, 4, 91, 1990.

117. **Massa, G., Verhaeghe, J., Vanderschueren-Lodeweyckx, M., et al.,** *Horm. Metab. Res.*, 25, 325, 1993.
118. **Ambler, G. R., Breier, B. H., Surus, A., et al.,** *Acta Endocrinol. (Copenh.)*, 126, 155, 1992.
119. **Cramer, S. D., Barnard, R., Engbers, C., et al.,** *Endocrinology (Baltimore)*, 130, 1074, 1992.
120. **Cramer, S. D., Barnard, R., Engbers, C., et al.,** *Endocrinology (Baltimore)*, 131, 876, 1992.
121. **Smith, W. C. and Talamantes, F.,** *Endocrinology (Baltimore)*, 123, 1489, 1988.
122. **Sadeghi, H., Wang, B. S., Lumanglas, A. L., et al.,** *Mol. Endocrinol.*, 4, 1799, 1990.
123. **Massa, G., Mulumba, N., Ketelslegers, J. M., et al.,** *Endocrinology (Baltimore)*, 126, 1976, 1990.
124. **Carmignac, D. F., Wells, T., Carlsson, L. M. S., et al.,** *J. Endocrinol.*, 135, 447, 1992.
125. **Amit, T., Barkey, R. J., Bick, T., et al.,** *Mol. Cell. Endocrinol.*, 70, 197, 1990.
125a. **Sadeghi, H. and Wang, B. S.,** *Endocrine J.*, 2, 153, 1994.
126. **Carmignac, D., Gabrielsson, B., and Robinson, I.,** *Endocrinology (Baltimore)*, 133, 2445, 1993.
127. **Tiong, T. S. and Herington, A. C.,** *Endocrinology (Baltimore)*, 129, 1628, 1991.
128. **Frick, G. P., Leonard, J. L., and Goodman, H. M.,** *Endocrinology (Baltimore)*, 126, 3076, 1990.
129. **Bick, T., Amit, T., Barkey, R. J., et al.,** *Endocrinology (Baltimore)*, 128, 181, 1991.
130. **Bick, T., Hochberg, Z., Amit, T., et al.,** *Endocrinology (Baltimore)*, 131, 423, 1992.
131. **Bick, T., Hochberg, Z., Amit, T., et al.,** *71st Annu. Meet. Endocr. Soc.*, Seattle, WA, 1989, p. 1307.
132. **Sanchez-Jimenez, F., Fielder, P., Martinez,R., et al.,** *Endocrinology (Baltimore)*, 126, 1270, 1990.
133. **Tobar-Dupres, E. T., Froman, D. P., and Davis, S. L.,** *Poultry Sci.*, 72, 2337, 1993.
134. **Spencer, S. A., Hammond, R. G., Henzel, W. J., et al.,** *J. Biol. Chem.*, 263, 7862, 1988.
135. **Vasilatos-Younken, R., Andersen, B. J., Rosebrough, R. W., et al.,** *J. Endocrinol.*, 130, 115, 1991.
136. **Hull, K. L., Fraser, R. A., and Harvey, S.,** *J. Endocrinol.*, 135, 459, 1992.
137. **Barnard, R., Bundesen, P. G., Rylatt, D. B., et al.,** *Biochem. J.*, 231, 459, 1986.
138. **Barnard, R., Bundesen, P. G., Rylatt, D. B., et al.,** *Endocrinology (Baltimore)*, 115, 1805, 1984.
139. **Smith, W. C., Kuniyoshi, J., and Talamantes, F.,** *Mol. Endocrinol.*, 3, 984, 1989.
140. **Haldosen, L. A. and Gustafson, J. A.,** *Mol. Cell Endocrinol.*, 68, 187, 1990.
141. **Spencer, S. A., Leung, D. W., Godowski, P. J., et al.,** *Recent Prog. Horm. Res.*, 46, 165, 1990.
142. **Barnard, R. and Waters, M. J.,** *Biochem. J.*, 237, 885, 1986.
143. **Smith, W. C., Kuniyoshi, L., and Talamantes, F.,** *71st Annu. Meet. Endocr. Soc.*, Seattle, WA, 1989, p. 258.
144. **Cramer, S. D. and Talamantes, F.,** *The Endocrinology of Growth, Development, and Metabolism in Vertebrates* (Eds. Schreibman, M. P., Scanes, C. G., and Pang, P. K. T.). Toronto, Academic Press, 1993, p. 117.
145. **Frick, G. P. and Goodman, H. M.,** *Endocrinology (Baltimore)*, 131, 3083, 1992.
146. **Tiong, T. S. and Herington, A. C.,** *Biochem. Biophys. Res. Commun.*, 180, 489, 1991.
147. **Bingham, B., Oldham, E. R., and Baumbach, W. R.,** *Proc. Soc. Exp. Biol. Med.*, 1994, in press.
148. **Oldham, E. R., Bingham, B., and Baumbach, W. R.,** *Mol. Endocrinol.*, 7, 1379, 1993.
149. **Bick, T., Amit, T., Barkey, R. J., et al.,** *Endocrinology (Baltimore)*, 128, 181, 1991.
150. **Walker, J. L., Moatsstaats, B. M., Stiles, A. D., et al.,** *Pediatr. Res.*, 31, 335, 1992.
151. **Tiong, T. S. and Herington, A. C.,** *Mol. Cell. Endocrinol.*, 83, 133, 1992.
152. **Garcia-Aragon, J., Lobie, P. E., Muscat, G. E. O., et al.,** *Development*, 114, 869, 1992.
153. **Hasegawa, O., Minami, S., Sugihara, H., et al.,** *Dev. Brain Res.*, 74, 287, 1993.
154. **Lobie, P. E., Garcia-Aragon, J., Lincoln, D. T., et al.,** *Dev. Brain Res.*, 74, 225, 1993.
155. **Mathews, L. S., Enberg, B., and Norstedt, G.,** *J. Biol. Chem.*, 264, 9905, 1989.
156. **Sotiropoulos, A., Goujon, L., Simonin, G., et al.,** *Endocrinology (Baltimore)*, 132, 1863, 1993.
156a. **Amit, T., Hacham, H., Daily, O., et al.,** *Mol. Cell. Endocrinol.*, 101, 29, 1994.
157. **Amit, T., Hartmann, K., Shoshany, G., et al.,** *Mol. Cell. Endocrinol.*, 94, 149, 1993.
158. **Lobie, P. E., Garcia-Aragon, J., Wang, B. S., et al.,** *Endocrinology (Baltimore)*, 130, 3057, 1992.
159. **Ymer, S. I., Stevenson, J. L., and Herington, A. C.,** *Endocrinology (Baltimore)*, 125, 516, 1989.

160. **Herington, A. C., Ymer, S., Roupas, P., et al.,** *Biochim. Biophys. Acta*, 881, 236, 1986.
161. **Barnard, R., Rowlinson, S. W., and Waters, M. J.,** *Biochem. J.*, 267, 471, 1990.
162. **Barnard, R. and Waters, M. J.,** *Biochem. J.*, 231, 459, 1985.
163. **Herington, A. C., Ymer, S. I., Roupas, P., et al.,** *Biochim. Biophys. Acta*, 88, 236, 1986.
164. **Ymer, S. I., Stevenson, J. L., and Herington, A. C.,** *Biochem. J.*, 221, 617, 1984.
165. **Fraser, R. A. and Harvey, S.,** *Endocrinology (Baltimore)*, 130, 3593, 1992.
166. **Hocquette, J. F., Postel-Vinay, M. C., Djiane, J., et al.,** *Endocrinology (Baltimore)*, 127, 1655, 1990.
167. **Hocquette, J. F., Postel-Vinay, M. C., Kayser, C., et al.,** *Endocrinology (Baltimore)*, 125, 2167, 1989.
168. **Ymer, S. I. and Herington, A. C.,** *Mol. Cell. Endocrinol.*, 41, 153, 1985.
169. **Harvey, S., Baumbach, W. R., Sadeghi, H., et al.,** *Endocrinology (Baltimore)*, 133, 1125, 1993.
170. **Lobie, P. E., Barnard, R., and Waters, M. J.,** *J. Biol. Chem.*, 265, 19947, 1991.
171. **Amit, T., Dirnfeld, M., Barkey, R. J., et al.,** *J. Clin. Endocrinol. Metab.*, 77, 33, 1993.
172. **Hattori, N., Shimatsu, A., Kato, Y., et al.,** *Kidney Int.*, 37, 951, 1990.
173. **Baumann, G., Amburn, K. D., and Buchanan, T. H.,** *J. Clin. Endocrinol. Metab.*, 64, 657, 1987.
174. **Baumann, G., Shaw, M. A., and Buchanan, T. A.,** *Metab. Clin. Exp.*, 38, 330, 1989.
175. **Veldhuis, J. D. and Johnson, M. L.,** *Excerpta Medica International Congress Series,* 1993, in press.
176. **Veldhuis, J. D., Johnson, M. L., Faunt, L. M., et al.,** *J. Clin. Invest.*, 91, 629, 1993.
177. **Veldhuis, J. D., Johnson, M. L., Faunt, L. M., et al.,** *J. Clin. Invest.*, 92, 1109, 1993.
178. **Gabrielsson, B., Fairhall, K. M., and Robinson, I. C. A. F.,** *J. Endocrinol.*, 124, 371, 1990.
179. **Jansson, J. O., Eden, S., and Isaksson, O.,** *Endocr. Rev.*, 6, 128, 1985.
180. **Minamitani, N., Chihara, K., Kaji, H., et al.,** *J. Neuroendocrinol.*, 1, 147, 1989.
181. **Roupas, P. and Herington, A. C.,** *Mol. Cell. Endocrinol.*, 61, 1, 1989.
182. **Baumann, G., Shaw, M. A., and Buchanan, T. A.,** *Metab. Clin. Exp.*, 38, 330, 1989.
183. **Mannor, D. A., Winer, L. M., Shaw, M. A., et al.,** *J. Clin. Endocrinol. Metab.*, 73, 30, 1991.
184. **Lim, L., Spencer, S. A., McKay, P., et al.,** *Endocrinology (Baltimore)*, 127, 1287, 1990.
185. **Hansen, B. S., Hjorth, S., Welinder, B. S., et al.,** *Endocrinology (Baltimore)*, 133, 2809, 1993.
185a. **Dattani, M. T., Hindmarsh, P. C., Brook, C. G. D., et al.,** *J. Endocrinol.*, 140, 445, 1994.
186. **Wang, B. S., Sadeghi, H., Fung, C., et al.,** *Mol. Cell. Endocrinol.*, 92, 161, 1993.
187. **Clark, R. G., Cunningham, B., and Moore, J. A.,** *73rd Ann. Meet. Endocrine. Soc.*, Washington, D.C., 1991, p. 1611 (Abstract).
188. **Martha, P. M., Reiter, E. O., Davila, N., et al.,** *J. Clin. Endocrinol. Metab.*, 75, 1464, 1992.
189. **Holl, R. W., Snehotta, R., Siegler, B., et al.,** *Horm. Res.*, 35, 190, 1991.
190. **Hochberg, Z., Amit, T., and Youdin, M. B. H.,** *Cell. Signal.*, 3, 85, 1991.
191. **Cramer, S. D., Wong, L., Kensinger, R. S., et al.,** *Endocrinology (Baltimore)*, 131, 2914, 1992.
192. **Postel-Vinay, M. C., Belair, L., Kayser, C., et al.,** *Proc. Natl. Acad. Sci. U.S.A.*, 88, 6687, 1991.
193. **Devolder, A., Renaville, R., Sneyers, M., et al.,** *J. Endocrinol.*, 138, 91, 1993.
194. **Mercado, M., Carlsson, L., Vitangcol, R., et al.,** *J. Clin. Endocrinol. Metab.*, 76, 1291, 1993.
195. **Mercado, M. and Baumann, G.,** *74th Annu. Meet. Endocr. Soc.,* San Antonio, TX,1992, p. 225. (Abstract)
195a. **Frick, G. P., Tai, L-R., and Goodman, H. M.,** *Endocrinology (Baltimore)*, 134, 1994.
196. **Lobie, P. E., Breipohl, W., Garcia-Aragon, J., et al.,** *Endocrinology (Baltimore)*, 126, 2214, 1990.
197. **Yoon, J., Berry, S. A., Seelig, S., et al.,** *J. Biol. Chem.*, 266, 19947, 1990.
198. **Saunders, J. C., Dickson, M. C., Pell, J. M., et al.,** *J. Mol. Endocrinol.*, 7, 233, 1991.
199. **Rajkovic, I. A., Valiontis, E., and Ho, K. Y.,** *J. Clin. Endocrinol. Metab.,* 70, 772, 1994.
200. **Bernardini, S., Spadoni, G. L., Povoa, G., et al.,** *Acta Endocrinol. (Copenh.)*, 127, 313, 1992.
201. **Rieu, M., Le Bouc, Y., Villares, S. M., et al.,** *J. Clin. Endocrinol. Metab.*, 76, 857, 1993.
202. **Hattori, N., Hino, M., Ikekubo, K., et al.,** *Clin. Endocrinol.*, 36, 351, 1992.
203. **Carlsson, L. M. S., Rowland, A. M., Clark, R. G., et al.,** *J. Clin. Endocrinol. Metab.*, 73, 1216, 1991.
204. **Ho, K. K. Y. and Weissberger, A. J.,** *J. Bone Min. Res.*, 7, 821, 1992.
205. **Carlsson, L. M. S., Attie, K. M., Compton, P. G., et al.,** *J. Clin. Endocrinol. Metab.*, 78, 1325, 1994.
206. **Zadik, Z., Amnon, Z., Altman, Y., et al.,** *Horm. Res.*, 40, 161, 1993.
207. **Postel-Vinay, M. C. and Fontoura, M.,** *Acta Pediatr. Scand.*, 79, 1991.

210. Zadik, Z., Landau, H., Chen, M., et al., *J. Clin. Endocrinol. Metab.*, 75, 412, 1992.
211. Baumann, G., Stolar, M. W., Amburn, K., et al., *J. Clin. Endocrinol. Metab.*, 62, 134, 1986.
212. Lauterio, T. J., Trivedi, B., Kapadia, M., et al., *Comp. Biochem. Physiol.*, 91A, 15, 1988.
213. Ymer, S. I. and Herington, A. C., *Mol. Cell. Endocrinol.*, 41, 153, 1985.

Chapter 15

Growth Hormone Metabolism

Stephen Harvey

CONTENTS

I. INTRODUCTION

The structure, secretion, and action of growth hormone (GH) are partly regulated by clearance mechanisms within target sites and sites of GH synthesis, distribution, and degradation. Circulating GH concentrations therefore reflect the dynamic rates of GH synthesis, release, and metabolic clearance and the products of proteolytic processing. The removal of GH from systemic circulation is largely due to its irreversible metabolism in tissues and the renal excretion of GH moieties. The heterogeneity of circulating and pituitary GH[1,2] partly reflects this metabolic processing, which may be a prerequisite for some biological activities of GH moieties in plasma.[3] This production of multiple forms of GH by the pituitary gland[4-8] may also provide a mechanism for controlling the posttranslational proteolytic modification of the hormone. Indeed, each of the multiple forms (different gene products, products of alternative mRNA processing and posttranslational modifications) may have a characteristic conformation and these spatial structures could expose different, specific regions to proteolytic degradation.[3,9-11]

II. INTRAPITUITARY GH DEGRADATION

In addition to GH synthesis, pituitary tissue is a site of GH degradation. This degradation is not due to humeral components of the pituitary gland, because proteolytic degradation of GH is minimal in the bloodstream.[12,13] The degradation of GH therefore occurs intracellularly and occurs within pituitary somatotrophs, in which newly synthesized GH appears to be more susceptible than older, stored GH.[14] Secreted GH is not, however, degraded by GH-secreting cells,[15,16] even though newly synthesized and secreted GH are readily degraded by extrapituitary tissues.[17]

The degradation of GH within the pituitary is achieved by lysosomal proteases with plasmin-, thrombin-, trypsin-, and chymotrysin-like activity.[13,18] The activity of these proteolytic enzymes is related to the rate of GH synthesis and regulated by regulators of GH release. Somatostatin (SRIF), for instance, not only inhibits GH release but increases the cytoplasmic volume density of somatotroph lysosomes and increases the intracellular degradation of stored hormone.[19-21] SRIF (and GH-releasing hormone [GHRH]) has no effect, however, on the degradation of GH released extracellularly from cultured pituitary cells.[14]

0-8493-8697-7/95/$0.00+$.50

In the pituitary gland, different GH moieties may be differentially degraded, resulting in the generation of moieties with different bioactivities, as suggested for deamidated GH ([Asp[152]]GH and [Glu[137]]GH).[22] In the human pituitary, three proteolytically cleaved forms of 22-kDa GH have been identified (see Chapter 1). Three acidic GH varieties (termed GH-C, GH-D, and GH-E by Yadley et al.,[23] and α_1, α_2, and α_3 by Singh et al.[24]) are derived by tryptic cleavage between Arg-135, Lys-140, and Lys-145 in 22-kDa hGH. These fragments are therefore two-chained forms because proteolytic cleavage occurs in the large disulfide loop and the disulfide bridge between Lys-53 and Lys-165 connects these chains. When this bridge is intact, the unfolding of the GH molecule artifactually increases its size to 24 kDa, although this moiety is likely to be subsequently degraded (Figure 1). The resulting fragments are thus composed of approximately 134 and 51 amino acids with molecular masses of 14 and 8 kDa, respectively.[25-27] The main product of plasmin-digested GH is thus composed of the NH_2-terminal portion, residues 1 to 134, and the COOH-terminal portion, residues 141 to 191.[26,28] These GH fragments have more potency than authentic 22-kDa GH in some bioassays[29-31] and plasmin digestion may activate the full-length molecule.[24] This proteolytic processing of GH is, however, unlikely to reflect posttranslational modifications of the GH molecule, because the fragments produced are not stored or secreted from the pituitary gland.[2,32]

A 5-kDa peptide corresponding to 22-kDa hGH_{1-43} has also been isolated from human pituitary glands and accounts for 0.9% of the GH isolated.[6] This amino-terminal fragment has insulin-potentiating effects in hypophysectomized rats and mice but no growth-promoting effect.[29,30] Cleavage of human GH (hGH) between amino acids 43 and 44 also generates an amino acid fragment (amino acids 44 to 191) of 17 kDa, with potent diabetogenic activity devoid of growth-promoting and insulin-like activity[2,8] (see Chapters 2 and 20). Proteolytic cleavage of pituitary GH also results in the formation of another truncated GH moiety (Δ4 GH) lacking the four-amino acid amino terminus of the full-length molecule.

III. EXTRAPITUITARY GH DEGRADATION

Although GH is stable in plasma,[12,13] proteolytic degradation of GH occurs in peripheral tissues and most (>90% of secreted GH) is degraded in the liver and kidney.[2,33] Hepatic or renal dysfunction therefore impairs GH clearance and elevates circulating GH concentrations.[34-37]

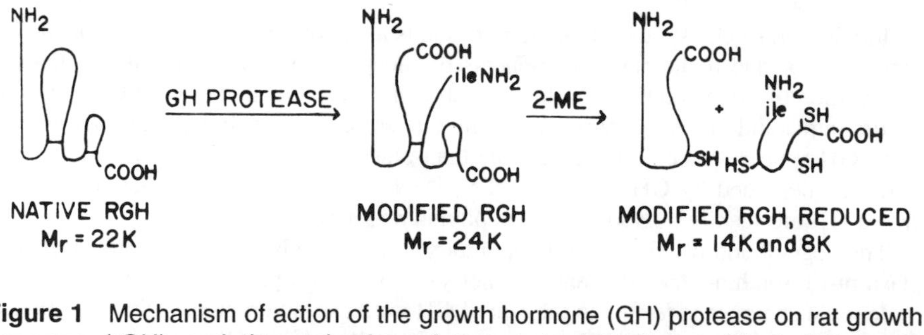

Figure 1 Mechanism of action of the growth hormone (GH) protease on rat growth hormone (rGH) and the reduction of the proteolytically modified rGH by 2-mercaptoethanol. Cleavage of native rGH by the GH protease occurs at an Arg_{132}-Ile_{133} peptide bond, resulting in the formation of a modified form of rGH that possesses a molecular weight of 24,000 by SDS gel electrophoresis. Reduction of the modified form of RGH with 2-mercaptoethanol results in the formation of the two peptides. (From Maciag T., Forand, R., Ilsley, S., et al., *J. Biol. Chem.*, 255, 6064, 1980.)

The majority of GH is removed irreversibly from blood by hepatic proteases.[38] In the liver, GH is incorporated into Kupffer cells and concentrated in peritoneal mast cells and macrophages.[39] Proteolytic degradation therefore occurs in endocytotic and lysosomal vesicles by proteases that differ in pH dependence and cleavage site specificity.[40] These proteases may completely degrade GH into component amino acids or generate smaller fragments that may[41] be returned to peripheral circulation[31,42] and augment the heterogeneity of circulating GH moieties.[2] Cleavage of GH also occurs in hepatic plasmalemma fractions,[43] and generates the 15-kDa GH moiety found in the bloodstream.[44] As a result of these protease actions, a single pass of blood through the liver results in the removal of approximately one-third of its GH content, at a rate of 2 to 3 ml of blood per minute per kilogram[38] (Table 1).

In the kidney renal tubular cells reabsorb GH from blood filtrates and degrade it in lysosomal compartments, from which catabolic fractions are returned to plasma.[2,33] Thrombin and collagenase-like proteases cleave GH between residues 132 and 135 and between 132 and 135, respectively, to produce other 14- and 15-kDa fragments. Chymotrypsin-like serine proteases that are not present in liver and kidney tissue also cleave GH in thyroid and skeletal tissues between its Tyr/Phe-Xaa bonds and generate two-chain GH forms by cleavage between des_{1-8}GH and residues 135–145.[45]

IV. RECEPTOR-MEDIATED GH CLEARANCE

Although some GH degradation occurs in the pituitary gland and occurs minimally in circulation and tissue plasmalemma fractions, GH is mostly degraded in intracellular compartments within target or clearance tissues.[46] The entry of GH into these compartments largely occurs through receptor-mediated endocytosis. Indeed, in the absence of GH receptors (as in Laron dwarfism), the metabolic clearance of GH appears to be markedly reduced.[47]

This endocytotic process is likely to result from the discrete aggregation of membrane GH-GH receptor complexes over coated pits[48-50] or from the constitutive internalization of membrane receptors.[51] This internalization of GH occurs as a function of time and temperature until the intracellular accumulation of GH reaches an equilibrium. This equilibrium occurs in human IM-9 lymphocyte cells and rat adipocytes when approximately half of the cell-associated GH is intracellular, at which time the rate of GH internalization is comparable to the rate of GH degradation and/or release.[48,52] Uptake of hGH by the rat liver occurs within 30 s of its intravenous injection and is maximal within 15 min, at which time 24% of the injected GH is present in hepatic cells.[53]

After binding to the receptors in the plasma membrane GH is transported in two successive endocytotic compartments (EI and EII), prior to arrival in the lysosomes that fuse with the endosomal vesicles.[40] The majority of the internalized GH (75%) is thought to be targeted for degradation by acidic endopeptidases[48,49,54] present in the endosomal and lysosomal fractions.[40] The translocation of GH into subcellular fractions is, however, temporally separated from its entry into degradative pathways, because only intact GH is recovered from the subcellular fractions of the rat liver during the first 15 min after exogenous GH administration.[53] The subsequent degradation of GH involves a non-trypsin-like serine protease that is present in plasma membranes and in endosomal and lysosomal vesicles. These proteases are instrumental in degrading plasma GH and an impairment of serine protease activity increases the half-life of GH in rat plasma.[17,43] The proteolytic cleavage of GH by serine proteases also generates a 15-kDa fragment that retains the ability to bind to membrane GH receptors[40,42] and may be more potent in inducing some biological effects.[3,10,24] This increased potency may reflect the greater resistance of cleaved GH to complete proteolysis by other protease fractions.[42] Proteolytic processing of GH is thus not only essential for its disposal but also its biological activity,

Table 1 **Metabolic Clearance of GH**

Species	Condition	GH	Half-Life[a]	Clearance Rate	Ref.
Human	Premature infant	Endogenous	20.0		158
	Term infant	Endogenous	24.0		158
		hGH	12.0		159
	Newborn infant	Endogenous	18.0		148
	Children	Endogenous	19.0		136
	Young men	rbhGH	13.8	2.9 ml/min/kg	124
	Boys	Endogenous	20.0		130
	Girls	Endogenous	11.0		128
	Pubertal boys	Endogenous	17.0		137
	Preadolescent	Endogenous	18.0		140
	Late adolescent	Endogenous	24.0		140
	Adults	hGH	19.0	2.99 ml/min/kg	35
	Adult men	rbhGH	12.8–19.4		100
		Endogenous	22.8	4.8 ml/h/m^2	133
		Endogenous	17		85
		Endogenous	18.9	3.07 ml/min/kg	161
		^{131}I-Labeled hGH		112 ml/min/m^2	88
		^{125}I-Labeled hGH		125 l/day/m^2	94
	Adult women	Endogenous	33.6	3.1 ml/h/m^2↓	133
		^{125}I-Labeled hGH		190 l/day/m^2	94
				89.4 l/day/m^2	88
	Elderly men	rbhGH	14.2	2.3 ml/min/kg	124
		Endogenous	32.4		133
	Postmenopausal women	^{125}I-Labeled hGH		161 l/day/m^2	94
Monkey	Lean ♂	rbhGH		12.7 l/day (1.2 l/day/kg)	162
	Obese ♂	rbhGH		1.2 l/day (1.2 l/day/kg)	162

Pig	3 month	pGH	8.7	3.4 ml/min/kg ↓	127
	30 month	pGH	8.4	1.9 ml/min/kg	127
	Lean ♂	pGH	13.4		163
	Obese	pGH	12.5		163
	15 week	pGH	7–12	253 ml/min	164
	Adult	pGH	20–30		165
	Boars	pGH		81.4 l/day	135
	Gilts	pGH		159.4 l/day	135
Cow	Fasted ♂ calves	bGH	31.9 ↑	4.5 l/h ↓	143
	Fed ♂ calves	bGH	20.3	7.5 l/h	143
		bGH	22–25		166
		bGH	26		167
	Growing steers	bGH	5.0/41.8	1.24 ml/min/kg	168
	Estrogenized steers	bGH	3.6/41.5	1.01 ml/min/kg	168
	3-month cows	bGH	9.4	17.4 ml/min	126
	9-month cows	bGH	7.9	341 ml/min ↑	126
	5-month ♂ ♀	Endogenous	14.7	2.17 ml/min/kg	125
	8-month ♂ ♀	Endogenous	14.8	2.67 ml/min/kg	125
	12-month ♂ ♀	Endogenous	14.5	2.17 ml/min/kg	125
	15-month ♂ ♀	Endogenous	14.2	2.33 ml/min/kg	125
	Bulls	Endogenous	15.0	2.33 ml/min/kg	125
	Steers	Endogenous	15.8	2.33 ml/min/kg	125
	Heifers	Endogenous	15.5	2.17 ml/min/kg	125
	Ovariectomized heifers	Endogenous	15.0	2.5 ml/min/kg	125
	Aging cattle	bGH	8.4	300 ml/min ↑	169
	Fasting cattle	bGH		↓	170
	Feed-restricted heifers	bGH	9.3	159 ml/min	126
	Lactating cows	bGH		260-370 ml/min/kg ↑	172
	Nonlactating cows	bGH		160 ml/min/kg	172

Table 1 **Metabolic Clearance of GH (Continued)**

Species	Condition	GH	Half-Life[a]	Clearance Rate	Ref.
	Cows	[Met^{1}, Leu^{127}]bGH	6.9/31.1	1.7 ml/min/kg	173
		[Ala^{1},Val^{127}]bGH	8.0/28.4	1.5 ml/min/kg	173
		[Ala^{1},Val^{127},His^{133}]bGH	9.5/27.9	1.4 ml/min/kg	173
	Heifers (warm environment)	bGH	6.3/24.0	2.95 ml/min/kg	174
	Heifers (cold environment)	bGH	8.2/40.8	2.23 ml/min/kg	174
	Newborn steers	bGH	6.86/44.1	1.60 ml/min/kg	175
	75-kg steers	bGH	7.45/66.6	1.19 ml/min/kg	175
	150-kg steers	bGH	7.53/40.5	0.94 ml/min/kg	175
	250-kg steers	bGH	3.67/24.6	0.94 ml/min/kg	175
	350-kg steers	bGH	4.62/23.9	1.07 ml/min/kg	175
	450-kg steers	bGH	2.94/19.5	0.94 ml/min/kg	175
Sheep	Adult	oGH	7–8		176
	Fed	oGH	9.2	13.5 l/h	143
	Fasting	oGH	13.4 ↑	8.2 l/h	143
	Normal ♂	^{125}I-Labeled oGH		61 l/d	177
	Nephrectomized	^{125}I-Labeled oGH		26 l/d ↓	177
		oGH	10.0		179
Dog		CGH	25		180
		^{125}I-Labeled oGH	6/20/64		181
Rabbit	♂	^{131}I-Labeled hGH	18		182
	Adult	Endogenous	6–15		183
		rGH	4		184
		hGH	9.3–20.8		12
Guinea pig	Normal	hGH	11–20		122

Mouse	Normal ♂	^{125}I-Labeled oGH	2.2/10.0/70		99
	bGH transgenic	^{125}I-Labeled oGH	1.2/10.0/160.0		99
Rat	Normal ♂	Endogenous	3.4/13.2		185
	Normal ♂	hGH (22K dimer)		0.75 ml/min/kg	89
	Normal ♂	hGH (20K dimer)		3.82 ml/min/kg	89
	Normal ♂	hGH (20K)		7.26 ml/min/kg	89
	Normal ♂	hGH (22K)		14.03 ml/min/kg	89
	Normal ♂	^{125}I-Labeled rGH	7.5	6.8 ml/min/kg	186
	Normal	^{125}I-Labeled rGH	5.0		36
	Normal ♂	hGH (22K complex)		2.3 ml/min/kg	106
	Normal ♂	hGH (free 22K GH)		14.0 ml/min/kg	106
	Normal ♂	rGH		10.7 ml/min/kg	121
	Normal ♂	^{125}I-Labeled rGH	8.5/51.6		95
	Dwarf ♀	^{125}I-Labeled rGH	8.5/56.6		95
	Adult ♂	rGH	5.7	50–70 ml/min/kg	93
	Adult ♀	rGH	5.8	26 ml/min/kg ↓	93
	Castrated ♂	rGH	5.0	50 ml/min/kg	93
	Castrated ♀	rGH	5.3	27 ml/min/kg ↓	93
	Castrated ♂ + estrogen	rGH	4.9	20 ml/min/kg ↓	93
	Fasted ♂	rGH		3.9 ml/min/kg ↓	121
	Underfed ♂	rGH		6.1 ml/min/kg ↓	121
	Diabetic ♂	rGH		4.2 ml/min/kg ↓	121
	Insulin-treated diabetic ♂	rGH		1.01 ml/min/kg ↓	121
	Uremic ♂	^{125}I-Labeled hGH	20–90 ↑		188
	Hypophysectomized ♂	^{125}I-Labeled rGH	7.7 ←	6.4 ml/min/kg ←	186
	4 day	^{125}I-Labeled rGH	8.64 ←	6.9 ml/min/kg ←	186
	15 day	^{125}I-Labeled rGH	7.26 ←	7.2 ml/min/kg ←	186
	Nephrectomized	^{125}I-Labeled rGH	11 ↑		36
		hGH	15.6	5.4 ml/min/kg	178

Table 1 **Metabolic Clearance of GH (Continued)**

Species	Condition	GH	Half-Life[a]	Clearance Rate	Ref.
Chicken	Immature Leghorn ♂	cGH	3/13/216		189
	Conscious Leghorn ♂	cGH		3.7 ml/min/kg	141
	Anaesthetized Leghorn ♂	cGH		8.1 ml/min/kg ↑	141
	Underfed Leghorn ♂	cGH		2.8 ml/min/kg	141
	Immature broiler ♂	cGH	8.4–13.9	2.75–5.0 ml/min/kg	190
Turkey	Intact (8 weeks)	rbcGH	10.5	2.66 ml/min	191
	Hypophysectomized (8 weeks)	rbcGH	15.9	1.17 ml/min	191
	Intact (13 weeks)	rbcGH	12.7	3.02 ml/min	191
	Hypophysectomized (13 weeks)	rbcGH	13.7	2.47 ml/min	191
	Intact (8 weeks)	tGH	9.9	3.24 ml/min	191
	Hypophysectomized (8 weeks)	tGH	14.8	1.00 ml/min	191
	Intact (13 weeks)	tGH	11.7	2.98 ml/min	191
	Hypohysectomized (13 weeks)	tGH	14.2	2.62 ml/min	191
Salmon	Transfer to seawater		↑	↑	192
	Transfer to freshwater	rbsGH	←	←	187

Note: (↑) Increased, (↓) decreased, (←) with respect to the corresponding control groups. hGH, human GH; pGH, porcine GH; bGH, bovine GH; cGH, canine CGH; oGH, ovine GH; gpGH, guinea pig GH; rGH, rat GH; RGH, rabbit GH; cGH, chicken GH; tGH, turkey GH; sGH, salmon GH; rb, recombinant.

[a] Single or multiphasic disappearance of GH from plasma (min).

because the potency of exogenous GH is reduced in the presence of serine protease inhibitors.[43]

Following proteolysis or cleavage, degraded GH fragments are then released from cells (e.g., adipocytes) with a $t_{1/2}$ of approximately 30 min, although the rate of processing may be tissue specific and attenuated in GH-deficient states.[48,55,56] A deficiency of tissue receptors, as in the livers of cirrhotic patients, may thus impair GH clearance and elevate the circulating GH concentration.[35]

In adipocytes and possibly other tissues, a small pool of internalized GH (25%) may alternatively enter a non-degradative pathway and be rapidly externalized in an intact form.[54,57,58] The externalization of intact GH occurs faster than the externalization of degraded GH fragments, indicating the presence of separate, independently regulated pathways for internalized GH[48,49,54] (Figure 2). The preferential entry of GH into this pathway could therefore modulate the circulating GH concentration and the rates of GH synthesis, release, and degradation and could serve as a reservoir of free, biologically active hormone.

V. RENAL EXCRETION

The metabolic clearance of GH is largely (about 70%) due to glomerular filtration[33] and hence plasma GH levels are increased and GH turnover is decreased in anephric animals[36] and in patients with renal insufficiency.[59-62] The small size of monomeric GH ensures that it is readily filtered at the glomerulus and (with a coefficient of 0.65[33] and an integrated plasma GH concentration of 2 μg/l), approximately 225 μg of hGH per day is filtered through human glomeruli.[63] However, as most of the filtered GH is reabsorbed and catabolized in the proximal tubules, very little is excreted intact in the urine (0.1% of the filtered load or 0.03% of the secretion rate).[63-65] The GH found in urine lacks oligomeric forms (which may be too large to be filtered), and proteolytically cleaved two-

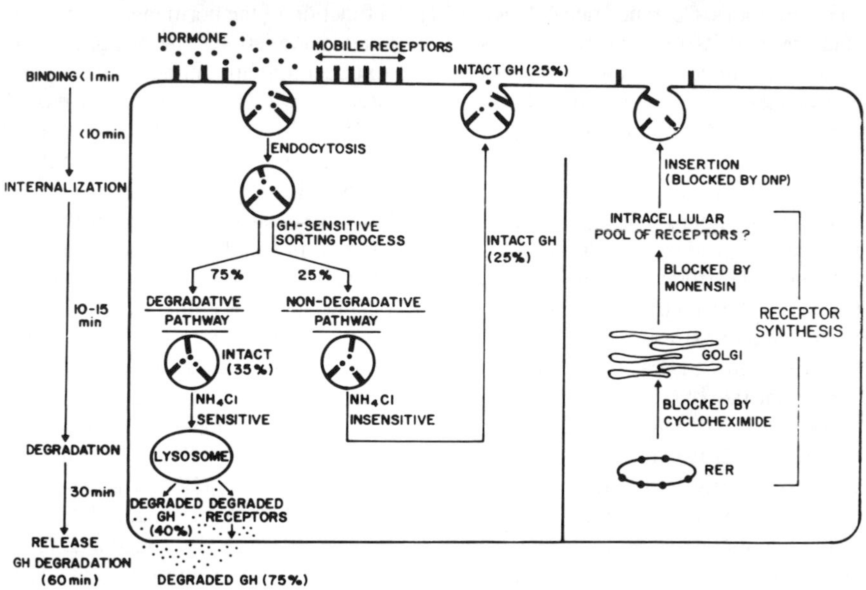

Figure 2 Model for the intracellular processing of GH and its receptor by rat adipocytes. (From Roupas, P. and Herington, A. C., *Mol. Cell. Endocrinol.*, 61, 1, 1989. Reprinted with permission.)

chain GH forms (hGHD and hGHE, with enhanced biological activity), but otherwise resembles monomeric plasma GH in the relative proportions of 22-kDa, 20-kDa, and acidic GH forms and is stable.[2,63-64,66,67] The renal clearance of GH would thus appear to be relatively nonspecific for the heterogeneous GH forms in plasma.

The amount of GH excretion in urine varies considerably among children and among adults (range, 0 to 37 ng/24 h)[68-73] and within individuals by 30 to 60%.[65,66,74] This variation is partly due to daily fluctuations in mean 24-h plasma GH concentrations,[75,76] although this only accounts for approximately 50% of the variation in urinary GH.[70,73,77,78] The correlation of urinary GH concentrations with urinary albumin and β_2-microglobin levels[79-81] also suggests that individual differences in glomerular and tubular function[82,83] also contribute to this variability in GH concentration. Variations in kidney function may also be related to variations in kidney size. The increased clearance of GH in some acromegalics may, for instance, result from the accompanying increase in kidney size and twofold doubling in renal tubule diameter.[84] The absorption of filtered GH by tubular epithelial cells requires the binding of GH to membrane receptors on the lumenal side of the cell and does not involve nonspecific engulfment of tubular fluid.[33] This reabsorptive process has a high capacity and is not impaired at supraphysiological loads of filtered GH. This lumenal route is responsible for almost all the accumulation, extraction, and catabolism of GH within the kidney and peritubular handling of GH is minimal.[33]

In the cells of the proximal tubule the receptor-mediated internalization[33] of GH delivers most of the internalized pool to phagolysosomes for irreversible and complete proteolytic degradation.[54] The catabolism of GH is almost entirely intracellular and is not performed by the surface proteases present at the lumenal cell surface.[33] The catabolic products from GH degradation do not accumulate in the tubular cells, but are returned to general circulation.

VI. METABOLIC CLEARANCE RATE

The metabolic clearance rate (MCR) of GH is a function of the hormone concentration in plasma, and its rates of synthesis, secretion, and degradation. The MCR of GH is therefore also dependent on the access of GH to extracellular and intracellular compartments and the volume of GH distribution, and consequently on the molecular size of GH and its degree of protein binding. The MCR of GH is also dynamically responsive to nutritional and endocrine factors and may differ between species and sexes and between physiological and pathophysiological states (see Tables 1 and 2).

In humans, approximately 180 to 200 l of blood is cleared of GH each day, at a clearance rate of about 3 ml/min/kg (Table 1), although in healthy individuals there are up to fourfold differences in the calculated MCR of GH.[38] The kinetics of this clearance is triphasic. The first phase represents distribution in plasma (approximately 4 to 8% body weight[85]) and in extravascular and extracellular compartments (V_0; 3 to 11%[85,86]) and has a $t_{1/2}$ of <5 min. Increasing the distribution volume decreases the half-life of GH and increases the rate of GH clearance.[87] The second phase represents GH metabolism and has a $t_{1/2}$ of <10 to 20 min in most species and accounts for approximately 67% of the GH removal.[38,88,89] However, as GH fragments may retain biological activity,[17,90,91] the half-life of biologically active GH may exceed that of immunoreactive GH.[92] The third phase is protracted and lasts several hours, during which GH fragments are cleared.

The rate of GH clearance is also dependent on the circulating GH concentration and its molecular composition. For instance, estrogens increase GH secretion in males and reduce the rate of GH clearance.[93] Similarly, increasing GH concentrations by GH administration increases the half-life of endogenous GH in rats, although as this also increases the volume of distribution the MCR of GH remains constant.[12] Owens et al.[35] similarly found that the MCR of GH in humans was relatively constant at plasma GH

Table 2 **Metabolic Clearance of Human GH**

Condition	Half-Life	Volume of Distribution	Clearance Rate	Ref.
Aging	↓		←	114
	←		←	124
Sexual maturation	↑			144
	↑			140
	←			145
Birth	←			146
	←			148
Time of day (AM VS. PM)	↓	←		110,111
Menstruation			←	94
Menopause			←	94
Estrogen therapy			↑	94
	←			136
	↓		↓	88
Dexamethasone therapy	←			149
Androgen receptor blockade	←			137
Androgen therapy	←			130
Fasting	←			142
GH therapy	↑			100
Acromegaly	↑		←	86
	←			152
	←			153
			←	98
			←	94
Panhypopituitarism	←		←	86
Hypopituitarism	←			35
GH-deficient	←		←	97
			←	98
Short stature	←			129
Laron dwarfism			↓	47
			↑	38,112
Obesity	↓			113
	↓			114
Diabetes mellitus (type I)	↓			140
	↑			100
Diabetes mellitus (type II)		←	←	35
	↑	↑		156
Turner's syndrome	↑			128
Hyperthyroidism			←	98
	←			155
Hypothyroidism			↑	98
	↓			157
Myxedema	↑	←	←	35
Renal failure	↑	↑	↓	35,37
Cushing's syndrome			←	98
Liver disease	↑	←	↓	35

Note: (↑) Increased, (↓) decreased, or (←) unchanged with respect to corresponding controls.

concentrations ranging between 5 and 50 µg/ml. Within individuals, the MCR of GH is also relatively constant over long periods of time.[94] Moreover, despite accompanying changes in GH secretion and circulating GH levels, GH clearance rates are not abnormal in many physiological (e.g., growth, fasting, menstruation) and pathophysiological states (e.g., hypothyroidism, hyperthyroidism, acromegaly and GH deficiency, Cushing's syndrome, Turner's syndrome).[38,94-98] Supraphysiological increases in circulating GH concentrations may, however, alter the distribution of GH between different compartments and/or saturate the internalization/degradative process of tissue GH receptors and redirect GH toward the extracellular space.[37] Indeed, this may account for the prolonged residence time of ovine GH in the plasma of mice transgenically expressing the bovine *GH* gene[99] and for the longer half-life of exogenous GH in patients infused with exogenous GH in comparison with patients receiving bolus GH injections.[100] Supraphysiological GH concentrations in chickens are, however, associated with increased GH clearance in domestic fowl.[101]

The clearance of GH from plasma normally involves the removal of heterogeneous proteins. The clearance of monomer (22-kDa) GH occurs faster than for other GH moieties, because it has greater affinity for tissue GH receptors.[2,89,102] Dimerized GH is thus protected from the lysosomal degradation that occurs after the internalization of receptor-bound GH and is cleared more slowly than 22-kDa (fivefold) GH and 20-kDa GH (twofold), leading to its accumulation in plasma.[89] The large size of the "big" GH also excludes it from glomerular filtration and the principal pathway involved in GH removal. Consequently, whereas high molecular weight GH aggregates account for 5 to 15% of pituitary GH, "big" GH comprises 15 to 50% of total plasma GH immunoreactivity (see Baumann[2] for review). The increased production of monomer GH by androgens and oligomeric GH by estrogens is also thought to be a causal factor in the sexual dimorphism in the MCR of humans[88] and rats,[93] as these gonadal steroids increase and decrease, respectively, GH clearance.[93]

The amount and avidity of the high-affinity GHBP are also thought to modulate the half-life of GH in plasma.[2,89,102-109] However, although GH clearance appears to be related to the plasma GHBP levels, neither the slightly longer nocturnal half-life of GH[110,111] nor the shorter GH half-life in obese patients[112-114] is accompanied by changes in the plasma GHBP concentration or avidity. The MCR of GH complexed to GHBP is, nevertheless, much less (tenfold) than that of free GH, which is not restricted to the intravascular compartment.[106,109] The clearance of GH in Laron dwarfs is, conversely, accelerated in the absence of GHBP.[38,47,112] The greater affinity of the GHBP in the plasma of female rats,[115] rabbits,[115] and humans[116,117] may similarly inhibit their MCR in comparison with male rats and rabbits[118,119] (Table 1). This sexual dimorphism in MCR may also reflect the stimulatory effect of estrogens and inhibitory effects of androgens on the concentrations of binding proteins in plasma.[93,117,120] The poor affinity of some GH fragments (e.g., "big" GH and 20-kDa GH) for tissue GH receptors and plasma binding proteins,[42,121] rather than their molecular size, also prolongs their metabolic clearance and accumulation in plasma. The 20-kDa GH variant therefore persists in plasma longer than 22-kDa GH, especially as it is more likely to form dimer associations.[89] The location and extent and rate of complex formation between GHBP and GH may thus be important determinants in the passage of GH between the intra- and extravascular compartments.[122,123]

In many species the half-life of GH is decreased and the metabolic clearance rate increased with increasing age and body size (e.g., in humans,[88,98,114,124] cattle,[125,126] pigs[127]). The increased GH clearance in adults partly reflects an increased blood volume and impaired renal reabsorption of filtered GH[35] and (together with a reduced amplitude and frequency of GH release) results in the low GH concentrations in adults. The GH half-life in girls with short stature and ovarian dysgenesis (Turner's syndrome) is similarly longer than that in healthy controls,[128] although the GH half-life is unchanged in boys with idiopathic short stature.[129] Aging and body size are, however, independent variables modifying GH clearance. For a normal body mass, each decade of increasing age attenuates the GH half-life by approximately 6%.[114] This age-related decline may reflect the decline in plasma androgen, because androgen levels are closely correlated with the GH half-life in normal humans.[114] Exogenous androgen administration to boys with constitutionally improved growth is, however, without effect on the GH half-life.[130]

Changes in body size and blood volume are also likely to be causally responsible for the aberrant GH clearance in obese patients. The GH distribution volume is increased in obese individuals (possibly from an increased permeability of blood vessels), the MCR is increased, and the half-life is decreased by 30% (in humans[113,114,131] and monkeys[132]). The reduced GH distribution volume in patients with renal failure (possibly due to reduced levels of GH receptors or GHBPs) may, conversely, account for their increased half-life of GH and decreased rate of GH clearance.[35,37]

The GH distribution volume in women is also less than that in men [38] and this may also account for their reduced rate of GH clearance.[88,133] The sexual dimorphism in GH clearance in some species (Table 1) may, however, also result from differential effects of gonadal steroids. The half-life of GH in humans correlates negatively with serum estradiol levels and positively with concentrations of testosterone and sex hormone-binding globulin (SHBG).[111,134] The aromatization of androgens and the androgenic inhibition of SHBG production would thus elevate free estradiol concentrations in men and this may account for the sexual dimorphism in the rate of GH clearance. Differential effects of estrogens and androgens on the production of molecular GH variants and on plasma GHBP concentrations may also contribute to this sexual dimorphism in GH clearance. The influence of gonadal steroids on GH clearance is, however, uncertain because the rate of GH clearance is not sexually dimorphic in other species (e.g., pigs[135]) and only minimal effects of sex steroids have been reported in most studies.[130,136,137] Moreover, although gonadal steroids have been considered to regulate GH clearance in rats and humans, castration has no effect on GH clearance in male or female rats[93] and elevated estrogen concentrations fail to modify the GH half-life in late pubertal boys.[137]

In addition to gonadal hormones, the half-life of GH may also be affected by changes in pancreatic function. The half-life of GH in some diabetic patients is, however, longer than that in nondiabetics, even though the distribution volume[100] and the renal clearance of GH[138] and monomeric plasma GH composition[139] are increased. The half-life of GH in other diabetic populations is, however, reported to be shorter than that in non-diabetic controls.[140] Moreover, although glucose might be expected to glycosylate protein and insulin might be expected to enhance receptor-mediated GH uptake in peripheral tissues,[100] the half-life of GH is not prolonged by exogenous insulin or exogenous glucose.[100] Diabetes in the rat is, however, accompanied by insulin-dependent reductions in the GH distribution volume and clearance rates of monomeric and oligomeric GH moieties.[121] Food restriction or acute periods of fasting similarly reduce the distribution volume and clearance rates of "little" and "big" GH in rats.[121] Nutritional restriction similarly reduced GH metabolic clearance in dairy heifers[126] but not in chickens[141] or normal men[142] and increases GH half-life in sheep and calves.[143]

The clearance of GH is also modified in other pathophysiological states (Table 2) and axiomatically is thus an important factor regulating the circulating GH concentration and GH biological activity.

REFERENCES

1. **Baumann, G.,** *Horm. Res.*, 36, 5, 1991.
2. **Baumann, G.,** *Endocr. Rev.*, 12, 424, 1991.
3. **Lewis, U. J., Singh, R. N. P., Lewis, L. J., and Abadi, N.,** *Basic and Clinical Aspects of Growth Hormone* (Ed. Bercu, B. B.). Plenum Press, New York, 1988, p. 43.
4. **Sinha, Y. N., Jacoobsen, B. P., and Lewis, U. J.,** *Biochem. Biophys. Res. Commun.*, 163, 386, 1989.
5. **Sinha, Y. N. and Jacobsen, B. P.,** *Biochem. Biophys. Res. Commun.*, 156, 171, 1988.
6. **Sinha, Y. N., Seavey, B. K., Lewis, L. J., et al.,** *J. Protein Chem.*, 2, 525, 1983.
7. **Yoyoka, S. and Friesen, H. G.,** *Endocrinology (Baltimore)*, 119, 2097, 1986.
8. **Lewis, U. J., Lewis, L. J., Salem, M. A. M., et al.,** *Mol. Cell Endocrinol.*, 78, 45, 1991.
9. **Baumann, G., Shaw, M. A., Merimee, T. J., et al.,** *Clin. Res.*, 36, 477A, 1988.
10. **Lewis, U. J.,** *Trends Endocrinol. Metab.*, 3, 117, 1992.
11. **Lewis, U. J.,** *Annu. Rev. Physiol.*, 46, 33, 1984.
12. **Lee, S. H., Yoon, W. H., Jang, S. H., et al.,** *Int. J. Pharmacol.*, 90, 81, 1993.
13. **Baumann, G.,** *J. Clin. Endocrinol. Metab.*, 43, 222, 1976.
14. **Fukata, J., Diamond, D. J., and Martin, J. B.,** *Endocrinology (Baltimore)*, 117, 457, 1985.
15. **Talamantes, F., Lopez, J., Lewis, U. J., et al.,** *Acta Endocrinol. (Copenh.)*, 98, 8, 1981.
16. **Cherington, P. V. and Tashjian, A. H.,** *Endocrinology (Baltimore)*, 13, 418, 1983.
17. **Vodian, M. A. and Nicoll, C. S.,** *J. Endocrinol.*, 80, 69, 1979.
18. **Ellis, S., Nuenke, J. M., and Grindeland, R. E.,** *Endocrinology (Baltimore)*, 83, 1029, 1968.
19. **Rene, E., Willoughby, J., and Brazeau, P.,** *Regul. Peptides*, 4, 325, 1982.
20. **Stachura, M. E., Costoff, A., and Tyler, T. M.,** *Neuroendocrinology*, 42, 383, 1986.
21. **Asa, S. L., Felix, I., Kovacs, K., et al.,** *Endocr. Pathol.*, 1, 228, 1990.
22. **Lewis, U. J., Singh, R. N. P., Bonewald, L. F., et al.,** *J. Biol. Chem.*, 256, 11645, 1981.
23. **Yadley, R. A., Rodbard, D., and Chrambach, A.,** *Endocrinology (Baltimore)*, 93, 866, 1973.
24. **Singh, R. N. P., Seavey, B. K., Rice, V. P., et al.,** *Endocrinology (Baltimore)*, 94, 883, 1974.
25. **Maciag, T., Forand, R., Ilsley, S., et al.,** *J. Biol. Chem.*, 255, 6064, 1980.
26. **Li, C. H. and Graf, L.,** *Proc. Natl. Acad. Sci. U.S.A.*, 71, 1197, 1974.
27. **Baumann, G. and MacCart, J.,** *J. Clin. Endocrinol. Metab.*, 55, 611, 1982.
28. **Graf, L. and Li, C. H.,** *Biochemistry*, 13, 5408, 1974.
29. **Frigeri, L. G., Teguh, C., Ling, N., et al.,** *Endocrinology (Baltimore)*, 122, 2940, 1988.
30. **Salem, M. A. M.,** *Endocrinology (Baltimore)*, 123, 1565, 1988.
31. **Mittra, I.,** *Cell*, 38, 347, 1984.
32. **Baumann, G., MacCart, J. G., and Amburn, K.,** *Endocrinology (Baltimore)*, 56, 946, 1983.
33. **Johnson, V. and Maack, T.,** *Am. J. Physiol.*, 233, F185, 1977.
34. **Cameron, D. P., Burger, H. G., Catt, K. J., et al.,** *Metab. Clin. Exp.*, 21, 895, 1972.
35. **Owens, D., Srivastava, M. C., Tompkins, M. V., et al.,** *Eur. J. Clin. Invest.*, 3, 284, 1973.
36. **Wallace, A. L. C. and Stacy, B. D.,** *Horm. Metab. Res.*, 7, 135, 1975.
37. **Garcia-Mayor, R. V. G., Perez, A. J., Gandara, A., et al.,** *Clin. Endocrinol. (Oxford)*, 39, 337, 1993.
38. **Iranmanesh, A. and Veldhuis, J. D.,** *Endocrinol. Metab. Clin. N. Am.* 21, 783, 1992.
39. **De Kretser, D. M., Catt, K. J., Burger, H. G., et al.,** *J. Endocrinol.*, 43, 105, 1969.
40. **Husman, B., Gustafson, J. A., and Andersson, G.,** *Mol. Cell Endocrinol.*, 59, 13, 1988.
41. **Baumann, G. and Hodgen, G.,** *J. Clin. Endocrinol. Metab.*, 43, 1009, 1976.
42. **Ingram, R. T., Afshari, N., and Nicoll, C. S.,** *Endocrinology (Baltimore)*, 130, 3085, 1992.
43. **Schepper, J. M., Hughes, E. F., Postel-Vinay, M. C., et al.,** *J. Biol. Chem.*, 259, 12945, 1984.
44. **Chawla, R. K., Parks, J. S., and Rudman, D.,** *Annu. Rev. Med.*, 34, 519, 1983.
45. **Wroblewski, V. J., Masnyk, M., and Becker, G. W.,** *Endocrinology (Baltimore)*, 129, 465, 1991.
46. **Hizuka, N., Gorden, P., Lesnick, M. A., et al.,** *Endocrinology (Baltimore)*, 111, 1576, 1982.
47. **Keret, R. A., Peretzelan, A., Zeharia, A., et al.,** *Isr. J. Med. Sci.*, 24, 75, 1988.
48. **Roupas, P. and Herington, A. C.,** *Endocrinology (Baltimore)*, 120, 2158, 1987.
49. **Roupas, P. and Herington, A. C.,** *Endocrinology (Baltimore)*, 121, 1521, 1987.
50. **Eshet, R., Peleg, S., and Laron, Z.,** *Acta Endocrinol. (Copenh.)*, 107, 9, 1984.
51. **Roupas, P. and Herington, A. C.,** *Mol. Cell. Endocrinol.*, 57, 93, 1988.
52. **Hizuka, N., Gorden, P., Lesniak, M. A., et al.,** *J. Biol. Chem.*, 256, 4591, 1981.
53. **Bullier-Picard, F., Postel-Vinay, M. C., and Kayser, C.,** *J. Endocrinol.*, 121, 19, 1989.

54. Roupas, P. and Herington, A. C., *Mol. Cell. Endocrinol.*, 61, 1, 1989.
55. Gorin, E., Grichting, G., and Goodman, H. M., *Endocrinology (Baltimore)*, 115, 467, 1984.
56. Grichting, G. and Goodman, H. M., *Endocrinology (Baltimore)*, 119, 847, 1986.
57. Ilondo, M., Smal, J., De Meyts, P., et al., *Endocrinology (Baltimore)*, 128, 1597, 1991.
58. Ilondo, M. M., Vanderschueren-Lodeweyckx, M., Courtoy, P. J., et al., *Endocrinology (Baltimore)*, 130, 2037, 1992.
59. Samaan, N. and Freeman, R. M., *Metab. Clin. Exp.*, 19, 102, 1970.
60. Maheshwari, H. G., Rifkin, I., Butler, J., et al., *Acta Endocrinol. (Copenh.)*, 127, 485, 1992.
61. Hattori, N., Kato, Y., Murakami, Y., et al., *J. Clin. Endocrinol. Metab.*, 66, 727, 1988.
62. Owens, D., Srivastava, M. C., Tompkins, C. V., et al., *Eur. J. Clin. Invest.*, 3, 2814, 1973.
63. Baumann, G. and Abramson, E. C., *J. Clin. Endocrinol. Metab.*, 56, 305, 1983.
64. Hattori, N., Shimatsu, A., Kato, Y., et al., *Kidney Int.*, 37, 951, 1990.
65. Hattori, N., Shimatsu, A., Kato, Y., et al., *Acta Endocrinol. (Copenh.)*, 121, 533, 1989.
66. Winer, L. M., Shaw, M. A., and Baumann, G., *J. Endocrinol. Invest.*, 12, 461, 1989.
67. Tanaka, T., Umezawa, S., Yano, H., et al., *Horm. Metab. Res.*, 21, 324, 1989.
68. Girard, J., Erb, T., Pampalone, A. N., et al., *Horm. Res.*, 28, 78, 1987.
69. Hashida, S., Ishikawa, E., Kato, Y., et al., *Clin. Chima Acta*, 162, 229, 1987.
70. Hattori, N., Shimatsu, A., Yamanaka, C., et al., *Acta Endocrinol. (Copenh.)*, 119, 113, 1988.
71. Albini, C. H., Quattrin, T., Vandlen, R. L., et al., *Pediatr. Res.*, 23, 89, 1988.
72. Skinner, A. M., Clayton, P. E., Price, D. A., et al., *Clin. Endocrinol. (Oxford)*, 39, 201, 1993.
73. Skinner, A. M., Clayton, P. E., Price, D. A., et al., *J. Endocrinol.*, 138, 337, 1993.
74. Sukegawa, I., Hizuka, N., Takano, K., et al., *Acta Endocrinol. (Copenh.)*, 121, 290, 1989.
75. Saini, S., Hindmarsh, P. C., Matthews, D. R., et al., *Clin. Endocrinol. (Oxford)*, 34, 455, 1991.
76. Albertsson-Wikland, K. and Rosberg, S., *Acta Endocrinol. (Copenh.)*, 126, 109, 1992.
77. Weissberger, A. J., Ho, K. Y., and Stuart, M. C., *Clin. Endocrinol. (Oxford)*, 30, 697, 1989.
78. Skinner, A. M., Price, D. A., Addison, G. M., et al., *Growth Regul.*, 2, 156, 1992.
79. Tanaka, T., Yoshizawa, A., Niki, Y., et al., *Acta Pediatr. Scand.*, 366, 155, 1990.
80. Pan, F. P., Stevenson, J. L., Donaldson, D. L., et al., *J. Clin. Endocrinol. Metab.*, 71, 611, 1990.
81. Skinner, A. M., Clayton, P. E., Price, D. A., et al., *Endocrinology (Baltimore)*, 138, 337, 1993.
82. Kusano, E., Suzuki, M., Asano, Y., et al., *Nephron*, 41, 320, 1985.
83. Moreira-Andres, M. N., Canizo, F. J., and Hawkins, F., *Acta Endocrinol. (Copenh.)*, 128, 197, 1993.
84. Mauri, M., Pico, A. M., Alfayate, R., et al., *Horm. Res.*, 39, 13, 1993.
85. Hartman, M. L., Faria, A. C. S., Vance, M. L., et al., *Am. J. Physiol.*, 260, E101, 1991.
86. Refetoff, S. and Sonksen, P. H., *J. Clin. Endocrinol.*, 30, 386, 1970.
87. Owens, D., Srivastava, M. C., Tompkins, C. V., et al., *Eur. J. Clin. Invest.*, 72, 51, 1991.
88. Rosenbaum, M. and Gertner, J. M., *J. Clin. Endocrinol. Metab.*, 69, 821, 1989.
89. Baumann, G., Stolar, M. W., and Buchanan, T. A., *Endocrinology (Baltimore)*, 119, 1497, 1986.
90. Mills, J. B., Gennick, S. E., and Kostyo, J. L., *Biochim. Biophys. Acta*, 742, 169, 1983.
91. Parks, J. S., *Acta Pediatr. Scand.*, 349, 127, 1989.
92. Lim, L., Spencer, S. A., McKay, P., et al., *Endocrinology (Baltimore)*, 127, 1287, 1990.
93. Badger, T. M., Millard, W. J., Owens, S. M., et al., *Endocrinology (Baltimore)*, 128, 1065, 1991.
94. Thompson, R. G., Rodriguez, A., Kowarski, A., et al., *J. Clin. Invest.*, 51, 3193, 1972.
95. Carmignac, D. F. and Robinson, I. C. A. F., *J. Endocrinol.*, 127, 69, 1990.
96. Iranmanesh, A., Lizarralde, G., Johnson, M. L., et al., *J. Clin. Endocrinol. Metab.*, 72, 108, 1991.
97. Jorgenson, J. O. L., Flyvjerg, A., and Christiansen, J. S., *Acta Endocrinol. (Copenh.)*, 120, 8, 1989.
98. MacGillivray, M. H., Frohman, L. A., and Doe, J., *J. Clin. Endocrinol.*, 30, 632, 1970.
99. Turyn, D. and Bartke, A., *Transgenic Res.*, 2, 219, 1993.
100. Mullis, P. E., Pal, B. R., Matthews, D. R., et al., *Clin. Endocrinol. (Oxford)*, 36, 255, 1992.
101. Harvey, S., *Avian Endocrinology* (Ed. Sharp, P.). Journal of Endocrinology, Ltd., Edinburgh, 1993, p. 11.
102. Baumann, G., Stolar, M. W., and Buchanan, T. A., *Endocrinology (Baltimore)*, 117, 1309, 1985.
103. Baumann, G., *Acta Endocrinol. (Copenh.)*, 124, 21, 1991.
104. Baumann, G., Vance, M. L., Shaw, M. A., et al., *J. Clin. Endocrinol. Metab.*, 71, 470, 1990.

105. Baumann, G., Amburn, K., and Shaw, M., *Endocrinology (Baltimore)*, 122, 976, 1988.
106. Baumann, G., Shaw, M. A., and Buchanan, T. A., *Metab. Clin. Exp.*, 38, 330, 1989.
107. Baumann, G., Stolar, M. W., Amburn, K., et al., *J. Clin. Endocrinol. Metab.*, 62, 134, 1986.
108. Baumann, G., *Proc. Soc. Exp. Biol. Med.*, 202, 392, 1993.
109. Baumann, G., Amburn, K. D. and Buchanan, T. H., *J. Clin. Endocrinol. Metab.*, 64, 657, 1987.
110. Holl, R. W., Schwartz, U., Schauwecker, P., et al., *J. Clin. Endocrinol. Metab.*, 77, 216, 1993.
111. Holl, R. W., Schwartz, U., Schauwecker, P., et al., *J. Clin. Endocrinol. Metab.*, 77, 216, 1993.
112. Veldhuis, J. D., Johnson, M. L., Faunt, L. M., et al., *J. Clin. Invest.*, 91, 629, 1993.
113. Veldhuis, J. D., Iranmanesh, A., Ho, K. K. Y., et al., *J. Clin. Endocrinol. Metab.*, 72, 51, 1991.
114. Iranmanesh, A., Lizarralde, G., and Veldhuis, J. D., *J. Clin. Endocrinol. Metab.*, 73, 1081, 1991.
115. Barnard, R. and Waters, M. J., *Biochem. J.*, 237, 885, 1986.
116. Barnard, R., Quirk, P., and Waters, M. J., *J. Endocrinol.*, 123, 327, 1989.
117. Baumann, G., Shaw, M. A., and Amburn, K., *Mol. Cell. Endocrinol.*, 38, 683, 1989.
118. Massa, G., Mulumba, N., Ketelslegers, J. M., et al., *Endocrinology (Baltimore)*, 126, 1976, 1990.
119. Amit, T., Barkey, R. J., Bick, T., et al., *Mol. Cell. Endocrinol.*, 70, 197, 1990.
120. Plymate, S. R., Leonard, J. M., Paulsen, C. A., et al., *J. Clin. Endocrinol. Metab.*, 57, 645, 1983.
121. Jolin, T., Gonzalez, C., and Gonzalez, T., *J. Endocrinol. Invest.*, 13, 209, 1990.
122. Fairhall, K., Carmignac, D., and Robinson, I., *Endocrinology (Baltimore)*, 131, 1963, 1992.
123. Poznansky, M. J., Halford, J., and Taylor, D., *FEBS Lett.*, 239, 18, 1988.
124. Sohmiya, M. and Kato, Y., *J. Clin. Endocrinol. Metab.*, 75, 1487, 1992.
125. Plouzek, C. A. and Trenkle, A., *Domest. Anim. Endocrinol.*, 8, 63, 1991.
126. Lapierre, H., Farmer, C., Girard, C., et al., *Domest. Anim. Endocrinol.*, 9, 199, 1992.
127. Farmer, C., Lapierre, H., Matte, J. J., et al., *Domestic Anim. Endocrinol.*, 10, 249, 1993.
128. Veldhuis, J., Sotos, J. F., and Sherman, B. M., *J. Clin. Endocrinol. Metab.*, 73, 71, 1991.
129. Veldhuis, J. D., Blizzard, R. M., Rogol, A. D., et al., *J. Clin. Endocrinol. Metab.*, 74, 766, 1992.
130. Ulloa-Aguirre, A., Blizzard, R. M., Garcia-Rubi, E., et al., *J. Clin. Endocrinol. Metab.*, 71, 846, 1990.
131. Hindmarsh, P. C., Matthews, D. R., Brain, C. E., et al., *Clin. Endocrinol.*, 30, 443, 1989.
132. Dudley, A. K., Ahanukoglu, A., Hansen, B. C., et al., *J. Clin. Endocrinol. Metab.*, 67, 1064, 1988.
133. Gupta, S. K., Krishnan, R. R., Ellinwood, E. H., et al., *Life Sci.*, 47, 1887, 1990.
134. Postel-Vinay, M. C., Tar, A., Hocquette, F., et al., *J. Clin. Endocrinol. Metab.*, 73, 197, 1991.
135. Arbona, J. R., Marple, D. N., Russell, et al., *J. Anim. Sci.*, 66, 3068, 1988.
136. Mauras, N., Rogol, A. D., and Veldhuis, J. D., *Pediatr. Res.*, 28, 626, 1990.
137. Metzger, D. L. and Kerrigan, J. R., *J. Clin. Endocrinol. Metab.*, 76, 1147, 1993.
138. Hourd, P., Edge, J. A., Dunger, D. B., et al., *Diabet. Med.*, 8, 237, 1991.
139. MacFarlane, I. A., Stafford, S., and Wright, A. D., *Acta Endocrinol. (Copenh.)*, 112, 547, 1986.
140. Nieves-Rivera, F., Rogol, A. D., Veldhuis, J. D., et al., *J. Clin. Endocrinol. Metab.*, 77, 638, 1993.
141. Lauterio, T. J. and Scanes, C. G., *Poultry Sci.*, 67, 120, 1988.
142. Hartman, M. L., Veldhuis, J. D., Johnson, M. L., et al., *J. Clin. Endocrinol. Metab.*, 74, 757, 1992.
143. Trenkle, A., *J. Anim. Sci.*, 43, 1035, 1976.
144. Gupta, D., *Neuroendocrinol. Lett.*, 14, 1, 1992.
145. Martha, P. M., Reiter, E. O., Davila, N., et al., *J. Clin. Endocrinol. Metab.*, 75, 1470, 1992.
146. Martha, P. M., Reiter, E. O., Davila, N., et al., *J. Clin. Endocrinol. Metab.*, 75, 1464, 1992.
147. Mauras, N., Blizzard, R. M., Link, K., et al., *J. Clin. Endocrinol. Metab.*, 64, 596, 1987.
148. de Zegher, F., Devlieger, H., and Veldhuis, J. D., *J. Clin. Endocrinol. Metab.*, 76, 1177, 1993.
149. Veldhuis, J. D., Lizarralde, G., and Iranmanesh, A., *J. Clin. Endocrinol. Metab.*, 74, 96, 1992.
150. Holl, R. W., Siegler, B., and Heinze, E., *73rd Ann. Meet. Endocr. Soc.*, Washington, D.C., 1991, p. 1548.
151. Holl, R. W., Siegler, B., Scherbaum, W. A., et al., *J. Clin. Endocrinol. Metab.*, 76, 165, 1993.
152. Yen, S. S. C., Siler, T. M., and DeVane, G. W., *N. Engl. J. Med.*, 290, 935, 1974.
153. Glick, S. M., Roth, J., and Lonergan, E. T., *J. Clin. Endocrinol.*, 24, 501, 1964.

154. **Hendricks, C. M., Eastman, R. C., Takeda, S., et al.,** *J. Clin. Endocrinol. Metab.*, 64, 51, 1985.
155. **Iranmanesh, A., Lizarralde, G., Johnson, M. L., et al.,** *J. Clin. Endocrinol. Metab.*, 72, 108, 1991.
156. **Hindmarsh, P. C., Matthews, D. R., Brain, C., et al.,** *Clin. Endocrinol. (Oxford)*, 32, 739, 1990.
157. **Chernausek, S. D., Underwood, L. E., and Utiger, R. D.,** *Clin. Endocrinol. (Oxford)*, 19, 337, 1983.
158. **Wright, N. M., Northington, F. J., Miller, J. D., et al.,** *Pediatr. Res.*, 32, 286, 1992.
159. **Cornblath, M., Parker, M. C., Reisner, S. H., et al.,** *J. Clin. Endocrinol. Metab.*, 25, 209, 1965.
161. **Faria, A. C. S., Veldhuis, J. D., Thorner, M. O., et al.,** *J. Clin. Endocrinol. Metab.*, 68, 535, 1989.
162. **Dubey, A. K., Hanokoglu, A., Hansen, B. C., et al.,** *J. Clin. Endocrinol. Metab.*, 67, 1064, 1988.
163. **Wangsness, P. J., Martin, R. J., and Gatchel, B. B.,** *Growth*, 44, 318, 1980.
164. **Althen, T. G. and Gerrits, R. J.,** *Endocrinology (Baltimore)*, 99, 511, 1976.
165. **Machlin, L. J., Horino, M., Hertelendy, F., et al.,** *Endocrinology (Baltimore)*, 82, 369, 1968.
166. **Yousef, M. K., Takahashi, Y., Robertson, W. D., et al.,** *J. Anim. Sci.*, 29, 341, 1969.
167. **Mitra, R., Christison, G. I., and Johnson, H. D.,** *J. Anim. Sci.*, 34, 776, 1972.
168. **Gopinath, R. and Kitts, W. K.,** *Growth*, 48, 499, 1984.
169. **Lapierre, H., Reynolds, C. K., Elsasser, T. H., et al.,** *J. Anim. Sci.*, 70, 742, 1992.
170. **Lapierre, H., Tyrrell, H. F., Reynolds, C. K., et al.,** *J. Anim. Sci.*, 70, 764, 1992.
172. **Hart, I. C., Bines, J. A., and Morant, S. V.,** *Life Sci.*, 27, 1839, 1980.
173. **Eppard, P. J., White, T. C., Birmingham, B. K., et al.,** *Endocrinology (Baltimore)*, 139, 441, 1993.
174. **Scott, S. L. and Christopherson, R. J.,** *Can. J. Anim. Sci.*, 73, 33, 1993.
175. **Mears, G. J. and Schaalje, G. B.,** *Can. J. Anim. Sci.*, 773, 277, 1993.
176. **Wallace, A. L. C. and Bassett, J. M.,** *J. Endocrinol.*, 47, 21, 1970.
177. **Wallace, A. L. C., Stacy, B. D., and Thorburn, G. D.,** *Pfluegers Arch.*, 331, 25, 1972.
178. **Thomsen, M. K., Friis, C., Hansen, B. S., et al.,** *J. Pediat. Endocrinol.*, 7, 93, 1994.
179. **Wagner, J. F., Veenhuizen, E. L., and Root, N. A.,** *J. Anim. Sci.*, 31, 232, 1970.
180. **Hampshire, J., Altzuler, N., Steele, R., et al.,** *Endocrinology (Baltimore)*, 96, 822, 1975.
181. **Kramer, M. W., Pearson, S. E., and Michaelson, S. M.,** *Horm. Metab. Res.*, 5, 470, 1973.
182. **Salmon, S., Utiger, R., Parker, M., et al.,** *Endocrinology (Baltimore)*, 70, 459, 1962.
183. **McIntyre, H. B. and Odell, W. D.,** *Neuroendocrinology*, 16, 8, 1974.
184. **Garcia, J. F. and Geschwind, L. L.,** *Growth Hormone* (Eds. Recile, A. and Muller, E. E.). Excerpta Medica Foundation, Amsterdam, 1968, p. 267.
185. **Chapman, I. M., Helfgott, A., and Willoughby, J. O.,** *J. Endocrinol.*, 128, 369, 1991.
186. **Strosser, M. T. and Mialhe, P.,** *Horm. Metab. Res.*, 7, 275, 1975.
187. **Sakamoto, T., Iwata, M., and Hirano, T.,** *Gen. Comp. Endocrinol.*, 82, 184, 1991.
188. **Rabkin, R., Pimstone, B. L., Marlis, T., et al.,** *Horm. Metab. Res.*, 4, 467, 1972.
189. **Harvey, S. and Scanes, C. G.,** *Horm. Metab. Res.*, 9, 340, 1977.
190. **Herremans, M., Buyse, J., Decuypere, E., et al.,** *Horm. Metab. Res.*, 25, 142, 1993.
191. **Proudman, J. A. and Opel, H.,** *Poultry Sci.*, 69, 1569, 1990.
192. **Sakamoto, T. and Hirano, T.,** *J. Endocrinol.*, 130, 425, 1991.

Chapter 16

Growth Hormone Action: Growth Hormone Receptors

S. Harvey and K. L. Hull

CONTENTS

I. INTRODUCTION

Unlike most pituitary hormones, growth hormone (GH) lacks specific target sites and has widespread effects in all tissue and organ systems, in which diverse biological responses to GH stimulation occur. These actions, *a priori* are mediated by GH receptors (GHRs), which have a heterogeneity of structure and intracellular distribution and may

0-8493-8697-7/95/$0.00+$.50

differ in their attendant signal transduction mechanisms. In this chapter the synthesis, structure, distribution, and localization of GHRs are reviewed. Physiochemical aspects of GHRs have also been reviewed by Kelly et al.,[1,2] Postel-Vinay et al.,[3] Cramer and Talamantes,[4] Spencer et al.,[5] Mendelsohn,[6] Waters et al.,[7,8] and Mathews.[9]

II. THE HEMATOPOIETIN RECEPTOR SUPERFAMILY

The growth hormone receptor belongs to the hematopoietic receptor (HR) superfamily, which also includes the receptors for prolactin, hematopoietin, erythropoietin, the β and γ chains of interleukin 2 (IL-2), IL-3 through IL-7, the interferons, gp130, and KH97/AIC2B (which are subunits of other receptors), granulocyte colony-stimulating factor, and granulocyte-macrophage colony-stimulating factor (GM-CSF)[10–12] (Figure 1). The structural characteristics of this family include two to four conserved cysteines in the extracellular, N-terminus domain, a type II fibronectin-like domain,[13] and, except for the IL-7 and GH receptors, a conserved WSXWS motif[10,12]. This region is conserved as an analogous aromatic-S/G-X-aromatic-S sequence in the GHR.[14] Although the homologies in primary sequence constitute only a minor part of the receptor, they provide for similarities in secondary and tertiary structure and are crucial for protein folding and/or formation of the binding pocket.[10] Moreover, all are type I glycoproteins with an extracellular N terminus and a single transmembrane domain. The majority of HR family members are also present as a truncated, soluble binding protein, which may arise from alternate splicing of the receptor gene (e.g., GHR[15] and GM-CSFR[16] or proteolysis of membrane receptors (e.g., GHR[17] and CNTFR[18]).

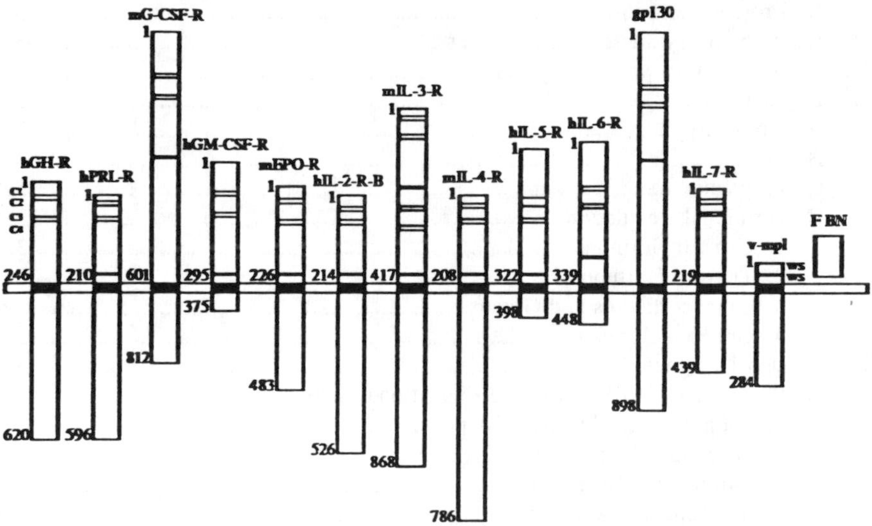

Figure 1 Schematic representation of members of the hematopoietin receptor family. The abbreviations used can be found in text. Homology with an approximately 90-amino acid type II domain of fibronectin (FBN) is also indicated. The first and last amino acids of the mature receptor, as well as the last amino acid of the extracellular region, are indicated. The transmembrane regions are shown as black boxes. The homologous pairs of cysteines are represented by thin black lines, and the WSxWS motif by thick black lines. (From Kelly, P. A., Ali, S., Rozakis, M., et al., The growth hormone/prolactin receptor family, *Recent Prog. Horm. Res.*, 48, 123–164, 1993. Copyright by Academic Press. Reprinted with permission.)

Receptors in this group may share mechanistic as well as structural similarities. For example, many HRs mediate cellular effects via ligand-induced receptor dimerization (e.g., GHR[19] and GM-CSFR[20]) and tyrosine kinase activation (e.g., GHR[21] and erythropoietin [ER][22]). This conservation of function across the family is surprising, as the length of the cytoplasmic, signal-mediating domain is not conserved, varying between 54 amino acids (GM-CSFR) and 568 amino acids (human IL-4R).[12] However, a conserved region containing two boxes of homology is located proximal to the transmembrane domain. Box 1, a proline-rich region, is present in at least 12 family members[20,23] and is necessary for induction of a proliferative response (e.g., GHR,[24,25] erythyropoietin,[26] GM-CSFβ[27]). Box 2, which is present in about half the family,[20,23] consists of a charge-conserved region and is necessary for (ER) or enhances (GHR, HMCSF) the proliferative response.[25,27]

Curiously, apart from GH and prolactin, the ligands for these receptors differ widely in primary structure. However, the tertiary structure of most of these hormones is organised as 4 antiparallel α helices.[12] This shared structure would therefore provide homologous regions to interact with the receptors. Indeed, the structure of the GH/GHR complex has been used to identify binding regions of the IL-4 molecule.[29]

III. CHARACTERIZATION

A. THE *GHR* GENE

The human GHR is encoded by a single gene localized on chromosome 5p13 to p14 or 5p13.1 to p.12.[30,31] This gene is composed of 10 exons ranging in size from 66 to 3400 bp, stretching over a distance of >87 kb.[32] Exon 2 encodes the signal peptide, exons 3 to 7 encode the extracellular domain, exon 8 encodes the transmembrane domain, and exons 9 and 10 encode the intracellular domain of the receptor [32] (Figure 2). A lengthy untranslated sequence of over 2 kb is located 3′ of exon 10 in the mature mRNA.[32] In chickens, the gene is located on the sex chromosome (chromosome Z) and may lack exon 3.[33]

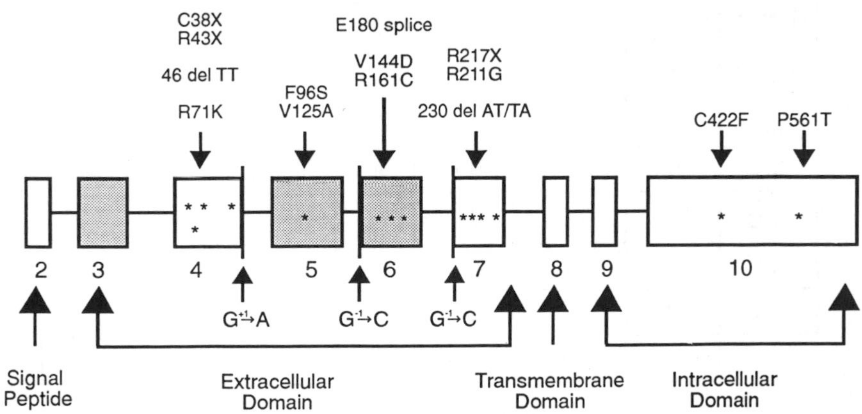

Figure 2 Structure of the *GHR* gene and mutations in Laron syndrome. The coding exons (boxes) are shown to approximate relative size; intron sizes are not to scale, and the noncoding part of exon 10 has been omitted. The structural protein domains encoded by the exons are indicated. The three shaded exons are involved in a deletion mutation. The locations (asterisks) and types of known point mutations within exons 4 to 7 are summarized above the diagram. The three splice site mutations are indicated by black bars at the exon 4/intron 4, intron 5/exon 6, and intron 6/exon 7 boundaries and are described below the diagram. (Based on Berg et al.[224])

The 5′ untranslated region (5′ UTR) in GHR cDNAs from several species is characterized by considerable heterogeneity. This is thought to indicate different promoters directing the expression of the gene from exons encoding 5′ UTR's which are alternatively spliced onto a common splice acceptor 11 bp upstream of the initiating AUG on exon 2. A liver-specific promoter for the ovine GHR gene has recently been identified[32a] in exon 1A, possessing a TATA box at –31, a CCAAT box at –88, and putative binding sites for several transcription factors involved in liver-specific gene expression. These include binding sites for the activated glucocorticoid receptor and C/EBP-like proteins.

The *GHR* cDNA has been cloned in sheep,[34] humans,[35] rabbits,[35] mice,[36] cattle,[37] pigs,[38,39] and chickens.[33] Multiple transcription products of the single *GHR* gene have been observed in most species, but may not be present in humans.[35] A 4.2- to 4.7-kb mRNA transcript encodes the full-length receptor, whereas a 1.0- to 1.4-kb transcript encodes the truncated binding protein in rat and mouse.[40–42] This smaller transcript is not present in humans, rabbits, or pigs,[33,39] suggesting the GH-binding protein (GHBP) in these species arises from proteolytic processing of membrane-bound receptors (see Chapter 14). Additional *GHR* transcripts of 1.7 to 1.9 kb in sheep[34] and 0.7 kb in chicken, quail, turkey, and duck[43] have also been detected; however, the proteins encoded by these transcripts have not been identified. A 2.1- to 2.6-kb transcript is also present in rat liver and adipose,[41,44] and in chickens[43] and sheep.[45] Larger transcripts of 8.0 kb have also been observed in mouse trophoblasts;[42] however, they may represent partially processed nuclear RNA.

The amino acid sequence predicted from these cDNAs is highly homologous in different mammals; however, the chicken GHR is less similar (Figure 3). The regions of high homology are also similar in the prolactin receptor and include the proximal cytoplasmic domain, the first two cysteine loops, and the distal extracellular domain.[38]

B. THE GHR PROTEIN

The human and rabbit GHR is a single-chain polypeptide of approximately 638 amino acids, composed of an 18-amino acids (aa) signal peptide, a 247-aa extracellular domain,

	Human	Rabbit	Sheep	Cow	Mouse	Rat	Chicken
Human							
Rabbit	84						
Sheep	81	77					
Cow	76	82	97				
Mouse	70	76	71	71			
Rat	69	74	71	71	85		
Chicken	59	60	58	57	56	56	

% GHR IDENTITIES

Figure 3 Percent amino acid sequence homology between growth hormone receptors. Percent identities between the indicated growth hormone receptors were determined after alignments were maximized. (From Cramer, S. and Talamantes, F., The growth hormone receptor and growth hormone-binding protein: structure, functions, and regulation, *The Endocrinology of Growth, Development, and Metabolism in Vertebrates* [Eds. Schreibman, M. P., Scanes, C. G., and Pang, P. K. T.]. Academic Press, Toronto, 1993, p. 117.)

a 24-aa transmembrane domain, and a 349-aa intracellular domain.[35] The ruminant (sheep and cow) GHR is highly similar; however, the extracellular domain is 242 aa in length and the intracellular domain is 350 aa in length.[34,37] In chickens, the cDNA sequence predicts a protein of 592 aa and a 16-aa signal peptide, a 221-aa extracellular domain, a 24-aa transmembrane domain and a 347-aa intracellular domain.[33] The receptor is extensively N-glycosylated at asparagine residues in the extracellular domain.[35,46] These glycosylated chains are converted to sialic acid containing complex-type chains rather than mannose chains.[46–48] However, the importance of glycosylation is not known, as transfection of *Escherichia coli* with the human (h)GHBP produces an unglycosylated GHBP capable of binding GH with normal affinity.[49] Approximately 50% of human GHRs are also covalently linked to to ubiquitin.[35]

1. Receptor Size

Analysis of GHR amino acid sequences predict a protein of 70 kDa; however, GH/GHR complexes of 122 to 142, 200, and/or 300 kDa have been purified from rat hepatocytes and adipocytes,[47,50,51] ovine hepatic microsomes,[52] human IM-9 cells and adipocytes,[46,53] porcine liver,[39] and rabbit liver[35,54,55] (Table 1). The 120-kDa moiety is probably the native form, corresponding to a 100-kDa receptor protein if only one 22-kDa GH is bound. Disulfide bonds between two or three 100-kDa subunits may account for the larger forms, as they resolve into the 100-kDa protein in the presence of reductants and have been shown to contain the 100 kDa form by two-dimensional electrophoresis (Table 1).[46,47] The additional 30-kDa discrepancy between the observed and computed GHR size is due to posttranslational modifications, such as glycosylation and covalent linkage to ubiquitin.[35] These modifications are tissue specific, as the ovine GHR/GH complex is smaller in ovine liver than when expressed in Chinese Hamster Ovary (CHO) cells (148 vs. 133 kDa).[52]

GH/GHR complexes containing a GHR equivalent to or smaller than the computed molecular weight have been isolated from sheep (80 kDa[52]), rat (66 kDa[53]), and rabbit (80 kDa[48,55]) liver and rabbit mammary glands (61 kDa[53]). These studies may be detecting the growth hormone-binding protein, which is approximately 55 kDa in size and is found in most tissues, or degradation products. Alternatively, the prolactin receptor (PRLR) may be responsible, as the 50- to 60-kDa PRLR is capable of binding hGH.[56–58] However, in the rabbit liver, the isolated receptor was capable of binding bGH (which is purely somatogenic) but not oPRL.[54]

2. Structure of the Intracellular Domain

Structural and functional aspects of the GHR have been recently reviewed by Beattie[58a] and by Beattie and Flint[58b] and Wells.[58c] The intracellular domain is essential for signal transduction[58d] and may be further subdivided into subdomains responsible for distinct GH actions. The proximal portion may mediate the proliferative effects of GH, as only the 54 amino acids adjacent to the membrane are required for GH-induced proliferation in a promyeloid cell line.[59] Moreover, the endogenous short form of the prolactin receptor (which lacks most of the cytoplasmic domain) is also active in proliferative assays.[60] This region includes the conserved proline-rich box that is necessary for the proliferative effects of most of the hematopoietic receptor family (see above). The importance of these prolines is underlined by a study by Billestrup et al.,[61] in which the conversion of Pro-300, -301, -303, and -305 to alanines eliminated the effect of GH on receptor internalization, metabolism, transcription, and mitogen-activated protein (MAP) kinase activity. Residues 249 to 381 are also necessary for metabolic and MAP kinase-mediated effects of GH, whereas residues 436 to 620 appear to play a role in transcriptional effects.[61] The carboxy terminus of the receptor is also necessary for GH-induced insulin synthesis in rat insulinoma cells.[63]

Table 1 **Identified Molecular Weights of GH/GHR Complexes under Reducing and Nonreducing Conditions**

Receptor Source	Native M_r(kDa)	Reduced M_r(kDa)	Ref.
Rat adipocyte	56	56	177
	108*	↑ 108	
	240*	↓ 230	
	310*	↓ 310	
	116–125	134	50
	230		
Rat liver	220	120	255
	300		
	130	100	47
	220		
	300		
	66	66	53
Rat hepatocyte	100	↑ 100	47
	220	↓ 220	
	300	↓ 300	
Rabbit mammary gland	61	61	51
Rabbit liver	70–80		48,53,54,55
	56		256
	68		
	76		
	57*	57*	257
	124	124	
Mouse liver		56	258
		62	
		125*	
Transgenic mouse liver		70	169,259
		82	
		122	
Human IM-9 cells	140	140	46
	270		
	130	130	53
Human skin	140		260
Human fibroblast	140		260
Human liver	124	↓ 124	261
	75	↑ 75	
Pig adipose	114		186
	75		
Sheep GHR cDNA	170	170	52
(expressed in CHO)	95*	95*	
Sheep liver		155	
		80	
Pig liver	140		39
Chicken liver and pituitary	80–86		262

GH, growth hormone; GHR, GH receptor; Mr, molecular weight; CHO, chinese hamster ovary cells; *, major form.

↓ increase; ↑ decrease.

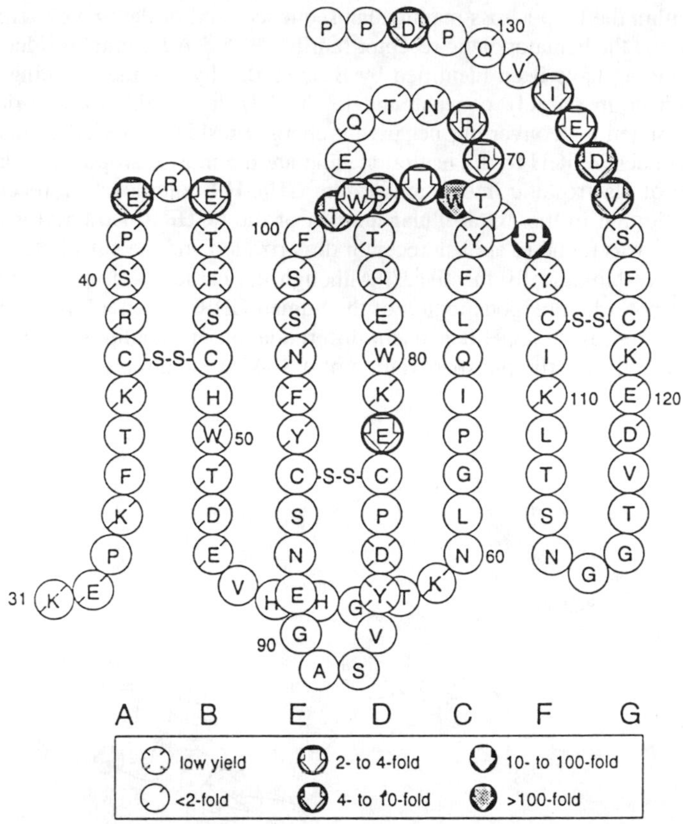

Figure 4 Location of residues in the hGHBP causing a twofold or greater effect on binding to hGH, mapped on an immunoglobulin-like folding diagram predicted for the cysteine-rich domain of the hGHBP. Seven antiparallel β strands (A–F) compose a β barrel by folding the G strand around the back until it hydrogen bonds to the A strand. Residues causing less than a 2-fold reduction, a 2- to 4-fold reduction, a 4- to 10-fold reduction, a 10- to 100-fold reduction, and >100-fold reduction in binding are indicated. Those residues that gave low expression yields are also indicated. Disulfides are shown by -S-S- and residues are numbered by tens. (From Bass, S. H., Mulkerrin, M. G., and Wells, J. A., A systematic mutational analysis of hormone binding determinants in the human growth hormone receptor, *Proc. Natl. Acad. Sci.*, U.S.A., 88, 4498–4502, 1991. Copyright by National Academy of Sciences. Reprinted with permission.)

3. Structure of the Extracellular Domain

The extracellular domain of the human GHR/GHBP is composed of two immunoglobulin-like domains, each consisting of seven β strands divided into two antiparallel β sheets (residues 1–123 and 128–238), separated by a four-residue connecting chain.[64] The two domains are linked by a salt bridge.[64] The NH_2 portion contains three disulfide loops linking six cysteines,[49] two of which link adjacent β strands and the last of which links the two β sheets.[64] These cysteines are widely conserved across species, including chickens[33] and rabbits.[65] The first disulfide loop of the GHR interacts with GH via electrostatic and hydrophobic interactions, and mutations within the loop eliminate GH binding in rabbits.[65] The importance of this loop is supported by the conservation of the residues within the loop across mammalian sequences and of the two cysteines forming the loop across the hematopoietic receptor family.[10,34–36,66] Additional residues important for hGH binding have been identified by Bass et al.[67] by alanine scanning (Figure 4). These residues are largely concentrated in the NH_2-terminal, disulfide-rich domain. Moreover, mutations converting negatively charged residues in GHBP and positively charged residues in hGH to the neutral alanine are the most disruptive, underlining the importance of electrostatic interactions in the GH/GHR complex.[67] In recent studies a single substitution in the extracellular domain of the hGHR (histidine for aspartate at position 152) was found to abolish receptor dimerization and the substituted monomeric receptors were biologically inactive.[67a] Although the carboxyl-terminal domain of the cytokine receptor homologous region of the human GHR does not have ligand binding activity, it enhances the hGH-dependent differentiation of preadipose 3T3-F442A cells into adipose cells and the phosphorylation of a 34-kDa membrane protein.[67b]

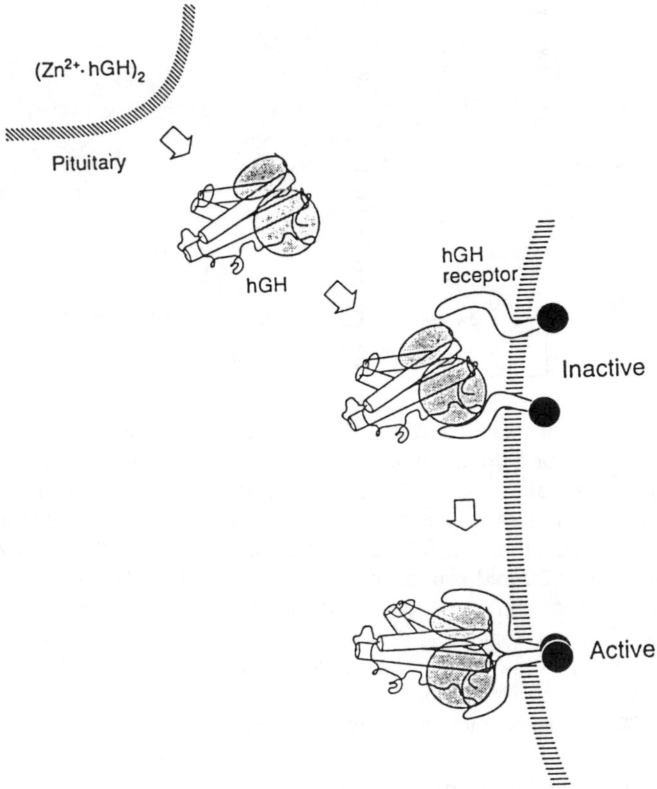

Figure 5 Summary of molecular endocrinology of hGH. hGH is released as a $(Zn^{2+}\text{-}hGH)_2$ complex from the pituitary. On dilution in the bloodstream, the dimer dissociates into a monomeric form, in which it is available to bind via site I to the hGH receptor. The membrane-bound hGH then complexes with a second receptor molecule using site II on hGH and binding determinants on the first receptor. Receptor dimerization by GH initiates signal transduction. (From Wells, J. A., Cunningham, B. C., Fuh, G., et al., The molecular basis for growth hormone-receptor interactions, *Recent Prog. Horm. Res.*, 48, 253, 1993. Copyright by Academic Press. Reprinted with permission.)

4. Mechanism of GH Binding

Although early binding studies assumed 1:1 stoichiometry in GH/GHR complexes, work has demonstrated that one hGH molecule binds two hGHBPs (thus, two receptors).[68,68a] The two binding sites on the growth hormone are distinct; however, the residues interacting with the two sites differ in only one residue (Asn-19)[68–70] (see Chapter 1). Binding occurs sequentially to the higher affinity site I and then to site II, as site I is larger and contains twice as many sites for the formation of hydrogen bonds and salt bridges.[64] Binding to site II is stabilized by interactions between the COOH tails of the two receptors; however, minimal interaction between receptors occurs in the absence of ligand.[64] Dimerization of the hGHR is rapidly induced by hGH, which induces its receptor to form a noncovalently associated complex, which then undergoes a rapid transition to a disulfide-linked form.[64a]

This interaction between the COOH tails is thought to initiate signal transduction, and consequently inhibition of dimerization by mutation of the GHBP or by immunoneutralization with a monoclonal antibody blocks GH action.[52,64] Moreover, excess GH results in the formation of inactive monomeric complexes, which may account for the biphasic effect of GH on cell proliferation[71] and lipogenesis.[71a] GH analogs containing only one binding site are thus inactive in cell-based[72] and rat weight gain bioassays.[72] However, this model of sequential dimerization does not apply to all species, as the binding of bovine GH to the bGHBP is disrupted by mutation at site II, and bGH:bGHBP monomers are not observed even under conditions of GH excess.[19]

The sequence of events between GH secretion from the pituitary and GH-induced dimerization of target tissues is summarized in Figure 5. The roles of zinc and the GHBP in this scheme are described in Chapters 5 and 6, and 14, respectively.

Calcium is necessary for the binding of hGH (but not oGH) to rabbit serum/cytosolic GHBPs[73] and hGH (but not rGH or bGH) to rabbit liver membranes.[74] Calcium serves to neutralize opposed negative charges in hGH and the GHR to permit binding.[73] The 20 kDa variant of hGH possesses an intermediate degree of calcium dependence, thus a critical negative charge is probably located between amino acids 32 and 46.[73] A possible candidate is Glu-33, as it is replaced by positively charged arginine in all characterized subprimate GHs.[75] Thus, the electrostatic interactions in this region would contain one interaction for subprimate GH, as the positive charge of the ligand would attract the negative charge of the receptor, and two interactions for human GH, as the calcium would interact with negative charges on both molecules. This additional attractant in the hGH molecule may account for the high affinity of hGH in heterologous radioreceptor assays. Alternatively, calcium may serve to increase the rigidity of the hGH molecule.[76]

5. Characteristics of GH Binding

The specificity and magnitude of GH binding are dependent on time, temperature, pH, and the presence of divalent and monovalent cations.[64a,77,78] Importantly, binding is reversible in fresh preparations but is increasingly irreversible due to the internalization of ligand (see below).[79,80] The binding affinity of GHRs is generally in the nanomolar range (Table 2). Commonly, the endogenous ligand is of lower affinity than hGH and/or bGH, which may be due to the effect of calcium described above. Although Scatchard analysis generally results in a linear graph, suggesting a single class of binding sites (e.g., in salmon liver[77]), high- and low-affinity binding sites have been identified in rabbit,[54] rat,[81] sheep,[82] pig,[83] bovine,[84] and human[85] hepatic membranes (Table 2).

IV. GHR HETEROGENEITY

The heterogeneity of growth hormone receptors has been well established by structural and binding criteria and may account for the tissue-specific actions of GH described in

Table 2 **Affinities and Capacities of Growth Hormone Receptors from Various Species and Tissues**

Species	Tissue	Ligand	Affinity	B_{max}[a]	Ref.
Tilapia	Liver	bGH	$K_d = 13\ nM$	180	263
		tGH	$K_d = 2.5\ nM$	390	263
Eel	Liver	eGH	$K_a = 1.1\ nM^{-1}$	105	264
Salmon	Liver	sGH	$K_a = 92\ nM^{-1}$	84	77
Trout	Liver	sGH	$K_a = 2.4\ nM^{-1}$	187 pmol/mg tissue	265
Rat	Adipocyte	rGH	$K_a = 1\ nM^{-1}$	15,000/cell	292
8-day ♂ or ♀	Liver	rGH	$K_a = 0.66\ nM^{-1}$	6.4	143
28-day ♂	Liver	rGH	$K_a = 0.66\ nM^{-1}$	30	143
28-day ♀	Liver	rGH	$K_a = 0.66\ nM^{-1}$	39	143
Adult ♀	Liver	bGH	$K_a = 0.5\ nM^{-1}$		81
			$K_a = 21.4\ nM^{-1}$		
Adult ♂	Liver	bGH	$K_a = 0.5\ nM^{-1}$		81
			$K_a = 12.1\ nM^{-1}$		
Adult ♀	Liver	bGH	$K_a = 0.8\ nM^{-1}$	35 pmol/mg tissue	164
Mouse	Liver	bGH	$K_a = 23.2\ nM^{-1}$		152
Guinea pig	Liver	oGH	$K_d = 2.34$ pmol/l	623	250
Human	Liver	hGH	$K_a = 2.0\ nM^{-1}$	14–53	85
			$K_a = 1.5\ nM^{-1}$	707	85
	Fibroblast	hGH	$K_a = 1.07\ nM^{-1}$	8305/cell	184
	Adipocyte	hGH	$K_a = 2.1\ nM^{-1}$	7.3/cell	266
	Lymphocyte	hGH	$K_a = 1.5\ nM^{-1}$	7.1/cell	267
Rabbit	Liver	rbGH	$K_a = 0.14\ nM^{-1}$	7–38	54
			$K_a = 14\ nM^{-1}$	990–480	
Adult	Lung	hGH	$K_a = 2.6\ nM^{-1}$	9.5	269
Fetus	Lung	hGH	$K_a = 1.85\ nM^{-1}$	27.6	269
Pig	Muscle	bGH	$K_a = 9–15\ nM^{-1}$	5–6	293
2-day	Liver	bGH	$K_a = 3.5\ nM^{-1}$	12	83
Adult	Liver	bGH	$K_a = 5.0\ nM^{-1}$	91	83
			$K_a = 0.35\ nM^{-1}$		83
Sheep	Liver	oGH	$K_a = 0.27nM^{-1}$		270
	Expressed	oGH	$K_a = 0.3\ nM^{-1}$		52
Chicken	Liver	cGH	$K_a = 0.12\ nM^{-1}$	1.91	271
	Hypothalamus	cGH	$K_d = 33.5\ nM$	2.14	272
Turkey	Liver		$K_a = 3.09\ nM^{-1}$	2.13	271

[a] The maximal binding capacity (B_{max}) values are provided as fmol/mg protein unless otherwise described.

other chapters. Variation may occur in the *GHR* transcript, posttranslational processing, and association with other proteins.

A. TRANSCRIPTIONAL VARIATION

1. Splicing Variants

Alterations in splice site usage within the coding region generally produce transcripts and proteins that vary in size. The best characterized variant of this type is the 1.2-kb transcript observed in rodents, which results from alternate splicing at the exon 7/8 boundary and contains a short hydrophilic tail in place of the transmembrane and intracellular domains (see Chapter 14). The GHBP mRNA in rodents also encodes a 50-kDa (processed to 38 or 42 kDa) short form of the GHR (GHR_s) in adipocytes that binds GH with high affinity and is regulated independently of the long form.[86] In contrast to the GHBP, the GHR_s is exclusively membrane bound and is not secreted.[87]

The existence of a novel binding protein is suggested by the identification of a rabbit GHR cDNA clone that diverges from the full-length GHR cDNA at the exon 8/9 boundary.[35] Translation of this sequence would result in a membrane-bound receptor with a truncated cytoplasmic domain of eight amino acids, somewhat homologous to the short form of the prolactin receptor. However, as yet there is no evidence supporting the existence of this protein. A similar membrane-bound GHBP would be generated by the chicken GHR splice variant identified by Bingham et al.[38] This variant uses an alternate splice acceptor site within exon 7, resulting in a 184-aa GHBP that includes the transmembrane domain and a portion of the intracellular domain.

Alternate splicing at the exon3/exon 4 junction results in production of human GHR mRNA lacking exon 3 (GHR3d).[32] Synthesis of GHRs containing or excluding exon 3 is regulated in a tissue-specific manner, as hGHR is the only form present in fat and fibroblasts but is excluded from fetal placenta cells.[88,89] GHR and GHR3d mRNA are equally abundant in total placenta, lymphocytes, brain, and perhaps liver, but hGHR predominates in prostate and lung and hGHR3d predominates in the stomach and some hepatoma cell lines.[88,89] The mouse placenta does not contain GHR/GHBP sequences with the exon 3 deletion,[90] but the only chicken GHR cDNA identified to date lacks exon 3.[33] This alternate splicing pattern is therefore also present in subprimates. Although binding proteins or membrane-bound receptors lacking the third exon bind hGH and hCS with normal affinity, this deletion eliminates a glycosylation site and may alter cellular trafficking or signal transduction of the receptor.[88,89] Conversely, GHR cDNA clones have been isolated from a porcine cDNA library containing two copies of exon 3, thus the presence, absence, or duplication of this exon is common.[38]

Conversely, multiple splice sites in the 5′ untranslated region (5′ UTR) produce transcripts of similar size that differ in the 5′ regulatory region. The use of multiple splice sites at the exonl/exon 2 junction is responsible for variation in the 5′ UTR of the human,[35,91] sheep,[34] rat,[66] and rabbit[35] GHR mRNA. This heterogeneity occurs 12 bp upstream of the initiating ATG codon.[34] Variant V1, the most abundant form in human liver,[91] also predominates in ovine liver but is totally absent from ovine nonhepatic tissues[34] and may not be present in rodents.[36,66,92] The less abundant V2, which was published as the cloned hGHR cDNA,[35] predominates in rodent liver,[93] whereas V7 is present in lower amounts in human and rat liver.[91,93] Additional human variants bearing no resemblance to cloned GHRs have also been observed, including GC-rich, upstream ATG-containing UTRs that may not be translated.[91]

Growth hormone receptor transcripts other than the full-length 4.4-kb mRNA may also contain variation in the 3′ region. For example, the 2.6-kb transcript present in rat liver contains the the GHBP hydrophilic tail present in the 1.2-kb transcript.[44] It has been suggested that the two transcripts differ only in the length and composition of the 3′ UTR and/or the length of the poly(A) tail.[44] The differential regulation of the two transcripts[41,44] is of interest, as it suggests that regulatory regions may be located in 3′ sequences. Although these regions were traditionally confined to the sequences upstream of the functional gene, response elements have been increasingly localized to structural and downstream sequences (e.g., *GH* gene). Alternatively, undetected variation in 5′ regulatory regions of the two transcripts may also be present.[41,44] Alterations in the 3′ untranslated region [including the poly(A) tail] of the 1.2-kb rodent GHBP mRNA are thought to result in a 2.6-kb transcript in rodents, but the derivation of this transcript in other species is presently unknown.[44]

2. Polyadenylation Variants

The use of stop and polyadenylation sites within the coding region of the GHR mRNA can result in drastic alterations in the size and composition of the resulting protein. In a study by Oldham et al.,[43] an alternate polyadenylation signal in the chicken (but not rat) cDNA has been identified that results in a 0.7-kb GHR transcript. This signal is located in exon 5 at residue 325, thus this transcript encodes a GHBP encompassing exons 2, 4, and 5. This site is located 10 to 30 nucleotides downstream of AATAAA and upstream of a GU-rich sequence, which is characteristic of poly(A) sites. Moreover, it is functional when inserted into the rat GHR gene [which contains only a single poly(A) site]. Usage of the two poly(A) sites is regulated in a tissue-specific manner, as full-length and 0.7-kb transcripts are present in equal amounts in liver but the larger transcript predominates in heart, muscle, and female kidney. Although a protein product of the 0.7-kb transcript has not been identified, translation of the mRNA is suggested by its close association with polyribosomes.

B. PROTEIN VARIATION

Variation can also occur in proteins translated from the same mRNA, owing to differential posttranslational processing and/or association with other proteins. For example, three subtypes of the rabbit liver GHR and a distinct adipose GHR have been differentiated by their reactivity with various monoclonal antibodies.[94,95] Binding to type 1 receptors can be inhibited by either MAb 7 and MAb 263, whereas binding to type 2 and type 3 receptors is sensitive only to MAb 7 or MAb 263, respectively.[94] This heterogeneity may be due to a single GHR subunit interacting with different membrane proteins, which would obscure certain epitopes and would permit receptors to be modified by tissue-specific factors. Indeed, the receptor subtypes are distributed in a tissue-specific manner, as only type 2 binding sites are present in clonal osteoblast-like cells and type 3 receptors are absent from GHR-transfected COS cells.[35,96] Moreover, cytosolic receptors, which would not interact with membrane proteins, are identical in rabbit liver and adipose tissue, using these criteria.[95]

Antibodies directed against different epitopes of the hGH molecule have identified further heterogeneity in the binding region of the GHR.[97] The ability of different antibodies to inhibit GH:GHR complex formation differs not only between species but also between normal and *lit/lit* mice and between normal, lactating, and pregnant rats.[97] Thus, as expected, the epitopes of hGH interacting with the GHR differ between species. However, this study also suggests the presence of several distinct populations of receptors within each species that are differentially sensitive to alterations in physiological state.

GH receptors also differ in their affinity for intact GH and GH variants. The high- and low-affinity binding sites for intact GH (Table 2) described above may correspond to the high- and low-affinity GHBPs in serum (see Chapter 14). These sites are also regulated independently. For instance, the high-affinity GH-binding site alone is involved in the GH-induced upregulation of hepatic GHRs in sheep[82] whereas the ontogenic rise in porcine hepatic GHRs is accompanied by the development of a low-affinity binding site.[83] Other examples of differential regulation are provided in Section VII, below. In addition, the binding kinetics of cleaved GH in liver are suggestive of multiple receptor subtypes; type 1 binds cleaved and intact GH equally, whereas type 2 binds intact GH with the same affinity as type 1, but binds cleaved GH poorly.[99]

GHR heterogeneity is further supported by the comparative binding activities and bioactivities of 22-kDa hGH (22K) and the 20-kDa hGH variant (20K). The two hGH forms are equally effective in growth-promoting assays but 20K competes poorly with labeled 22-kDa GH (hGH) in female rat and rabbit liver (3% vs. 20%) and rabbit mammary gland (22% vs. 53%).[100,101] Thus, 20K may bind a specific receptor. This possibility is supported by the ability of 20K to compete equally or more effectively for labeled rGH or 20K binding sites in rat liver (respectively).[102] On the basis of these and other studies, Hughes et al.[102] have proposed two GHR subtypes: GHR-1, a low-capacity receptor that binds rGH and the two hGH variants equally, and GHR-2, which binds 22K with high affinity but rGH and 20K with low affinity. GHR-1 may be the sole variant present on IM-9 lymphocytes, as 20K inhibits the binding of 22K as effectively as 22K itself,[103] whereas both variants are present in liver membranes. Moreover, only a portion of hepatic 20K binding is calcium dependent.[73]

V. RECEPTOR BIOSYNTHESIS AND TURNOVER

The biosynthetic pathway of the GHR is common to most peptides, in that the protein is synthesized on the rough endoplasmic reticulum, processed and glycosylated by the Golgi apparatus, and subsequently inserted into the plasma membrane. The receptors may also require some form of activation, as Gorin et al.[104] suggested that a cAMP-dependent process regulated the availability of membrane GHRs. The appearance of new receptors on the plasma membrane is dependent on protein synthesis, normal vesicular traffic, normal microtubule and/or microfilament systems, and cellular energy but is independent of glycosylation state and pH.[105,106] Growth hormone receptor biosynthesis is extremely rapid, as adipocytes require only 2 h to restore GH-binding levels to normal following the destruction of existing receptors by trypsinization.[105]

The rapidity of GHR synthesis is necessitated by the rapid turnover of GHRs in adipocytes ($t_{1/2}$ = 45 min),[106] hepatocytes ($t_{1/2}$ = 30 to 40 min),[107] and fibroblasts ($t_{1/2}$ = 75 min).[108] Indeed, in the presence of translational blockers, GH-binding activity in adipocytes is entirely blocked within 3 h. Curiously, the early study of Lesniak and Roth[109] reported a slow turnover rate in IM-9 lymphocytes ($t_{1/2}$ = 8 h), which is similar to the $t_{1/2}$ of insulin and insulin-like growth factor I (IGF-I). However, this study failed to observe GHR internalization, a well-accepted phenomenon responsible for the short GHR half-life in other tissues (see below), thus the conclusions are somewhat suspect. The rate of GHR turnover is thought to be increased by receptor occupancy, as the GHR $t_{1/2}$ was decreased in the presence of ligand in adipocytes, fibroblasts, and lymphocytes.[108,110,111] However, Gorin and Goodman[106] failed to observe any effect of receptor occupancy on receptor turnover in freshly isolated adipocytes.

This loss of membrane binding activity is due to internalization by both constitutive and ligand-induced pathways.[105] The ligand-induced pathway is the best understood, in which GH binding results in GHR aggregation in coated pits followed by phosphorylation-independent endocytosis.[112,113] The composition of the resulting vesicle is then modified by fusion with acidic prelysosomes and subsequently protein-degrading lysosomes. The GH:GHBP complex does not dissociate at pH 5.5, the environment of the acidic prelysosomal endosomes, thus the hormone and receptor are processed together. This finding contrasts with many peptide hormones that dissociate from their receptors in the endosome, permitting receptor recycling.[105]

The fate of the internalized GH:GHR complex is still somewhat controversial. Degradation of internalized GH is suggested by the inhibitory effect of lysosomal enzyme inhibitors on release of radiolabeled GH.[114] However, internalized radiolabeled hGH recovered from rat hepatic microsomes and lysosomes was trichloroacetic acid (TCA)-precipitable and capable of rebinding to membranes, suggesting that GH remains intact throughout processing. Moreover, the internalized hormone released from IM-9 cells in the study of Ilondo et al.[115] was comparable in size to intact GH (22 kDa). In accordance with these observations, kinetic studies suggest that internalized GH is released from two vesicular compartments, the endosome and the lysosome,[115,116] which would permit both GH/GHR degradation and GH/GHR release intra- or extracellularly. Accordingly, Roupas and Herington[117] claimed that 75% of GHRs are degraded and the remaining 25% may be recycled to the membrane intact or released into the intracellular space. Growth

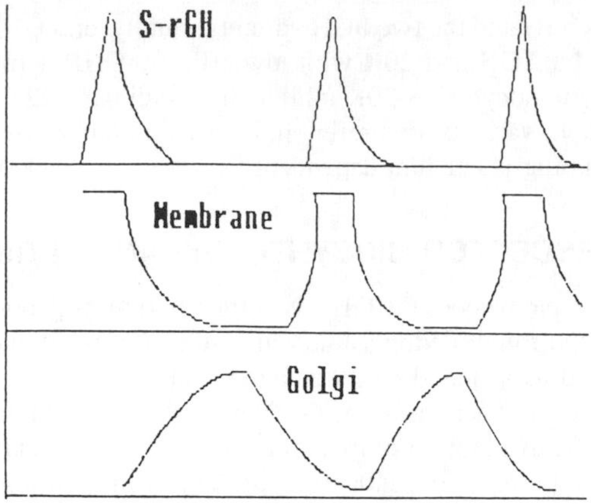

Figure 6 Schematic model of the spontaneous changes of the hGH receptors and their internalization to the Golgi membranes. GH secretion is followed by an immediate downregulation of the plasma membrane receptors to GH. This is followed by a spontaneous, slow upregulation, which prepares for another surge at 3-h intervals. The receptors that disappeared from the plasma membrane are internalized to the Golgi membranes. An unknown fraction of the Golgi receptors is recycled to the plasma membrane, awaiting the next GH pulse. (From Bick, T., Yardin, M. B. H., and Hochberg, Z., Adaptations of liver membrane somatogenic and lactogenic growth hormone (GH) binding to the spontaneous pulsation of GH secretion in the male rat, *Endocrinology*, 125(3), 1711–1722, 1989. Copyright by the Endocrine Society. Reprinted with permission.)

hormone may also be modified prior to release, as endosomal and lysosomal fractions contain a serine protease capable of cleaving GH into a 15-kDa form, which is capable of binding the GH receptor and is biologically active.[118] However, it is not known whether this internalized receptor and/or ligand is involved in signal transduction, as internalization inhibitors did not lessen the bioactivity of GH.[114]

The rapid turnover and internalization of the GHR are thought to enable recognition of specific GH pulses and to permit responses on the basis of the frequency of GH pulses[119,120] (Figure 6). (See Hochberg et al.[138a] for recent review). In the normal male rat, GH is secreted in a pulsatile fashion with 3 h between pulses. Both somatogenic and lactogenic receptors are present at higher levels immediately after a GH pulse than during a GH trough, thus enabling the maximum sensitivity.[121] Within the hour following a pulse, 65 to 80% of the plasma membrane-bound receptors are internalized to the Golgi fraction, due to increased receptor occupancy. Plasma membrane receptors are restored to the original level within 2 h for somatogenic receptors and within 2.75 h for lactogenic receptors; thus they are available for interaction with the subsequent GH pulse.

The internalization of GH/GHR complexes ensures that the interaction between GH and the GHR does not demonstrate simple equilibrium kinetics. Instead, there is a time-dependent increase in apparently irreversible binding. For example, GH binding to fresh lymphocytes is essentially reversible, as 100% of bound hormone is dissociable by trypsin, whereas this acid-soluble fraction drops to 50% and 70% after 30 min and 5 h, respectively [116]. Partially irreversible GH binding has also been demonstrated for adipocytes and hepatocytes. Thus, Scatchard analysis is not strictly applicable for the GHR as for other receptors, but still provides an indication of the relative affinity and capacity of the GHR in different species and tissues (Table 2).

VI. LOCALIZATION

Growth hormone receptors are almost ubiquitously present in rodents (rats and mice), ruminants (sheep and bovines), rabbits, humans, fish, and birds (chickens) (Table 3). Growth hormone-binding activity has also been localized in undomesticated mammals (porcupines[28]), reptiles (turtles[122]), and amphibians (frogs[123]). In most species, the adult liver is the primary GH target site, as hepatic GHR mRNA is three- to tenfold greater than that in other tissues.[124] However, GH binding and/or GHR mRNA is highest in the kidney of frogs,[123] porcupines[28] and fetal sheep,[45] whereas thymus and spleen GHR mRNA levels are higher than those in the liver of fetal cows.[124] The variable roles of GH in different species and physiological states may thus be mediated in part by altered receptor levels in different target tissues.

The intracellular distribution of GHRs is equally widespread, as GHR/GHBP immunoreactivity and binding activity have been localized to the plasma membrane, secretory vesicles, Golgi apparatus, and nucleus in liver and pituitary cells.[85,125,125a] Putative roles of the intracellular receptors are discussed in Chapters 14 and 18.

Despite the ubiquitous nature of the GHR in most species, the intratissue localization of GHRs is often quite specific and delineates spheres under direct and indirect (IGF-I-mediated) GH influence. For example, hepatic GHRs are located on hepatocytes and the hematopoietic Kupffer cells,[126] reflecting GH effects on the synthesis of IGF-I and hepatic enzymes and on immune function, respectively.

The relative distributions of GHR mRNA and IGF-I receptor (IGF-IR) mRNA have been used to distinguish between direct and IGF-I-mediated effects in the kidney.[127] For instance, the proximal straight tubule was identified as a GH-responsive structure due to the presence of GHR mRNA but not IGF-IR mRNA, whereas the glomerulus, distal tubule, and collecting system contained only IGF-IR mRNA. The medullary thick ascending limb (MTAL), the site of renal IGF-I synthesis, was also rich in GHR mRNA, suggesting that GH plays a direct role in renal IGF-I synthesis but does not directly affect the glomerulus, distal tubule, or collecting system.

Growth hormone receptor expression in bone appears to be limited to specific developmental stages. In human infants and newborn rabbits, GHR immunoreactivity is intense in proliferative zones of the growth plate, whereas increased staining in the less mature reserve zone is present in young rabbits.[98,128] The closure of the epiphyseal growth plate in rabbits is associated with the loss of GHRs, suggesting GH is primarily involved in the differentiation of prechondrocytes into mature chondrocytes rather than in bone maintenance.[98,128] However, GHR expression is observed in mature rats, in which the epiphyseal plate does not close.[98,128]

Growth hormone receptors are extensively localized in epithelial cells and their derivatives throughout the reproductive and gastrointestinal systems and in skin. In skin, immunoreactive GHRs are present in all epidermal layers except the stratum corneum but are less abundant in dermal layers.[129,130] GHRs are also abundantly present in sweat and sebaceous glands as well as the smooth muscle cells and endothelium of arteries.[129] These receptors may be responsible for the oily skin and increased incidence of atherosclerosis (respectively) associated with acromegaly.[62,129]

The responsiveness of the gastrointestinal tract (GIT) to the mitogenic effects of GH is well established; however, many groups were unable to detect GH-binding sites in this tissue. These negative results may, however, be due to receptor degradation by gut enzymes, as GHR mRNA[130a] and immunoreactivity is present nearly ubiquitously throughout the GIT[131] and GH binding has been identified in gut extracts from fish.[132,133] In general, immunoreactivity is more intense and heterogeneous in the epithelial cells and their derivatives than in mesenchymal cells, suggesting epithelial cells are the primary GH target sites within the GIT. Curiously, despite the well-known mitogenic effect of GH

Table 3 **Distribution of GHR mRNA and Immunoreactive and/or Bioactive GHRs**

Tissue	Humans		Rodents[a]		Ruminants[b]		Rabbits		Pigs		Fish		Birds[c]	
	GHR mRNA	GHR	GHR mRNA	GHR	GHR mRNA	GHR	GHR mRNA	GHR	GHR mRNA	GHR	GHR mRNA	GHR	GHR mRNA	GHR
Liver	Low[273*]		+[40,41] +[139*,140*]	+[126] Low[139*]	+[34,37,124] Low[45,124*]	+[45]	+[148]	+[123]	+[38]	+[293]		+[77,265,275,264]	+[33,278]	+[262,279]
Kidney		+[280*]	+[40,41,127] +[140*]	+[139*]	+[34,37] Low[124*]		+[135,148]		+[38]			+[77,265,275] –[264]	+[33]	
Adrenal	+[273*]	–[280*]	+[40,41]	+[139*]	+[34]		+[148]	+[123]					+[33]	
Heart	+[273*]		+[40,41]		+[34]		+[135,148]		+[38]			–[275]	+[33]	
Lung	+[273*]	+[280*]	+[140] +[140*]	+[139*]	Low[34]				+[38]				+[33]	
Skin	+[260,281]	+[128,260,281,280*]		+[128,139*]				+[128]				+[265,282] –[275]		
Fibroblasts	+[283*]			+[139*]										
Muscle	+[273*]	–[280*]	+[40,41]	+[139*]	+[34] +[45*]		+[148]		+[38]			+[265] –[28]	+[33,278]	
Chondrocytes /bone	+[283*]	–[280*] +[128]		+[139*]				+[98]						
GIT	+[273*]	–[280*]	+[40,41] +[140*]	+[139*] +[131]	+[124*]							–[28,265] +[275,277]		
Fat		+[129]	+[40]	+[128]	+[37]			+[128]	+[38]			+[265]	+[278]	
Mammary gland			–[40]		+[37,135]		+[135]		+[135]					
Thymus			–[40]	+[284]	+[124*]		+[148]		+[38]				+[285]	

Spleen	+[273*]		–[40,41]		+[124*]			+[38]	+[265]		
Hematopoietic cells		+[267]									
Anterior Pituitary	+[286]		+[286]	+[139*]	+[37]					+[33,262]	+[262]
Posterior Pituitary	+[273*]				Low[37]						
Brain	+[273*]	+[280*]	–[40]		Low[37]	+[135]	+[135]		+[77,265,275]	+[33,262]	
			+[287–289]	+[139*]	+[45,124*]				–[264]		
Tongue			+[41]								
Gall bladder	–[273*]										
Pancreas	+[273*]			+[139*]	Low[34]						
Urinary Tract	+[273*]			+[139]							
Male reproductive tract				+[130]		+[148]					
Ovary	+gonad[273*]		+[40,41]	+[130]		+[135,148]			+[265,275]		
Granulosa	+[290]	+[290]	+[142]	+[130]							
Corpus lutea			+[142]	+[130]	+[124]						
Testis	–[273*]		–[40]	+[130]		+[148]			+[28,265]	+[33]	
Endometrium				+[130]	+[124]						
Placenta	+[273,283]	+[280,283]			Low[37]						
					+[45]						
Teeth				+[130]							
Gill									+[77,265,276]		
									+[264]		

Note: +, GHR mRNA or protein is present; –, GHR mRNA or protein is absent; *, studies perfomed in fetal animals.

[a] Rodents: rats and mice.

[b] Ruminants: sheep and bovines.

[c] Birds: chickens.

[d] Fish.

on gastric mucosa, the most intense immunoreactivity was associated with the chief cells of the exocrine gastric gland rather than the proliferating cells of the gastric isthmus.[134] Thus, GH may induce gastric proliferation via gastric or systemic IGF-I or gastrin, and the chief cell receptors may mediate GH-induced synthesis of intrinsic factor.[134] Direct GH involvement in exocrine gut functioning is also supported by the presence of GHRs in pancreatic acini, islets, and ducts, and in intestinal enteroendocrine cells.[131] Conversely, GHRs localized in columnar and crypt cells of the small intestine and colon may mediate GH-induced stimulation of mucosal proliferation and intestinal ion transport or may act to increase IGF-I in these tissues.[131]

Lobie and colleagues[130] have determined that GHR immunoreactivity in the male and female reproductive tracts also follows a ubiquitous epithelial/endothelial distribution. In accordance with functional studies, GHRs in the endometrium and the epithelium of the oviduct in the female and the vas deferens, seminal vesicles, and epididymis in the male may be important in epithelial maintenance. The endometrial receptors may also be a source of the GHBP in uterine fluid. Moreover, GHRs are present in the prostatic epithelium during the secretory but not the nonsecretory phase, thus suggesting a secretory role for GH. A role for GH and IGF-I in the maturation of gametes is supported by the intense staining in immature granulosa cells and spermatozoa precursors that is absent or decreased in their mature counterparts, and GH influences on gonadal steroid synthesis may be mediated by GHRs on Leydig and Sertoli cells in the male and granulosa and luteal cells in the female.

The importance of GH in mammary development is well accepted, but conventional binding assays were unable to detect GH-binding activity in the mammary gland. The proliferative effects of GH were thought to be mediated by lactogenic receptors as a result of these studies. However, GHR mRNA is present in this tissue[135] and a study by Feldman et al.[136] identified rGH binding sites in rat mammary gland membranes that did not cross-react with prolactin. The difficulty in receptor detection may be due to the heterogeneous composition of the gland. Alternatively, these receptors may be particularly sensitive to GH synthesized within the mammary gland, as this GH may be different from pituitary GH.[137] The mammary gland may also be the source of the GHBP found in milk.[138]

The localization of GHRs in immune tissues and the central nervous system also has important functional implications; however, these tissues are discussed in Chapters 24 and in 27 and 28, respectively.

VII. REGULATION

The regulation of the GHR occurs at transcriptional and translational levels by tissue-specific mechanisms that may differentially or coordinately modulate GHR and GHBP activity (see Hochberg et al. for recent review). Variation in both GH-binding activity and GHR mRNA is observed throughout ontogeny and pregnancy of most species; however, the magnitude and characteristics of the changes are highly variable. These changes are mediated in part by other hormones, as GH, estrogen, thyroid hormones, insulin, and adrenal steroids are all regulators of GHR expression and/or translation.

A. ONTOGENY

In rodents, expression of both the 4.4- and 1.2-kb GHR mRNAs has been observed as early as embyonic day 16 (e.16) in the fetal rat liver and rises exponentially during fetal development.[139,140] One or both of the transcripts is also translated, as GHR and/or GHBP immunoreactivity is detectable in hematopoietic cells and endothelium by e.12 and is ubiquitously present by e.18.[139] However, in the study of Tiong and Herington,[141] only the GHBP mRNA was detected prenatally, whereas the GHR mRNA and GHR peptide were not detected before postnatal day 20. Growth hormone receptor and GHBP mRNA were also detected prenatally in kidney, lung, and ileum of fetal rats on the 19th day of gestation.

The ontogeny of post-natal GHR expression is both transcript and tissue specific. Both transcripts increase three- to fourfold in liver, reaching adult levels by day 42.[140] In kidney and lung the two transcripts are differentially regulated, as the receptor increases three- to fourfold or twofold, whereas the GHBP mRNA increases twofold or not at all (respectively). Both transcripts decrease twofold in ileum by day 43 and one or both transcripts show a similar ontogenic decrease in the ovary.[142] GHR mRNA in the liver, kidney, muscle, and heart also increased with age in one study[92]; however, the solution hybridization assay used did not permit distinctions between the two transcripts. The 2.6-kb transcript is not developmentally regulated.[44]

Age-related changes in hepatic GH-binding levels are slightly more complex, as the rat liver contains somatogenic and lactogenic binding sites. Lactogenic (prolactin) receptors increase in females with age, but remain constant or decrease in males.[143,144] The abundance of somatogenic (bGH) receptors is similar in newborn male and female rats, and a pubertal surge of similar magnitude is observed in both sexes.[143] However, in females, levels continue to rise steadily until day 120, reaching a final concentration of 33.5 fmol/mg protein, whereas in males receptor levels reach a maximum of 13.3 fmol/mg protein at day 50.[143] The opposite changes in somatogenic and lactogenic binding sites in male rats may explain the lack of an age-related increase in hGH binding in male rats[145] and mice.[146]

The ontogeny of GHR mRNA in fetal sheep is differentially controlled in different tissues. GHR transcripts appear in the kidney, liver, brain, and muscle by embryonic day 51 and increase in abundance between e.51 and e.120.[45] Growth hormone binding is detectable in fetal liver by e.51[45] but has not been investigated in the other tissues. The magnitude of the GHR mRNA increase is highly tissue specific, as it is greatest in muscle (1 to 43%), approximately equal in kidney (32 to 57%) and liver (13 to 28%), and less in brain (7 to 11%)(levels are expressed as the percentage of maternal hepatic GHR mRNA). The relative distribution of GHR mRNA is also of interest, as kidney GHR mRNA predominates at e.51 and kidney and muscle GHR mRNA are both quite abundant at e.120, suggesting prenatal actions of GH may be primarily restricted to these organs. However, the high fetal levels of GH and the abundance of GHR transcripts in extrahepatic tissues suggest that GH may play a vital role in fetal metabolism, fine structural development, and/or water balance rather than in fetal growth. The liver plays a greater role in the somatotropic axis in the postnatal period, as hepatic GHR mRNA levels peak 1 week following birth but then decline until adulthood.[34]

An ontogenic rise in hepatic GHRs is also observed in steers; however, this increase in binding capacity is confined to the high-affinity GH-binding site.[84] An age-related increase in hepatic GHR mRNA and GH-binding activity is also observed in weaned pigs that correlates inversely with plasma GH levels.[38,83,147] A portion of the augmented binding capacity in older pigs is due to the development of a low-affinity, high-capacity binding site.[83] Transcript levels also rise in muscle, fat, and kidney; however, GHR mRNA levels decline in porcine brain, thymus, heart, diaphragm, lung, pancreas, spleen, and stomach with advancing age.[38]

The 4.4-kb hepatic GHR transcript is of low abundance in fetal and early neonatal rabbit liver, heart, and skeletal muscle but increases markedly by postnatal day 16 to reach adult levels between 2 and 6 months of age.[148] In contrast, kidney levels (largely the cortex) of the GHR mRNA are high in fetuses and neonates and do not alter dramatically during development.[148] Somatogenic and lactogenic membrane-binding sites and cytosolic GHRs also increase with age,[145,148] but the magnitude of the increase is six times greater than the increase in mRNA levels. In contrast, GHRs in bone are observed only in younger, actively growing rabbits.[98]

In male chickens, hepatic GHR mRNA levels are detectable by e.15, peak transiently prior to hatching, then rise to adult levels between 2 and 20 weeks of age.[43,149] Muscle and perhaps hepatic GHR mRNA levels also show a postnatal rise in female chickens.[43] GH-binding activity in domestic fowl (chickens and turkeys) is also low until sexual maturity but then rises in both males and females.[150] Generally, plasma GH levels are inversely correlated with GHR expression postnatally.[149]

B. SEX

Hepatic GHR mRNA and GH-binding levels appear to be sexually dimorphic in rats,[143,144,151] but not in rabbits,[145] mice,[152] cows,[37] or turkeys.[150] This dimorphism is observed after puberty, resulting in significantly higher levels of GHR/GHR mRNA in adult female rats. The increased GH binding results from both increased GHR binding capacity and increased receptor affinity (Table 2).[81] The pattern of GH secretion may play a significant role in the higher levels of GHRs in female rats. Females are characterized by having relatively continuous GH secretion, which has been demonstrated to be highly effective in inducing GHR upregulation.[153] In contrast, multiple daily injections of GH mimicking the highly pulsatile GH secretion observed in males are less effective in inducing GHR upregulation.[153] Autoregulation of the GHR is further discussed below.

C. PREGNANCY

In rat, the smaller, 1.2-kb GHR gene transcript is significantly increased in late pregnancy[40] but levels of the 2.6- or 4.4-kb transcripts remain constant.[44] Accordingly, GHBP, but not hepatic GHR, levels are also increased.[40] Conversely, Moldrup et al.[154] observed a significant increase in both the receptor and the binding protein mRNA in the rat liver and pancreas at midpregnancy. Kelly et al.[145] and Herington et al.[155] also observed a significant increase in hepatic hGH binding in pregnant rats; however, this may reflect lactogenic binding sites. Pregnancy also increases GHR transcripts in the liver, but not adipose tissue, of heifers.[37] In mouse, pregnancy increases hepatic and serum mGH and bGH binding activity and hepatic GHR/GHBP mRNA levels.[152,156] This rise is dependent on adequate GH concentrations,[156] the feto-placental unit, and the maternal adrenal gland, and correlates with the number of conceptuses.[152,157] In rabbits, hepatic hGH and bGH binding is increased by day 20 of pregnancy[145] and bGH receptor levels increase markedly at parturition.[158] Conversely, hepatic GHR mRNA and GH-binding levels in sheep remain constant throughout pregnancy; however, sheep have a distinct placental lactogen receptor that may increase in abundance during gestation.[45,159]

D. HORMONAL REGULATION

1. Growth Hormone

Studies investigating regulatory effects of GH on the GHR are often contradictory. The use of heterologous ligands in binding assays and different regimens of GH administration have contributed to the confusion. However, some species- and tissue-specific conclusions can be drawn.

In the absence of pituitary GH, in hypophysectomized rats, GHR mRNA and GH binding are reduced in females[155,160] but unchanged[92,121,161,162] or increased[163,164] in adult and juvenile males. However, the effects of chronic GH administration are not as sexually dimorphic, as somatogenic and lactogenic binding sites are increased in the livers of GH-deficient, hypophysectomized or intact male and/or female rats following GH therapy.[121,151,165,166] Indeed, exposure to high, chronic levels of GH from GH_3 or MT/W15 tumors[155,167,168] (which may also indicate prolactin-induced upregulation) or the bGH transgene[169] also correlates with increased somatogenic and lactogenic binding sites and increased GHR mRNA[170] in both sexes. This upregulation is also dose dependent, as incubation of rat hepatocytes with physiological levels of GH upregulate GHR mRNA and somatogenic receptors for over 48 h, whereas GHR mRNA/GHR levels are acutely downregulated by pharmacological GH levels.[171,172] However, chronic bGH administration downregulates hepatic GHR mRNA levels in male GH-deficient dwarf rats[173] and has no effect in hypophysectomized male rat liver membranes.[155]

The pattern of GH administration determines the alterations in rat hepatic GH-binding levels, as intermittent injections are less effective[121] or ineffective[160,174] stimulants than the continuous infusion of GH. Curiously, the livers of GH infused rats contain higher transcript levels for the GHBP but normal levels for the GHR,[160] suggesting that increased binding levels may correspond to the truncated binding protein or that the receptor is regulated posttranscriptionally. In contrast, a single injection of GH downregulates hepatic GHRs by approximately 50%[174] for nearly 12 h, suggesting differential effects of acute and chronic exposure to GH. This acute downregulation is also observed following GH secretory pulses in male rats; however, in this case the downregulation is transitory (3 h) and due to receptor internalization.[119]

Growth hormone receptor autoregulation in other species and tissues could also be considered to be bidirectional. In general, GH upregulates its own receptors, as hypophysectomy decreases GH binding in rabbit, sheep, fish, and eel liver[175,176,189] and rat adipocytes[177,178] by up to 75% and also decreased GHR mRNA levels in rat ovary, kidney, and adipose tissue.[127,142,160,179–181] The partial restoration of GH binding and/or GHR mRNA levels by treatment with exogenous GH observed in these studies supports this conclusion. The mode of adminstration also determines the efficacy of GH treatment, as GH injections four times, but not two times, daily reverse the GHR loss in rat adipocytes.[178] Chronic GH administration also increases hepatic GH binding in sheep,[82] infant and pubertal pigs,[147,182] and male lambs[183] and increases the levels of GHR mRNA in isolated rat epiphyseal chondrocytes.[184] The effect of acute GH administration on its receptor may also be tissue-specific. For instance, a single GH injection increases GHR levels in the lymphocytes of GH-deficient children 2.4 to 5 h after the injection.[184a] However, the low levels of GHR in the fetuses of most species, despite the high levels of circulating GH, suggest that GH-mediated upregulation of tissue GHRs is restricted to the postnatal period.

Although GH has been observed to upregulate tissue GHRs, substantial evidence supporting an inhibitory effect of GH on GHR levels also exists. For example, hypophysectomy increases GH binding in rat muscle[162] and chick liver.[185] Moreover, chronic exposure to GH down-regulates the GHR in porcine adipose tissue,[186] salmon liver,[175] fibroblasts[187] and IM-9 cells.[109] However, since some of these *in vitro* studies measured binding to intact cells, ligand-induced internalization may have been responsible for the apparent downregulation observed. Nevertheless, free, but not total, hepatic binding sites are also downregulated by chronic GH exposure in chickens.[188]

Finally, in many situations, GH status is not correlated with GHR levels, as hypophysectomy does not alter GH-binding levels in rat adipose cells or in rabbit and sheep hepatic cells.[92,189] Twice-daily injections of GH similarly have no effect in chicks.[185]

In summary, the majority of studies suggest that GH maintains its own receptors in most species but can cause some degree of downregulation. This conclusion is supported by the concordant loss of GH and GHRs in cultured rat hepatocytes.[171] However, the critical GH concentration required for GHR maintenance is less than that required for growth in rodents, as Little mice and GH-deficient rats with subnormal GH concentrations and growth have relatively normal ontogenic patterns of GHR development.[144,165] Moreover, GHR levels do not always correlate negatively or positively with serum GH concentrations. For instance, a negative[143] or zero[144] correlation has been observed between ontogenic changes in GH and GHR levels in rats, and the correlation between GHR and GH levels in chickens and turkeys varies between positive, 0 and negative values over ontogeny.[190,191] Thus, the GH-GHR regulatory interaction is highly complex and dependent on the physiological status of the animal.

2. Thyroid Hormone

Thyroid status appears to be directly correlated with hepatic GHR expression in mammals. For instance, triiodothyronine (T_3), alone or in conjunction with bGH, upregulates GH-binding sites in hepatocytes isolated from dwarf mice[192] or normal rats[171] or hGH binding sites in hypophysectomized rat adipocytes.[177,178] Moreover, in the study of Hochberg et al.,[193] somatogenic and lactogenic receptor levels in purified plasma membranes were low in female hypothyroid rats, restored by exogenous thyroxine (T_4), and high in hyperthyroid rats. A previous study by Chernausek et al.[194] observed high levels of GH-binding activity in hypothyroid male rat liver. However, in contrast to the Hochberg study, endogenous ligand was not removed, thus only free binding sites were detected. Alternatively, the effect of thyroid status may be species-dependent. For instance, thyroid hormone status is inversely related to hepatic GHR activity in chickens[195] but not in fish.[175]

3. Glucocorticoids and Glucose

Glucocorticoids and/or glucose may be responsible for the reduced GHR levels observed in fasted rats, as levels of both are decreased with nutritional deprivation. For instance, in starved rats, the decrease in plasma glucose levels parallels the loss of GHRs,[196] suggesting a role for glucose in GHR maintenance. Accordingly, incubation with glucose partially counteracts the loss of GH-binding sites in cultured hepatocytes by stimulating *GHR* gene transcription, and glucose plus glucocorticoids (dexamethasone) restores GHR expression to 70% of the level in freshly isolated hepatocytes.[197] In the same system, dexamethasone and glucagon increase GHR levels to 25% of initial levels, and this enhancement is inhibited by epidermal growth factor (EGF) and insulin.[198] The dexamethasone effect is also dependent on cell density, suggesting the possibility of paracrine interactions between dexamethasone, the GHR, and an as-of-yet unidentified hepatic paracrine factor. Dexamethasone also increases GHR levels and GHBP mRNA in newborn rat pancreatic islets and RIN-5AH cells[154] and in human hepatocytes,[172] but not in rat adipocytes,[177] suggesting that the GHR is upregulated only in specific tissues in response to glucocorticoids. In addition, the stimulatory effect of glucocorticoids may also be important in pregnancy, as both glucocorticoid and GHR levels are increased in pregnant rats.[154]

4. Gonadal Hormones

The influence of gonadal hormones on GHR synthesis may be responsible for increased GHR expression during pregnancy and female puberty. Estradiol upregulates the capacity of the low- and the high-affinity GH-binding sites in bovine liver[84] and increases hGH binding to hypophysectomized rat liver.[155] However, in the study by Carmignac et al.,[198a] estradiol was effective only in rats with adequate GH levels, suggesting that estradiol effects are GH dependent. This conclusion is also supported by the ability of GH or estradiol to restore GHR levels in rat liver following ovariectomy.[198a] In contrast, testosterone reduces GH binding in diabetic and normal rats,[199] which may in part explain the lower levels of GHRs in male rats.

VIII. GROWTH HORMONE RESISTANCE

GH resistance is observed in numerous physiological and pathological situations, including fasting, diabetes, liver disease, and renal disorders, and is often associated with a deficiency in the quantity of GHRs. Moreover, genetic models of GH resistance have also been identified in humans and domesticated animals.

A. PHYSIOLOGICAL GH RESISTANCE (NUTRITIONAL STATUS)

Growth hormone receptor levels generally correlate well with nutritional status, and GHR deficiency has been implicated in the poor growth, GH resistance and low IGF-I levels associated with nutritional deprivation and fasting.[196,200,201] Short-term (3 to 4 days) but not long-term fasting severely inhibits GHR mRNA and/or somatogenic receptor levels in rat[81,196,200–202] and chicken liver[203] and in rat adipose, heart, and diaphragm,[200] but not in rat kidney.[200] This effect is reversible, as receptor levels are usually restored by refeeding.[201,203] Thus, in total nutritional deprivation, a GHR defect could account for the observed GH resistance.

However, despite a similar degree of GH resistance in protein- or calorically deprived animals, GH-binding levels are depressed in young animals (steers and rats) but are relatively normal in adult sheep or rats.[84,183,202,204,205] A postreceptor defect is thus thought to be responsible for the decreased IGF-I levels in adult protein-restricted animals, as GH is inactive despite normal GHR levels.[206] Incidentally, the situation in young bovines is somewhat unique, as the decreased binding capacity in underfed steers is associated with the loss of the high-affinity GH-binding site and decreased affinity for the low-affinity binding site.[84] Inversely, obesity in Zucker rats is associated with increased hepatic GH binding.[207]

B. PATHOPHYSIOLOGICAL GH RESISTANCE

Hepatic GH-binding levels are severely decreased in nephrectomized rats.[208] Moreover, as GH levels remain normal, this downregulation is due to the deficiency or excess of factors other than GH.[208] Low GHR levels may thus be partially responsible for the growth retardation in children with renal tubular acidosis, especially as nonuremic acidosis in rats is associated with decreased hepatic GH binding.[209] However, GHR downregulation in this disorder may be mediated by alterations in feeding and pituitary GH secretion, as both are reduced in acidotic rats.[209] A decrease in liver function due to partial hepatectomy in rats or by cirrhosis in humans is also accompanied by hepatic GHR downregulation.[210,211] In rats, this effect is posttranslational and independent of GH, as a decrease of equal magnitude is observed in hypophysectomized rats following hepatectomy.[210]

Diabetes is also associated with GH resistance, as chronic, but not acute, streptozocin-induced diabetes in rats results in decreased GHR mRNA levels in the diaphragm and heart but normal levels in the liver and kidney.[200] Moreover, these changes are probably the result of insulin deficiency rather than streptozotocin itself, as GHR mRNA levels are restored by exogenous insulin in diabetic rats[81,212] and insulin enhances hGH binding to adipocytes in hypophysectomized rats.[177] The effect of insulin may be mediated by adrenal steroids, as streptozocin-treated rats have increased levels of adrenal androgens and adrenalectomy restores both GHR and testosterone levels to normal.[199] However, as exogenous testosterone does not eliminate the restorative effects of adrenalectomy, additional adrenal factors may also be involved.[199]

C. INHERITED GH RESISTANCE

The etiology of dwarfism is often linked to a GH deficiency or to a GH mutation, although dwarfism can also be associated with supranormal GH levels and GH resistance.

Laron-type dwarfism in humans and sex-linked dwarfism in chickens are both growth hormone-resistant states associated with GHR dysfunctions. GH-resistant strains of guinea pigs, dogs, cattle, and pigs have also been identified; however, the lesion in these animals is not as well understood.

1. Laron-Type Dwarfism

Laron-type dwarfism (LTD) is an autosomal recessive disorder characterized by severe growth retardation that is clinically indistinguishable from isolated GH deficiency (reviewed in Ref. 212a). Patients with LTD, however, have no abnormalities of *GH* gene expression and have normal or elevated serum concentrations of biologically active GH. Patients with LTD also have normal *IGF-I* genes, but are IGF-I deficient. These patients respond dramatically to IGF-I but are unresponsive to exogenous GH, which is unable to restore growth or IGF-I production. Laron-type dwarfism is therefore a genetic form of GH resistance, suggesting a dysfunction in the *GHR* gene may be the underlying defect. Indeed, linkage analysis has demonstrated that LTD and *GHR* gene polymorphisms in intron 9 were coinherited in two Mediterranean families in which the parents were consanguinous and had some children with LTD. The children with LTD were homozygous for the same DNA markers for which the parents were heterozygous.

A large number of LTD patients lack hepatic GH-binding activity[213] and have deficient GH-binding activity in their serum.[214,215] This dysfunction may arise at the level of the gene itself, as exons 3, 5, and 6 are deleted but exon 4 is retained in the *GHR* gene of 2 patients (Figure 2).[32] Exon 4 is syntenic and linear with the other exons.[216,217] The loss of noncontiguous exons suggests a gene mutation resulting from two independent deletion events or from a complex deletion and rearrangement. The deletion of exons 5 and 6 causes a frameshift, generating a stop codon in exon 7.[216,217] Thus the resulting GHR of 53 residues would lack a portion of the extracellular domain and the entire transmembrane and cytoplasmic domains,[217] which would account for the GH resistance and growth defect in these patients. However, as this defect is not observed in other LTD individuals, a heterogeneity of gene defects is likely to be responsible for the LTD phenotype.

Smaller deletions of only two residues can be equally deleterious, as they shift the reading frame of the gene. Dinucleotide deletions of TA/AT or TT have been identified in South African and Spanish LTDs at residue 46 or 230 (respectively).[218] The deleted sequences are both within regions of direct repeats (TTTTT or TAT), which are common regions for short deletions in other genes.[218]

The molecular defects in LTD also include nonsense mutations within the GH receptor gene (Figure 2). Amselem et al.[219] identified a nonsense mutation at residue 43 in exon 4 that consists of an arginine (CGA) to stop (TGA) conversion. Unlike most LTD lesions, this mutation arose independently in patients from several different ethnic backgrounds,[219–221] underlining the hypermutability of CpG dinucleotides.[222] Indeed, a nonsense mutation in the other CGA codon in the coding region of the *GHR* gene (residue 217 in exon 7) has also been detected by several investigators in unrelated LTD populations.[220,221] A further nonsense mutation has been identified in residue 38 that does not involve a CpG dinucleotide.[219] These three defects would result in GH receptors of 38, 43, or 271 amino acids, in which a portion of the extracellular GH-binding domain and the entire transmembrane and cytoplasmic domains would be deleted. Thus, patients with these genetic defects are completely GH resistant and lack the GH-binding protein.

The Laron syndrome has also been linked to point mutations in the GHR gene that create or destroy splice donor or acceptor sequences. The "594 A to G" point mutation in codon 180 in Equadorean LTDs generates a new and efficiently used 5′ splice consensus sequence within exon 6, resulting in an eight-amino acid deletion in the extracellular domain of the GHR.[218,220] The destruction of crucial splicing residues may result in the retention of introns, frameshifts, and/or the use of alternate splice sites, thereby totally altering the sequence of the GHR. A splice donor site is destroyed by the G-to-A transversion at the first nucleotide of intron 4 (nucleotide 266+1 or residue 71+1),[220,221,223] whereas splice acceptor sites are destroyed by a G-to-C transversion at the last nucleotide of intron 5 (nucleotide 440–1)[220] or the last nucleotide of intron 6 (codon 189–1).[221]

Point mutations at residues important in GH binding or in GHR trafficking can also result in the LTD phenotype (Figure 2). A thymidine-to-cysteine substitution has been identified by Amselem et al.[223a] that generated a serine in place of a highly conserved phenylalanine at residue 96 of the extracellular domain. This mutation was not observed in the polymorphic *GHR* gene sequences of seven unrelated subjects with LTD who belonged to diferent population groups. Duquesnoy et al.[224] subsequently investigated the effect of this point mutation on GH-binding activity by expressing the total human GHR cDNA and the mutant form in eukaryotic cells. The wild-type phenotype expressed was able to bind GH, but no plasma membrane binding was detectable on cells transfected with the mutant cDNA or with a Phe96-Ala mutant cDNA, suggesting the lack of binding activity was not due to a posttranslational modification of serine. It was hypothesized that this mutation may impair the intracellular trafficking of the GHR and its insertion into the plasma membrane, because some GH-binding activity was detected in lysosomal fractions and immunoreactive mutant proteins were present in cytosol. Moreover, this mutant protein also had the same GH affinity as the wild-type binding protein when expressed in bacteria, in which the GHR is not glycosylated.[67] However, in studies by Edery et al.,[225] the phenylalanine-serine substitution totally eliminated GH-binding activity in hGHR or rabGHR-transfected COS-7 cells. This substitution would result in the replacement of a nonpolar residue with a polar residue and is adjacent to a glycosylation site and thus may disrupt electrostatic interactions with the ligand and the tertiary form of the receptor. Moreover, these patients were totally deficient in serum GHBP activity, which is evidently independent of receptor trafficking. Thus, an LTD mutation affecting the subcellular localization of the GHR rather than GH binding has not yet been identified.

Additional point mutations, that is, R71K (arginine to lysine), V125A (valine to alanine), R161C (arginine to cysteine), R211G (arginine to glycine), or V144D (valine to aspartic acid), have also been detected in the *GHR* gene of various subpopulations of patients with LTD.[220] However, only one of these mutations occurs at a hypermutable CpG dinucleotide (R161C). In contrast, these mutations are thought to involve residues necessary for GH binding, as residues 71, 125, and 211 are totally conserved in the *GHR* gene from different species. The importance of residues 71 and 125 has already been demonstrated by Bass et al.[67] by *in vitro* mutagenesis. Curiously, conversion of residues 161 and 211 to alanines does not eliminate binding in *in vitro* studies. However, the *GHR* gene of the LTD patients contains cysteine or glycine substitutions, which may disrupt disulfide bridge formation, electrostatic interactions, and/or tertiary structure more severely than alanine substitution.

The molecular defects described above result in complete GH insensitivity. However, populations with low or normal levels of serum GHBP and/or exhibiting only partial GH insensitivity have also been identified.[226,226a] The etiology in these patients could relate to defects in the regulatory regions of the *GHR* gene and/or in the intracellular, signal-mediating regions. The former possibility may be responsible for the etiology of an Equadorean population of LTDs, as they possess low levels of functional GHBP.[227] Moreover, adult African pygmies are characterized by a lesser degree of growth impairment and GH resistance than LTDs and low but detectable levels of GHBP.[228,229] Indeed, GHBP levels are relatively normal in young pygmies, but do not rise at puberty, as in normally growing individuals. A defect in the regulatory regions of the *GHR* gene has thus been implicated in this defective ontogenic regulation; however, primary analysis of the pygmy *GHR* gene has revealed only an intronic defect of unknown significance.[230]

A possible defect in the intracellular domain of the GHR was also detected in an LTD patient by Kou et al.[231] Clones prepared from fibroblast mRNA from this patient were heterozygous for two mutations in the cytoplasmic GHR domain: a cysteine-to-phenylalanine conversion at residue 422 and a proline-to-threonine conversion at residue 561. The mother was also heterozygous for these mutations and was not growth deficient. Moreover, this patient was also defective in plasma GHBP, but no mutations were detected in the mRNA region encoding the extracellular domain. The true nature of this GHR defect remains unknown.

2. Sex-Linked Dwarfism

Sex-linked dwarfism (SLD) in poultry is due to a recessive gene (*dw*) on the Z chromosome that results in mature body weights 30% to 40% smaller than that of normal chickens.[232,233] This mutation occurs in a number of different populations of domestic fowl, in which it is thought to arise independently.[233–235]

The pituitary *GH* gene in SLD chickens is normally transcribed and, as in patients with LTD, the GH molecule is immunologically, electrophoretically, and bioactively normal[236] and present at normal or supranormal concentrations in peripheral plasma.[237,238] These birds are, however, abnormally fat[234] and are deficient in plasma T_3 and IGF-I.[237,238] As lipolysis, T_3 production and IGF-I synthesis are GH dependent, a GH receptor dysfunction may therefore be causal in the expression of the dwarf phenotype in birds. This possibility is supported by the inability of exogenous GH to restore IGF-I production in SLD chickens,[239] in which cartilaginous responses to plasma IGF-I activity are normal or supranormal.[233] Sex-linked dwarfism, like LTD, is therefore an inherited form of GH resistance.

As in patients with LTD, dwarf birds are deficient in hepatic GH-binding sites.[240] The remaining GH-binding capacity is highly variable, ranging from 1% to 50% in broiler and egg-laying SLDs. Both the affinity and the capacity of hepatic GHRs are reduced by 50% in a small group of broiler SLDs, but only the number of binding sites are reduced in a population of Leghorn SLDs.[240] The low level of liver GH-binding in dwarf chickens may therefore indicate a deficiency or defect in the expression of the GHR gene. This conclusion is supported by Northern analysis of dwarf broiler fowl mRNA , in which a 1.0- to 1.7-kb deletion in the GHR mRNA was identified.[33] However, sex-linked dwarfism, like the Laron syndrome, is a heterogeneous disorder, as the GHR mRNA is of normal length in a European population of Leghorn SLDs.[241] The *GHR* gene in these birds is characterized by a G-to-T transversion that converts the serine at residue 199 into an isoleucine. This missense mutation occurs within the critically conserved WS-like motif close to the junction between the extracellular and transmembrane domains and results in a drastic reduction in hepatic GH binding and expression of the receptor at the plasma membrane.

The *GHR* gene is also of normal length in the Georgia strain of *dw/dw* birds. However, in contrast to the major 4.3-kb GHR mRNA observed in normal birds, the truncated 0.8-kb GHR mRNA is the major *GHR* transcript in Georgia dwarfs and small amounts of an intron 5-containing transcript are also detected.[242] This shift in transcription is due to a T→C conversion at the exon 5/intron 5 junction that eliminates a splicing consensus site in the full-length transcript. The retained intron contains four AUUUA sequences, a consensus motif for transcript destabilization; thus inappropriate splicing of intron 5 would result in an unstable message. However, the 0.8-kb transcript is truncated prior to intron 5 and would be unaffected.

The source of the GHR dysfunction remains unknown in certain populations of SLDs. For example, in a closed strain of SLDs, *GHR* transcripts of appropriate size are present in liver, hypothalamus, brain, muscle, and fat.[243] Moreover, amplification of GHR cDNA encoding large portions of the extracellular, transmembrane, and intracellular domain and subsequent sequencing failed to identify any deletions or missense mutations.

3. Additional Models

Historically, the guinea pig has been considered to be GH resistant, as guinea pig pituitary extracts or bovine or monkey GH were incapable of stimulating bone growth, body weight, or ornithine carboxylase synthesis.[244,245] The normal bioactivity of guinea pig pituitary extract in rats is therefore suggestive of GHR deficiency.[246] A defect in GHR synthesis is supported by the lack of ^{125}I-labeled GH-binding sites in guinea pig serum[244,247] and the significantly lower concentrations of hepatic binding sites.[244] Hypophysectomy, nevertheless, does not affect the growth rate or IGF-I concentrations in these animals.[245,248] Moreover, guinea pigs are only partially GH resistant, as exogenous GH is capable of stimulating estrogen receptor synthesis in the uterus.[249]

However, studies have revealed the presence of GHR mRNA and GH-binding activity in the guinea pig liver,[250] and immunoreactive GHBPs have been identified by radioimmunoassay and Western blotting in guinea pig serum.[251] The failure of earlier studies to detect this GHBP by radioreceptor assay may be due to a mutated GHBP that is incapable of binding GH. Alternatively, the iodination of GH employed in these studies may interfere with GH binding. Preliminary data also suggest that these binding proteins may be present at extremely high concentrations,[251a] which may account for the GH resistance observed in these animals. Thus, the guinea pig provides an interesting model of partial GH resistance, but the etiology of their GH resistance is not yet understood.

Toy poodles, miniature pigs, and miniature Brahmin cattle are additional examples of animals with greatly impaired IGF-I concentrations and growth retardation despite normal or supranormal GH concentrations.[248] The plasma GHBP concentrations in poodles and small cattle are normal, suggesting a defect in the intracellular or transmembrane domains of the GHR or in postreceptor mechanisms.[252–254] However, the decreased affinity of the plasma GHBP in these pigs is suggestive of a dysfunction in the extracellular domain of the GHR.[253]

Thus, inherited GH resistance is not associated with a single GHR defect, but rather a spectrum of dysfunctions at the level of the structural GHR and perhaps at the level of gene regulation and signal transduction. Further elucidation of the lesions in these animal models should provide further insight into the structure and function of the GHR.

REFERENCES

1. **Kelly, P. A., Ali, S., Rozakis, M., et al.,** *Rec. Prog. Horm. Res.*, 48, 123, 1993.
2. **Kelly, P. A., Djiane, J., Postel-Vinay, M. C., et al.,** *Endocr. Rev.*, 12, 235, 1991.
3. **Postel-Vinay, M. C., Finidori, J., Goujon, L., et al.,** *Ann. d'Endocrinol.*, 52, 469, 1991.
4. **Cramer, S. and Talamantes, F.,** *The Endocrinology of Growth, Development, and Metabolism in Vertebrates* (Eds. Schreibman, M. P., Scanes, C. G., and Pang, P. K. T.). Academic Press, Toronto, 1993, p. 117.
5. **Spencer, S. A., Leung, D. W., Godowski, P. J., et al.,** *Rec. Prog. Horm. Res.*, 46, 165, 1990.
6. **Mendelsohn, L. G.,** *Life Sci.*, 43, 1, 1988.
7. **Waters, M. J., Barnard, R. T., Lobie, P. E., et al.,** *Acta Paediatr. Scand.*, 366(Suppl.), 60, 1990.
8. **Waters, M. J., Spencer, S. A., Leung, D., et al.,** *Biotechnology in Growth Regulation*, (Ed. Heap, B.). Butterworths Scientific, Surrey, 1989, p. 15.
9. **Mathews, L. S.,** *Trends Endocrinol. Metab.*, 2, 176, 1991.
10. **Bazan, J. F.,** *Biochem. Biophys. Res. Commun.*, 164, 788, 1989.
11. **Hochberg, Z., Amit, T., and Youdin, M. B. H.,** *Cell. Signal.*, 3, 85, 1991.

11a. **Ihle, J. N.,** *Trends Endocrinol. Metab.*, 5, 137, 1994.

12. **Cosman, D., Lyman, S. D., Idzerda, R. L., et al.,** *Trends Biochem. Sci.*, 15, 265, 1990.
13. **Patthy, L.,** *Cell*, 61, 13, 1990.
14. **Rozakis-Adcock, M. and Kelly, P. A.,** *J. Biol. Chem.*, 267, 7428, 1992.
15. **Sadeghi, H., Wang, B. S., Lumanglas, A. L., et al.,** *Mol. Endocrinol.*, 4, 1799, 1990.
16. **Fukunaga, R., Seto, Y., Mizushima, S., et al.,** *Proc. Natl. Acad. Sci. U.S.A.*, 87, 8702, 1990.

17. **Trivedi, B. and Daughaday, W. H.,** *Endocrinology (Baltimore)*, 123, 2201, 1988.
18. **Davis, S., Aldrich, T. H., Valenzuela, D. M., et al.,** *Science*, 253, 59, 1991.
19. **Staten, N. R., Byatt, J. C., and Krivi, G. G.,** *J. Biol. Chem.*, 268, 18467, 1993.
20. **Fukunaga, R., Ishizaka-Ikeda, E., Pan, C. X., et al.,** *EMBO J.*, 10, 2855, 1991.
21. **Stred, S. E., Stubbart, J. R., Argetsinger, L., et al.,** *Endocrinology (Baltimore)*, 130, 1626, 1992.
22. **Dusanter-Fourt, I., Casadevall, N., Lacombe, C., et al.,** *J. Biol. Chem.*, 267, 10670, 1992.
23. **Murakami, M., Narazaki, M., Hibi, M., et al.,** *Proc. Natl. Acad. Sci. U.S.A.*, 88, 11349, 1991.
24. **Colosi, P., Wong, K., Leong, S. R., et al.,** *J. Biol. Chem.*, 268, 12617, 1993.
25. **Cosman, D.,** *Cytokine*, 5, 95, 1993.
26. **Chiha, T., Kishi, A., Sugiyama, M., et al.,** *Biochem. Biophys. Res. Commun.*, 186, 1236, 1992.
27. **Sakamaki, K., Miyajima, I., Katmura, T., et al.,** *EMBO J.*, 11, 3541, 1992.
28. **Ng, T. B., Cheng, C. H. K., Woo, N. Y. S., et al.,** *Comp. Biochem. Physiol.*, 196A, 483, 1993.
29. **Wlodawer, A., Pavlovsky, A., and Gustchina, A.,** *Protein Sci.*, 2, 1373, 1993.
30. **Barton, D. E., Foellmer, B. E., Wood, W. I., et al.,** *Cytogenet. Cell. Genet.*, 50, 137, 1989.
31. **Arden, K. C., Boutin, J. N., Kelly, P. A., et al.,** *Cytogenet. Cell. Genet.*, 53, 161, 1990.
32. **Godowski, P. J., Leung, D. W., Meacham, L. R., et al.,** *Proc. Natl. Acad. Sci. U.S.A.*, 86, 8083, 1989.
32a. **O'Mahoney, J. V., Brandon, M. R., and Adams, T. E.,** *Mol. Cell. Endocrinol.*, 101, 129, 1994.
33. **Burnside, J., Liou, S. S., and Cogburn, L. A.,** *Endocrinology (Baltimore)*, 128, 3183, 1991.
34. **Adams, T. E., Baker, L., Fiddes, R. J., et al.,** *Mol. Cell. Endocrinol.*, 73, 135, 1990.
35. **Leung, D. W., Spencer, S. A., Cachianes, G., et al.,** *Nature (Lond.)*, 330, 537, 1987.
36. **Smith, W. C., Linzer, D. I. H., and Talamantes, F.,** *Proc. Natl. Acad. Sci. U.S.A.*, 85, 9576, 1988.
37. **Hauser, S. D., McGrath, M. F., Collier, R. J., et al.,** *Mol. Cell. Endocrinol.*, 72, 187, 1990.
38. **Bingham, B., Oldham, E. R., and Baumbach, W. R.,** *Proc. Soc. Exp. Biol. Med.*, 1994 (in press).
39. **Wang, X., Cioffi, J. A., Kelder, B., et al.,** *Mol. Cell. Endocrinol.*, 94, 89, 1993.
40. **Tiong, T. S. and Herington, A. C.,** *Endocrinology (Baltimore)*, 129, 1628, 1991.
41. **Carlsson, B., Billig, H., Rymo, L., et al.,** *Mol. Cell. Endocrinol.*, 73, R1, 1990.
42. **Barnard, R., Edens, A., Thordarson, G., et al.,** *74th Annu. Meet. Endocr. Soc.*, San Antonio, TX, 1992, p. 17.
43. **Oldham, E. R., Bingham, B., and Baumbach, W. R.,** *Mol. Endocrinol.*, 7, 1379, 1993.
44. **Tiong, T. S. and Herington, A. C.,** *Biochem. Biophys. Res. Commun.*, 180, 489, 1991.
45. **Klempt, M., Bingham, B., Breier, B. H., et al.,** *Endocrinology (Baltimore)*, 132, 1071, 1993.
46. **Asakawa, K., Hedo, J. A., McElduff, A., et al.,** *Biochem. J.*, 238, 379, 1986.
47. **Yamada, K., Lipson, K. E., and Donner, D. B.,** *Biochemistry*, 26, 4438, 1987.
48. **Waters, M. J. and Friesen, H. G.,** *J. Biol. Chem.*, 254, 6815, 1979.
49. **Fuh, G., Mulkerrin, N. G., Bass, S., et al.,** *J. Biol. Chem.*, 265, 3111, 1990.
50. **Schwartz, J. and Carter-Su, C.,** *Horm. Metab. Res.*, 19, 242, 1987.
51. **Carter-Su, C., Schwartz, J., and Kikuchi, G.,** *J. Biol. Chem.*, 259, 1099, 1984.
52. **Fiddes, R. J., Brandon, M. R., and Adams, T. E.,** *Mol. Cell Endocrinol.*, 86, 37, 1992.
53. **Hughes, J. P., Simpson, J. S. A., and Friesen, H. G.,** *Endocrinology (Baltimore)*, 112, 1980, 1983.
54. **Hughes, J. P.,** *Endocrinology (Baltimore)*, 105, 414, 1979.
55. **Tsushima, T., Murakami, H., Wakai, K., et al.,** *FEBS Lett.*, 147, 49, 1982.
56. **Crumeyrolle-Arias, M., Latouche, J., James, M. M., et al.,** *Neuroendocrinology*, 57, 457, 1993.
57. **Borst, D. W. and Sayare, M.,** *Biochem. Biophys. Res. Commun.*, 105, 194, 1982.
58. **Liscia, D. S. and Vonderhaar, B. K.,** *Proc. Natl. Acad. Sci. U.S.A.*, 79, 5930, 1982.
58a. **Beattie, J.,** *Biochem. Biophys. Acta*, 1203, 1, 1993.
58b. **Beattie, J. and Flint, D. J.,** *Endocrinology (Baltimore)*, 139, 349, 1993.
58c. **Wells, J. A.,** *Curr. Opinion Cell Biol.*, 6, 163, 1994.
58d. **Goujon, L., Allevato, G., and Simonin, G., et al.,** *Proc. Natl. Acad. Sci. (USA)*, 91, 957, 1994.
59. **Colosi, P., Wong, K., Leong, S. R., et al.,** *J. Biol. Chem.*, 268, 12617, 1993.
60. **Ali, S., Edery, M., Pellegrini, I., et al.,** *Mol. Endocrinol.*, 6, 1242, 1992.
61. **Billestrup, N., Allevato, G., and Nielsen, J. H.,** *2nd Cell. Growth Forum, Israel,* 1993, p. 9.
62. **Motlich, M. E.,** *Endocrinol. Metab. Clin. N. Am.*, 21, 597, 1992.
63. **Moldrup, A., Allevato, G., Dyrberg, T., et al.,** *J. Biol. Chem.*, 266, 17441, 1991.
64. **DeVos, A. M., Ultsch, M., and Kossiakoff, A. A.,** *Science*, 255, 306, 1991.

64a. **Frank, S. J., Gilliand, G., and Van Epps, C.,** *Endocrinology (Baltimore)*, 135, 148, 1994.
65. **Gobius, K. S., Rowlinson, S. W., Barnard, R., et al.,** *J. Mol. Endocrinol.*, 9, 213, 1992.
66. **Baumbach, W. R., Horner, D. L., and Logan, J. S.,** *Gene Dev.*, 3, 1199, 1989.
67. **Bass, S. H., Mulkerrin, M. G., and Wells, J. A.,** *Proc. Natl. Acad. Sci. U.S.A.*, 88, 4498, 1991.
67a. **Duquesnoy, P., Sobrier, M-L., Duriez, B., et al.,** *EMBO J.*, 13, 1386, 1994.
67b. **Asakura, A., Kikuchi, M., Uchida, E., et al.,** *Biomed. Pharmacother.*, 48, 35, 1994.
68. **Ultsch, M., de Vos, A., and Kossiakoff, A. A.,** *J. Mol. Biol.*, 222, 865, 1991.
68a. **Baumann, G., Lowman, H. B., Mercado, M.,** *J. Clin. Endocrinol. Metab.*, 78, 1113, 1994.
69. **Demeyts, P.,** *Trends Biochem. Sci.*, 17, 169, 1992.
70. **Cunningham, B. C., Ultsch, M., de Vos, A., et al.,** *Science*, 254, 821, 1991.
71. **Fuh, G., Cunningham, B. C., Fukunaga, R., et al.,** *Science*, 256, 1677, 1992.
71a. **Ilondo, M., Damholdt, A. B., Cunningham, B. A., et al.,** *Endocrinology*, 134, 2397, 1994.
72. **Wells, J. A., Cunningham, B. C., Fuh, G., et al.,** *Rec. Prog. Horm. Res.*, 48, 253, 1993.
73. **Barnard, R. and Waters, M. J.,** *Biochem. J.*, 250, 533, 1988.
74. **Fix, J. A., Leppert, P., and Moore, W. V.,** *Horm. Metab. Res.*, 13, 508, 1981.
75. **Nicoll, C. S., Mayer, G. L., and Russell, S. M.,** *Endocr. Rev.*, 7, 169, 1986.
76. **Moore, W. V., Wong. K. P., and Rawitch, A. B.,** *60th Ann. Meet. Endocr. Soc.*, Miami, abst. 603, 1978.
77. **Gray, E. S., Young, G., and Bern, H. A.,** *J. Exp. Zool.*, 256, 290, 1990.
78. **Krishnan, K. A., Proudman, J. A., and Bahr, J. M.,** *Mol. Cell Endocrinol.*, 66, 125, 1989.
79. **Gorden, P., Carpentier, J. L., Hizuka, N., et al.,** *Hormone Receptors in Growth and Reproduction* (Eds. Saxema, B. B., Catt, K. J., Bernbaumer, L., et al.), Raven Press, New York, pp. 317–329.
80. **Rosenfeld, R. G. and Hintz, R. L.,** *J. Clin. Endocrinol. Metab.*, 51, 368, 1980.
81. **Baxter, R. C., Bryson, J. M., and Turtle, J. R.,** *Endocrinology (Baltimore)*, 107, 1176, 1980.
82. **Sauerwein, H., Breier, B. H., Bass, J. J., et al.,** *Acta Endocrinol. (Copenh.)*, 124, 307, 1991.
83. **Breier, B. H., Gluckman, P. D., Blair, H. T., et al.,** *J. Endocrinol.*, 123, 25, 1989.
84. **Breier, B. H., Gluckman, P. D., and Bass, J. J.,** *J. Endocrinol.*, 116, 169, 1988.
85. **Hocquette, J. F., Postel-Vinay, M. C., Kayser, C., et al.,** *Endocrinology (Baltimore)*, 125, 2167, 1989.
86. **Frick, G. P. and Goodman, H. M.,** *Endocrinology (Baltimore)*, 131, 3083, 1992.
87. **Goodman, H. M., Frick, G. P., and Tai, L.,** *2nd Cell. Growth Forum,* Israel, 1993, p. 35.
88. **Sobrier, M. L., Duquesnoy, P., Duriez, B., et al.,** *FEBS Lett.*, 319, 16, 1993.
89. **Urbanek, M., Macleod, J. N., Cooke, N. E., et al.,** *Mol. Endocrinol.*, 6, 279, 1992.
90. **Barnard, R., Southard, J. N., Edens, A., et al.,** *Endocrinology (Baltimore)*, 133, 1474, 1993.
91. **Pekhletsky, R. I., Chernov, B. K., and Rubtsov, P. M.,** *Mol. Cell Endocrinol.*, 90, 103, 1992.
92. **Mathews, L. S., Enberg, B., and Norstedt, G.,** *J. Biol. Chem.*, 264, 9905, 1989.
93. **Pekhletskii, R. I., Chernov, B. K., and Rubtsov, P. M.,** *Mol. Biol.-Engl. Tr.*, 25, 1114, 1991.
94. **Barnard, R., Bundesen, P. G., Rylatt, D. B., et al.,** *Biochem. J.*, 231, 459, 1986.
95. **Barnard, R., Rowlinson, S. W., and Waters, M. J.,** *Biochem. J.*, 267, 471, 1990.
96. **Barnard, R., Ng, K. W., Martin, T. J., et al.,** *Endocrinology (Baltimore)*, 128, 1459, 1991.
97. **Thomas, H., Green, I. C., Wallis, M., et al.,** *Biochem. J.*, 243, 365, 1978.
98. **Barnard, R., Haynes, K. M., Werther, G. A., et al.,** *Endocrinology (Baltimore)*, 122, 2562, 1988.
99. **Ingram, R. T., Afshari, N., and Nicoll, C. S.,** *Endocrinology (Baltimore)*, 130, 3085, 1992.
100. **Wohnlich, L. and Moore, W. V.,** *Horm. Metab. Res.*, 14, 138, 1982.
101. **Sigel, M. B., Thorpe, N. A., Kobrin, M. S., et al.,** *Endocrinology (Baltimore)*, 108, 1600, 1981.
102. **Hughes, J. P., Tokuhiro, E., Simpson, S. A., et al.,** *Endocrinology (Baltimore)*, 113, 1904, 1983.
103. **Smal, J., Closset, J., Hennen, G., et al.,** *Biochem. Biophys. Res. Commun.*, 134, 159, 1986.
104. **Gorin, E., Honeyman, T. W., Tai, L. R., et al.,** *Endocrinology (Baltimore)*, 123, 328, 1988.
105. **Roupas, P. and Herington, A. C.,** *Mol. Cell. Endocrinol.*, 57, 93, 1988.
106. **Gorin, E. and Goodman, H. M.,** *Endocrinology (Baltimore)*, 116, 1796, 1985.
107. **Amit, T., Hartmann, K., Shoshany, G., et al.,** *Mol. Cell. Endocrinol.*, 94, 149, 1993.
108. **Murphy, L. J. and Lazarus, L.,** *Endocrinology (Baltimore)*, 115, 1625, 1984.
109. **Lesniak, M. A. and Roth, J.,** *J. Biol. Chem.*, 251, 3720, 1976,
110. **Hizuka, N., Gorden, P., Lesniak, M. A., et al.,** *J. Biol. Chem.*, 256, 4591, 1981.
111. **Barazzone, P., Lesniak, M. A., Gorden, P., et al.,** *J. Cell Biol.*, 87, 360, 1980.
112. **Eshet, R., Peleg, S., and Laron, Z.,** *Acta Endocrinol. (Copenh.)*, 107, 9, 1984.

113. **Asakawa, K., Grunberger, G., McElduff, A., et al.,** *Endocrinology (Baltimore)*, 117, 631, 1985.
114. **Weyer, B. and Sonne, O.,** *Mol. Cell. Endocrinol.*, 41, 85, 1985.
115. **Ilondo, M. M., Vanderschueren-Lodeweyckx, M., Courtoy, P. J., et al.,** *Endocrinology (Baltimore)*, 130, 2037, 1992.
116. **Rosenfeld, R. G. and Hintz, R. L.,** *J. Clin. Endocrinol. Metab.*, 51, 368, 1980.
117. **Roupas, P. and Herington, A. C.,** *Endocrinology (Baltimore)*, 121, 1521, 1987.
118. **Husman, B., Gustafson, J. A., and Andersson, G.,** *Mol. Cell. Endocrinol.*, 59, 13, 1988.
119. **Bick, T., Youdim, M. B. H., and Hochberg, Z.,** *Endocrinology (Baltimore)*, 125, 1711, 1989.
120. **Bick, T., Youdim, M. B. H., and Hochberg, Z.,** *Endocrinology (Baltimore)*, 125, 1718, 1989.
121. **Bick, T., Hochberg, Z., Amit, T., et al.,** *Endocrinology (Baltimore)*, 131, 423, 1992.
122. **Marques, M., Silva, R. S. M., Turyn, D., et al.,** *Gen. Comp. Endocrinol.*, 37, 487, 1979.
123. **Posner, B. I., Kelly, P. A., Shiu, R. P. C., et al.,** *Endocrinology (Baltimore)*, 95, 521, 1974.
124. **Scott, P., Kessler, M. A., and Schuler, L. A.,** *Mol. Cell. Endocrinol.*, 89, 47, 1992.
125. **Fraser, R. A. and Harvey, S.,** *Endocrinology (Baltimore)*, 130, 3593, 1992.
125a. **Mertani, H. C., Waters, M. J., and Jambou, R.,** *Neuroendocrinology*, 59, 483, 1994.
126. **Kover, K. and Moore, W. V.,** *Horm. Metab. Res.*, 16, 193, 1984.
127. **Chin, E., Zhou, J., and Bondy, C. A.,** *Endocrinology (Baltimore)*, 131, 3061, 1992.
128. **Werther, G. A., Haynes, K. M., Barnard, R., et al.,** *J. Clin. Endocrinol. Metab.*, 70, 1725, 1990.
129. **Lobie, P. E., Breipohl, W., Lincoln, D. T., et al.,** *J. Endocrinol.*, 126, 467, 1990.
130. **Lobie, P. E., Breipohl, W., Garcia-Aragon, J., et al.,** *Endocrinology (Baltimore)*, 126, 2214, 1990.
130a. **Delehaye-Zervas, M. C., Mertani, H., Martini, J. F.,** *J. Clin. Endocrinol. Metab.*, 78, 1473, 1994.
131. **Lobie, P. E., Breipohl, W., and Waters, M. J.,** *Endocrinology (Baltimore)*, 126, 299, 1990.
132. **Sakamoto, T. and Hirano, T.,** *J. Endocrinol.*, 130, 425, 1991.
133. **Ng, T. B., Kwan, L., and Cheng, C. H. K.,** *Biochem. Mol. Biol. Int.*, 29, 695, 1993.
134. **Lobie, P. E., Garcia-Aragon, J., and Waters, M. J.,** *Endocrinology (Baltimore)*, 130, 3015, 1992.
135. **Jammes, H., Gaye, P., Belair, L., et al.,** *Mol. Cell. Endocrinol.*, 75, 27, 1991.
136. **Feldman, M., Ruan, W., Cunningham, B. C., et al.,** *Endocrinology (Baltimore)*, 133, 1602, 1993.
137. **Helman, P. J., Mol, J. A., Rutteman, G. R., et al.,** *Endocrinology (Baltimore)*, 134, 287, 1994.
138. **Devolder, A., Renaville, R., Sneyers, M., et al.,** *J. Endocrinol.*, 138, 91, 1993.
138a. **Hochberg, Z., Phillip, M., Youdim, M. B. H., et al.,** *Metabolism*, 42, 1617, 1993.
139. **Garcia-Aragon, J., Lobie, P. E., Muscat, G. E. O., et al.,** *Development*, 114, 869, 1992.
140. **Walker, J. L., Moatsstaats, B. M., Stiles, A. D., et al.,** *Pediatr. Res.*, 31, 335, 1992.
141. **Tiong, T. S. and Herington, A. C.,** *Mol. Cell. Endocrinol.*, 83, 133, 1992.
142. **Carlsson, B., Nilsson, A., Isaksson, O. G. P., et al.,** *Mol. Cell. Endocrinol.*, 95, 59, 1993.
143. **Maes, M., De Hertogh, R., Watrin-Granger, P., et al.,** *Endocrinology (Baltimore)*, 113, 1325, 1983.
144. **Herington, A. C.,** *Horm. Metab. Res.*, 14, 422, 1982.
145. **Kelly, P. A., Posner, B. I., Tsushima, T., et al.,** *Endocrinology (Baltimore)*, 95, 532, 1974.
146. **Sorrentino, R. N. and Florini, J. R.,** *Exp. Ageing Res.*, 2, 191, 1970.
147. **Ambler, G. R., Breier, B. H., Surus, A., et al.,** *Acta Endocrinol. (Copenh.)*, 126, 155, 1992.
148. **Ymer, S. I. and Herington, A. C.,** *Mol. Cell. Endocrinol.*, 83, 39, 1992.
149. **Burnside, J. and Cogburn, L. A.,** *Mol. Cell. Endocrinol.*, 89, 91, 1992.
150. **Vasilatos-Younken, R., Gray, K. S., Bacon, W. L., et al.,** *J. Endocrinol.*, 126, 131, 1990.
151. **Baxter, R. C. and Zaltsman, Z.,** *Endocrinology (Baltimore)*, 115, 2009, 1984.
152. **Sasaki, N., Imai, Y., Tsushima, T., et al.,** *Acta Endocrinol. (Copenh.)*, 101, 574, 1982.
153. **Maiter, D., Underwood, L. E., Maes, M., et al.,** *Endocrinology (Baltimore)*, 123, 1053, 1988.
154. **Moldrup, A., Petersen, E. D., and Nielsen, J. H.,** *Endocrinology (Baltimore)*, 133, 1165, 1993.
155. **Herington, A. C., Phillips, L. S., and Daughaday, W. H.,** *Metab. Clin. Exp.*, 25, 341, 1976.
156. **Sanchez-Jimenez, F., Fielder, P., Martinez, R. R., et al.,** *Endocrinology (Baltimore)*, 126, 1270, 1990.
157. **Cramer, S. D., Wong, L., Kensinger, R. S., et al.,** *Endocrinology (Baltimore)*, 131, 2914, 1992.
158. **Gerasimo, P., Djiane, J., and Kelly, P. A.,** *Mol. Cell. Endocrinol.*, 13, 11, 1979.
159. **Chan, J. S., Robertson, H. A., and Friesen, H. G.,** *Endocrinology (Baltimore)*, 102, 632, 1978.
160. **Maiter, D., Walker, J. L., Adam, E., et al.,** *Endocrinology (Baltimore)*, 130, 3257, 1992.

161. **Domene, H., Krishnamurti, K., Eshet, R., et al.,** *Endocrinology (Baltimore),* 133, 675, 1993.
162. **Frick, G. P., Leonard, J. L., and Goodman, H. M.,** *Endocrinology (Baltimore),* 126, 3076, 1990.
163. **Messini, J. L., Eden, S., and Kostyo, J. L.,** *Am. J. Physiol.,* 249, E56, 1985.
164. **Picard, F. and Postel-Vinay, M. C.,** *Endocrinology (Baltimore),* 114, 1328, 1984.
165. **Carmignac, D. F., Robinson, I. C. A. F., Enberg, B., et al.,** *J. Endocrinol.,* 138, 267, 1993.
166. **Bick, T., Amit, T., Barkey, R. J., et al.,** *Endocrinology (Baltimore),* 128, 181, 1991.
167. **Furuhashi, W. and Fang, V. S.,** *Endocrinology (Baltimore),* 103, 2053, 1978.
168. **Baxter, R. C., Zaltsman, Z., and Turtle, J. R.,** *Endocrinology (Baltimore),* 111, 1020, 1982.
169. **Aguilar, R. C., Fernandez, H. N., Dellacha, J. M., et al.,** *Life Sci.,* 50, 615, 1992.
170. **Orian, J. M., Snibson, K., Stevenson, J. L., et al.,** *Endocrinology (Baltimore),* 128, 1238, 1991.
171. **Barash, I. and Posner, B. I.,** *Mol. Cell. Endocrinol.,* 62, 281, 1989.
172. **Mullis, P. E., Lund, T., Patel, M. S., et al.,** *Mol. Cell. Endocrinol.,* 76, 125, 1991.
173. **Singh, K., Ambler, G. R., Breier, B. H., et al.,** *Endocrinology (Baltimore),* 130, 2758, 1992.
174. **Maiter, D., Underwood, L. E., Maes, M., et al.,** *Endocrinology (Baltimore),* 122, 1291, 1988.
175. **Gray, E. S., Kelley, K. M., Law, S., et al.,** *Gen. Comp. Endocrinol.,* 88, 243, 1992.
176. **Mori, I., Sakamoto, T., and Hirano, T.,** *Gen. Comp. Endocrinol.,* 85, 385, 1992.
177. **Gause, I. and Eden, S.,** *J. Endocrinol.,* 105, 331, 1985.
178. **Gause, I. and Staffan, E.,** *Endocrinology (Baltimore),* 118, 119, 1986.
179. **Vikman, K., Carlsson, B., Billig, H., et al.,** *Endocrinology (Baltimore),* 129, 1155, 1991.
180. **Frick, G. P., Leonard, J. L., and Goodman, H. M.,** *Endocrinology (Baltimore),* 126, 3076, 1990.
181. **Frick, G. P. and Goodman, H. M.,** *71st Ann. Meet. Endocr. Soc.,* Seattle, WA, 1307, 1991.
182. **Chung, C. S. and Etherton, T. D.,** *Endocrinology (Baltimore),* 119, 780, 1986.
183. **Bass, J. J., Oldham, J. M., Hodgkinson, P. J., et al.,** *J. Endocrinol.,* 128, 181, 1992.
184. **Nilsson, A., Carlsson, B., Mathews, L., et al.,** *Mol. Cell. Endocrinol.,* 70, 237, 1990.

184a. **Stewart, C., Clejan, S., Fugler, T., et al.,** *Arch. Biochem. Biophys.,* 220, 309, 1983.

185. **Vanderpooten, A., Darras, V. M., Huybrechts, L. M., et al.,** *J. Endocrinol.,* 129, 275, 1991.
186. **Sorensen, M. T., Chaudhuri, S., Louveau, I., et al.,** *Domest. Anim. Endocrinol.,* 9, 13, 1992.
187. **Murphy, J., Vrhovsek, E., and Lazarus, L.,** *J. Clin. Endocrinol. Metab.,* 57, 1117, 1983.
188. **Leung, F. C.,** *Control and Manipulation of Animal Growth* (Eds. Buttery, P. J., Lindsay, D. B., and Haynes, N. B.), Butterworths, London, 1986, pp. 233–248.
189. **Posner, B. I., Patel, B., Vezinhet, A., et al.,** *Endocrinology (Baltimore),* 107, 1954, 1980.
190. **Burnside, J., and Cogburn, L.,** in *Avian Endocrinology* (Ed. Sharp, P.), Journal of Endocrinology, Bristol, 1993, p. 161.
191. **Vasilatos-Younken, R., Gray, K. S., Bacon, W. L., et al.,** *J. Endocrinol.,* 126, 131, 1990.
192. **Fouchera-Peron, C., Broer, Y., and Rosselin, G.,** *Biochim. Biophys. Acta,* 677, 445, 1981.
193. **Hochberg, Z., Bick, T., and Harel, Z.,** *Endocrinology (Baltimore),* 126, 325, 1990.
194. **Chernausek, S., Underwood, L. E., and Van Wyk, J.,** *Endocrinology (Baltimore),* 111, 1534, 1982.
195. **Burnside, J. and Cogburn, L. A.,** *Poult. Sci.,* 66(Suppl. 1), 28, 1989.
196. **Maes, M., Underwood, L. E., and Ketelslegers, J. M.,** *J. Endocrinol.,* 97, 243, 1983.
197. **Niimi, S., Hayakawa, T., Tanaka, A., et al.,** *Endocrinology (Baltimore),* 129, 2734, 1991.
198. **Niimi, S., Hayakawa, T., and Tanaka, A.,** *Endocrinology (Baltimore),* 127, 688, 1990.

198a. **Carmignac, D. F., Gabrielsson, B. G., and Robinson, T. C. A. F.,** *Endocrinology (Baltimore),* 133, 2445, 1993.

199. **Bryson, J. M. and Baxter, R. C.,** *Diabetologica,* 29, 106, 1986.
200. **Bornfeldt, K. E., Arnqvist, H. J., Enberg, B., et al.,** *J. Endocrinol.,* 122, 651, 1989.
201. **Postel-Vinay, M. C., Cohen-Tanugi, E., and Charrier, J.,** *Mol. Cell. Endocrinol.,* 28, 657, 1982.
202. **Triest, S., Thissen, J., Mauerhoff, T., et al.,** *73rd Annu. Meet. Endocr. Soc.,* Washington, D.C., 1991, p. 1655.
203. **Vanderpooten, A., Dewil, E., Decuypere, E., et al.,** *Med. Sci. Res.,* 17, 1031, 1989.
204. **Maes, M., Underwood, L. E., Gerard, G., et al.,** *Endocrinology (Baltimore),* 115, 786, 1984.
205. **Fliesen, T., Maiter, D., Gerard, G., et al.,** *Pediatr. Res.,* 26, 415, 1989.
206. **Thissen, J. P., Triest, S., Underwood, L. E., et al.,** *Endocrinology (Baltimore),* 126, 908, 1990.
207. **Postel-Vinay, M. C., Durand, D., Lopez, S., et al.,** *Horm. Metab. Res.,* 22, 7, 1990.
208. **Finidori, M. C., Postel-Vinay, M., and Kleinknecht, C.,** *Endocrinology (Baltimore),* 106, 1960, 1980.
209. **Challa, A., Chan, W., Krieg, R. J., et al.,** *Kidney Int.,* 44, 1224, 1993.

210. **Husman, B. and Andersson, G.,** *J. Mol. Endocrinol.*, 10, 289, 1993.
211. **Chang, T. C., Yu, S. C., and Chang, T. J.,** *Hepatology*, 11, 123, 1989.
212. **Maes, M., Ketelslegers, J. M., and Underwood, L. E.,** *Diabetes*, 32, 1060, 1983.
212a. **Laron, Z. and Parks, J. S. (Eds.),** *Lessons from Laron Syndrome 1966–1992*, Karger, Basel, 1992.
213. **Eshet, R., Laron, Z., Pertzelan, A., et al.,** *Israel J. Med. Sci.*, 20, 8, 1984.
214. **Aguirre, A., Donnadieu, M., and Job, J. C.,** *Horm. Res.*, 34, 4, 1990.
215. **Baumann, G., Shaw, M. A., and Winter, R. J.,** *J. Clin. Endocrinol. Metab.*, 65, 814, 1987.
216. **Meacham, L. R., Brown, M. R., Murphy, T. L., et al.,** *J. Clin. Endocrinol. Metab.*, 77, 1379, 1993.
217. **Brown, M. R., Meacham, L. R., Pfaffle, R. W., et al.,** *Lessons from Laron Syndrome 1966–1992* (Eds. Laron, Z. and Parks, J. S.), Karger, Basel, 1992, p. 127.
218. **Berg, M. A., Guevara-Aguirre, J., Rosenbloom, A. L., et al.,** *Hum. Mutat.*, 1, 24, 1992.
219. **Amselem, S., Sobrier, M. L., Duquesnoy, P., et al.,** *J. Clin. Invest.*, 87, 1098, 1991.
220. **Amselem, S., Duquesnoy, P., Duriez, B., et al.,** *Hum. Mol. Genet.*, 2, 355, 1993.
221. **Berg, M. A., Argente, J., Chernausek, S., et al.,** *Am. J. Hum. Genet.*, 52, 998, 1993.
222. **Duncan, B. K. and Miller, J. H.,** *Nature (Lond.)*, 287, 560, 1980.
223. **Amselem, S., Duquesnoy, P., Sobrier, M., et al.,** *Ped. Adolesc. Endocrinol.*, 24, 132,1992.
223a. **Amselem, S., Duquesnoy, P., Attree, O., et al.,** *N. Engl. J. Med.*, 321, 989, 1989.
224. **Duquesnoy, P., Sobrier, M. L., Amselem, S., et al.,** *Proc. Natl. Acad. Sci. U.S.A.*, 88, 10272, 1991.
225. **Edery, M., Rozakisadcock, M., Goujon, L., et al.,** *J. Clin. Invest.*, 91, 838, 1993.
226. **Maheshwari, H. G., Clayton, P. E., Mughal, et al.,** *Ped. Adolesc. Endocrinol.*, 24, 160, 1992.
226a. **Oakes, S. R., Haynes, K. M., and Batch,** *J. Mol. Cell. Endocrinol.*, 99, 125, 1994.
227. **Fielder, P. J., Guevara, J., Rosenbloom, A. L., et al.,** *J. Clin. Endocrinol. Metab.*, 743, 750, 7492.
228. **Merimee, T. J., Baumann, G., and Daughaday, W. H.,** *J. Clin. Endocrinol. Metab.*, 71, 1183, 1990.
229. **Baumann, G., Shaw, M. A., and Merimee, T. J.,** *N. Engl. J. Med.*, 230, 1705, 1989.
230. **Merimee, T.,** *Lessons from Laron Syndrome 1966–1992* (Eds. Laron, Z. and Parks, J. S.). Karger, Basel, 1992, p. 266.
231. **Kou, K., Lajara, R., and Rotwein, P.,** *J. Clin. Endocrinol. Metab.*, 76, 54, 1993.
232. **Decuypere, E., Huybrechts, L. M., Kuhn, E. R., et al.,** *Crit. Rev. Poultry Biol.*, 2, 191, 1991.
233. **Tixier-Boichard, M., Huybrechts, L. M., Kuhn, E. R., et al.,** *Genet. Select. Evol.*, 21, 217, 1989.
234. **Guillaume, J.,** *World Poultry Sci. J.*, 32, 285, 1976.
235. **Leung, F. C., Gillett, J., Lilburn, M. S., et al.,** *J. Steroid Biochem.*, 20, 1557, 1984. (Abstract)
236. **Harvey, S., Scanes, C. G., and Marsh, J. A.,** *Gen. Comp. Endocrinol.*, 55, 493, 1984.
237. **Huybrechts, L. M., Kuhn, E. R., Decuypere, E., et al.,** *Reprod. Nutr. Dev.*, 27, 547, 1987.
238. **Scanes, C. G., Marsh, J., Decuypere, E., et al.,** *J. Endocrinol.*, 97, 127, 1983.
239. **Bowen, S., Huybrechts, L., Marsh, J., et al.,** *Comp. Biochem. Physiol.* 86A, 137, 1987.
240. **Leung, F. C., Styles, W. J., Rosenblum, C. I., et al.,** *Proc. Soc. Exp. Biol. Med.*, 184, 234, 1987.
241. **Duriez, B., Sobrier, M. L., Duquesnoy, P., et al.,** *Mol. Endocrinol.*, 7, 806, 1993.
242. **Burnside, J., Liou, S. S., Zhong, C., et al.,** *Gen. Comp. Endocrinol.*, 88, 20, 1992.
243. **Hull, K. L., Harvey, S., Marsh, J. A., et al.,** *Pediatr. Adolesc. Endocrinol.*, 24, 294, 1993.
244. **Daughaday, W. H., Tobin, G. S., Lenny, N. T., et al.,** *Clin. Res.*, 34, 950A, 1986.
245. **Mitchell, M. L., Guilleman, R., and Selye, H.,** *Endocrinology (Baltimore)*, 54, 111, 1954.
246. **Knobil, E. and Greep, R. O.,** *Recent Prog. Horm. Res.*, 15, 1, 1959.
247. **Fairhall, K. M., Carmignac, D. F., and Robinson, I.,** *Endocrinology (Baltimore)*, 131, 1963, 1992.
248. **Daughaday, W. H.,** *Lesson from Laron Syndrome* (Eds. Laron, Z. and Parks, J. S.), Karger, Basel, 1992, p. 282.
249. **Bezecny, I., Bartova, J., and Skarda, J.,** *J. Endocrinol.*, 134, 5, 1992.
250. **Harvey, S. and Fraser, R. A.,** *J. Endocrinol.*, 133, 357, 1992.
251. **Harvey, S., Hull, K., Janssens, W., et al.,** *2nd Cell. Growth Forum,* Israel, 1993, p. C10.
251a. **Harvey, S.,** unpublished observations, 1994.
252. **Eigenmann, J. E., Patterson, D. F., Zapf, J., et al.,** *Acta Endocrinol. (Copenh.)*, 105, 294, 1984.
253. **Lauterio, T. J., Trivedi, B., Kapadia, M., et al.,** *Comp. Biochem. Physiol.*, 91A, 15, 1988.

254. **Hammond, A. C. E. and Olson, T. H.**, *Proc. Soc. Exp. Biol. Med.*, 197, 450, 1991.
255. **Donner, D. B.**, *J. Biol. Chem.*, 258, 2736, 1983.
256. **Blossey, H. C.**, *Horm. Metab. Res.*, 11, 616, 1979.
257. **Ymer, S. I. and Herington, A. C.**, *Endocrinology (Baltimore)*, 114, 1732, 1984.
258. **Smith, W. C. and Talamantes, F.**, *J. Biol. Chem.*, 262, 2213, 1987.
259. **Orian, J. M., Snibson, K., Stevenson, J. L., et al.**, *Endocrinology (Baltimore)*, 128, 1238, 1991.
260. **Tavakkol, A., Elder, J. T., Griffiths, C. E. M., et al.**, *J. Invest. Dermatol.*, 99, 343, 1992.
261. **Hocquette, J. F., Postel-Vinay, M. C., Djiane, J., et al.**, *Endocrinology (Baltimore)*, 127, 1655, 1990.
262. **Hull, K. L., Fraser, R. A., and Harvey, S., et al.**, *J. Endocrinol.*, 135, 459, 1992.
263. **Ng, T. B., Leung, T. C., Cheng, C. H. K., et al.**, *Gen. Comp. Endocrinol.*, 86, 111, 1992.
264. **Hirano, T.**, *Gen. Comp. Endocrinol.*, 81, 383, 1991.
265. **Yao, K., Niu, P., LeGac, F., et al.**, *Gen. Comp. Endocrinol.*, 81, 72, 1991.
266. **DiGirolamo, M., Eden, S., Enberg, G., et al.**, *FEBS Lett.*, 205, 15, 1986.
267. **Kiess, W. and Butenandt, O.**, *J. Clin. Endocrinol. Metab.*, 60, 740, 1985.
268. **Herington, A. C. and Veith, N. M.**, *Endocrinology (Baltimore)*, 101, 984, 1977.
269. **Labbe, A., Delcros, B., Dechelotte, P., et al.**, *Biol. Neonate*, 61, 179, 1992.
270. **Freemark, M., Kirk, K., Pihoker, C., et al.**, *Endocrinology (Baltimore)*, 133, 1830, 1993.
271. **Gray, K. S., Vasilatos-Younken, R., Bacon, W. L., et al.**, *Gen. Comp. Endocrinol.*, 81, 313, 1991.
272. **Fraser, R. A., Attardo, D., and Harvey, S.**, *J. Endocrinol.*, 5, 231, 1990.
273. **Miettinen, P. J., Ilvesmaki, V., and Voutilainen, R.**, *J. Mol. Endocrinol.*, 31, 28, 1992.
274. **Martinoli, M. G., Oullet, J., Rheaume, E., and Pelletier, G.**, *Neuroendocrinology*, 54, 607, 1991.
275. **Sakamoto, T. and Hirano, T.**, *J. Endocrinol.*, 130, 425, 1991.
276. **Ikuta, K., Hirano, T., and Aida, K.**, *11th Int. Symp. Comp. Endocrinol.*, Spain, 1989, p. 688.
277. **Ng, T. B., Kwan, L., and Cheng, C. H. K.**, *Biochem. Mol. Biol. Int.*, 29, 695, 1993.
278. **Hull, K. L., Fraser, R. A., Marsh, J. A., and Harvey, S.**, *J. Endocrinol.*, 137, 91, 1993.
279. **Gray, K. S., Vasilatos-Younken, R., Bacon, W. L., et al.**, *Gen. Comp. Endocrinol.*, 81, 313, 1991.
280. **Hill, D. J., Riley, S. C., Bassett, N. S., et al.**, *J. Clin. Endocrinol. Metab.*, 75, 646, 1992.
281. **Oakes, S. R., Haynes, K. M., Waters, M. J., et al.**, *J. Clin. Endocrinol. Metab.*, 75, 1368, 1992.
282. **Ng, T. B., Cheng, C. H., Hui, S. T., et al.**, *Biochem. Int.*, 27, 939, 1992.
283. **Frankenne, F., Alsat, E., Scippo, M. L., et al.**, *Biochem. Biophys. Res. Commun.*, 182, 481, 1992.
284. **Ban, E., Gagnerault, M. C., Jammes, H., et al.**, *Life Sci.*, 48, 2141, 1991.
285. **Marsh, J. A.**, *Poultry Sci. Rev.*, 4, 129, 1992.
286. **Fraser, R. A., Siminoski, K., and Harvey, S.**, *J. Endocrinol.*, 128, R9, 1991.
287. **Burton, K. A., Kabigting, E. B., Clifton, D. K., et al.**, *Endocrinology (Baltimore)*, 130, 958, 1992.
288. **Hasegawa, O., Minami, S., Sugihara, H., et al.**, *Dev. Brain Res.*, 74, 287, 1993.
289. **Lobie, P. E., Garcia-Aragon, J., Lincoln, D. T., et al.**, *Dev. Brain Res.*, 74, 225, 1993.
290. **Carlsson, B., Bergh, C., Bentham, J., et al.**, *Hum. Reprod.*, 7, 1205, 1992.
291. **Zhang, C. Z., Young, C. G., and Waters, M. J.**, *Arch. Oral Biol.*, 37, 77, 1992.
292. **Gavin, J. R., Sattman, R. J., and Tollefsen, S. E.**, *Endocrinology (Baltimore)*, 110, 637, 1982.
293. **Louveau, I. and Etherton, T. D.**, *J. Anim. Sci.*, 70, 1801, 1992.

Chapter 17

Growth Hormone Action: Intracellular Mechanisms

C. G. Scanes

CONTENTS

I. SIGNAL TRANSDUCTION AND GH ACTION

This chapter considers the possible role of tyrosine kinase(s), calcium, phosphatidylinositol/protein kinase C, cAMP/protein kinase A, and protooncogenes in the mechanism of growth hormone (GH) action.

II. TYROSINE KINASES

A. INTRODUCTION

There is growing evidence that tyrosine kinase may be part of the signal transduction mechanism for GH,[1–3] as is the case with insulin and insulin-like growth factor I (IGF-I). In studies with preadipocytes (3T3-F442A fibroblasts), it would appear that binding of GH to the GH receptor (GHR) stimulates phosphorylation of tyrosine residues in the intracellular domain of the GHR[4–5] or to a closely associated protein with a molecular weight similar to the GHR. This is based on a series of ingenious studies in which the GH-GHR complex was found to bind to an affinity column with antisera to phosphorylated tyrosine and in which [^{32}P]phosphate is found to be incorporated into GHR[1–5] or closely associated protein. The endogenous tyrosine kinase appears to employ ATP rather than other GTP, CTP, or UTP.[2] A GHR-associated tyrosine kinase is found in IM-9 lymphocytes, preadipocytes, and adipocytes.[2,3]

The GHR receptor does not appear to be a tyrosine kinase itself. The sequence of the GHR does not show any homology with any known tyrosine kinase.[6] Moreover, when the

0-8493-8697-7/95/$0.00+$.50

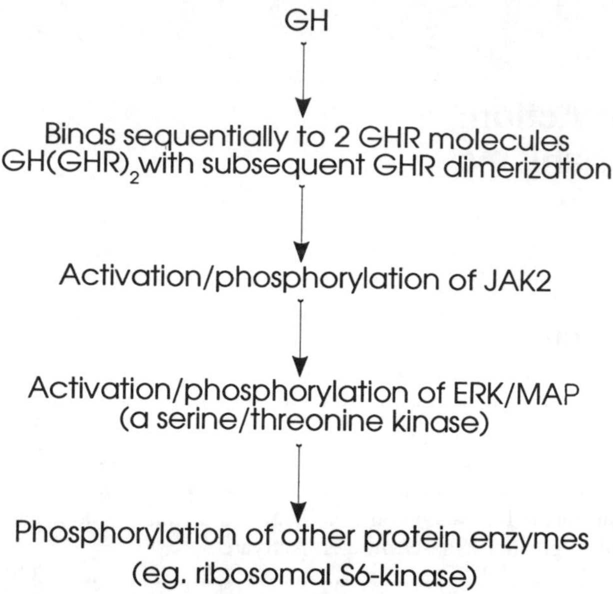

Figure 1 Putative mechanism of GH signal transduction involving JAK2 and ERK/MAP.

liver GHR is transfected into different cell types (e.g., COS-7 cells), then various and much reduced tyrosine phosphorylation of 120- to 130-kDa proteins is observed.[7,8] This is consistent with phosphorylation of the GH receptor (or an associated protein), depending on "the amount of GHR associated tyrosine kinase being cell type-specific"[7] or not GHR itself.

The available evidence suggests that GH acts via a tyrosine kinase, probably Janus kinase 2 (JAK2), associated with the GHR (for details see below and Figure 1). There are, in addition, other proteins that are tyrosine phosphorylated in response to GH. It is probable that a GHR-associated tyrosine kinase (JAK2) phosphorylates other kinases, including the extracellular signal-regulated kinase (ERK)/mitogen-activated protein (MAP) kinase. There is evidence that exposure of at least some cell types to GH results in the phosphorylation of proteins distinct from the GHR-associated 130-kDa protein. For instance, the antilipolytic (inhibition of norepinephrine-stimulated lipolysis) effect of GH on rat adipocytes is associated with phosphorylation of a 46-kDa protein in the membrane[9] in which 14% of the phosphate was found with tyrosine residues.[9] In addition, in 3T3-H442A adipocytes, GH stimulates tyrosine phosphorylation of three cytoplasmic proteins[9,10] (Janus kinase, extracellular signal-regulated kinases [ERK]/mitogen-activated protein [MAP] kinase, and S6 kinase).

B. JANUS KINASE

1. Introduction

The Janus kinase (JAK) family of protein tyrosine kinases includes JAK1, JAK2, and tyk2.[11–13] These have the unusual structural feature of two protein kinase-related domains; one is structurally similar to other tyrosine kinases and the other is unique. The GHR is a member of the cytokine receptor superfamily of proteins including the receptors for GH, prolactin, erythropoietin, interleukins, granulocyte-macrophage colony-stimulating factor (GM-CSF), granulocyte colony-stimulating factor (G-CSF), and interferons α, β, and γ.[14–16] There is increasing evidence that signal transduction for many members of this superfamily involves a tyrosine kinase closely associated with the receptor; for example the tyrosine kinase JAK2 is associated with the erythropoietin receptor, the G-CSF receptor (G-CSFR),[17] and the interleukin 3 receptor,[18] while tyk2 is associated with the interferon α receptor.[12,19] GH also acts via binding to the GHR and thence activation of the closely associated JAK2.[20] Consequently, in view of the common signal transduction pathways, it is therefore not surprising that a chimeric G-CSFR (intracellular and transmembrane domain) and GHR (extracellular domain) can mediate the effects of GH.[21]

2. JAK2 and GH Action

The tyrosine kinase JAK2 consists of 993 amino acid residues[13] with a molecular weight of approximately 130 kDa.[20] A series of studies by Carter-Su and colleagues[20] have provided strong evidence that GH rapidly promotes tyrosyl phosphorylation (within 30 s and maximally at 5 min) of JAK2 in 3T3-F442A fibroblasts. Moreover, this phosphorylated JAK2 is associated, but not covalently bound to the GH:GHR complex, as evidenced by the ability of antisera to either GH or the GHR to immunoprecipitate JAK2.[20] Association of GH with the GHR leads to both phosphorylation and activation of JAK2.[18]

3. Structural Requirements for Activation

Studies with a human lymphocyte cell line provide evidence for the requirements of GH binding for activation of tyrosine kinases.[22] With as little as 1 min exposure to GH, three proteins undergo tyrosine phosphorylation, these having molecular masses of 93, 120, and 134 kDa.[22] It is not clear which of these is JAK2 but, based on molecular weight, either of the latter may be synonymous with JAK2. It has been concluded that binding of GH site 1 to the GHR is required for tyrosine phosphorylation[22] of each of the three proteins. Evidence of this came from the lack of tyrosine phosphorylation following exposure to a mutant form of GH (hGH K172A, F176A) which has greatly reduced binding for the GHR at site I.[22] It is thought that GH binds to GHR at site I, followed by binding to a second GHR molecule at site II followed by GHR dimerization[16] (Figure 2). If GHR dimerization is required for GH signal transduction/tyrosine phosphorylation of proteins, then a GH analog that has no binding site II should be inactive. Such a GH analog is the mutant hGH G120R. This mutant is biologically inactive and acts as a GH antagonist.[23] Not only does this GH analog not induce tyrosine phosphorylation of the 93-kDa protein but its presence also prevents (antagonizes) GH induction of phosphorylation of this protein. Thus, dimerization of the GHR following GH binding appears to be required for GH to induce phosphorylation of the 93-kDa protein. The effects of the mutant hGH G120R on tyrosine phosphorylation of the other proteins were somewhat equivocal. Thus, it is not possible to conclude, as yet, that dimerization is required for GH/GHR phosphorylation of JAK2.

C. EXTRACELLULAR SIGNAL-REGULATED KINASE/MITOGEN-ACTIVATED PROTEIN KINASE

Extracellular signal-regulated kinases (ERK) are both serine/threonine and tyrosine protein kinases with molecular masses between 42 and 45 kDa.[24] In view of the activation of the ERK during mitogenic stimulation, the ERKs are also known as mitogen-activated protein (MAP) kinase. This acronym is also applicable owing to their ability to phosphorylate microtubule-associated protein 2 (MAP2). An alternative name is MBP kinase, in view of the ability of the kinase to phosphorylate myelin basic protein (MBP).

In studies with 3T3-F442A, a mouse preadipocyte cell line, GH rapidly induces the phosphorylation of tyrosine residues in two proteins. These have molecule masses of 42 and 45 kDa and are immunologically related to ERK1.[10,25] The time course of GH induction of the phosphorylation of the three proteins, such that both maximal phosphorylation and MAP/MBP enzyme activity are observed, requires between 5 and 10 min of

GH exposure,[25–27] with a second peak occurring at 120 min.[26] These peaks are considerably delayed relative to the phosphorylation of JAK2[20]. This is consistent with the model shown in Figure 1 with JAK2 activated prior to ERK/MAP kinase.

There is evidence both for and against GH activating MAP kinase(s) via protein kinase C. Such an effect would represent a separate or parallel mechanism of GH-induced kinase activation. However, the effects of GH on the activation of MAP kinase are not consistently influenced by protein kinase C inhibitors. For instance, H7 is without effect whereas staurosporin blocks the GH effect.[26,27] GH can induce tyrosine phosphorylation/activation of ERK/MAP kinase/MBP kinase not only in 3T3-F442A cells but also in Chinese hamster ovary cells transfected with GHR cDNA.[28] However, in studies with a human lymphocyte cell line there is no evidence for tyrosine phosphorylation of proteins with molecular weights similar to ERK.[22] Thus, ERK/MAP kinase may not be required for all cellular actions of GH in all cell types. The phosphorylated ERKs in turn can activate other serine/threonine kinases such as the ribosomal S6 kinase II (reviewed by Crew et al.[24]).

D. S6 KINASE

In studies again with 3T3-F442A preadipocytes, GH stimulates the phosphorylation of ribosomal S6 kinase.[10,27,29] As might be expected were this being mediated by ERK/MAP kinase (Figure 1), the time course of GH induction of S6 kinase activity is delayed relative to that of MAP kinase activity.[27] Maximal increases in MAP kinase activity are observed after a 10 min exposure to GH whereas those of S6 kinase were observed after 20 min.[27,29] The S6 kinase can be resolved into two components distinguishable by size: respectively, $p70^{S6k}$ and $p90^{rsk}$ with molecular masses of 70 and 90 kDa. Exposure of 3T3-F442A preadipocytes to GH induces phosphorylation of both components but with a markedly different time course. The $p70^{S6K}$ was maximally phosphorylated after exposure to GH for 20 min,[29] with phosphorylation maintained for at least 1 h. On the other hand, $p90^{rsk}$ was transitorily phosphorylated in the presence of GH, with peaks at 10 min and a return to basal by 1 h.[29] On the basis of the ability of chelerythrine, a protein kinase C inhibitor, to prevent GH-induced phosphorylation of the S6 kinase,[29] a role for protein kinase C in the scheme proposed in Figure 1 is possible. However, it it not known whether this protein kinase C inhibitor influences either ERK/MAP kinase or JAK2 activity.

III. CALCIUM AND GH ACTION

There is evidence for a role of calcium in some but not all the actions of GH. A role for calcium in the mechanism of GH action was first suggested from studies examining the lipolytic effect of GH on chicken adipose tissue *in vitro*. The lipolytic effect of GH is not observed if the calcium concentration in the medium is reduced or the calcium antagonist, verapamil, is present.[30] Evidence that GH may be acting via increased intracellular calcium concentrations comes from the observation that GH induction of lipolysis could be mimicked by a calcium ionophore.[30] The insulin-like effect of GH on chicken adipose cells (inhibiting glucagon-induced lipolysis) is, however, unaffected by the concentration of calcium in the medium or the presence of verapamil.

Other studies provide strong, direct evidence that GH induces refractoriness to its insulin-like effect by increasing the intracellular concentration of calcium.[31,32] Parenthetically, it might be noted that this effect, like the lipolytic effect of GH on chicken adipose tissue, can be blocked by inhibitors of RNA/protein synthesis. The refractoriness-inducing effect of GH is prevented by verapamil (a calcium antagonist) or trifluoperazine or calmidazolium (calmodulin antagonists).[31] These agents do not, however, influence the insulin-like effects of GH.[31] If GH is inducing refractoriness by increasing the intracellular calcium concentration, we would expect agents that increase intracellular calcium

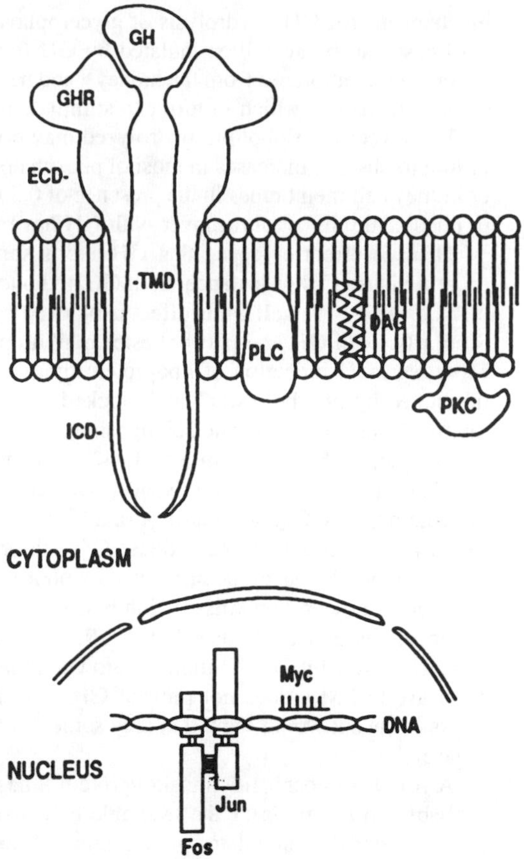

Figure 2 GH signal transduction pathway. GH brings about receptor dimerization by binding first to the extracellular domain (ECD) of one receptor molecule and then to that of another receptor molecule. The signal of GH binding is probably transmitted via the transmembrane domain (TMD) to bring about conformational changes in the intracellular domain (ICD). Following this, several early biochemical events have been recorded including phospholipase C (PLC)-mediated DAG production and protein kinase C (PKC)-induced phosphorylation of the GHR and possibly some cellular proteins. Induction of *fos*, *jun*, and *myc* genes is also observed. (From Maharajan, P. and Maharajan, V., Growth hormone signal transduction, *Experientia*, 49, 980, 1993. Reprinted with permission.)

concentrations to mimic this effect of GH. Indeed, a calcium ionophore (A23187) alone induces refractoriness.[31] There is also direct evidence that GH increases intracellular calcium [as indicated by fluorescence of cells loaded with Fura-2 hexakis(acetoxymethyl) ether] in adipocytes under conditions in which they are developing refractoriness but not when GH is exerting an insulin-like effect.[32,33] As is the case with the involvement of polyamines, there is also evidence that the mechanism of action of prolactin, a hormone closely related to GH, also requires calcium.[34] The increase in intracellular calcium (Ca^{2+}) concentration does not occur if actinomycin D is present.[33] This would suggest that GH induces increased intracellular Ca^{2+} concentrations by a mechanism requiring RNA/protein synthesis.

In summary, the available evidence suggests that calcium is involved in mediating some of the actions of GH that require RNA/protein synthesis — the lipolytic effect on chicken adipose and the induction of refractoriness to the insulin-like effect of GH on rat adipose tissue. However, insulin-like effects of GH do not appear to require calcium. The induction of *c-fos* mRNA by GH in mouse osteoblasts is also calcium-independent.[35]

IV. PHOSPHATIDYLINOSITOL/PROTEIN KINASE C

There is strong evidence that GH influences diacylglycerol formation and/or protein kinase C under some circumstances (Figure 2). This may be one of the signal transduction mechanisms for GH. Hydrolysis of glycerophospholipid in the cell membrane by phospholipase can be acutely stimulated by GH (e.g., in preadipose Ob1771 cells[36] and in basolateral membranes from the kidney[37] and hepatocytes[38]). This results in the formation of diacylglycerol, which in turn can stimulate protein kinase C.

The glycerophospholipid hydrolyzed may not be phosphatidylinositol in view of the failure to observe increases in inositol phosphate formation in GH-treated Ob1771 cells[36] or kidney cell membranes in the presence of 0.3 to 1.0 m*M* calcium.[37] However increases in phosphatidylinositol turnover with GH have been observed with adipocytes.[39]

There is direct evidence that GH exerts some of its effects via diacylglycerol and protein kinase C. For instance, both GH and phorbol esters stimulate expression of *c-fos* mRNA in Ob1771 cells; this effect is blocked by H7, an inhibitor of protein kinase C.[40] Similarly, both GH and phorbol ester inhibit epidermal growth factor (EGF) binding.[36] Moreover, GH induction of lipoprotein lipase by GH in Ob1771 preadipocytes can be mimicked by phorbol ester[40] and blocked by inhibitors of protein kinase C (H7, polymyxin B, staurosporine, and sphingosine).[40] Similarly, GH stimulates ERK/MAP kinase and ribosomal S6 kinase in 3T3-F442A preadipocytes. This can be mimicked by the phorbol ester, phorbol 12-myristate 13-acetate (PMA), and inhibited by some inhibitors of protein kinase C (e.g., staurosporine[26,27,41] or chelerythrine[29]) but not by H7 or H8[26] or by downregulation of protein kinase C by chronic exposure to PMA.[29,41] All the effects of GH cannot, however, be mimicked by phorbol esters (stimulation of protein kinase C) or prevented by protein kinase C inhibitors or by downregulation of protein kinase C. For instance, while GH increases IGF-I mRNA, this is not observed with phorbol ester and is not blocked by H7.[36] Similarly, downregulation of protein kinase C by prolonged exposure to PMA does not prevent GH from inducing *c-fos* mRNA in mouse osteoblasts.[35] Thus it would seem that only some of the effects of GH are mediated via protein kinase C.

A role for phosphatidylinositol/protein kinase C in the insulin-like effects of GH is probable on the basis of the available evidence. Preincubation of rat adipocytes with phorbol ester downregulates protein kinase C and reduces lipogenesis (incorporation of radioactive glucose into lipid) in the presence of either GH or insulin.[41] Similarly, both GH- and insulin-induced lipogenesis is inhibited by acridine orange, an inhibitor of protein kinase C.[42] (The specificity of this effect may be questioned, however, as acridine orange also inhibits basal lipogenesis.[41]) Moreover, GH-induced glucose oxidation is unaffected by the presence of H7, an inhibitor of protein kinase C.[41] This is unexpected if protein kinase C is part of the GH signal transduction mechanism. The involvement of phosphatidylinositol/protein kinase C in the insulin-like action of GH is, nevertheless, supported by another insulin-like effect, the inhibition of norepinephrine-induced lipolysis. In this case, GH increases phosphoinositide turnover in a concomitant manner.[38]

Phosphatidylinositol/protein kinase C may also be involved in mediating the lipolytic and/or diabetogenic effects of GH. Inhibitors of protein kinase C (sphingosine and staurosporine) inhibit GH/dexamethasone-induced lipolysis by rat adipose tissue *in vitro*.[43,44] Similarly, GH/theophylline-induced lipolysis is reduced in the presence of inhibitors of protein kinase C (sphingosine, H7, and H8).[43,44] As with all pharmacological studies, the specificity of the effects of blockers may be questioned. For instance, H7 inhibits RNA/protein synthesis[43] whereas sphingosine inhibits both isoproterenol- and forskolin-stimulated lipolysis[44] — effects that are not normally associated with protein kinase C. The chronic diabetogenic effect of GH is associated with a reduced response to insulin. Chronic administration of GH (to *ob/ob* mice) has been found to inhibit both the activation of the acute phosphatidylinositol/phospholipase C by insulin[45] and the insulin-induced increase in diacylglycerol (DAG) in adipocytes.[46] In summary, some of the effects of GH therefore appear to be mediated via a protein kinase C pathway.

V. G PROTEINS

The mechanism by which GH decreases phosphatidylinositol/phospholipase C and DAG in adipose tissue from *ob/ob* mice appears to involve a pertussis toxin-sensitive G protein (possibly G_i), which may similarly be involved in mediating some of the effects of insulin[45,47] and prolactin.[48]

Pertussis toxin, for instance, can block the effect of prolactin on lactose production but not on ornithine decarboxylase activity in mouse mammary tissue incubated *in vitro*.[48] The protein dose-response curve for the proliferation of Nb_2 lymphoma cells is also shifted to the right in the presence of pertussis toxin.[49] It should also be noted that exposure of Nb_2 cells to lactogens (including hGH) increases pertussis toxin stimulation of ADP ribosylation of G_i proteins.[50]

Thus, pertussin toxin-sensitive G proteins, presumably G_i proteins, may be related to at least some of the mechanisms of actions for both GH and prolactin, because they are members of a common gene family.

VI. cAMP

In the presence of GH, the lipolytic effect of epinephrine is reduced, at least in adipose tissue from hypophysectomized rats.[51] Similarly with chicken adipose tissue, GH partially suppresses the lipolytic effect of glucagon,[52] glucagon being the major lipolytic hormone in this species.[53] In view of the common second messenger (cAMP) for both epinephrine and glucagon stimulation of lipolysis in (respectively) rat[54] and chicken[55] adipose tissue, it is reasonable to suggest that GH in some way is disrupting cAMP signal transduction. However, it is still not clear at which point GH is exerting its antilipolytic effect and also whether the antilipolytic effects of GH on adipose tissue from hypophysectomized rats and chicken represent the same overall phenomenon. The antilipolytic effect of GH may involve (1) depression in the intracellular concentration of cAMP (via decreased adenylate cyclase activity or increased phosphodiesterase activity), (2) a decrease in protein kinase A activity (via decreased cAMP), and (3) a reduction in the phosphorylation of hormone-sensitive lipase (via decreased protein kinase A and/or increased phosphatase activity). There is evidence that GH can decrease phosphorylation of hormone-sensitive lipase induced by norepinephrine in adipocytes from hypophysectomized rats.[56] Birnbaum and Goodman[57] report that GH reduces the accumulation of cAMP in the presence of epinephrine in hypophysectomized rat adipose tissue. Moreover, GH reduces, albeit very slightly, protein kinase A activity in the absence of cAMP.[57] An increase in the activity of low K_m cyclic AMP phosphodiesterase is also observed. It is unlikely that GH activation of the phosphodiesterase represents the only aspect of the antilipolytic effect of GH, as in short incubations GH can inhibit lipolysis before any detectable activation of phosphodiesterase is observed.[57] The antilipolytic effect of GH on chicken adipose tissue appears to be mediated downstream from cAMP in view of the ability of GH to reduce lipolysis induced by glucagon, 8-bromo-cAMP, and a phosphodiesterase inhibitor.[52]

In summary, the relationship between GH and cAMP is therefore still unclear.

VII. PROTOONCOGENES

The expression of protooncogenes is thought to be an early step in the mechanism by which many growth factors stimulate proliferation or differentiation of specific cell types. GH has been demonstrated to induce c-*fos* expression in Ob1771 cells[36,58] and c-*fos* and c-*jun* expression in 3T3 cells (preadipocytes)[59,60] and mouse osteoblasts.[35] In the former study with Ob1771 cells, the induction of c-*fos* expression may have been subsequent to activation of protein kinase C.[36] However, whereas the protein kinase C activator (PMA) induces superinduction of the c-*fos* gene in the presence of cycloheximide, this is not observed with GH.[58] This is not consistent with the notion that GH is inducing c-Fos mRNA via a protein kinase C mechanism.

Although GH can increase c-*fos* and c-*jun* expression in two cell lines, this does not automatically mean that in all tissues responding to GH an induction of c-*fos* and/or c-*jun* is occurring. Moreover, just because GH induces expression of these protooncogenes, it does not necessarily follow that the Fos-Jun protein complex is mediating the effects of GHs. Schwartz and colleagues[60] employed an elegant approach to examine this question. Differentiation of some but not all 3T3 cell lines can be induced by GH. However, GH induces expression of c-*fos* and c-*jun* irrespective of whether differentiation is also induced (as in 3T3-F442A cells) or not (as is the case with NIH-3T3 cells). Thus, although GH induces c-*fos* and c-*jun*, this does not appear to be sufficient to evoke its full biological response.

VIII. POLYAMINES AND GH ACTION

Polyamines may be part of the signal transduction mechanism for at least some of the actions of GH. The polyamines putrescine, spermidine, and spermine are sequentially synthesized from ornithine by ornithine decarboxylase (ODC), *S*-adenosylmethionine decarboxylase (SAMD), spermidine synthetase, and SAMD spermidine synthetase, respectively.

In vivo studies have demonstrated that GH can increase greatly the activity of ornithine decarboxylase in the liver.[61,62] This effect is dose dependent and of a greater magnitude in young than adult rats.[61] The effect of GH on ornithine decarboxylase can be blocked by the administration of inhibitors of RNA (actinomycin) or protein (cycloheximide) synthesis.[61] This suggests that GH is increasing ornithine decarboxylase activity by stimulation of transcription and translation. However, it is also possible to envisage GH as activating ornithine decarboxylase, if this enzyme were extremely short lived[62] and with a presumed high rate of *de novo* synthesis. In another system with accumulated T lymphocytes, ornithine decarboxylase is activated during phosphatidylinositol breakdown.[63] In studies with sheep adipose tissue *in vitro*, GH also increases adenosylmethionine decarboxylase.[64] Moreover, whereas GH alone does not influence ornithine decarboxylase activity, GH does overcome the reduction in ornithine decarboxylase activity accompanying dexamethasone treatment.[64]

The ability of GH to change the activities of enzymes involved in polyamine synthesis does not *a priori* indicate that the anticipated changes in polyamines mediate the effects of GH. Pharmacological studies provide support for the notion that polyamines are obligatory for (although not necessarily mediating) at least some of the effects of GH. Difluoromethylornithine, an inhibitor of ornithine decarboxylase, prevents the insulin-like effect of GH on chicken adipose tissue explants (GH reducing lipolysis) in the presence of glucagon.[30] This insulin-like effect of GH is restored by spermidine.[30] Inhibition of polyamine synthesis does not appear to influence the lipolytic effect on GH, at least in chicken adipose tissue.[30] This would suggest that the mechanisms by which GH exerts different effects are not identical.

There is also evidence that polyamines may be involved in the mechanism of action of prolactin. Not only can prolactin increase ornthine decarboxylase in mouse mammary tissue *in vitro* but pharmacological inhibitors of polyamine synthesis can suppress the effects of prolactin.[65,66]

REFERENCES

1. **Foster, C. M., Shafer, J. A., Rozsa, F. W., et al.,** *Biochemistry*, 27, 326, 1988.
2. **Stred, S. E., Stubbart, J. R., Argetsinger, L. S., et al.,** *Endocrinology*, 127, 2506, 1990.
3. **Stubbart, J. R., Barton, D. F., Tai, P.-K. K., et al.,** *Endocrinology*, 129, 1659, 1991.
4. **Carter-Su, C., Stubbart, J. R., Wang, X., et al.,** *J. Biol. Chem.*, 264, 18654, 1989.
5. **Stred, S. E., Stubbart, J. R., and Argetsinger, L. S.,** *Endocrinology*, 130, 1626, 1992.
6. **Leung, D. W., Spenser, S. A., Cachianes, G., et al.,** *Nature (Lond.)*, 330, 537, 1987.
7. **Wang, X., Uhler, M. D., Billestrup, N., et al.,** *J. Biol. Chem.*, 267, 17390, 1992.
8. **Wang, X., Moller, C., Norstedt, G., et al.,** *J. Biol. Chem.*, 268, 3573, 1993.
9. **Donner, J., Eriksson, H., and Belfrage, P.,** *FEBS Lett.*, 208, 269, 1986.
10. **Anderson, N. G.,** *Biochem. J.*, 284, 649, 1990.
11. **Wilks, A. F., Harpur, A. G., Kurban, R. R., et al.,** *Mol. Cell. Biol.*, 11, 2057, 1991.
12. **Firmbach-Kraft, I., Byers, M., Shows, T., et al.,** *Oncogene*, 5, 1329, 1990.
13. **Harpur, A. G., Andres, A.-C., Ziemiecki, A., et al.,** *Oncogene*, 7, 1347, 1992.
14. **Bazan, J. F.,** *Proc. Natl. Acad. Sci. U.S.A.*, 87, 6934, 1990.
15. **Bazan, J. F.,** *Immunol. Today*, 10, 350, 1991.
16. **De Vos, A. M., Ultsch, M., and Kossiakoff, A. A.,** *Science*, 255, 306, 1992.
17. **Witthuhn, B. A., Quelle, F. W., Silvennoinen, O., et al.,** *Cell*, 74, 227, 1993.
18. **Silvennoinen, O., Witthuhn, B. A., Quelle, F. W., et al.,** *Proc. Natl. Acad. Sci. U.S.A.*, 90, 8429, 1993.
19. **Fu, X.-Y.,** *Cell*, 70, 323, 1992.
20. **Argetsinger, L. S., Campbell, G. S., Yang, X., et al.,** *Cell*, 74, 237, 1993.
21. **Ishizaka-Ikeda, E., Fukunaga, R., Wood, W. I., et al.,** *Proc. Natl. Acad. Sci. U.S.A.*, 90, 123, 1993.
22. **Silva, C. M., Weber, M. J., and Thorner, M. O.,** *Endocrinology*, 132, 101, 1993.
23. **Cunningham, B. C. and Wells, J .A.,** *Proc. Natl. Acad. Sci. U.S.A.*, 88, 3407, 1991.
24. **Crew, C. M., Alessandrini, A. A., and Erikson, R. L.,** *Proc. Natl. Acad. Sci. U.S.A.*, 88, 8845, 1991.
25. **Winston, L. A. and Bertics, P. J.,** *J. Biol. Chem.*, 267, 4747, 1992.
26. **Campbell, G. S., Pang, L., Miyasaka, T., et al.,** *J. Biol. Chem.*, 267, 6074, 1992.
27. **Anderson, N. G.,** *Biochem. J.*, 284, 649, 1992.
28. **Moler, C., Hansson, A., Emberg, B., et al.,** *J. Biol. Chem.*, 267, 23403, 1992.
29. **Anderson, N. G.,** *Biochem. Biophys. Res. Commun.*, 193, 284, 1993.
30. **Campbell, R. M. and Scanes, C. G.,** *Proc. Soc. Exp. Biol. Med.*, 188, 177, 1988.
31. **Schwartz, Y. and Goodman, H. M.,** *Endocrinology*, 126, 170, 1990.
32. **Schwartz, Y., Goodman, H. M., and Yamaguchi, H.,** *Proc. Natl. Acad. Sci. U.S.A.*, 88, 6790, 1991.
33. **Schwartz, Y., Yamaguchi, H. and Goodman, H. M.,** *Endocrinology*, 131, 772, 1992.
34. **Murphy, P. R., DiMattia, G. E., and Friesen, H. G.,** *Endocrinology*, 122, 2476, 1988.
35. **Slootweg, M. C., vanGenesen, S. T., Otte, A. P., et al.,** *J. Mol. Endocrinol.*, 4, 265, 1990.
36. **Doglio, A., Dani, C., Grimaldi, P., and Ailhaud, G.,** *Proc. Natl. Acad. Sci. U.S.A.*, 86, 1148, 1989.
37. **Rogers, S. A. and Hammerman, M. R.,** *Proc. Natl. Acad. Sci. U.S.A.*, 86, 6363, 1989.
38. **Johnson, R. M., Napier, M. A., Cronin, M. J., et al.,** *Endocrinology*, 127, 2099, 1990.
39. **Eriksson, H., Sundler, R., and Donner,** *J. Mol. Cell. Biochem.*, 97, 181, 1990.
40. **Pradines-Figueres, A., Barcellini-Couget, S., Dani, C., et al.,** *J. Lipid Res.*, 31, 1283, 1990.
41. **Smal, J. and De Meyts, P.,** *Biochem. Biophys. Res. Commun.*, 147, 1232, 1987.
42. **Smal, J., Kathuria, S., and De Meyts, P.,** *FEBS Lett.*, 244, 465, 1989.
43. **Goodman, H. M., Tai, L.-R., and Chipkin, S. R.,** *Endocrinology*, 126, 441, 1990.
44. **Gorin, E., Tai, L.-R., Honeyman, T. W., and Goodman, H. M.,** *Endocrinology*, 126, 2973, 1990.
45. **Chou, S. Y., Kostyo, J. L., and Adamafio, N. A.,** *Endocrinology*, 126, 62, 1990.
46. **Towns, R., Kostyo, J. L., Martin, D., et al.,** *Endocrinology*, 132, 1671, 1993.
47. **Roupas, P., Chou, S. Y., Towns, R. J., et al.,** *Proc. Natl. Acad. Sci. U.S.A.*, 88, 1691, 1991.
48. **Rillema, J. A. and Koduri, P. B.,** *Proc. Soc. Exp. Biol. Med.*, 203, 424, 1993.
49. **Larsen, J. L. and Dufau, M. L.,** *Endocrinology*, 123, 438, 1988.
50. **Larsen, J. L.,** *J. Biol. Chem.*, 267, 10583, 1992.

51. **Goodman, H. M.,** *Metabolism*, 19, 849, 1970.
52. **Campbell, R. M. and Scanes, C. G.,** *Proc. Soc. Exp. Biol. Med.*, 184, 456, 1987.
53. **Goodridge, A. G.,** *Am. J. Physiol.*, 214, 902, 1968.
54. **Butcher, R. W., Ho, R. J., Meng, H. C., et al.,** *J. Biol. Chem.*, 240, 4515, 1965.
55. **Boyd, T. A., Weiser, P. B., and Fain, J. N.,** *Gen. Comp. Endocrinol.*, 26, 243, 1975.
56. **Bjorgell, P., Rosberg, S., Isaksson, O., et al.,** *Endocrinology (Baltimore)*, 15, 1154, 1984.
57. **Birnbaum, R. S. and Goodman, H. M.,** *Endocrinology*, 99, 1336, 1976.
58. **Barcellini-Couget, S., Doglio, A., and Dani, C.,** *Endocrinology (Baltimore)*, 132, 1875, 1993.
59. **Gurland, G., Ashcom, G., Cockran, B., et al.,** *Endocrinology (Baltimore)*, 127, 3187, 1990.
60. **Sumantran, V. N., Tsai, M.-L., and Schwartz, J.,** *Endocrinology (Baltimore)*, 130, 2016, 1992.
61. **Russell, D. H., Snyder, S. H., and Medina, V. J.,** *Endocrinology (Baltimore)*, 86, 1414, 1970.
62. **Martin, J. V., Wyatt, R. J., and Mendelson, W. B.,** *Life Sci.*, 44, 1891, 1989.
63. **Mustelin, T., Poso, H., Lapinjoki, S. P., et al.,** *Cell*, 49, 171, 1987.
64. **Snoswell, A. M., Finley, E., and Vernon, R. G.,** *Horm. Metab. Res.*, 22, 650, 1990.
65. **Rillema, J. A.,** *Endocrinol. Res. Commun.*, 2, 296, 1976.
66. **Oppat, C. A. and Rillema, J. A.,** *Am. J. Physiol.*, 257, 328, 1989.

Chapter 18

Growth Hormone Action: Genomic Mechanisms

S. Harvey and K. L. Hull

CONTENTS

I. INTRODUCTION

Genomic actions of growth hormone (GH), resulting in increased protein synthesis, are well established and are thought to largely reflect the activation of signal transduction cascades and second messengers with *trans*-acting roles in gene transcription. Genomic actions of GH may, however, also result from direct effects of GH within the nucleus. This possibility is supported by the presence of nuclear GH receptors and the identification of GH response elements in individual genes.

II. NUCLEAR GH RECEPTORS

Immunoreactive growth hormone receptors and GH-binding activity have been observed in nuclear fractions of the rat, human, and chicken liver,[1-4] in the chicken and rat pituitary gland,[4-6] and in the central nervous system (CNS), reproductive tract, alimentary tract, and skin.[7-10] GH-binding activity is greater in the nucleus than in microsomal fractions in the rat pituitary and liver, although less than in the cytosol.[2,5] Within the nucleus of rat pituitary cells, GH-binding protein (GHBP) immunoreactivity is most intense in heterochromatin and the nucleolus, but is also present in inner and outer nuclear membranes.[5] In contrast, hepatic GH-binding activity is most abundant in the outer nuclear membrane, constituting 60% of total nuclear binding, whereas nucleoplasm, chromatin, and inner nuclear membranes are responsible for 25, 10, and 2%, respectively.[2] The binding characteristics and antigenic determinants of the hepatic nuclear high-affinity, low-capacity binding sites are almost identical to the well-characterized cytosolic/plasma GHBP.[2] The molecular size of the hepatic chromatin-associated GHBP (68 kDa) corresponds in size to the glycosylated soluble GHBP rather than the full-length receptor; however, the size of the hepatic nuclear membrane-bound GHBP/GHR has not been determined.[3] The intense nuclear labeling in rats, using monoclonal antibody (MAb) 4.3 (which recognizes the unique GHBP tail), also suggests that most, if not all, hepatic nuclear GH-binding activity corresponds to the alternatively spliced GHBP.[3,6] Studies have demonstrated that both the full-length receptor and the binding protein are present in the nucleus of rat hepatic cells; however, only the GHR contains the intracellular residues (amino acids 255 to 454) enabling nuclear anchorage.[11] Conversely, in the chicken, pituitary nuclear GH-binding moieties corresponded in size to membrane-bound receptors rather than to the smaller plasma/cytosolic binding proteins.[4]

0-8493-8697-7/95/$0.00+$.50

III. MECHANISM OF NUCLEAR ENTRY

Although nuclear receptors in the pituitary may mediate intracrine actions of GH, endocytosis of extracellular GH is necessary for nuclear GH action in extrapituitary sites. The existence of this process is indicated by the nuclear accumulation of radiolabeled human and rat GH[12] as well as prolactin[13] and other growth factors (e.g., insulin[14] and NGF[15]). This uptake process is likely to be receptor mediated, because Schepper et al.[16] have demonstrated rapid translocation of GHR immunoreactivity from hepatic membranes to the nucleus in response to GH.[17] Moreover, unlike most hormones, GH does not dissociate from its receptor in the acidic prelysosomal endosomes,[18,19] facilitating receptor-mediated translocation within the cell. Although 75% of GH/GHR complexes are targeted for lysosomal degradation, 25% are recycled to the plasma membrane and/or delivered to intracellular sites such as the nucleus.[18,19] This mode of nuclear translocation has previously been demonstrated for EGF/EGFR complexes, which then act at the nucleus to stimulate nucleoplasmic transport and DNA synthesis.[20,21] Passage across nuclear membranes may also be facilitated by the hydrophilic tail of the GHBP, which contains four amino acids that may interact with the nuclear pore complex.[22]

IV. PUTATIVE MECHANISMS OF ACTION

Nuclear growth hormone receptors may affect transcription directly by acting as DNA-binding proteins or may be linked to a nuclear signal transduction system. The former possibility is supported by the close association between nuclear GHRs and chromatin[2] and by the identification of GH-responsive regions in the *Spi2.1* and *IGF-I* genes.[23,24] The association of the GH receptor with DNA is not, however, mediated by zinc fingers, leucine zippers, or helix-turn-helices, which permit the binding of steroid receptors to DNA, because the GH receptor lacks these consensus sequences.[25,26] The transcriptional efficiency of GH-responsive genes could therefore be altered by protein kinase cascades inititiated by protein kinase C (PKC), as a prolactin- and hGH-sensitive PKC isoform is present in whole and membrane-deprived rat hepatic nuclei[27-30] (Figure 1). This PKC isoform shares some characteristics with the well-characterized GH-sensitive PKC in adipocytes,[31] in that it is independent of phosphatidylinositol.[27,30] However, oGH was ineffective in the hepatocyte system, suggesting a lactogenic rather than a somatogenic effect.[28] c-Fos is another likely mediator of nuclear GH action, as c-Fos is involved in GH effects in adipocytes and possibly the hypothalamus[31,32] and is a well-established transcription factor.

V. TRANSCRIPTIONAL EFFECTS

The presence of GHR/BPs in the nucleus suggests that the nuclear GHR may constitute a significant component of the GH signaling system. It is now clear that GH can influence the expression of specific genes encoding receptors, growth factors, enzymes, and transcription factors (Table 1), and GH also stimulates DNA synthesis and cell proliferation.[33] Although some of these effects may be indirectly mediated by IGF-I, the *Spi2.1*, c-*fos*/c-*myc,* and *IGF-I* genes appear to respond directly to GH. Of particular interest is the GH-responsive region of the serine protease inhibitor 2.1 gene (*Spi2.1*), which is located upstream of the transcription start site (–174/–102).[24] This sequence is bound by an unidentified hepatic nuclear factor that is activated in response to GH stimulation. The GH-responsive region in the *IGF-I* gene has also been mapped,[23] although the DNA-binding factor(s) involved have not been identified.

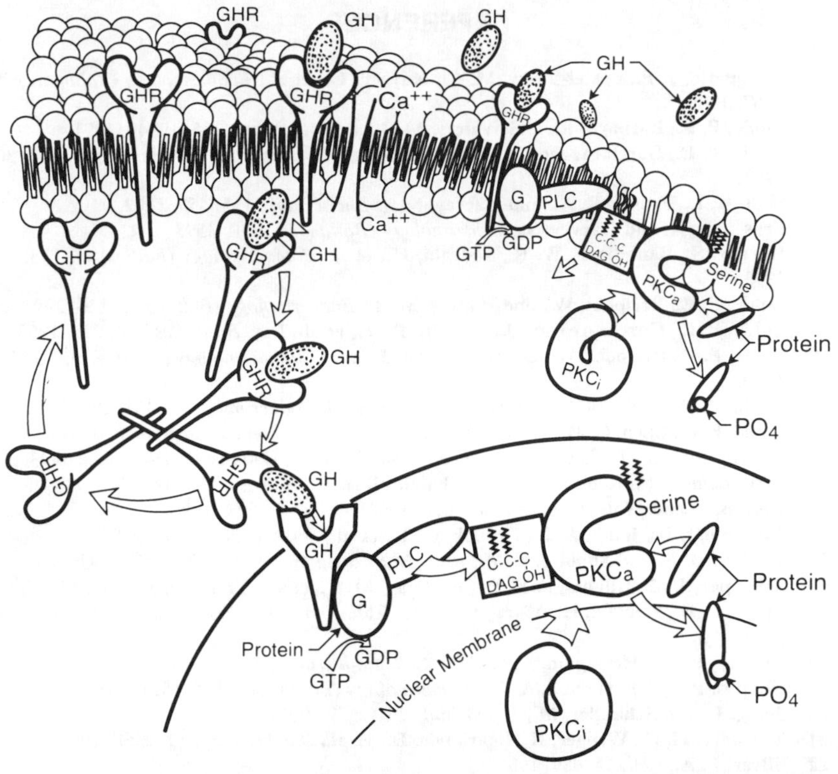

Figure 1 Schematic representation of GH receptor coupling to the activation of protein kinase C (PKC). This may occur at both the plasma and nuclear membranes. (Based on Russell.[27])

Table 1 **Pretranslational Regulation of GH-Responsive Genes**

Gene	Tissue	Ref.
IGF-I	Liver, adipose, pituitary	23,34–37
Albumin	Liver	38
HGF	Liver	39
P-450 15β-hydroxylase	Liver	40
α_2-microglobulin	Liver	41
EGFR	Liver	42
GHR	Hypothalamus, liver	43
Spi2.1	Liver	24
SRIF	Hypothalamus	44
GHRH	Hypothalamus	45
Myosin heavy chain	Muscle	46
c-*myc*	Liver	34
c-*fos*	Fat, hypothalamus	31,32
Prolactin receptor	Liver	47

Note: IGF-I, Insulin-like growth factor I; HGF, hepatocyte growth factor; EGFR, epidermal growth factor receptor; GHR, growth hormone receptor; Spi2.1, serine protease inhibitor 2.1; SRIF, somatostatin; GHRH, growth hormone-releasing hormone.

REFERENCES

1. **Hocquette, J. F., Postel-Vinay, M. C., Kayser, C., et al.,** *Endocrinology (Baltimore)*, 125, 2167, 1989.
2. **Lobie, P. E., Barnard, R., and Waters, M. J.,** *J. Biol. Chem.*, 265, 19947, 1991.
3. **Lobie, P. E., Garcia-Aragon, J., Wang, B. S., et al.,** *Endocrinology (Baltimore)*, 130, 3057, 1992.
4. **Hull, K. L., Fraser, R. A., and Harvey, S.,** *J. Endocrinol.*, 135, 459, 1992.
5. **Fraser, R. A. and Harvey, S.,** *Endocrinology (Baltimore)*, 130, 3593, 1992.
6. **Harvey, S., Baumbach, W. R., Sadeghi, H., et al.,** *Endocrinology (Baltimore)*, 133, 1125, 1993.
7. **Lobie, P. E., Breipohl, W., and Waters, M. J.,** *Endocrinology (Baltimore)*, 126, 299, 1990.
8. **Lobie, P. E., Garcia-Aragon, J., Lincoln, D. T., et al.,** *Dev. Brain Res.*, 74, 225, 1993.
9. **Lobie, P. E., Breipohl, W., Garcia-Aragon, J., et al.,** *Endocrinology (Baltimore)*, 126, 2214, 1990.
10. **Lobie, P. E., Breipohl, W., Lincoln, D. T., et al.,** *J. Endocrinol.*, 126, 467, 1990.
11. **Lobie, P. E., Chen, C. M., Waters, M. J., et al.,** *Workshop on the Superfamily of Receptors for Growth Hormone, Prolactin, Erythropoietin and Cytokines,* Haifa Israel, November, 1993, p. C8.
12. **Bonifacino, J. S., Rougin, L. P., and Paladini, A. C.,** *Biochem. J.*, 214, 121, 1983.
13. **Giss, B. J. and Walker, A. M.,** *Mol. Cell. Endocrinol.*, 42, 259, 1985.
14. **Goldfine, I. D., Jones, A. L., Hradek, G. T., et al.,** *Science*, 202, 760, 1978.
15. **Yanker, B. A. and Shooter, E. M.,** *Proc. Natl. Acad. Sci. U.S.A.*, 76, 1269, 1979.
16. **Schepper, J. M., Hughes, E. F., Postel-Vinay, M. C., et al.,** *J. Biol. Chem.*, 259, 12945, 1984.
17. **Lobie, P. E., Mertani, H., Morel, G., et al.,** *76th Ann. Meet. Endocrine Soc.,* Anaheim, CA, p. 432 (Abst. 927).
18. **Roupas, P. and Herington, A. C.,** *Mol. Cell. Endocrinol.*, 57, 93, 1988.
19. **Roupas, P. and Herington, A. C.,** *Endocrinology (Baltimore)*, 121, 1521, 1987.
20. **Jiang, L. and Schindler, M.,** *J. Cell Biol.*, 110, 559, 1990.
21. **Schroder, H. C., Wenger, R., Ugarkovic, D., et al.,** *Biochemistry*, 29, 2368, 1990.
22. **Silver, P. A.,** *Cell*, 64, 489, 1991.
23. **Saunders, J. C., Dickson, M. C., Pell, J. M., et al.,** *J. Mol. Endocrinol.*, 7, 233, 1991.
24. **Yoon, J., Berry, S. A., Seelig, S., et al.,** *J. Biol. Chem.*, 266, 19947, 1990.
25. **Schwabe, J. W. R. and Rhodes, D.,** *Trends Biochem. Sci.*, 16, 291, 1991.
26. **Leung, D. W., Spencer, S. A., Cachianes, G., et al.,** *Nature (Lond.)*, 330, 537, 1987.
27. **Russell, D. H.,** *TIPS*, 10, 40, 1989.
28. **Buckley, A. R., Crowe, P. D., and Russell, D. H.,** *Proc. Natl. Acad. Sci. U.S.A.*, 85, 8649, 1988.
29. **Capitani, S., Girard, P. R., Mazzei, G. J., et al.,** *Biochem. Biophys. Res. Commun.*, 142, 367, 1987.
30. **Masmoudi, A., Labourdette, G., Mersel, M., et al.,** *J. Biol. Chem.*, 264, 1172, 1989.
31. **Doglio, A., Dani, C., Grimaldi, P., et al.,** *Proc. Natl. Acad. Sci. U.S.A.*, 86, 1148, 1989.
32. **Minami, S., Kamegai, J., Sugihara, H., et al.,** *Endocrinology (Baltimore)*, 131, 247, 1992.
33. **Isaakson, O. G. P., Eden, S., and Jansson, J. O.,** *Annu. Rev. Physiol.*, 47, 483, 1985.
34. **Murphy, L. J., Bell, G. I., and Friesen, H. G.,** *Endocrinology (Baltimore)*, 120, 1806, 1987.
35. **Fagin, J. A., Brown, A., and Melmed, S.,** *Endocrinology (Baltimore)*, 122, 2204, 1988.
36. **Roberts, C. T., Brown, A. L., Graham, D. E., et al.,** *J. Biol. Chem.*, 261, 10025, 1986.
37. **Wolverton, C. K., Azain, M. J., Duffy, J. Y., et al.,** *Am. J. Physiol.*, 263, E637, 1992.
38. **Keller, G. H. and Taylor, J. M.,** *J. Biol. Chem.*, 254, 276, 1979.
39. **Ekberg, S., Luther, M., Nakamura, T., et al.,** *J. Endocrinol.*, 135, 59, 1992.
40. **MacGeouch, C., Morgan, E. T., Cordell, B., et al.,** *Biochem. Biophys. Res. Commun.*, 143, 782, 1987.
41. **Roy, A. K., Chaterjee, B., Demyan, W. F., et al.,** *J. Biol. Chem.*, 257, 7834, 1982.
42. **Johansson, S., Husman, B., Morstedt, G., et al.,** *J. Endocrinol.*, 3, 113, 1989.
43. **Mullis, P. E., Lund, T., Patel, M. S., et al.,** *Mol. Cell Endocrinol.*, 76, 125, 1991.
44. **Wood, T. L., Berelowitz, M., Gelato, M. C., et al.,** *Neuroendocrinology*, 53, 298, 1991.
45. **Bloch, B., Gaillard, R. C., Brazeau, P., et al.,** *Regul. Peptides*, 8, 21, 1984.
46. **Fong, Y., Rosenbaum, M., and Tracy, K. J.,** *Proc. Natl. Acad. Sci. U.S.A.*, 86, 3371, 1989.
47. **Robertson, J. A., Haldosen, L., Wood, T. J. J., et al.,** *Mol. Endocrinol.*, 4, 1235, 1990.

Chapter 19

Growth Hormone Action: Growth

C. G. Scanes and W. H. Daughaday

CONTENTS

I. INTRODUCTION

The presence of a principle in the anterior pituitary gland that maintains and/or stimulates animal growth has been recognized since the 1920s.[1–3] This activity was determined to be localized in a single hormone, growth hormone (GH), by the combination of protein chemistry and biological assays.[4–6]

Effects of GH on growth in agricultural animals are considered in detail in Chapter 29, and effects in humans are covered in Chapter 30. Thus, these areas are not discussed in detail in this chapter.

0-8493-8697-7/95/$0.00+$.50

As a prelude to discussion of the role of GH in the control of growth, it is necessary to consider what is meant by growth. A number of growth-related parameters are employed. These include: body weight or body weight gain (this has the potential disadvantage of hepatomegaly or obesity being interpreted as growth); indices of skeletal growth, including height (human, horses), crown-rump length (pigs and sheep), tail length or tail length gain (rats and mice), or the width/thickness of the epiphyseal cartilage plate in long bones (e.g., tibia); and also nitrogen retention (indicating muscle growth):

$$\text{Nitrogen retention} = \text{net } N_2 \text{ absorbed} - N_2 \text{ in urine as g/day}$$

where net N_2 absorbed equals the intake of feed – N_2 in feces.

The most widely used biological assay for GH is the tibia test,[7] which employs the width (thickness) of the cartilage plate of tibia from hypophysectomized rats as the response to GH (Figure 1). A similar assay has also been developed using hypophysectomized mice.[8]

II. EFFECTS OF GH IN GH-DEFICIENT MAMMALS

A. INTRODUCTION

There is abundant evidence that GH is required for growth in all species examined, except guinea pigs, in at least the latter stages of growth. Growth is reduced or completely suppressed in the absence of GH (e.g., following hypophysectomy or the administration of antisera against GH or in genetic models in which GH is not secreted and restored by GH replacement therapy; for details see below).

B. HYPOPHYSECTOMIZED MAMMALS

1. Rats

During development, growth becomes progressively more GH dependent (reviewed by Gluckman et al.[9]). For instance, in rats hypophysectomized after 25 days of age, there is a complete cessation of growth irrespective of whether body weight gain or tail growth is considered.[9–13] Moreover, growth is restored by GH administration to the hypophysectomized rat.[10–13] In neonatal rats (6 days old) growth rate and circulating concentrations of insulin-like growth factor I (IGF-I) (a mediator of GH action) are decreased by approximately 50% by hypophysectomy.[12,14,15] Growth is partially restored by GH injections[13,15] (also see Table 1). Full restoration of skeletal growth (tail growth) requires both GH and thyroxine.[14]

In the fetal rat, however, neither hypophysectomy nor GH therapy appears to influence growth.[15–17] The growth of fetal rat tissues transplanted under the kidney capsule of young rats nevertheless appears to be GH dependent.[18,19] The growth rate of transplanted paws or small intestines from fetal rats is reduced in hypophysectomized hosts but increased in hypophysectomized hosts receiving GH replacement therapy.[16,17] This is likely to be due to GH increasing IGF-I in the host as injections of IGF-I to host hypophysectomized rats increases the growth rate of transplanted fetal tissues.[20] It might be noted that GH administration to pregnant rats reduces the weight of both the placenta and fetuses.[21]

In the postnatal rat, GH exerts a major role in the control of growth. Not only does hypophysectomy completely suppress growth and GH restore growth but there are qualitative changes following manipulation of GH availability. For instance, with muscles, hypophysectomy has been found to reduce type 2 but not type 1 muscle fibers while inducing the so-called hypophysectomized-induced fibers.[22,23] The administration of GH to hypophysectomized rats eliminates the hypophysectomized-induced muscle fibers and increases both type 1 and type 2 muscle fibers.[22,23] The cessation of GH availability in hypophysectomized rats leads to marked changes in the growth plate of the long bone,

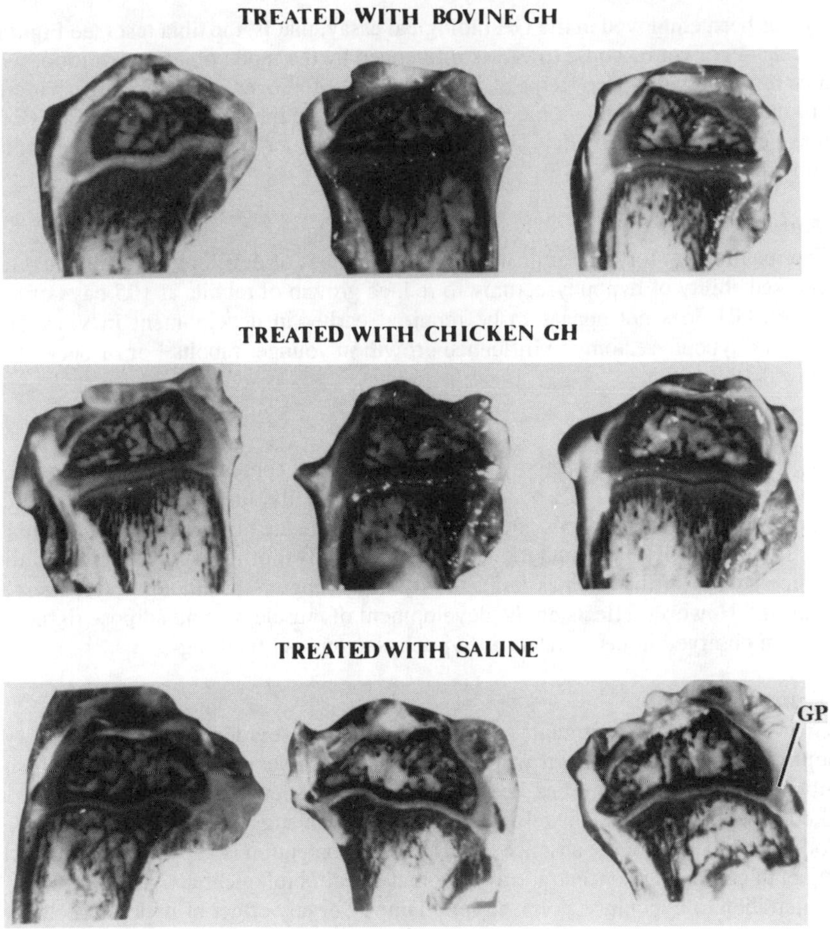

Figure 1 The effect of GH on tibial growth plate (GP) width in hypophysectomized rats receiving bovine or chicken GH in comparison with controls receiving the 0.9% NaCl vehicle. (From Harvey, S., Purification, radioimmunoassay and physiology of chicken growth hormone, Ph.D. thesis, University of Leeds.)

Table 1 **Effect of Hypophysectomy (at Day 6) and Replacement Therapy (7 to 15 Days of Age) on Growth in Young Rats**

	Growth+	
	Body Weight Gain	**Tail Length Gain**
Hypophysectomy + vehicle	100%[a]	100%[a]
Hypophysectomy + GH	168%[a]	121%[b]
Hypophysectomy + prolactin	114%[a]	110%[ab]
Sham operated	179%[b]	161%[c]

[+] Data expressed as a percentage of hypophysectomy + vehicle.

[a,b,c] Values with different superscript differ $p < 0.05$.

Source: Data from Glasscock et al.[13]

which has been employed in the GH biological assay, that is, the tibia test (see Figure 1). The skeletal growth response to GH is influenced by the mode of administration, with a greater response with more frequent injections.[24,25] Moreover, if the pulsatile pattern of GH in the circulation of intact rats[26] is mimicked by the infusion of pulses of GH[27] then both skeletal and body growth is stimulated to a far greater extent than with the continuous infusion of GH.[27]

2. Rabbits

The requirement for GH for growth in the latter part of development is attested to by the marked ability of hypophysectomy to reduce growth of rabbits at 105 days of age.[28] However, GH does not appear to be required earlier in development in view of the inability of hypophysectomy to influence growth in younger rabbits[28] or of decapitation to affect fetal growth.[16]

3. Pigs

In growing pigs, hypophysectomy reduces but does not completely suppress growth.[29–31] Replacement therapy with GH overcomes, albeit partially, the reduction in growth.[30,31] The growth of the fetal pig does not require GH, despite the high endogenous circulating concentrations of GH.[32] Decapitation, and hence removal of the pituitary gland, fails to influence growth of the fetus as indicated by body length,[33,34] although body weight is decreased.[35] However, effects on the development of muscle[36–38] and adipose tissue[35,39,40] have been observed in decapitated or hypophysectomized fetal pigs.

4. Ruminants

Growth of young ruminants requires GH, as is evidenced by the ability of hypophysectomy to decrease growth in young cattle (calves),[41] goats (kids),[42] and sheep (lambs).[43] Hypophysectomy has been found to reduce fetal growth in sheep[44–46] with marked retardation of bone development.[46] It has been suggested that this reflects fetal hypothyroidism and not the absence of GH.[47] This contention is supported by the lack of changes in circulating concentrations of somatomedin biological activity following GH administration to hypophysectomized fetal lambs[45] or any effect of hypophysectomy on circulating concentrations of IGF-I or IGF-II.[47] The lack of a role from GH in fetal growth is understandable in view of the lack of specific binding of ^{125}I-labeled GH in the fetal liver[48,49] (despite the expression of GHR mRNA[49]). An effect of GH on adipose development, as in the fetal pig, is considered possible in view of changes in adipose development following hypophysectomy.[50]

5. Other Species

In all other mammals examined with the exception of the guinea pig, hypophysectomy reduces growth and GH restores, or partially restores, growth (e.g., in monkey[51] and hamster[52]). Unlike the situation in other mammals, GH appears not to be required for growth, or capable of stimulating growth, in guinea pigs. Evidence for this unusual situation comes from the inability of hypophysectomy or GH replacement therapy to influence growth rate.[51,53,54] GH is, however, present in the guinea pig[51,55] and thus is biologically active in rat bioassays.[51] Moreover, secretion of GH in the guinea pig can be stimulated by physiological GH secretagogues.[55] Furthermore, although guinea pigs are GH-resistant, GH receptors are present in guinea pig tissue,[55a] implying that post-receptor defects are responsible for GH resistance in this species.

In summary, most postnatal growth appears to be GH dependent in almost all mammals examined.

Table 2 **Effect of GH or Prolactin on Growth and Circulating "Somatomedin" Biological Activity in Dwarf Mice**

	Body Weight Gain[a]	Somatomedin Biological Activity[a]
Vehicle	100%	100%
GH		
5 μg/day	354%	168%
20 μg/day	508%	168%[b]
Prolactin		
5 μg/day	315%	169%
20 μg/day	398%	180%

[a] Data expressed as a percentage of vehicle-treated dwarf mice.

[b] Administration of 15 μg of GH per day increases serum concentrations of IGF-I as determined by RIA by approximately eightfold (Bates and Pell[59]).

Source: Data from Wallis and Dew,[56] Holder and Wallis,[57] and Holder et al.[58]

C. GENETIC DWARF MODELS

Considerable progress has been made in our understanding of the physiology of GH by using genetic dwarf models, predominantly dwarf rodents (mice and rats). Three genetic mouse models exist, namely the Snell (*dw*), Ames (*df*), and Little (*lit*) mice. The Snell and Little dwarf mice have been used extensively in studying the role of GH in growth[56–60] (see Table 2) but the Snell dwarf mouse,[61] like the Ames dwarf mouse, exhibits panhypopituitarism (lacking GH, prolactin, and thyrotropin).[62,63]

The administration of GH to Snell dwarf mice increases growth rate and circulating concentrations of IGF-I (Table 2) compared to control lines. Moreover, the Snell mouse shows marked differences in the tibial epiphysis (or growth plate), with reductions in the size of the ossification center, and metaphysis, and in the number of capillaries in the metaphysis but an increase in the size of the resting zone.[64] Administration of GH increases the size of the ossification center and the metaphysis, and the number of capillaries in the metaphysis.[65]

Protein metabolism in Snell dwarf mice is influenced by GH administration. Not only does GH increase the weight of skeletal muscle but also the rate of net accretion of protein in muscle. This is due to GH stimulating both the rates of protein synthesis (muscle and whole body) and degradation (muscle only), the latter albeit to a less extent (see Table 3).[59,60] This is similar to the situation with GH-deficient children receiving GH therapy,[66] in which protein metabolism was examined prior to and following hGH administration, using changes in the metabolism of the nonradioactive isotope [^{15}N]glycine as the indicator of protein metabolism. Growth hormone therapy markedly improved protein balance with an increase in whole body rate of protein synthesis, in protein degradation, and in protein accretion/gain[66] (see Table 3).

The Little (*lit*) mouse has isolated GH deficiency[67,68] with considerably reduced circulating somatomedin biological activity[69] and circulating concentrations of IGF-I[70] and insulin-like growth factor binding protein (IGF-BP).[70] Despite the obvious advantages of this dwarf model, it has not been employed as much as the Snell dwarf mouse. A biological assay depending on the growth response of the Little dwarf mouse to GH has been proposed.[71]

Table 3 **Effect of GH on Protein Metabolism in Snell Dwarf Mice and GH-Deficient Children**

Treatment	Protein Synthesis[a]	Protein Degradation[a] or Breakdown	Protein Accretion/ Gain[a]
	Whole Body		
Mice			
+ Vehicle	21.3	19.8	1.45
+ GH	24.3	18.1	6.20
Increase with GH or Δ GH	+3.0[b]	−1.7	+4.75[b]
Children			
Pretreatment	2.4	1.8	0.6
+ GH	4.2	3.1	1.1
ΔGH	+1.8[b]	+1.3[b]	+0.5[b]
	Skeletal Muscle		
Mice			
+ Vehicle	6.72	6.44	0.28
+ GH	9.28	7.67	1.60
ΔGH	+2.56[b]	+1.13[b]	+1.32[b]

[a] Data for mice[59] are expressed as a percentage per day whereas those from children are expressed as grams per kilogram per day.[66]

[b] Difference with GH, $p < 0.05$.

Genetically, GH deficient models have also been described for the rat (e.g., see Refs. 72 and 73) in which GH deficiency is thought to be due to a reduction by >90% in the number of GH-producing cells.[74] Although marked effects of GH on growth[27,73] or IGF-I[75] on growth are evident, this model has not been utilized to its potential.

The above-described rodent dwarf lines are used as biological models for human GH deficiency (see Chapter 30). In addition, a clinical situation, Laron dwarfism,[76,77] exists in which circulating concentrations of GH are normal. This is thought to be due to a defect in the GH receptor (GHR). The circulating concentrations of the GH-binding protein are low or absent[78,79] due to *GHR* gene deletions, exon deletions, or point mutations in the *GHR* gene.[80] A biological model for Laron's dwarfism is the sex-linked dwarf chicken. As in Laron's dwarfism, the sex-linked dwarf chicken shows normal circulating levels of GH[81] but greatly reduced plasma concentrations of IGF-I.[82] Moreover, the GHR is absent or defective as evidenced by low or nonexistent binding of ^{125}I-labeled GH to membrane preparations from sex-linked dwarf chickens[83,84] and the abnormal pattern of expression of GHR mRNA.[85,86] Other models for Laron dwarfism undoubtedly exist. For instance, there are dwarf lines of both pigs and dogs with normal GH concentrations but deficiencies in IGF-I.[87–89] However, there is little evidence that the GHR is missing in either dwarf dogs or pigs although the level of expression may be reduced in the micropig.[89]

In summary, in a variety of genetic models, the absence of GH reduces growth whereas exogenous GH stimulates growth.

D. GH IMMUNIZATION

Administration of polyclonal antisera against GH has been demonstrated to decrease growth in a number of species including rats,[90–95] mice,[96] and chickens.[97] In rodent models, growth (as indicated by body weight and tail length) is reduced within 3 weeks of anti-GH administration (e.g., see Ref. 95). This reduction in growth is accompanied by marked decreases in circulating concentrations of IGF-I[93,94] (which is to be expected as IGF-I mediates the growth stimulating effects of GH). It might be noted that in mice, antiserum to prolactin also depresses growth.[97] This might suggest that prolactin may also have a role in the control of growth and/or that passive immunization may have nonspecific effects.

While the administration of polyclonal GH antisera consistently depresses growth, there are marked differences with studies employing antisera either raised against specific fragments (epitopes) of GH or some monoclonal antibodies recognizing specific epitopes of GH. For instance, monoclonal antibodies to GH have been found to enhance the growth response to GH in hypophysectomized rats.[98] Similarly, monoclonal antibodies to GH have been found to magnify the increase in circulating concentrations of IGF-I induced by GH in hypophysectomized rats[98] and intact sheep.[99] The growth-promoting effects of GH (increases in both body weight gain and incorporation of [^{35}S]sulfate in cartilage) in dwarf mice are also accentuated by the administration of either antisera raised against GH fragment 35–53 or a specific monoclonal antibody against GH.[100] Some monoclonal antibodies to GH also enhanced the anabolic response to GH (muscle protein synthesis) in dwarf mice.[101] These reports are consistent with the specific antibodies either influencing the three-dimensional structure of GH to facilitate receptor interaction, or sterically impeding GH binding to the circulating GH-binding protein (GHBP), and hence making more GH available at the GHR. In the case of the latter, there is now direct evidence that a monoclonal antibody, which enhances growth, decreases the ability of GH to bind to GHBP.[102]

E. PROLACTIN AND GROWTH

In view of the structural similarities between prolactin and GH (see Chapter 1), it might be questioned whether prolactin can stimulate growth, or whether prolactin replacement therapy can substitute for GH in restoring growth in GH-deficient animals. Prolactin appears to be capable of enhancing growth in a few, unusual and perhaps not representative, mammalian models (and also in some lower vertebrates [see Chapter 2]).

The administration of either prolactin or GH to dwarf mice will stimulate growth and elevate circulating concentrations of bioassayable somatomedins (reflecting IGF-I and IGF-II concentrations and availability of free IGF-I and -II, depending on their binding to IGF-BPs) (see Table 2). Moreover, in one study, increased growth was observed in intact rats receiving prolactin or GH injections.[103] In contrast, however, Glasscock and Nicoll conducted a definitive series of studies that compared the effects of GH and prolactin (in the presence or absence of thyroxine) on growth in young hypophysectomized rats. Whereas GH stimulated both body weight gain and tail growth, no effect was observed with prolactin[14] (see Table 1). Similarly, prolactin injections do not influence growth in intact lambs.[104]

If prolactin is involved in the control of growth, depression of circulating prolactin concentrations would be expected to reduce growth rate. This is supported by the reduced growth of intact mice treated with prolactin antisera.[97] Similarly, bromocriptine (which suppresses endogenous prolactin secretion) reduces growth in rat pups[105] but not in lambs.[106] However, these depletion studies may have confounding problems related to antisera specificity or effects of bromocriptine unrelated to suppression of prolactin release.

In some early studies, prolactin was reported to increase bioassable somatomedins (Table 2), for instance, in the circulation of dwarf mice and in a liver perfusate.[107,108] However there is little evidence that prolactin stimulates release of IGF-I *in vivo* or *in vitro*. Moreover, whereas GH markedly increases hepatic IGF-I mRNA, prolactin is without effect.[109]

In summary, the available evidence does not support the contention that prolactin can substitute for GH in maintaining growth via an IGF-I mechanism. Some role for prolactin in the control of growth cannot be precluded, however.

III. EFFECTS OF GH IN INTACT ANIMALS

A. INTRODUCTION

This section considers the effect of increasing circulating concentrations of GH in normal intact animals, which should be envisaged as having sufficient GH. Effects of GH in humans and in agricultural animals are considered elsewhere (respectively, in Chapters 29 and 30).

B. RODENTS

It has long been recognized that GH can stimulate growth (body weight gain, increases in bones and tail length) in mature, plateaued rats,[13,27,103,110] that is, rats in which the growth curve has achieved a plateau. Not only is growth stimulated but also there are increases in the circulating concentrations of IGF-I.[111] Although the body weight gain in response to GH is a specific effect of GH, bovine placental lactogen (evolutionarily derived from prolactin [see Chapters 1 and 2]) appears to be equipotent with GH.[111]

The growth rate of young growing female rats can be increased considerably by GH, which concomitantly enhances circulating concentrations of IGF-I.[13] The administration of GH to young male rats may, however, have equivocal effects on growth,[23] despite increasing type I muscle fibers.[23] In old rats, GH can still stimulate growth,[112] increase circulating concentrations of IGF-I,[112] and increase the rate of protein synthesis in muscle to match that in younger rats.[113] GH similarly influences protein metabolism in young men by reducing leucine oxidation.[114]

C. TRANSGENIC ANIMALS

Transgenic animals expressing high levels of GH have been widely used in GH research (reviewed by Wanke et al.[115] and Pursel et al.[116]). This section considers only GH-transgenic rodents (see Chapter 29 for a discussion of GH-transgenic livestock).

Dramatically increased growth rates have been observed in mice carrying GH transgenes and expressing high levels of GH.[117,118] Not only is the growth rate increased (body weight, bone growth, etc.)[117–119] but also the duration of growth is markedly extended.[119] Expression of GH appears to exert a greater effect on growth (increase) and body fat (decrease) in female than in male mice.[120] Along with the increase in growth, there are concomitantly higher circulating concentrations of IGF-I in GH-transgenic mice.[121–123]

In transgenic mice expressing a mutant form of bovine GH (Glu-117→Leu, Gly-119→Arg, and Ala-122→Asp), both the growth rate and circulating concentrations of IGF-I are reduced.[123–125] This is explicable in terms of the mutant form of GH being expressed and acting as an antagonist to the endogenous GH.

An increase in circulating concentrations of IGF-I is observed in GH-transgenic pigs[126] despite the lack of a pronounced increase in growth. This is thought to be due to the pathological effects of excessive concentrations of GH. Frequently the level of GH transgene expression is high, depending on gene dosing and/or induction of the transgene metallothionein promoter of the fusion genes by heavy metals.[127,128] Glucocorticoids are reported to be able to influence GH expression independently of the metallothionein promoter in transgenic mice.[127]

D. GH-PRODUCING TUMORS

When rats are implanted with GH-producing tumors, marked effects on growth are observed. For instance, young rats implanted with GH_3 cells (producing both GH and prolactin) show a substantial increases in both growth rate[129] and liver IGF-I expression.[130] Similarly, growth is enhanced if rats are implanted with StW5 tumor cells, which produce only GH.[131]

E. CONCLUSIONS

In virtually all of the systems discussed above, GH is required for postnatal growth. In many cases, growth in normal intact animals can also be enhanced by elevating GH concentrations in the circulation. Not only does GH consistently stimulate growth, it also elevates the circulating concentrations of IGF-I. The relationship between GH, IGF-I, and growth is considered in the next section.

IV. MECHANISM BY WHICH GH INDUCES GROWTH

A. THE SOMATOMEDIN HYPOTHESIS

1. Introduction

The development of the somatomedin hypothesis is an example of how clinical research leads to major advances in basic research. This was summarized in a review[132] on the history of the somatomedin hypothesis as follows: "modest observation prompted by clinical questions have provided the clues that have led to fundamental insights into important physiologic mechanism."

In 1957, Salmon and Daughaday[120] advanced the somatomedin hypothesis. This envisioned that a peripheral factor(s) mediated the effect of GH on growth. Obviously, subsequent research allowed a refinement of our understanding of the nature of the factor(s) involved and what it does. In a review, Holley and Wass[134] summarized the original hypothesis as follows: **"GH... exerts its effects by stimulating IGF-I release from the liver which then mediates the somatogenic actions in the target tissues."**

The evidence for this somatomedin hypothesis is considered in terms of the following questions: (1) What is the identity of somatomedin(s) (Section IV.A.2)? (2) Does GH increase circulating concentrations of IGF-I and IGF-I production (Section IV.A.3)? (3) Does GH increase hepatic IGF-I synthesis (Section IV.A.4)? (4) Does IGF-I stimulate growth (Section IV.A.4)?

Undoubtedly, GH influences bone/cartilage and muscle growth by maintaining optimal circulating concentrations of IGF-I. In addition, it is now recognized that GH can also act directly on target tissues such as cartilage and perhaps also muscle. This somatogenic or anabolic effect of GH is thought to be largely mediated by GH inducing IGF-I release, which then mediates the effect of GH locally in a paracrine or autocrine manner.[135,136] Whether GH can act directly on target tissues (e.g., cartilage) is considered in Section IV.A.5. This section not only addresses evidence for direct effects of GH and their mediation by IGF-I but also the extent to which GH acts via circulating and via locally produced IGF-I. It should be stressed that some IGF-I release by the liver and by target tissues is autonomous, that is, independent of GH. The extent of this is briefly addressed in Sections IV.A.3 and IV.A.5.

Much of the IGF-I in circulation is bound to specific IGF-binding proteins (IGF-BPs), which influence the biological activity of the IGF-I, principally by inhibiting the activity of IGF-I. Synthesis and release of specific IGF-BPs can be controlled by GH. This is considered below in Section IV.B.

2. Somatomedin Identification

The first indication that the action of GH on skeletal growth is mediated, at least in part, by an intermediate factor arose from studies of isolated cartilage from hypophysectomized rats (for review see Daughaday[132]). After hypophysectomy, there is a marked decrease in the ability of cartilage to incorporate [^{35}S]sulfate into chondroitin sulfate. This defect could be corrected within 24 h by the administration of GH *in vivo*. When, however, GH was added *in vitro* to segments of costal cartilage from hypophysectomized rats, little or no stimulation of [^{35}S]sulfate occurred. In contrast, when serum from normal rats was added *in vitro* to such cartilage segments, a 100 to 200% increase in [^{35}S]sulfate uptake occurred, but the serum of hypophysectomized rats was virtually inactive.[133] This factor, initially called *sulfation factor*, was renamed somatomedin on the basis of its being the mediator for GH (somatotropin).[135] Several somatomedins were purified, including somatomedins A and C.[137]

Independent of the research on somatomedin(s), other investigators (e.g., E. R. Froesch and R. E. Humbel) focused on the insulin-like activity in plasma that was not influenced by the presence of antisera to insulin. This was referred to as nonsuppressible insulin-like activity.[138,139] The circulating concentrations of this nonsuppressible insulin-like activity were found to be GH dependent, as was the case with somatomedin activity.[140,141] Nonsuppressible insulin-like activity in human serum was determined, following purification and biological and radioreceptor assay, to be due to two peptides, respectively insulin-like growth factor I (IGF-I) and IGF-II.[142–144] Subsequent work by many investigators established that somatomedin C and IGF-I are identical[145–147] and IGF-I is now the conventionally employed term.

Both IGF-I and IGF-II have been chemically characterized by conventional techniques of protein chemistry and/or their sequences have been deduced from the cDNA nucleotide sequence in many mammalian species (considered below). IGF-I and IGF-II show homology with each other and with pro-insulin. The three peptides make up the insulin family of peptides and are thought to have evolved from a common precursor molecule (reviewed by Hintz *et al.*[146]). An additional IGF-I like entity was purified from conditioned media from Buffalo rat liver cells.[148] This was named multiplication stimulating activity (MSA)[148–150] but MSA is now known to be rat IGF-II.

3. GH Dependence

Circulating concentrations of IGF-I are predominantly under GH control, with GH being thought to act on the liver to stimulate IGF-I release. Evidence for this comes from the depression in circulating concentrations of IGF-I in hypophysectomized animals and in dwarf rodents and the ability of GH administration to increase circulating IGF-I levels (in hypophysectomized rats,[12,14,15] dwarf rodents,[59,70,82] intact animals,[13,151] and GH-transgenic animals[121–123]). Not only does GH increase circulating concentrations of IGF-I,[21] it also stimulates IGF-I release and production from liver tissues *in vitro*.[152–155] In addition, GH increases hepatic IGF-I mRNA *in vivo*, for instance in rats,[156,157] pigs,[158] and Little mice,[149] and *in vitro*[109] (Figure 2).

It is pertinent to consider whether the liver is the major site of IGF-I production. There is considerable evidence that the liver is the major source of plasma IGF-I.[160–162] For instance, hepatectomy reduces circulating IGF activity whereas liver regeneration following partial hepatectomy restores circulating IGF activity.[163] However, increases in IGF-I production following GH treatment are also found in other tissues (see below).

4. IGF-I Stimulation of Body Growth

The somatomedin hypothesis predicted that somatomedins should be able to substitute for GH and stimulate growth. In view of the high levels of IGF-I postnatally, when GH

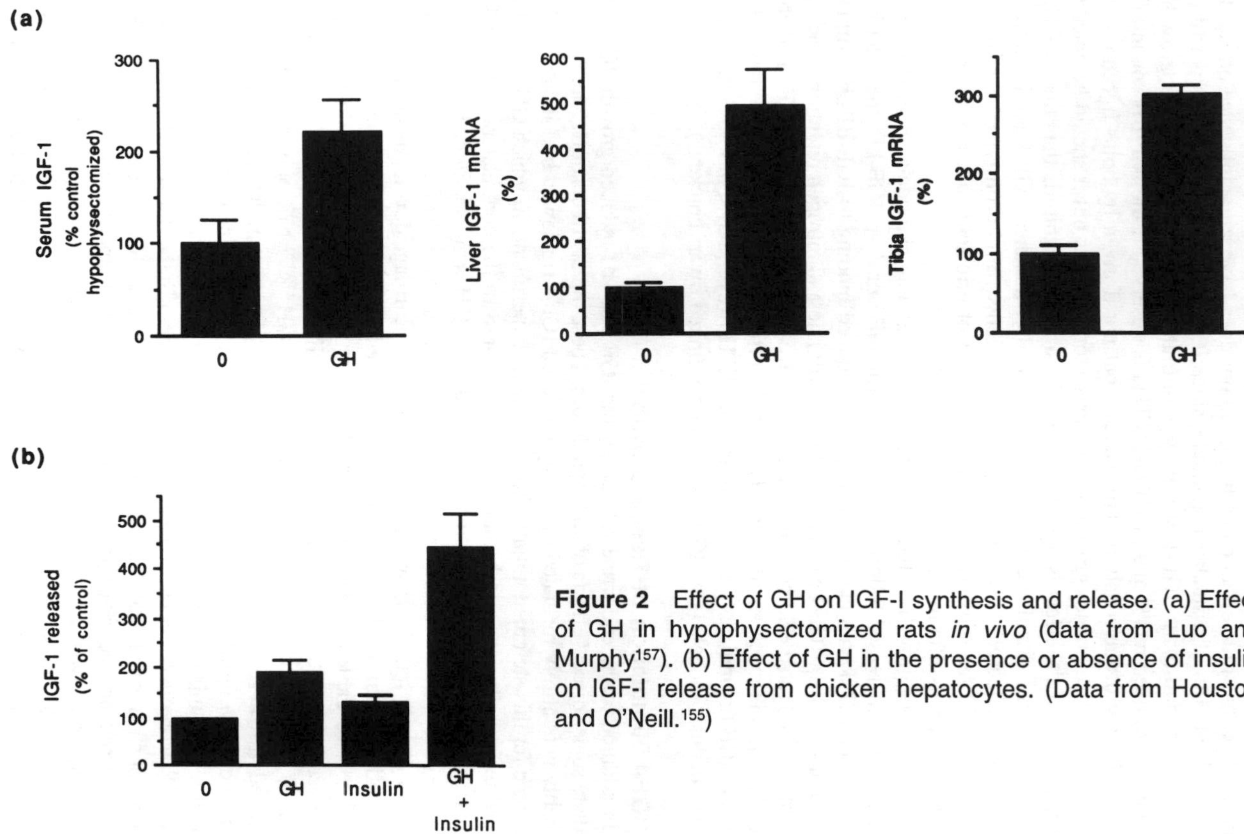

Figure 2 Effect of GH on IGF-I synthesis and release. (a) Effect of GH in hypophysectomized rats *in vivo* (data from Luo and Murphy[157]). (b) Effect of GH in the presence or absence of insulin on IGF-I release from chicken hepatocytes. (Data from Houston and O'Neill.[155])

stimulates growth and the ability of GH to stimulate IGF-I production is high (see above for details), it was appropriate to examine whether IGF-I can stimulate growth. Purified human IGF-I, administered as a constant infusion subcutanously, has been shown to mimic the effects of GH in increasing growth (body weight gain, tibia epiphyseal width and [^{3}H]thymidine incorporation into cartilage) in hypophysectomized rodents.[164,165] In contrast, human IGF-II has little effect on body growth but does have some influence on cartilage growth, albeit both of a smaller magnitude, since IGF-II is considerably less potent than IGF-I.[165] Recombinant IGF-I similarly stimulates body and cartilage growth in hypophysectomized rats[166] and dwarf rats.[75] This is observed if the IGF-I is continuously infused either subcutaneously or intravenously but not if given by bolus injections.[166] In summary, IGF-I stimulates growth in hypophysectomized rats but it should be noted that the magnitude of the effect is predominantly of a smaller magnitude than that of GH.

Growth is also stimulated by IGF-I in humans. For instance, IGF-I administration is reported to be effective in stimulating growth in patients with Laron-type dwarfism.[167,168] Moreover, IGF-I reduces circulating concentrations of uric acid and creatinine.[169] Arguably this reflects shifts in nitrogen metabolism.[169]

Stimulation of growth by IGF-I is observed in other animal models. In gut-resectioned rats, growth is increased following treatment with IGF-I or des$_{1-3}$IGF-I,[170] the latter being an analog of IGF-I with elevated potency and reduced binding to IGF-BPs.[171] Similarly, either IGF-I or des$_{1-3}$IGF-I can increase the growth rate and nitrogen retention in partially nephrectomized rats.[172] Some stimulation of growth can also be evoked by IGF-I in diabetic rats,[173] but this is much smaller than the anabolic effect of insulin in diabetic rats.[173,174] In addition, IGF-I stimulates growth in dwarf mice.[175] However, IGF-I has failed to influence growth in other models (e.g., IGF-I-deficient sex-linked dwarf chickens).[176]

5. IGF-I Stimulation of Tissue Growth

In addition to the endocrine role of hepatic IGF-I in GH-induced growth, IGF-I may mediate some "direct" effects of all target tissues. These direct effects of GH are largely but probably not entirely mediated by locally produced IGF-I (reviewed by Lindahl et al.[177]). Evidence for this stemmed initially from ingenious experiments in which GH was administered unilaterally into one leg and growth compared to the contralateral limb. A local stimulation of cartilage proliferation is observed when GH is infused directly into the tibial cartilage plate.[178] Similarly, infusion of GH into one limb, by way of a catheterized femoral artery or directly into the growth plate, results both in cartilage development in the growth plate and in enhanced linear growth of the infused limb compared to the contralateral limb[179,180] (see Table 4). The view that GH exerts this effect by the local release of IGF-I (reviewed in Ref. 181) is supported both by the ability of antibodies to IGF-I to block the effect of GH and of IGF-I to mimic the effect of GH.[179,180] Not only can GH evoke a localized growth response in cartilage *in vivo,* but GH also stimulates the proliferation of chondrocytes *in vitro*[182,183] (as does IGF-I). All the key components for a local GH-IGF-I response mechanism are present in growth plate cartilage, including GH and IGF-I receptors.[184,185] Moreover, GH can induce local increases both in the IGF-I mRNA[186–188] and the number of chondrocytes containing IGF-I,[189] and changes in IGF-I receptor number.[190] Not all GH effects are, however, mediated by IGF-I. It was found that local infusion of GH stimulated the *in vivo* incorporation of [^{3}H]thymidine into the germinal layer of the rat tibial epiphyseal plate, whereas IGF-I was inactive.[191] The same selective action of GH on isolated rabbit prechondrocytes was demonstrated by Lindahl and colleagues.[192] Accordingly, these workers have hypothesized that GH action "is limited to a precursor cell population 3X and when the cells have started to differentiate IGF-I promotes a limited clonal expansion of these cells." It is also possible that GH directly influences the growth of muscle and perhaps other tissues. For instance, GH has been found to increase IGF-I

Table 4 **Effect of Growth Hormone Administration into the Right Proximal Tibial Growth Plate of Hypophysectomized Rats**

Treatment		Tibial Plate Width		
Left	Right	Left	Right	Δ
Saline	Saline	155 ± 9	154 ± 14	—
Saline	Rat GH (5 μg)	160 ± 8	190 ± 8[a]	30 ± 4
Saline	Human GH (5 μg)	167 ± 19	211 ± 19[a]	44 ± 5

[a] $p < 0.05$.

Source: Data from Russell and Spencer.[179]

mRNA in the muscles of rats leg (e.g., in gastrocnemius, heart, and diaphragm),[130,193] but not in pigs,[158] and to elevate IGF-I mRNA levels in adipose tissue.[194]

These findings do not refute the central premise of the somatomedin hypothesis, that IGF-I is an essential mediator of the somatogenic actions of GH. Moreover, the experimental observations cannot be explained by an exclusive endocrine or autocrine/paracrine model of IGF-I action. In summary, GH exerts its effects on growth by hepatic production of IGF-I together with local release of IGF-I in at least some target tissues. The balance between these two models of action undoubtedly differs in various tissues and at various ages of the animal. The relative importance of these two modes is considered below.

6. Endocrine, Paracrine, and Autocrine Roles of IGF-I

A major role for locally produced IGF-I in GH-induced cartilage growth is supported by direct evidence. This includes the *in vitro* effects of GH on cartilage and the ability of unilaterally administered GH to enhance linear growth of limbs compared to the control contralateral limbs (see above). Moreoever, there is also some indirect evidence favoring a significant role of locally produced IGF-I in mediating some of the effects of GH on growth. Increases in growth in response to GH have been observed in the absence of changes in circulating concentrations of IGF-I. For instance, whereas GH administration to hypophysectomized rats equally increases growth whether given as a bolus injections (daily or four times daily) or as a continuous infusion, these treatments have markedly different effects on circulating IGF-I concentrations.[195,196] Similarly, GH increased growth in both selenium-treated or control rats without influencing circulating concentrations of IGF-I relative to the appropriate controls.[197] Moreover, GH has been observed to markedly enhance linear growth of hypophysectomized lambs without affecting circulating concentrations of IGF-I.[198]

Arguments that GH is exerting its principal effects by increasing circulating concentrations of IGF-I have been advanced.[132] These include the following: (1) IGF-I is present in the circulation at much higher concentrations than in any tissue. For instance, adult human serum contains 200 to 300 ng of IGF-I per milliliter whereas rat serum has even higher levels;[199] (2) the liver has a greater ability to synthesize IGF-I than do cartilage or muscle;[132] and (3) GH, even at very high concentrations, provokes some cartilage growth but to a much lesser degree than IGF-I.[132] Indeed, GH has little or no direct effect on cartilage uptake of radioactive sulfate.[132]

B. IGFS AND IGF-BINDING PROTEINS

1. Introduction

A brief account is presented here on the chemistry and physiology of the IGFs and the IGF-BPs, which bind both IGF-I and IGF-II. For detailed reviews the reader is referred elsewhere.[200–204]

2. Chemistry of IGF-I and IGF-II

There are two IGFs: IGF-I and IGF-II. These are both single-chain polypeptides, with marked homologies to each other and to proinsulin. On the basis of these homologies it is thought that insulin, IGF-I, and IGF-II were derived during evolution from a common ancestral gene. This gene is thought to have undergone duplication followed by point mutations.

3. IGF-Binding Proteins

It is well established that both IGF-I and IGF-II in the circulation are predominantly bound to a 150-kDa IGF-BP complex. A subunit of this complex is one of six specific IGF-BPs (reviewed by Holly and Martin[202a] and Baxter[203]). Six circulating IGF-BPs have been characterized. The amino acid sequences of these forms have been deduced from cDNA (IGF-BP-1,[205,206] IGF-BP-2,[207,208] IGF-BP-3,[209] and IGF-BP-4, -5, and -6[209a]). Four of these BPs are glycosylated (IGF-BP-3, -4, -5, and -6).

There is strong evidence that production of IGF-BP-3 from the liver is GH dependent (or IGF-I dependent, which itself is GH dependent).[202,203,209] For instance, the isolated GH-deficient *lit/lit* mouse has considerably reduced circulating IGF-BP-3 levels.[70] Moreover, in pigs, a strong positive correlation exists with levels of IGF-BP-3 and either body weight gain or circulating concentrations of IGF-I.[204] In rats, both circulating levels of IGF-BP-3 and of IGF-BP-3 mRNA in the liver are reduced following hypophysectomy and this is overcome by GH replacement therapy.[210,211] Circulating IGF-BP-3 levels are also increased in transgenic mice with high expression of IGF-I.[212] There is evidence that GH increases IGF-BP-3 indirectly via elevated IGF-I.[212,213] However, the administration of IGF-I to hypophysectomized rats fails to influence hepatic levels of IGF-BP-3 mRNA.[211]

Whereas levels of IGF-BP-1 are unaffected by GH status (e.g., see Ref. 70), there is evidence that IGF-BP-2 may be influenced by GH. Hypophysectomy depresses IGF-BP-2 in rodents[214,215] but IGF-BP-2 levels are increased in GH-deficient humans.[216,217] The expression and release of IGF-BP-1 and -2 appear to be affected by development[218] and are markedly increased by nutrient protein/energy deprivation.[219,200]

The presence of IGF-BPs with either IGF-I or IGF-II predominantly acts to reduce the activity of the IGF.[221–223] However, there is evidence that a complex of IGF-I to at least one IGF-BP enhances the activity of the growth factor.[224,225]

4. Control of IGF-I and IGF-II Expression and Release

There is abundant evidence that expression of IGF-I in the liver and other tissues can be stimulated by GH (see Section II.A above). Some IGF-I expression is, however, independent of GH.[226,227] In the chick embryo, for instance, significant expression of IGF-I is observed in the absence of circulating GH.[228–230] Evidence that this IGF-I has a physiological role comes from the reduction in circulating concentrations of IGF-I in growth-retarded chick embryos.[228] Moreover, IGF-I has been demonstrated to influence embryonic growth and development in the chick.[232–233]

Circulating concentrations of IGF-II appear to be developmentally controlled. For instance, circulating concentrations of IGF-II are high prenatally but fall at the time of birth (e.g., in rat,[234] and sheep[235]) or postnatally (in pigs[236]). There is little evidence that circulating concentrations of IGF-II are controlled by GH. In some cases, circulating concentrations of IGF-II are in fact elevated by hypophysectomy[237] but are unaffected by GH.[237] Moreover, circulating concentrations of IGF-II are unaffected by hypophysectomy in fetal lambs despite marked reductions in growth.[237] Surprisingly, if IGF-II is having a physiological role, growth retardation of mammalian fetuses is associated with elevations in circulating concentrations of IGF-II.[238] The lack of consistent effects of GH on circulating concentrations of IGF-II is not surprising if the liver is the major source of

IGF-II. Radical increases in the circulating concentrations of GH do not affect liver IGF-II mRNA levels[130] but do increase muscle IGF-II mRNA levels (e.g., in gastrocnemius and heart[130]).

5. IGF Receptors

In mammals, IGF-I and IGF-II act after binding to the IGF-I receptor. This receptor shows similarities to the insulin receptor; it binds IGF-I > IGF-II > proinsulin > insulin (reviewed by Rechler and Nissley[239]). In addition, a second receptor for IGF-II exists, this being the IGF-II/mannose 6-phosphate receptor.[239] This receptor is unlikely to be involved in many IGF-II actions but acts as a "sink" for excess IGF-II. Instead, IGF-II acts via the IGF-I receptor despite its lower potency than IGF-I.[239]

In birds, the mannose 6-phosphate receptor does not bind IGF-II[240] and hence a specific IGF-II receptor is absent.[241] Moreover, IGF-I and IGF-II are equipotent in avian biological assays.

REFERENCES

1. **Evans, H. M. and Long, J. A.,** *Proc. Natl. Acad. Sci. U.S.A.*, 8, 38, 1922.
2. **Evans, H. M. and Long, J. A.,** *Anat. Rec.*, 19, 1922.
3. **Evans, H. M. and Simpson, M. E.,** *Am. J. Physiol.*, 98, 511, 1931.
4. **Simpson, M. E., Asling, C. W., and Evans, H. M.,** *Yale J. Biol. Med.*, 23, 1, 1950.
5. **Cotes, P. M., Reid, E., Young, F. G., et al.,** *Nature (Lond)*, 164, 209, 1949.
6. **Lewis, U. J., Singh, R. N. P., Tutwiler, G. F., et al.,** *Recent Prog. Horm. Res.*, 36, 377, 1988.
7. **Greenspan, F. S., Li, C. H., Simpson, M. E., et al.,** *Endocrinology*, 45, 455, 1949.
8. **Lostroh, A. J. and Li, C. H.,** *Endocrinology*, 69, 308, 1957.
9. **Gluckman, P. D., Grumbach, M. M., and Kaplan, S. L.,** *Endocrine Rev.*, 2, 363, 1981.
10. **Simpson, M. E., Evans, H. M., and Li, C. H.,** *Growth*, 13, 151, 1949.
11. **Simpson, M. E., Asling, C., and Evans, H. M.,** *Yale J. Biol. Med.*, 23, 1, 1951.
12. **Walker, D. G., Simpson, M. E., Asling, C. W., et al.,** *Anat. Res.*, 106, 539, 1950.
13. **Groesbeck, M. D., Parlow, A. F., and Daughaday, W. H.,** *Endocrinology*, 120, 1963, 1987.
14. **Glasscock, G. L. and Nicoll, C. S.,** *Endocrinology*, 1089, 179, 1981.
15. **Glasscock, G. L., Gelber, S. E., Lamson, G., et al.,** *Endocrinology*, 127, 1792, 1990.
16. **Jost, A.,** *Recent Prog. Horm. Res.*, 22, 541, 1966.
17. **Jost, A.,** *J. Physiol. Paris*, 73, 877, 1977.
18. **Cooke, P. S., Russell, S. M., and Nicoll, C. S.,** *Endocrinology*, 112, 806, 1983.
19. **Cooke, P. S., Yonemura, C. U., Russell, S. M., et al.,** *Biol. Neonate*, 49, 211, 1986.
20. **Liu, L., Greenberg, S., Russell, S. M., et al.,** *Endocrinology*, 124, 3077, 1989.
21. **Chiang, M. H. and Nicoll, C. S.,** *Endocrinology*, 129, 2491, 1991.
22. **Ayling, C. M., Moreland, B. H., Zanelli, J. M., et al.,** *J. Endocrinol.*, 123, 429, 1989.
23. **Ayling, C. M., Zanelli, J. M., Moreland, B. M., et al.,** *Growth Regulat.*, 2, 133, 1992.
24. **Thorngren, K.-G. and Hansson, L. I.,** *Acta Endocrinol.*, 84, 497, 1977.
25. **Jansson, J.-O., Albertsson-Wikland, K., Eden, S., et al.,** *Acta Physiol. Scand.*, 114, 261, 1982.
26. **Tannenbaum, G. S. and Martin, J. B.,** *Endocrinology*, 98, 562, 1976.
27. **Skottner, A., Clark, R. G., Fryklund, L., et al.,** *Endocrinology*, 124, 2519, 1989.
28. **Vezinhet, M. A.,** *C.R. Acad. Sci. (D), Paris*, 266, 2348, 1968.
29. **Ford, J. J. and Anderson, L. L.,** *J. Endocrinol.*, 37, 347, 1967.
30. **Anderson, L. L., Feder, J., and Bohnker, C. R.,** *J. Endocrinol.*, 68, 345, 1976.
31. **Anderson, L. L., Bohnker, C. R., Parker, R. O., et al.,** *J. Anim. Sci.*, 53, 1981.
32. **Atinmo, T., Baldijao, C., Pond, W. G., et al.,** *J. Nutr.*, 106, 940, 1976.
33. **Stryker, J. L. and Dzuik, P. J.,** *J. Anim. Sci.*, 40, 282, 1975.
34. **Colebrander, B., Van Rossum-Kok, C. M. J. E., Van Straaten, H. W. M., et al.,** *Biol. Reprod.*, 20, 198, 1979.
35. **Hausman, G. J. and Thomas, G. B.,** *J. Anim. Sci.*, 58, 1540, 1984.
36. **Campion, D. R., Hausman, G. J., and Richardson, R. L.,** *Biol. Neonate*, 39, 253, 1980.
37. **Hausman, G. J., Campion, D. R., and Thomas, G. B.,** *J. Anim. Sci.*, 55, 1330, 1982.
38. **Hausman, G. J.,** *J. Anim. Sci.*, 67, 1367, 1989.

39. **Hausman, G. J., Campion, D. R., Richardson, R. L., et al.,** *J. Anim. Sci.*, 53, 1634, 1981.
40. **Hausman, G. J., Hentges, E. J., and Thomas, G. B.,** *J. Anim. Sci.*, 64, 1255, 1987.
41. **Anderson, L. L.,** *Am. J. Physiol.*, 232, E497, 1977.
42. **Tindal, J. S. and Yokoyama, A.,** *J. Endocrinol.*, 31, 45, 1964.
43. **Courot, M.,** *J. Reprod. Fert. Suppl.*, 2, 89, 1967.
44. **Liggins, G. C. and Kennedy, P. C.,** *J. Endocrinol.*, 40, 371, 1968.
45. **Barnes, R. J., Comline, R. S., and Silver, M.,** *J. Physiol.*, 264, 429, 1977.
46. **Parkes, M. J. and Hill, D. J.,** *J. Endocrinol.*, 104, 193, 1985.
47. **Mesiano, S., Young, I. R., Baxter, R. C., et al.,** *Endocrinology*, 120, 1821, 1987.
48. **Gluckman, P. D., Butler, J. H., and Elliot, T. B.,** *Endocrinology*, 112, 1607, 1983.
49. **Klempt, M., Bingham, B., Breier, B. H., et al.,** *Endocrinology*, 132, 1077, 1993.
50. **Stevens, D. and Alexander, G.,** *J. Dev. Physiol.*, 8, 139, 1986.
51. **Knobil, E. and Greep, R. O.,** *Recent Prog. Horm. Res.*, 15, 1, 1959.
52. **Li, C. H. and Yang, W. H.,** *Growth*, 39, 417, 1975.
53. **Mitchell, M. L., Guillemin, R., and Selye, H.,** *Endocrinology*, 54, 111, 1954.
54. **Clayton, B. E. and Worden, J. A.,** *J. Endocrinol.*, 20, 30, 1960.
55. **Gabrielsson, B., Fairhall, K. M., and Robinson,** *J. Endocrinol.*, 24, 371, 1990.
55a. **Harvey, S. and Fraser, R. A.,** *J. Endocrinol.*, 133, 357, 1992.
56. **Wallis, M. and Dew, J. A.,** *J. Endocrinol.*, 56, 235, 1973.
57. **Holder, A. T. and Wallis, M.,** *J. Endocrinol.*, 74, 223, 1977.
58. **Holder, A. T., Wallis, M., Biggs, P., et al.,** *J. Endocrinol.*, 85, 35, 1980.
59. **Bates, P. C. and Pell, J. M.,** *Br. J. Nutr.*, 65, 115, 1991.
60. **Bates, P. C. and Holder, A. T.,** *J. Endocrinol.*, 119, 31, 1988.
61. **Snell, G. D.,** *Proc. Natl. Acad. Sci. U.S.A.*, 15, 733, 1929.
62. **Sinha, Y. N., Salocks, C. B., and Vanderlaan, W. P.,** *Proc. Soc. Exp. Biol. Med.*, 150, 207, 1975.
63. **Slabaugh, M. B., Lieberman, M. E., Rutledge, J. J., et al.,** *Endocrinology*, 109, 1040, 1981.
64. **Smeets, T. and van Buul-Offers, S.,** *Growth*, 47, 145, 1983.
65. **Smeets, T. and van Buul-Offers, S.,** *Growth*, 47, 160, 1983.
66. **Zeisel, H. J., Willgerodt, H., Richter, I., et al.,** *Horm. Res.*, 37(Suppl. 2), 14, 1992.
67. **Beamer, W. G. and Eicher, E. M.,** *J. Endocrinol.*, 71, 37, 1976.
68. **Phillips, J. A. I., Beaker, W. G., and Bartke, A.,** *J. Endocrinol.*, 92, 405, 1982.
69. **Nissley, S. P., Knazel, R. A., and Wolff, G. L.,** *Horm. Metab. Res.*, 12, 158, 1980.
70. **Donahue, L. R. and Beamer, W. G.,** *J. Endocrinol.*, 136, 91, 1993.
71. **Bellini, M. H. and Bartolini, P.,** *Endocrinology*, 132, 2051, 1993.
72. **Charlton, H. M., Clark, R. G., Robinson, I. C. A. F., et al.,** *J. Endocrinol.*, 119, 51, 1988.
73. **Nogami, H., Watanabe, T., and Takeuchi, T.,** *Horm. Metab. Res.*, 24, 300, 1992.
74. **Kineman, R. D., Chen, T. T., and Frawley, L. S.,** *Endocrinology*, 125, 2035, 1989.
75. **Skottner, A., Clark, R. G., Fryklund, L., et al.,** *Endocrinology*, 124, 2519, 1989.
76. **Laron, Z., Pertzelan, A., and Karp, M.,** *Isr. J. Med. Sci.*, 4, 883, 1968.
77. **Laron, Z., Konadlo-Silbergeld, A., Eshet, R., et al.,** *Ann. Clin. Res.*, 12, 269, 1980.
78. **Baumann, G., Shaw, M. A., and Winter, R. J.,** *J. Clin. Endocrinol. Metab.*, 65, 814, 1987.
79. **Daughaday, W. H. and Trivedi, B.,** *Proc. Natl. Acad. Sci. U.S.A.*, 84, 4636, 1987.
80. **Godowski, P. J., Leung, D. W., Meacham, L. R., et al.,** *Proc. Natl. Acad. Sci. U.S.A.*, 86, 8083, 1989.
81. **Scanes, C. G., Marsh, J., Decuypere, E., et al.,** *J. Endocrinol.*, 97, 127, 1983.
82. **Huybrechts, L. M., King, D. B., Lauterio, T. J., et al.,** *J. Endocrinol.*, 104, 233, 1985.
83. **Leung, F. C., Styles, W. J., Rosenblum, C. I., et al.,** *Proc. Soc. Exp. Biol. Med.*, 184, 234, 1987.
84. **Vanderpooten, A., Huybrechts, L. M., Decuypere, E., et al.,** *Reprod. Nutr. Dev.*, 31, 47, 1991.
85. **Burnside, J., Liou, S. S., and Cogburn, L. A.,** *Endocrinology*, 128, 3183, 1991.
86. **Burnside, J., Liou, S. S., Zhong, C., et al.,** *Gen. Comp. Endocrinol.*, 88, 20, 1992.
87. **Eigenmann, J. E., Zanesco, S., Arnold, U., et al.,** *Acta Endocrinol.*, 105, 289, 1984.
88. **Buonomo, F. C., Lauterio, T. J., Baile, C. A., et al.,** *Domest. Anim. Endocrinol.*, 4, 23, 1987.
89. **Lauterio, T. J., Trivedi, B., Kapadia, M., et al.,** *Comp. Biochem. Physiol.*, 91A, 15, 1988.
90. **Grindeland, R. E., Smith, A. T., Evans, E. S., et al.,** *Endocrinology*, 95, 793, 1974.
91. **Duquesnoy, R. J. and Good, R. A.,** *J. Endocrinol.*, 48, 465, 1970.
92. **Gause, I., Eden, S., Jansson, J.-O., et al.,** *Endocrinology*, 112, 1559, 1983.
93. **Flint, D. J. and Gardner, M. J.,** *J. Endocrinol.*, 122, 79, 1989.
94. **Gardner, M. J. and Flint, D. J.,** *J. Endocrinol.*, 124, 381, 1990.

95. **Flint, D. J. and Gardner, M. J.,** *J. Endocrinol.*, 137, 203, 1993.
96. **Scanes, C. G., Harvey, S., and Chadwick, A.,** *Growth and Poultry Meat Production* (Eds. K. M. Boorman and B. J. Wilson). British Poultry Science, Edinburgh, 1977, p. 79.
97. **Sinha, Y. N. and Vanderlaan, W. P.,** *Endocrinology*, 110, 1871, 1982.
98. **Wallis, M., Daniels, M., Ray, K. P., et al.,** *Biochem. Biophys. Res. Commun.*, 149, 187, 1987.
99. **Pell, J. M., Johnsson, I. D., Pullar, R. A., et al.,** *J. Endocrinol.*, 120, R15, 1989.
100. **Bomford, R. and Aston, R.,** *J. Endocrinol.*, 125, 31, 1990.
101. **Bates, P. C., Aston, R., and Holder, A. T.,** *J. Endocrinol.*, 132, 369, 1992.
102. **Wang, B. S., Sadeghi, H., Fung, C., et al.,** *Mol. Cell. Endocrinol.*, 92, 161, 1993.
103. **Bates, R. E., Milkovic, G., and Garrison, M. M.,** *Endocrinology*, 74, 714, 1964.
104. **Eisemann, J. H., Bauman, D. E., Hogue, D. E., et al.,** *J. Anim. Sci.*, 59, 86, 1984.
105. **Nicoll, C.S.,** *Progress in Prolactin Physiology and Pathology* (Eds. C. Rozyn and M. Hartger). Elsevier, Amsterdam, 1978, p. 175.
106. **Johnsson, I. D., Hart, I. C., and Butler-Hogg, B. W.,** *Anim. Prod.*, 41, 207, 1985.
107. **Francis, M. J. O. and Hill, D. J.,** *Nature (Lond.)*, 255, 167, 1975.
108. **Bala, R. M., Bohnot, H. G., Carter, J. N., et al.,** *Can. J. Physiol. Pharmacol.*, 56, 984, 1978.
109. **Norstedt, G. and Moller, C.,** *J. Endocrinol.*, 115, 135, 1987.
110. **Evans, H. M., Simpson, M. E., and Li, C. H.,** *Growth*, 12, 15, 1948.
111. **Byatt, J. C., Staten, N. R., Schmuke, J. J., et al.,** *J. Endocrinol.*, 130, 11, 1991.
112. **Ullman, M., Ullman, A., Sommerland, H., et al.,** *Acta Physiol. Scand.*, 140, 521, 1990.
113. **Sonntag, W. E., Hylka, V. W., and Meites, J.,** *J. Gerontol.*, 689, 1985.
114. **Yamasheski, K. E., Campbell, J. A., Smith, K., et al.,** *Am. J. Physiol.*, 262, E261, 1992.
115. **Wanke, R., Wolf, E., Hermanns, W., et al.,** *Horm. Res.*, 37(Suppl. 3), 74, 1992.
116. **Pursel, V. G., Pinkert, C. A., Miller, K. F., et al.,** *Science*, 244, 1281, 1989.
117. **Palmiter, R. D., Brinster, R. L., Hammer, R. E., et al.,** *Nature (Lond.)*, 300, 611, 1982.
118. **Palmiter, R. D., Norstedt, F. M., Gelinas, R. E., et al.,** *Science*, 222, 809, 1983.
119. **Oberbauer, A. M., Currier, T. A., Nancurrow, C. D., et al.,** *Am. J. Physiol.*, 262, E936, 1992.
120. **Seale, T. W., Murray, J. D., and Baker, P. J.,** *J. Endocrinol.*, 132, 285, 1992.
121. **Quaife, C. J., Mathews, L. S., Pinkert, C. A., et al.,** *Endocrinology*, 124, 40, 1989.
122. **Mathews, L. S., Hammer, R. E., Brinster, R. L., et al.,** *Endocrinology*, 123, 433, 1988.
123. **Chen, W. Y., White, M. E., Wagner, T. E., et al.,** *Endocrinology*, 129, 1402, 1991.
124. **Chen, W. Y., Wright, D. C., Wagner, T. E., et al.,** *Proc. Natl. Acad. Sci. U.S.A.*, 87, 5061, 1990.
125. **Chen, W. Y., Wight, D. C., Chen, N. Y., et al.,** *J. Biol. Chem.*, 266, 2252, 1991.
126. **Miller, K. F., Bolt, D. J., Pursel, V. G., et al.,** *J. Endocrinol.*, 120, 481, 1989.
127. **Pavlakis, G. M. and Hamer, D. H.,** *Proc. Natl. Acad. Sci. U.S.A.*, 80, 397, 1983.
128. **Durnam, D. M., Hoffman, J. S., Quafe, C. J., et al.,** *Proc. Natl. Acad. Sci. U.S.A.*, 81, 1053, 1984.
129. **Turner, J. D., Novakofski, J., and Bechtel, P. J.,** *Growth*, 50, 402, 1986.
130. **Turner, J. D., Rotwein, P., Novakofski, J., et al.,** *Am. J. Physiol.*, E513, 1988.
131. **Coyne, M. D., Alpert, L. C., Harter, K. C., et al.,** *Horm. Res.*, 14, 36, 1981.
132. **Daughaday, W. H.,** *Perspect. Biol. Med.*, 32, 194, 1989.
133. **Salmon, W. D., Jr. and Daughaday, W. H.,** *J. Lab. Clin. Med.*, 49, 825, 1957.
134. **Holly, J. M. P. and Wass, J. H. H.,** *J. Endocrinol.*, 122, 611, 1989.
135. **Underwood, L. E., D'Eriole, A. J., Clemmons, D. R., et al.,** *Clin. Endocrinol. Metab.*, 15, 89, 1986.
136. **Daughaday, W. H., Hall, K., Raben, M. S., et al.,** *Nature (Lond.)*, 235, 107, 1972.
137. **Fryklund, L., Skottner, A., Sievertsson, H., et al.,** *Growth Hormone and Related Peptides* (Eds. E. Pecile and E. E. Muller). Excerpta Medica, Amsterdam, 1976, p. 156.
138. **Froesch, E. R., Burgi, H., Ramseier, E. R., et al.,** *J. Clin. Invest.*, 42, 1816, 1963.
139. **Zapf, J., Mader, M., Waldvogel, M., et al.,** *Arch. Biochem. Biophys.*, 168, 630, 1975.
140. **Schlumpf, U., Heimann, R., Zapf, J., et al.,** *Acta Endocrinol.*, 81, 28, 1976.
141. **Heinrich, U. E., Schalch, D. S., Koch, J. G., et al.,** *J. Clin. Endocrinol. Metab.*, 46, 672, 1978.
142. **Rinderknecht, E. and Humbel, R. E.,** *Proc. Natl. Acad. Sci. U.S.A.*, 73, 4379, 1976.
143. **Rinderknecht, E. and Humbel, R. E.,** *Proc. Natl. Acad. Sci. U.S.A.*, 73, 2365, 1976.
144. **Rinderknecht, E. and Humbel, R. E.,** *J. Biol. Chem.*, 253, 2769, 1978.
145. **Van Wyk, J. J., Svoboda, M. E., and Underwood, L. E.,** *J. Clin. Endocrinol. Metab.*, 50, 206, 1980.
146. **Hintz, R. L., Liu, F., and Richerknecht, E.,** *J. Clin. Endocrinol. Metab.*, 51, 672, 1980.

147. **Klapper, D. G., Svoboda, M. E., and Van Wyk, J. J.,** *Endocrinology*, 112, 2215, 1983.
148. **Dulak, N. and Temin, H.,** *J. Cell Physiol.*, 81, 153, 1973.
149. **Smith, G. L. and Temin, H. M.,** *J. Cell Physiol.*, 84, 181, 1974.
150. **Nissley, S. P. and Rechler, M. M.,** *Natl. Center Inst. Monogr.*, 48, 167, 1978.
151. **Bick, T., Amit, T., Barkey, R. J., et al.,** *Endocrinology*, 126, 1914, 1990.
152. **Schalch, D. S., Heinrich, O. E., Draznin, B., et al.,** *Endocrinology*, 104, 1143, 1979.
153. **Scott, C. D., Martin, J. L., and Baxter, R. C.,** *Endocrinology*, 116, 1102, 1985.
154. **Norstedt, G. and Moller, C.,** *J. Endocrinol.*, 115, 135, 1987.
155. **Houston, B. and O'Neill, I. E.,** *J. Endocrinol.*, 128, 389, 1991.
156. **Roberts, C. T., Brown, A. L., Graham, D. E., et al.,** *J. Biol. Chem.*, 261, 10025, 1986.
157. **Luo, J. and Murphy, L. J.,** *Endocrinology*, 125, 165, 1989.
158. **Grant, A. L., Helferich, W. G., Kramer, S. A., et al.,** *J. Endocrinol.*, 130, 331, 1991.
159. **Matthews, L. S., Norstedt, G., and Palmiter, R. D.,** *Proc. Natl. Acad. Sci. U.S.A.*, 83, 9343, 1986.
160. **Froesch, E. R., Schmid, C., Schwander, J., et al.,** *Annu. Rev. Physiol.*, 47, 443, 1985.
161. **Schwander, J., Hauri, C., Zapf, J., et al.,** *Endocrinology*, 113, 297, 1983.
162. **Schalch, D. S., Heinrich, U. E., Draznin, B., et al.,** *Endocrinology*, 104, 1143, 1979.
163. **Uthne, K. and Uthne, T.,** *Acta Endocrinol.*, 71, 255, 1972.
164. **Schoele, E., Zapf, J., Humbel, R. E., et al.,** *Nature (Lond.)*, 296, 252.
165. **Schoenle, E., Zapf, J., Hauri, C., et al.,** *Acta Endocrinol.*, 108, 167, 1985.
166. **Skottner, A., Clark, R. G., Robinson, I. C. A. F., et al.,** *J. Endocrinol.*, 112, 123, 1987.
167. **Laron, Z., Klipper-Aurbach, Y., and Klinger, B.,** *Lancet*, 339, 1258, 1992.
168. **Walker, J. L., Van Wyk, J. J., and Underwood, L. E.,** *J. Pediatr.*, 121, 641, 1992.
169. **Takano, K., Hizuka, N., Shizume, K., et al.,** *Growth Regulat.*, 1, 23, 1991.
170. **Lemmey, A. B., Martin, A. A., Read, L. C., et al.,** *Am. J. Physiol.*, E213, 1991.
171. **Francis, G. L., Upton, F. M., Ballard, F. J., et al.,** *Biochem. J.*, 251, 95, 1988.
172. **Martin, A. A., Tomas, F. M., Owens, P. C., et al.,** *Am. J. Physiol.*, 261, F626, 1991.
173. **Tomas, F. M., Knowles, S. E., Owens, P. C., et al.,** *Biochem. J.*, 276, 547, 1991.
174. **Carlsson, L. M. S., Clark, R. G., Skottner, A., et al.,** *J. Endocrinol*, 122, 661, 1989.
175. **Van Buuls-Offers, S., Uleda, I., and Van den Brande, J. L.,** *Pediatr. Res.*, 20, 825, 1986.
176. **Tixier-Boichard, M., Huybrechts, L. M., Decuypere, E., et al.,** *J. Endocrinol.*, 133, 101, 1992.
177. **Lindahl, A., Isgaard, J., and Isaksson, O.,** *Bailliere's Clin. Endocrinol. Metab.*, 5, 671, 1991.
178. **Isaksson, O. G. P., Jansson, J.-O., and Gause, I. A. M.,** *Science*, 216, 1237, 1982.
179. **Russell, S. M. and Spencer, E. R.,** *Endocrinology*, 116, 2563, 1985.
180. **Schechter, N. L., Russell, S. M., Greenberg, S., et al.,** *Am. J. Physiol.*, 250, E231, 1986.
181. **Isaksson, O. G. P., Lindahl, A., Nilsson, A., et al.,** *Endocrine Rev.*, 8, 426, 1987.
182. **Madsen, K., Friberg, U., Roos, P., et al.,** *Nature (Lond.)*, 304, 545, 1983.
183. **Ohlsson, C., Nilsson, A., Isaksson, O. G. P., et al.,** *J. Endocrinol.*, 133, 291, 1992.
184. **Eden, S., Isaksson, O. G., Madsen, K., et al.,** *Endocrinology*, 112, 1127, 1983.
185. **Bentham, J., Ohlsson, C., Lindahl, A., et al.,** *J. Endocrinol.*, 137, 361, 1993.
186. **Isgaard, J., Moller, C., Isaksson, O. G. P., et al.,** *Endocrinology*, 122, 1515, 1988.
187. **Isgaard, J., Moller, C., Isaksson, O. G. P., et al.,** *Endocrinology (Baltimore)*, 122, 1515, 1988.
188. **Isgaard, J., Carlsson, L., Isaksson, O. G. P., et al.,** *Endocrinology (Baltimore)*, 123, 2605, 1988.
189. **Nilsson, A., Isgaard, J., Lindahl, A., et al.,** *Science*, 233, 571, 1986.
190. **Watanabe, N., Rosenfeld, R. G., Hintz, R. L., et al.,** *J. Endocrinol.*, 107, 275, 1985.
191. **Ohlsson, C., Nilsson, A., Isaksson, O., et al.,** *Proc. Natl. Acad. Sci. U.S.A.*, 9826, 1992.
192. **Lindahl, A., Nilsson, A., and Isaksson, O. G. P.,** *J. Endocrinol.*, 115, 263, 1987.
193. **Isgaard, J., Nilsson, A., Vikman, K., et al.,** *J. Endocrinol.*, 120, 107, 1989.
194. **Wolverton, C. K., Azain, M. J., Duffy, J. Y., et al.,** *Am. J. Physiol.*, 263, E637, 1992.
195. **Orlowski, C. C. and Chernausek, S. D.,** *Endocrinology (Baltimore)*, 123, 44, 1988.
196. **Maiter, D., Underwood, L. E., Maes, M., et al.,** *Endocrinology (Baltimore)*, 123, 1059, 1988.
197. **Thorlacius-Ussing, O., Flyvbjerg, A., Damm Jorgensen, K., et al.,** *Acta Endocrinol.*, 117, 65, 1988.
198. **Young, I. R., Mesiano, S., Hintz, R., et al.,** *J. Endocrinol.*, 121, 563, 1989.
199. **Hizuka, N., Takano, K., Asakawa, K., et al.,** *Growth Regul.*, 1, 55, 1991.
200. **Rechler, M. M. and Nissley, S. P.,** *Handbook of Experimental Pharmacology* (Eds. M. B. Sporm and A. B. Roberts), Vol. 95. Springer-Verlag, Berlin, 1990, p. 263.
201. **de Pablo, F., Perez-Villamil, B., Serna, J., et al.,** *Mol. Reprod. Dev.*, 35, 427, 1993.
202. **Holly, J. M. P. and Martin, J. L.,** *Growth Regulat.*, 4(Suppl. 1), 20, 1994.

203. **Baxter, R. C.,** *Modern Concepts of Insulin-Like Growth Factors* (Ed. E. M. Spencer). Elsevier, New York and Amsterdam, 1991, p. 371.
204. **Dauncey, M. J., Rudd, B. T., White, D. A., et al.,** *Growth Regulat.*, 3, 198, 1993.
205. **Brinkman, A., Groffen, C., Kortleve, D. J., et al.,** *EMBO J.*, 7, 2417, 1988.
206. **Murphy, L. J., Seneviratne, C., Ballejo, G., et al.,** *Mol. Endocrinol.*, 4, 329, 1990.
207. **Brown, A. L., Chiariotti, L., Orlowski, C. G., et al.,** *J. Biol. Chem.*, 264, 5148, 1989.
208. **Brown, A. L., Chiariotti, L., Orlowski, C. G., et al.,** *J. Biol. Chem.*, 264, 5148, 1989.
209. **Wood, W. I., Cachianes, G., Henzel, W. J., et al.,** *Mol. Endocrinol.*, 2, 1176, 1988.
209a. **Shimasaki, S., Shimonaka, M., Zhang, H.-P., and Ling, N.,** in *Modern Concepts of Insulin-Like Growth Factors* (Ed. E. M. Spencer). Elsevier, New York, 1991, 343.
210. **Albiston, A. L. and Herington, A. C.,** *Endocrinology (Baltimore)*, 130, 497, 1992.
211. **Domene, H., Krishnamurthe, K., Eshet, R., et al.,** *Endocrinology (Baltimore)*, 133, 675, 1993.
212. **Camacho-Hubner, C., Clemmons, D. R., and D'Ercole, A. J.,** *Endocrinology (Baltimore)*, 129, 1201, 1991.
213. **Zapf, J., Hauri, C., Waldvogel, M., et al.,** *Proc. Natl. Acad. Sci. U.S.A.*, 86, 3813, 1989.
214. **Yang, W.-H., Wang, J.-F., Orlowski, C. C., et al.,** *Endocrinology (Baltimore)*, 125, 1540, 1989.
215. **Fielder, P. J., Thordarson, G., Talamantes, F., et al.,** *Endocrinology (Baltimore)*, 127, 2270, 1990.
216. **Drop, S. L. S., Kortleve, D. J., Guyda, H. J., et al.,** *J. Clin. Endocrinol. Metab.*, 59, 908, 1984.
217. **Hardouin, S., Hossenlopp, P., Segovia, B., et al.,** *Eur. J. Biochem.*, 170, 121, 1987.
218. **Lee, C. Y., Bazer, F. W., Etherton, T. D., et al.,** *Endocrinology (Baltimore)*, 128, 2336, 1991.
219. **Takenaka, A., Hirosawa, M., Mori, M., et al.,** *Br. J. Nutr.*, 69, 73, 1993.
220. **Tseng, L. Y.-H., Ooi, G. T., Brown, A. L., et al.,** *Mol. Endocrinol.*, 6, 1195, 1992.
221. **Chochinov, R. H., Mariz, I. K., Hajek, A. S., et al.,** *J. Clin. Endocrinol. Metab.*, 44, 902, 1977.
222. **Knauer, D. J. and Smith, G. I.,** *Proc. Natl. Acad. Sci. U.S.A.*, 77, 7252, 1980.
223. **Burch, W. H., Correa, J., Shively, J. E., et al.,** *J. Clin. Endocrinol.*, 70, 173, 1990.
224. **Elgin, R. G., Busby, W. H., and Clemmons, D. R.,** *Proc. Natl. Acad. Sci. U.S.A.*, 84, 3254, 1987.
225. **DeMellow, J. S. M. and Baxter, R. C.,** *Biochem. Biophys. Res. Commun.*, 156, 199, 1988.
226. **D'Ercole, A. J., Applewhite, G. T., and Underwood, L. E.,** *Dev. Biol.*, 75, 315, 1980.
227. **D'Ercole, A. J., Hill, D. J., Strain, A., et al.,** *Pediatr. Res.*, 20, 253, 1986.
228. **Robcis, H. L., Caldes, T., and de Pablo, F.,** *Endocrinology (Baltimore)*, 128, 1895, 1991.
229. **Serrano, J., Shuldiner, A. R., Roberts, C. T., et al.,** *Endocrinology (Baltimore)*, 127, 1547, 1990.
230. **Alemany, J., Borras, T., and de Pablo, F.,** *Proc. Natl. Acad. Sci. U.S.A.*, 87, 3353, 1990.
231. **de Pablo, M., Girbau, M., Gomez, J. A., et al.,** *Diabetes*, 34, 1063, 1985.
232. **Girbau, M., Gomez, J. A., Lesniak, M. A., et al.,** *Endocrinology (Baltimore)*, 121, 1477, 1987.
233. **Girbau, M., Lesniak, M. A., Gomez, J. A., et al.,** *Biochem. Biophys. Res. Commun.*, 153, 142, 1988.
234. **Moses, A. C., Nissley, S. P., Short, P. A., et al.,** *Proc. Natl. Acad. Sci. U.S.A.*, 77, 3649, 1980.
235. **Mesiano, S., Young, I. R., Hey, A. W., et al.,** *Endocrinology (Baltimore)*, 124, 1485, 1989.
236. **Owens, P. C., Conlon, M. A., Campbell, R. G., et al.,** *J. Endocrinol.*, 128, 439, 1991.
237. **Mesiano, S., Young, I. R., Baxter, R. C., et al.,** *Endocrinology (Baltimore)*, 120, 1821, 1987.
238. **Jones, C. T., Gu, W., Harding, J. E., et al.,** *J. Dev. Physiol.*, 10, 179, 1988.
239. **Rechler, M. M. and Nissley, S. P.,** *Annu. Rev. Physiol.*, 47, 425, 1985.
240. **Yang, Y. W.-H., Robbins, A. R., Nissley, S. P., et al.,** *Endocrinology (Baltimore)*, 126, 1177, 1991.
241. **Bassas, L., Lesniak, M. A., Serrano, J., et al.,** *Diabetes*, 37, 637, 1988.
242. **Harvey, S.,** Ph.D. thesis, University of Leeds, 1977.

Chapter 20

Growth Hormone Action: Carbohydrate Metabolism

C. G. Scanes

CONTENTS

I. INTRODUCTION

Growth hormone (GH) has diverse effects on both carbohydrate and lipid metabolism (for reviews see Refs. 1 and 2). As Davidson notes,[2] "unfortunately no unified picture has emerged. Both insulin-like and anti-insulin-like effects have been documented." The effects of GH on carbohydrate metabolism are considered in this chapter. It is obvious that GH may also influence carbohydrate metabolism by influencing the secretion of other metabolic (e.g., insulin) hormones and by its effects on lipid and protein metabolism. Endocrine effects of GH are considered in Chapter 26, whereas effects of GH on lipid and protein metabolism are discussed, respectively, in Chapters 21 and 22.

II. GROWTH HORMONE AND HYPERGLYCEMIA

A. THE DIABETOGENIC EFFECT OF GH

It has been known since the 1940s that GH can exert a diabetogenic effect; that is, under some circumstances GH can elevate the circulating concentration of glucose. As early as 1949, daily administration of various preparations of GH was found to induce glycosuria after 4 to 6 days in cats.[3] Similar diabetogenic effects were observed in adult dogs but not consistently in rabbits, rats, or mice.[3] In the same year GH was found to reduce the sensitivity of normal or diabetic rats to insulin.[4]

Administration of GH alone has been found to influence circulating concentrations of glucose and/or insulin. The reports are, however, not entirely consistent even within a given species but these differences can be explained. For instance, chronic administration of GH to growing or nonlactating cattle has been reported to increase plasma concentrations of glucose but not insulin,[5] insulin but not glucose,[6] and both glucose and insulin.[7] This apparent lack of consistency is readily explained in terms of GH increasing plasma

0-8493-8697-7/95/$0.00+$.50

concentrations of glucose, which in turn stimulates insulin secretion in a homeostatic manner. Thus, if we assume GH to be stimulating net glucose output by liver, then a new balance of glucose to insulin or set point is reached following GH administration. In conscious chickens, the acute administration of GH is followed rapidly by hyperglycemia.[8] However, GH does not appear to influence circulating concentrations of glucose in anesthetized chickens irrespective of whether a glucose load is also given.[9]

B. BIOLOGICAL ASSAY SYSTEMS

A response widely used to examine the diabetogenic effects of GH is the change in circulating concentrations of glucose following a glucose load in *ob/ob* mice (see Figure 1). In this system GH is administered daily for 4 days prior to the glucose load.[11] This bioassay system has been employed to examine, for instance, both the activities of GH fragments and of lower vertebrate GH preparations.[12–14] Prolonged treatment of growing pigs with porcine GH (pGH) also has a diabetogenic effect. Chronic administration of native (n) or recombinant (r) pGH to growing pigs increases serum concentrations of both glucose and insulin in a dose-dependent manner.[15] These effects require about 10 days of GH treatment before they become manifest.[16] Young pigs show no acute response to pGH but if they have previously received daily injections of pGH then both insulin and glucose are acutely increased by pGH.[16] Further evidence that chronic GH administration exerts a diabetogenic effect on young pigs comes from the increases in both peak height and area of the insulin response to a glucose challenge[17] (Figure 2, top). Moreover, tissue sensitivity to insulin is reduced by chronic GH administration, as evidenced by the marked attenuation in the decline in plasma concentrations of glucose in response to insulin infusion in GH-treated pigs (Figure 2, bottom).[17] Similarly, in sheep, either native or recombinant bovine GH reduces the ability of insulin to lower circulating concentrations of glucose.[18] In this case, GH was administered twice daily for 2 days prior to the insulin tolerance tests.

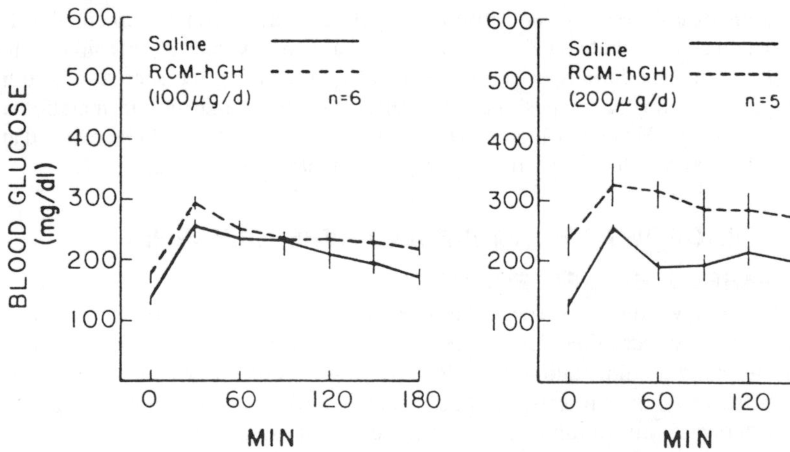

Figure 1 Effect of chronic GH administration (three daily injections) on the response of fasted *ob/ob* mice to an oral glucose load. Vertical lines indicate 2 SEM. (From Cameron, C. M., Kostyo, J. L., Rillema, J. A., and Gennick, S. E., Reduced and S-carboxymethylated human growth hormone: a probe for diabetogenic action, *Am. J. Physiol.*, 247, E639, 1984. Reprinted with permission.)

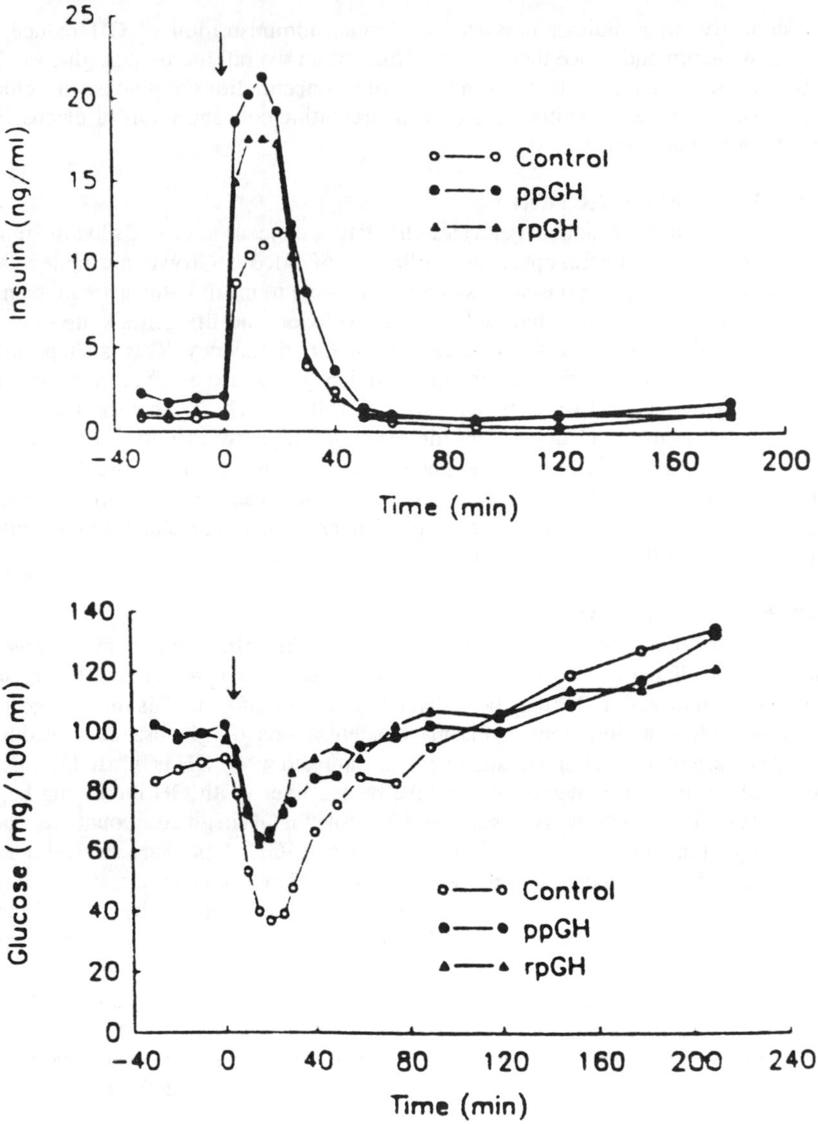

Figure 2 Effect of chronic administration of GH on insulin and glucose dynamics in pigs. *Top*: Effect of chronic porcine GH injections on the insulin secretory response (circulating concentrations of insulin) to a glucose challenge (infused at 1 g/kg). *Bottom*: Effect of chronic porcine GH treatment on the response to insulin (infused at 0.125 U/kg^{75} body weight), as indicated by the decrease in circulating concentration of glucose. (From Gopinath, R. and Etherton, T. D., Effects of porcine growth hormone on glucose metabolism of pigs. II. Glucose tolerance, peripheral tissue sensitivity and glucose kinetics, *J. Anim. Sci.*, 67, 689–697, 1989. Copyright by *Journal of Animal Science*. Reprinted with permission.)

In summary, in a number of systems, chronic administration of GH reduces the response to insulin and hence there is a net effect of an overall increase in glucose. This may be reflected as a greater increase in circulating concentrations of glucose in a glucose load study or a smaller magnitude decrease in circulating concentrations of glucose with an insulin load paradigm.

C. MECHANISM OF ACTION

The mechanism by which GH elevates circulating concentrations of glucose appears to involve reduced peripheral uptake and utilization of glucose. Growth hormone reduces the apparent sensitivity/responsiveness of adipose tissue to insulin, downstream from the insulin receptor. Both insulin-induced glucose oxidation and lipogenesis are either reduced by chronic exposure to GH or increased by GH deficiency. This is supported by clinical studies in which there is either a GH deficiency or an excess,[19] or in rats in which GH deficiency is induced by injection of antisera to GH.[20,21] The mechanism for this does not involve GH inducing changes in insulin receptor number or affinity.[22] Effects of GH on adipose tissue glucose transport and lipogenesis are considered in more detail below. It might be noted that the diabetogenic effect of GH is not observed following chronic GH deficiency (e.g., following hypophysectomy). Under these circumstances GH exerts an acute insulin-like effect (see section below).

D. CLINICAL ASPECTS

Growth hormone exerts an initial, transitory insulin-like effect in humans followed by an anti-insulin (diabetogenic) effect.[23] The initial effect is considered later this chapter. The anti-insulin effect of GH may be a physiological role of GH. This may suggest that GH has pathological importance. Plasma concentrations of glucose and insulin are elevated postprandially in humans and this is accentuated when GH is infused.[24] This GH effect is due to decreased net glucose uptake by the liver (with GH increasing hepatic glucose release and decreasing glucose uptake[24]). A role for GH in glucose counterregulation (during hypoglycemia) is indicated[25] as, for instance, GH deficiency reduces hepatic glucose production and increases glucose utilization.[26] Hence, GH deficiency accentuates hypoglycemia.[26] Moreover, in adults, GH deficiency is accompanied by reduced fasting circulating concentrations of glucose.[27] The situation is even more marked in young children with GH deficiency, in whom a 24-h fast induces considerable hypoglycemia.[28,29] In insulin-dependent diabetics, there is no effect of GH on circulating concentrations of glucose or the requirements of insulin to maintain euglycemia, this being observed either when GH release was suppressed by somatostatin (SRIF) infusion or when endogenous GH concentrations are maintained by the administration of both SRIF and GH.[30]

III. GROWTH HORMONE AND GLUCOSE TRANSPORT

A. ANTI-INSULIN, DIABETOGENIC, OR GLUCOSE TRANSPORT-DEPRESSING EFFECTS OF GH

Similar to its actions *in vivo*, GH has been found to exert an acute insulin-like effect and a chronic anti-insulin effect *in vitro*. Studies on the chronic effect of GH on 3T3-F442A adipocytes demonstrated that prolonged incubation with GH can inhibit both basal and insulin stimulated uptake of 2-deoxy[^{3}H]glucose[31,32] and 3-*O*-methyl-D-glucose uptake.[32] The mechanism for this would seem to be that GH reduces the synthesis of one of the glucose transporters, glut 1,[32] glut 1 being the major glucose transporter in the plasma membranes of adipocytes. Exposure of 3T3 - F442A adipocytes to GH reduces immunoreactive glut 1 and glut 1 mRNA but has no discernible effect on glut 4 or glut 4 mRNA.[32]

B. INSULIN-LIKE OR GLUCOSE TRANSPORT-STIMULATING EFFECTS OF GH

Growth hormone has long been known to acutely increase the metabolism of [U-^{14}C]-glucose to lipid and carbon dioxide by adipose tissue.[33] This is probably mediated by increased substrate availability due to increased glucose transport and also by changes in the content/activity of key enzymes in the metabolic pathways. It has been demonstrated that GH can stimulate glucose uptake by chicken adipose tissue,[34] rat adipocytes differentiated *in vitro*,[35] 3T3-F422A adipocytes, and muscle tissue (hemidiaphragms from hypophysectomized rats).[36] The mechanism by which GH increases glucose uptake is identical to that of insulin. This involves translocations of glut 4, and to a less extent glut 1, from an intracellular pool to the plasma membrane.[35] It may be suggested that GH and insulin are exerting this common effect by the same signal transduction mechanism.

It is generally viewed that the insulin-like effect of GH is best observed in cells not previously challenged by GH exposure. Studies on this effect of GH therefore frequently use tissue from hypophysectomized rats,[33] or cell lines that have not been exposed to GH.[32,35] Preincubation of adipose tissue from normal rats (such that the tissue is not exposed to GH) allows an insulin-like effect (e.g., glucose oxidation) *in vitro* to become manifest.[37] An exception to this general rule is the GH-induced increase in [U-^{14}C]glucose oxidation and incorporation of ^{14}C into fatty acids in adipose tissue freshly taken from young rats.[38]

C. OTHER EFFECTS OF GH ON GLUCOSE TRANSPORT

There are effects of GH that are difficult to categorize as insulin-like or diabetogenic/anti-insulin. One of these is glucose transport, as indicated by ^{14}C-labeled 3-*O*-methylglucose efflux from adipocytes, which are chronically affected by GH status. Basal glucose efflux is maximal in adipocytes from hypophysectomized rats and cannot be stimulated further by insulin.[39] This can be envisaged in terms of GH reducing glucose transport, that is, a diabetogenic effect. Addition of GH to adipocytes *in vitro* reduces the rate of glucose (^{14}C-labeled 3-*O*-methylglucose) efflux in a manner that might be considered the opposite to an insulin effect.[40] However, in doing so GH is restoring sensitivity to insulin.[39]

Although GH has little acute effect on glucose output from hepatocytes from intact rats, GH increases glucose output from hepatocytes from hypophysectomized rats (due to increased glycogenolysis and gluconeogenesis).[41] This effect is opposite to that found with insulin.[41] However, as with the insulin-like effect of GH, this effect requires GH to be absent or deficient prior to GH exposure.

IV. METABOLIC EFFECTS OF GROWTH HORMONE FRAGMENTS

An intriguing series of reports suggest that specific GH fragments can exert either hypoglycemic or hyperglycemic effects. The biological activities of synthesized peptides have been examined. The rationale for these studies was either to map the functional domains on the GH molecule or to provide information on the biological activity of GH fragments that are known to be present in the pituitary gland.[42]

Peptides corresponding to the N terminus of human GH have been consistently found to have effects that can be variously referred to as hypoglycemic or insulin-like or insulin potentiating. For instance, hGH_{1-15} enhances the hypoglycemic effect of insulin in both control and streptozocin-treated (diabetic) rats.[43] Similarly, hGH_{4-15} accentuates the decrease in circulating glucose concentrations following insulin challenge in either young or adult rats.[44] This appears to be due, at least in part, to hGH_{4-15} increasing insulin receptors in adipose tissue.[44] Potentiation of insulin-induced hypoglycemia in rats occurs with hGH_{1-43} in diabetic rats[45] and obese yellow mice and also with a constrained analog of hGH_{6-13} λ-lactam[11]-hGH_{6-13}.[46] Similarly, hGH_{1-43} decreases the increment in circulating concentrations of glucose following a glucose load in hypophysectomized AS^{av}/A mice[47] or following a glucose load together with insulin treatment in either obese yellow A^{vy}/A or C57BL/6J *ob/ob* mice.[48]

It is reasonable to suggest that the effects of the peptides corresponding to the N terminal of GH are due to shifts in glucose uptake and/or metabolism. Studies to examine this have involved the paradigm of *in vivo* administration of the peptide and then determination of tissue changes or tissue metabolism *in vitro*. These N-terminal peptides of GH exert effects that mimic or accentuate those of insulin. For instance, *in vivo* administration of λ-lactam-hGH_{6-13} increases both diaphragm glycogen and [^{14}C]glucose incorporation into lipid in the presence or absence of insulin in rats.[46] Similarly, *in vivo* treatment of rats with hGH_{4-15} affects several aspects of metabolism in rat adipose tissue *in vitro*. These stimulatory effects of hGH_{4-15} are observed in the presence or absence of insulin and can be enumerated: (1) glucose uptake (as indicated by 2-deoxy[^{14}C]glucose), (2) [^{14}C]glucose oxidation to $^{14}CO_2$, (3) [^{14}C]glucose incorporation into lipid, and (4) glycogen synthase activity.[49]

Conversely, $hGH_{172-191}$ has been reported to exert a hyperglycemic effect.[50] This is explicable in terms of the anti-insulin/antilipogenic effect of the GH fragment, $hGH_{172-191}$ having been shown to inhibit acetyl-CoA carboxylase in both adipocytes and hepatocytes as does hGH itself.[51]

It may be questioned whether these effects suggest that a specific domain on the N terminal of the GH molecule is responsible for this hypoglycemic/insulin-like effect. Indeed, it is tempting to suggest that the sequence of these peptides corresponds to part of the second GH receptor-binding site on GH.[52] Studies with N-terminal GH peptides frequently do not include a comparison with full-length GH. However, even when this is done, GH appears to be either less effective or completely ineffective. For instance, one comprehensive study compared the effectiveness of GH and GH fragments in increasing the insulin-induced glucose oxidation by adipose tissue from yellow A^{vy}/A obese mice. Whereas hGH_{1-15}, hGH_{1-43}, and hGH_{1-139} were active, this was not the case with native hGH, $hGH_{146-191}$, or hGH_{32-46}.[42]

It may be questioned whether the effects of the peptides corresponding to the N terminal of GH are pharmacological or of doubtful physiological significance, due to the peculiar or pathological models that need to be employed. Alternatively, it may well be the case that endogenously produced GH fragments have biological activities distinct from GH itself and that the studies with the N-terminal peptides may have provided the initial thrust that leads to this being considered.

REFERENCES

1. **Goodman, H. M.,** *The Endocrinology of Growth, Development and Metabolism in Vertebrates* (Eds. M. P. Schreibman, C. G. Scanes, and P. K. T. Pang). Academic Press, San Diego, 1993, p. 93.
2. **Davidson, M. B.,** *Endocr. Rev.*, 8, 115, 1987.
3. **Cotes, P. M., Reid, E., and Young, F. G.,** *Nature (Lond.)*, 164, 209, 1949.
4. **Milman, A. E. and Russell, J. P.,** *Fed. Proc.*, 8, 111, 1949.
5. **Crooker, B. A., McGuire, M. A., Cohick, W. S., et al.,** *J. Nutr.*, 120, 1256, 1990.
6. **Eisemann, J. H., Hammond, A. C., Bauman, D. E., et al.,** *J. Nutr.*, 116, 2504, 1986.
7. **Byatt, J. C., Eppard, P. J., Veenhuizen, J. J., et al.,** *J. Endocrinol.*, 132, 185, 1992.
8. **Hall, T. R., Cheung, A., and Harvey, S.,** *Comp. Biochem. Physiol.*, 86A, 29, 1987.
9. **Scanes, C. G.,** *Comp. Biochem. Physiol.*, 101A, 871, 1992.
10. **Cameron, C. M., Kostyo, J. L., Rillema, J. A., et al.,** *Am. J. Physiol.*, 247, E639, 1984.

11. **Kostyo, J. L., Gennick, S. E., and Sauder, S. E.,** *Am. J. Physiol.*, 246, E356, 1984.
12. **Cameron, C. M., Kostyo, J. L., and Papkoff, H.,** *Endocrinology (Baltimore)*, 116, 1501, 1985.
13. **Mills, J. B., Kostyo, J. L., Reagan, C. R., et al.,** *Endocrinology (Baltimore)*, 107, 391, 1980.
14. **Reagan, C. R., Kostyo, J. L., Mills, J. B., et al.,** *Endocrinology (Baltimore)*, 102, 1377, 1978.
15. **Evock, C. M., Etherton, T. D., Chung, C. S., et al.,** *J. Anim. Sci.*, 66, 1928, 1988.
16. **Gopinath, R. and Etherton, T. D.,** *J. Anim. Sci.*, 67, 682, 1989.
17. **Gopinath, R. and Etherton, T. D.,** *J. Anim. Sci.*, 67, 689, 1989.
18. **Hart, I. C., Chadwick, P. M. E., Boone, T. C., et al.,** *Biochem. J.*, 224, 93, 1984.
19. **MacGorman, L. R., Rizza, R. A., and Gerich, J. E.,** *J. Clin. Endocrinol. Metab.*, 53, 556, 1981.
20. **Schwartz, J.,** *Endocrinology (Baltimore)*, 107, 877, 1980.
21. **Gause, I., Eden, S., Janssen, J. O., et al.,** *Endocrinology (Baltimore)*, 112, 1559, 1983.
22. **Schwartz, J. and Eden, S.,** *Endocrinology*, 116, 1806, 1985.
23. **MacGorman, L. R., Rizza, R., and Gerich, J.,** *J. Clin. Endocrinol. Metab.*, 53, 556, 1981.
24. **Butler, P., Kryshak, E., and Rizza, R.,** *Am. J. Physiol.*, 260, E513, 1991.
25. **Boyle, P. J. and Cryer, P. E.,** *Am. J. Physiol.*, 260, E395, 1991.
26. **DeFeo, P., Perrielo, G., Torlon, T., et al.,** *Am. J. Physiol.*, 256, E835, 1989.
27. **Merimel, T. J., Felig, P., Marliss, E., et al.,** *J. Clin.*, F4, 1971.
28. **Hopwood, N. J., Foprman, P. J., Kenny, F. M., et al.,** *Am. J. Dis. Child.*, 129, 918, 1975.
29. **Wolfsdorf, J. I., Sadeghi-Nejad, A., and Senior, B.,** *Metabolism*, 32, 457, 1983.
30. **Skor, D. A., White, N. H., Thomas, L., et al.,** *Diabetes*, 34, 135, 1985.
31. **Silverman, M. S., Mynarcek, D. C., Corin, R. E., et al.,** *Endocrinology (Baltimore)*, 125, 2600, 1989.
32. **Tar, P.-K., Liao, J.-F., Chen, E. H., et al.,** *J. Biol. Chem.*, 521–828, 1990.
33. **Goodman, H. M.,** *Endocrinology (Baltimore)*, 114, 431, 1984.
34. **Rudas, P. and Scanes, C. G.,** *Poultry Sci.*, 62, 1838, 1983.
35. **Tanner, J. W., Leingang, K. A., and Mueckler, M. M.,** *Biochem. J.*, 282, 99, 1992.
36. **Cameron, C. M., Kostyo, J. L., and Adamafio, P.,** *Endocrinology (Baltimore)*, 122, 471, 1988.
37. **Eden, S., Schwartz, J., and Kostyo, J. L.,** *Endocrinology*, 111, 1505, 1982.
38. **Goodman, H. M. and Corro, V.,** *Endocrinology (Baltimore)*, 109, 2046, 1981.
39. **Schoenle, E., Zapf, J., and Froesch, E. R.,** *Am. J. Physiol.*, 242, E368, 1982.
40. **Schoenle, E., Zapf, J., and Froesch, E. R.,** *Endocrinology (Baltimore)*, 112, 384, 1983.
41. **Blake, W. L. and Clarke, S. D.,** *J. Endocrinol.*, 457, 1989.
42. **Frigeri, L. G., Teguh, C., and Ling, N.,** *Endocrinology (Baltimore)*, 122, 2940, 1988.
43. **Ng, F. M. and Bornstein, J.,** *Diabetes*, 28, 1126, 1979.
44. **Ng, F. M. and Bornstein, J.,** *Diabetologia*, 23, 534, 1982.
45. **Salem, M. A. M.,** *Endocrinology*, 123, 1565, 1988.
46. **Lim, N., Ng, F. M., and Wu, Z. M.,** *Endocrinology (Baltimore)*, 131, 835, 1992.
47. **Salem, M. A. M., Lewis, U. J., and Haro, L. S.,** *Proc. Soc. Exp. Biol. Med.*, 191, 408, 1989.
48. **Salem, M. A. M. and Wolff, G. L.,** *Proc. Soc. Exp. Biol. Med.*, 191, 113, 1989.
49. **Ng, F. M. and Harcourt, J. A.,** *Diabetogia*, 29, 882, 1986.
50. **Wade, J. D., Pullin, C. D., Ng, F. M., and Bornstein, J.,** *Biochem. Biophys. Res. Commun.*, 78, 827, 1977.
51. **Bornstein, J., Ng, F. M., and Heng, D. R.,** *Acta Endocrinol.*, 103, 479, 1983.
52. **deVos, A. M., Ultsch, M., and Kossiakoff, A. A.,** *Science*, 255, 306, 1992.

Chapter 21

Growth Hormone Action: Lipid Metabolism

C. G. Scanes

CONTENTS

I. INTRODUCTION

Profound effects on body composition can be attributed to growth hormone (GH) in both growing and adult humans and in a range of animal species. For instance, in men over 60 years old, hGH administration for 6 months resulted in a 14% decrease in adipose tissue mass.[1] Similarly, body fat is reported to be reduced following hGH administration in obese women.[2] In growing rats, the administration of antisera to GH has different effects on the weight of adipose tissue depending on sex, the fat depot, and whether adipose tissue is expressed as weight, weight relative to body weight, mean adipocyte volume, or adipocyte number.[3,4] Prolonged administration of antisera to rat GH decreases the number of adipocytes in parametrial, perirenal, and subcutaneous depots while not influencing adipocyte volume.[4] Concomitant injections of bovine GH overcome this depression in adipocyte number.[4] These observations are consistent with GH being required for adipocyte proliferation. It might be noted that the bovine GH injections are followed by decreases in mean adipocyte volume in perirenal depot, presumably due to effects of GH on lipolysis and/or lipogenesis. In growing farm animals (pigs, cattle, and sheep), GH also markedly decreases the proportion of adipose tissue/fat in the carcass (discussed in detail in Chapter 29).

II. LIPOLYTIC EFFECT OF GH

A. INTRODUCTION

Growth hormone exerts an overall lipolytic effect (hydrolysis of triglyceride to free fatty acids and glycerol). However, there is controversy as to whether GH per se is influencing lipolysis.

B. *IN VIVO* STUDIES

The administration of GH *in vivo* increases the circulating concentrations of free fatty acids in many species, including an acute effect in humans[5] (see Figure 1), sheep,[6,7]

0-8493-8697-7/95/$0.00+$.50

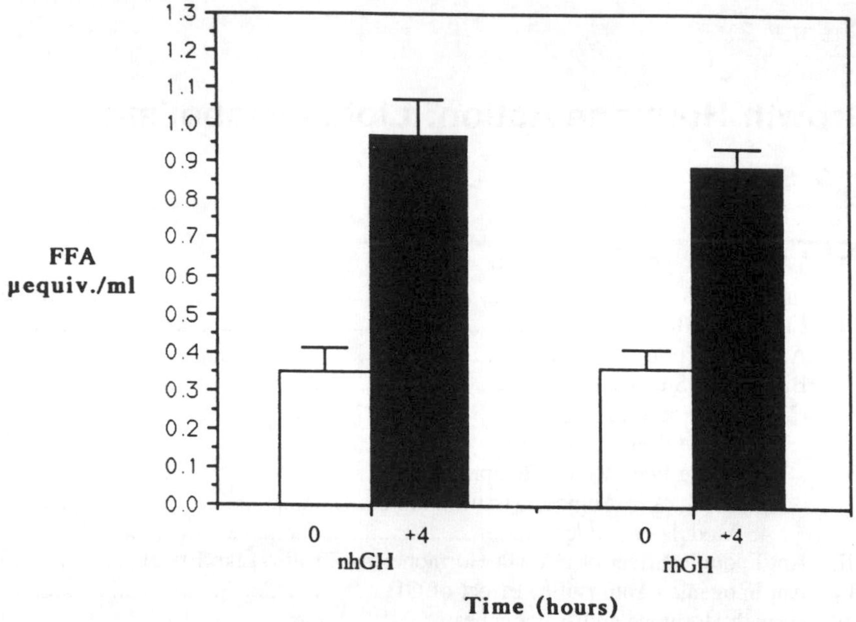

Figure 1 Effect of recombinant (r) and native (n) hGH on circulating concentrations of FFA in adult men. (Data from Takano et al.[5])

cattle,[8] and chickens[9,10] but not in other species (e.g., rabbit).[11] Further evidence that GH influences lipolysis in humans includes the following. Insulin administration *in vivo* induces an increase in circulating concentrations of free fatty acids (after a transient decrease[12,13]). It can be presumed that this effect is indirect, being either prolonged or completely mediated by GH, as no such effect is found in GH-deficient hypopituitary individuals[12] or when GH secretion is suppressed by somatostatin (SRIF).[13] Moreover, GH administration restores the effect of insulin in these GH-deficient states. Chronic administration of GH has also been reported to elevate circulating concentrations of free fatty acids in growing cattle.[8,14] However, in lactating cattle, GH has been reported to have little effect[15,16] or to increase circulating concentrations of free fatty acids.[15,17–19] This is explicable in terms of a third fate for circulating free fatty acids in a lactating animal, that is, in the synthesis of milk components.

The increases in circulating concentrations of free fatty acids observed following GH administration are explicable in terms of increased lipolysis and/or decreased fatty acid oxidation. The available evidence favors GH influencing lipolysis; this may be achieved either directly or indirectly (considered in detail below). There is also some contrary evidence for this contention. For instance, chronic administration of bovine GH to growing female cattle actually increases both irreversible loss and oxidation of circulating free fatty acids.[14] Similarly, GH increases palmitate uptake by ovine hepatocytes *in vitro*.[20]

The stimulation of lipolysis by GH may involve one or more of the following possibilities: (1) GH may act indirectly by increasing the release of endogenous hormones that affect lipolysis. This may involve increased release of lipolytic hormones (e.g., epinephrine, norepinephrine, and glucagon) or decreased release of antilipolytic hormone (e.g., insulin); (2) GH may directly increase lipolysis; or (3) GH may affect the sensitivity and/or responsiveness of adipose tissue to hormones (increasing lipolytic and/or decreasing antilipolytic response).

An indirect effect of GH acting via other lipolytic agents is, however, unlikely. This is because circulating concentrations of insulin are frequently elevated by GH administration (for details see Chapter 26), which would be expected to reduce lipolysis. However, the opposite is observed. There is little information on the effects of GH on circulating concentrations of epinephrine, norepinephrine, or glucagon. This is in part due to the absence, until recently, of assay systems to distinguish between glucagon, glucagon-like peptides, oxyntomodulin, and glycentin.

The second possible mechanism by which GH may evoke a *de facto* lipolytic effect is by directly acting on adipose tissue and stimulating lipolysis. For GH to act directly on lipolysis, GH needs to be shown to stimulate lipolysis by adipose tissue *in vitro.*

C. *IN VITRO* STUDIES

Studies indicate that GH may have both direct lipolytic and antilipolytic actions on adipose tissue. The latter will be considered under the insulin-like/antilipolytic actions of GH. Direct stimulation of lipolysis by GH has been observed *in vitro* with adipose tissue from some species, for example, rats[21–24] and chickens,[25,26] but not from other species (e.g., sheep[27] and rabbits[28]). There are five systems (two with rats, two mouse, and one with chickens) in which GH has been reported to stimulate lipolysis (glycerol release). These are referred to as *in vitro* GH stimulation of lipolysis in presence of glucocorticoid (rat; also in mice[29]), *in vitro* GH stimulation of lipolysis in the presence of theophylline (rat), and *in vitro* GH stimulation of lipolysis (chicken).

1. In the Presence of Glucocorticoids

In the presence of the glucocorticoid dexamethasone, GH stimulates lipolysis *in vitro* with rat adipose tissue explants, as indicated by increased release of glycerol;[21,22,24,27] the effect is observed with either native[21,22] or recombinant human GH[24] and increases in magnitude in four sequential 1-h incubations.[24] On the other hand, it should be noted that some investigators have reported highly purified GH or recombinant GH failed to stimulate lipolysis.[30,31] It was argued that the lipolytic effect of GH was due to a contamination of GH preparations previously employed. However, the specificity of this lipolytic effect comes from the ability of a GH analog with GH antagonist activity to inhibit the GH effect.[32] Moreover, the GH effect can be blocked by inhibitors of RNA and protein synthesis,[22,23] which do not inhibit the effect of other lipolytic agents.[33]

The mechanism by which GH exerts a lipolytic effect (in the presence of the glucocorticoid dexamethasone) is not well established. Obviously, for GH to exert a direct effect on adipose tissue, there must be GH binding to GH receptors and indeed the presence of GH receptors in adipose tissue is well established.[34–37] In view of the ability of inhibitors of RNA and protein synthesis to block the lipolytic effect of GH,[22,33] it is presumed that newly synthesized protein is required for the GH effect. This protein may be mediating the GH effect directly, or its presence may be obligatory. Other studies, with pharmacological blockers, have provided some insight into the mechanism of action of GH. The isoquinoline sulfonamides H8 and H7 have been reported to inhibit the lipolytic effects of GH while not influencing the lipolytic effect of either dibutyryl cAMP or isoproterenol.[33] The interpretation of these data is not straightforward; however, it is generally considered that H8 is an inhibitor of protein kinase A and to a lesser extent protein kinase C, whereas H7 is an inhibitor of protein kinase C and to a lesser extent protein kinase A. In view of the ability of the same dose of H7 to suppress phorbol ester (and hence protein kinase C)-induced lipolysis in the same system, a role for protein kinase C in the lipolytic effect of GH can be postulated. However, both H7 and H8 have also been shown to inhibit RNA synthesis[33] and this effect may explain their ability to reduce the lipolytic effect of GH. Another pharmacological agent, swainsonine, also suppresses the lipolytic effect of GH.[38] This alkaloid interferes with Golgi functioning, specifically the processing of carbohydrate chains of glycoproteins by mannosidase A. The inhibition of the lipolytic effect of GH by swainsonine is suggested to be due to a requirement for the synthesis of new glycosylated GH receptors,[38] implying a rapid turnover of glycosylated GH receptors. However, a requirement for other glycosylated proteins cannot be ruled out.

A simple system for investigating the lipolytic effect of GH involves preincubation of adipose tissue explants (pieces) with varying doses of GH together with dexamethasone, and has utilized tissue from mice.[29] In this case, lipolysis was stimulated (by two- to fourfold) by mouse GH with tissue from either virgin or pregnant mice.

While the presence of dexamethasone is commonly thought to be necessary in *in vitro* incubations of rat adipose tissue for GH to exert a lipolytic effect,[23,24] a marked lipolytic response to GH alone has been reported with adipose tissue from young rats (20 day old and to a lesser extent also at 24 days old).[39] It is also unclear whether the magnitude of this GH effect with young rat tissue increases with time of exposure to GH.[39] The magnitude of incremental release of glycerol over that of control was found to be increased with repeated 1-h exposures to GH but no changes with time were observed if lipolytic rates in the presence of GH are expressed as a percentage of the control.[39]

2. In the Presence of Theophylline

A frequently employed system for examining the lipolytic effect of GH involves preincubating rat adipose tissue explants (segments) for 3 h in the presence or absence of GH. Lipolysis, as indicated by glycerol released into the incubation medium, is determined in a subsequent 1-h incubation period in which theophylline is present.[39,40] Evidence of the specificity of the GH effect comes from the ability of both native and recombinant (human) GH to evoke the same lipolytic response.[23]

The mechanism by which the lipolytic effect of GH is expressed in this system (with theophylline) is not established. It does not appear that RNA and protein synthesis are obligatory.[33] Similarly, in view of the inability of high doses of H8 or HA-4004 (inhibitors of protein kinase A and to a less extent protein kinase C) to decrease the lipolytic response to GH,[33] a role for either cAMP/protein kinase A or Ca^{2+}/protein kinase C is mitigated.

An analogous lipolytic response to theophylline after GH pretreatment has been observed with adipose tissue from virgin mice.[29] Unlike the ability of GH to stimulate lipolysis in the presence of dexamethasone in adipose tissue from virgin or pregnant mice, no such response with theophylline was observed with tissue from pregnant mice.[29] This would suggest that the two lipolytic *in vitro* rodent systems (the dexamethasone model versus the theophylline model) are not indices of the same inherent phenomenon. Evidence supporting this contention comes from the inability of inhibitors of RNA or protein synthesis to interfere with the lipolytic response (theophylline system), while completely preventing lipolysis (dexamethasone model).[40] There is also some evidence that the lipolytic response to GH in the presence of theophylline differs from the insulin-like effect of GH. An analog of human GH (Da1, a noncovalent complex of residues 1–134 and 141–191 from plasmin digests of reduced and *S*-carbamidomethylated GH) is equipotent with human GH in evoking lipolysis (with theophylline) but has a much reduced potency in bioassays with an insulin-like end point.[41]

3. In Chicken Adipose Tissue

A simple system exists in which GH exerts a direct lipolytic effect with chicken adipose tissue explants *in vitro*.[26] If adipose tissue from either young or adult chickens is incubated for 1-h in the presence of GH, lipolysis (release of glycerol) is increased.[26] The magnitude of the lipolytic effect of GH increases if adipose tissue is repeatedly exposed for 1-h periods to GH.[26] This lipolytic effect of GH has been observed in the absence of glucocorticoids and with a variety of GH preparations, including native or recombinant bovine GH,[26] human GH,[42] and chicken GH[26] and also with the 20K variant of human GH and reduced and carboxymethylated GH.[42] However, GH from lower vertebrates (reptiles, amphibians, or fish) and human placental lactogen fail to exert a lipolytic effect on chicken adipose tissue.[42,43] The specificity of the lipolytic effect of GH on chicken adipose tissue is further indicated by the ability of a GH antagonist to inhibit GH-induced lipolysis.[44]

It is reasonable to consider the possibility that the lipolytic effect of GH in chickens shows similarities to the GH/dexamethasone rat lipolytic model and to GH alone in the young rat systems. As with the former, the lipolytic effect of GH on chicken adipose tissue is abolished if RNA and/or protein synthesis is inhibited.[45] However, the ability of GH to stimulate lipolysis within 1 h *in vitro* and in the absence of glucocorticoid appears to be similar to the latter system with adipose tissue from young rats *in vitro*.[39]

Pharmacological studies have been conducted on the mechanism by which GH exerts its lipolytic effect on chicken adipose tissue. As with the lipolytic effects of GH in the presence of dexamethasone in rat adipose tissue, the lipolytic effect of GH is not observed if inhibitors of RNA or protein synthesis are added to the incubation medium.[45] In view of the requirement for high concentrations of glucose for GH to stimulate lipolysis,[46] it would seem reasonable to suggest that the lipolytic effect of GH is an energy-dependent process. The lipolytic effect of GH does not appear to involve cAMP/protein kinase A, as their effects are not additive.[47] In fact, GH depresses the lipolytic effect of 8-bromo-cAMP.[47] A role for calcium and perhaps also protein kinase C may be indicated from pharmacological studies. If the concentration of calcium in the medium is reduced, the lipolytic effect of GH is lost.[45] This may be due to a calcium requirement for GH binding to the receptor or to an intracellular role for calcium in mediating the GH effect. The latter would appear to be the case in view of the ability of the calcium channel blocker, verapamil, to block the lipolytic effect of GH[45] and the ability of a calcium ionophore to stimulate lipolysis.[45] There is also some evidence for a role of protein kinase C in the lipolytic effect of GH, since phorbol ester stimulates lipolysis *de novo* by a mechanism additive with GH.[45]

To summarize, GH appears to exert its lipolytic effect on chicken adipose tissue by a process that is energy dependent, requires calcium entry into the adipocytes, and RNA and protein synthesis.

4. Modulatory Roles

There is evidence that GH can influence the sensitivity and/or responsiveness of adipose tissue to agents that influence lipolysis. If adipocytes from either control or diabetic (streptozocin-treated) rats are preincubated with GH, both the sensitivity and responsiveness to a subsequent *in vitro* lipolytic challenge with epinephrine are markedly improved.[48] Similar *in vivo* treatment of lactating rats with GH is also followed by an increase in *in vitro* lipolytic response to norepinephrine.[49] Conversely, *in vivo* treatment with antisera to GH reduces the *in vitro* response to norepinephrine.[49] These effects might be explainable in terms of GH increasing β-adrenergic receptor number or reducing inhibitory influences. In view of the ability of adenosine to reduce the lipolytic response to β-adrenergic agonists,[48] adenosine is a good candidate for such an inhibitory factor. There is indeed evidence that the enhanced response to norepinephrine following GH treatment is lost if adenosine cannot exert an effect, for instance, if adenosine is destroyed by adenosine deaminase (ADA)[49] or antagonized by theophylline.[39,40]

In sheep, GH alone does not influence lipolysis *in vitro* either in short-term or prolonged (48-h) incubations of adipose tissue.[50] However, exposure of ovine adipose tissue to GH does increase the lipolytic responses to both norepinephrine and isoproterenol,[50] this being mediated by increases in the number of β-adrenergic receptors.[50] Similarly, *in vivo* in GH-treated cattle, there is a greater response (circulating concentrations of free fatty acids) to lipolytic agents, for example, epinephrine.[51,52]

It might be argued that one of the *in vitro* lipolytic effects of GH in rat adipose tissue (preincubation for 3 h in the presence of GH and lipolysis determined in a subsequent 1-h incubation with theophylline) is in fact due to GH influencing the response to another agent. Theophylline acts both to inhibit phosphodiesterase activity and as an adenosine antagonist. Thus the lipolytic effect of GH *in vitro* may in fact be due to GH either increasing the responsiveness to cAMP or to lack of an inhibitory effect of endogenous adenosine.

III. ANTILIPOLYTIC EFFECT OF GROWTH HORMONE: AN INSULIN-LIKE EFFECT?

In addition to its direct lipolytic effects, GH can also inhibit lipolysis induced by some lipolytic agents. This antilipolytic effect of GH can be observed with rat adipose tissue extracts[53,54] or adipocytes[55] and with chicken adipose tissue explants[47] *in vitro*.

An antilipolytic effect of (either native or recombinant) GH is found with adipose tissue from either hypophysectomized rats[54] or from normal rats following prolonged preincubation *in vitro*.[55] It would seem that, like the other insulin-like actions of GH, this is observed only in tissue that has not been exposed to GH prior to attempting to examine the antilipolytic effects. The antilipolytic effect of GH appears to involve GH reducing the activation and phosphorylation of hormone-sensitive lipase.[55] This is mediated in part by a reduction in cAMP production.[56]

Growth hormone also exerts an antilipolytic effect with chicken adipose tissue *in vitro*, in this case depressing glucagon-induced lipolysis.[47] Unlike the situation in the rat, in which a response to GH is observed predominantly in hypophysectomized rat tissues, the GH effect in the chicken is observed with adipose tissue from intact animals.[47] Moreover, as GH also partially suppresses the lipolytic effect of a cAMP analogue (8-bromo-cAMP), it might be suggested that the locus of the GH effect is subsequent to adenylate cyclase.[47]

IV. ANTILIPOGENIC (ANTIINSULIN) EFFECT OF GH

There is considerable evidence that GH can act directly on adipose tissue to reduce lipogenesis, that is, the synthesis of fatty acids. *In vitro*, GH reduces lipogenesis in adipose tissue incubated with insulin (and in some cases also glucocorticoids) for a prolonged period (24 or 48 h). This has been observed with adipose tissue from pigs,[57,58] cattle,[59,60] and sheep.[61,62] In studies with ovine adipose tissue explants, GH reduces the activity of both acetyl-CoA carboxylase and glucose-6-phosphate dehydrogenase, with also some tendency for depressed fatty-acid synthase, 6-phosphogluconate dehydrogenase, and isocitrate dehydrogenase activities.[63] The mechanism by which GH exerts its effect on lipogenesis does not appear to involve changes in the insulin receptor, based on studies with porcine adipocytes.[58] Moreover, this antilipogenic effect of GH is not due to stimulation of adipose insulin-degrading (protease) activity. In fact, GH has been reported to reduce insulin-degrading enzyme activity in bovine adipose tissue *in vitro*.[59]

The intracellular mechanism by which GH exerts an antilipogenic/antiinsulin effect is also unknown. There is no information of involvement of either cAMP/protein kinase A or Ca^{2+}/protein kinase C as the signal transducer for the antilipogenic/antiinsulin effect of GH. However, a possibility exists for polyamines being involved in the mechanisms of

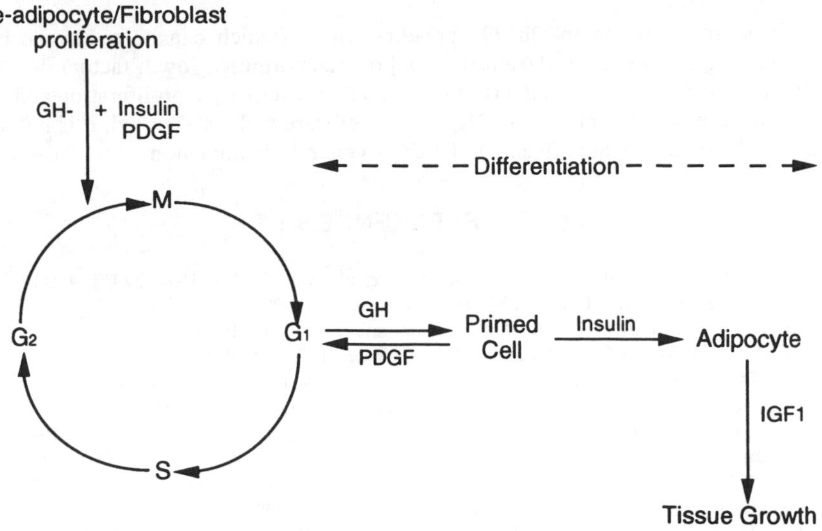

Figure 2 Model for the role of GH and other hormones/growth factors in the adipogenesis of 3T3 cells. (Based on Refs. 65 through 69.)

both insulin and GH in ovine adipose tissue. Insulin has been observed to reduce ornithine decarboxylase[64] and hence presumably decreases synthesis of polyamines. On the other hand, GH increases *S*-adenosylmethionine decarboxylase[64] and, thus, would be expected to increase polyamine synthesis.

In summary, GH exerts an antilipogenic/antiinsulin effect in reducing fatty acid synthesis in the presence of insulin or insulin and dexamethasone.[61,62] Moreover, GH also reduces the insulin-stimulated glucose utilization by ovine adipose tissue.[58] Other antiinsulin effects of GH include effects on glucose transport. These are considered under the influence of GH on glucose transport elsewhere in this chapter. Not only can GH can exert an antilipogenic effect but also it would appear that this is one of the direct physiological actions of GH, in view of the ability of GH to reduce adiposity in pigs, sheep, and cattle.

V. GROWTH HORMONE AND ADIPOGENESIS

A useful model for investigating the mechanism of GH action is its ability to stimulate differentiation of rodent 3T3 preadipocytes[65] (as indicated by the marker enzymes, e.g., glycerophosphate dehydrogenase). This effect appears to be specific for GH, as it is not observed with other pituitary hormones[65] and serum adipogenic activity is completely suppressed by antisera to GH.[66] Although GH induces differentiation, it also reduces proliferation in the presence of insulin.[67,68]

The full induction of adipocyte differentiation of at least 3T3 preadipocytes requires both GH and insulin. In the presence of insulin, GH evokes complete adipocyte differentiation, as indicated by fat accumulation (oil red 0 staining) in 3T3(F442A) cells.[69] A series of differentiational changes in 3T3 preadipocytes is provoked by GH. Among these are changes in the cellular morphology, with a different shape and a reduction in the cell volume.[70] These are presumably mediated by the cytoskeleton, with GH inducing increases in both B-tubulin[68] and vinculin.[71] GH may also be affecting the extracellular matrix in which adipose differentiation occurs. GH reduces the synthesis of the matrix proteins fibronectin and collagen in 3T3 adipocytes.[72] In addition, GH increases lipoprotein lipase transcription in Ob1771 preadipocytes[73] (which can be overcome by the presence of glucocorticoid [dexamethasone] or transforming growth factor).[74]

Figure 2 summarizes the effects of GH on 3T3 preadipocyte proliferation and differentiation. Although the effects of GH on 3T3 cells are well established, it is not known whether these adipogenic effects of GH are a general phenomenon.

REFERENCES

1. **Rudman, D., Feller, A. G., Nagraj, H. S., et al.,** *New Engl. J. Med.*, 323, 1, 1990.
2. **Skaggs, S. R. and Crist, D. M.,** *Horm. Res.*, 35, 19, 1991.
3. **Gardner, M. J. and Flint, D. J.,** *J. Endocrinol.*, 124, 381, 1990.
4. **Flint, D. J. and Gardner, M. J.,** *J. Endocrinol.*, 137, 203, 1993.
5. **Takano, K., Hizuka, N., Shikume, K., et al.,** *Endocrinol. Jpn.*, 30, 79, 1983.
6. **Bassett, J. M. and Wallace, A. L. C.,** *Metab. Clin. Exp.*, 15, 933, 1966.
7. **Hart, I. C., Chadwick, P. M. E., Boone, T. C., et al.,** *J. Endocrinol.*, 224, 93, 1984.
8. **Byatt, J. C., Eppard, P. J., Veenhuzen, J. J., et al.,** *J. Endocrinol.*, 132, 185, 1992.
9. **Hall, T. R., Cheung, A., and Harvey, S.,** *Comp. Biochem. Physiol.*, 86A, 29, 1987.
10. **Scanes, C. G.,** *Comp. Biochem. Physiol.*, 101A, 871, 1992.
11. **Bowden, C. R., White, K. D., Lewis, J. J., et al.,** *Metabolism*, 4, 237, 1985.
12. **Boyle, P. J. and Cryer, P. E.,** *Am. J. Physiol.*, 260, E395, 1991.
13. **DeFeo, P., Perriello, G., Torlane, E., et al.,** *Am. J. Physiol.*, 256, E835, 1989.
14. **Eisemann, J. H., Hammond, A. C., Bauman, D. E., et al.,** *J. Nutr.*, 116, 2504, 1986.
15. **Peel, C. J., Bauman, D. E., Gorewit, R. C., et al.,** *J. Nutr.*, 111, 1662, 1981.
16. **McDowell, G. H., Hart, I. C., Bines, J. A., et al.,** *Aust. J. Biol. Sci.*, 40, 191, 1987.
17. **Peel, C. J., Frank, T. J., Bauman, D. E., et al.,** *J. Dairy Sci.*, 66, 776, 1983.
18. **Pocius, P. A. and Herbein, J. H.,** *J. Dairy Sci.*, 69, 713, 1986.
19. **Lough, D. S., Muller, L. D., Kensinger, R. S., et al.,** *J. Dairy Sci.*, 72, 1469, 1989.
20. **Emmison, N., Agius, L., and Zammit, V. A.,** *Biochem. J.*, 274, 21, 1991.
21. **Fain, J. N., Kovacev, V. P., and Scow, R. O.,** *J. Biol. Chem.*, 240, 3522, 1965.
22. **Fain, J. N.,** *Adv. Enzyme Regulat.*, 5, 39, 1967.
23. **Goodman, H. M.,** *Endocrinology*, 114, 131, 1984.
24. **Goodman, H. M. and Grichting, G.,** *Endocrinology*, 113, 1697, 1983.
25. **Harvey, S., Scanes, C. G., and Howe, T.,** *Gen. Comp. Endocrinol.*, 33, 322, 1977.
26. **Campbell, R. M. and Scanes, C. G.,** *Proc. Soc. Exp. Biol. Med.*, 180, 513, 1985.
27. **Duquette, P. F., Scanes, C. G., and Muir, L. A.,** *J. Anim.Sci.*, 58, 1191, 1984.
28. **Barenton, B., Batifol, V., Combarnous, Y., et al.,** *Biochem. Biophys. Res. Commun.*, 122, 197, 1984.
29. **Fielder, P. J. and Talamantes, F.,** *Endocrinology*, 121, 493, 1989.
30. **Frigeri, L. G., Robel, G., and Stebbing, N.,** *Biochem. Biophys. Res. Commun.*, 104, 1041, 1982.
31. **Frigeri, L. G.,** *Endocrinology*, 107, 738, 1980.
32. **Okada, S., Chen, W. Y., Wiehl, P., et al.,** *Endocrinology*, 130, 2284, 1992.
33. **Goodman, H. M., Tai, L. R., and Chipkin, S. R.,** *Endocrinology*, 126, 441, 1990.
34. **Fagin, K. D., Lackey, S. L., Reagan, C. R., et al.,** *Endocrinology*, 107, 608, 1980.
35. **Gavin, J. R., Saltman, R. J., and Tollefsen, S. E.,** *Endocrinology*, 110, 637, 1982.
36. **Eden, S., Schwartz, J., and Kostyo, J. L.,** *Endocrinology*, 111, 1505, 1982.
37. **Grichting, G., Levy, L. K., and Goodman, H. M.,** *Endocrinology*, 113, 1111, 1983.
38. **Chipkin, S. R., Szecowka, J., Tai, L.-R., et al.,** *Endocrinology*, 125, 450, 1989.
39. **Goodman, H. M. and Coiro, V.,** *Endocrinology*, 109, 2046, 1981.
40. **Goodman, H. M.,** *Endocrinology*, 82, 1027, 1968.
41. **Goodman, H. M. and Kostyo, J. L.,** *Endocrinology*, 1080, 553, 1981.
42. **Campbell, R. M., Kostyo, J. L., and Scanes, C. G.,** *Proc. Soc. Exp. Biol. Med.*, 193, 269, 1990.
43. **Campbell, R. M., Kawauchi, H., Lewis, U. J., et al.,** *Proc. Soc. Exp. Biol. Med.*, 197, 409, 1991.
44. **Campbell, R. M., Chen, W. Y., Wiehl, P., et al.,** *Proc. Soc. Exp. Biol. Med.*, 203, 311, 1993.
45. **Campbell, R. M. and Scanes, C. G.,** *Proc. Soc. Exp. Biol. Med.*, 188, 177, 1988.
46. **Scanes, C. G., Aramburo, C., Campbell, R. M., et al.,** *Avian Endocrinology* (Ed. Sharp, P. J.). Journal of Endocrinology, London, 1994 (in prcss).

47. **Campbell, R. M. and Scanes, C. G.,** *Proc. Soc. Exp. Biol. Med.*, 184, 456, 1987.
48. **Solomon, S. S., Schwartz, Y., and Kawlinson, T.,** *Endocrinology*, 121, 1056, 1987.
49. **Vernon, R. G., Finley, E., and Flint, D. J.,** *Biochem. J.*, 242, 931, 1987.
50. **Watt, P. W., Finley, E., Cork, S., et al.,** *Biochem. J.*, 273, 39, 1991.
51. **Sechen, S. J., Dunshea, F. R., and Bauman, D. E.,** *Am. J. Physiol.*, E582, 1990.
52. **Sechen, S. J., McCutcheon, S. N., and Bauman, D. E.,** *Domest. Anim. Endocrinol.*, 6, 141, 1989.
53. **Goodman, H. M.,** *Metabolism*, 19, 849, 1970.
54. **Goodman, H. M.,** *Endocrinology*, 114, 131, 1984.
55. **Bjorgell, P., Rosberg, S., Isaksson, O., et al.,** *Endocrinology*, 115, 1151, 1984.
56. **Birnbaum, R. S. and Goodman, H. M.,** *Endocrinology*, 99, 1336, 1976.
57. **Walton, P. E., Etherton, T. D., and Evock, C. M.,** *Endocrinology*, 118, 2577, 1986.
58. **Magri, K. A., Adamo, M., Leroith, D., et al.,** *Biochem. J.*, 266, 107, 1990.
59. **Etherton, T. D., Evock, C. M., and Kensinger, R. S.,** *Endocrinology*, 121, 699, 1987.
60. **Marinchenko, G. V., McNamara, J. P., and Becker-Khaleel, B.,** *Proc. Soc. Exp. Biol. Med.*, 200, 57, 1992.
61. **Vernon, R. G.,** *Int. J. Biochem.*, 14, 255, 1987.
62. **Vernon, R. G. and Finley, E.,** *Biochem. J.*, 256, 873, 1988.
63. **Vernon, R. G., Barber, M. C., and Finley, E.,** *Biochem. J.*, 274, 1990, 1991.
64. **Snoswell, A. M., Finley, E., and Vernon, R. G.,** *Horm. Metab. Res.*, 22, 650, 1990.
65. **Morikawa, N., Nixon, T., and Green, H.,** *Endocrinology*, 114, 527, 1984.
67. **Guller, S., Corin, R. E., Myhnarcik, D. C., et al.,** *Endocrinology*, 122, 2084, 1988.
68. **Guller, S., Sonenberg, M., Wu, K. Y., et al.,** *Endocrinology*, 125, 2360, 1989.
69. **Guller, S., Sonenberg, M., and Corin, R. E.,** *Endocrinology*, 124, 325, 1989.
70. **Guller, S., Corin, R. E., Wu, K.-Y., et al.,** *Biochem. Biophys. Res. Commun.*, 163, 895, 1989.
71. **Guller, S., Corin, R. E., Wu, K.-Y., et al.,** *Endocrinology*, 129, 527, 1991.
72. **Guller, S., Allen, D. L., Corin, R. E., et al.,** *Endocrinology*, 130, 2609, 1990.
73. **Pradines-Figueres, A., Barcillini-Couget, S., Dani, C., et al.,** *J. Lipid Res.*, 31, 1283, 1992.
74. **Pradines-Figueres, A., Barcillini-Couget, S., Dani, C., et al.,** *Biochem. Biophys. Res. Commun.*, 166, 1118, 1990.

Chapter 22

Growth Hormone Action: Protein Metabolism

C. G. Scanes

CONTENTS

I. INTRODUCTION

Prolonged *in vivo* treatment with growth hormone (GH) increases muscle weight and improves nitrogen retention (see Chapters 19, 28, and 29). The mechanism by which these effects are mediated includes indirect effects on various tissues by hepatic and locally produced insulin-like growth factor I (IGF-I) and direct effects of GH on the use of amino acids for either gluconeogenesis or oxidation.[1] These effects are considered in this chapter.

II. AMINO ACID TRANSPORT IN MUSCLE TISSUE

Growth hormone is reported to exert a marked short-term effect on amino acid transport by muscle tissue *in vitro*. In studies with hemidiaphragms from hypophysectomized rats, GH stimulates the uptake of amino acids, as does insulin.[2–4] This phenomenon has been most extensively studied using transport of a nonmetabolizable amino acid, amino isobutyl acid (AIB).[3,4] As with other insulin-like effects, the 20-kDa variant of human GH (hGH) shows only about 20% of the activity of the 22K hGH.[2] There are reports that GH can increase AIB uptake by diaphragms from intact young rats.[4] This effect of GH is also observed in intact, fasted rats[5] and in thyroxine-treated dwarf mice.[6]

III. MUSCLE PROTEIN SYNTHESIS

Growth hormone increases net muscle protein synthesis (protein accretion) (Chapters 19 and 29). This is probably largely an indirect effect of GH mediated via IGF-I. In addition, GH can also directly stimulate muscle protein synthesis. For instance, GH at high but physiological concentrations increases the incorporation of [^{14}C]phenylalanine into proteins in hemidiaphragms from hypophysectomized rats incubated *in vitro*.[2,7] Similarly, GH, albeit at very high (possibly pharmacological) concentrations, enhances [^{3}H]leucine incorporation into proteins in hemidiaphragms from hypophysectomized rats,[3,5] fasted young rats,[4] and dwarf mice[6] but not generally from intact rodents.

0-8493-8697-7/95/$0.00+$.50

The effects of GH on muscle protein synthesis may be due to GH enhancing amino acid transport and hence the availability of amino acids for protein synthesis. *In vitro* studies in which the effect of GH on the rate of protein synthesis is determined relative to intracellular amino acid precursor-specific activity (e.g., [^{14}C]phenylalanine) have not been reported. It is considered unlikely that the direct effect of GH on muscle protein synthesis is of major physiological significance.

IV. INSULIN-LIKE EFFECT ON AMINO ACID UTILIZATION BY ADIPOSE TISSUE

Growth hormone stimulates the *in vitro* oxidation of [1-^{14}C]leucine to CO_2 by adipose tissue from hypophysectomized rats.[8] This effect is observed in the presence or absence of glucose in the incubation medium.[8] In addition, GH increases the incorporation of [2-^{14}C]leucine as a substrate into lipid.[8] As with other insulin-like actions of GH, this effect on leucine oxidation is transitory, with the adipose tissue becoming refractory with prolonged exposure to GH. The induction of refractoriness can be blocked by inhibitors of RNA/protein synthesis (respectively, actinomycin b and cycloheximide).[8]

V. PROTEIN DIGESTION AND AMINO ACID ABSORPTION

The available evidence indicates that GH has little, if any, effect on gut functioning. Although hypophysectomy in rats results in reduced growth of the gut mucosa and decreased levels of some digestive enzymes (alkaline phosphate, maltase, and sucrase),[9] these effects are not reversed by GH injections.[10] No changes in net nitrogen absorption are observed when GH is administered to cattle.[1] The only studies in which a direct effect of GH on amino acid transport across the gut was measured directly were in fish (Coho salmon) and in these GH failed to have any effect on proline transport.[11]

VI. BONE PROTEIN SYNTHESIS

Growth hormone stimulates cartilage and bone growth by increasing circulating and also local concentrations of IGF-I (discussed in detail in Chapter 19). GH injection into intact rats has been found to increase rapidly (i.e., within 50 min of GH administration) the fractional synthetic rate of tibial protein.[12] This occurs despite no discernable changes in circulating concentrations of IGF-I.[12] Local increases in IGF-I production cannot be precluded, however.

VII. COLLAGEN SYNTHESIS BY FIBROBLASTS

Although it is undoubtedly the case that changes in the skin can be induced by GH administration, these may represent indirect effects via increased circulating concentrations of IGF-I. For instance, hGH increases the strength and collagen content of rat skin.[13] Until recently, the direct effect of GH on collagen synthesis by fibroblasts had not been investigated. At physiological concentrations and *in vitro*, GH reduces both the synthesis and appearance of extracellular collagen and also the level of type 1 procollagen mRNA in chicken fibroblasts.[14] In view of the ability IGF-I to induce similar effects,[14] it is not possible to preclude GH acting via paracrine IGF-I.

VIII. WOUND HEALING

Wound healing has been reported to be improved by GH administration.[14–19] This is thought to be due in part to increased collagen formation by fibroblasts and hence improved mechanical strength of the wound.[14–20] Moreover, local injections of GH into wounds (specifically wound cylinders) increased the weight of ingrown granulation tissue and the level of IGF-I mRNA in the wound tissue.[17] There is no effect of GH on wound tissue GH receptor mRNA or IGF-I receptor mRNA or on the concentration of IGF-I in the wound fluid.[20] It is thus unclear if GH is exerting a direct effect or is increasing local IGF-I production.

IX. ERYTHROPOIESIS

Growth hormone may have some effect on erythropoiesis. In the presence of erythropoietin, GH has been observed to increase erythroid colony formation from bone marrow *in vitro*.[21] This may be viewed as GH potentiation of erythropoiesis. The erythropoietic effect of GH appears to be specific. Bovine GH stimulates erythroid colony formation in mouse bone marrow. Human GH enhances erythropoiesis in human bone marrow, whereas bGH has no effect.[21] It might be argued that the hGH effect is mediated via prolactin receptors. However, human placental lactogen, which has high lactogenic activity and low somatogenic effect, has little if any effect *in vitro* on erythropoiesis.[22] The absence of an effect of bGH in human marrow cells is not surprising in view of the species specificity of the human GH receptor (GHR).

X. ARCTIC FISH "ANTIFREEZE" PROTEIN

Some arctic fish have "antifreeze" polypeptides in their plasma that help them survive in waters containing ice. The synthesis of these polypeptides occurs in the winter and is under pituitary control. In hypophysectomized winter flounders, the synthesis of antifreeze protein continues throughout the year.[23,24] The pituitary hormone responsible for suppressing "antifreeze" polypeptide synthesis may well be GH. This is supported by the observed reduction of antifreeze polypeptide in hypophysectomized or intact flounders receiving injections of salmon GH.[25]

REFERENCES

1. **Eisemann, J. H., Hammond, A. C., Bauman, D. E., et al.,** *J. Nutr.*, 116, 2504, 1986.
2. **Cameron, C. M., Kostyo, J. L., Adamafio, P., et al.,** *Endocrinology*, 122, 471, 1988.
3. **Kostyo, J. L.,** *Ann. N.Y. Acad. Sci.*, 148, 389, 1968.
4. **Nutting, D. F.,** *Endocrinology*, 98, 1273, 1976.
5. **Nutting, D. F. and Coats, L. J.,** *Proc. Soc. Exp. Biol. Med.*, 156, 446, 1977.
6. **Nutting, D. F.,** *Endocrinology*, 99, 1423, 1976.
7. **Cameron, C. M., Kostyo, J. L., Rillema, J. A., et al.,** *Am. J. Physiol.*, 247, E639, 1984.
8. **Goodman, H. M.,** *Endocrinology*, 102, 210, 1978.
9. **Yeh, K. and Moog, F.,** *Dev. Biol.*, 47, 156, 1975.
10. **Yeh, K. and Moog, F.,** *Dev. Biol.*, 47, 173, 1975.
11. **Collie, N. L. and Stevens, J. J.,** *Gen. Comp. Endocrinol.*, 59, 399, 1985.
12. **Martinez, J. A., Del Barrio, A. S., and Larraldo, J.,** *Biochim. Biophys. Acta*, 93, 111, 1991.
13. **Jorgensen, P. H., Andreassen, T. T., and Jorgensen, K. D.,** *Acta Endocrinol.*, 120, 767, 1989.
14. **Granot, I., Halevy, O., Hurwitz, S., et al.,** *Mol. Cell. Endocrinol.*, 80, 1, 1991.
15. **Barbul, A., Rettura, G., Levenson, S. M., et al.,** *Am. J. Clin. Nutr.*, 37, 786, 1983.
16. **Hollander, D. M., Devereux, D. F., Marafino, B. J., et al.,** *Surg. Forum*, 35, 612, 1984.
17. **Pessa, M. E., Bland, K. I., Sitren, H. S., et al.,** *Surg. Forum*, 36, 6, 1985.
18. **Waago, H.,** *Lancet*, 1485, 1987.
19. **Jorgensen, P. H. and Andreassen, T. T.,** *Endocrinology*, 121, 1637, 1987.
20. **Zaizen, Y., Ford, E. G., Costin, G., et al.,** *J. Pediatr. Surg.*, 25, 70, 1990.
21. **Steenfos, H. H. and Jansson, J.-O.,** *J. Endocrinol.*, 132, 293, 1992.
22. **Golde, D. W., Bersch, N., and Li, C. H.,** *Science*, 196, 1112, 1977.
23. **Hew, C. L. and Fletcher, G. L.,** *FEBS Lett.*, 99, 337, 1979.
24. **Fourney, R., Fletcher, G. L., and Hew, C. L.,** *Gen. Comp. Endocrinol.*, 54, 392, 1984.
25. **Idler, D. R., Fletcher, G. L., Belkhode, S., et al.,** *Gen. Comp. Endocrinol.*, 74, 327, 1989.

Chapter 23

Growth Hormone Action: Reproductive Function

C. G. Scanes and S. Harvey

CONTENTS

I. INTRODUCTION

Ogilvy-Stuart and Shalet[1] concluded that "... the action of GH on the reproductive axis is more akin to "fine tuning" than that of a "major player ..." Nevertheless, such a role may have physiological importance. Accordingly, growth hormone (GH) has been termed a "cogonadotropin," and may serve a modulatory role in both male and female reproduction.

0-8493-8697-7/95/$0.00+$.50

II. MALE REPRODUCTION

A. GH DEFICIENCY MODELS

The effects of GH on reproductive functioning have been examined in four GH deficiency models, namely hypophysectomized rats, genetically GH deficient rats, rats passively immunized against GH-releasing hormone (GHRH) and in the clinical (human) setting.

1. Hypophysectomized Rats

Ablation/replacement therapy studies suggest some role for GH in the control of reproduction in the male. For instance, GH administration to hypophysectomized rats increases both testicular ^{125}I-labeled human chorionic gonadotropin (HCG) binding (reflecting more luteinizing hormone [LH] receptors) and circulating testosterone concentrations.[2] Although GH replacement therapy alone was not observed to influence the weight of the ventral prostate or seminal vesicle in hypophysectomized rats, GH did augment the effect of LH.[2] Moreover, *in vivo* GH treatment of hypophysectomized rats has been found to increase insulin-like growth factor I (IGF-I) receptors in Leydig cells,[3] together with androgen receptors, androgen receptor mRNA, IGF-I mRNA, and IGF-I receptor mRNA in the prostate.[4]

2. Mutant Dwarf Rats

Testicular functioning is normal in mutant dwarf rats with isolated GH deficiency. There are no differences in either androgen production (as indicated by both serum testosterone and epididymal weights) or spermatogenesis (by histological/morphometric analysis of the seminiferous tubules) compared to a control strain of rats.[5-7] Testicular weight is reduced, however, in these dwarf rats.[5-7] This may reflect some relationship between testicular weight and either body size or circulating concentrations of GH and/or IGF-I. In other rodent (mouse) genetic models, GH deficiency is accompanied by abnormalities in other pituitary hormones. Thus studies in which testicular functioning is modified in these models may be excluded from consideration, albeit cautiously.

3. GHRH Immunization

Passive immunization of young male rats with antisera against rat GHRH depresses growth and delays sexual maturation.[8] For instance, in these GH-deficient rats, testicular size is reduced due to decreased plasma concentrations of follicle-stimulating hormone (FSH) while spermatogenesis is delayed.[8] It is not clear, however, if these effects on reproduction are a consequence of the GH deficiency, the reduced size of the rats, or other (specific or nonspecific) effects of the antisera against GHRH.

4. Clinical Studies

There is evidence that GH exerts a modulatory effect on reproduction in human males. This comes from studies providing GH therapy to GH-deficient individuals. For instance, increasing the dose of human GH administered to preadolescent boys was reported to be followed by a faster progression through puberty.[9] Moreover, GH treatment potentiated the increase in both testosterone and androstenedione evoked by HCG.[10]

B. GH-REPLETE MAMMALS

There is little evidence to suggest that GH can evoke consistent changes in the reproductive physiology of the normal male. Examples of studies in which plasma concentrations of GH are artifically elevated are considered below.

1. Pigs

Exogenous GH has little or no effect on reproductive development in male pigs. For instance, the administration of porcine GH to growing pigs had no effect on genital tract development or testicular androgen biosynthesis or the production of the pig pheromone, 16-androstene.[11] This and related steroids give meat from male pigs the peculiar flavor, which is known as boar taint or boar taint odor. However, in view of the ability of GH to reduce lipid content of meat and because 16-androstene is fat soluble, it is not surprising that GH reduces boar taint content of meat.[11]

2. Transgenic Mice

There have been a number of studies examining reproduction in transgenic male mice expressing hybrid genes including human *GH* and bovine *GH*. In view of the prolactin-like activity of human GH, caution is needed in drawing conclusions from transgenic mice expressing human GH. However, transgenic mice expressing either human GH or bovine GH show increases in both testicular and seminal vesicle weights[12] (although this is not apparent if the weights of these organs are expressed relative to body weight).[12,13] Nevertheless, transgenic mice expressing bovine GH show no differences in either fertility or spermatozoa production[13] but they have a higher content of androgen receptors in the seminal vesicles.[13]

III. GH AND FEMALE REPRODUCTION

A. *IN VITRO* EFFECTS

There is increasing evidence that GH can influence functioning of mammalian ovarian cells *in vitro*. Although the data do not per se indicate a major role for GH, they at least are supportive of GH exerting a modulatory role in female reproduction.

1. Rats

There is considerable evidence that GH can affect rat ovarian cells. Studies with granulosa cells from estrogen-treated hypophysectomized immature rats provide strong evidence that GH can markedly change the response of the cells to gonadotropins. In 3-day cultures, GH increases the production of both progesterone (P_4) and 20α-progesterone in the presence of FSH or cAMP or the adenylate cyclase stimulator, forskolin, but GH is not effective alone.[14] Moreover, GH enhances FSH-induced differentiation of the granulosa cells as indicated by increases in the number of LH receptors.[14] It is likely that the effects of GH on rat granulosa cells are mediated by stimulating ovarian IGF-I production, which in turn influences granulosa cell maturation. The administration of GH to hypophysectomized rats has been shown to increase ovarian IGF-I concentrations[15]. Moreover, IGF-I has been demonstrated to be capable of profoundly influencing rat ovarian cell functioning; these effects are similar, if not identical, to those evoked by GH.[16-18]

2. Pigs

Effects of GH have been demonstrated in granulosa cells from prepubertal pig ovaries. In this model, GH elevates basal progesterone production and markedly increases P_4 synthesis in the presence of FSH.[19] It might be noted that GH also increases release of IGF-I from the pig granulosa cells.[19] It is therefore possible that these effects of GH on ovarian cells may be mediated by local production of IGF-I.

3. Humans

Interpretations of investigations of the ability of hGH to influence human ovarian functioning are complicated by two factors: (1) human GH has both GH and prolactin activity; and (2) non-primate GHs, which lack prolactin activity, are inactive in man (see Chapters 1 and 2). Thus, it is not possible to determine whether GH is affecting a particular function via a GH receptor (GHR) unless human prolactin is used in tandem in the studies as a negative control. Human GH has been found to enhance both basal and FSH-stimulated aromatization of testosterone to estradiol (E_2) by human granulosa cells *in vitro*.[20] Moreover, hGH increased both basal and induced progesterone synthesis by small but not large cells derived from human corpora lutea.[21]

4. Rabbits

Studies in which rabbit ovaries were perfused *in vitro* demonstrated that bGH alone can influence a number of aspects of ovarian function normally thought to be controlled by gonadotropins. For instance, GH increases follicular diameter, estrogen release, and oocyte meiosis maturation;[22] in the latter case GH increases geminal vesicle breakdown from less than 10% to over 70%.[22]

B. *IN VIVO* EFFECTS

Although there is considerable evidence that exogenous GH can influence reproduction *in vivo*, the effects are not consistent across different species. It is, therefore, difficult to draw overall conclusions. Moreover, ready interpretation of data may be confounded by effects of GH on body mass and/or body fat, parameters that may be linked directly or indirectly to sexual maturation and reproductive functioning. Finally, it should be noted that GH may be exerting its effects on reproduction at the levels of the hypothalamus, the pituitary gland, or the gonad. Direct effects of GH on ovarian cells have been demonstrated in vitro (see above).

1. Primates

The available evidence suggests that GH has a stimulatory modulatory role in primates. For instance, chronic peripheral administration of hGH to rhesus monkeys accelerates sexual maturation as indicated by an earlier onset of first ovulation and premature increases in circulating concentrations of estradiol.[23] The administration of hGH to normal cycling females does not markedly influence the menstrual cycle. For instance, no changes in circulating concentrations of LH, FSH, E_2, or P_4 are observed irrespective of whether women are receiving hGH injections or not.[24] Similarly, the administration of recombinant GH to normally cycling women does not markedly influence the menstrual cycle. For instance, hGH injections administered to women (also receiving hCG and HMG treatment to induce ovulation for *in vitro* fertilization) do not influence circulating concentrations of LH, FSH, testosterone, sex-steroid binding globulin, or follicular fluid levels of E_2, P_4, or IGF-I.[25] However, there is some increase in the circulating concentrations of E_2 and P_4, in the follicular concentration of testosterone, and in granulosa 3β-hydroxysteroid dehydrogenase and aromatase mRNA following GH administration.[25] In some studies in which hGH treatment is combined with HMG/HCG therapy, more oocytes are collected that have increased sensitivity to HMG,[26,27] although this is a controversial finding.[25,28]

2. Rodents

In rats, there is evidence that high levels of GH inhibit reproduction but basal levels of GH are required for reproductive development. The administration of GH appears to delay puberty. Chronic peripheral administration of rat GH (rGH) to intact young rats delays puberty (vaginal opening) while increasing growth.[29] Similarly, implanting bovine GH (bGH) into the median eminence of young female rats delays puberty (vaginal opening).[30] Rats infected with plerocercoid larvae of the tapeworm (*Spirometra mansonoides*) also show delayed puberty.[31] These larvae produce a GH-like factor that depresses plasma concentrations of GH.[31] This delay in reproduction is reversed by exogenous GH.[31]

3. Pigs

The available evidence would suggest that GH has little effect on reproduction except when present at very high (pharmacological or pathological) levels. Daily injections of porcine GH (pGH) to young female pigs have no consistent effect on the age of puberty[32] or on maturation of reproductive organs.[33,34] For instance, pGH administration does not consistently influence either ovarian or uterine weights.[35,36] Injections of pGH have little discernable effects on the ability of PMSG/HCG to induce ovulation in gilts.[35,36] Although chronic treatment of young pigs (gilts) with GH does not influence circulating concentrations of LH or FSH, there are shifts in the secretory profile of LH (pulse frequency increased, amplitude decreased).[37] In transgenic gilts expressing bovine GH at high levels, plasma concentrations of LH but not FSH are reduced.[38] Porcine GH has no discernible effect on reproduction in adult pigs. The administration of pGH to pregnant pigs does not influence number of live or dead neonates, or the birth weight or survival rate of piglets.[39]

4. Cattle and Sheep

Growth hormone would appear to have some slight stimulatory effect on reproduction in cattle. Evidence for an important, if not essential, role for GH in puberty in cattle comes from studies involving active immunization of heifers against GHRH. This regimen results in decreases in circulating concentrations of GH and growth rate with an increase in adiposity, as might be expected.[40] In addition, only about 40% of the immunized cattle go through puberty compared with 100% of controls.[40] Similarly, a role for GH in the return to cycling of postpartum beef cattle is suggested, since cycling is delayed if the heifers are immunized against GHRH.[41] However, while these results may be attributable to GH deficiency, they may also reflect shifts in body composition.

In cattle, chronic administration of bovine GH has little effect on reproduction apart from some modulatory effects on the estrous cycle. No changes in the ovulation rate occur in heifers receiving bovine GH.[42] Similarly, GH does not affect the concentrations of circulating reproductive hormones (LH, FSH, E_2, and P_4) in heifers.[42] In addition, bovine GH has little effect on the estrous cycle of lactating cows, including no changes in the length of follicular or luteal phases of the cycle, or in follicular diameter,[43] or in the diameter of the preovulatory follicle, or in the circulating concentrations of E_2.[43] However, bGH treatment elevates circulating concentrations of P_4[43,44] and increases the LH secretory resonse to gonadotropin-releasing hormone (GnRH).[44] Although bovine GH does not appear to influence the number of large follicles (>5 mm in diameter),[42,45] GH does appear to stimulate follicular recruitment, as indicated by increased numbers of small follicles (2 to 5 mm).[45,46] There is little evidence of any effect of GH on reproduction in sheep; there is, for instance, no change in ovulation rate.[46]

C. GH AND VITELLOGENESIS

As opposed to the situation in viviparous mammals, many female vertebrates produce an egg with substantial quantities of yolk. The most obvious example is the yolks in the eggs of chickens and other birds. Yolk-containing eggs are also found in reptiles, amphibians, and fish. Yolk is deposited in the ovum in the ovary but yolk precursors are synthesized (vitellogenesis) in the liver. Vitellogenesis is primarily induced by estrogen. The major yolk precursor is vitellogenin, a phospholipoprotein.

Growth hormone would appear to be required for estrogen stimulation of vitellogenesis. This is supported by the absence of a full vitellogenic response (liver RNA for phosphoproteins and circulating concentrations of phosphoprotein or lipid) in hypophysectomized reptiles (lizards)[47] and the marked restoration of vitellogenesis following bGH replacement therapy.[47] Similarly, estrogen-induced synthesis of yolk precursors (incorporation of ^{32}P into phosphoprotein) is suppressed in hypophysectomized turtles and restored in bGH-treated hypophysectomized turtles.[48] Furthermore, estrogen can stimulate [^{35}S]methionine incorporation into vitellogenin *in vitro* in liver explants only if ovine GH (oGH) is also present in the incubation media.[49] In summary, GH appears to be required for estrogen induction of vitellogenesis. This is perhaps not surprising in view of the effect of GH on estrogen receptor concentrations (see the next section). Moreover, GH per se appears capable of stimulating vitellogenin synthesis in at least some amphibians. This is supported by *in vitro* studies with frog (*Rana esculenta*) liver explants incubated in the presence of varying doses of mammalian or frog GH.[50] Under these circumstances, vitellogenin synthesis was consistently increased by GH.[50]

D. ESTROGEN RECEPTORS

1. Liver

The induction and/or maintenance of estrogen receptors in the liver appears to require GH. This is supported by studies in rats and also in the turtle. Hypophysectomy reduces liver estrogen receptor concentrations in the livers of ovariectomized rats and this effect is overcome by replacement therapy with either hGH or bGH if infused,[51] but not if given by intermittent administration.[52] Similarly, hGH infusion into male rats increases the total hepatic estrogen receptor concentration[52] while decreasing the concentration of the hepatic atypical steroid-binding protein.[52] Moreover, rGH (rat GH) depresses the concentration of a nonreceptor estrogen-binding protein in rat liver.[51] In turtle *(Chrysemys picta)*, GH is required for the maintenance (induction) of hepatic estrogen receptors with decreases in estrogen receptor concentrations following hypophysectomy and partial restoration with oGH replacement therapy.[53] Similarly, estrogen and GH administration synergistically increased the number of "occupied" nuclear estrogen receptors in the liver of an oviparous lizard.[54]

2. Other Organs

The administration of GH does not affect body growth or the growth of the mammary glands or uterus in guinea pigs.[55] Moreover, GH does not influence mammary estrogen receptor. In marked contrast, GH increases uterine estrogen receptors (cytosolic by 10-fold; nuclear by 4.9-fold).[55]

IV. SEX STEROID METABOLISM

Growth hormone can influence the metabolism of sex steroids in male rats, such that an apparent female or feminized pattern of circulating GH concentrations emerges.[56] For instance, infusion of GH into male rats reduces both 3α-hydroxysteroid dehydrogenase and 3β-hydroxysteroid dehydrogenase activity while increasing 5α-reductase activity in the liver.[57] Infusion of hGH into male rats also feminizes steroid hormone metabolism in rat liver with a considerable reduction in microsomal 2-hydroxylation of estradiol to catechol estrogens.[58] *In vitro* recombinant human GH has been found to depress both 6β-hydroxylase and 16α-hydroxylase activity induced by glucocorticoids in hepatocytes from either male or female rats.[59]

Studies with the dwarf rat with an isolated GH deficiency unexpectedly revealed a normal maturation pattern of liver microsome P-450 enzymes and 5α-reductase in the female.[60] It had previously been thought that endogenous GH (continuously released) in prepubertal female rats was acting to induce steroid metabolizing enzymes. Thus no firm conclusions can be reached as to the physiological significance of the feminizing effect of GH on the rat liver.

V. REPRODUCTION IN FISH

There is a growing body of evidence that GH is required for normal reproductive functioning in fish. It might have been possible to dismiss the results of some studies in which mammalian GH was observed to influence gonadal function in fish. For instance, ovarian growth in Coho salmon *in vivo* is stimulated by the administration of bovine GH.[61,62] Moreover, treatment of hypophysectomized killifish (*Fundulus heteroclitus*) with this GH partially restores testicular 3β-hydroxysteroid dehydrogenase levels.[63] These data might be explicable in terms of gonadotropin contamination of the GH preparation, or as being artifactual due to the heterologous hormone preparation (due perhaps to effects via the prolactin receptor).[64] However, studies have now been reported with recombinant fish GH. This is an obvious advantage in that an evolutionarily distant GH is not being employed and that contamination with gonadotropin(s) is not possible. It is clear that GH does, indeed, promote gonadal functioning in fish (for details, see below).

In vivo studies with hypophysectomized adult male killifish unequivically demonstrated that fish GH replacement therapy increased testicular weight (as indicated by the gonadosomatic index), the plasma concentration of testosterone, and testosterone synthesis *in vitro*.[64] GH administration was accompanied by elevated plasma concentrations of estradiol in hypophysectomized female killifish.[64] Similarly, the production of the major sex steroid *in vitro* (testosterone from the testes, E_2 from the ovary) is increased in the presence of fish GH per se with either killifish or trout gonadal tissue.[64] The situation is not, however, as clearcut as might be envisaged. For instance, with trout testicular cells incubated *in vitro*, fish GH decreases both basal and gonadotrophin-induced production of 11-ketotestosterone while increasing release of 17α,20β-hydroxyprogesterone.[65] In the goldfish, GH alone has no effect on steroidogenesis *in vitro*, but does potentiate the stimulatory influence of gonadotropins on ovarian follicles (testosterone or E_2 release).[66] In summary, GH appears to be capable of directly influencing gonadal function in fish.

VI. PLACENTAL GROWTH HORMONE

The *hGH-N* gene is expressed in the pituitary gland and encodes human GH. *hGH-V*, another member of the GH gene family (additionally comprising *hCS-A*, *hCS-B*, and *hCS-L*) (see Chapter 4), is at least 92% homologous to the *hGH-N* gene,[67] and is expressed in syncytiotrophoblasts[68,69] and epithelium[70,71] of term placentas[72-74] and in choriocarcinoma cells,[75] leading to the production of placental GH. The synthesis of this GH moiety may also occur in hematopoietic cells of the syncytiotrophoblast, because the *GH* gene is transcribed within cells of the immune system (see Chapter 24).

Placental GH is a 191-amino acid, 22-kDa protein that differs from pituitary GH by 13 amino acid residues. It is more basic and contains a unique N-linked glycosylation site at ASN-140.[73,76-78] The glycosylation of this site results in a protein with a molecular mass of 25 kDa.[79] Both GH-N and GH-V are encoded by an 800-bp mRNA.[80] The primary transcript of the *hGH-V* gene can, however, be alternatively spliced (unlike the *hGH-N* gene) to generate a 1250-bp mRNA that retains intron D between the fourth and fifth exons.[81,82] This splice variant, hGH-V2, accounts for approximately 15% of the hGH-V mRNA and encodes a third placental GH protein with a molecular mass of 26 kDa.[70]

A placental GH (mGH-V) is also produced in rhesus monkeys[83] and its mRNA is detectable from day 18 of pregnancy, approximately 9 days after implantation. The abundance of mGH-V increases from the first to the second trimester and then remains relatively constant, similar to hGH-V mRNA.[81] In contrast, the placentas of rodents appear to express placental lactogens and prolactin-like proteins but not placental GH proteins.[84,85]

A. TISSUE-SPECIFIC EXPRESSION OF THE *HGH-V* GENE

Several mechanisms appear to play a role in the tissue-specific regulation of the *GH* genes. DNase 1 hypersensitivity sites in chromatin often signal the presence of *cis*-acting elements involved in the control of gene transcription.[86] These sites are distributed in a complex pattern along the *hGH* locus, and most appear to be restricted to cells of placental origin. Placental-specific hypersensitive sites within the first intron of the *hGH-N* and *hGH-V* genes have been identified.[87] It has been suggested that extensive alteration in chromatin structure of the *hGH* locus may, therefore, be required for transcriptional activation and control in placental cells.

In contrast to the *hGH-N* gene, *hGH-V* appears to have only a single site for pit-1/GHF-1 binding. This is the distal site described in *hGH-N*, located –140/–107 bases upstream of the mRNA start site,[88] and the presence of only a single binding site may account for the weaknesses of the *GH-V* promoter in transfection assays. The distal binding site has only a single base difference compared to hGH-N, whereas the proximal pit-1/GHF-1 site contains seven base changes, which clearly prevent efficient pit-1/GHF-1 binding. Other members of the placental *GH* family do not share these sequence changes. The chorionic somatomammotropin gene (*CS*) shares sequence homology with *hGH-N* at its proximal pit-1/GHF-1-binding site, and pit-1/GHF-1 has been shown to bind to this region but not the distal region in the *CS* gene.[89] These data indicate that the pattern of binding is distinct for the placental members of the GH family. The *GH-V* gene may also contain a T_3-responsive region, as this hormone has been shown to increase hGH-V mRNA levels in placental carcinoma cells,[88] but not in pituitary tumor cells.[90]

B. PITUITARY-SPECIFIC REPRESSION OF PLACENTAL *GH* GENES

Negative regulation is implicated in the transcriptional control of several genes, including *hGH-N*.[67] Highly conserved regions in the distal 5′ flanking sequences of the placental (*CS* and *GH-V*) genes have been identified that differ from those in the *GH-N* gene. These DNA elements occur in a 263-bp fragment located about 2 kb upstream of the placental genes and are able to repress placental promoter activity >98% in pituitary cells after gene transfer, but they permit efficient promoter activity in placental cells.[91–93] These sequences, PSF-A and PSF-B, appear to bind a single protein factor (pituitary-specific factor), which is probably absent from placenta but appears to be ubiquitous otherwise. This mechanism may, therefore, play a role in tissue-specific expression/repression of *GH-V*; however, the *GH-V* gene has been shown to be weakly expressed in the pituitary,[94] indicating other factors may play a more important role in expression.

C. SYNTHESIS AND RELEASE IN PREGNANCY

Coordinated expression of the *hCS-A*, *hCS-B*, *hCS-L*, and *hGH-V* genes occurs in the placenta during gestation. A 5- to 10-fold increase in hCS-B, hCS-L, and hGH-V mRNA levels per unit of villus tissue occurs between 8 and 39 weeks of gestation, whereas a 30-fold increase in hCS-A occurs during the same period.[81] In each case, gene expression is maximally increased between 12 and 20 weeks of gestation and then plateaus through term. The relative increase in the abundance of the *hCS-A* and *hGH-V* gene transcripts is, however, at least tenfold less than the increase in circulating hCS and hGH-V concentrations during the same period. The developmental profiles of the hormones in maternal sera are thus likely to reflect increases in both placental mass and transcriptional activity, although the stability in processing or transport of the gene transcripts may also be modified during fetal development.[81] A developmental difference in the processing of the hGH-V primary transcript is indicated by the relative increase in the proportion of hGH-V2 mRNA during fetal development, being 5 to 7% of the total hGH-V mRNA in the first trimester and 15 to 20% at term.[81] This is likely to reflect developmental control over splice site selection or increased stability of hGH-V2 mRNA. In either case, the finding that levels of hGH-V2 mRNA are controlled independently from hGH-V mRNA suggests different functional roles for hGH-V and hGH-V2 during gestation.

During early pregnancy, pituitary GH is the only measurable GH in maternal serum and is secreted in a highly pulsatile pattern.[95,96] However, from 15 to 17 weeks of gestation until term, pituitary GH in serum is progressively replaced by increasing levels of placental GH (Figure 1), reaching the highest serum GH concentrations found in females.[74] This increase in maternal GH concentrations is directly related to the size of the

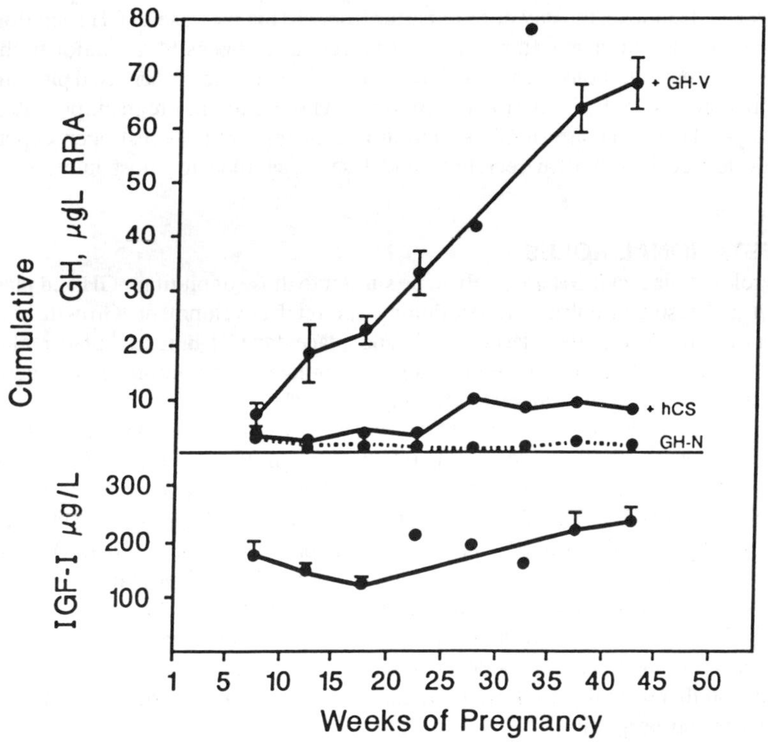

Figure 1 Serum growth hormone (GH) and insulin-like growth factor I (IGF-I) levels in 66 women during pregnancy. Placental GH (GH-V) concentrations were determined by a human liver radioreceptor assay whereas immunoreactive human chorionic somatomammotropin (hCS) and pituitary GH (GH-N) concentrations were determined by immunoradiometric assays. (From Daughaday, W. H., Trivedi, B., Winn, H. N., and Yan, H., Hypersomatotropism in pregnant women, as measured by a radioreceptor assay, *J. Clin. Endocrinol. Metab.*, 70(1), 215–221, 1990. Reprinted with permission.)

fetoplacental unit and is reduced in pregnancies with intrauterine growth retardation.[97] This increase in placental GH release accompanies an elevation in peripheral plasma GH-releasing hormone (GHRH) levels,[98] which may reflect transcriptional activation of the *GHRH* gene in placental or peripheral tissues.[99] Placental GH levels subsequently fall with the onset of labor, probably as a result of decreased uteroplacental blood flow or increased metabolism, by increased placental protease activity.[95,97] These developmental changes in maternal GH concentrations are not, however, paralleled in the fetus, because placental GH does not cross the fetal-placental barrier.[100] The secretion of placental GH is not episodic (unlike pituitary GH), and maternal serum GH concentrations remain relatively constant during a 24-h period. This increase in placental GH secretion is causally related to the simultaneous suppression of pituitary GH release, which becomes unresponsive to hypothalamic GHRH,[100] arginine, and insulin[101,102] and to other provocative tests of GH release.[103,104] These diminished responses may be mediated by the accompanying increase in serum IGF-1 concentrations,[97] which are not pituitary GH dependent in late pregnancy.[105] This diminution of pituitary GH secretion does not occur, however, in acromegalic women,[105] because adenomatous somatotrophs are resistant to IGF-I feedback.

Maternal serum GH concentrations also progressively increase during pregnancy in rats[106,107] and decline on the day prior to parturition.[108] This increase in GH secretion may be causally related to increased secretion of placental lactogens.[108] Gestational changes in serum GH concentrations in rats are, however, a consequence of increased pituitary GH secretion, contrary to the inhibition of pituitary GH secretion in pregnant primates. The pregnancy-induced increase in GH secretion in rats appears to be estrogen dependent, inversely related to prolactin secretion, and less susceptible to stress-induced inhibition.[109]

D. GESTATIONAL ROLES

The roles of placental GH are likely to be similar to those of pituitary GH and placental lactogen and result in nutrient repartitioning for fetal development. Growth hormone receptors are present in the placenta[110,111] and placental GH binds to placental membranes.[112] Placental GH may therefore also have endocrine, paracrine, and/or autocrine actions within the placenta, mediated through IGF-I.[68]

In comparison with hGH, placental GHs selectively bind to somatogenic (rather than lactogenic) receptors[112,113] and to plasma GH-binding proteins.[114] mGH-V also contains two of the four residues shown to be critical for high-affinity hGH binding to the GH receptor.[83] Placental GHs, therefore, have potent growth-promoting properties in transgenic mice[115] and hypophysectomized rats[113] but are poorly lactogenic.[116] hGH-V is also a potent somatogen in humans and some of the manifestations of late pregnancy, such as the coarsening of features, are probably related to its high circulating level in maternal serum.[114] These physical manifestations of GH excess are, however, relatively mild and subtle, reflecting the relatively modest rise in circulating IGF-I levels. Tissue resistance to GH action therefore results during pregnancy, probably as a result of the high levels of estrogens and progestins.[117,118]

Despite these marked changes in the GH axis in pregnancy, it is remarkable that a number of cases have been reported in which maternal serum GH-V fails to rise and this has been shown to result from homozygous deletion of both *hCS-A* and *hCS-B* genes, and therefore, likely includes the *GH-V* gene.[119] It is surprising that fetal growth is unimpaired in this condition and that abnormalities of maternal metabolism have not been recognized. It is therefore likely that maternal GH secretion is not suppressed in this condition and that it compensates for the loss of placental GHs.

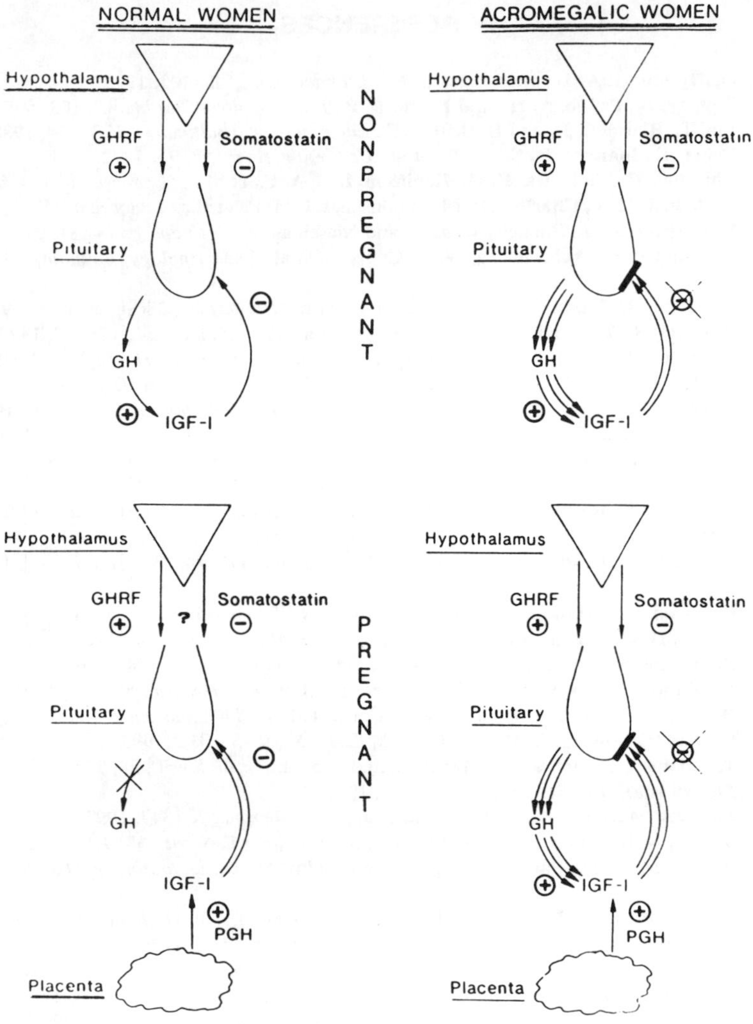

Figure 2 Proposed model for regulation of the secretion of pituitary and placental GH in normal and acromegalic women (From Beckers, A., Stevenaert, A., Foidart, J. M., et al., *J. Clin. Endocrinol. Metab.*, 71, 725, 1990. Reprinted with permission.)

VII. MAMMARY GROWTH HORMONE

In addition to its presence in the placenta, GH has been reported to be produced in the dog mammary gland.[120] In this species, progestin induces GH excess, which is present even after hypophysectomy. Normal levels of circulating GH are, however, observed after removal of the mammary gland. High levels of GH immunoreactivity have been demonstrated in this tissue, primarily in hyperplastic ductular epithelial elements and neoplastic epithelium of mammary tumors. These data, therefore, indicate that GH may have endocrine, paracrine, or autocrine actions within the mammary gland, in which GH receptors and GH actions have also been documented (see Chapters 16, 29, and 30).

REFERENCES

1. Ogilvy-Stuart, A. L., and Shalet, S. M., *J. Endocrinol.*, 135, 405, 1992.
2. Zipf, W. B., Payne, A. H., and Ketch, R. P., *Endocrinology (Baltimore)*, 103, 595, 1978.
3. Lin, T., Blaisdell, J., and Haskell, J. F., *Endocrinology (Baltimore)*, 123, 134, 1988.
4. Reiter, E., Bonnett, P., Sente, B., et al., *Cell. Endocrinol.*, 88, 77, 1992.
5. Charlton, H. M., Clark, R. G., Robinson, I. C. A. F., et al., *J. Endocrinol.*, 119, 51, 1988.
6. Bartlett, J. M. S., Charlton, H. M., Robinson, I. C. A. F., et al., *J.Endocrinol.*, 126, 193, 1990.
7. Spiteri-Grecht, J., Bartlett, J. M. S., and Nieschlag, E., *J. Endocrinol.*, 131, 279, 1991.
8. Arsenijevic, Y., Wehrenberg, W. B., Conz, A., et al., *Endocrinology (Baltimore)*, 124, 3050, 1989.
9. Stanhope, R., Albanese, A., Hindmarsh, P., et al., *Horm. Res.*, 38(Suppl. 1), 9, 1992.
10. Balducci, R., Toscano, V., Mangiantini, A., et al., *Acta Endocrinol.*, 128, 19, 1993.
11. Bonneau, M., Meadus, M. J., and Squires, E. J., *Can. J. Anim. Sci.*, 72, 537, 1992.
12. Bartke, A., Naar, E. M., Johnson, L., et al., *J. Reprod. Fertil.*, 95, 109, 1992.
13. Prins, G. S., Cecim, M., Birch, L., et al., *Endocrinology (Baltimore)*, 131, 2016, 1992.
14. Jia, X.-C., Kalmijn, J., and Hsueh, A. J. W., *Endocrinology (Baltimore)*, 118, 1401, 1986.
15. Davoren, J. B. and Hsueh, A. J. W., *Endocrinology (Baltimore)*, 118, 888, 1986.
16. Davoren, J. B., Hsueh, A. J. W., and Li, C. H., *Am. J. Physiol.*, 249, E26, 1984.
17. Adashi, E. Y., Resnick, C. E., Svoboda, M. E., et al., *Endocrinology (Baltimore)*, 116, 2369, 1985.
18. Adashi, E. Y., Resnick, C. E., Svoboda, M. E., et al., *Endocrinology (Baltimore)*, 116, 2135, 1985.
19. Hsu, C.-J., and Hammond, J. H., *Endocrinology (Baltimore)*, 121, 1343, 1987.
20. Mason, H. D., Martikainen, H., Beard, R. W., et al., *J. Endocrinol.*, 126, R1, 1990.
21. Di Simone, N., Caruso, A., Castellani, R., et al., *Fertil. Steril.*, 60, 47, 1993.
22. Yoshimura, Y., Iwashita, M., Makamura, Y., et al., *Fertil. Steril.*, 59, 917, 1993.
23. Wilson, M. E., Gordon, T. P., Rudman, C. G., et al., *J. Clin. Endocrinol. Metab.*, 68, 69, 1989.
24. Ovesen, P., Moller, N., Moller, J., et al., *Fertil. Steril.*, 59, 311, 1993.
25. Tapanainen, J., Orava, M., Martikainen, H., et al., *Fertil. Steril.*, 58, 726, 1992.
26. Jacobs, H., *Horm. Res.*, 38(Suppl. 1), 14, 1992.
27. Volpe, A., Artini, P. G., Barreca, A., et al., *Hum. Reprod.*, 7, 1347, 1992.
28. Shaker, A. G., Yates, R. W. S., Fleming, R., et al., *Fertil. Steril.*, 58, 919, 1992.
29. Groesbeck, M. D., Parlow, A. F., and Daughaday, W. D., *Endocrinology (Baltimore)*, 120, 1963, 1987.
30. Advis, J. P., Smith White, S., and Ojeda, S. R., *Endocrinology (Baltimore)*, 108, 1343, 1981.
31. Ramaley, J. A. and Phares, C. K., *Endocrinology*, 106, 1989, 1980.
32. Andres, C. J., Green, M. L., Clapper, J. A., et al., *J. Dairy Sci.*, 69, 3754, 1991.
33. Bryan, K. A., Hammond, J. M., Canning, S., et al., *J. Anim. Sci.*, 67, 196, 1989.
34. Bryan, K. A., Clark, A. M., and Haagen, D. R., *J. Anim. Sci.*, 68, 2357, 1990.
35. Kirkwood, R. N., Thacker, P. A., Korchinski, R.S., et al., *Domest. Anim. Endocrinol.*, 5, 317, 1988.
36. Kirkwood, R. N., Thacker, P. A., and Laarveld, B., *Can. J. Anim. Sci.*, 69, 265, 1989.
37. Gilbertson, J., Kirkwood, R. N., and Thacker, P. A., *Can. J. Anim. Sci.*, 71, 717, 1991.
38. Guthrie, H. D., Pursel, V. G., Miller, K. F., et al., *Domest. Anim. Endocrinol.*, 8, 423, 1991.
39. Kveragas, C. L., Seerley, R. W., Martin, R. J., et al., *J. Anim. Sci.*, 63, 1877, 1986.
40. Simpson, R. B., Armstrong, J. D., Harvey, R. W., et al., *J. Anim. Sci.*, 69, 4914, 1991.
41. Simpson, R. B., Armstrong, J. D., and Harvey, R. W., *J. Anim. Sci.*, 70, 1478, 1992.
42. Gong, J. G., Bramley, T., and Webb, R., *Biol. Reprod.*, 45, 941, 1991.
43. Schemm, S. R., Deaver, D. R., Griel, L. C., et al., *Biol. Reprod.*, 42, 815, 1990.
44. Gallo, G. F. and Block, E., *Can. J. Anim. Sci.*, 71, 343, 1991.
45. Gong, J. G., Bramley, T. A., and Webb, R., *J. Reprod. Fertil.*, 97, 247, 1993.
46. Davis, S. R., Smith, J. F., and Gluckman, P. D., *Reprod. Fertil. Dev.*, 2, 173, 1990.
47. Callard, I. P., Banks, S. J., and Banks, W. L., *Comp. Biochem. Physiol.*, 41B, 503, 1972.
48. Ho, S.-M., Taylor, S., and Callard, I. P., *Gen. Comp. Endocrinol.*, 48, 254, 1982.
49. Ho, S.-M., Wangh, L. J., and Callard, I. P., *Comp. Biochem. Physiol.*, 81B, 467, 1985.
50. Carnevali, O., Mosconi, G., Yamamoto, K., et al., *Gen. Comp. Endocrinol.*, 88, 466, 1992.
51. Norstedt, G., Wrange, O., and Gustafsson, J.-A., *Endocrinology (Baltimore)*, 108, 1190, 1981.

52. **Norstedt, G.,** *Endocrinology,* 40, 2107, 1982.
53. **Riley, D., Heisermann, G. J., MacPherson, R., et al.,** *J. Steroid Biochem.,* 26, 41, 1987.
54. **Paulucci, M., et al.,** *Mol. Cell. Endocrinol.,* 66, 101, 1989.
55. **Bezecny, I., Bartova, J., and Skarda, J.,** *J. Endocrinol.,* 134, 5, 1992.
56. **Mode, A., Norstedt, G., Simic, B., et al.,** *Endocrinology (Baltimore),* 108, 2103, 1991.
57. **Lax, E. R., Rumstadt, F., Plasczyk, H., et al.,** *Endocrinology,* 113, 1043, 1983.
58. **Jellinck, P. H., Quail, J. A., and Crowley, C. A.,** *Endocrinology (Baltimore),* 117, 2274, 1985.
59. **Vind, C., Dich, J., and Grunnet, N.,** *Biochem. Pharmacol.,* 44, 1523, 1992.
60. **Bullock, P., Gemzik, B., Johnson, D., et al.,** *Proc. Natl. Acad. Sci. U.S.A.,* 88, 5227, 1991.
61. **Higgs, D. A., Donaldson, E. M., Dye, H. M., et al.,** *J. Fish. Res. Board Can.,* 33, 1585, 1976.
62. **Higgs, D. A., Fagerlund, U. H. M., McBride, J. R., et al.,** *Can. J. Zool.,* 55, 1048, 1977.
63. **Pickford, G. E., Lofts, B., Bara, G., et al.,** *Biol. Reprod.,* 7, 370, 1972.
64. **Singh, H., Griffith, R. W., Takahashi, A., et al.,** *Gen. Comp. Endocrinol.,* 72, 144, 1988.
65. **LeGac, F., Ollitrault, M., Lori, M., et al.,** *Biol. Reprod.,* 46, 949, 1992.
66. **Vander Kraaf, G., Rosenblum, P. M., and Peter, R. E.,** *Gen. Comp. Endocrinol.,* 79, 233, 1990.
67. **Parks, J. S.,** *Acta Paediatr. Scand.,* 349, 127, 1989.
68. **Scippo, M. L., Frankenne, F., Hooghe-Peters, E. L., et al.,** *Mol. Cell. Endocrinol.,* 92, 7, 1993.
70. **Liebhaber, S. A., Urbanek, M., Ray, J., et al.,** *J. Clin. Invest.,* 83, 1985, 1989.
71. **Jara, C. S., Salud, A. T., Bryant-Greenwood, G. D., et al.,** *J. Clin. Endocrinol. Metab.,* 69, 1069, 1989.
72. **Frankenne, F., Rentier-Delrue, F., Scrippo, M. L., et al.,** *J. Clin. Endocrinol. Metab.,* 64, 635, 1987.
73. **Frankenne, F., Scrippo, M. L., Beeuman, J. V., et al.,** *J. Clin. Endocrinol. Metab.,* 71, 15, 1990.
74. **Cooke, N. E., Ray, J., Watson, M. A., et al.,** *J. Clin. Invest.,* 82, 270, 1988.
75. **Nickel, B. E., Bock, M. E., Nachtigal, M. W., et al.,** *Mol. Cell. Endocrinol.,* 91, 159, 1993.
76. **Chen, E. Y., Liao, Y. C., Smith, S. H., et al.,** *Genomics,* 4, 479, 1989.
77. **Igout, A., Scippo, M. L., Frankenne, F., et al.,** *Arch. Inst. Physiol. Biochem.,* 96, 63, 1988.
78. **Evain-Brion, D., Alsat, E., Mirlesse, V., et al.,** *Horm. Res.,* 33, 256, 1990.
79. **Ray, J., Jones, B. K., Liebhaber, S. A., et al.,** *Endocrinology,* 125, 566, 1989.
80. **Igout, A., Van Beeumen, J., Frankenne, F., et al.,** *Biochem. J.,* 295, 719, 1993.
81. **Macleod, J. N., Lee, A. K., Liebhaber, S. A., et al.,** *J. Biol. Chem.,* 267, 14219, 1992.
82. **Cooke, N. E., Ray, J., Watson, M. A., et al.,** *J. Biol. Chem.,* 263, 9001, 1988.
83. **Golos, T. G., Durning, M., Fisher, J. M., et al.,** *Endocrinology (Baltimore),* 133, 1744, 1993.
84. **Soares, M. J., Faria, T. N., Roby, K. F., et al.,** *Endocr. Res.,* 12, 402, 1991.
85. **Talamantes, F., Ogren, L., Thordarson, G., and Southard, J. N.,** *Prolactin Gene Family and Its Receptors* (Ed. Hoshino, K.). Elsevier Science Publishers, Amsterdam, 1988, p. 145.
86. **Gross, D. S. and Garrad, W. T.,** *Annu. Rev. Biochem.,* 57, 159, 1988.
87. **Jimenez, G., Ford, A. M., Enver, T., et al.,** *J. Clin. Invest.,* 83, 1985, 1989.
88. **Nickel, B. E., Nachtigal, M. W., Bock, M. E., et al.,** *Mol. Cell. Biol.,* 106, 181, 1991.
89. **Nickel, B. E., Kardami, E., and Cattini, P. A.,** *Biochem. J.,* 267, 653, 1990.
90. **Nickel, B. E. and Cattini, P. A.,** *74th Annu. Meet. Endocr. Soc.,* San Antonio,TX, 1992, p. 36.
91. **Treacy, M. N., Ryan, F., and Martin, F.,** *J. Steroid Biochem. Mol. Biol.,* 38, 1, 1991.
92. **Nachtigal, M. W., Nickel, B. E., and Cattini, P. A.,** *J. Biol. Chem.,* 268, 8473, 1993.
93. **Nachtigal, M. W., Nickel, B. E., and Cattini, P. A.,** *75th Annu. Meet. Endocr. Soc.,* Las Vegas, NV, 1993, p. 1256.
94. **Scrippo, M., Frankenne, F., Igout, A., et al.,** *73rd Annu. Meet. Endocr. Soc.,* Washington, D.C., 1991, p. 923.
95. **Frankenne, F., Closset, J., Gomez, F., et al.,** *J. Clin. Endocrinol. Metab.,* 66, 1171, 1988.
96. **Eriksson, L., Frankenne, F., Eden, S., et al.,** *Br. J. Obstet. Gynaecol.,* 96, 949, 1989.
97. **Mirlesse, V., Frankenne, F., Alsat, E., et al.,** *Pediatr. Res.,* 34, 439, 1993.
98. **Mazlan, M., Spence-Jones, C., Chard, T., et al.,** *J. Endocrinol.,* 125, 161, 1990.
99. **Frohman, L. A., Downs, T. R., and Chomczynski, P.,** *Front. Neuroendocrinol.,* 13, 344, 1992.
100. **de Zegher, F., Vanderschueren-Lodeweyckx, M., Spitz, B., et al.,** *J. Clin. Endocrinol. Metab.,* 71, 520, 1990.
101. **Artenisio, A. C., Volpe, A., Rayonese, F., et al.,** *Horm. Metab. Res.,* 12, 205, 1980.
102. **Yen, S. S. C., Vela, P., and Tsai, C. C.,** *J. Clin. Endocrinol. Metab.,* 31, 29, 1970.
103. **Grumbach, M. M., Kaplan, S. L., Sciarra, J. J., et al.,** *Ann. N.Y. Acad. Sci.,* 148, 501, 1968.

104. **Spellacy, W. N., Buhi, W. C., and Birk, S. A.,** *Obstet. Gynecol.*, 36, 238, 1970.
105. **Beckers, A., Stevenaert, A., Foidart, J. M., et al.,** *J. Clin. Endocrinol. Metab.*, 71, 725, 1990.
106. **Blazquez, E., Simon, F. A., Blazquez, M., et al.,** *Proc. Soc. Exp. Biol. Med.*, 147, 780, 1974.
107. **Carlsson, L., Eden, S., and Jansson, J. O.,** *J. Endocrinol.*, 124, 191, 1990.
108. **Kishi, K., Hirashiba, M., and Hasegawa, Y.,** *Endocrinol. Jpn.*, 38, 589, 1991.
109. **Jahn, G. A., Rastrilla, A. M., and Deis, R. P.,** *J. Reprod. Fertil.*, 98, 327, 1993.
110. **Urbanek, M., Russell, J. E., Cooke, N. E., et al.,** *J. Biol. Chem.*, 268, 19025, 1993.
111. **Frankenne, F. J., Closset, J., Gomez, F., et al.,** *J. Clin. Endocrinol. Metab.*, 66, 1171, 1988.
112. **Ray, J., Okamura, H., Kelly, P. A., et al.,** *J. Biol. Chem.*, 265, 7939, 1990.
113. **Macleod, J. N., Worsley, I., Ray, J., et al.,** *Endocrinology*, 128, 1298, 1991.
114. **Baumann, G., Davila, N., Shaw, M. A., et al.,** *J. Clin. Endocrinol. Metab.*, 73, 1175, 1991.
115. **Selden, R. F., Wagner, T. E., Blethen, S., et al.,** *Proc. Natl. Acad. Sci. U.S.A.*, 85, 8241, 1988.
116. **Igout, A., Frankenne, F., Lhermitte-Balleriaux, M., et al.,** *74th Annu. Meet. Endocr. Soc.*, San Antonio, TX, 1992, p. 1308.
117. **Caufriez, A., Frankenne, F., Englert, Y., et al.,** *Am. J. Physiol.*, 258, E1014, 1990.
118. **Daughaday, W. H., Trivedi, B., Winn, H. N., et al.,** *J. Clin. Endocrinol. Metab.*, 70, 215, 1990.
119. **Wurzel, J. M., Parks, J. S., Herd, J. E., et al.,** *DNA*, 1, 251, 1982.
120. **Selman, P. J., Mol, J. A., Rutteman, G. R., et al.,** *Endocrinology*, 134, 287, 1994.

Chapter 24

Growth Hormone Action: Immune Function

C. G. Scanes

CONTENTS

I. INTRODUCTION

There is considerable evidence for the importance of growth hormone (GH) in the maintenance, control, and modulation of the immune system (reviewed by Kelley[1]). These roles for GH are discussed below by addressing the following questions: is GH required for the maintenance of immune tissues and their functioning? Does GH directly influence immune tissues? For instance, can GH act directly on immune cells *in vitro* and are there GH receptors (GHRs) on immune cells? Might immune cells themselves be a source of GH?

II. *IN VIVO* EFFECTS ON IMMUNE FUNCTION

A. INTRODUCTION

A role for GH in the maintenance of immune function can be summarized by the following quotation: "Growth hormone is necessary to maintain lymphatic tissue populated with lymphocytes since the removal of ... GH... results in the atrophy of the thymus and the secondary lymphoid tissues."[2]

B. IMMUNE FUNCTION IN GH-IMMUNIZED ANIMALS

The administration of antisera against bovine GH to mice results in decreased growth and large reductions in the weights of the thymus and of the spleen;[3,4] there is also a decrease in the weights of thymus and spleen relative to body weight.[3,4] These effects of antisera against bovine GH appear to be specific and due to the immunoneutralization of endogenous GH. The effects are not observed with nonimmune sera. Moreover, the antiserum effect can be reversed by the concomitant administration of GH.[3]

0-8493-8697-7/95/$0.00+$.50

C. IMMUNE FUNCTION IN GH-DEFICIENT MODELS

The hypopituitary Snell dwarf mouse has been extensively employed in the study of GH effects on the immune system. Snell dwarf mice are hypopituitary and GH deficient. Both the thymus and spleen are reduced in size in Snell dwarf mice relative to normal Snell mice.[5,6] Administration of GH to dwarf mice consistently increases the weights of the thymus and spleen[6,7] (also see Table 1), but only in some studies does GH increase the weights of these organs relative to body weight.[6] In Snell dwarf mice, GH administration also stimulates the proliferation of immune cells as indicated by higher rates of [^{3}H]thymidine labeling in the cells of the thymus and lymph node.[6] In addition, GH improves the antibody titer response to a challenge with sheep erythrocytes.[6] It might be mentioned that only with this latter effect does thyroxine administration have a synergistic action with GH.[6] Growth hormone administration to GH-deficient puppies (dogs) has also been found to improve immune function, including enhanced thymus development.[8] Conversely, hypophysectomy depresses immune function. For instance, the antibody titer in response to a sheep erythrocyte challenge is reduced in hypophysectomized rats.[4] Replacement therapy with GH partially restores the primary antibody response.[4] If the rate of cell proliferation following sheep erythrocyte challenge is estimated by [^{3}H]thymidine incorporation, this too is found to be reduced following hypophysectomy and increased by GH administration.[4] Similarly in young chickens, hypophysectomy results in depressed weights for both the thymus and bursa of Fabricius (where in birds B lymphocytes originate) (Table 2).[9-11] Replacement therapy with GH consistently exerts a marked stimulating effect on thymus weights in hypophysectomized chicks (Table 2).[9-11] It might be noted that the administration of bovine GH (bGH) was reported to increase bursal weight and the primary antibody response to sheep erthrocytes in sex-linked dwarf chickens,[12] which are thought to have deficiencies and/or abnormalities associated with their GH receptors.

D. CLINICAL ASPECTS

Clinical studies do not provide a clear picture of the effect of GH in immune function in humans. There are few differences in immune parameters between GH-deficient and aged-matched children. For instance, there are normal circulating levels of IgG, IgA, and IgM, expected levels of T and B rosette activity, and normal mitogen induced proliferation of lymphocytes (in response to phytohemagglutinin [PHA][13]). Moreover, GH-deficient children have normal levels of T helper (OKT4$^+$) cells and total T cells although B

Table 1 **Effect of hGH on Body, Thymus, and Spleen Weights in Snell Dwarf Mice**

Treatment Weight	Body Weight (g ± SEM)	Thymus Weight mg ± SEM	Spleen Weight mg ± SEM
Vehicle for 4 weeks (*N* = 16)	11 ± 0.1	15 ± 2	11 ± 0.7
hGH for 4 weeks (*N* = 16)	12 ± 0.1[a]	25 ± 2[a]	17 ± 0.7[a]
Vehicle for 10 weeks (*N* = 7)	10 ± 0.5	11 ± 0.1	18 ± 0.8
hGH for 10 weeks (*N* = 10)	14 ± 0.7[a]	13 ± 0.2[a]	23 ± 1.0[a]

[a] $p < 0.05$ compared to respective vehicle-injected control.

Source: Data from Van Buul-Offers and Van den Brande.[7]

Table 2 **Effect of GH on Immune Tissue Weights in Hypophysectomized Chicks**

Treatment	Thymus Weight mg ± SEM	Bursa Weight mg ± SEM
Sham operated ($N = 20$)	4.65 ± 0.27^{e}	3.51 ± 0.24^{d}
Hypophysectomized (Hx) ($N = 38$)	1.22 ± 0.07^{b}	1.80 ± 0.09^{b}
Hx + rcGH ($N = 19$)	1.89 ± 0.13^{c}	1.75 ± 0.11^{b}
Hx + ncGH ($N = 4$)	1.67 ± 0.16^{c}	—
Hx + oGH ($N = 7$)	2.55 ± 0.24^{b}	2.23 ± 0.16^{c}
Hx + corticosterone (B) ($N = 25$)	0.51 ± 0.02^{a}	1.27 ± 0.06^{a}
Hx + B + bGH ($N = 11$)	0.94 ± 0.07^{b}	1.65 ± 0.10^{ab}

Note: rcGH, Recombinant chicken GH; ncGH, natural chicken GH; oGH, ovine GH; bGH, bovine GH.

[a,b,c,d,e] Different superscript letter indicates difference ($p < 0.05$) by ANOVA and LSD.

Source: Data from Scanes *et al.*,[9] King and Scanes,[10] and Johnson *et al.*[11]

cell and T suppressor (OKT8^{+}) cell numbers are somewhat elevated.[11] Growth hormone therapy to GH-deficient children has no effect on T or B rosette activity or on circulating concentrations of IgG, IgA, or IgM but does appear to enhance the ability of lymphocytes to proliferate in response to PHA.[13]

E. AGING

It is possible that the decline in immune functioning in the elderly may be due to GH insufficiency and hence be amenable to GH therapy. There is some evidence for this from animal studies in both dogs and rats and also from clinical experimentation. Administration of bGH to old dogs partially restores thymic weight with an elevated ratio of cortex to medulla.[15] Moreover, bGH injections result in higher circulating thymulin levels in old but also in young dogs.[15] In aged rats, GH treatment improves immune functioning. Implanting rats with GH_3 cells (which secrete GH) stimulates immune function, with increases in thymus weight, in the number of peripheral blood leukocytes, and in the mitotic response of lymphocytes to either PHA or concanavalin A (ConA).[16] Similarly, the injection of ovine GH (oGH) into aged rats improves the proliferative response of T cells (splenocytes) to either PHA or ConA and increases natural killer cell activity.[17] However, this treatment regimen does not influence interleukin 2 (IL-2) production by splenocytes, thymic size, or lymphocyte number.[17]

In humans, aging reduces immune functioning and, on the basis of the scant evidence available, GH may ameliorate this. Middle-aged women have been found to have reduced natural killer and cytotoxic activity compared to young women.[18] Chronic administration of GH has been found to enhance natural killer cell cytotoxic activity.[18]

III. *IN VITRO* EFFECTS ON IMMUNE FUNCTION

A. INTRODUCTION

Growth hormone exerts some marked effects on immune cells *in vitro*. This is considered under the following subtopics: GH and cell differentiation, GH and macrophage functioning, GH receptors (GHRs) on immune cells, and GH synthesis by immune tissues.

B. IMMUNE CELL DIFFERENTIATION

The differentiation of lymphocytes can be influenced by GH *in vitro*. For instance, both bGH and human GH (hGH) increase both the extent of blastogenesis and the percentage of lymphocytes transformed to macrophages in a 5-day culture of human peripheral blood lymphocytes.[19] This effect of bGH may, however, be questioned as it is generally thought that bGH does not bind to the hGHR. Human GH but not bGH has been found to enhance colony formation by human T cells (either Ficoll gradient-separated peripheral blood mononuclear leukocytes or a T cell lymphoblastic cell line).[20] In studies with a mixed lymphocyte culture (derived from mouse spleen), GH increases the number of cytotoxic T lymphocytes generated in 5 days but has no effect on the blastogenic response.[19] Even at very high doses insulin fails to influence the generation of cytotoxic cells.[19] This would suggest that the GH effect is not mediated by insulin-like growth factor I (IGF-I). In contrast, differentiation of myeloid progenitors into granulocytes *in vitro* is enhanced by either IGF-I or GH but only in the presence of granulocyte/macrophage colony-stimulating factor (GM-CSF).[21] It would seem in this case that GH may be enhancing local IGF-I production to exert its effect.

C. GH AND MACROPHAGE FUNCTION

Growth hormone can influence macrophage function. Macrophages can be activated to produce reactive oxygen intermediates, including the superoxide anion (O_2^-). These reactive oxygen intermediates participate in the defense against foreign organisms. *In vitro* GH increases the production of superoxide anions by macrophages in response to opsonized zymosan (see Table 3).[22] Moreover, the administration of porcine GH (pGH) to hypophysectomized rats (*in vivo*) potentiates opsonized zymosan-induced O_2 production.[22]

D. GH RECEPTORS ON IMMUNE CELLS

The presence of specific binding sites for ^{125}I-labeled GH on immune tissues argues strongly for the presence of GHRs. Specific binding of ^{125}I-labeled bGH has been demonstrated with bovine and murine thymocytes, and with lymph node lymphocytes.[23] Similarly, ^{125}I-labeled hGH binds to membrane preparations from cultured human lymphocytes.[24] It is not surprising that binding of ^{125}I-labeled hGH is inhibited by hGH but not by bGH, pGH, or oGH in view of the species specificity of the hGHR.[24]

Table 3 **Effect of GH on Reactive Oxygen Intermediate (Superoxide Anion) Production by Porcine Alveolar Macrophages *In Vitro***

Treatment (*In Vitro*)	**Release of Superoxide Anion from Pig Macrophage (nmol O_2^-/mg/h ± SEM)**
Unstimulated	28 ± 14^a
Opsonized zymosan (o.z.)$^+$	199 ± 48^b
o.z. + npGH (500 ng/ml)	430 ± 90^c
o.z. + rpGH (500 ng/ml)	431 ± 81^c
o.z. + rpGH + antisera to GH	48 ± 20^a

Note: npGH, Natural porcine Gh; rpGH, recombinant porcine GH.

$^+$ Effect can be blocked by superoxide dismutase.

^{a,b,c} Different superscript letter indicates difference ($p < 0.05$).

Source: Adapted from Edwards *et al.*[22]

Table 4 **Effect of *In Vivo* GH Treatment on ^{125}I-labeled hGH Binding to Human Lymphocytes in GH-Deficient Children**

Treatment	Specific Binding (% ± SEM)
Pretreatment	3.5 ± 1.5
2.5 h following hGH	14.8 ± 4.2[a]
5 h following hGH	8.7 ± 1.8

[a] Increase compared with pre-treatment $p < 0.01$.

Source: Data from Stewart.[25]

The GHR on human lymphocytes appear to be upregulated by GH. Not only is specific ^{125}I-labeled hGH binding to fresh circulating lymphocytes low in GH-deficient children[25] but specific binding of ^{125}I-labeled hGH is markedly increased following acute hGH treatment in GH-deficient children[25] (see Table 4).

Downregulation (internalization) of GHRs on human lymphocytes can be induced by activators of protein kinase C and inhibited by inhibitors of protein kinase C.[23] This supports a role for protein kinase C in GHR downregulation. This is thought to involve phosphorylation of a 55-kDa protein that co-isolates with GHR.[26]

E. GH SYNTHESIS IN IMMUNE TISSUES

Immune tissues are not only GH target tissues, but sites of GH synthesis, because GH mRNA is present in human and rat peripheral blood mononuclear leukocytes[27] and rat spleen, thymus, and bone marrow leukocytes.[27,28] This message is translated in immune cells, as shown by the incorporation of ^{3}H-labeled amino acids into immunoreactive GH.[27] However, immunocytochemical studies reveal that only about 10% of leukocytes (either human or rat) are GH immunoreactive.[27] Leukocytes also release immunoreactive and biologically active GH,[27] with molecular weights of 22 and 300 kDa.[28] Although GH mRNA and GH in immune tissues have been reported to differ from pituitary GH in molecular size,[29–31] this conclusion is not supported by the studies of Weigent et al.,[27,31] in which immune GH has antigenic similarity with pituitary GH and its activity can be blocked by antibodies against pituitary GH.[32] The nucleotide sequence of the immune *GH* gene is also very similar to that in the pituitary[33,34] and its transcription is also likely to be pit-1 dependent, since the *pit*-1 gene is transcribed in immune tissues.[35] The number of GH secreting cells and release of immunoreactive GH from human lymphocytes are

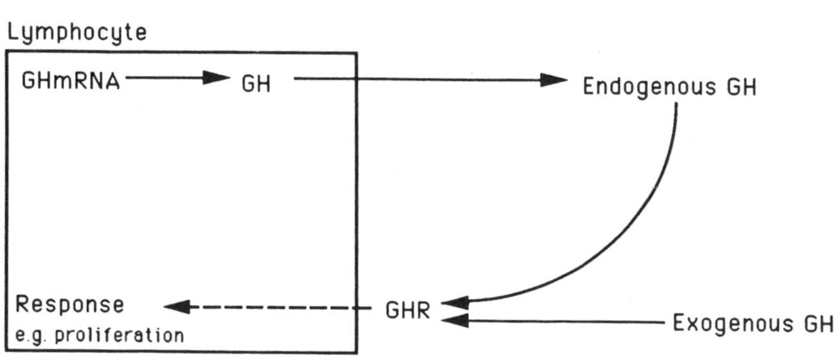

Figure 1 Endogenous GH secreted by lymphocytes exerts an autocrine effect. (Based on Weigent et al.[27])

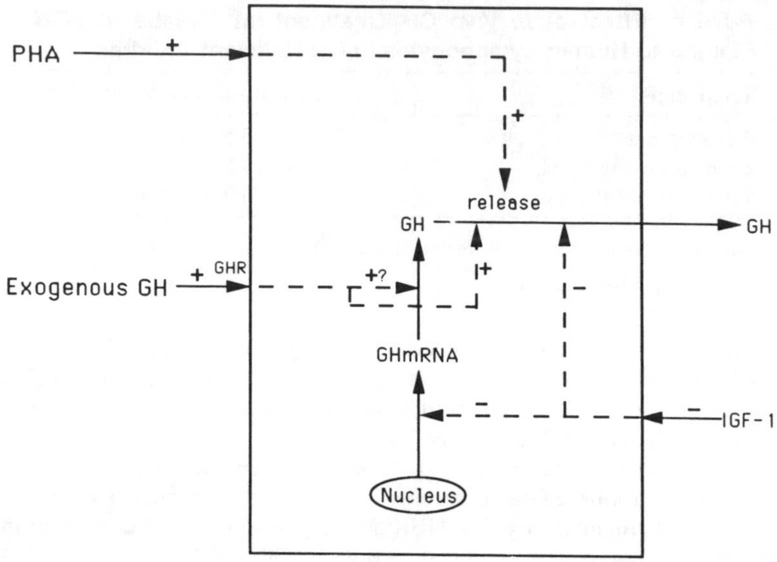

Figure 2 Influence of PHA, exogenous GH, and IGF-I on secretion of GH by lymphocytes. (Based on Weigent et al.[27,31,32])

stimulated by mitogens (e.g., PHA)[33,36] and exogenous GH,[36] but GH mRNA and GH immunoreactivity are suppressed in IGF-I-treated leukocytes[28] (see Figure 1). The production of GH in spleen, thymus, and bone marrow is reduced in rats following hypophysectomy, although increased in peripheral blood leukocytes.[37] The stimulation of GH synthesis and/or release in immune tissues may be induced by autocrine, paracrine, or endocrine actions of GH-releasing hormone (GHRH)[38] produced in immune cells,[27,39,40] although some studies were unable to demonstrate a stimulatory effect of GHRH on GH synthesis. GHRH also influences immune function by increasing the migration of human peripheral blood lymphocytes but reduces the chemotaxic response[41] and depresses natural killer activity *in vitro*.[36] The release of GH from immune cells may also be suppressed by endocrine, paracrine, or autocrine actions of somatostatin (SRIF), which is also present in immune cells,[42] although Hattori et al. again found no effect of SRIF on GH synthesis in human lymphocytes.[36]

This locally produced GH appears to exert an autocrine effect, because IGF-I production by leukocytes is reduced in the presence of GH antibodies.[32] Similarly, addition of GH antisense oligonucleotides, which would block GH synthesis, reduces the proliferation ([^{3}H]thymidine incorporation) of rat lymphocytes *in vitro*.[31] This blockade can be overcome by either GH sense oligonucleotides or exogenous GH.[31] It would, therefore, seem that some immune responses are mediated by GH in an autocrine manner (Figure 2). The relative importance of endogenous and exogenous GH to lymphocyte functioning remains to be elucidated. It is also unclear whether locally produced IGF-I exerts effects on lymphocytes, or if the effects of GH on immune tissues are mediated via IGF-I produced locally or systemically.

REFERENCES

1. **Kelley, K. W.,** *Biochem. Pharmacol.*, 38, 705, 1989.
2. **Snow, E. C., Feldbush, T. L., and Oaks, J. A.,** *J. Immunol.*, 126, 161, 1981.
3. **Pierpauli, W. and Sorkin, E.,** *J. Immunol.*, 101, 1036, 1968.
4. **Pandian, M. R. and Talwar, G. P.,** *J. Exp. Med.*, 134, 1095, 1971.
5. **Baroni, C.,** *Experientia*, 23, 282, 1967.
6. **Baroni, C., Fabris, N., and Bertoli, G.,** *Immunology*, 17, 303, 1969.
7. **Van Buul-Offers, S. and Van den Brande, J.,** *Acta Endocrinol.*, 96, 46, 1981.
8. **Roth, J. A., Kaeberle, M. L., Grier, R. L., et al.,** *Am. J. Vet. Res.*, 45, 1151, 1984.
9. **Scanes, C. G., Duyka, D. R., Lauterio, T. J., et al.,** *Growth*, 50, 12, 1986.
10. **King, D. B. and Scanes, C. G.,** *Proc. Soc. Exp. Biol. Med.*, 182, 201, 1986.
11. **Johnson, B. E., Scanes, C. G., King, D. B., et al.,** *Dev. Comp. Immunol.*, 17, 331, 1993.
12. **Marsh, J. A., Gause, W. C., Sandhu, S., et al.,** *Proc. Soc. Exp. Biol. Med.*, 175, 351, 1984.
13. **Abbassi, V. and Bellanti, J. A.,** *Pediatr. Res.*, 19, 299, 1985.
14. **Gupta, S., Fikrig, S. M., and Novol, M. S.,** *Clin. Exp. Immunol.*, 54, 87, 1983.
15. **Goff, B. L., Roth, J. A., Arp, L. H., et al.,** *Clin. Exp. Immunol.*, 68, 580, 1987.
16. **Kelley, K. W., Brief, S., Westly, H. J., et al.,** *Proc. Natl. Acad. Sci. U.S.A.*, 83, 5663, 1986.
17. **Davila, D. R., Brief, S., Simon, J., et al.,** *J. Neurosci. Res.*, 18, 108, 1987.
18. **Crist, D. M., Peak, G. T., MacKinnon, L. T., et al.,** *Metabolism*, 36, 1115, 1987.
19. **Astaldi, A., Yalcin, B., Meardi, G., et al.,** *Blut*, 26, 74, 1973.
20. **Mercola, K. E., Cline, M. J., and Golde, D. W.,** *Blood*, 58, 337, 1981.
21. **Merchav, S., Tatarsky, I., and Hochberg, Z.,** *J. Clin. Invest.*, 81, 791, 1988.
22. **Edwards, C. K., Ghiasuddin, S. M., Schepper, J. M., et al.,** *Science*, 239, 769.
23. **Arrenbrecht, S.,** *Nature (Lond.)*, 252, 255, 1974.
24. **Lesniak, M. A., Gorden, P., Roth, J., et al.,** *J. Biol. Chem.*, 249, 1661, 1974.
25. **Stewart, C., Clejan, S., Fugler, L., et al.,** *Arch. Biochem. Biophys.*, 220, 309, 1983.
26. **Suzuki, K., Suzuki, S., Saito, Y., et al.,** *J. Biol. Chem.*, 265, 11320, 1990.
27. **Weigent, D. A., Baxter, J. B., Wear, W. E., et al.,** *FASEB J.*, 2, 2812, 1988.
28. **Weigent, D. A., Riley, J. E., Galin, F. S., et al.,** *Proc. Soc. Exp. Biol. Med.*, 198, 643, 1991.
29. **Baglia, L. A., Cruz, D., and Shaw, J. E.,** *Endocrinology*, 130, 2446, 1992.
30. **Hiestand, P. C., Mekler, P., Nordmann, R., et al.,** *Proc. Natl. Acad. Sci. (USA)*, 83, 2599, 1986.
31. **Weigent, D. A., Blalock, J. E., and LeBoeuf, R. D.,** *Endocrinology*, 128, 2053, 1991.
32. **Baxter, J. B., Blalock, J. E., and Weigent, D. A.,** *Endocrinology*, 129, 1727, 1991.
33. **Baxter, J. B., Blalock, J. E., and Weigent, D. A.,** *J. Neuroimmunol.*, 33, 43, 1991.
34. **Weigent, D. A. and Blalock, J. E.,** *Cell. Immunol.*, 135, 55, 1991.
35. **Delhase, M., Vergani, P., Malur, A., et al.,** *Eur. J. Immunol.*, 23, 951, 1993.
36. **Hattori, N., Shimatsu, A., Sugita, M., et al.,** *Biochem. Biophys. Res. Commun.*, 168, 396, 1990.
37. **Weigent, D. A., Baxter, J. B., and Blalock, J. E.,** *Brain Behav. Immun.*, 6, 365, 1992.
38. **Guarcello, V., Weigent, D. A., and Blalock, J. E.,** *Cell. Immunol.*, 136, 291, 1991.
39. **Weigent, D. A. and Blalock, J. E.,** *J. Neuroimmunol.*, 29, 1, 1990.
40. **Stephanou, A., Knight, R. A., and Lightman, S. L.,** *Neuroendocrinology*, 53, 628, 1991.
41. **Zelazowski, P., Dohler, K. D., Stepien, H., et al.,** *Neuroendocrinology*, 50, 236, 1989.
42. **O'Dorsio, M. S.,** *The Neuroendocrine-Immune Network* (Ed. Freier, S.). CRC Press, Boca Raton, FL, 1990, p. 187.

Chapter 25

Growth Hormone Action: Osmoregulation

C. G. Scanes

CONTENTS

I. INTRODUCTION

Growth hormone (GH) can influence renal function in mammals and also affects water and mineral balance in salmonid fish during transfer from freshwater to seawater. These distinctly different actions of GH are discussed in this chapter.

II. RENAL FUNCTION IN MAMMALS

The first reports that GH influences renal function came in the late 1940s. In a study reported in 1949, GH administration was found to increase both the glomerular filtration rate (GFR) (inulin clearance) and renal plasma flow (RPF) (sodium *p*-aminohippurate clearance) in dogs.[1] Hypophysectomy, on the other hand, had the opposite effect.[1] In the same year, it was reported that plasma concentrations of phosphate were elevated in hypophysectomized rats receiving GH injections,[2] presumably reflecting increased renal phosphate reabsorption. The effects of GH on the GFR, RPF, and water and sodium balance are considered prior to a discussion of the role of GH related to phosphate reabsorption.

Chronic administration of GH increases the GFR in dogs[1] (see Table 1)[3] and in other species, including humans (see below) and hypophysectomized rats.[4] Similar actions are also induced in isolated functioning kidneys from intact rats.[4] Growth hormone exerts another effect on kidneys, that is, it influences renal sodium potassium-dependent ATPase activity. Chronic administration of GH to hypophysectomized rats increases (by 2.1-fold) the activity of ATPase (i.e., that ATPase inhibitable by ouabain).[5]

In humans, high circulating concentrations of GH appear to influence renal function. For instance, acromegalic patients show elevated plasma volume (see Table 2).[9] Similarly, extracellular water is increased relative to intracellular water or exchangeable potassium.[11] There are also marked changes in the GFR and RPF (Table 2). These changes can be mimicked by the chronic administration of human GH (hGH) to normal subjects (for details see Table 2). Administration of GH does not affect the volume of the kidney or the excretion of albumin or β_2-macroglobulin or sodium or potassium.[9] Chronically high circulating concentrations of GH influence renal functioning. This may be a direct effect of GH or may be mediated by insulin-like growth factor I (IGF-I), either because of increases in the circulating concentrations of IGF-I or by locally (kidney) produced IGF-I. Certainly circulating concentrations of IGF-I are elevated in acromegaly and following GH administration.[9] Moreover, kidney tissue can synthesize IGF-I in a GH-dependent manner (reviewed by Hammerman[12]). The acute administration of hGH does not have the same effect as that observed chronically. For instance, infusion of hGH

0-8493-8697-7/95/$0.00+$.50

Table 1 **Effect of Chronic GH Treatment on Renal Phosphate Reabsorption in Dogs**

Parameter	Intact Dogs			Thyroidectomized/ Parathyroidectomized Dogs			Ref.
	Pre-GH	Post-GH	Δ	Pre-GH	Post-GH	Δ	
Study 1: Daily GH Injections for 8 Days							
GFR (ml/min)	70.5 ± 12.0	93.6 ± 15.1[a]	23.1 ± 4.9	70.4 ± 4.3	86.0 ± 5.5[a]	15.6 ± 4.0	Data calculated from Corvilain and Abramow[3]
Fasting plasma phosphate concentration (mg/ml)	2.9 ± 0.3	3.8 ± 0.36[a]	0.8 ± 0.3	4.8 ± 0.4	5.8 ± 0.5[a]	1.0 ± 0.2	Data calculated from Corvilain and Abramow[3]
$TmPO_4$[b] (mg/min)	2.8 ± 0.5	4.9 ± 0.8	2.1 ± 0.47	4.6 ± 0.6	6.6 ± 1.0[a]	2.0 ± 0.6	Data calculated from Hammerman et al.[6]
$TmPO_4/GFR \times 100$	3.9 ± 0.16	5.25 ± 0.2[a]	1.3 ± 0.3	6.4 ± 0.8	7.5 ± 0.9[a]	1.1 ± 0.4	Data calculated from Hammerman et al.[6]
Study 2: Daily GH Injections for 3 Days							
Plasma calcium (mg/ml)	9.7 ± 0.8	9.5 ± 0.3	—	6.3 ± 0.3	6.8 ± 0.5	—	Hammerman et al.[6]
Plasma phosphate (mg/ml)	4.2 ± 0.2	6.3 ± 0.3[b]	—	5.3 ± 0.4	5.5 ± 0.2	—	Hammerman et al.[6]

Reabsorption of phosphate (*in vivo*)	1.1 ± 0.1	2.2 ± 0.2[a]	—	2.4 ± 0.2	2.7 ± 0.5	—	Hammerman et al.[6]
Sodium gradient-dependent phosphate uptake by brush border membrane (vesicles *in vitro* (nmol/20 s/mg protein)	1.6 ± 0.02	2.1 ± 0.10[a]	—	2.5 ± 0.14	3.0 ± 0.16[a]	—	Hammerman et al.[6]

[a] $p < 0.05$ compared to respective control.

[b] Maximal tubular reabsorption of phosphate.

Table 2 **Effect of Chronic Elevation of Circulating Concentrations of GH on Kidney Function in Adult Humans**

Index	Control (Normal Adults) (Mean ± SEM)	High GH (Acromegalic) (Mean ± SEM)	Ref.
In acromegalic patients with elevated plasma levels			
Plasma volume (ml/kg body weight)	37.9 ± 0.6	54.7 ± 1.7[a]	7
Blood pressure (mm Hg)	90.6 ± 1.7	105 ± 3[a]	7
Glomerular filtration rate (ml/min/1.73 m[2])	107 ± 1.16	136 ± 6.6[a]	8
Renal plasma flow (ml/min/1.73 m^2)	531 ± 15.4	687 ± 55.7[b]	8
Filtration Fraction	0.21 ± 0.00	0.21 ± 0.01	8
In patients receiving exogenous GH	**Pre-treatment** (Mean ± SEM)	**Post-GH** (Mean ± SEM)	
1. hGH injected twice daily for 7 days in normal adults			
Glomerular filtration rate (ml/min/1.73m^2)	114 ± 5	125 ± 4[c]	9
Renal plasma flow (ml/min/1.73m^2)	554 ± 30	601 ± 36[a]	9
Filtration fraction	0.21 ± 0.00	0.21 ± 0.01	9
Heart rate	55 ± 1	60 ± 1[c]	9
2. hGH injected in normal adults (sampling after 48 h)			
Glomerular filtration rate (ml/min/1.73 m^2)	100 ± 3	130 ± 5	10
Renal plasma flow (ml/min/1.73 m^2)	546 ± 19	713 ± 21[b]	11

[a] $p < 0.001$ compared to respective control.
[b] $p < 0.01$ compared to respective control.
[c] $p < 0.05$ compared to respective control.

in normal adults has no effect on the GFR, but depressed the RPF and increased the filtration fraction.[13] Similarly, the acute injection of bovine GH (bGH) had no acute effect on the GFR, RPF, or the clearance of sodium, chloride, or phosphate in either intact or thyroidectomized/parathyroidectomized dogs.[14]

Chronic GH treatment increases plasma concentrations of phosphate[3] by increasing renal phosphate reabsorption (for examples, see Table 1). This effect is not observed if GH is given acutely.[14] Moreover, *in vitro* studies involving microperfusion of rabbit proximal convoluted tubules indicate that GH does not influence phosphate transport.[15] It is likely that the chronic effects of GH on phosphate excretion are mediated by IGF-I. This is supported by the ability of IGF-I to stimulate phosphate transport by proximal convoluted tubules *in vitro* and the ability of GH to increase renal IGF-I concentrations both by elevating circulating concentrations of IGF-I and by local production. This is similar to the effects of GH on the GFR and RPF. Growth hormone also exerts effects on renal acid excretion, which reflects the amount of bicarbonate ions reabsorbed and the amount of ammonium ions secreted.

In studies in rats, GH acutely increases both reabsorption of bicarbonate ions and excretion of ammonium ions. This is observed irrespective of the model employed: isolated functioning kidneys (from either hypophysectomized or intact rats) or *in situ* kidneys (from hypophysectomized rats).[4] *In vitro* studies with canine renal proximal tubule segments indicate that GH stimulates ammonia formation from glutamine directly.[16] This effect is pH sensitive and requires an extracellular-intracellular sodium concentration gradient.[16] Moreover, GH appears to stimulate the ouabain-sensitive Na^+,K^+-ATPase pump.[16] The ammonia effect of GH may be mediated via increased transport of glutamine into the renal cells (sodium-dependent L-glutamine transport).

III. OSMOREGULATION IN SALMONIDS

The life history of salmonid fish includes migrations from freshwater to seawater and vice versa. It is, therefore, apparent that these transitions require extraordinary homeostatic mechanisms to facilitate adaptation and the ultimate achievement of homeostasis. Although cortisol appears to be the major hormone involved in this,[18] there is substantial evidence that GH can exert considerable effects on osmoregulation during the transition from freshwater to seawater. For instance, GH administration considerably improves the rate of survival of young salmonid fish transferred experimentally from freshwater to seawater (Table 3). Moreover, the administration of GH to salmonid fish during a transfer from freshwater to seawater also reduces the magnitude of physiological changes induced by exposure to high salinity. Plasma concentrations of both sodium and chloride ions are increased by exposure of salmonids to seawater, as is the plasma osmolality, whereas muscle water content decreases. The magnitude of these changes is attenuated if GH treatment accompanies the transfer from freshwater to seawater (see Table 4 for details).

It would appear that the GH effect of blunting the increase in plasma sodium concentrations in salmonid fish represents a true effect of GH and not GH exerting an effect via prolactin receptors. A sodium-lowering effect has been observed with mammalian GH[18,19,23,24] and also with preparations of fish GH, namely *Tilapia* GH[23] and salmon GH.[21] Although ovine prolactin exerts a similar effect,[23] no such effect is observed with either *Tilapia* prolactin[23] or salmon prolactin.[21] These data would argue strongly for the effect of ovine prolactin being artifactual, owing to the use of a heterologous hormone preparation and perhaps ovine prolactin binding to fish GH receptors.

The mechanism by which GH exerts its osmoregulatory effect is not clear. It is possible that GH is acting by increasing cortisol production by fish interrenal tissue. A

Table 3 **Effect of GH on Survival of Fish Transferred from Freshwater to Seawater**

Species	Stage	Treatment	Death Rate (Percentage Mortality[a])	Ref.
Atlantic Salmon (*Salmo salar*)	Parr	Control	>50[b]	17
		GH	0	
Sea trout (*Salmo trutta*)	Parr	Control	46	18
		GH	4	
Amago salmon (*Oncorhynchus rhodurus*)	Parr	Control	73	19
		GH	36	
		T4	82	
		GH + T_4	100	

[a] Mortality after a 1-week exposure for Atlantic salmon and sea trout, but 1 day for Amago salmon.

[b] No mortality was observed in another study[20] of Atlantic salmon parr (in Sweden).

steroidogenic effect of GH on fish interrenal tissue has been demonstrated (see Chapter 26). Alternatively GH may be acting directly on osmoregulatory organs such as the gills. There is specific binding of ^{125}I-labeled GH to both salmon kidney and gills.[25] Administration of GH has been found to have marked effects on gills of salmonid fish with changes in Na^+,K^+-ATPase and chloride cells (see Table 4 for details).

Because GH can influence water/mineral balance in salmonid fish during the transition from freshwater to saltwater, this does not per se indicate that GH is in fact playing a physiological role at this time. If GH were to be involved in the homeostatic regulating mechanism, the changes in plasma concentrations of sodium would be expected to be greater in hypophysectomized fish transferred from freshwater to seawater. Although this is the case,[22] neither GH nor cortisol replacement therapy fully restores plasma concentrations of sodium to those in sham-operated controls.[22] A second line of reasoning favoring an osmoregulatory role for GH comes from studies in which circulating concentrations of GH were measured in young fish undergoing a transition from freshwater to seawater. As might be anticipated were GH to have an osmoregulatory role, plasma concentrations of GH are increased following the transition from freshwater to seawater in young Coho salmon and sexually mature Atlantic salmon parr.[22,24,26] Conversely, transfer from seawater to freshwater transitorily depresses circulating concentrations of GH in juvenile chum salmon.[27] It might be noted that other pituitary hormones (prolactin and possibly somatolactin) are probably of importance during the transition from seawater to freshwater, this effect occurring prior to reproduction.

Table 4 Effects of GH on Osmoregulatory Indices in Salmonid Fish Parr during Transfer from Freshwater to Seawater

Index	Effect	Species	Ref.
Plasma osmolality[a]	↓	Amago salmon (*Oncorhynchus rhodurus*) (parr)	19
Plasma sodium[a]	↓[b]	Sea trout (*Salmo trutta*) (parr)	18
	↓	Amago salmon (*Oncorhynchus rhodurus*) (parr)	
	↓	Rainbow trout (*Salmo gairdneri*) (juveniles)	21
	↓	(Hypophysectomized) Coho salmon (juveniles)	22
		Amago Salmon (*Oncorhynchus kisutch*) (hypophysectomized)	23
	↓	Sockeye salmon (*Oncorhynchus nerka*) (juvenules)	18
Plasma chloride[a]	↓[b]	Sea trout (*Salmo trutta*) (parr)	21
Plasma calcium	↓	Rainbow trout (*Salmo gairdneri*)	21
Plasma magnesium	↓	Rainbow trout (*Salmo gairdneri*)	18
Muscle water[c]	↑[b]	Sea trout (*Salmo trutta*) (parr)	18
Gill Na^+,K^+-ATPase[a]	↑[b]	Sea trout (*Salmo trutta*) (parr)	19
	↑[d]	Amago salmon (*Oncorhynchus rhodurus*)	24
	↑	Coho salmon (*Oncorhynchus kisutch*) (presmolt)	24
	↑	Coho salmon (*Oncorhynchus kisutch*) (desmolt)	18
Gill chloride cell density	↑	Sea trout (*Salmo trutta*) (parr)	24
	→	Coho salmon (*Oncorhynchus kisutch*) (presmolt)	18
Gill chloride cell length	↑	Sea trout (*Salmo trutta*) (parr)	18

[a] Increased by exposure to seawater per se.

[b] Decreased by exposure to seawater per se.

[c] Effect in presence of GH or cortisol treatments alone but marked effect (additive or greater) in presence of both GH and cortisol.

[d] Effect observed only if GH and T_4 administered together.

REFERENCES

1. **White, H. L., Heinbecker, P., and Rolf, D.,** *Am. J. Physiol.*, 157, 47, 1949.
2. **Li, C. H., Geschwind, I., and Evans, H. M.,** *Endocrinology*, 44, 67, 1949.
3. **Corvilain, J. and Abramow, A.,** *J. Clin. Invest.*, 43, 1608, 1964.
4. **Welbourne, T. C. and Cronin, M. J.,** *Am. J. Physiol.*, 260, R1036, 1991.
5. **Shimomura, M., Lee, M., Oku, J., et al.,** *Metabolism*, 31, 213, 1982.
6. **Hammerman, M. R., Karl, I. E., and Hruska, K. A.,** *Biochim. Biophys. Acta*, 603, 322, 1980.
7. **Deray, G., Rieu, M., Devynch, M. A., et al.,** *New Engl. J. Med.*, 316, 575, 1987.
8. **Ikkos, D., Ljunggren, H., and Luft, R.,** *Acta Endocrinol.*, 21, 226, 1956.
9. **Sandahl Christiansen, J., Grammelgaard, J., Orskov, H., et al.,** *Eur. J. Clin. Invest.*, 11, 487, 1981.
10. **Hirschberg, R., Rabb, H., Bergemo, R., et al.,** *Kidney Int.*, 35, 865, 1989.
11. **Ikkos, D., Ljunggren, H., and Luft, R.,** *Acta Endocrinol.*, 21, 211, 1956.
12. **Hammerman, M. R.,** *Am. J. Physiol.*, 257, F503, 1989.
13. **Parving, H.-H., Noer, I., Mogensen, C. E., et al.,** *Acta Endocrinol.*, 89, 796, 1978.
14. **Westby, G. R., Goldfarb, S., Goldberg, M., et al.,** *Metabolism*, 26, 525, 1977.
15. **Quigley, R. and Baum, M.,** *J. Clin. Invest.*, 88, 368, 1991.
16. **Chobanian, M. C., Julin, C. M., Molteni, K. H., et al.,** *Am. J. Physiol.*, 262, F878, 1992.
17. **Komourdjian, M. P., Saunders, R. L., and Fenwick, J. C.,** *Can. J. Zool.*, 54, 531, 1976.
18. **Madsen, S.,** *Gen. Comp. Endocrinol.*, 791, 1990.
19. **Miwa, S. and Inui, Y.,** *Gen. Comp. Endocrinol.*, 58, 436, 1985.
20. **Rydevik, M., Borg, B., Haux, C., et al.,** *Gen. Comp. Endocrinol.*, 80, 9, 1990.
21. **Bolton, J. P., Collie, N. L., Kawauchi, H., et al.,** *J. Endocrinol.*, 112, 63, 1987.
22. **Richman, N. H., Nishioka, R. S., Young, G., et al.,** *Gen. Comp. Endocrinol.*, 67, 194, 1987.
23. **Clark, W. C., Farmer, S. W., and Hartwell, K. M.,** *Gen. Comp. Endocrinol.*, 33, 174, 1977.
24. **Richman, N. H. and Zaugg, W. S.,** *Gen. Comp. Endocrinol.*, 65, 189, 1987.
25. **Fryer, J. N. and Bern, H. A.,** *J. Fish Biol.*, 15, 527, 1979.
26. **Sweeting, R. M. and McKeown, B. A.,** *Gen. Comp. Endocrinol.*, 88A, 147, 1987.
27. **Ogasawara, T., Hirano, T., Akiyama, T., et al.,** *Fish Physiol. Biochem.*, 7, 309, 1989.

REFERENCES

Chapter 26

Growth Hormone Action: Endocrine Function

C. G. Scanes

CONTENTS

I. INTRODUCTION

There is abundant evidence that growth hormone (GH) influences the release/degradation and hence the circulating concentrations of other peripheral hormones. The effects of GH on insulin-like growth factors (IGFs) and on gonadal hormones are considered in Chapters 19 and 23, respectively, and are not included in this chapter.

II. GH EFFECTS ON RELEASE OF PANCREATIC HORMONES

A. INSULIN RELEASE

A strong case can be made for GH (or perhaps a specific GH variant) being an insulin secretagogue. In the "classic" ablation-replacement therapy, insulin release from the pancreatic islets of hypophysectomized rats is considerably less than that released by tissue from control rats.[1] Moreover, daily injections of native ovine GH (oGH) into hypophysectomized rats restores the ability of islet tissue to release insulin.[1] Similarly the administration of native bovine GH (bGH) acutely increases insulin release into the portal vein of either normal or alloxan-diabetic dogs.[2] Native bGH also increases release of insulin into the portal vein of perfused isolated rat pancreas.[3] However, studies on the effects of GH on pancreatic islets incubated *in vitro* have yielded inconsistent results. At low glucose concentrations, native bGH stimulates basal insulin release but inhibits insulin release at high prevailing glucose concentrations.[4] Clinical-grade human GH (hGH) also stimulates insulin release *in vitro*, but this is not the case with highly purified hGH.[5] This might suggest that a contaminant of the GH preparation is the insulin secretagogue. The identity of this is not known, although it may well be a cleaved GH variant. Indeed, proteolytic cleavage of purified hGH engenders it with insulin secretagogue activity,[5] and this may also account for the ability of recombinant hGH to increase blood insulin levels in humans.[5a]

0-8493-8697-7/95/$0.00+$.50

In prolonged culture of neonatal islets of Langerhans, rat GH increases insulin secretion together with stimulating proliferation of islet cells, predominantly β cells.[6] These effects are, however, of a considerably smaller magnitude than those evoked by prolactin.[6] In summary, GH and/or a cleaved GH have the net effect of increasing insulin secretion.

B. GLUCAGON RELEASE

Growth hormone stimulates the release of immunoreactive glucagon from perfused isolated rat pancreas/duodenum[3] and from the dog pancreas *in vitro*.[2] It is possible that GH is stimulating glucagon release per se. However, possible effects of GH on enteroglucagon (oxytomodulin, glicentin) release from the gut (e.g., duodenum) cannot be precluded.

III. GH AND THYROID HORMONE PRODUCTION

A. INTRODUCTION

Growth hormone can, under some circumstances, influence circulating concentrations of thyroxine (T_4) and triiodothyronine (T_3). Predominantly, it would seem that the administration of GH increases the circulating concentrations of T_3 while decreasing those of T_4. This would suggest that GH is not affecting thyroidal secretion of T_4 and T_3, but rather influencing peripheral 5- and/or 5′-monodeiodinase activity and the metabolism of the thyroid hormones (Figure 1) and hence conversion of T_4 to T_3 (the active form of thyroid hormone). The major locus for the production of circulating T_3 is predominantly the liver but other organs (e.g., kidney) also participate.

B. THYROID HORMONE PRODUCTION IN HUMANS

There is evidence that GH can exert an acute effect on the metabolism of T_4 to T_3. GH administration to hGH-deficient children is followed by increased circulating concentrations of T_3 but decreased concentrations of T_4.[7] Recombinant hGH similarly increases plasma T_3 levels in normal adult men[5a] and women.[7a] This is consistent with GH enhancing 5′-monodeiodination.

C. THYROID HORMONE PRODUCTION IN RATS

In thyroidectomized rats receiving T_4 replacement therapy, infusion of hGH enhances the plasma appearance rate of T_3. There is, in addition, increased (net) T_3 production in the liver and to a lesser extent also in the kidney and pituitary gland.[8] The authors concluded that GH increased T_4 to T_3 conversion "probably by stimulation of monodeiodination."[8] Circulating concentrations of T_4 and T_3 are only slightly, if at all, affected in GH-deficient rats.[8] This suggests that any such GH effect on T_4/T_3 monodeiodination is not obligatory.

D. THYROID HORMONE PRODUCTION IN PIGS

Daily injections of porcine GH (pGH) into young pigs exert marked effects on circulating concentrations of T_3 (increased)[9,10] and T_4 (decreased).[9] These reports are also consistent with GH increasing 5′-monodeiodination of T_4.

E. THYROID HORMONE PRODUCTION IN CATTLE AND SHEEP

There is contradictory evidence as to whether GH influences thyroid hormone metabolism in cattle and sheep. Chronic administration of bGH does not change circulating concentrations of either T_3 or T_4 in lactating dairy cattle,[11,12] growing cattle,[13] or growing sheep.[14,15] Moreover, bGH injected daily for 5 days does not affect 5′-monodeiodination of T_4 to T_3 at its major locus, the liver.[12] However, an increase in 5′-monodeiodinase is

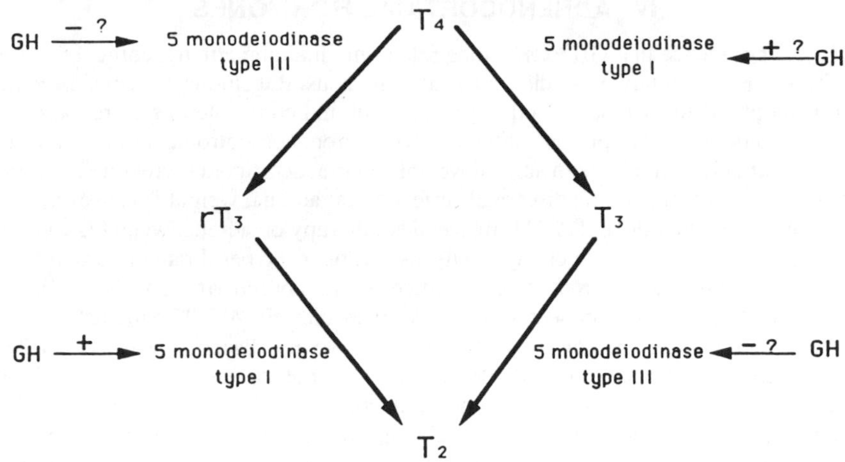

Figure 1 Effect of GH on the metabolism of thyroid hormones in the chicken liver. (Based on work by Kuhn and colleagues.[102–111])

observed in mammary tissue.[12] It might be noted that effects of GH on thyroid hormones have been observed in some studies. For instance, chronic administration of bGH is reported to increase circulating concentrations of T_3 significantly, albeit slightly.[16] Moreover, the acute administration of oGH to newborn lambs has been observed to increase circulating concentrations of T_3.[17]

F. THYROID HORMONE PRODUCTION IN CHICKENS

A single injection of either oGH or chicken GH (cGH) increases circulating concentrations of T_3 and stimulates liver 5′-monodeiodinase activity in both chick embryos[18,19] and adult chickens.[20–22] However, this effect is not observed in young chickens, in which GH receptors are downregulated,[19] or in sex-linked dwarf chick embryos, in which GH receptors are either not present or are aberrant.[19] Recently more specific assays for type I and type III deiodinase have been employed to reexamine the effects of GH on thyroid hormone metabolism. Type III deiodinase activity is depressed following GH injection into chick embryos, whereas type I was unaffected (Figure 1).[19] It would appear, therefore, that the increase in circulating concentrations of T_3 and 5′-monodeiodinase activity previously observed is due in fact to a reduction in the catabolism (monodeiodination) of T_3. The physiological nature of this effect is supported by the "mirror image" changes in type III deiodinase activity and circulating concentrations of GH during embryonic and early posthatch development of the chick.[20]

These data provide strong evidence that GH can influence the peripheral metabolism of T_4. It might be noted that GH also influences circulating concentrations of T_4[21,23] and appears to potentiate the biological response (circulating concentrations of T_4) to thyroid-stimulating hormone (TSH) in a 1-day-old chick bioassay.[24] These effects are presumed to be mediated by GH influencing metabolism of T_4 but a direct thyrotropic effect of GH cannot be precluded.

G. THYROID HORMONE PRODUCTION IN FISH

In fish, GH can increase peripheral conversion of T_4 to T_3 *in vivo*.[25–27] However, *in vitro*, GH does not affect 5′-monodeiodinase activity by trout hepatocytes.[28]

IV. ADRENOCORTICAL HORMONES

There is evidence that GH exerts some role in maintaining adrenal cortical functioning. This is based largely on studies in rats and chickens. Rat studies are considered here first. Hypophysectomy reduces both adrenal weight and corticosterone secretion *in vivo* and *in vitro* in rats.[29,30] Replacement therapy with adrenocorticotropic hormone (ACTH) restores, but only partially, both adrenal weight and glucocorticoid secretion.[29,30] Administration of bGH alone has no discernible effect on rat adrenal weight.[29] However, in the presence of GH, the effect of ACTH replacement therapy on adrenal weight is considerably augmented.[29] Hypophysectomy in rats also reduces the basal rate of corticosterone secretion *in vitro*.[30] Basal production of corticosterone both *in vitro* and *in vivo* is only partially restored in hypophysectomized rats given chronic ACTH administration.[29,30] Whereas chronic GH therapy alone appears to have no discernable effect on basal corticosterone secretion,[29,30] GH (bGH) consistently potentiates the effect of chronic ACTH treatment on basal corticosterone production.[29,30] It might be noted that these studies estimated net corticosterone production, and, in fact, bGH also augments the ACTH-induced decline in adrenal 5α-reductase,[29] an enzyme involved in corticosterone catabolism.

In chickens, hypophysectomy decreases adrenal weight and adrenal weight relative to body weight.[31] Replacement therapy with cGH alone increases relative adrenal weight and also influences adrenal corticosteroidogenesis.[31] Adrenal cortical cells from hypophysectomized chicks or cGH-treated hypophysectomized chicks show little difference in basal corticosterone release. However, marked differences are observed in ACTH-induced corticosterone production, with hypophysectomy reducing responsiveness to ACTH *in vitro* and cGH increasing responsiveness to ACTH.[31] This effect of GH appears to be downstream from both ACTH activation of adenylate cyclase and cholesterol side-chain cleavage as GH also increases the responsiveness (corticosterone production) to 8-bromo-cAMP or pregnenolone.[31]

Circulating concentrations of glucocorticoids can be changed by exogenous GH in intact animals. The acute administration of cGH has been reported to increase plasma concentrations of corticosterone in chickens.[32] Implanting rats with GH-producing tumors results in elevated adrenal weights and ACTH-induced corticosterone production *in vitro* and *in vivo*.[33] Similarly, pretreatment of Coho salmon with GH enhances cortisol secretion (*in vivo* and *in vitro*) in response to both ACTH and pregnenolone.[34] However, GH does not influence circulating cortisol concentrations uniformly in all species. For instance, in pigs, chronic administration of GH fails to affect the circulating concentrations of cortisol.[35]

In summary, GH exerts some stimulating effects on the adrenal cortex and on glucocorticoid production in some species under some circumstances.

V. GROWTH FACTOR RECEPTORS

Growth hormone can be envisaged as potentially exerting profound effects by influencing the number and/or affinity of receptors for growth factors or other hormones. One well-characterized example of this is the stimulating influence of GH on the hepatic epidermal growth factor (EGF) receptor, which is also the receptor for transforming growth factor α (TGF-α). The number of EGF receptors is reduced in hypophysectomized rats[36,37] and *lit/lit* dwarf mice (which lack GH)[36] and, furthermore, GH therapy partially restores EGF receptor number.[36,37] The pattern of GH secretion/administration is thought to be a major influence on EGF receptor number. There are considerably higher numbers of EGF receptors in male rats, in which GH secretion is episodic, than in females, in which GH secretion is more or less continuous.[36,37] If the male or female endogenous secretory pattern of GH is mimicked (by GH infusion or injection) in the opposite sex then EGF receptor number is shifted to that appropriate to the imposed GH profile.[36,37] The GH effect would appear to be at the pretranslational level, as the hepatic EGF receptor mRNA level changes in a manner similar to EGF receptor number.[36,37] Growth hormone similarly increases steady-state mRNA levels of the IGF-I receptor gene and the androgen receptor gene in the prostate of immature rats.[38] "Masked" or "cryptic" insulin receptors are also increased in abundance in the livers of transgenic mice expressing the bovine GH gene.[39]

REFERENCES

1. **Malaisse, W. J., Malaisse-Lagae, F., King, S., et al.,** *Am. J. Physiol.*, 215, 423, 1968.
2. **Sirek, A., Vranic, M., Sirek, O. V., et al.,** *Am. J. Physiol.*, 237, E107, 1979.
3. **Tai, T.-Y. and Pek, S.,** *Endocrinology*, 99, 669, 1976.
4. **Kawabe, T. and Morgan, C. R.,** *Endocrinology*, 32, 728, 1983.
5. **Larson, B. A., Williams, T. L., Lewis, U. J., et al.,** *Diabetogia*, 15, 129, 1978.
5a. **Brixen, K., Nielsen, H. K., Bouillon, R., et al.,** *Acta Endocrinol.*, 127, 331, 1992.
6. **Brelje, T. C. and Sorenson, R. L.,** *Endocrinology*, 28, 45, 1991.
7. **Sato, T., Suzuki, Y., Taketani, T., et al.,** *J. Clin. Endocrinol. Metab.*, 45, 324, 1977.
7a. **Moller, J., Jorgensen, J. O. L., Moller, N., et al.,** *Metabolism*, 41, 728, 1992.
8. **Gelhoed-Durjvestyn, P. H. L. M., Roelfsema, F., Schroder-van der Elst, J., et al.,** *J. Endocrinol.*, 133, 45, 1992.
9. **Kirkwood, R. N., Thacker, P. A., and Laarveld, B.,** *Domest. Anim. Endocrinol.*, 6, 167, 1989.
10. **Etherton, T. D., Wiggins, J. P., Evock, C. M., et al.,** *J. Anim. Sci.*, 64, 433, 1987.
11. **Peel, C. J., Frank, T. J., Bauman, D. E., et al.,** *J. Dairy Sci.*, 66, 776, 1983.
12. **Capuco, A. V., Keys, J. E., and Smith, J. J.,** *J. Endocrinol.*, 121, 205, 1989.
13. **Eisemann, J. H., Hammond, A. C., Bauman, D. E., et al.,** *J. Nutr.*, 116, 2504, 1986.
14. **Rosenberg, E., Thonney, M. L., and Butler, W. R.,** *J. Anim. Sci.*, 67, 3300, 1989.
15. **Pell, J. M., Elcock, C., Harding, R. L., et al.,** *Br. J. Nutr.*, 63, 431, 1990.
16. **Byatt, J. C., Eppard, P. J., Veenhuizen, J. J., et al.,** *J. Endocrinol.*, 132, 185, 1992.
17. **Kuhn, E. R., Van Osselaer, P., Siau, O., et al.,** *ICRS Med. Sci.*, 13, 425, 1985.
18. **Kuhn, E. R., Huybrechts, L. M., Decuypere, E., et al.,** *Proc. 7th Eur. Poultry Conf.*, 1986, p. 965.
19. **Darras, V. M., Berghman, L. R., Vanderpooten, A., et al.,** *FEBS Lett.*, 310, 5, 1992.
20. **Darras, V. M., Visser, T. J., Berghman, L. R., et al.,** *Comp. Biochem. Physiol.*, 103A, 131, 1992.
21. **Darras, V. M., Huybrechts, L. M., Berghman, L. R., et al.,** *Gen. Comp. Endocrinol.*, 77, 212, 1990.
22. **Kuhn, E. R., Verheyen, G., Decuypere, E., et al.,** *IRCS Med. Sci.*, 14, 479, 1986.
23. **Kuhn, E. R., Verheyen, G., Chiasson, R. B., et al.,** *IRCS Med. Sci.*, 13, 451, 1988.
24. **MacKenzie, D. S.,** *Poultry Sci.*, 60, 2136, 1981.
25. **deLuze, A. and Leloup, J.,** *Gen. Comp. Endocrinol.*, 56, 308, 1984.
26. **deLuze, A., Leloup, J., Papkoff, H., et al.,** *Gen. Comp. Endocrinol.*, 186, 1989.
27. **MacLatchy, D. L. and Eales, J. G.,** *Gen. Comp. Endocrinol.*, 78, 164, 1990.
28. **Sweeting, R. M. and Eales, J. G.,** *Gen. Comp. Endocrinol.*, 88, 169, 1992.
29. **Colby, H. D., Caffrey, J. L., and Kitay, J. I.,** *Endocrinology*, 188, 1973.
30. **Kramer, R. E., Greiner, J. W., and Colby, H. D.,** *Endocrinology*, 101, 297, 1977.
31. **Carsia, R. V., Weber, H., King, D. B., et al.,** *Endocrinology*, 117, 928, 1985.
32. **Cheung, A., Hall, T. R., and Harvey, S.,** *Gen. Comp. Endocrinol.*, 69, 128, 1988.
33. **Coyne, M. D., Alpert, L. C., Harter, K. C., et al.,** *Horm. Res.*, 14, 36, 1981.
34. **Young, G.,** *Gen. Comp. Endocrinol.*, 71, 85, 1988.
35. **Kirkwood, R. N., Thacker, P. A., and Laarveld, B.,** *Domest. Anim. Endocrinol.*, 6, 167, 1989.
36. **Johansson, S., Husman, B., Norstedt, G., et al.,** *J. Mol. Endocrinol.*, 3, 113, 1989.
37. **Kashima, K., Hiramatsu, M., and Minami, N.,** *J. Endocrinol.*, 123, 75, 1989.
38. **Reiter, E., Bonnet, P., Sente, B., et al.,** *Mol. Cell. Endocrinol.*, 88, 77, 1992.
39. **Balbis, A., Dellacha, J. M., Calandra, R. S., et al.,** *Life Sci.*, 51, 771, 1992.

REFERENCES

Chapter 27

Growth Hormone Action: Neuroendocrine Function

S. Harvey

CONTENTS

I. INTRODUCTION

Neuroendocrine actions of growth hormone (GH) are indicated by the presence of GH receptors in hypophysiotropic regions of the hypothalamus and within cells of the neurohypophysis and adenohypophysis. Neuroendocrine roles for GH have also been demonstrated by its autoregulation of secretion and by its actions at central nervous system (CNS) and pituitary sites in modulating the secretion of other pituitary hormones.

II. HYPOTHALAMIC REGULATION

Autoregulatory effects of GH, mediated through the hypothalamus, are well-established.[1] Systemic or intraventricular injections of GH reduce the pituitary GH content and reduce basal, sleep-related, episodic, or pharmacologically induced pituitary GH release,[2–4] as does the implantation of GH-secreting tumors or GH pellets into the hypothalamus or third ventricle.[3,5–12] These autoregulatory actions may be mediated through opioidergic interneurons[13] or by insulin-like growth factor I (IGF-I),[14] which results in an increase in the content and release of SRIF in hypothalamic tissue.[15–22] The localization of GH receptors in hypophysiotropic neurons secreting SRIF[23] and GH-releasing hormone (GHRH)[24,25] also indicates direct effects of GH in hypothalamic regulation, especially as exogenous GH induces neuronal activity and the expression of the c-*fos* protooncogene.[26] Exogenous GH or chronic GH hypersecretion also stimulates the synthesis and release of SRIF,[7,18,27–33] while inhibiting the synthesis and release of GHRH.[28,29,34–38] Actions of GH within the CNS are also thought to be responsible for the inhibition of gonadotropin secretion and enhanced prolactin secretion in birds receiving intraventricular injections of GH[3,39] and for the modulation of gonadotropin and prolactin secretion in mice transgenically expressing *GH* genes.[40] Hypothalamic actions of GH are also responsible for the downregulation of pituitary TRH (thyrotropin-releasing hormone) receptors in the chicken pituitary gland,[7,41] which are insensitive to GH *in vitro*.[41]

0-8493-8697-7/95/$0.00+$.50

III. PITUITARY REGULATION

The possibility that GH has direct effects on pituitary function is supported by the presence of GH receptors in adenohypophysial and neurohypophysial tissue.

A. GH RECEPTOR GENE: EXPRESSION

Transcription of the GH receptor gene (which encodes GH receptors and GH-binding proteins) in pituitary glands was demonstrated by the abundance of GH receptor mRNA in extracts of rat, rabbit, guinea pig, human, bovine, and chicken pituitary glands.[42–48] Multiple transcripts of the GH receptor gene of approximately 4.2–4.5, 2.4–2.7, and 1.0–1.2 kb occur, although the larger moiety predominates. These transcripts are therefore comparable in size to those in the liver and other peripheral tissues. The larger transcript is thus likely to encode the full-length receptor, whereas the smaller transcripts may be degradation products or splice variants that encode truncated receptors or binding proteins. The large transcripts in rat, rabbit, guinea pig, and chicken pituitary glands have sequences and endonuclease cleavage sites (determined by polymerase chain reaction) homologous to sequences in the liver that encode intracellular, transmembrane, and extracellular domains of the receptor. Southern blotting of pituitary cDNA also suggests that GH receptor gene transcripts in the rabbit pituitary are identical to those in the rabbit liver.[49]

Transcripts of the GH receptor gene are widely distributed throughout the adenohypophysis of the rat and rabbit pituitary gland and also occur in the neural and intermediate lobes.[44] They are also present in both the somatotroph-rich caudal lobe and the somatotroph-poor cephalic lobe of the chicken pituitary gland,[45] and in GH- and non-GH-secreting cells (including lactotrophs) of the rat and rabbit pituitary.[46,47] Transcription of the GH receptor gene nevertheless occurs only in a proportion of the somatotrophs in the rat adenohypophysis[46] but is reported to be absent in thyrotrophs and corticotrophs.[47] Within pituitary cells the GH receptor RNA is localized near the endoplasmic reticulum and in the cytoplasmic matrix.[47]

B. GH RECEPTOR GENE: TRANSLATION

Translation of the GH receptor gene in chicken pituitary glands was indicated by the specific binding of radiolabeled GH to plasma membrane proteins with a molecular mass of 60 kDa.[45] Although this is smaller than the predicted (70 kDa) and actual (130 kDa) size of the rat GH receptor[50] the GH receptor detected in the avian pituitary may have lacked ubiquitin or carbohydrate linkage or may have been partially degraded prior to chromatographic separation. This smaller size may also partially reflect the deletion of exon 3 in the avian GH receptor gene.[51] These studies also showed that GH-binding proteins of comparable mass to those in plasma membranes were also present in nuclear membranes of the chicken adenohypophysis. In addition, smaller GH-binding proteins of 50 kDa, comparable in size to those in chicken plasma, were identified in cytosolic fractions of the chicken pituitary gland.[45]

Translation of the GH receptor gene in the rat pituitary gland has also been indicated by the binding of labeled GH to crude solubilized preparations of the plasma and nuclear membranes and to pituitary cytosol.[46] The binding of the tracer to these fractions was inhibited by prior exposure to a monoclonal antibody (MAb 263), which binds to epitopes in the ligand-binding domain of the rat liver GH receptor.[52] Translation of the GH receptor gene has also been indicated by electron microscopy and immunogold staining of rat and rabbit pituitary sections.[46,47] Monoclonal antibodies that recognize the extracellular domain of the rat (MAb 263) or rabbit (MAb 7) GH receptor showed GH immunoreactivity for the GH receptor or GH-binding protein (GHBP) throughout the rat and rabbit adenohypophysis.[46] Immunostaining occurred in some somatotroph and some

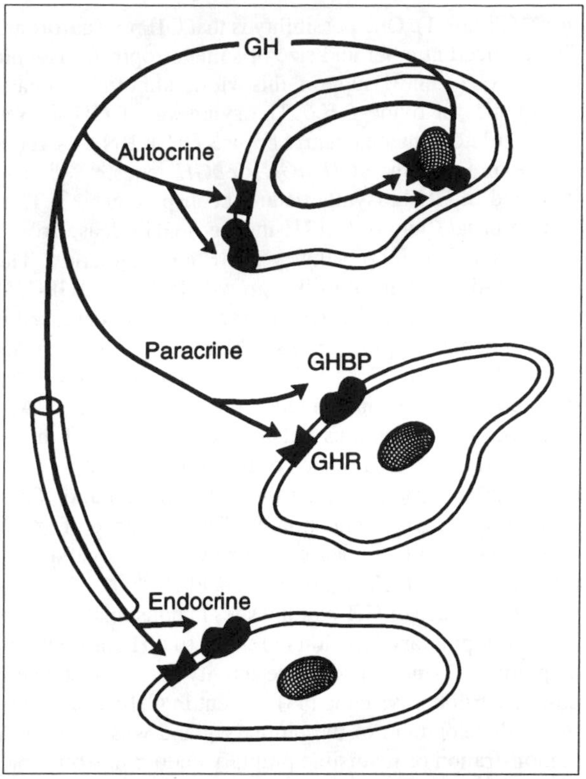

Figure 1 Putative sites of growth hormone (GH) action within the pituitary gland. The presence of GH receptors (GHRs) in somatotrophs suggests the release of GH may have endocrine, paracrine, autocrine, or intracrine autocrine actions on somatotroph function, which may be modulated by competitive binding to GH-binding proteins (GHBPs). The presence of GH receptors in other pituitary cell types indicates other endocrine or paracrine actions of GH within the pituitary gland. (Based on Harvey, S., Hull, K. L., and Fraser, R. A., *Growth Regul.*, 3, 1, 1993.)

nonsomatotroph cells, and was widely distributed throughout the intracellular and nuclear compartments. In GH-secreting cells gold particles were specifically accumulated in the secretory granules and heterochromatin, but were also present in mitochondria, Golgi, endoplasmic reticulum, cytoplasm, nucleoplasm, and in cellular and nuclear membranes.[46,47] This labeling was specific and was not observed with nonimmune serum or phosphate-buffered saline (PBS) or when the antibody was preabsorbed by incubation with liver membranes. A similar distribution of GH receptor immunoreactivity was also observed within rat and rabbit hepatocytes. Subsequent studies with monoclonal (MAb 4.3) and polyclonal antibodies raised against the GHBP clearly demonstrate the presence of the GHBP in adenohypophysial cells.[53] In all immunoreactive cells the labeling was most intense in secretory granules and, in contrast to MAb 263 and MAb 7 (which recognize both the receptor and the GHBP), was not in the nucleus or associated with heterochromatin.[53,54] Moreover, whereas staining with MAb 263 appeared to be ubiquitous, some pituitary cells were not labeled by the polyclonal GHBP antiserum or by MAb 4.3. GHBP immunoreactivity was, however, co-localized with hormones in all pituitary cell types, although in some GH-, prolactin-, luteinizing hormone (LH)-, follicle-stimulating hormone (FSH)-, thyrotropin (TSH)-, and adrenocorticotropin (ACTH)-secreting cells, GHBP immunoreactivity was not present in all secretory granules. These results contrast, however, with those of Mertani et al.,[55] which indicated an absence of GHR/GHBP immunoreactivity in corticotrophs and thyrotrophs.

C. GH ACTIONS

The presence of GHR mRNA and GH-binding sites in the pituitary gland suggests endocrine, paracrine, autocrine, or intracrine roles for GH in regulating pituitary function[49] (Figure 1). One possibility is that GH may autoregulate its own synthesis or release. The reduced number and size of somatotrophs transgenically expressing the human[56] or bovine *GH* gene[57] support this view, although somatotrophs are unaffected in mice transgenic for ovine *GH*.[58] The synthesis of GH also appears to be suppressed by an autoregulatory mechanism, because GH mRNA is reduced in pituitaries of transgenic mice expressing the *oGH*, *hGH*, or *bGH* genes,[56,59,60] in which the cytoplasmic organelles involved in GH biosynthesis are inconspicuous.[56,57] The inhibitory action of exogenous GH on basal (but not GHRH-induced) GH release from bovine pituitary glands[61,62] also suggests a short-loop pathway in GH autoregulation. This pathway may be mediated by pituitary-derived insulin-like growth factor I (IGF-I),[63] as *IGF-I* gene expression is increased, whereas *GH* gene transcription is decreased in the pituitaries of rats bearing somatomammotrophic tumors.[64] Moreover, exogenous human GH evokes a dose-dependent induction of IGF-1 mRNA in tumorous (thyroid hormone-deficient) somatotrophs (GH_3 cells), and a concomitant reduction in GH mRNA.[65,66] Endocrine or local paracrine or autocrine interactions between GH and IGF-1 in the pituitary gland may also be facilitated by the localization of IGF-I receptors in somatotroph cells.[14,67–69] An inhibitory autocrine or paracrine factor distinct from GH and IGF-I is also produced by GH_3 cells in response to GH stimulation.[70,71] The existence of an ultrashort-loop feedback pathway in the control of GH release is, however, controversial and not supported by studies with rat, murine, or chicken pituitary glands.[72–76]

The presence of GH-binding sites in nonsomatotroph pituitary cells also suggests roles for GH in pituitary function unrelated to GH autoregulation. A stimulatory action of GH on pituitary 5′-monodeiodinase activity has, for instance, been suggested by the increased pituitary triiodothyronine (T_3) content in GH-treated rats.[77] Indeed, a pivotal role for GH in the development of the pituitary gland was recently suggested by Flint et al., by the demonstration of reversible pituitary endocrine abnormalities and greatly impaired pituitary growth in neonatal rats injected with GH antibodies.[78,79] The results of these studies indicate a facilitatory role for GH in regulating the activity of gonadotropic, thyrotropic, and corticotropic cells and in the maintenance of pituitary size. This possibility is supported by the increased number and size of corticotrophs and gonadotrophs in the pituitaries of mice transgenically expressing the human or bovine *GH* gene,[56] in which lactotroph cells are scarce and immature. An inhibitory action of GH on lactotroph function is also indicated by the pronounced hypoprolactinemia in GH transgenic animals,[80] even though Gardner and Flint[79] found no diminution in the pituitary prolactin content after passive immunoneutralization of endogenous GH.

In hGH transgenic animals, plasma and pituitary LH levels are elevated and luteinizing hormone-releasing hormone (LHRH)-stimulated LH release from incubated pituitaries and steady state levels of LH-β mRNA are increased.[40,81–83] Chronic stimulation of pituitary gonadotrophs is also indicated by the intense immunocytochemical staining for LH, increased cytoplasmic volume, and extensive vacuolation of gonadotrophs in hGH transgenic mice.[56] The postcastration increase in peripheral plasma LH levels and the ability of exogenous testosterone to suppress plasma LH levels is, however, reduced in these transgenic animals[83,84] in which prolactin suppresses, rather than stimulates, LH secretion.[40] The concentrations of FSH in mice transgenically expressing the *hGH* or *bGH* genes are normal or suppressed in comparison with LH.[40,80] These gonadotropic effects of GH overproduction therefore indicate divergent GH actions in the hypothalamus-pituitary-gonadal axis that modulate the release of pituitary LH and FSH. Because the *in vitro* release of LH and FSH from normal mouse pituitaries transplanted into transgenic hosts expressing the *hGH* gene is similar to that from the *in situ* pituitary glands of the transgenic hosts,[40,81,82] these actions of GH may be exerted directly at the pituitary level. The possibility that GH modulates gonadotropin function is also supported by the sexual dysfunction in patients with acromegaly[85] and the suppression of fertility in transgenic mice and pigs overexpressing GH.[81,82,86] The influence of GH on gonadotropin secretion and action is considered in further detail in Chapter 23.

D. PITUITARY GH-BINDING PROTEINS

In addition to authentic receptors, the GH-binding sites in the pituitary gland may include GHBPs not coupled to signal transduction systems, and the binding of GH to these proteins may inhibit GH action or protect against degradation or have roles in the intracellular trafficking and secretion of pituitary GH.

1. Pituitary GH-Binding Protein: Secretion

The extensive tissue and cellular distribution of GHBP within the pituitary gland suggests the pituitary may be a source of the circulating GHBP, especially as GHBPs are abundantly present in the secretory vesicles of all adenohypophysial cells.[46,53] The possibility that the pituitary gland may be a source of the circulating binding protein was first suggested by Ymer and Herington,[87] because large molecular variants of GH were found in plasma, pituitary glands, and pituitary incubation media.[88,89] The large GH variants were thought to be protein-bound moieties. The appearance of specific and abundant GHBP immunoreactivity in the secretory granules of adenohypophysial cells[53] strongly suggests the quantum release of GHBP from pituitary glands, some of which may already be bound to its ligand. However, although GHBP immunoreactivity is located in secretory granules, it is possible that GHBPs are not released if the immunoreactivity is associated with membrane proteins of the secretory granules. During exocytosis the plasma and secretory granules fuse[90] and this may provide a mechanism for the insertion of binding proteins into the plasma membrane. Membrane bound binding proteins could thus modulate GH action by competing with authentic membrane receptors for the ligand. The presence of GHBPs in plasma membranes is indicated by the studies of Frick and Goodman,[91] who demonstrated GHBPs associated with membrane proteins in rat adipocytes.

2. Pituitary GH-Binding Protein: Roles

The abundance of the GHBP in pituitary cells also suggests that it has function(s) at the cellular/tissue level independent of GH binding in plasma (Figure 1). The binding of GH to cellular GHBPs may facilitate the internalization and intracellular transport of GH within the pituitary gland, by analogy to the internalization of GH bound to membrane proteins in liver, kidney, and adipose cells.[92] This possibility is supported by the uptake and accumulation of radioactivity in the pituitaries of rats systemically injected with radiolabeled hGH.[47,93] The intracellular association of GH and GHBP may also facilitate GH translocation to the nucleus and induction of genomic actions, especially as the unique hydrophilic tail of the GHBP includes a sequence of four basic amino acids thought to facilitate movement through the nuclear envelope by interaction with the nuclear pore complex.[94] The complexing of GH to GHBP may also enhance the stability of GH stores, becuase the intracellular degradation of GH in the pituitary gland appears to be mainly of newly synthesized hormone rather than of older, stored GH,[95] which may be protein bound. This possibility is also supported by the finding that secreted GH is less stable than purified (stored) GH when the two forms are incubated with slices of peripheral tissues *in vitro*.[96]

The presence of GHBPs in somatotrophs does not, however, imply a functional role in GH storage and release, because GHBPs are also widely distributed in pituitary cells secreting hormones with little or no affinity for the GHBP. It is, therefore, also possible that the abundance and widespread distribution of the GHBP in the pituitary may provide a buffering capacity that normally protects pituitary cells from excess GH stimulation. Putative paracrine, autocrine, or intracrine actions of GH within the pituitary gland may, therefore, be modulated by the competitive binding of GH to its binding protein rather than to its receptor. Indeed, such a role could explain the inability of numerous investigators to demonstrate autoregulatory actions of GH in the rat pituitary gland.[72–74,76] Pituitary GHBPs may not, however, be sufficiently abundant to inactivate the supraphysiological GH excess in transgenic mice expressing heterologous *GH* genes, in which novel pituitary actions of GH have been indicated.[56,80]

REFERENCES

1. **Muller, E. E.,** *Horm. Res.*, 33(Suppl. 2), 90, 1990.
2. **Krulich, L. and McCann, S. M.,** *Proc. Exp. Biol. Med.*, 121, 1114, 1966.
3. **Lea, R. W. and Harvey, S.,** *J. Endocrinol.*, 125, 409, 1990.
4. **Sakuma, M. and Knobil, E.,** *Endocrinology (Baltimore)*, 86, 890, 1970.
5. **Tannenbaum, G. S.,** *Endocrinology (Baltimore)*, 107, 2117, 1980.
6. **Advis, J. P., White, S. S., and Ojeda, S. R.,** *Endocrinology (Baltimore)*, 108, 1343, 1981.
7. **Lea, R. W., Ahene, C., Marsh, J. A., et al.,** *J. Endocrinol.*, 126, 237, 1990.
8. **MacLeod, R. M., DeWitt, G. W., and Smith, D.,** *Endocrinology (Baltimore)*, 79, 1149, 1966.
9. **Peake, C. T., Mariz, L. K., and Daughaday, W. H.,** *Endocrinology (Baltimore)*, 83, 714, 1968.
10. **Yamashita, S., Slanina, S., Kado, H., et al.,** *Endocrinology (Baltimore)*, 118, 915, 1986.
11. **Katz, S. H., Molitch, M., and McCann, S. M.,** *Endocrinology (Baltimore)*, 85, 725, 1969.
12. **Voogt, J. L., Clemens, J. A., Negro-Vilar, A., et al.,** *Endocrinology (Baltimore)*, 88, 1363, 1971.
13. **Ganzetti, I., Petraglia, F., Capuano, I., et al.,** *J. Endocrinol. Invest.*, 10, 241, 1987.
14. **Walker, D. A., Hogg, A., Haynes, K., et al.,** *J. Neuroendocrinol.*, 2, 305, 1990.
15. **Berelowitz, M., Szabo, M., Frohman, M., et al.,** *Science*, 212, 1279, 1981.
16. **Berelowitz, M., Firestone, S. I., and Frohman, L. A.,** *Endocrinology (Baltimore)*, 109, 714, 1981.
17. **Aguila, M. C. and McCann, S. M.,** *Brain Res.*, 623, 89, 1993.
18. **Chihara, K., Minamitani, N., Kaji, G., et al.,** *Endocrinology (Baltimore)*, 109, 2279, 1981.
19. **Molitch, M. E. and Hlivyak, L. E.,** *Horm. Metab. Res.*, 12, 519, 1980.
20. **Sheppard, M. C., Kronheim, S., and Pinstone, B. L.,** *Clin. Endocrinol.*, 9, 583, 1978.
21. **Hoffman, D. L. and Baker, B. L.,** *Proc. Soc. Exp. Biol. Med.*, 156, 265, 1977.
22. **Patel, Y. C.,** *Life Sci.*, 24, 1589, 1979.
23. **Burton, K. A., Kabigting, E. B., Clifton, D. K., et al.,** *Endocrinology (Baltimore)*, 130, 958, 1992.
24. **Burton, K. A., Kabigting, E., Clifton, D., et al.,** *73rd Annu. Meet. Endocr. Soc.*, Washington, D.C., 1991, p. 419.
25. **Muccioli, G., Ghe, C., and Di Carlo, R.,** *Neuroendocrinology*, 53, 47, 1991.
26. **Minami, S., Kamegai, J., Sugihara, H., et al.,** *Endocrinology (Baltimore)*, 131, 247, 1992.
27. **Sato, M., Chihara, K., Kita, T., et al.,** *Neuroendocrinology*, 50, 139, 1989.
28. **Conway, S., McCann, S. M., and Kulich, L.,** *Endocrinology (Baltimore)*, 117, 2284, 1985.
29. **Minami, S., Kamegari, J., Hasegawa, O., et al.,** *J. Neuroendocrinol.*, 5, 691, 1993.
30. **Rogers, K. V., Vician, L., Steiner, R. A., et al.,** *Endocrinology (Baltimore)*, 122, 586, 1988.
31. **Robbins, R. J., Leidy, J. W., Jr., and Landon, R. M.,** *Endocrinology (Baltimore)*, 117, 538, 1985.
32. **Bertherat, J., Timsit, J., Bluet-Pajot, M. T., et al.,** *J. Clin. Invest.*, 91, 1783, 1993.
33. **Kanatsuka, A., Makino, H., Matsushima, Y., et al.,** *Neuroendocrinology*, 29, 186, 1979.
34. **Chomczynski, D., Downs, T. R., and Frohman, L. A.,** *Mol. Endocrinol.*, 2, 236, 1988.
35. **De Gennaro Colonna, V., Cattaneo, E., Cocchi, D., et al.,** *Peptides*, 9, 985, 1988.
36. **Miki, N., Ono, M., Miyoshi, H., et al.,** *Life Sci.*, 44, 469, 1985.
37. **Cella, S. G., De Gennaro Collona, V., Locatelli, M., et al.,** *J. Endocrinol.*, 199, 1990.
38. **Sato, M. and Frohman, C. A.,** *Endocrinology (Baltimore)*, 133, 793, 1993.
39. **Buntin, J. D., Lea, R. W., and Figge, G. R.,** *J. Endocrinol.*, 118, 33, 1988.
40. **Bartke, A., Chandrashekar, V., Tang, K., et al.,** *Neuroendocrinol. Lett.*, 15, 89, 1993.
41. **Harvey, S. and Baidwan, J. S.,** *J. Mol. Endocrinol.*, 4, 13, 1990.

42. **Harvey, S. and Fraser, R. A.,** *J. Endocrinol.*, 133, 357, 1992.
43. **Fraser, R. A., Attardo, D., and Harvey, S.,** *J. Mol. Endocrinol.*, 5, 231, 1990.
44. **Fraser, R. A., Siminoski, K., and Harvey, S.,** *J. Endocrinol.*, 128, R9, 1991.
45. **Hull, K. L., Fraser, R. A., and Harvey, S.,** *J. Endocrinol.*, 135, 459, 1992.
46. **Fraser, R. A. and Harvey, S.,** *Endocrinology (Baltimore)*, 130, 3593, 1992.
47. **Mertani, H., Jambou, R., Waters, M. J., et al.,** *J. Endocrinol. Invest.*, 15, 61, 1992.
48. **Hauser, S. D., McGrath, M. F., Collier, R. J., et al.,** *Mol. Cell. Endocrinol.*, 72, 187, 1990.
49. **Harvey, S., Hull, K. L., and Fraser, R. A.,** *Growth Regul.*, 3, 1, 1993.
50. **Kelly, P. A., Djiane, J., Postel-Vinay, M. C., et al.,** *Endocr. Rev.*, 12, 235, 1991.
51. **Burnside, J., Liou, S. S., and Cogburn, L. A.,** *Endocrinology (Baltimore)*, 128, 3183, 1991.
52. **Barnard, R., Bundesen, P. G., Rylatt, D. B., et al.,** *Biochem. J.*, 231, 459, 1986.
53. **Harvey, S., Baumbach, W. R., Sadeghi, H., et al.,** *Endocrinology (Baltimore)*, 133, 1125, 1993.
54. **Harvey, S., Baumbach, W. R., Sadeghi, H., et al.,** *75th Annu. Meet. Endocr. Soc.*, Las Vegas, NV, Abstr. 467, p. 167.
55. **Mertani, H. C., Waters, M. J., Jambou, R., et al.,** *Neuroendocrinology*, 59, 483, 1994.
56. **Stefaneanu, L., Kovacs, K., Horvath, E., et al.,** *Endocrinology (Baltimore)*, 126, 608, 1990.
57. **Stefaneanu, L., Kovacs, K., Bartke, A., et al.,** *Lab. Invest.*, 68, 584, 1993.
58. **Orian, J. M., Weiss, L. M., and Braundon, M. R.,** *Endocrinology (Baltimore)*, 126, 608, 1989.
59. **Sotelo, A. I., Bartke, A., and Turyn, D.,** *Acta Endocrinol.*, 129, 446, 1993.
60. **Stefaneanu, L., Kovacs, K., and Bartke, A.,** *Endocr. Pathol.*, 4, 73, 1993.
61. **Glenn, K. C.,** *Endocrinology (Baltimore)*, 118, 2450, 1986.
62. **Rosenblum, S. M., Silverman, B. L., and Wehrenberg, W. B.,** *Neuroendocrinology*, 53, 597, 1991.
63. **Ceda, G. P., Hoffman, A. R., Silver, G. D., et al.,** *J. Clin. Endocrinol. Metab.*, 600, 1204, 1985.
64. **Fagin, J. A., Brown, A., and Melmed, S.,** *Endocrinology (Baltimore)*, 122, 2204, 1988.
65. **Fagin, J. A., Fernandez-Megia, C., and Melmed, S.,** *Endocrinology (Baltimore)*, 125, 2385, 1989.
66. **Ezzat, S. and Melmed, S.,** *J. Endocrinol. Invest.*, 13, 691, 1990.
67. **Oerantt, I., Valentino, K. L., Hoffman, A. R., et al.,** *Neuroendocrinology*, 49, 248, 1989.
68. **Rosenfeld, R. G., Pham, H., Oh, Y., et al.,** *Endocrinology (Baltimore)*, 124, 2867, 1989.
69. **Rosenfeld, R. G., Ceda, G., Cuttler, C. W., et al.,** *Endocrinology (Baltimore)*, 117, 2008, 1985.
70. **Lapp, C. A., Tyler, J. M., Lee, Y. S., et al.,** *In Vitro Cell. Dev. Biol.*, 25, 528, 1989.
71. **Stachura, M. E., Lapp, C. A., Tyler, J. M., et al.,** *In Vitro Cell. Dev. Biol.*, 25, 482, 1990.
72. **Richman, R. A., Weiss, J. P., Hochberg, Z., et al.,** *Endocrinology (Baltimore)*, 108, 2287, 1981.
73. **Kraicer, J., Lussier, B., Moor, B. C., et al.,** *Endocrinology (Baltimore)*, 122, 1511, 1988.
74. **Becker, K. and Conway, S.,** *Brain Res.*, 578, 107, 1992.
75. **Goodyer, C. G., De Stephano, L., Guyda, H. J., et al.,** *Endocrinology (Baltimore)*, 115, 1568, 1984.
76. **de Zegher, F., Bettendorf, M., Grumbach, M. M., et al.,** *Neuroendocrinology*, 52, 429, 1990.
77. **Geelhoed-Duijvestijn, F., Roelfsema, F., Schroder-van der Elst, J., et al.,** *J. Endocrinol.*, 133, 45, 1992.
78. **Flint, D. J., Gardner, M. J., Akinsanya, K., et al.,** *J. Endocrinol.*, 132(Suppl.), abstract 60, 1992.
79. **Gardner, M. J. and Flint, D. J.,** *J. Endocrinol.*, 124, 381, 1990.
80. **Chandrashekar, V., Bartke, A., and Wagner, T. E.,** *Endocrinology (Baltimore)*, 130, 1802, 1992.
81. **Tang, K., Bartke, A., Gardiner, C. S., et al.,** *Biol. Reprod.*, 49, 346, 1993.
82. **Tang, K., Bartke, A., Gardiner, C., et al.,** *Endocrinology (Baltimore)*, 132, 2518, 1993.
83. **Steger, R. W., Bartke, A., Parkening, T. A., et al.,** *Neuroendocrinology*, 52, 106, 1990.
84. **Chandrashekar, V., Bartke, A., and Wagner, T. E.,** *Endocrinology (Baltimore)*, 123, 2717, 1988.
85. **Jadresic, A., Banks, L. M., Child, D. F., et al.,** *Q. J. Med.*, 202, 189, 1982.
86. **Pursel, V. G., Pinkert, C. A., Miller, K. F., et al.,** *Science*, 244, 1281, 1989.
87. **Ymer, S. I. and Herington, A. C.,** *Mol. Cell. Endocrinol.*, 41, 153, 1985.
88. **Talamantes, F., Lopez, J., Lewis, U. J., et al.,** *Acta Endocrinol. (Copenh.)*, 98, 8, 1981.
89. **Baumann, G.,** *Proc. Soc. Exp. Biol. Med.*, 202, 392, 1993.
90. **Burgess, T. L. and Kelly, R. B.,** *Annu. Rev. Cell Biol.*, 3, 243, 1987.

91. **Frick, G. P. and Goodman, H. M.,** *Endocrinology (Baltimore)*, 131, 3083, 1992.
92. **Roupas, P. and Herington, A. C.,** *Mol. Cell Endocrinol.*, 61, 1, 1989.
93. **Van Houten, H., Posner, B. I., and Walsh, R. J.,** *Exp. Brain. Res.*, 38, 455, 1980.
94. **Silver, P. A.,** *Cell*, 64, 489, 1991.
95. **Fukata, J., Diamond, D. J., and Martin, J. B.,** *Endocrinology (Baltimore)*, 117, 457, 1985.
96. **Vodian, M. A. and Nicoll, C. S.,** *J. Endocrinol.*, 80, 69, 1979.
97. **Vaisman, N., Zadik, Z., Shamai, Y., et al.,** *Metab. Clin. Exp.*, 41, 483, 1992.

Chapter 28

Growth Hormone Action: Neural Function

S. Harvey

CONTENTS

I. INTRODUCTION

Neural actions of growth hormone (GH) were first documented more than 50 years ago,[1] but these were largely ignored and GH has rarely been considered as a neuropeptide or neuromodulator (see Harvey et al.[2] for review). It is, however, now clear that GH receptors and binding proteins are widely distributed within the central nervous system (CNS), in which GH has roles in neurotransmission, central behavior, growth, and development. These may be ancestral roles of GH, because GH is present in the brains of primitive vertebrates lacking pituitary somatotrophs[3] and in the nervous tissue of invertebrates lacking pituitary glands,[4,5] and is present in the mammalian brain prior to its ontogenic appearance in the pituitary gland.[6] The possibility that GH has neural sites of action is suggested by its presence within the brain, the permeability of the blood-brain barrier to circulating GH, and the presence of neural GH receptors.

II. NEURAL GH

GH-like immunoreactivity at concentrations greater than those in blood[6-12] has been identified in telencephalic, diencephalic, mesencephalic, metencephalic, and myelencephalic regions of the brain (Table 1). The GH content is greatest in the amygdala, hippocampus, and hypothalamus, although at concentrations <1% of those in the anterior pituitary gland. Immunoreactivity is present in perikarya, nerve fibers and axon terminals and localized in neurons in association with somatostatin (SRIF) and thyrotropin-releasing hormone (TRH).[12-14] It is also present in granular secretory vesicles in terminal boutons and may be colocalized with classic neurotransmitters.[12,13] Immunoreactive GH tracts originating in the preoptic hypothalamus (POA) and projecting through the infundibulum to the posterior pituitary gland or through the wall of the preoptic recess to the third ventricle have also been seen in the fish brain.[15]

0-8493-8697-7/95/$0.00+$.50

Table 1 **Distribution of Growth Hormone, GH Receptor, and GH-Binding Protein, and of GH and GHR mRNA, in Neural Tissues***

Region	GH	GH mRNA	GHR/GHBP	GHR mRNA
Whole brain	11		37,53,165,166	35,38,43,45
Telencephalon			46	28
Corpus striatum		18		
Caudate nucleus	6,8,9	18		
Putamen	6			
Stria terminalis	12,13			
Amygdala	6–8,12	18	38	37
Globus pallidus	6		38	37
Hippocampus	6–8,12	18		
Dentate gyrus			38,49,165	37,48
Cortex	6–8	18	38	37
Pyramid cells	6		49,165	
Olfactory bulb			38,49,165	
Septum	12,13		49,165	37
Diagonal band of Broca	12			37
Diencephalon	8			
Hypothalamus	6–9 13	18	30,50	35,37
Periventricular nucleus	12,13	19	30,165	37,39,44,48
Paraventricular nucleus	12,13		30	37,39,44,48
Arcuate nucleus	12,13		30	37,39,44,48
Preoptic area	13		30,49,165	37,44,48
Mamillary nucleus	12		30	
Dorsomedial nucleus	12,13		30	37,44
Lateral hypothalamus	12,13		30	
Median eminence	12,13		30	
Medial tuberal nucleus				37
Ventromedial nucleus				39
Supraoptic nucleus				39
Circumventricular organs				
Subfornical organ	6			
Choroid plexus	6,12		167,168	27
OVLT	13,30		30,49,165	
Third ventricle (lining)				37,44,48
Septohippocampal tract	6		49,165	
Thalamus	6,8,12	18		
Sub-thalamus	6,12	18	49,165	
Midbrain	12			
Reticular formation	12			
Superior colliculus				
Inferior colliculus			38,49,165	

Table 1 **(Continued)**

Region	GH	GH mRNA	GHR/GHBP	GHR mRNA
Metencephalon				
Pons	12			37,44,48
Parabrachial nucleus	12			
Trigeminal nucleus	12			
Cerebellum	8	57	38	
Nuclei	12	57	57	
Purkinje cells				
Myencephalon				
Cranial ganglia			38	37,44,48
Miscellaneous				
Pineal gland	17			
Posterior pituitary	3,12			
Spinal cord	12			
Brain stem	12		38	
Gray matter			49,57,165	
Peripheral ganglia	12		49,57,165	
Ventral horn				
Retinal ganglia			49,57,165	
Astrocytes			38,49,57,165	
Oligodendrocytes/Schwann cells				49,57,165
Pia mater			49,57,165	
Cerebrospinal fluid	16			

Note: OVLT, Organum vasculosum lamina terminalis.

* Numbers refer to Reference citations.

Source: Based on Harvey et al.[2]

The GH-like protein in most of the rat brain is chromatographically identical to monomeric pituitary GH and has a comparable molecular size (approximately 22 kDa).[6,8,11] It is also immunologically similar to pituitary GH and to placental lactogen[6,8,11,12,16] but is more akin to the central portion of human GH than to rat pituitary GH.[12,13] Moreover, whereas hypothalamic nuclei contain cells that cross-react with human GH antisera,[13] the amygdaloid nucleus cross-reacts only with antisera against rat GH.[7,12] It is therefore possible that a number of GH-like proteins are expressed in the brain, some of which are likely to differ from pituitary GH in amino acid sequence or conformational structure. Indeed, GH-like proteins of 55 and 25 kDa have also been identified in the ovine pineal gland.[17] However, despite these differences, brain GH has the same biological activity and potency as pituitary GH.[7,8,11] The structural difference between brain GH and pituitary GH is, nevertheless, indicative of separate sites of synthesis rather than the accumulation of pituitary GH in the brain. This possibility is also supported by the presence of GH immunoreactivity in neuronal perikarya and the increase in immunoreactive GH staining after colchicine blockade of axonal transport.[13] The persistence of GH in the brain following hypophysectomy[7,8,13] also indicates its neuronal synthesis, as does the continuous release of immunoreactive GH from cultured amygdaloid, hippocampal, and hypothalamic cells[7,8] and the depletion of GH in the median eminence following electrical stimulation of the paraventricular nucleus (PVN).[13] It is therefore surprising that Gossard et al.[18] failed to detect GH mRNA in the amygdaloid nucleus or the hypothalamus, although they found GH mRNA in the caudate putamen, basal cortex, striatum, central thalamus, formation reticularis, and the outer layer of the hippocampus.[18,19] Expression of the *GH* gene also occurs in large cell bodies of the periventricular nucleus[19] (Table 1).

The synthesis of GH within the brain appears to occur prior to that in the pituitary gland, because it is present in the fetal brain of rats on the 10th day of gestation whereas it is not present in the pituitary gland until the 12th day.[6] Moreover, whereas the GH content of the brain peaks prenatally and is only slightly higher in adults than in neonates, the GH content in the pituitary gland increases during ontogeny.[2,6] *GH* gene transcription in the brain is therefore likely to differ from that in the pituitary gland. Indeed, whereas ovariectomy and thyroidectomy reduce GH concentrations in the pituitary gland they increase those in the brain[9]. Similarly, although GH release from the pituitary gland is not autonomous, GH is continuously released from cultured brain cells.[7,8] The release of GH from amygdaloid, hippocampal, and hypothalamic cells *in vitro* is, however, suppressed by SRIF, as is the release of pituitary GH.[7,8]

The synthesis of GH within the brain may also be related to the synthesis of GH within the pituitary gland. The GH concentration in the brain may be regulated by pituitary GH, because hypophysectomy augments GH levels in the amygdala, hypothalamus, and thalamus, while lowering those in the caudate nucleus, hippocampus, and cerebral cortex.[7-9,12,13] Because GH acts within the hypothalamus to inhibit the synthesis and release of pituitary GH,[2,20,21] the synthesis of GH within hypophysiotropic regions of the brain may therefore participate in the short-loop regulation of pituitary GH synthesis.

III. PASSAGE ACROSS THE BLOOD-BRAIN BARRIER

The presence of GH-like immunoreactivity in cerebrospinal fluid (CSF)[16] suggests GH synthesis within the brain and/or the passage of GH through the blood-brain barrier. The permeability of the blood-brain barrier to GH is, however, controversial. Although Linfoot et al.[16] found that acromegalics with excessive pituitary GH secretion had a CSF GH concentration 740- to 800-fold higher than normal, GH concentrations in systemic circulation are not always mirrored by changes in CSF GH concentrations.[22-24] Thus whereas Pacold et al.[7] and Hojvat et al.[6,8] found negligible uptake of radiolabeled GH from the systemic circulation into the brain stem, Stern et al.[25] found GH radioactivity in the telencephalon, diencephalon, midbrain, pons-medulla, and cerebellum. A carrier-mediated transport system for GH has also been suggested by Banks and Kastin[26] by their demonstration that aluminum specifically enhances the permeability of the blood-brain barrier to labeled GH, but not to labeled thyrotropin stimulating hormone (TSH), luteinizing hormone (LH), or iodine. This possibility is also supported by the demonstration of GH receptor mRNA in the choroid plexus of adult[27] and fetal[28] rats, suggesting GH entry through circumventricular organs. Moreover, as the blood-brain barrier is incomplete in the median eminence, the presence of GH in hypophysial portal blood (as a result of retrograde flow) at concentrations far higher than those in peripheral plasma[29] may also facilitate GH entry into the brain. Indeed, systemically administered radiolabeled GH has been shown to localize in the median eminence and other hypothalamic nuclei.[30] Systemic injections of GH have also been found to have rapid effects on brain neurochemistry,[31-34] suggesting poor impedance by the blood-brain barrier. Movement through the blood-brain barrier may also be bidirectional, because the transplantation of fibroblasts containing the *hGH* transgene into the cerebrum[23] results in the presence of hGH in rat peripheral plasma.

IV. NEURAL GH RECEPTORS

Neural sites of GH action are indicated by the presence of GH receptor mRNA and GH-binding sites within the CNS. Expression of the GH receptor gene has been demonstrated in the brains of rats,[35-40] guinea pigs,[41] rabbits,[42,43] pigs,[44] cattle,[45] chickens,[43] and fish.[46] In the rat the small (1.2-kb) transcript of the GH receptor gene encoding the GH binding protein is expressed in the hypothalamus, although the large (4.5-kb) transcript coding for the full-length receptor predominates.[39,40] Splicing of the primary GH receptor gene transcript in the hypothalamus is thus different from that in the liver, in which the 1.2-kb transcript predominates.[42,47] Moreoever, although both transcripts are expressed in the rat hypothalamus from embryonic day 15,[40] after the ontogenic appearance of brain GH,[6] the large transcript is preferentially processed neonatally.[40] The ratio between the 4.5- and 1.2-kb transcripts was less than 2.0 from embryonic day 15 to 3 days after birth, but increased toward puberty in male and female rats. This developmental pattern of GH receptor gene expression does not, therefore, appear to be coordinately regulated by GH. Plasma GH levels are known to peak in and around the time of parturition, when GH receptor transcripts are maximal, and then decline until puberty, when GH secretion increases in a sexually dimorphic way. Although the GH receptor mRNA levels in the hypothalamus also decline neonatally[38,40] they are comparable in males and females and plateau much earlier than the onset of puberty. This age-related decline in the expression of the GH receptor in hypothalamic tissue[38,40] contrasts, however, with its increased expression in peripheral tissues and ontogenic increase in the rabbit and chicken hypothalamus.[43] In the rat the level of GH receptor message in the brain is approximately fivefold lower than that in the liver but comparable to that in other peripheral tissues.[35,38]

In the rat, GH receptor mRNA-containing cells are present in the arcuate nucleus (AN), periventricular nucleus (PEVN), ventrolateral region of the ventromedial nucleus, paraventricular nucleus (PVN), supraoptic nucleus,[39] dorsomedial nucleus, and the medial tuberal nucleus.[37] In the rat and rabbit hypothalamus the localization of GH receptor mRNA is reported to occur in SRIF neurons in the PEVN and PVN and to be coexpressed with GH-releasing hormone (GHRH) in the AN,[39,42,43] suggesting their participation in GH autoregulation. This possibility is also supported by the 66% decrease in the number of cells in the rostral PEVN expressing the GH receptor gene following hypophysectomy.[48] The increase in the relative abundance of the 4.5-kb GH receptor transcript in the rat hypothalamus from 7 days of age to puberty may also participate in the autofeedback inhibition of GH, because plasma GH levels are reduced during this period of development.[40] However, according to Minami et al.,[39] the localization of GH receptor mRNA in the AN of the rat hypothalamus does not correspond to the localization of cells in the dorsomedial region containing SRIF mRNA or cells in the ventrolateral region containing GHRH mRNA, even though GHRH and SRIF expression in the AN respond to hypophysectomy and exogenous GH therapy. In extrahypothalamic regions of the brain, GH receptor mRNA has also been located in the bed nucleus of the stria terminalis, the diagnonal band of Broca, the thalamus, septal region, hippocampus, and amygdala, and in cells not expressing the *SRIF* gene.[37] This wide distribution suggests other actions of GH within the brain that may be unrelated to GH autoregulation. Because the distribution of GH receptor mRNA is concordant with the distribution of IGF-I mRNA and IGF-I immunoreactivity,[38] these roles may include the regulation of brain growth and maturation, especially as both genes are primarily expressed in the neonatal period.

The widespread distribution of GH receptor message within the brain is also similar to the distribution of the translated protein. Immunoreactivity for the GH receptor/GH-binding protein in the rat occurs strongly in layers 2, 3, 5, and 6 of the cerebral cortex, in neurons of the thalamus and hypothalamus, in Purkinje cells of the cerebellum, in neurons of the trapezoid body of the brain stem, and in retinal ganglion cells. Glial cells, notably astrocytes, are also strongly reactive, along with ependyma of the choroid plexus, the ventricular lining, and pia mater. Lesser amounts of GH receptor/GH-binding protein immunoreactivity occur in the hippocampus, dentate gyrus, olfactory bulb, inferior colliculus, and spinal cord[38,49] (Table 1). Immunoreactivity for the receptor/binding protein is also present in the rabbit brain, particularly in layers 2 and 3 of the cerebral cortex, in pyramidal cells of the hippocampus, and in neurons of the inferior and superior colliculi, brain stem reticular formation, dorsal thalamus, and hypothalamus.[38] As with GH receptor mRNA in the rat, this immunoreactivity in the brain is more prominent in the neonate and declines with postnatal age.[38] In addition to being widely distributed throughout the brain, immunoreactivity for the GH receptor/GH-binding protein also occurs ubiquitously within brain cells. Immunoreactive staining occurs on cellular and nuclear membranes, in secretory granules and other intracellular membranes, and in the cytoplasm and nucleoplasm of neural tissues. It is, therefore, possible that some of the central actions of GH may be independent of signal transduction systems and may involve intracrine or genomic mechanisms.[2]

This immunoreactivity for the GH receptor/binding protein is also likely to reflect the distribution of ligand-binding proteins, because specific high-affinity, low-capacity membrane binding sites for ^{125}I-labeled GH are present in rabbit,[43] rat,[30] and chicken[50] hypothalami and extrahypothalamic brain regions and in the human choroid plexus[27] (Table 2). These binding moieties have estimated molecular weights of 51 kDa in the human brain[27] and 50 to 55 kDa in the chicken hypothalamus.[50a] These binding moieties are smaller than liver GH receptors but of a size comparable to the circulating GH-binding protein.[51,52] It is, however, possible that these moieties are proteolytic fragments of larger receptors or differ from the liver receptor in their degreee of glycosylation or association with ubiquitin. Specific GH-binding sites also occur in the telencephalon, thalamus-midbrain, cerebellum-medulla, and hypothalamus of teleost fish.[46,53-55] In each species the affinity of these GH-binding sites is comparable to those in the liver, although they are of lower (15- to 30-fold less) abundance (Table 2). The abundance of these binding sites is inversely related to circulating GH levels in fish during starvation[46] and in chickens and rabbits during growth,[43] similar to the downregulation and upregulation of liver GH receptors by elevated and suppressed GH concentrations.[36,56]

In addition to neurons of the CNS, neurons of peripheral ganglia (sympathetic, parasympathetic, and dorsal root ganglia) also possess immunoreactivity for the GH receptor and for the GH-binding protein.[57] Satellite cells of the axons also contain GH receptor/GH-binding protein immunoreactivity.[57]

Table 2 **Growth Hormone-Binding Sites in Neural Tissues**

Species	Tissue	K_d (nM)	B_{max}	Ref.
Human	Choroid plexus	0.63	0.67 pmol/mg protein	27
Rabbit	Hypothalamus	0.19–0.20	30–43.9 fmol/g tissue	43
Rabbit	Hypothalamus		0.75–1.1 fmol/g protein	43
Rabbit	Hypothalamus	4.45	135.2 fmol/g tissue	43
Rabbit	Hypothalamus		3.38 fmol/mg protein	43
Chicken	Hypothalamus	26.1		50
Chicken	Hypothalamus	0.04	2.14 fmol/mg protein	50
Salmon	Telencephalon	10.0	62.3 fmol/g tissue	46
Salmon	Thalamus (mid brain)	8.2	53.2 fmol/g tissue	46
Salmon	Cerebellum-medulla	8.2	42.1 fmol/g tissue	46
Salmon	Hypothalamus	9.8	25.8 fmol/g tissue	46

V. NEURAL GH ACTIONS

A. BRAIN GROWTH

It is unknown if GH has a mitogenic effect in the human brain[58] but it is now well established that GH stimulates neuronal and glial proliferation and increases brain and cranial size in young animals.[59-61] Intracisternal injections of GH increase ornithine decarboxylase (ODC) activity (a marker for myelination) in neonatal rats,[62] in which isolated cerebral nuclei have increased rates of ribosomal RNA transcription.[63] The synthesis of RNA in rat cerebral slices is directly stimulated by GH *in vitro*,[59] as is the ODC activity in fetal rat brain cells.[64] The activity of CNPase (2′, 3′-cyclic-nucleotide 3′-phosphodiesterase, an enzyme involved in myelin synthesis) is also increased by GH in cultures of fetal rat telencephalon cells.[65] This increase in myelination occurs in the absence of increased DNA synthesis, indicating that GH may modulate the differentiation of oligodendrocytes.

A neurotrophic role for GH is also indicated by the severe deficits in brain development and RNA/DNA synthesis, and impaired proliferation and myelination of neural and glial cells that occur in GH-deficient states.[66,67] These neuronal defects are reversed by GH administration[68] and may be thyroid hormone dependent.[69] Exogenous GH similarly corrects the altered glucose metabolism, nucleic acid and protein synthesis, lipid content, and enzyme activity in the brains of GH-deficient cretinoid rats[63] and GH-deficient Snell dwarf mice,[68] although only prior to a critical period of neonatal development.[70] These actions on myelinogenesis and neural proliferation may be mediated by gangliosides, because myelin gangliosides and some neuron-maturation gangliosides are reversibly reduced in GH-deficient animals.[71] Glial cell development in rats is similarly dependent on a transient period of neonatal development.[69]

The possibility that GH is an agent promoting the maturation of the CNS is also supported by the transient elevation in brain GH during the late fetal-early neonatal stage,[6] at a time when brain GH receptors are maximally expressed[38] and interconnections of neuronal circuitry and formation of glial cells are maximal.[38] Gray cell volume and dentritic length are also increased in the newborn progeny of GH-injected mothers.[73–74] GH thus appears to stimulate neuronal and/or glial proliferation and development, thereby increasing brain DNA and size.[73-77] These effects may be dependent on thyroid hormone interactions and may occur only during critical periods of perinatal development.[69]

The actions of endogenous GH on brain development may be mediated by IGF-I[38,58,78] and probably reflect endocrine actions of pituitary GH rather than actions of GH of neural origin. This possibility is supported by the increased brain weight in mice bearing *IGF-I* transgenes, in which plasma GH levels are suppressed,[79] because mice with *GH* transgenes have normal-sized brains but enlarged peripheral tissues.[80] Moreover, if neural GH were responsible for the neural development of the brain, the intellectual or visuomotor disturbances associated with isolated (pituitary) GH deficiency syndromes[81,82] would not occur. The impaired growth and function of the brain following neonatal hypophysectomy[83,84] would similarly not occur if neural GH is an important regulator of neural development.

B. NEUROTRANSMISSION

Neuromodulatory actions of GH occur in discrete brain regions. For instance, in the rat[31-33,85] and chicken[34] hypothalamus the content and turnover of dopamine (DA) and norepinephrine (NE) are reduced by exogenous GH. These actions occur within 2 to 4 h of systemic GH injections in intact and hypophysectomized animals. The transgenic expression of *hGH* or the *hGH-B* trangenes or of the bovine *GH* (*bGH*) gene in female (but not male) mice is similarly associated with reduced DA and NE levels in the medial basal hypothalamus.[86] The *hGH* transgene also nullifies the castration-induced rise in median eminence NE levels observed in nontransgenic mice.[87] This may result in the disinhibition of SRIF release and provide negative feedback for somatotroph regulation.[32]

In contrast to the hypothalamus, the production, release, and reuptake of DA metabolites in striatal and septal tissue are reduced in GH-deficient Snell dwarf mice[88] and Ames dwarf mice,[89] indicating that GH may stimulate DA transmission in discrete extrahypothalamic brain regions.[90] Indeed, tissue stores of NE and epinephrine in the tuberculum olfactorium, caudate nucleus, telencephalon and pons medulla are not affected by exogenous GH.[31-33,85] Similarly, whereas GH increases the affinity of β-adrenergic receptors in the rat hypothalamus,[91] and thereby promotes SRIF release, it reduces the affinity of these receptors in the cerebral cortex.[91] Catecholaminergic actions of GH have also been observed in mice transgenically expressing the human or bovine *GH* gene and correlated with increased pituitary gonadotropin secretion.[92] Increased dopamine turnover in the median eminence of transgenic animals expressing hGH has also been correlated with a suppression of prolactin synthesis and release,[93–96] although this may result from the activation of lactotrophic receptors. Overexpression of hGH in transgenic mice is accompanied by an increased content of pituitary prolactin and an increase in the plasma prolactin level,[86] associated with a reduction in the turnover of dopamine in the median eminence.[86]

The number and affinity of serotonin (5-HT) receptors in the rat hypothalamus, but not in the cerebral cortex, are also reduced by exogenous GH,[96] which suppresses 5-HT synthesis and turnover in the hypothalamus[90] and diencephalon,[31,32,48,91,97] but increases 5-HT concentrations in the pons-medulla. These actions may also contribute to GH autoregulation, because 5-HT stimulates GH release in mammals[98] and inhibits SRIF secretion.[99] In GH-deficient Snell dwarfs, concentrations of 5-HT and its precursors and metabolites are elevated in the striatum and hippocampus[88] and similar findings are observed in the whole brain of hypophysectomized animals.[97,100] Because exogenous GH does not alter tryptophan hydroxylase activity, GH may inhibit 5-HT metabolism by altering the availability of tryptophan,[97,100,101] especially as GH decreases brain tryptophan levels in hypophysectomized and Snell dwarf mice.[100] The memory and motor control deficit of these dwarf animals may thus result from abnormal neurotransmission, particularly the altered balance between dopamine and 5-HT in the septohippocampal tract and nigrostriatal loop, respectively.[88]

The deficiency of choline acetyltransferase (ChAT) activity and choline uptake in the hippocampus, olfactory tubules, and striatum of Snell dwarfs also indicates a role for GH in cholinergic regulation, especially as normal ChAT activity is restored by exogenous GH.[102,103] Autoregulatory actions of GH on GHRH and SRIF neurons are also thought to be mediated through cholinergic neurons.[104–106]

C. CENTRAL BEHAVIOR

1. Feeding

The localization of GH in the ventromedial nucleus (VMN) of the hypothalamus[9] suggests GH involvement in feeding, because the VMN is the primary regulatory center for appetite control. Moreover, under certain conditions GH injected centrally or peripherally stimulates food intake in mammalian, avian, and reptilian species,[107-109] whereas suppression of GH secretion inhibits food intake.[110] This stimulation of food intake may serve a physiological role, as the rise in GH levels several hours after a meal may stimulate hunger.[111] The induction of IGF-I production,[112] which inhibits feeding, may also protect against excessive food intake, especially as the abnormal food consumption of VMN-lesioned rats is corrected by GH administration.[113] Normal levels of GH are therefore thought to be required for normal feeding behavior.[110]

2. Learning and Memory

The prenatal administration of GH increases the number and length of dendrites, brain weight, DNA content, cortical cell density, and the ratio of neuron to glia, and enhances performance in conditional discrimination tasks.[114,115] The administration of GH to pregnant female animals also improves the cortically mediated but not innate or automatic behavior of their offspring, suggesting a specific action on higher learning.[73,116,117] However, as the placenta may be impermeable to maternal GH this action may indirectly result from a prolongation of gestation and from an enhanced frequency and duration of nursing episodes. Learning ability may, nevertheless, be stimulated by GH, as GH-deficient Snell dwarf mice and hydrocortisone-intoxicated rats have impaired abilities[68,118] that can be corrected by exogenous GH.[68] The poor performance of Snell dwarfs in cortically mediated behaviors is similar to that of mice with hippocampal lesions or with deficiencies in the septal-hippocampal tract or in cholinergic neurotransmission in the basal ganglia.[88,103,118] Thus, GH may influence memory by affecting cholinergic transmission and the septal-hippocampal tract or by generally enhancing brain development. Suboptimal brain development is also thought to be responsible for the learning disabilities of some Laron dwarfs,[82,119] characterized by tissue resistance to GH action. Indeed, the administration of hGH to GH-deficient children has been shown to increase head circumference, intellectual quotient, and school performance, particularly if GH treatment is commenced before 5 years of age.[119-121] Adults with GH deficiency are, however, reported to have some improved psychcometric tests after GH therapy.[122] An improved sense of well-being, mental alertness, and vitality have also been observed in such patients.[123,124] A facilitatory role for exogenous GH in memory acquisition has also been demonstrated[125] and may reflect a physiological role for GH during sleep. Memory and learning are enhanced by rapid eye movement (REM) sleep, which may be induced by the surges in GH secretion during sleep.[126] However, studies have identified subpopulations of intellectually gifted Laron dwarfs, indicating only a minor role, if any, for GH in learning and memory processes.[127,128]

3. Sleep

Sleep and GH secretion are temporally correlated in humans and animals.[129,130] Normal sleep patterns, particularly the maintenance of REM sleep and slow-wave sleep (SWS) may thus be dependent on GH, which is released with the onset of sleep.[131] A major burst of GH secretion occurs soon after sleep onset, in association with non-REM sleep. This GH burst is thus unrelated to SWS but correlates with sleep onset.[132] This burst may, therefore, be masked by circadian bursts of GH release before or after sleep onset, depending on bed time,[133] although most nocturnal bursts of pulsatile GH release closely correlate with sleep stages in humans.[134] GH release and sleep deprivation therefore results in decreased GH secretion, which is elevated during recovery sleep. The relationship between sleep onset and GH secretion is also shown by the reduced GH levels in the plasma of patients with sleep disorders.[130] This relationship may be causal, because abnormally high or low GH secretion may modulate sleep duration and/or intensity.[126] Indeed, psychosocial dwarfs and young adults with severe isolated GH deficiency have abnormal sleep patterns,[135,136] because less time is spent in REM sleep and SWS, even though total sleep time (TST) is increased.[137-140] These abnormalities can be corrected by GH administration[135,140,141] or normalization of GH secretion.[142] This action of GH may be mediated by IGFs, cholinergic transmission, or opioid sleep-modifying peptides.[58,138,140] Increased protein synthesis is probably also involved, because GH can reverse the inhibition of REM sleep induced by actinomycin.[143] The sleep-inducing action of GH may, however, be restricted to GH-deficient states, because GH has minimal effects on REM sleep in normal individuals[126,131,144] and a pathological excess of GH may induce abnormal sleep patterns.[144]

The sleep patterns of acromegalics (characterized by hypersecretion of GH) are characterized by decreased REM sleep time, more night-time awakenings, more day-time somnolence, less total sleep time and decreased SWS in the presence or absence of sleep apnea,[126] and increased REM and delta sleep energy.[145] The normalization of GH concentrations in these patients by adenectomy is associated with restoration of REM sleep, increased SWS and TST, reduced night-time awakenings and day-time sleepiness, and a normalization of sleep energy.[145]

The relationship between GH and sleep duration/intensity may not, however, be causal (reviewed by Krueger and Obal).[130] The dissociation of sleep and GH secretory pattern in human pathophysiological conditions,[146] in elderly individuals,[147] and by pharmacological drugs[130,148,148a] supports this view. Moreover, changes in GH secretion correlate only with changes in SWS, because non-REM sleep is unaffected,[149] increased,[150] or decreased[151] by hypophysectomy, whereas exogenous GH has inconsistent effects on non-REM sleep in rats, cats, and human subjects.[126,139,152-154] It has, therefore, been suggested[130] that the stimulation of SWS and GH secretion soon after the onset of sleep results from a common subcortial mechanism, rather than reflecting cause and effect relationships.[154,156] This mechanism may be mediated through GHRH, which independently promotes both sleep (particularly non-REM sleep) and GH secretion,[130,156,157] both of which are suppressed by GHRH immunoneutralization[158] and GHRH antagonists.[157]

4. Breathing

The central control of breathing during sleep may also be related to GH action. Central sleep apnea occurs in a large number (37%) of acromegalic patients, in which its severity closely correlates with the degree of GH hypersecretion.[145,159] The presence of GH receptor in the hypothalamus, which controls breathing, supports this possibility.

5. Motor Activity

Growth hormone may play a role in the development of motor activity, as Snell dwarf mice have impaired motor activity[103] that can be normalized by GH therapy.[160] The postulated defect in the 5-HT/DA balance in the nigrostriatal tract of these animals[88] or their deficiency in cholinergic transmission[103] have also been proposed as factors responsible for these motor deficits. This chemical finding is consistent with the localization of GH receptors in Purkinje cells of the cerebellum, in brain stem nuclei and motor neurons, and in retinal ganglion cells.[38] Alternatively, GH may facilitate locomotion by stimulating neuronal myelination.[160] Motor development is also characteristically slow in GH-resistant Laron dwarfs,[161] which characteristically have visuomotor disturbances.[162,163]

Inhibitory effects of GH on locomotor activity have, conversely, been observed in wheel-running[164] and open-field[115] rats, in which total movement and jumping frequency are reduced following GH injections. These inhibitory effects of GH on locomotor activity are, however, observed with submaximal and suprathreshold doses of GH, because these behaviors show a biphasic dose relationship with exogenous GH.[115]

REFERENCES

1. **Zamenoff, S.,** *Growth*, 5, 123, 1941.
2. **Harvey, S., Hull, K. L., and Fraser, R. A..,** *Growth Regul.*, 3, 1, 1993.
3. **Wright, G.,** *Cell Tiss. Res.*, 246, 23, 1986.
4. **Inestrosa, N. C., Labarca, R., Perelman, et al.,** *Arch. Biol. Med. Exp. Santiago*, 23, 179, 1990.
5. **Tsushima, T., Friesen, H. G., Chang, T. W., et al.,** *Biochem.Biophys.Res.Commun.*, 59, 1062, 1974.
6. **Hojvat, S., Emanuele, N., Baker, G., et al.,** *Dev. Brain Res.*, 4, 427, 1982.
7. **Pacold, S. T., Kirsteins, L., Hojvat, S., et al.,** *Science*, 199, 804, 1978.
8. **Hojvat, S., Baker, G., Kirsteins, L., et al.,** *Brain Res.*, 239, 543, 1986.

9. **Hojvat, S., Emanuele, N. V., Kirsteins, L., et al.,** *Neuroendocrinology*, 44, 355, 1986.
10. **Yon, L., Feuilloley, M., Kobayashi, T., et al.,** *Gen. Comp. Endocrinol.*, 83, 142, 1991.
11. **Kyle, C. V., Evans, M. C., and Odell, W. D.,** *J. Clin. Endocrinol. Metab.*, 53, 1138, 1981.
12. **Lechan, R. M., Molitch, M. E., and Jackson, I. M. D.,** *Endocrinology (Baltimore)*, 112, 877, 1983.
13. **Lechan, R. M., Nestler, J. L., and Molitch, M. E.,** *Endocrinology (Baltimore)*, 109, 1950, 1981.
14. **Krieger, D. T.,** *Fed. Proc.*, 39, 2937, 1980.
15. **Hansen, B. L. and Hansen, G. N.,** *Cell Tiss. Res.*, 222, 615, 1982.
16. **Linfoot, J. A., Garcia, J. F., Wei, W., et al.,** *J. Clin. Endocrinol. Metab.*, 31, 230, 1970.
17. **Noteborn, H. P. J. M., Vanbalen, P. P., Vandergugten, A. A., et al.,** *J. Pineal Res.*, 14, 11, 1993.
18. **Gossard, F., Dihl, F., Pelletier, G., et al.,** *Neurosci. Lett.*, 79, 251, 1987.
19. **Martinoli, M. G., Oullet, J., Rheaume, E., et al.,** *Neuroendocrinology*, 54, 607, 1991.
20. **Melmed, S.,** *The Brain as an Endocrine Organ* (Eds. Cohen, M. P. and Foa, P. P.). Springer-Verlag, Berlin, 1992, p. 193.
21. **Tannenbaum, G. S.,** *Endocrinology (Baltimore)*, 107, 2117, 1980.
22. **Belchetz, P. E., Ridley, R. M., and Baker, H. F.,** *Brain Res.*, 239, 310, 1982.
23. **Doering, L. C. and Chang, P. L.,** *J. Neurosci. Res.*, 29, 292, 1991.
23a. **Hashimoto, N., Handa, H., and Nishi, S.,** *J. Neurosurg.*, 64, 140, 1986.
24. **Tamasawa, N., Kurahashi, K., Babe, T., et al.,** *J. Endocrinol. Invest.*, 11, 429, 1988.
25. **Stern, W. C., Miller, M., Resnick, O., et al.,** *Am. J. Anat.*, 144, 503, 1976.
26. **Banks, W. A. and Kastin, A. J.,** *Neuropharmacology*, 24, 407, 1985.
27. **Lai, Z., Emnter, M. R. P., and Nyberg, F.,** *Brain Res.*, 546, 222, 1991.
28. **Garcia-Aragon, J., Lobie, P. E., Muscat, G. E. O., et al.,** *Development*, 114, 869, 1992.
29. **Sato, M., Chihara, K., Kita, T., et al.,** *Neuroendocrinology*, 50, 139, 1989.
30. **Van Houten, H., Posner, B. I., and Walsh, R. J.,** *Exp. Brain. Res.*, 38, 455, 1980.
31. **Stern, W. C., Miller, M., Jalowiec, J. E., et al.,** *Pharmacol. Biochem. Behav.*, 1992.
32. **Andersson, K., Fuxe, K., Eneroth, P., et al.,** *Neurosci. Lett.*, 5, 83, 1977.
33. **Andersson, K., Fuxe, K., Eneroth, P., et al.,** *Eur. J. Pharmacol.*, 95, 271, 1983.
34. **Lea, R. W. and Harvey, S.,** *J. Endocrinol.*, 136, 245, 1993.
35. **Mathews, L. S., Enberg, B., and Norstedt, G.,** *J. Biol. Chem.*, 264, 9905, 1989.
36. **Kelly, P. A., Djiane, J., Postel-Vinay, M. C., et al.,** *Endocr. Rev.*, 12, 235, 1991.
37. **Burton, K. A., Kabigting, E. B., Clifton, D. K., et al.,** *Endocrinology (Baltimore)*, 130, 958, 1992.
38. **Lobie, P. E., Garcia-Aragon, J., Lincoln, D. T., et al.,** *Dev. Brain Res.*, 74, 225, 1993.
39. **Minami, S., Kamegari, J., Hasegawa, O., et al.,** *J. Neuroendocrinol.*, 5, 691, 1993.
40. **Hasegawa, O., Minami, S., Sugihara, H., et al.,** *Dev. Brain Res.*, 74, 287, 1993.
41. **Harvey, S. and Fraser, R. A.,** *J. Endocrinol.*, 133, 357, 1992.
42. **Tiong, T. S. and Herington, A. C.,** *Endocrinology (Baltimore)*, 129, 1628, 1991.
43. **Fraser, R. A., Attardo, D., and Harvey, S.,** *J. Mol. Endocrinol.*, 5, 231, 1990.
44. **Bingham, B., Oldham, E. R., and Baumbach, W. R.,** *72nd Annu. Meet. Endocr. Soc.*, Atlanta, GA, 1991, p. 871.
45. **Hauser, S. D., McGrath, M. F., Collier, R. J., et al.,** *Mol. Cell Endocrinol.*, 72, 187, 1990.
46. **Sanchez, J. P., Smal, J., and Le Bail, P. Y.,** *Growth Regul.*, 1, 145, 1991.
47. **Baumbach, W. R., Horner, D. L., and Logan, J. S.,** *Gene Dev.*, 3, 1199, 1989.
48. **Burton, K. A., Kabigting, E., Clifton, D. K., et al.,** *73rd Annu. Meet. Endocr. Soc.,* Washington, D.C., 1991, p. 419.
49. **Waters, M. J., Barnard, R. T., Lobie, P. E., et al.,** *Acta Pediatr. Scand.*, 366(Suppl.), 60, 1990.
50. **Attardo, D. and Harvey, S.,** *J. Mol. Endocrinol.*, 4, 123, 1990.
50a. **Harvey, S. and Hull, K. L.,** unpublished observations, 1994.
51. **Baumann, G.,** *Endocr. Rev.*, 12, 424, 1991.
52. **Baumann, G.,** *Proc. Soc. Exp. Biol. Med.*, 202, 392, 1993.
53. **Sakamoto, T. and Hirano, T.,** *J. Endocrinol.*, 130, 425, 1991.
54. **Gray, E. S., Young, G., and Bern, H. A.,** *J. Exp. Zool.*, 256, 290, 1990.
55. **Yao, K., Niu, P., LeGac, F., et al.,** *Gen. Comp. Endocrinol.*, 81, 72, 1991.
56. **Kelly, P. A., Ali, S., Rozakis, M., et al.,** *Recent Prog. Horm. Res.*, 48, 123, 1993.
57. **Lobie, P. E., Garcia-Aragon, J., Wang, B. S., et al.,** *Endocrinology (Baltimore)*, 130, 3057, 1992.
58. **Hoffman, A. R., Lieberman, S. A., and Ceda, G. P.,** *Psychoneuroendocrinology*, 17, 327, 1992.
59. **Zamenhof, S., Mosley, J., and Schuller, E.,** *Science*, 152, 1396, 1966.

60. **Zamenhof, S., Van Marthens, E., and Grauel, L.,** *Science*, 174, 954, 1971.
61. **Oyhenart, E. E. and Pucciarelli, H. M.,** *Growth Dev. Aging*, 56, 179, 1992.
62. **Roger, L. J., Schanberg, S. M., and Fellows, R. E.,** *Endocrinology (Baltimore)*, 95, 904, 1974.
63. **Berti-Mattera, L. N., Gomez, C. J., and Krawiec, L.,** *Horm. Metab. Res.*, 15, 286, 1983.
64. **Yang, J. W., Raizada, M. K., and Fellows, R. E.,** *J. Neurochem.*, 36, 1050, 1981.
65. **Almazan, G., Honegger, P., Matthieu, J. M., et al.,** *Dev. Brain Res.*, 21, 257, 1985.
66. **Pelton, E. W., Grindeland, R. E., Young, E., et al.,** *Neurology*, 27, 282, 1977.
67. **Pelton, E. W., Young, E., and Bass, N. H.,** *Neurology*, 24, 377, 1974.
68. **Noguchi, T., Sugisaki, T., Takamatsu, K., et al.,** *J. Neurochem.*, 39, 1693, 1982.
69. **Sugisaki, T. and Noguchi, T.,** *J. Neurochem.*, 59, 2005, 1992.
70. **van Buul-Offers, S. and Van den Brande, J. L.,** *Horm. Metab. Res.*, 17, 20, 1985.
71. **Noguchi, T. and Sugisaki, T.,** *J. Neurochem.*, 47, 1785, 1986.
72. **Clendinnen, B. G. and Eayrs, J. T.,** *J. Endocrinol.*, 22, 183, 1961.
73. **Ginalska-Malinowska, M. and Romer, T. E.,** *Endokrinologie*, 77, 341, 1981.
74. **Porterfield, S. P. and Hendrich, C. E.,** *Endocrinology (Baltimore)*, 111, 406, 1982.
75. **Sara, V. R. and Lazarus, L.,** *Science*, 186, 446, 1974.
76. **Sara, V. R. and Lazarus, L.,** *Prog. Brain Res.*, 42, 368, 1974.
77. **Sara, V. R., King, T. L., Stuart, M. C., et al.,** *Endocrinology (Baltimore)*, 99, 1512, 1976.
78. **Hynes, M. A., Van Wyk, J. J., Brooks, P. J., et al.,** *Mol. Endocrinol.*, 1, 233, 1987.
79. **Matthews, L. S., Hammer, R. E., Behringer, R. R., et al.,** *Endocrinology (Baltimore)*, 123, 2827, 1988.
80. **Hurley, D. L. and Phelps, C. J.,** *Endocrinology (Baltimore)*, 130, 1809, 1992.
81. **Pollitt, E. and Morey, J.,** *J. Pediatr.*, 64, 415, 1964.
82. **Rosenbloom, A. L., Aguirne, J. S., Rosenfeld, R. G., et al.,** *New Engl. J. Med.*, 323, 1367, 1990.
83. **Asling, C. W., Walker, D. G., Simpson, M. E., et al.,** *Anat. Rec.*, 114, 49, 1952.
84. **Diamond, M. C.,** *Brain Res.*, 7, 407, 1968.
85. **Andersson, K., Juxe, K., Balke, C. A., et al.,** *Prolactin and Prolactinomas* (Ed. Tolis, G.). Raven Press, New York, 1983, p. 43.
86. **Steger, R. W., Bartke, A., Parkening, T. A., et al.,** *Neuroendocrinology*, 53, 365, 1991.
87. **Steger, R. W., Bartke, A., Parkening, T. A., et al.,** *Neuroendocrinology*, 52, 106, 1990.
88. **Kempf, E., Fuhrmann, G., Thiriet, G., et al.,** *Neurochem. Res.*, 7, 969, 1985.
89. **Morgan, W. W. and King, T. S.,** *Neuroendocrinology*, 42, 351, 1986.
90. **Pan, J. T., Kow, L. M., and Pfaff, D. W.,** *Neuroendocrinology*, 43, 189, 1986.
91. **Popova, J., Robeva, A., Iavorska, N., et al.,** *Comp. Biochem. Physiol.*, 100C, 543, 1991.
92. **Bartke, A., Chandrashekar, V., Tang, K., et al.,** *Neuroendocrinol. Lett.*, 15, 89, 1993.
93. **Bartke, A., Steger, R. W., Hodges, S. L., et al.,** *J. Exp. Zool.*, 248, 121, 1988.
94. **Chandrashekar, V., Bartke, A., and Wagner, T. E.,** *Endocrinology (Baltimore)*, 123, 2717, 1988.
95. **Milton, S., Cecim, M., Li, V. S., et al.,** *Endocrinology (Baltimore)*, 131, 536, 1992.
96. **Popova, J., Ivanova, E., Tosheva, T., et al.,** *Gen. Pharmacol.*, 22, 1143, 1991.
97. **Cocchi, D., Giulio, A., Groppetti, A., et al.,** *Acta Vit. Enzymol.*, 29, 90, 1975.
98. **Muller, E. E.,** *Physiol. Rev.*, 67, 962, 1987.
99. **Richardson, S. B., Hollander, C. S., Prasad, J. A., et al.,** *Endocrinology (Baltimore)*, 109, 602, 1981.
100. **Cocchi, D., DiGiulio, A., Groppetti, A., et al.,** *Experientia*, 31, 384, 1975.
101. **Cocchi, D., DiGiulio, A., Groppetti, A., et al.,** *Br.J. Pharmacol.*, 52, 441P, 1974.
102. **Fuhrmann, G., Durkin, T., Thiriet, G., et al.,** *J. Neurosci. Res.*, 13, 417, 1985.
103. **Fuhrmann, G., Kempf, E., and Ebel, A.,** *J.Neurosci.Res.*, 16, 527, 1986.
104. **Ross, R. J. M., Tsagarakis, S., Grossman, A., et al.,** *Clin. Endocrinol.*, 27, 727, 1987.
105. **Torsello, A., Panzeri, G., Cermenati, P., et al.,** *J. Endocrinol.*, 117, 273, 1988.
106. **Kelijman, M. and Frohman, L. A.,** *J. Clin. Endocrinol. Metab.*, 72, 1081, 1991.
107. **Bray, G. A.,** *Brain Res. Bull.*, 14, 505, 1985.
108. **Buntin, F. and Figge, G. R.,** *Pharmacol. Biochem. Behav.*, 31, 533, 1989.
109. **Kimwele, C. N., Kanui, T. I., and Aulie, A.,** *Comp. Biochem. Physiol.*, 102A, 553, 1992.
110. **Tannenbaum, G. S., Huyda, H. J., and Posner, B. I.,** *Science*, 77, 1983.
111. **Panksepp, J., Bishop, P., and Rossi, J.,** *Psychoneuroendocrinology*, 4, 89, 1979.
112. **Holly, J. M. P. and Wass, J. A. H.,** *J. Endocrinol.*, 122, 611, 1989.
113. **York, D. A. and Bray, G. A.,** *Endocrinology (Baltimore)*, 90, 885, 1972.

114. **Sara, V. R. and Lazarus, L.,** *Nature (Lond.)*, 250, 257, 1974.
115. **Alvarez, X. A. and Cacabelos, R.,** *Peptides*, 14, 707, 1993.
116. **Block, J. B. and Essman, W. B.,** *Nature (Lond.)*, 205, 1136, 1965.
117. **Croskerry, P. G., Smith, G. K., Shepard, B. J., et al.,** *Brain Res.*, 52, 413, 1973.
118. **Bouchon, R. and Will, B.,** *Physiol. Behav.*, 30, 213, 1983.
119. **Laron, Z. and Galatzer, A.,** *Brain Dev.*, 7, 559, 1985.
120. **Laron, Z. and Galatzer, A.,** *Early Hum. Dev.*, 5, 211, 1981.
121. **Laron, Z., Roitman, A., and Kauli, R.,** *Clin. Endocrinol.*, 10, 393, 1979.
122. **Almqvist, O., Thoren, M., Saaf, M., et al.,** *Psychoneuroendocrinology*, 11, 347, 1986.
123. **Mcgauley, G. A.,** *Acta Pediatr. Scand.*, 356(Suppl.), 70, 1989.
124. **Rosen, T., Wiren, L., Wilhelmsen, L., et al.,** *Clin. Endocrinol. (Oxford)*, 40, 111, 1994.
125. **Hoddes, E. S.,** *Sleep*, 1, 287, 1979.
126. **Mendelsohn, W. B., Slater, S., Gold, P., et al.,** *Biol. Psychiatry*, 15, 613, 1980.
127. **Guevera-Aguirre, J., Rosenbloom, A. L., Fielder, P. J., et al.,** *J. Clin. Endocrinol. Metab.*, 76, 417, 1993.
128. **Meyer-Bahlburg, H. F. L., Feinman, J. A., MacGillivray, M. H., et al.,** *Psychiatr. Med.*, 9, 187, 1979.
129. **Oswald, I.,** (Eds. Obal, F., Jr., Shultz, H., and Visser, P. G.) Fischer, New York, 1988, p. 23.
130. **Krueger, J. M. and Obal, F.,** *FASEB J.*, 7, 645, 1993.
131. **Kupfer, D. J., Jarrett, D. B., and Ehlers, C. L.,** *Psychoneuroendocrinology*, 17, 37, 1992.
132. **Born, J., Muth, S., and Tehm, H. L.,** *Psychoneuroendocrinology*, 13, 233, 1988.
133. **Van Cuter, E.,** *Horm. Res.*, 34, 45, 1990.
134. **Holl, R. W., Hartman, M. L., and Veldhuis, J. D.,** *J. Endocrinol. Metab.*, 72, 854, 1991.
135. **Guilhaume, A., Benoit, O., Gourmelen, M., et al.,** *Pediatr. Res.*, 16, 299, 1982.
136. **Wolff, G. and Money, J.,** *Psychiatr. Med.*, 3, 18, 1973.
137. **Astrom, C. and Lindholm, J.,** *Neuroendocrinology*, 51, 82, 1990.
138. **Astrom, C. and Trojaborg, W.,** *Clin. Endocrinol.*, 36, 241, 1992.
139. **Wu, R. H. and Thorpy, M. J.,** *Sleep*, 11, 425, 1988.
140. **Astrom, C., Pedersen, S. A., and Lindholm, J.,** *Clin. Endocrinol.*, 33, 495, 1990.
141. **Guilhame, A., Benoit, O., Gournelen, M., et al.,** *Pediatr. Res.*, 16, 299, 1982.
142. **Brown, G. M., Scggie, J. A., Chambers, J. W., et al.,** *Psychoneuroendocrinology.*, 3, 131, 1978.
143. **Drucker-Colin, R. R., Spanis, C. W., Hunyadi, J., et al.,** *Neuroendocrinology*, 18, 1, 1975.
144. **Grunstein, R. R., Ho, K. Y., and Sullivan, C. E.,** *Ann. Int. Med.*, 115, 527, 1991.
145. **Astrom, C., Gjerris, F., Tojaborg, W., et al.,** *Neuroendocrinology*, 53, 328, 1991.
146. **Weill, J., Dherbomez, M., Fialdes, P., et al.,** *Horm. Res.*, 34, 9, 1990.
147. **Carlson, H. E., Gillin, J. G., Gorden, P., et al.,** *J. Clin. Endocrinol.*, 34, 1102, 1972.
148. **Martin, J. B.,** *Front. Neuroendocrinol.*, 4, 129, 1976.
148a. **Colle, M., Rosenzweig, P., Bianchetti, G., et al.,** *Horm. Res.*, 35, 30, 1991.
149. **Sallanon, M., Buda, C., Puymartin, M., et al.,** *Neurol. Sci.*, 88, 173, 1988.
150. **Zhang, J. Y., Valaty, J. L., and Jouvet, M.,** *Brain Res. Bull.*, 21, 897, 1988.
151. **Valaty, J. L., Chouvet, G., and Jouvet, M.,** *Prog. Brain Res.*, 42, 115, 1975.
152. **Drucher-Colin, R., Spanis, C. W., Hunyadi, J., et al.,** *Neuroendocrinology*, 18, 1, 1975.
153. **Stern, W. C., Jalowiecz, J. E., Shabshelowitz, H., et al.,** *Horm. Behav.*, 6, 189, 1975.
154. **Buzi, F., Zanotti, P., Tiberti, A., et al.,** *J. Clin. Endocrinol. Metab.*, 77, 1495, 1993.
155. **Parker, D. C., Sassin, J. F., Mace, J. W., et al.,** *J. Clin. Endocrinol.*, 29, 871, 1969.
156. **Steiger, A., Guldner, J., Hemmeter, U., et al.,** *Neuroendocrinology*, 56, 566, 1992.
157. **Obal, F., Payne, L., Kapas, L., et al.,** *Brain Res.*, 557, 149, 1991.
158. **Obal, F., Payne, L., Opp, M., et al.,** *Am. J. Physiol.*, 263, R1078, 1992.
159. **Perks, W., Horrocks, P. M., Cooper, R. A., et al.,** *Br. Med. J.*, 1980.
160. **Noguchi, T., Sugisaki, T., and Tsukada, Y.,** *J. Neurochem.*, 38, 257, 1982.
161. **Laron, Z. and Galatzer, A.,** *Brain Dev.*, 7, 559, 1985.
162. **Frankel, J. J. and Laron, Z.,** *Israel J. Med. Sci.*, 4, 953, 1968.
163. **Shurka, E. and Laron, Z.,** *Israel J. Med. Sci.*, 11, 352, 1975.
164. **Kelly, P. H.,** *Horm. Behav.*, 17, 163, 1983.
165. **Lobie, P. E., Barnard, R., and Waters, M. J.,** *J. Biol. Chem.*, 265, 19947, 1991.
166. **Mustafa, A., Adem, A., Roos, P., et al.,** *Neurosci. Res.*, 19, 93, 1994.
167. **Lai, Z., Roos, P., Zhai, Q. Z., et al.,** *Brain Res.*, 621, 260, 1993.
168. **Mangurian, L. P., Lewis, R., and Walsh, R. J.,** *J. Anat.*, 184, 425, 1994.

Chapter 29

Growth Hormone Action: Agricultural Significance

C. G. Scanes

CONTENTS

0-8493-8697-7/95/$0.00+$.50

I. INTRODUCTION

There are considerable possibilities for growth hormone (GH) to be employed to enhance the efficiency of animal agricultural production. It should be noted that there is a voluminous literature in this area, of which much is in abstract form. Although every effort has been made to review the available information objectively, data that have been presented only in abstract form or in nonrefereed papers are not included.

II. GROWTH HORMONE AND LACTATION IN DAIRY CATTLE

A. INTRODUCTION

It is now well established that GH stimulates milk production (galactopoiesis) in dairy cattle. As early as the 1930s, extracts of anterior pituitary glands were shown to increase the amount of milk produced by cows.[1,2] Studies by Folley and Young went on to demonstrate that this (galactopoietic) effect was not due solely to prolactin but to a fraction with high antiinsulin/diabetogenic activity.[3] This was identified as GH in 1949.[4] The galactopoietic effect of GH in dairy cows has been repeatedly confirmed (e.g., see Refs. 5 through 7 and Table 1). The contention that bovine GH (bGH) can affect milk production only in low-yielding cows was refuted initially by studies of Dale Bauman and colleagues at Cornell University (e.g., see Ref. 8). Examples of the effects of bGH on milk production in dairy cows are shown in Table 1.[6,7,9-12] It should be noted parenthetically that throughout this volume, GH is the term of choice (following the terminology of the Endocrine Society) and hence bGH is used in this section, rather then bovine somatotropin (bST).

It now is almost axiomatic that bGH increases milk production in dairy cattle (see Table 1 and Figure 1[13] for examples). This effect is observed when either native or biosynthetic bGH is employed[12] or whether bGH is administered as a daily injection, as intravenous (i.v.) pulses every 2 h, or as a continuous subcutaneous (s.c.) infusion (see Table 1).[10]

The ability of bGH to influence milk production is markedly affected by the adequacy of the cow's diet.[15] For instance, injections of biosynthetic bGH increase milk production by 7.7 kg/day (31%) in cows on a high-energy, high-protein diet, but only by 4.8 kg/day (24%) in cows receiving 80% of estimated energy and protein requirements[15] (see Table 2). Further deterioration of nutrition will reduce milk production both in the presence and absence of bGH; indeed, the incremental increase in production evoked by GH will be expected to be greatly depressed.

B. MILK PRODUCTION

Consistently, bGH improves feed efficiency (milk produced/feed intake) whether expressed as kilograms of milk produced per unit feed consumed or as milk energy as a percentage of consumed energy.[8,9,10,13] The improvement in feed efficiency is obviously of potential advantage to dairy farmers. Moreover, there is additional improvement in economic efficiency with greater production relative to fixed costs. This is true irrespective of whether we consider the cow or the worker or the farm as the fixed cost.

C. MILK COMPOSITION/HUMAN SAFETY CONSIDERATIONS

Not only does bGH increase milk production but there are negligible or no changes in the composition of the milk in terms of solids, fat, protein, lactose, and water (e.g., see Refs. 8–10, 13, 16; and Table 3). The administration of GH induces a small or no increase in the concentration of insulin-like growth factor I (IGF-I) (e.g., Δ IGF-I = 0.39 n*M*) in the milk[15] (also reviewed by Juskevich and Guyer[17]). It is generally viewed that bGH has no impact on the safety of milk for human consumption.[17]

D. HEALTH OF DAIRY CATTLE

Does prolonged bGH treatment influence either the health or reproductive performance of dairy cows? Consistently chronic bGH administration to dairy cows increases milk production but does not adversely influence the health of the animals. For instance, the administration of biosynthetic methionyl-bGH does not affect the incidence of clinical mastitis or subclinical mastitis as determined by subsequent reproductive performance.[18] On the basis of the effectiveness of GH in stimulating milk production, the human safety data, and the lack of problems to the health of dairy cows, the U.S. Food and Drug Administration approved the commercial use of GH in 1994.

E. MAMMARY ANATOMY

Lactating cows injected daily (5 days) with bGH show increased mammary gland weight but no change in mammary gland DNA.[14] In contrast, in prepubertal heifers, GH reduces mammary gland volume while increasing parenchymal tissue. This has been demonstrated in a definitive study with 8-month-old heifers, using pairs of monozygous twins.[19] Daily injections of bGH for 16 weeks reduce mammary weight due to a decrease in extraparenchymal tissue but an increase in parenchymal tissue as determined by dissection and weighing or by computer-assisted X-ray tomography.[19] Similarly, in prepubertal heifers, GH-releasing hormone (GHRH) reduces mammary gland volume owing to a decrease in extraparenchymal tissue.[16] Also, in prepubertal sheep, GH tends to reduce total mammary gland volume and that of the mammary fat pad.[20]

Relatively few biochemical changes have been observed in the mammary glands of dairy cattle treated with bGH. For instance, no changes in lipoprotein lipase *in vivo*[21] or triglyceride formation *in vitro*[22] are reported with bGH treatment. Bovine GH does, however, reduce mammary fatty acid synthesis *in vitro*[22] and tends to exert the same effect *in vivo*.[21]

F. METABOLIC EFFECTS

Changes in energy balance in cows receiving bGH (for 9 to 14 days) have been examined by respiration calorimetry.[23] Oxygen consumption increases by 4.1% with bGH treatment whereas carbon dioxide production is little affected (1.2% tendency to increase). There is little effect of GH on energy intake.[23] Increases in heat energy (8.6%) and milk energy (28.4%) are observed with a concomitant decrease in tissue energy.[23] The changes in heat energy reflect the energy needs for the synthesis of milk components and ultimately milk production.

To accommodate the presumed increases in mammary gland metabolism required for milk synthesis with bGH treatments, there is an increase in blood flow to the mammary gland.[24] Moreover, GH treatment increases glucose extraction from mammary arterial blood[24] and irreversible glucose loss from circulation.[25]

Table 1 **Examples of the Effect of Bovine GH Treatment on Milk Production in Dairy Cattle**

Milk Production (kg/day)		Duration of GH	Source and	Mode of	Breed of	
Pretreatment	GH Treatment	Treatment (days)	Dose of GH	Administration	Cattle	Ref.
13.3	16.6	10	Native bGH, 33 mg/day	Daily injection	Holstein	Machlin (1973)[6]
17.4	19.6	7	Native bGH, 30 mg/day	Daily injection	Friesian	Bines et al. (1980)[7]
4.76	5.6	7	Native bGH 30 mg/day	Daily injection	Hereford	Bines et al. (1980)[7]
34.4	37.7	11	Native bGH, 40 mg/day	Daily injection	Holstein	Peel et al. (1980)[8]
28.3	32.6	10	Native bGH, 40 mg/day	Daily injection	Holstein	Peel et al. (1983)[9]
16.8	16.7	10	Native bGH, 40 mg/day	Daily injection	Holstein	Peel et al. (1983)[9]
13.4	17.4	10	Native bGH, 40 mg/day	Daily injection	Holstein	Fronk et al. (1983)[10]
13.4	17.9	10	Native bGH, 40 mg/day	i.v. injection every 2 h	Holstein	Fronk et al. (1983)[10]
13.4	17.5	5	Native bGH, 40 mg/day	Continuous infusion s.c.	Holstein	Fronk et al. (1983)[10]

26.7	28.6	10	Native GH, 10 mg/day	Daily injection	Holstein	Eppard et al. (1985)[11]
26.7	31.5	10	25 mg/day	Daily injection	Holstein	Eppard et al. (1985)[11]
26.7	34.3	10	50 mg/day	Daily injection	Holstein	Eppard et al. (1985)[11]
26.7	35.2	10	100 mg/day	Daily injection	Holstein	Eppard et al. (1985)[11]
27.9[a]	32.5[a]	188	Native bGH, 27 mg/day	Daily injection	Holstein	Bauman et al. (1985)[12]
27.9[a]	34.4[a]	188	Biosynthetic bGH, 13.5 mg/day	Daily injection	Holstein	Bauman et al. (1985)[12]
27.9[a]	38.0[a]	188	27.0 mg/day	Daily injection	Holstein	Bauman et al. (1985)[12]
27.9[a]	39.4[a]	188	40.5 mg/day	Daily injection	Holstein	Bauman et al. (1985)[12]
26.8	29.5	252	Biosynthetic bGH, 500 mg/14 day	Prolonged release	Holstein	Bauman et al. (1985)[12]
15.9	18.3	5	Biosynthetic bGH, 40 mg/day	Daily injection	Holstein	Bauman et al. (1985)[12]

[a] Fat-corrected milk.

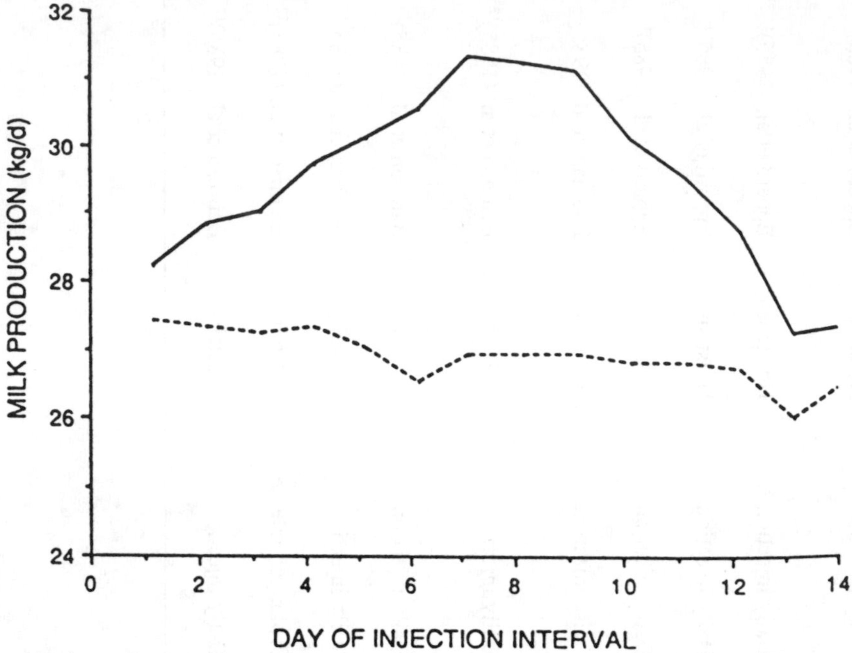

Figure 1 Effect of recombinant bovine GH on milk production. Cows received bovine GH (solid line) (500 mg of somatotribove in a slow release formulation at 14-day intervals) (n = 40) or vehicle control (dashed line) (n = 39). Treatment was initiated on day 0. (From Bauman, E. E., Hard, D. L., Crooker, B. A., et al., *J. Dairy Sci.*, 72, 642, 1989. Reprinted with permission.)

Table 2 **Effect of Nutrition on the Milk Production and IGF Response of Dairy Cows to bGH**

Diet[a]	Increase in Milk Production (kg/day)	% Increase	Change in Circulating IGF-I (ΔIGM-I [n*M*])	Change in Circulating IGF-II (ΔIGF-II [n*M*])
High crude protein, high net energy	7.7	31	17.7	27
High crude protein, low net energy	5.8	28	11.3	13
Low crude protein, high net energy	4.4	19	11.1	15
Low crude protein, low net energy	4.8	24	11.6	10

[a] High. 120% of requirements; low, 80% of requirements.

Source: Based on McGuire et al.[15]

Table 3 **Effect of Prolonged bGH on Milk Production in Dairy Cows**

	Milk Production (kg/d)	Milk Composition					
		3.5% Fat Corrected Milk (kg/d)	Fat %	Protein %	Lactose %	Calcium %	Phosphorus %
Study 1							
Daily injection[a]							
Vehicle ($n = 6$)	—	27.9	3.6	3.4	4.8	—	—
Native bGH (27 mg/day) ($n = 6$)	—	32.5	3.3	3.4	4.8	—	—
Biosynthetic bGH							
13.5 mg/day ($n = 6$)	—	34.4[b]	3.8	3.4	4.9	—	—
27 mg/day ($n = 6$)	—	38.0[b]	3.6	3.4	4.8	—	—
40.5 mg/day ($n = 6$)	—	39.4[b]	3.6	3.4	4.9	—	—
Study 2							
Prolonged release formulation[c]							
Vehicle ($n = 39$)	26.8	27.3	3.7	3.2	4.8	0.11	0.11
Biosynthetic bGH ($n = 40$)	29.5[b]	30.4[b]	3.8	3.3[b]	4.8	0.09	0.09

[a] Daily i.m. injections for 188 days (data from Ref. 12).

[b] $p < 0.05$.

[c] Formulations of 500 mg for 14 days with regular reimplantation. Length of study, 252 days.

Source: Data from Bauman et al.[13]

In ruminants, all of the circulating glucose is synthesized by gluconeogenesis in the liver. Thus the need for increased milk sugar (lactose) requires a marked rise in gluconeogenesis. It is not surprising that *in vitro* glucose synthesis from [^{14}C]propionate is greater (63%) in liver slices from bGH-treated cattle than in vehicle-injected cows.[26] In dairy cattle, chronic GH administration has little effect on nitrogen excretion.[8,23] This is in contrast to the situation in growing cattle receiving bGH, in which nitrogen excretion is reduced.[27,28] However, an increase in milk nitrogen output is observed.[23] Moreover, some decreases in circulating concentrations of urea have been observed in some studies.[28]

G. GALACTOPOIETIC MECHANISMS

The available evidence does not allow a firm conclusion as to whether GH exerts its effect directly on the mammary gland or whether the effect is mediated predominantly by effects on the other organs, for example, liver and adipose tissue. Evidence against a direct effect of GH on the mammary gland includes the failure of bGH to stimulate milk production in sheep and goats when GH is infused into the mammary artery.[29] Conversely, demonstration of the presence of GH receptors in bovine mammary tissue would support a direct galactopoietic effect of GH (see below).

1. IGF-I Mediation

The galactopoietic effect of GH may be mediated or potentiated by peripheral or locally produced IGF-I. The roles of IGF-I and IGF-II in the galactopoietic effect of GH are not fully established. Administration of bGH is followed by increases in the concentration of both IGF-I[15,30,31] and IGF-II in the circulation[15] (also see Table 2). It seems likely that these changes in circulating concentrations of IGF-I and possibly also IGF-II contribute to the galactopoietic effect of GH. Indeed, IGF-I has been found to increase mammary DNA synthesis *in vitro*[32] and thus is involved, presumably, in stimulating mammary cell proliferation and tissue growth. Moreover, either IGF-I or IGF-II infused directly into one mammary gland of goats has recently been found to rapidly increase milk production.[32a]

The rise in the circulating concentrations of IGF-I and IGF-II reflects increased synthesis/release from the liver and perhaps other organs. Significant local mammary production of IGF-I is unlikely. The argument for this contention is that if mammary IGF-I production were high following GH administration, then local concentrations of IGF-I would be expected to be high. If we assume that the concentration of IGF-I in the milk reflects mammary IGF-I levels, there is no evidence for local production of IGF-I. Milk IGF-I concentrations are calculated to be 6.1% of circulating concentrations prior to GH treatment, but rather than increasing, they fall to 4.3% of circulating concentrations following GH administration.[15] However, IGF-mRNA has been detected in stromal/blood components of bovine mammary tissue.[33]

2. GH Mediation

There is now convincing evidence that bovine mammary tissue has GH receptors. GH receptor mRNA has been detected in mammary tissue by either Northern blot analysis[33,34] or solution hybridization-nuclease protection assay.[33] The GH receptor mRNA has been localized in the alveolar epithelial cells by *in situ* hybridization with anti-RNA probes.[34]

3. Thyroid Hormone Mediation

Administration of bGH does not consistently influence circulating concentrations of either thyroxine (T_4) or triiodothyronine (T_3) (with little effect, e.g., see Refs. 14 and 16). However, a local increase in the concentration of mammary T_3 is a distinct possibility; increased mammary but not liver or kidney T_4 5′-monodeiodinase activity is observed following daily injections of bGH.[14]

4. Other Hormones

Chronic administration of bGH to dairy cows is associated with changes in the sensitivity/responsiveness of tissue to hormones. Hepatic gluconeogenic (and perhaps also glycogenolytic) responses to glucagon may be increased with chronic GH treatment. The hypoglycemic response to glucagon is greater in bGH-treated cows in some studies[35] but not others.[36] Moreover, the lipolytic effect of epinephrine on adipose tissue is elevated in GH-treated cows compared to control cows as evidenced by the greater increases in circulating concentrations of free fatty acids.[36,37]

5. Summary

It can be tentatively concluded that bGH increases milk production (and the efficiency of milk production) by a combination of direct effects on the mammary gland (include increases in galactopoietic tissue) and indirect effects. These indirect effects including increases in circulating concentrations of IGF-I and IGF-II, elevated gluconeogenesis, and greater sensitivity/responsiveness of various tissues to homeostatic stimulation. The latter is consistent with the original homeorhetic hypothesis of D. E. Bauman — *homeorhesis* being "orchestrated changes for the priorities of a physiological state, i.e., coordination of metabolism in various tissues to support a physiological state," namely lactation.[38]

III. GH AND GROWTH, PERFORMANCE, AND CARCASS COMPOSITION

A. INTRODUCTION

As would be expected from studies of GH in rodent models (see Chapter 19) and in clinical situations with children (see Chapter 30), GH stimulates postnatal growth in agricultural animals. Chronic administration of GH increases growth rate in many meat-producing animals. Moreover, this GH treatment improves feed-to-gain efficiency (weight gained per unit feed consumed) and enhances carcass quality (reduced fat, increased protein). Table 4[39-49] summarizes studies on the effects of chronic GH treatment in pigs, cattle, sheep, chickens, and fish. It is apparent that GH enhances performance in livestock mammals (pigs > sheep > cattle) and fish.

B. GROWTH/PERFORMANCE CHARACTERISTICS IN PIGS

Considerable potential exists for the application of porcine GH (porcine somatotropin, pST) to enhance the efficiency of pork production. This is based on the marked improvement in growth rate and feed efficiency coupled with reduced carcass fat and increased leanness (muscle, protein) (see Table 4 for examples). These performance-enhancing effects of chronic pGH are consistently observed (see Table 4), being found with either native pituitary (npGH)-derived or biosynthetic/recombinant derived pGH (rpGH),[44] and are dose dependent[43,44,48] (also see Figure 2).

The effects of pGH on pig performance can be summarized as increasing net protein synthesis (protein accretion) and decreasing fat accretion in adipose tissue. This leads to more lean meat. The effects of pGH are observed qualitatively irrespective of sex (see Table 5). The magnitude of the pGH effect is greater in females and castrates. Chronic pGH treatment increases protein accretion while decreasing fat accretion and, thereby, abolishes sexual dimorphism in growth rate.[46] There is a carryover effect of chronic pGH-treatment with the rate of protein accretion continuing to be elevated in pGH treated pigs even during a prolonged withdrawal period.[47]

It would seem that pGH is increasing the efficiency of pork production; there is, for instance, improved growth, feed efficiency, and carcass quality. The increase in effi-

Table 4 **Summary of Studies Examining the Effect (as Percentage Increase over Control) of Chronic GH Treatment on Performance Characteristics in Livestock, Poultry, and Fish**

Species	ADG %	FG %	Carcass Fat %	Carcass Protein %	Ref.
Pigs	+3.2	−19.3[b]	−25.4[b]	+25.0[b]	39
	+12.4[a]	−16.0[b]	−11.2[a]	+16.7[a]	40
	+9.9[b]	−4.0[a]	+3.7	+2.2	41
	+11.1[a]	−18.9[a]	−18.0[a]	+8.2[a]	42
	+14.4[a]	−17.2[a]	−24.7[a]	+18.2[a]	43
	+19.3[a]	−24.2[a]	−68.1[a]	+36.9[a]	44
	+16.2[a]	−23.7[a]	−28.6[a]	+11.0[a]	45
	+17.5[a]	−28.4[a]	−30.9[a]	+13.7[a]	46
	+13.6[b]	−28.2[b]	−33.0[b]	+6.1[a]	47
	+26.2[a]	−25.4	—	—	48
	+12.6[a]	−6.1	—	—	49
Arithmetic mean:	+14.2	−19.2	−26.2	+14.7	
Cattle					
Heifers	+8.6[b]	−2.1	−17.4	+0.5	57
Steers	−6.3	—	−32.2[b]	+22.4[b]	58
Steers	+15.4[a]	—	−7.7	+4.8	59,60
Steers	+5.3	−3.8	—	—	61
Bull calves	+17.8[a]	0.0	−15.2[a]	+1.7	62
Veal calves	+7.1	−1.2	—	—	63
Arithmetic mean:	+8.0	−1.7	−18.1	+7.3	
Sheep	+3.7	−7.4[a]	−9.2[a]	+7.5	64
	+22.2[a]	−12.2[a]	−12.5[a]	+23.9[a]	65
	+1.4	−0.4	−10.0	+4.2	66
	+40.9[b]	—	−10.9[b]	+9.4[a]	67
	+45.0[a]	−29.1[a]	−19.8	+29.1[b]	68
	+1.3	−7.2	−14.5	+5.0	69
Arithmetic mean:	+19.1	−11.3	−12.8	+13.1	
Chickens					
1–7 weeks old (broiler aged)					
	+1.5	−1.4	+7.9%	−0.0%	76
	+2.9	+3.4	+6.7	+0.7%	77
	−5.4	−4.4	+18.2	−2.9	78,79
	+1.5	+2.0	−3.3	−0.4	80
	+1.0	+2.0	—	—	81
7 weeks old	+17.1	−25.8[a]	−31.3	+0.7	82,83
Fish					
Oncorhynchus kisutch (Coho salmon)	+68.0[a]				89

Table 4 **(Continued)**

Species	ADG %	FG %	Carcass Fat %	Carcass Protein %	Ref.
O. kina (chum salmon)	+85.3[a]				89
O. tchawytscha (chinook salmon)	+18.3[a]				89
O. mykiss (rainbow trout)	+90.5[a]				89

[a] $p < 0.05$ compared to controls.

[b] $p < 0.01$ compared to controls.

ciency is true if we are considering production of meat (pork, ham, etc). However, on an energetic basis, this is not true. Less energy is retained in pGH-treated pigs[49] and more energy is utilized (measured as heat production, presumably, for protein synthesis).[49]

1. GH and Nutrition

The ability of pGH to influence accretion of either protein (net protein synthesis) or of fat is considerably influenced by diet. This is illustrated in Figure 3. For instance, if energy consumption is limited by as little as 20%, the pGH effect on fat accretion is attenuated.[45] Similar reductions in dietary protein also exert modifying effects on protein accretion. It is clear that pGH has little effect on protein accretion when dietary lysine intake is low.[50] As the intake of this limiting amino acid and other amino acids is increased, pGH progressively exerts a greater effect on protein accretion.

2. GH and Protein Metabolism

In GH-treated pigs, there is a reduction in the circulating concentration of urea.[41,44] This is consistent with GH decreasing catabolism/deamination of amino acids and hence increasing the availability of amino acids for the synthesis of proteins. As might be expected with the improvements of growth and carcass protein, the rates of both protein accretion (see Table 1) and nitrogen retention (e.g., see Ref. 49) are increased. However, there is little effect on protein (nitrogen) digestibility.[49]

3. Mechanisms of Action

The mechanism by which GH increases growth, improves feed efficiency, and modifies carcass composition is not completely understood. It is likely to involve (1) the growth-promoting effects of GH being mediated by IGF-I (the somatomedin hypothesis), (2) direct effects of GH on carbohydrate/lipid metabolism, and (3) effects of GH on the secretion of other hormones.

a. GH and GH receptors

It may be questioned whether chronic GH administration affects GH receptors in pigs. If we assume that specific binding of ^{125}I-labeled pGH to liver microsomes (in the presence of magnesium chloride) reflects GH receptor numbers and/or affinity, then chronic treatment of pigs with pGH increases GH receptor (GHR) number and/or affinity.[51] The increase in specific binding is dependent on the dose of GH administered *in vivo*.[51] The physiological importance of this upregulation of the GHR is underscored by the high correlation ($r^2 = 0.50$) between circulating concentrations of IGF-I and liver ^{125}I-labeled pGH specific binding.[51]

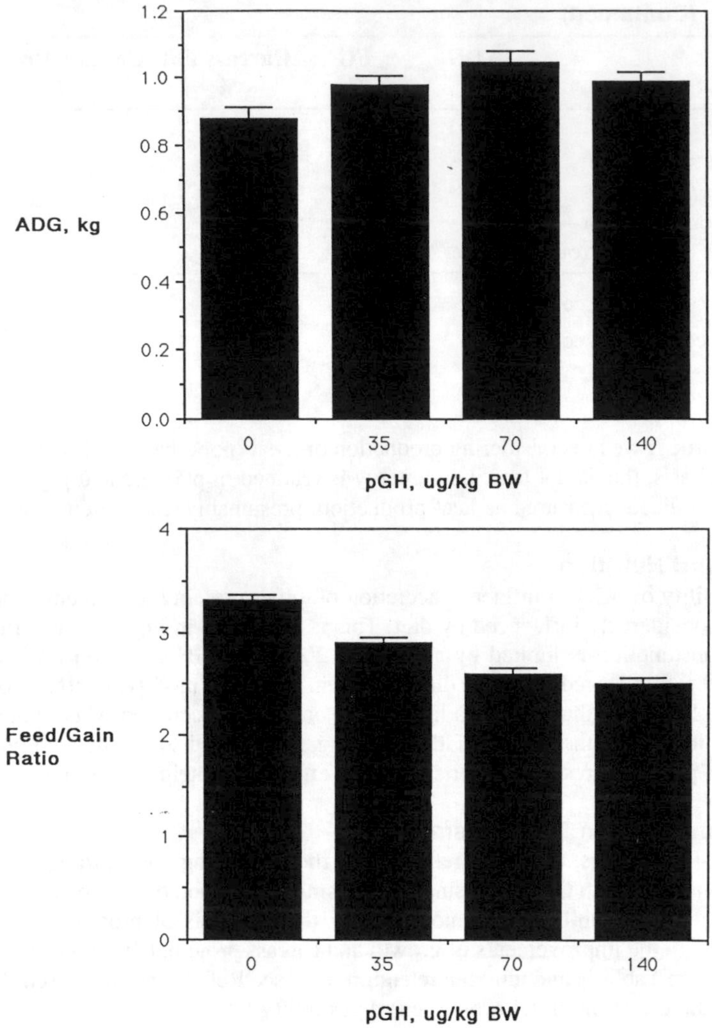

Figure 2 Effect of recombinant porcine GH on growth performance in pigs. *Top:* Growth (average daily gain [ADG]). *Bottom:* Feed: gain efficiencies. Recombinant pGH treatment was initiated at 27-kg body weight and continued for 11 weeks. Vertical bars indicate SEM. (Data from Evock et al.[44])

b. GH and IGF-I

The mechanism by which GH enhances growth and muscle protein accretion is probably mediated by somatomedin (IGF-I). Circulating concentrations of IGF-I have been consistently observed to be increased by chronic administration of pGH (e.g., see Ref. 44). Furthermore the circulating concentrations of IGF-I are increased by GH in a dose-dependent manner.[43] Thus, *in vivo* treatment of pigs with pGH results in a threefold increase in the relative content of IGF-I mRNA in the liver, together with a 54% elevation in total hepatic RNA.[52] There is, however, considerably more IGF-I mRNA available for translation. No significant changes in muscle IGF-I mRNA are found with GH treatment[52] but some increase in adipose IGF-I mRNA is reported.[53] The elevated circulating concentration of IGF-I inhibit GH release and/or synthesis in a negative feedback manner. The

Table 5 **Influence of Sex on the Response of Pigs to pGH**

Parameter Measured	Control	pGH (100 μg/kg/day)
Protein accretion (g/day)		
Boar (♂)	196	238
Gilt (♀)	148	235
Barrow (♂)	139	225
Protein accretion (as % of control)		
Boar (♂)	100	121
Gilt (♀)	100	159
Barrow (♂)	100	162
Fat accretion (g/day)		
Boar (♂)	317	203
Gilt (♀)	411	185
Barrow (♂)	462	223
Fat accretion (as % of control)		
Boar (♂)	100	64
Gilt (♀)	100	46
Barrow (♂)	100	48

Source: Based on Campbell et al.[46]

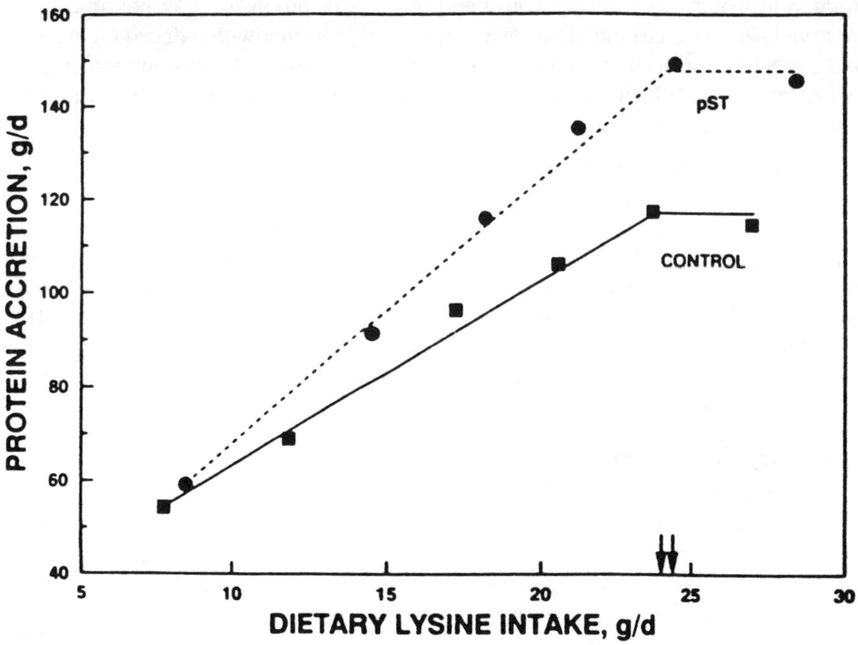

Figure 3 Effect of dietary amino acid (lysine) on protein accretion in control and pGH (150 μg/kg)-treated growing pigs (between 20 and 55 kg in body weight). (From Boyd, R. D., Bauman, D. E., Fox, D. G., et al., *J. Anim. Sci.*, 69(Suppl. 2), 56, 1991. Copyright by the *Journal of Animal Science*. Reprinted with permission.)

decreases in pituitary GH that have been reported[46] are likely to be due to negative feedback by IGF-I at the hypothalamus and/or pituitary levels.

c. Direct Effects of GH

The reduction in carcass fat/adipose tissue in pGH-treated pigs is primarily due to a decrease in lipogenesis.[52] This is likely to be due to direct effects of GH. *In vitro*, both native and recombinant pGH suppress insulin-induced lipogenesis (fatty acid synthesis) in porcine adipose tissue.[6,44,54,55] This is consistent with pGH *in vivo* reducing triglyceride accumulation by decreasing the availability of the precursor fatty acids. GH does not, therefore, appear to influence lipolysis. Even chronic administration of pGH does not elevate circulating concentrations of free fatty acids in this species.[41,43]

d. GH and Other Hormones

It is conceivable that some of the effects of pGH on performance in pigs are related to GH influencing the secretion of other hormones and hence their circulating concentrations. For instance, circulating concentrations of insulin are elevated by as much as 6.8-fold in pigs receiving pGH (40 μg/kg/day),[44] whereas those of T_3 are also somewhat elevated.[45] However, chronic pGH treatment has little, if any effect, on circulating concentrations of cortisol.[45]

e. Human Health Considerations

A case can be made that the application of pGH to enhance pig performance would have a positive impact on human health. Treatment of growing pigs with pGH reduces the amount of fat in the carcass and hence in the resulting meat, pork, ham, and so on. Moreover, pGH treatment also influences the relative level of saturated fatty acids.[56] A reduction in dietary fats and particularly saturated fatty acids in our diets could improve human health. However, it is unclear what impact the introduction of pGH to pig production would have on the consumption of pork products. It is possible that the improved performance (and likely decrease in price) together with reduced fat in the meat (and probable enhanced consumer demand) may increase pork consumption (replacing beef or poultry?). Whether this would result in a net decrease or increase in human fat consumption is not clear.

f. Animal Health Considerations

At high doses, pGH can induce adverse, even toxic, effects in pigs. In an early study by Machlin,[40] high mortality was observed in pigs receiving a very high dose of native pGH (1.1 mg/kg/day). These data may be explicable in terms of contamination of the GH preparation or the pharmacological dose of pGH or of pGH per se. Some increase in mortality or disease (e.g., osteochondrosis) has been observed in studies in which very high doses of either native or recombinant pGH were administered to pigs.[44,48] It is also possible that GH influences reproductive development. This is not likely to be important as pigs go to market (slaughter) prior to reproductive maturation.

C. GROWTH IN CATTLE AND SHEEP

Growth hormone has considerable potential as a performance enhancer (improving growth rate and/or feed:gain efficiency and/or carcass composition) for the production of cattle (beef) and, perhaps, also sheep (lamb) and veal. This section considers the ability of GH to modify the growth rate/efficiency and carcass composition in both cattle and sheep together, both being domesticated ruminants and ungulates. Another rationale for discussing the work on these two species together is that sheep are frequently employed as a model for cattle, sheep being much more cost efficient for experimentation due to their smaller size.

Table 4 summarizes a number of studies in which GH was administered to cattle or sheep. It is readily apparent that the effects of GH on growth or other performance characteristics are less consistent and of a smaller magnitude than those observed in pigs (see Table 4). Nevertheless, a general pattern might be discerned in which GH administration increases growth (average daily gain [ADG]), improves feed:gain efficiency, and reduces carcass fat in cattle and sheep. Moreover, there may also be some tendency for increased carcass protein.

1. Effect on Nitrogen Metabolism

Effects of GH on whole body nitrogen metabolism have been determined in cattle. The indices of nitrogen whole body metabolism are as follows:

1. Absorbed N_2^a = intake N_2^b – fecal N_2^b
2. Retained N_2^a = absorbed N_2 – urinary N_2^b
3. Digestibility[a] = absorbed[a]/intake[b]

[a] Calculated index.
[b] Directly determined.

GH administration to young cattle increases nitrogen retention in cattle (summarized in Table 6). There is little or no effect of GH on nitrogen intake or digestibility.[27,70–72] Administration of GH reduces both urinary nitrogen (Table 6) and circulating concentrations of urea.[27,70,71] Similarly, effects of GH on nitrogen retention have also been observed in sheep.[73] Moreover, GH has been found to reduce both the circulating concentrations of one amino acid, leucine, and also its oxidation in young cattle.[74] Net protein synthesis[74] also tends ($p < 0.1$) to be increased by GH.

Net protein synthesis = protein synthesis – protein degradation

Thus, GH may be increasing protein synthesis and/or decreasing protein degradation.

In summary, GH improves the efficiency of utilization of nitrogen. It is not clear to what the extent GH exerts its effects of nitrogen metabolism directly or via IGF-I.

2. Effect on Lipid Metabolism

The decreases in carcass lipid and improved feed efficiency observed in GH-treated cattle and sheep are probably due to direct effects of GH on carbohydrate and lipid metabolism (see Chapters 20 and 21).

Changes in carcass fat may be due to:

- Reduced adipose triglyceride synthesis
- Increased adipose triglyceride degradation (lipolysis)
- Decreased fatty acid synthesis (lipogenesis)
- Increased fatty acid oxidation

There is evidence, albeit scant, for each of these possibilities. For instance, in a comprehensive study in open-circuit respiration calorimeters, GH administration to growing heifers increased circulating free fatty acids (indicating increased lipolysis) and also fatty acid oxidation.[79] It should be noted that increases in circulating concentrations of free fatty acids are not consistently observed in GH-treated cattle.[61] This apparent lipolytic effect is likely to be due to GH enhancing the sensitivity of adipose tissue to lipolytic hormones (e.g., epinephrine) and/or decreasing the responsiveness to antilipolytic agonists (adenosine). Growth hormone also decreases insulin-induced lipogenesis *in vitro*.

3. GH and IGF-I

It is probable that the mechanism by which GH enhances growth and nitrogen retention is due to increases in the circulating concentrations of IGF-I. Increases in circulating concentrations of IGF-I have been consistently observed in both GH-treated cattle[61,62,71,72] and sheep.[67]

4. GH and GH Receptors

Chronic GH treatment in sheep upregulates liver GHR numbers. Although there is no effect of GH administration on dissociation constants of either low- or high-affinity binding sites, the concentration/number of both low- and high-affinity ^{125}I-labeled bGH-binding sites in the liver increases with GH treatment.[67]

5. Interactions with Performance Enhancers

There have been some studies in which the effects of GH administration are compared in a factorial manner with those of other metabolism/growth modifiers. For instance, both estradiol and GH have been found to stimulate growth rate (in terms of ADG), feed conversion efficiency, and plasma IGF-I concentrations. Together these effects are additive, with no interaction being observed.[59] Similar effects with GH and the estrogen analog, zeranol, on feed-to-gain efficiency have been reported in sheep.[75] However, on the basis of a single report, it would seem that GH in the presence of thyroxine has (or tends to have) a greater effect on growth, feed:gain, individual muscle weights, and carcass protein in sheep.[66] β-Adrenergic agonists and GH exert their performance-

Table 6 **Effect of GH on Nitrogen Balance in Cattle**

	Nitrogen Balance (g/day)				
	Intake	**Absorbed**	**Urinary**	**Retained**	**Ref.**
Study 1: Native bGH to Growing Hereford Heifers on a Maintenance Diet					
Vehicle	112	84[a]	81	2	70
bGH (30 mg/day)	112	85	72[a]	12	70
Study 2 Recombinant bGH to Growing Holstein Heifers					
Vehicle	100	71	44	27	71
bGH					
33 μg/kg/day	100	72	41	32	71
67 μg/kg/day	99	70	39	31	71
100 μg/kg/day	100	70	37	32	71
Study 3: Recombinant bGH to Growing Steers					
Vehicle	123	81	52[a]	29[a]	27
bGH (200 μg/kg/day)	121	80	42[a]	39[a]	27
Abomasal casein infusion and vehicle	144	106	70[a]	35[a]	27
Abomasal casein infusion and bGH	144	102	52[a]	51[a]	27

[a] $p < 0.05$.

enhancing actions by different mechanisms. It is, therefore, not surprising that additive effects of these agents are observed.[63]

D. POULTRY PRODUCTION

Growth hormone does not have the dramatic effect on growth, feed efficiency, or carcass composition in chickens as that seen in other livestock (mammals) (see Table 4). GH is, however, required for normal growth in chickens. This is supported by hypophysectomy/replacement therapy studies.[84,85] However, administration of either mammalian GH[86] or chicken GH (Table 4) has no consistent effect on growth or performance indices in young chickens, at least up to the age (~7 week old) when "broiler" chickens go to market. A possible explanation for this absence of an effect is GH down-regulation of GH receptors.

The administration of GH to chickens of other ages may, however, influence growth/ performance. For instance, injection of GH into chick embryos has been found to enhance posthatch growth and modify adipose tissue development.[87] Moreover, cGH treatment has been found to exert marked effects on growth in chickens older than 7 weeks of age. For instance, administration of GH in a pulsatile manner increases growth, improves feed efficiency, and decreases carcass fat[82] (Table 4). Conversely, when GH is given as a continuous infusion, adipose growth is enhanced[82,83] with either stimulatory[83] or no effects[82] on muscle growth. The practicality or applicability of cGH as a growth modifier/ performance enhancer in poultry is therefore dubious, especially in view of its ineffectiveness in young birds and in view of the relatively small proportion of chickens raised for more than 7 weeks.

1. GH and GH Receptors

The chronic administration of GH to chickens does not have the dramatic effect observed in other livestock (see Table 4). It is possible that the lack of a robust growth response to GH is a consequence of GH downregulating the GH receptors. Indeed, chronic administration of cGH has been reported to reduce specific ^{125}I-labeled cGH binding to liver membrane fractions.[74] Moreover, hypophysectomy increases both specific ^{125}I-labeled cGH binding and GH receptor number. These effects in chicken livers are reversed by cGH administration.[87] It may be argued that in intact young chickens the number of GH receptors is limiting to growth and that, if it were possible to manipulate GH receptor numbers, GH (either endogenous or exogenous) may exert a marked performance-enhancing effect in poultry.

E. FISH GROWTH

Normal growth throughout the vertebrates requires adequate secretion of GH. It, therefore, should not be surprising that exogenous GH can accelerate growth in fish as is the case in mammals. Table 4 summarizes the average of the maximal stimulation of growth evoked by GH in a number of studies in salmon and related fish. The growth rate of these fish can be greatly increased by GH administration (reviewed in Ref. 89). This is apparent, irrespective of whether mammalian, avian, or porcine GH is administered and whether GH is given by regular intraperitoneal (i.p.) or intramuscular (i.m.) injection, or by sustained release implants. Some encouraging initial results have also been obtained with GH being delivered by immersion of the fish in a GH-containing medium.[89]

The magnitude of the improvements to growth and feed efficiency, coupled with both the high price of fish paid by consumer and the availability of agriculture/aquaculture technology, support the notion that GH may be utilized for commercial fish production. Certainly further research is warranted. Moreover, particular attention should be focused on methods of delivering GH, including transgenic fish expressing high levels of GH and the use of GH secretagogues.

IV. MISCELLANEOUS EFFECTS

A. INTRODUCTION

As GH stimulates milk production, the efficiency of meat production, and the quality of meat (increased protein/reduced fat), it might be speculated that GH might enhance the production of other agricultural products. This is considered below in reference to the production of wool and chicken eggs.

B. WOOL PRODUCTION

Wool production involves the synthesis of proteins by sheep. In view of the profound effect of GH on the synthesis of milk proteins and muscle (meat) proteins, it is reasonable to consider whether GH might influence wool production. There is not, however, consistent data to support this possibility. Whereas in one study GH was reported to increase clean wool weight,[67] this stimulatory effect has not been observed by other investigators.[68]

C. EGG PRODUCTION

In view of the established effects of GH on ovarian function and yolk synthesis in lower vertebrates (see Chapter 23), it might be questioned whether GH enhances egg production and/or egg quality. The available information suggests that this is not likely. An early report concluded that GH depressed egg production.[90] However, cGH has been found to have no effect on egg production.[91] Similarly, no effect of cGH on egg quality was noted with the exception of some, albeit small, improvement in egg shell quality.

V. ALTERNATE METHODS TO INCREASE GROWTH HORMONE AVAILABILITY AND/OR EFFECTIVENESS

A. INTRODUCTION

The administration of exogenous GH results in marked increases in milk production in dairy cattle and in growth in pigs, sheep, and beef cattle (see above). An alternative to treatment with exogenous GH is to increase the secretion of endogenous GH or to elevate the effective level or availability of the GH. Increases in GH and/or availability may be achieved by a variety of means. These are considered below.

B. GROWTH HORMONE-RELEASING HORMONE

In view of the ability of GHRH to stimulate GH release and of bGH to increase milk production in cattle, it is not surprising that the effect of GHRH on lactation in dairy cattle has been investigated.[92] Infusion of bGHRH (i.v.) increases milk yield and also the production of milk solids (protein, lactose, and fat).[92] It is, therefore, reasonable to suggest that GHRH is an alternative to the use of exogenous bGH to enhance milk production in dairy cattle.

It is now well established that GH stimulates growth while improving feed and carcass composition efficiency in pigs (Table 4). Consequently, since GHRH stimulates GH release and has been shown to augment growth in humans,[93] GHRH would be expected to stimulate growth of livestock, especially because a GHRH deficiency is associated with growth retardation in rats[94–96] and steers.[97] The results from these studies are encouraging. For instance, GHRH increases carcass weight while reducing carcass lipid.[98] Moreover, a potent analog of hGHRH elevates growth rate and feed:gain efficiency while improving carcass composition (reduced fat, increased protein).[99,100] Kahn et al.[101] also found that injections of GHRH to pregnant ewes during days 136–146 of pregnancy induced higher birth weights and faster growth of the lambs.

C. INSULIN-LIKE GROWTH FACTORS

Many of the actions of GH on growth are mediated by IGF-I. Exogenous IGF-I might therefore be expected to improve growth and livestock performance. The effects of exogenous IGF-I on the growth of rats and dogs and on lactation in goats are, however, less than those of GH.[102–104] This may reflect the requirement of binding proteins in the induction of IGF-I bioactivity, and the rapid clearance of free IGF-I in peripheral plasma.

D. GENETIC SELECTION

Lactation in dairy cattle and growth/feed efficiency in pigs and cattle are stimulated by GH. It may be questioned whether genetic selection for GH would improve milk yield or growth? Conversely, does selection for maximal growth or milk yield influence circulating concentrations of GH? There is some evidence for this second possibility, with somewhat greater GH secretory responses to secretagogues in immature cattle in lines selected for high milk yield than in lower merit cattle.[100] The pulsatility of GH release precludes single blood samples being useful for understanding the GH status of an animal. Selection for GH as one of a spectrum of parameters may prove to be advantageous to animal agriculture, as may knowledge of the molecular control of GH synthesis. If circulating concentrations of GH "downregulate" GH receptors then selection for circulating concentrations of GH could be disadvantageous. Under these circumstances it is argued that selection for both circulating concentration of GH and the number of GH receptors should be considered.

E. GENETICALLY ENGINEERED CELLS

It is possible that GH-secreting cells could be introduced into a host. If these GH-producing cells can be maintained or preferably made to proliferate (as would be the case if the cells were embryonic/fetal cells from the same species), this could be a method to deliver GH and other proteins. The potential for this has been demonstrated. A recombinant gene encoding hGH has been introduced into cultured myoblasts. Following introduction of this into mice, the myoblasts become fused into preexisting multinucleated myofibers that are vascularized and then they release GH.[105]

F. GH IMMUNIZATION

Immunological approaches may be applicable to improving growth efficiency in agricultural animals and/or lactation in dairy cattle if the biological activity of either endogenous or exogenous GH can be increased. This would be expected to evoke a superior effect on production parameters in agricultural animals.

There is now a reasonable body of evidence showing that antibodies to specific GH epitopes potentiate growth responses to GH.[106] For instance, administration of monoclonal antibodies to GH enhances the effect of exogenous GH on $^{35}SO_4$ incorporation into rodent cartilage,[107,108] on growth (body weight gain) and circulating IGF-I concentrations in hypophysectomized rats,[106] on growth in Snell dwarf mice,[110] and on milk production, circulating concentrations of IGF-I, and the diabetogenic effect of GH in sheep.[111] Moreover, monoclonal antibodies to GH can potentiate the effect of endogenous GH. For instance, monoclonal antibodies to GH stimulate growth in the marmoset[110] and, in sheep, elevate circulating concentrations of IGF-I and induce a diabetogenic effect.[111] It is possible to envisage the production of large quantities of specific monoclonal antibodies to GH by recombinant technology. However, the practicality of the approach of passive immunization with mouse monoclonal antibodies can be questioned. This is based on the problem of the recipient animal developing an immune response to the foreign proteins and, at the extreme, anaphylactic shock. If homologous monoclonal antibodies against GH can be produced, this approach may offer a real possibility to enhance the effectiveness of GH in agricultural species.

It is possible to envisage antibodies to specific epitopes on the GH molecule, produced by immunization against specific amino acid residue sequences within GH. This has been done successfully with sequences within bGH.[112] Passive administration of rabbit antiserum to $bGH_{110-118}$ alone has no effect on the growth of hypophysectomized rats but does potentiate the effect of GH.[112] This sequence corresponds to part of α-helix 3 of GH, a region important to the second site on GH that binds to the GHR (see Chapter 16). Similarly, sheep antiserum against GH_{35-53} enhances the effect of GH on $^{35}SO_4$ incorporation in dwarf mouse cartilage.[112] It is, therefore, reasonable to suggest that either passive or active immunization to peptides corresponding to specific sequences of GH could be a practical means of enhancing the activity of either exogenous or endogenous GH in agricultural animals. If active immunization were to be a suitable method, this potentially has the advantage of animals receiving the peptide during conventional vaccination and booster shots.

The mechanism by which provocation of antibodies against specific epitopes on GH enhances the response to GH is not established. It might be speculated that antiserum to GH epitopes might reduce GH binding to a circulating GH-binding protein and hence increase GH availability. In addition, GH bound to antibodies may have an extended half-life (this is consistent with the observations that antibodies to some but not other epitopes enhance the activity of GH). An alternative explanation is that epitope-specific antibodies change the structure of the GH molecule such that there is enhanced biological activity. Indeed monoclonal antibodies to bGH have recently been shown to enhance GH binding to hepatic somatogenic receptors.[113] However, this needs to be examined in *in vitro* biological assays, in which the confounding influences of GH-binding proteins and clearance rates are not present.

An alternative approach to increasing the effective concentration of GH may be to produce anti-idiotypic antibodies to GH antibodies, which could be introduced by passive immunization. Anti-idiotypic antibodies to rat GH antibodies have been produced that not only bind to GH receptors but also are capable of stimulating growth (weight gain) in hypophysectomized rats.[111] Growth hormone receptor antibodies with GH-like activity also occur spontaneously in human subjects with acromegaly.[115]

G. SRIF IMMUNIZATION

Since SRIF inhibits GH secretion (Chapter 8), GH secretion and growth might be expected to be enhanced by SRIF immunoneutralization. Indeed, a number of studies have utilized this approach to enhance growth rate in poultry and livestock.[116–121] These finding are, however, controversial and have not been confirmed by others.[97,104] This may reflect the short-lived effects of both active and passive SRIF immunoneutralization on GH secretion, nonspecific effects on other endocrines involved in growth, and deleterious effects of active immunization on animal welfare.

H. TRANSGENIC ANIMALS

Tremendous increases in growth rate are observed in transgenic mice with additional *GH* (*hGH* or *bGH*) genes.[122,123] Such was the magnitude of the effect (increase in growth rate) in mice that photographs of the transgenic mice graced the covers of both *Science* and *Nature*.[122,123] The obvious potential application of this technology to agriculture has

Table 7 **Growth Performance of Transgenic Pigs (Metallothionein-bGH)**

Animals Tested	Control[a]	Transgenic[a]
	Body Weight Gain (ADG ± (N=) SEM)	
Founder animals	743 ± 32 (6)	690 ± 65 (6)
Progeny	815 ± 17 (30)	905 ± 21[b] (20)
	Feed: Gain Efficiency	
Founder animals	3.12 ± 0.15 (6)	2.62 ± 0.16[c] (6)
Progeny	2.99 ± 0.12 (8)	2.46 ± 0.16[c] (6)

[a] Numbers in parentheses represent number of pigs surveyed.

[b] $p < 0.001$ compared to control.

[c] $p < 0.05$ compared to control.

Source: Data from Pursel et al.[124]

received considerable attention and financial support. Transgenic livestock with *GH* transgenes have been produced (reviewed, e.g., in Refs. 124 through 127).

The introduction of transgenes (e.g., metalliothionein promoter-GH) has been achieved in pigs and sheep by microinjection of DNA into the egg nucleus.[124,126] The rate of success of this procedure is low in livestock (0.6% integration efficiency in pigs) and the percentage efficiency is even lower if expression of GH is considered.[124] Nevertheless, if *GH* transgenes can be introduced into the gene line of a few or even one founder animal, many transgenic progeny can be produced. A major question for the agricultural application of transgenic technology is, "do transgenic livestock exhibit superior growth and other production criteria?" The answer to this question is at present equivocal. Transgenic pigs (founder and progeny) have been produced with increases in growth rate and feed efficiency (Table 7); there were considerable problems of mortality and disease in the transgenic pigs. These may be a dose phenomenon related to the high (pharmacological?) circulating concentrations of GH in the transgenic pigs. Transgenic sheep (again with *GH* transgenes) show no improvement in growth rate[125,128] but a marked reduction in body fat.[129]

In poultry, it is not possible to inject DNA into the nucleus of the 2.5-cm wide ovum. Transgenes, however, have been introduced by injection of infectious nonreplicating retrovirus vectors (carrying the transgene) beneath the blastoderm of the very early chick embryo.[127] Chick embryos with insertion of GH genes have been produced by this procedure.[127] An alternative approach would be to introduce the transgene into poultry germ cells, producing germline chimeras.[130]

REFERENCES

1. **Asimov, G. J. and Krouze, N. K.,** *J. Dairy Sci.*, 20, 289, 1937.
2. **Folley, S. J. and Young, F. G.,** *Proc. R. Soc. B.*, 126, 45, 1938.
3. **Folley, S. J. and Young, F. G.,** *J. Endocrinol.*, 2, 226, 1940.
4. **Cotes, P. M., Crichton, J. A., Folley, S. J., et al.,** *Nature (Lond.)*, 164, 992, 1949.
5. **Brumby, P. and Hancock, J.,** *New Zealand J. Sci. Tech.*, A36, 417, 1955.
6. **Machlin, L. J.,** *J. Dairy Sci.*, 56, 575, 1973.
7. **Bines, J. A., Hart, I. C., and Morant, S. V.,** *Br. J. Nutr.*, 43, 179, 1980.

8. **Peel, C. J., Bauman, D. E., Gorewit, R. C., et al.,** *J. Nutr.*, 111, 1662, 1981.
9. **Peel, C. J., Fronk, T. J., Bauman, D. E., et al.,** *J. Dairy Sci.*, 66, 776, 1983.
10. **Fronk, T. J., Peel, C. J., and Bauman, D. E.,** *J. Anim. Sci.*, 57, 699, 1983.
11. **Eppard, P. J., Bauman, D. E., and McCutcheon, S. N.,** *J. Dairy Sci.*, 68, 1109, 1985.
12. **Bauman, D. E., Eppard, P. J., DeGeeter, M. J., et al.,** *J. Dairy Sci.*, 68, 1352, 1985.
13. **Bauman, D. E., Hard, D. L., Crooker, B. A., et al.,** *J. Dairy Sci.*, 72, 642, 1989.
14. **Capuco, A. V., Keys, J. E., and Smith, J. J.,** *J. Endocrinol.*, 121, 205, 1989.
15. **McGuire, M. A., Bauman, D. E., Miller, M. A., et al.,** *J. Nutr.*, 122, 128, 1982.
16. **Ringuet, H., Petitclerc, D., Sorensen, M. T., et al.,** *J. Dairy Sci.*, 72, 2928, 1989.
17. **Juskevich, J. C. and Guyer, C. G.,** *Science*, 249, 875, 1990.
18. **Eppard, P. J., Bauman, D. E., Curtis, C. R., et al.,** *J. Dairy Sci.*, 70, 582, 1987.
19. **Sejrsen, K., Foldager, J., Sorensen, M. T., Akers, R. M., et al.,** *J. Dairy Sci.*, 69, 1528, 1986.
20. **McFadden, T. B., Daniel, T. E., and Akers, R. M.,** *J. Anim. Sci.*, 68, 316, 1990.
21. **Fekry, A. E., Keys, J. E., Capuco, A. V., et al.,** *Domest. Anim. Endocrinol.*, 6, 87, 1989.
22. **Lough, D. S., Muller, L. D., Kensinger, R. S., et al.,** *J. Dairy Sci.*, 72, 1469, 1989.
23. **Tyrrell, H. F., Brown, A. C. G., Reynolds, P. J., et al.,** *J. Nutr.*, 118, 1024, 1988.
24. **McDowell, G. H., Gooden, J. M., Leenanuruksa, D., et al.,** *Aust. J. Biol. Sci.*, 40, 295, 1987.
25. **McDowell, G. H., Hart, I. C., Bines, J. A., et al.,** *J. Biol. Sci.*, 40, 191, 1987.
26. **Pocius, P. A. and Herbein, J. H.,** *J. Dairy Sci.*, 69, 713, 1986.
27. **Houseknecht, K. L., Bauman, D. L., Fox, D. G., et al.,** *J. Nutr.*, 122, 1717, 1992.
28. **Byatt, J. C., Eppard, P. J., Veenhuizen, J. J., et al.,** *J. Endocrinol.*, 132, 185, 1992.
29. **McDowell, G. H., Hart, I. C., and Kirby, A. C.,** *Aust. J. Biol. Soc.*, 40, 181, 1987.
30. **Davis, S. R., Gluckman, P. D., Hart, I. C., et al.,** *J. Endocrinol.*, 114, 17, 1987.
31. **Vicini, J. L., Buonomo, F. C., Veenhuizen, J. J., et al.,** *J. Nutr.*, 121, 1656, 1991.
32. **Baumrucker, C. R. and Stemberger, B. H.,** *J. Anim. Sci.*, 67, 3503, 1989.
32a. **Prosser, C. G., Davis, S. R., Farr, V. C., et al.,** *J. Endocrinol.*, 142, 93, 1994.
33. **Hauser, S. D., McGrath, M. F., Collier, R. J., et al.,** *Mol. Cell. Endocrinol.*, 72, 187, 1990.
34. **Glimm, D. R., Baracos, V. E., and Kennelly, J. J.,** *J. Endocrinol.*, R5, 1990.
35. **DeBoer, G. and Kennelly, J. J.,** *J. Dairy Sci.*, 72, 427, 1989.
36. **Sechen, S. J., McCutcheon, S. N., and Bauman, D. E.,** *Domest. Anim. Endocrinol.*, 6, 141, 1989.
37. **Sechen, S. J., Dunshea, F. R., and Bauman, D. E.,** *Am. J. Physiol.*, 258, E582, 1990.
38. **Bauman, D. E. and Currie, W. B.,** *J. Dairy Sci.*, 63, 514, 1980.
39. **Turman, E. J. and Andrews, F. N.,** *J. Anim. Sci.*, 14, 7, 1955.
40. **Machlin, L. J.,** *J. Anim. Sci.*, 35, 794, 1972.
41. **Chung, C. S., Etherton, T. D., and Wiggins, J. P.,** *J. Anim. Sci.*, 60, 118, 1985.
42. **Etherton, T. D., Wiggins, J. P., Chung, C. S., et al.,** *J. Anim. Sci.*, 63, 1389, 1986.
43. **Etherton, T. D., Wiggins, J. P., Evock, C. M., et al.,** *J. Anim. Sci.*, 64, 433, 1987.
44. **Evock, C. M., Etherton, T. D., Chung, C. S., et al.,** *J. Anim. Sci.*, 66, 1928, 1988.
45. **Campbell, R. G., Steele, N. C., Caperna, T. J., et al.,** *J. Anim. Sci.*, 66, 1643, 1988.
46. **Campbell, R. G., Steele, N. C., Caperna, T. J., et al.,** *J. Anim. Sci.*, 67, 177, 1989.
47. **Campbell, R. G., Steele, N. C., Caperna, T. J., et al.,** *J. Anim. Sci.*, 67, 1265, 1989.
48. **McLaren, D. G., Bechtel, P. J., Grebner, G. L., et al.,** *J. Anim. Sci.*, 68, 640, 1990.
49. **Verstegen, M. W. A., van der Hel, W., Henken, A. M., et al.,** *J. Anim. Sci.*, 68, 1008, 1990.
50. **Boyd, R. D., Bauman, D. E., Fox, D. G., et al.,** *J. Anim. Sci.*, 69(Suppl. 2), 56, 1991.
51. **Chung, C. S. and Etherton, T. D.,** *Endocrinology*, 119, 780, 1986.
52. **Grant, A. L., Helferich, W. G., Kramer, S. A., et al.,** *J. Endocrinol.*, 130, 331, 1991.
53. **Wolverton, C. K., Azain, M. J., Duffy, J. Y., et al.,** *Am. J. Physiol.*, 263, E637, 1992.
54. **Walton, P. E. and Etherton, T. D.,** *J. Anim. Sci.*, 62, 1584, 1986.
55. **Walton, P. E., Etherton, T. D., and Evock, C. M.,** *Endocrinology*, 118, 2577, 1986.
56. **Clark, S. L., Wander, R. C., and Hu, C. Y.,** *J. Anim. Sci.*, 70, 3435, 1992.
57. **Sejrsen, K., Foldager, J., Sorensen, M. T., et al.,** *J. Dairy Sci.*, 69, 1528, 1986.
58. **Peters, J. P.,** *J. Nutr.*, 116, 2490, 1986.
59. **Early, R. J., McBride, B. W., and Ball, R. O.,** *J. Anim. Sci.*, 68, 4134, 1990.
61. **Enright, W. J., Quirke, J. F., Gluckman, P. D., et al.,** *J. Anim. Sci.*, 68, 2345, 1990.
62. **Groenewegan, P. P., McBride, B. W., Burton, J. H., et al.,** *Domest. Anim. Endocrinol.*, 7, 43, 1990.
63. **Maltin, C. A., Delday, M. I., Hay, S. M., et al.,** *Br. J. Nutr.*, 63, 535, 1990.
64. **Muir, L. A., Wien, S., Duquette, P. F., et al.,** *J. Anim. Sci.*, 56, 1315, 1983.

65. **Johnsson, I. D., Hart, I. C., and Butler-Hogg, B. W.,** *Anim. Prod.*, 41, 207, 1985.
66. **Rosemberg, E., Thonney, M. L., and Butler, W. R.,** *J. Anim. Sci.*, 67, 3300, 1989.
67. **Pell, J. M., Elcock, C., Harding, R. L., et al.,** *Br. J. Nutr.*, 63, 431, 1990.
68. **Zainur, A. S., Tassell, R., Kellaway, R. C., et al.,** *Aust. J. Agric. Res.*, 40, 195, 1990.
69. **Johnsson, I. D., Halthorn, D. J., Wilde, R. M., et al.,** *Anim. Prod.*, 44, 405, 1987.
70. **Eisemann, J. H., Tyrell, H. F., Hammond, A. C., et al.,** *J. Nutr.*, 116, 157, 1986.
71. **Crooker, B. A., McGuire, M. A., Cohick, W. S., et al.,** *J. Nutr.*, 120, 1256, 1990.
72. **Moseley, W. M., Krabill, L. F., and Olsen, R. F.,** *J. Anim. Sci.*, 55, 1062, 1982.
73. **Davis, S. L., Garrigus, U. S., and Hinds, F. C.,** *J. Anim. Sci.*, 30, 236, 1969.
74. **Eisenmann, J. H., Hammond, A. C., Bauman, D. E., et al.,** *J. Nutr.*, 116, 2504, 1986.
75. **Olivares, V. H. and Hallford, D. M.,** *J. Anim. Sci.*, 68, 1971, 1990.
76. **Leung, F. C., Taylor, J. E., Wien, S., et al.,** *Endocrinology*, 118, 1961, 1986.
77. **Burke, W. H., Moore, J. A., Ogez, J. R., et al.,** *Endocrinology*, 120, 651, 1987.
78. **Bowen, S., Huybrechts, L. M., Marsh, J. A., et al.,** *Comp. Biochem. Physiol.*, 86A, 137, 1987.
79. **Cogburn, L. A., Liou, S. S., Rand, A. L., et al.,** *J. Nutr.*, 119, 1213, 1989.
80. **Cravener, T. L., Vasilatos-Younken, R., and Wellenreiter, R. H.,** *Poultry Sci.*, 68, 1133, 1989.
81. **Cogburn, L. A.,** *Crit. Rev. Poultry Biol.*, 3, 283, 1991.
82. **Vasilatos-Younken, R., Cravener, T. L., Cogburn, L. A., et al.,** *Gen. Comp. Endocrinol.*, 71, 268, 1988.
83. **Scanes, C. G., Peterla, T. A., Kantor, S., et al.,** *Growth Dev. Aging*, 54, 95, 1990.
84. **King, D. B. and Scanes, C. G.,** *Proc. Soc. Exp. Biol. Med.*, 182, 201, 1986.
85. **Scanes, C. G., Duyka, D. R., Lauterio, T. J., et al.,** *Growth*, 50, 12, 1986.
86. **Scanes, C. G., Harvey, S., Marsh, J. A., et al.,** *Poultry Sci.*, 63, 2062, 1984.
87. **Hargis, P. S., Dean, C. E., and Hargis, B. M.,** *Crit. Rev. Poultry Biol.*, 3, 307, 1991.
88. **Vanderpooten, A., Darras, V. M., Huybrechts, L. M., et al.,** *J. Endocrinol.*, 129, 275, 1991.
89. **McLean, E. and Donaldson, E. M.,** *The Endocrinology of Growth, Development, and Metabolism in Vertebrates* (Eds. Schreibman, M. P., Scanes, C. G., and Pang, P. K. T.,). Academic Press, San Diego, p. P43, 1992.
90. **Carter, R. D., Risner, R. N., and Yacowitz, H.,** *Poultry Sci.*, 34, 1407, 1955.
91. **Donoghue, D. J., Campbell, R. M., and Scanes, C. G.,** *Poultry Sci.*, 69, 1813, 1990.
92. **Daul, G. E., Chapin, L. T., Zinn, S. A., et al.,** *J. Dairy Sci.*, 73, 2444, 1990.
93. **Thorner, M. O., Reschke, J., Chitwood, J., et al.,** *N. Engl. J. Med.*, 312, 4, 1985.
94. **Wehrenberg, W. B.,** *Endocrinology*, 118, 489, 1986.
95. **Cella, S. G., Locatelli, V., Mennini, T., et al.,** *Endocrinology*, 127, 1625, 1990.
96. **Wehrenberg, W. B., Block, B., and Phillips, B. J.,** *Endocrinology*, 115, 1218, 1984.
97. **Trout, W. E. and Schanbacher, B. D.,** *J. Endocrinol.*, 125, 123, 1990.
98. **Dubreuil, P., Petitclerc, D., Pelletier, G., et al.,** *J. Anim. Sci.*, 68, 1254, 1990.
99. **Pommier, S. A., Dubreuil, P., Pelletier, E., et al.,** *J. Anim. Sci.*, 68, 1291, 1990.
100. **Lovendaul, P., Angus, K. D., and Woolliams, J. A.,** *J. Endocrinol.*, 128, 419, 1991.
101. **Kahn, G., Perier, A., and Martinet, J.,** *J. Anim. Sci.*, 67(Suppl. 1), 332, 1989.
102. **Guler, H. P., Zapf, J., Binz, K., et al.,** *Biotechnology in Growth Regulation* (Ed. Heap, R. B., Prosser, C. G., and Lamming, G. E.). Butterworths, London, p. 119, 1989.
103. **Robinson, I. C. A. F. and Clark, R. G.,** *Biotechnology in Growth Regulation* (Ed. Heap, R. B., Prosser, C. G., and Lamming, G. E.). Butterworths, London, p. 129, 1989.
104. **Scanes, C. G. and Baile, C.,** *The Endocrinology of Growth, Development, and Metabolism in Vertebrates* (Eds. Schreibman, M. P., Scanes, C. G., and Pang, P. K. T.). Academic Press, New York, p. 541, 1993.
105. **Dhawan, J., Pan, L. C., Pavlath, G. K., et al.,** *Science*, 254, 1509, 1991.
106. **Aston, R., Holder, A. T., Ivanyi, J., et al.,** *Mol. Immunol.*, 24, 143, 1987.
107. **Aston, R., Holder, A. T., Preece, M. A., et al.,** *J. Endocrinol.*, 110, 381, 1986.
108. **Wallis, M., Daniels, M., Ray, K. P., et al.,** *Biochem. Biophys. Res. Commun.*, 149, 187, 1987.
109. **Holder, A. T., Aston, R., Preece, M. A., et al.,** *J. Endocrinol.*, 107, R9, 1985.
110. **Pell, J. M., Johnsson, I. D., Pullar, R. A., et al.,** *J. Endocrinol.*, 120, R15, 1989.
111. **Wang, B. S., Swewczyk, E., Shieh, H. M., et al.,** *J. Endocrinol.*, 127, 481, 1990.
112. **Bomford, B. and Aston, R.,** *J. Endocrinol.*, 125, 31, 1990.
113. **Massart, S., Maiter, D., Portetelle, D., et al.,** *J. Endocrinol.*, 139, 383, 1993.
114. **Gardner, M. J., Morrison, C. A., Stevenson, L. Q., et al.,** *J. Endocrinol.*, 125, 53, 1990.
115. **Campino, C., Szecowka, J., Lopez, J. M., et al.,** *J. Clin. Endocrinol. Metab.*, 74, 751, 1992.

116. **Vicini, J. L., Clark, J. H., Hurley, et al.,** *Domest. Anim. Endocrinol.*, 5, 35, 1988.
117. **Van Kessel, A. G., Korchinski, R. S., Hampton, C. H., et al.,** *Domest. Anim. Endocrinol.*, 7, 217, 1990.
118. **Wehrenberg, W. B., Bergman, P. J., Stagg, L., et al.,** *Endocrinology*, 127, 2705, 1990.
119. **Chaplin, R. K., Kerr, D. E., and Laarvald, B.,** *Can. J. Anim. Sci.*, 64(Suppl.), 312, 1984.
120. **Spencer, G. S. G.,** *J. Roy. Soc. Med.*, 77, 496, 1984.
121. **Spencer, G. S. G., Harvey, S., Audsley, A. R. S., et al.,** *Comp. Biochem. Physiol.*, 85A, 553, 1986.
122. **Palmiter, R. D., Brinster, R. L., Hammer, R. E., et al.,** *Nature (Lond.)*, 300, 611, 1982.
123. **Palmiter, R. D., Norstedt, G., Gelinas, R. E., et al.,** *Science*, 222, 809, 1983.
124. **Pursel, V. G., Pinkert, C. A., Miller, K. F., et al.,** *Science*, 244, 1281, 1989.
125. **Rexroad, C. E., Hammer, R. E., Behringer, R. R., et al.,** *J. Reprod. Fert.*, Suppl. 41, 419, 1990.
126. **Steele, M. C. and Pursel, V. G.,** *Annu. Rev. Nutr.*, 10, 213, 1990.
127. **Bosselman, R. A., Hsu, R.-Y., Briskin, M. J., et al.,** *J. Reprod. Fert.*, Suppl. 41, 183, 1990.
128. **Rexroad, C. E., Hammer, R. E., Bolt, D. J., et al.,** *Mol. Reprod. Dev.*, 1, 164, 1989.
129. **Ward, K. A., Nancarrow, C. D., Murray, J. D., et al.,** *J. Cell. Biochem.*, 13B, 164, 1989.
130. **Petitte, J. N., Clark, M. E., Liu, G., et al.,** *Development*, 108, 185, 1990.

Chapter 30

Growth Hormone Action: Clinical Significance

W. H. Daughaday and S. Harvey

CONTENTS

0-8493-8697-7/95/$0.00+$.50

I. INTRODUCTION

Since the introduction of recombinant growth hormone (GH) in 1985 there has been extensive research into its possible use in conditions other than classic growth hormone deficiency. It is now clear that slightly higher doses of GH can stimulate skeletal growth in many children with proportionate short stature. Particularly promising results have been obtained in Turner's syndrome and in the growth failure of children with chronic renal failure. The anabolic effects of GH may also have clinical applications in adults, as the treatment of adult patients with hypopituitarism increases muscle bulk and strength. In elderly individuals, a decrease in GH secretion occurs and GH treatment may therefore be of benefit. There is also evidence that GH may be a useful adjuvant to nutritional therapy in the hypermetabolic state induced by clinical malnutrition and trauma. The identification of the immune and reproductive systems as GH target sites suggests novel therapeutic uses for GH in immune dysfunction and in infertility. The use of other components in the GH axis (growth hormone-releasing hormone [GHRH] and insulin-like growth factor I [IGF-I]) as alternatives to GH therapy is also discussed. Therapeutic uses of GH have also been reviewed by other investigators.[1-9]

II. GH THERAPY

The proper dose of GH and the optimal treatment regimen for growth promotion have not been conclusively established, largely due to the inconsistent bioactivity of extracted GH used in early studies. However, biosynthetic preparations of GH, recombinant human GH (rhGH) (with and without an amino-terminal methionine), are now used extensively[8] and are of consistent potency, thus facilitating comparisons between studies. The dose employed in the various therapeutic uses of GH is variable and described under individual headings, but optimal doses have rarely been established. Ideally, the dose should produce the greatest growth or metabolic response without adverse effects, but the high cost of the biosynthetic hormone constrains the use of the higher doses.

The route of GH administration has important implications in the kinetics of GH absorption and for patient compliance. Initially, physicians employed intramuscular (i.m.) injections; however, subcutaneous (s.c.) injections are more acceptable to the patient, as they are less painful and can be self-administered.[7] Moreover, the growth response may be greater with s.c. than with i.m. injections, as the slower reabsorption from a s.c. injection produces a more physiological GH peak.[7,10] The abdomen rather than the thigh appears to be the optimal site of injection, due to increased absorption.[10]

The frequency of GH injections is an equally important response determinant as the dosage. In all species studied thus far, pituitary GH is secreted in a pulsatile fashion, and the pattern of GH secretion is an important determinant of not only the magnitude of the GH response but also the existence of the response. Thus, the optimal GH treatment regimen would also be pulsatile, thus maximizing the desired physiological effects of GH and avoiding unwanted, unphysiological effects. Unfortunately, the high frequency of injections required to mimic pulsatile secretion is expensive and inconvenient, although the future development of nasal GH sprays or improved pumps may increase the practicality of artificial pulsatile GH secretion. Presently, the induction of a partially pulsatile, partially continuous administration pattern by daily GH injections is the optimal alternative, as it provides a significantly greater growth response than thrice-weekly injections.[7,10,11] Attempts at precisely mimicking the normal pattern of serum GH with exogenous GH are, however, inconvenient and expensive and likely to result only in minor economies in GH usage. Although some clinicians claim that night-time GH injections induce the maximal metabolic response,[7,11] other investigators have observed no difference between the efficacy of night-time and day-time injections.[12,13]

The pattern of GH administration is especially important in GH-replete or partially GH-deficient patients, as GH negatively affects its own secretion and exogenous GH may thus interfere with endogenous GH pulses. Optimization of the therapy regimen may avoid this inhibitory effect, as by analogy GH administration to conscious rats at 3-h intervals enhances endogenous GH peaks, whereas administration at 1.5-h intervals suppresses GH pulses.[14] Moreover, initial fears that the acute inhibitory effect of exogenous GH on somatotroph function may be irreversible following chronic GH administration have not been justified (reviewed by Muller[15]).

Growth hormone-releasing hormone (GHRH) is the primary secretogogue for pituitary GH, and may provide an alternative to GH therapy in patients with normal pituitary responsiveness, although its effects have so far proved inconsistent. The main advantage of GHRH is that the increase in plasma GH levels follows a physiological pattern, as GHRH administration increases the magnitude of naturally occurring pulses and does not disrupt the normal secretory pattern or downregulate GH secretion.[16,17] Moreover, the pituitary does not become desensitized following chronic GHRH administration, but maintains its sensitivity to GHRH stimulation.[16,17] GHRH therapy may thus be preferable to GH for therapeutic applications in GH-replete patients. However, continuous GHRH infusion, or s.c. or intravenous (i.v.) injections every 3 h, can also stimulate endogenous GH secretion in some GH-deficient (GHD) children and adults and in normal men,[18-20] thus expanding the possible recipients of GHRH therapy. This increase in GH secretion is also associated with indices of GH action, including increased linear growth, nitrogen retention, and insulin-like growth factor I (IGF-I).[18,19] The major disadvantage of GHRH treatment is the requirement for frequent injections, as once-daily injections of GHRH are only minimally effective.[21] Possible alternative methods of GHRH delivery include implants,[22,23] intranasal sprays,[24,25] or oral supplements.[26,27] However, intranasal (i.n.) administration requires a 300-fold increase in dose compared to infusion[24,25] and GHRH and peptidergic GHRH analogs are only minimally active when administered orally, due to digestion by intestinal proteases.[26,27] Thus, an inexpensive, oral system of GHRH therapy must await the development of nonpeptidergic GHRH analogs. L692,429 is a possible candidate, as it is a potent, nonpeptidyl GH secretogogue, and may be orally active.[28]

A large proportion of the metabolic and anabolic effects of GH are mediated by IGF-I, thus IGF-I is a further alternative to GH therapy. However, IGF-I doses of over 75 μg/kg (bolus i.v.) or over 25 μg/kg/h (infusion) cause acute hypoglycemia and hypoinsulinemia, and thus the dose must be carefully controlled.[29-31] The efficacy of IGF-I varies between different therapeutic applications, and is discussed below under the individual headings. However, apart from GH-insensitivity syndromes, IGF-I has proved to be considerably less potent than GH in stimulating skeletal growth.

III. GROWTH

Growth hormone is axiomatically involved in growth. A deficiency in GH during childhood would therefore result in serious growth failure, but growth failure can also occur as a result of diseases unrelated to the GH axis or as a result of unknown causes. GH replacement is the obvious therapy for GHD-related growth failure, but recent studies indicate that GH can effectively increase growth in most children. The effectiveness of GH therapy in these different conditions and the ethical dilemmas raised by widespread GH therapy are discussed below.

A. CLASSIC GH DEFICIENCY

Severe GHD results in a marked retardation of growth with height more than 3 standard deviations below the norm for age and sex (–3 SDs). There is also retardation of bone age behind the chronological age. The definition of GHD has been the object of some dispute; however, a failure of serum GH concentrations to rise above 5 μg/l in the presence of provocative stimuli is indicative of severe GH deficiency whereas concentrations of 5 to 10 μg/l are indicative of partial GH deficiency.

Current dosage recommendations for growth acceleration in GH-deficient children vary between 0.3 to 0.6 mg rhGH/kg/week.[7,10] The response to GH therapy is greatest during the first year of GH treatment, as growth velocity usually doubles or triples.[32,33] However, growth velocities usually decrease in subsequent years, and an increase in the dose of GH may be required.[33] Despite the dramatic early response to GH treatment, the final height achieved seldom reaches the mean height of normal individuals of the same sex and ethnicity, or the midparental height.[34] The failure to attain normal stature is attributable to delay in instituting treatment or to the use of inadequate dose regimens. For instance, the growth response to GH therapy is frequency dependent, as a greater response is observed with daily injections (10.6 cm/year) than with three- or five-times-weekly injections (8.6 cm/year).[33] In addition, GH treatment should be instituted as early as the diagnosis can be established, as delay in starting GH treatment limits the final height that can be achieved. Early (before 5 years of age) intervention is also required for normalization of head circumference in some GHD patients.[35] Thus, early GH replacement therapy may also improve mental well-being, as head circumference has been hypothesized to be correlated with cerebral development and some GHD children are mentally impaired.

Higher doses of GH in GHD children may accelerate sexual maturation and reduce the duration of puberty by about 6 months.[36] Evidently, this decreases the duration of the pubertal growth spurt and consequently limits the final growth attainment. The pubertal growth spurt can be maximized by low doses of GH, as the detrimetal effect of GH on pubertal length may be dose related,[37] and by coadministration of gonadotropin-releasing hormone (GnRH), which delays puberty.[36]

Although GH replacement therapy is indicated as therapy for correcting symptoms associated with genetic defects in the *GH* gene,[38] some cases of GH deficiency are due to defects in GHRH (growth hormone neurosecretory dysfunction [GHND]). Growth retardation in these patients can be effectively treated with twice-daily GHRH administration (20 to 40 μg/kg/day)[39] or with six-times weekly GH administration (0.5 IU/kg/week).[40]

B. PARTIAL GH DEFICIENCY AND IDIOPATHIC SHORT STATURE

In the past, when supplies of GH derived from human pituitaries were limited, only short children with the most severe GH deficiency were selected for treatment. It is now recognized that GHD is seldom complete, but the definition of partial GHD remains somewhat ambiguous. In some studies, partial GHD has been diagnosed in patients with normal GH responses to provocative tests, but impaired spontaneous GH secretion as determined by frequent monitoring for 24 h or overnight.[41-43] However, nights of low GH secretion are occasionally observed in normally growing children, and the test is not very repeatable.[44] Measurements of serum IGF-I[45] or serum IGF-binding protein 3 (IGFBP-3) may be better indicators of relative GH deficiency.[46]

There have now been over 18 reports of GH treatment of short children without classic GH deficiency (as reviewed by Cara and Johnson[47] and Lippe and Nakamoto[5]). Most investigators have given GH doses somewhat greater than those used in classic GH deficiency and have reported successful growth acceleration in 40 to 100% of treated, normally proportioned, well-nourished children. In the large Genentech Collaborative Study, GH was administered three times per week at a total dosage of 0.3 kg/week[48-50] to short children (average, –2.9 SDs) with normal GH secretion. After 1 year of treatment, the average growth velocity rose from 4.6 to 8.0 cm/year, and continued at 6.9 cm/year even after 3 years of treatment, with few adverse side effects. Although the degree of growth improvement in different children could not be predicted by pretreatment variables, the growth response in the first year was reasonably predictive of future growth reponses. The final height prediction of the GH-treated patients in the Genentech study was improved by 1.2 SDs, but smaller studies by the Dutch Growth Hormone Working Group[51] and Albertsson-Wikland and Karlberg[52] observed smaller increases of 0.5 SDs or 0.1 to 0.7 SDs, respectively.

The degree of growth acceleration is determined by the dosage,[53,54] the frequency of administration,[48] and the timing of treatment.[53] The growth response in the first year of treatment also appears to be inversely related to the initial growth rate.[55] Daily administration of high doses of GH 2 years prior to puberty appears to be the optimal treatment regimen. In the study of Lesage et al.,[53] administration of high (1.05 mg/kg/week) doses of rhGH to short children 2 years prior to puberty increased the mean growth velocity *z* scores to +6.9 and +3.1 at 12 and 24 months, respectively. These high doses of GH appear to be well tolerated, with only a slight, transient increase in blood pressure, body weight, and plasma aldosterone caused by the water-retaining effects of GH.[56] Moreover, chronic GH therapy does not interfere with normal pulsatile release.[15]

Growth failure in some children may be due to GH resistance rather than inadequate GH, as growth hormone-binding protein (GHBP) levels (which are often an indicator of hepatic growth hormone receptor [GHR] status) were decreased in a population of children with idiopathic short stature.[57] High doses of GH significantly increased GHBP levels and improved growth velocity in comparison with untreated controls; thus GH therapy appears to be beneficial in short stature caused by GH resistance and partial GH deficiency.

GH treatment can also cause a transition from slightly subnormal growth rates to above average. For instance, in the study of Hindemarsh and Brook,[58] the growth rate in GH-treated children improved from –0.44 to +2.2 mean growth rate *z* score (SDs). This finding is in agreement with findings in clinical and experimental states of GH excess, as pituitary tumors in children[59] and the *GH* transgene in mice[60] can increase growth to beyond the genetic potential.

C. DISEASES ASSOCIATED WITH GROWTH FAILURE

1. Turner's Syndrome

The phenotype of Turner's syndrome in girls includes short stature, primary amenorrhea, webbed neck and other physical abnormalities.[61] It is caused by total or partial loss of one sex chromosome. In many cases there is mosaicism of normal and abnormal X chromosome-containing cells.[61] Curiously, although the final height of Turner's patients is very low owing to decreased intrauterine and postnatal growth (an average of 143.5 cm in one study),[62,63] no consistent abnormality of GH secretion or serum IGF-I has been recognized. However, early studies suggested that height could be increased by GH treatment, but the number of cases involved and the differences in treatment schedules made these studies hard to interpret.[64] There is presently a large collaborative study investigating the effects of GH alone or in combination with the androgen oxandrolone.[65,66] After 5 years, GH and oxandrolone increased height over predicted values by 15.3 cm, whereas GH alone resulted in a 9.3-cm increase, and final adult height was increased by 8.6 cm in some girls. Clinical studies are also ongoing in a variety of other centers, and generally report a significant increase in growth velocity during 2 years of treatment and an increase in final height prediction, although the subjects in these studies have not yet finished growing.[67-70] It has been found that six injections per week of doses up to 1.0 IU/kg/week provide the greatest growth acceleration, but any treatment regimen is of limited effectiveness in patients over 16 years of age.[70] The usefulness of GH in Turner's syndrome is now generally accepted, in contrast to treatment in idiopathic short stature. This distinction is not due to the increased responsiveness of Turner's patients, but rather to the homogeneity of this population in contrast to idiopathic short stature.

The dosages employed in Turner's syndrome are generally higher than in classic GH deficiency, raising some concern about adverse side effects. However, when the growth in the left middle fingers of GH-treated Turner's patients was compared to investigate the possibility of acromegaloid-like hand development, no difference was observed in the GH-treated group.[71] In general, treatment produces few side effects; however, edema, increased number of nevi, and fatigue were noted in some patients.[67-70]

2. Down's Syndrome

Down's syndrome, which is caused by trisomy of chromosome 21, results in short stature, mental retardation, and various skeletal and visceral malformations. Like Turner's syndrome, no consistent abnormalities of GH secretion or serum IGF-I levels have been established, but GH secretion and IGF-I plasma levels are decreased in some patients.[72] GH treatment results in a substantial increase in growth velocity and head circumference but no improvement in mental function.[73] Unless it can be shown that GH treatment significantly improves the quality of life of Down's syndrome patients it cannot be recommended.[74]

3. Prader-Willi Syndrome

Prader-Willi syndrome is characterized by short stature, hyperphagia, obesity, cryptorchidism, hypotonia, and occasionally mental retardation. Deletions of the long arm of chromosome 15 are present in about half the patients.[75,76] As is true of any obese subject, the spontaneous and provoked secretion of GH is suppressed.[77] Short-term GH administration to a small cohort of these patients decreased weight gain and improved growth velocity,[78] but long-term results of GH treatment have not been reported.

4. Noonan's Syndrome

Noonan's syndrome is characterized by short stature in both sexes, webbed neck and ptosis, but the underlying genetic lesion is unknown.[79] IGF-I levels are low despite normal GH responses to provocative testing.[79,80] Initial increases in growth velocity have usually, but not always, followed GH treatment, and the growth response is greatest with daily, high-dosage therapies.[79,80]

5. Intrauterine Growth Retardation and Silver-Russell Syndrome

Intrauterine growth retardation (IUGR) results in a reduction in birth length and a decreased adult height to approximately –3.6 SDs.[81] IUGR is also a factor in the growth failure associated with Silver-Russell syndrome, which is also characterized by skeletal asymmetry and triangular facies.[82] A role for GH is suggested by the decreased IGF-I levels in IUGR infants, and the correlation between IGF-I levels and postnatal catch-up growth.[83] Indeed, IUGR and partial GH deficiency share some clinical characteristics, as GH responses to provocative stimuli are normal but spontaneous pulsatile GH secretion is abnormal.[82] In mixed populations of IUGR and Silver-Russell patients, GH treatment resulted in an initial increase in growth velocity SDs.[82,84,85] However, this growth response progressively decreased and height SDs for bone age did not improve,[84] thus casting some doubt on the utility of GH therapy in IUGR patients.

6. Thalassemia

β-Thalassemia major, a hemoglobin disorder, is often accompanied by short stature and sexual immaturity. Studies in different populations of thalassemic patients have observed normal[86,87] or impaired[88,89] GH responses to provocative stimuli and low levels of IGF-I.[90] However, these patients possess hepatic GH-binding activity[91] and their erythroid precursors are normally responsive to GH; thus GH resistance is not responsible for the short stature. In patients with impaired GH secretion, GH therapy significantly increased growth velocity after 1 year of treatment; however, the eventual height obtained by these patients was not determined.[89]

7. Genetic Skeletal Disorders

Hypochondroplasia ("short-leggedness") results in short stature and abnormal proportions, as growth of leg bones is considerably impaired. A polymorphism has been identified in the *IGF-I* gene of affected individuals but serum IGF-I levels are normal, suggesting the possibility of a subtle structural defect or a defect in locally produced IGF-I.[92] The genetic defect appears to be partially dominant, as leg length was increased by GH treatment in individuals that were heterozygous, but not homozygous, for the gene polymorphism.[92] Growth hormone effectively increased the growth velocity (primarily by increased growth of the lower limbs) in two additional clinical trials.[93,94]

Impaired renal phosphate absorption and abnormal vitamin D_3 synthesis are thought to be responsible for the reduced linear bone growth and height in children with X-linked hypophosphatemic rickets.[95] Initial studies suggest that GH can increase height velocity in these patients,[96] but long-term data on adult height attainment and possible adverse effects are not available.

8. Neural Tube Defects

Neural tube defects such as spina bifida or myelomeningocele are also associated with short stature, and GH treatment in one study increased the growth rate from 1.7 to 7.9 cm/year without affecting bone age.[97]

9. Chronic Renal Disease

The slow growth rates in children with chronic renal failure are due in part to alterations in mineral metabolism, including defective vitamin D synthesis and parathyroid hormone (PTH)-induced bone disease.[98,99] Correction of mineral abnormalities does not always correct the growth deficit, and a defect in the GH axis has been implicated. However, basal serum GH concentrations in these patients may actually be elevated and are not easily suppressed by glucose ingestion.[100] GH secretion and IGF-I responses to GH stimulation are essentially normal despite occasional abnormalities in IGFBP levels, suggesting some degree of GH/IGF-I resistance.[101,102] Sufficient doses of GH can partially offset this resistance to GH and IGF-I action, as GH therapy in uremic children significantly improves growth velocity.[103-107] The best results are obtained in children with stable renal failure. Children on dialysis or receiving corticosteroids posttransplantation responded poorly to therapy in some studies; however, the age and the severity of renal impairment are often greater in these patients. GH treatment in younger patients receiving dialysis has resulted in a dramatic increase in height velocity SDs, from –2.5 to +1.5,[108] and some posttransplantation patients responded positively to treatment.[109] Thus, GH therapy is recommended for treatment of short stature associated with chronic renal failure.[110]

10. Corticosteroid Growth Inhibition

Many allergic and chronic inflammatory conditions require prolonged treatment with corticosteroids, often to the detriment of skeletal growth. This growth impairment can be mitigated in some cases by a reduction in steroid dose or its administration on alternate days. Initial attempts at restoring normal growth in children receiving chronic steroid administration for rheumatoid arthritis,[111] nephrosis,[112] or inflammatory bowel disease[113] with GH therapy have been disappointing. This resistance to GH action may be due to the multiple sites of corticosteroid interference in the GH axis, as steroids may inhibit not only GH release but also IGF-I secretion and action.[114,115] Nevertheless, as the studies described above used relatively low GH doses, higher, more frequent GH administration may provide a better growth response. This possibility is supported by experimental data in rats, as high doses of rhGH were able to partially reverse the growth deficit caused by corticosteroid therapy of uremic rats.[116] However, the edema and abnormal glucose metabolism associated with high-dose GH therapy raise some doubts as to the advisability of GH treatment for this condition.

11. Psychosocial Growth Failure

In addition to the myriad of physical causes described above, growth failure can also arise from emotional disturbances. The defective growth of these children is thought to be due to inadequate nutrition and GH neurosecretory dysfunction. Despite increased basal GH levels, the GH response to provocative stimuli is blunted,[117] which is suggestive of partial GH deficiency. Accordingly, psychosocial growth failure (PSGF) patients also show a similar growth response to GH therapy, as growth velocity SD scores were increased from –2.32 to +4.66 after 6 months of daily GH administration.[117] However, this increased growth response is also due in part to the improvement in social environment that often accompanies therapeutic intervention.

12. Cancer Patients

Prepubertal patients with brain tumors, craniopharyngioma, and leukemia who are treated with cranial irradiation or total body irradiation usually develop some degree of GH deficiency and growth failure.[118-122] Bozzola et al.[119] observed that GH therapy could reverse the growth failure in five of seven cancer patients to a similar extent as in GHD patients. Untreated children do not spontaneously recover from irradiation-induced GH deficiency and there is no period of catch-up growth; thus GH treatment of these patients is certainly warranted.[119] Bone marrow transplantation in conjunction with low doses of radiation therapy has a similar deleterious effect on GH secretion and linear growth, and in neoplastic, but not hematological, patients, no catch-up growth is observed.[123,124] Studies in a small group of children with neuroblastoma suggest that GH can also be an effective growth stimulant in these patients.[124] However, certain combinations of chemotherapy drugs are thought to induce GH resistance,[124,125] and thus alternate therapies, such as IGF-I, should also be considered. Concerns over the possible oncogenic potential of GH therapy itself are discussed later in this chapter.

In summary, administration of GH to children with proportionate short stature of many different etiologies often results in an initial stimulation of skeletal growth. The dose required may be larger than customarily used in the treatment of pituitary dwarfism. Except for Turner's syndrome, increases in final height have rarely been established and the risks of GH treatment remain to be determined.

IV. MODIFICATION OF METABOLISM

GH therapy was invariably restricted to severely growth-impaired, GHD children when the only GH available was from cadaver pituitaries. However, with the advent of virtually unlimited supplies of biosynthetic GH, the issue of GH replacement therapy in adults received increased attention. It is now known that GH can beneficially affect various metabolic parameters in GHD adults and children and in the relatively GH-deficient elderly.

A. GH-DEFICIENT ADULTS

GH-deficient adults often complain of muscle weakness, fatigue, and a loss of well-being. Objective measurements have confirmed a decrease in muscle mass and relative increase in body fat. The benefits of GH have been reported in a number of double-blind, placebo-controlled studies and uncontrolled studies and have been extensively reviewed.[126] The doses used are higher than those used in GH-deficient children (0.91 to 1.82 mg/kg/week), and should be optimized to produce normal IGF-I levels but minimal fluid retention.[127,128] Specific effects of GH on lipid, glucose, and protein metabolism and general well-being are detailed below.

1. Lipids

GH-deficient patients have a relative increase in adipose tissue mass,[129,130] and this excess body fat is significantly reduced by GH treatment.[130-133] GH also promotes a healthier fat distribution, as adipose mass is primarily lost from visceral adipose stores.[134] Changes in fat metabolism reflect a direct lipolytic action on adipose cells, which results in an increased use of fatty acids for metabolic fuel and increased amino acid availability for protein synthesis. This increased lipid oxidation is an acute GH effect, as it is reduced in GH-treated patients that miss 1 day of treatment.[135]

The increase in body fat is associated with a high incidence of hypercholesterolemia in GHD adults and children and may be a contributing factor.[136-139] Generally, plasma levels of low-density lipoprotein (LDL) cholesterol, very-low-density lipoprotein (VLDL) cholesterol, and triglycerides are increased whereas high-density lipoprotein (HDL) cholesterol levels are reduced by GH deficiency. These opposing effects may be responsible for the normal cholesterol levels detected in some studies of GHD patients.[132,140] GH therapy can correct the hypercholesterolemia and high LDL levels in some, but not all, patients,[136] and the success rate may relate to the familial history of hyperlipidemia (as summarized by Keller and Miles[141] and Eden et al.[142]). The effect of GH on HDL is equally inconsistent, as serum levels were increased,[143] decreased,[144] or unaffected.[145] The increased concentrations of free fatty acids in GH-treated patients are also indicative of increased lipolysis.[130]

However, despite inhibitory effects on plasma cholesterol, the pharmacological amounts of GH observed in acromegaly and high-dose GH therapy are associated with increased plasma triglycerides, especially in the very-low-density fraction.[138,139,144,146-149] This apparently lipogenic effect may be due to GH-induced inhibition of lipoprotein lipase activity,[144] which would impair VLDL clearance. The increase in nonesterified fatty acids caused by lipolysis may also contribute to increased triglyceride formation by increasing substrate availability.[141] VLDL cholesterol is an important risk factor in atherosclerosis, thus necessitating close monitoring of VLDL levels in high-dose GH therapy.

2. Protein/Muscle

GH therapy in adults with childhood- or adult-onset hypopituitarism reverses the decreased lean body mass associated with GHD, and increases thigh muscle volume, sub maximal and maximal exercise capacity, and occasionally muscle strength.[130-134,150-152] This anabolic effect of GH is mediated largely by increased protein synthesis, and stable isotope studies have confirmed that protein synthesis is increased whereas leucine oxidation (which is indicative of protein degradation) is decreased in GH-treated patients.[151,153] Other markers of GH-induced protein synthesis are also directly correlated with GH status. For instance, GH administration increases retention of the constituents of anabolism (nitrogen, phosphorus, magnesium, calcium, sodium, and chloride),[154] and nitrogen retention (as measured by blood urea levels) is decreased in GHD and enhanced by GH treatment. Moreover, GH also provides the energy for protein synthesis, presumably by induction of lipolysis,[34,155] and several studies have observed a stimulatory effect of GH on energy expenditure in GHD patients.[135,156]

Although GH treatment undeniably enhances muscle physiology, the anabolic effects of GH are transitory, and daily, life-long therapy is necessary for maximal benefit. The omission of GH therapy for a single day results in increased protein breakdown (as demonstrated by increased nitrogen excretion).[138] Moreover, cessation of GH therapy in growing GH-deficient patients reduces muscle size and strength by about 5 to 10%.[157,158]

3. Glucose

GH deficiency is traditionally thought to be associated with hypoglycemia; however, evidence appears to support a defect in the recovery of blood glucose levels following hypoglycemia rather than a defect in the nadir levels of blood glucose (reviewed by Sonksen et al.[159]). This "hypoglycemia unresponsivenss" may account for the preventive effect of GH therapy on starvation-induced hypoglycemia.[160]

The traditional view of consistent insulin hypersensitivity in GHD patients is equally at variance with evidence, as GHD associated with obesity can result in insulin resistance.[161] However, increased insulin sensitivity is observed in some GHD patients, and GH administration to normal or GH-deficient people induces insulin resistance.[135] Serum glucose is increased but plasma insulin is decreased.[137] However, insulin and glucose concentrations were only transiently increased in the study of Fowelin et al.,[162] and levels returned to pretreatment values by 26 weeks. In contrast, insulin sensitivity was chronically reduced.

4. Psychological Well-Being

Patients with GH deficiency that arose in childhood or adulthood have demonstrated a decreased quality of life.[134,163-165] Mental health questionaires have revealed that GH-deficient adults are generally more socially isolated, anxiety ridden, and less energetic and self-controlled. They are also less likely to be married and more likely to be unemployed than the general population. GH therapy improves a number of indices of mood and well-being as quantified by the Comprehensive Psychological Rating Scale and the Symptom Check List-90.[134,137,163,165-167] The relationship between GH and psychological well-being may be mediated in part by the effects GH on protein and carbohydrate metabolism, which would enhance strength, energy, and general physical well-being.

However, direct effects of GH on the brain are also likely, as the central nervous system is also a target site for GH (see Chapters 27 and 28).

B. AGING

There is a progressive decrease in GH secretion in elderly adults (Chapter 12), which results in a decrease in serum IGF-I (reviewed by O'Neill,[168] Blackman et al.,[169] and Corpas et al.[170]). The magnitude of this decrease is such that a considerable number of the elderly have serum IGF-I levels that would indicate GH deficiency in younger adults. Indeed, the body composition of the elderly is suggestive of partial GH deficiency.[171] Lean body mass declines by up to 50% between 20 and 80 years of age, and total body fat (especially in the visceral fat compartment of the abdomen) increases by 18% in men and by 12% in women.[172-175] Skin thickness and organ volume are also decreased.[176] These findings have prompted clinical trials of GH therapy in elderly people selected on the basis of low IGF-I concentrations. The seminal studies by Rudman et al.[171] and Marcus et al.[177] have shown that short-term (7 day) GH therapy increased nitrogen and phosphorus retention[177] and longer term (6 to 12 months) GH administration increased lean body mass, skin thickness, and muscle and organ (spleen and liver) volume but decreased total body fat content.[171,178] Adverse side effects were noted in 13 of 26 patients (including carpal tunnel syndrome and/or gynecomastia) but disappeared within 3 months after therapy was discontinued. The incidence of these side effects is closely related to IGF-I levels, and can be lessened by keeping plasma IGF-I concentrations between 0.5 and 1.0 U/ml.[179] However, the improvement in lean body mass and skin thickness disappears 3 months after GH therapy was discontinued, thus indicating the need for lifelong GH administration.

GHRH therapy may provide a viable alternative to GH in the elderly, as GHRH injections reverse the age-related decreases in GH and plasma IGF-I while preserving physiological GH release.[180] GHRH is also advantageous in that it does not induce hypoglycemia. However, frequent GHRH injections are necessitated by its short half-life; thus the therapeutic use of GHRH must await the development of novel administration methods such as nasal sprays, implantable pellets, or superactive GHRH analogs (see above).[169]

C. CHILDREN

1. GH Deficiency/Idiopathic Short Stature

GHD in children is also associated with decreased lean muscle mass and nitrogen retention and increased liposity. GH therapy effectively increases lean muscle mass at the expense of adipose tissue and increases nitrogen retention[181-183] and skin-fold thickness.[184,185] A similar beneficial effect on body composition and skin-fold thickness was also observed in GH-treated children with partial GH deficiency.[186] GH treatment in short normal children can decrease fat mass by up to 76% and increase lean muscle mass by 25%.[187] GH therapy in short children can also produce a healthier fat distribution, as adipose tissue is primarily lost from central (abdominal) fat stores and is deposited in peripheral (gluteal) fat stores.[188] GH therapy is also of psychological benefit, as GH-treated patients with GHD, Turner's syndrome, or chronic renal failure are generally satisfied with treatment despite the inconvenience of daily injections.[189]

The role of GH in maintaining glucose homeostasis is important in children, and hypoglycemic episodes occur in young children on less-than-daily GH therapeutic regimens.[190,191] Appropriate GH treatment restores glucose levels to normal in these children and prevents hypoglycemic incidents.[192]

Moreover, the detrimental effects of GH therapy on lipid metabolism and water balance are not as prolonged or as widespread in GH-treated children as in GH-treated adults.[193-195]

2. Turner's Syndrome

Hyperglycemia and noninsulin dependent diabetes are increased in older girls with Turner's syndrome and may be aggravated by GH treatment.[196] Moreover, insulin levels were increased in some[197,198] but not all[199,200] clinical studies using varying doses of GH. Thus, although the beneficial effect of GH therapy on growth in Turner's syndrome is undisputed, the dose must be optimized to produce the minimal disruption in glucose homeostasis.

Turner's patients exhibit deficits in both cognitive and social/emotional functioning, the latter being due in part to short stature.[201,202] The growth-promoting effects of GH therapy may therefore be of some benefit in the psychosocial development of these patients, as improved growth is often correlated with an improved self-concept and/or improved social and emotional functioning.[203,204]

D. OBESITY

The improvement in body composition in GH-treated GHD patients suggests that GH treatment might be a useful adjunct in the treatment of obesity.[205] Indeed, GH administration increases lipolysis during short-term fasting.[206,207] In contrast, in placebo-controlled cross-over studies of obese women on a restricted diet, GH treatment for 3 to 12 weeks did not accelerate the loss of body fat.[208,209] However, GH therapy (especially at higher doses) may be beneficial in the maintenance of lean body mass.[210,211] A minor, 2.3% loss of body fat has been observed in women ingesting a weight maintenance diet,[212] but these limited studies of GH treatment of obesity do not indicate an important role for GH in the treatment of this often refractory condition.

E. WEIGHT TRAINING

The increased availability of GH has also resulted in its use as an anabolic drug by athletes and weight lifters,[213,214] and early studies suggested that GH could further enhance the body composition of highly conditioned athletes.[215] However, in double-blind placebo-controlled studies, GH did not improve body composition or increase muscle size or strength in athletes engaged in a weight training program.[216,217] Thus, the metabolic effects of GH appear to be minimal in individuals with near-optimal body composition and muscle development.

V. ANABOLIC ACTIONS OF GH IN TRAUMA AND DISEASE

Conditions such as malnutrition, surgery, severe infection, or burns induce a hypermetabolic stress response. This response is due in part to the release of "stress hormones" such as catecholamines and corticosteroids and is characterized by protein wasting and catabolism. In the posttrauma period, this catabolic state cannot always be reversed by nutritional therapy alone, thus prompting the search for anabolic therapeutic agents. The availability of GH and its minimal side effects have prompted extensive research into its suitability as an anabolic agent, and results have generally been promising.

A. PULMONARY DISEASE

Patients with chronic pulmonary obstructive disease (CPOD) are often characterized by low body weight despite a normal caloric intake, and the decreased muscle mass further exacerbates their pulmonary dysfunction. Studies in CPOD patients on a normocaloric diet have demonstrated that GH therapy increases weight gain and nitrogen retention and consequently improves lung functioning.[218] However, GH administration to patients with pulmonary failure receiving parenteral nutrition did not improve pulmonary function despite a similar increase in nitrogen retention.[219]

B. BURNS

Patients with thermal injuries can benefit twofold from GH therapy, as GH enhances burn healing (described below) and attenuates the trauma-induced protein catabolism.[220] rhGH administration to these patients reduces the loss of nitrogen and body weight[220] and stimulates whole-body and leg protein synthesis.[221-223] Surprisingly, rhGH treatment exacerbates the trauma-induced rise in catecholamine and glucagon levels, which contribute to the catabolic state.[224] Evidently, the effects of these hormones are overcome by direct anabolic effects of GH and the GH-induced rise in insulin.

C. MALNUTRITION

The possible use of GH as an adjunct to therapy of clinical malnutrition in the elderly or following surgery or infection has been the subject of a large number of studies, and initial results are promising.[225] For instance, 4 to 8 days of GH administration to elderly individuals with recent weight loss of different etiologies increased nitrogen and phosphorus retention and serum IGF-I levels.[219,226] A longer 3-week treatment period resulted in a sustained improvement in these parameters of anabolism and a significant increase in body weight.[227] IGF-I alone was relatively inactive in patients with severe catabolism, but when combined with GH, nitrogen retention was even greater than with GH alone.[225] The synergistic effect of IGF-I may be due to a decreased ability of the malnourished liver to secrete adequate IGF-I in response to GH. Coadministration of insulin with GH may also improve whole-body and skeletal muscle protein balance.[228]

D. INFECTION

Chronic infection is often associated with progressive weight loss due to a loss of body cell mass. The wasting observed in patients infected with the human immunodeficiency virus (HIV) is particularly debilitating, and can be as lethal as the direct effects of the infection itself.[229] GH therapy may be beneficial in this condition, as GH administration (in combination with a controlled metabolic diet) significantly increased body weight, nitrogen retention, and lipid oxidation and decreased protein oxidation to a similar extent as in normal controls.[230,231] However, GH therapy in wasted individuals with chronic sepsis produced inconsistent results.[232]

E. SURGERY

Major surgery is another condition associated with a marked increase in catabolism, particularly when full oral nutrition cannot be resumed. A 20% loss in body protein following surgery is common, and postoperative feeding generally results in increased deposition of fat and body water but minimal protein accretion.[233,234] Beneficial effects of GH treatment on protein metabolism have been noted in patients receiving subcaloric parenteral nutrition after major gastrointestinal surgery and other postoperative conditions.[235-238] These patients appear to be highly sensitive to the anabolic effects of GH, as GH therapy increased nitrogen, potassium, and phosphorus retention in postsurgical but not normal individuals.[239] GH therapy also enables the optimal use of nutrients in patients receiving hypercaloric nutrient therapy, as excess nutrients are generally stored as fat in patients not receiving GH but as protein in patients receiving GH.[238]

F. CORTICOSTEROIDS

Corticosteroids are highly catabolic, and thus their administration in autoimmune diseases or organ transplantation is associated with muscle wasting.[240] Moreover, they are often used in experimental studies to mimick the hypermetabolic states observed in catabolic illness.[241] For instance, administration of the corticosteroid, prednisone, to normal volunteers dramatically increases protein proteolysis and results in a negative nitrogen balance.[241] GH therapy partially nullifies these catabolic effects by stimulating whole-body protein synthesis, but does not affect proteolysis.[241] A similar anabolic effect of GH was observed in patients with lung disease receiving therapeutic corticosteroids.[219] However, the study in normal volunteers also revealed the negative aspects of GH therapy, as insulin resistance and hyperglycemia were observed in some patients.[242,243] Thus, on the basis of experimental and clinical observations, GH therapy is an effective anabolic adjuvant to nutritional therapy in catabolic states; however, carbohydrate metabolism in these patients must be closely monitored.

G. IGF-I THERAPY

The hypermetabolic state involves some degree of GH resistance, as GH levels are generally increased but serum IGF-I levels are decreased in catabolic patients.[244] Although the studies cited above demonstrate that pharmacological doses of GH can partially overcome this resistance, high levels of GH can also provoke hyperglycemic reactions. IGF-I therapy may thus provide a beneficial addition to the treatment regimen, as IGF-I infusion was as effective as GH at limiting nitrogen loss in fasting individuals.[245] Indeed, Clemmons and Underwood[246] observed that a combination of GH and IGF-I provided the optimal increase in nitrogen retention, and normal serum glucose levels were maintained owing to the opposing hyperglycemic actions of GH and the hypoglycemic actions of IGF-I.

VI. OSTEOPOROSIS

The dependence of linear bone growth in children on GH is well accepted, and GH is commonly used in GH-deficient children to correct deficient growth. However, studies have suggested that GH may be necessary for bone function in postpubertal patients. The actions of GH on bone have been reviewed by Bouillon.[247] In brief, GH stimulates linear bone growth by increased proliferation of chondrocytes in the growth plate.[248] In growing bones and bones that have completed linear growth, GH also increases osteoblast proliferation and activity, as indicated by increased concentrations of osteocalcin and alkaline phosphatase in GH-treated osteoblast cultures.[248] The resulting stimulation of bone formation is shown by the increased bone strength and bone mineral content in GH-treated animals (reviewed in Ref. 247). GH also enhances bone function by increasing intestinal uptake of calcium,[249] which may be mediated via increased activation of vitamin D_3.[250,251]

A. GH DEFICIENCY/RESISTANCE

The importance of growth hormone in proper bone formation is emphaiszed by the low bone mineral content (BMC), total bone mass, and bone density in adult GH-deficient and hypopituitary patients.[128,252-254] These changes in bone function are not only due to the defective bone growth before puberty, as osteoporotic dysfunction is observed in both childhood- and adult-onset GH deficiency.[253] Moreover, GH resistance in female Laron-type dwarfs[255] and Turner's syndrome[256] is also associated with defective bone formation, as these women are characterized by low bone density resulting in an increased incidence of osteoporotic fractures.[255,257,258]

Several studies suggest that GH therapy may provide an effective treatment for bone dysfunction in GH-deficient patients. Daily GH injections (0.125 IU/kg/week for 4 weeks, 0.25 IU/kg/week for 1 year) significantly increased proximal and distal forearm BMC in a small group ($n = 12$) of GH-deficient men and women.[259] A similar effect of GH therapy was also observed on spinal and/or forearm BMC in other studies of GH-deficient adults[128,150,260,262] and children.[186] Indeed, collagen synthesis (the major component of bone and connective tissue) is stimulated after only 14 days of GH treatment[263] and can be used as an indicator of the efficiency of GH therapy.[1,264] Short-term GH administration (less than 3 months) can cause an apparent decrease in bone mineral content that is a result of increased bone remodeling (see below), but by 1 year the BMC is considerably higher than pretreatment levels.[128] GH also increases intestinal uptake of calcium and phosphate via stimulation of vitamin D_3 activation.[134,137,265-268] GH therapy does not, however, improve the stimulatory effects of vitamin D_3 on osteocalcin levels in GH-deficient children.[265]

B. SENILE OSTEOPOROSIS

The decreased BMC and increased fractures in the elderly are thought to be due in part to the age-related decrease in plasma GH. Thus, it was originally thought that GH therapy might efficiently reverse both changes. Accordingly, GH therapy for 6 months in 21 elderly men with lower IGF-I levels increased spinal BMD but not forearm BMC by 1.6%.[178] However, an extended study of 45 men for 21 months did not detect any change in lumbar, forearm, or hip bone density,[178] and thus widespread use of this therapy must await further study. In some respects this decline in GH secretion may actually be beneficial, as the incidence of radiographic osteoarthritis was reduced in elderly GH-deficient patients.[269]

C. POSTMENOPAUSAL OSTEOPOROSIS

Initial studies suggest that age-related decreases in GH secretion may contribute to the development of osteoporosis, as peak GH levels were lower in osteoporotic than in osteoarthritic elderly patients.[247,269] Unfortunately, GH therapy has not been successful in reversing the bone loss associated with osteoporosis. Indeed, daily GH therapy alone or in conjunction with calcitonin for 12 months decreased bone mineral content of the distal radius and increased osteoclast markers.[270] However, alternation of GH with calcitonin (which inhibits bone resorption) was modestly successful at increasing total body calcium, but still resulted in a decrease in radial bone density.[271] Despite the apparent unsuitability of GH as a treatment for osteoporosis, it may be useful as a treatment for the "low-turnover" form of osteoporosis.[5] Alternatively, GH could be coadministered with inhibitors of bone resorption, such as diphosphonates.

D. FRACTURES

GH-induced improvements in bone density may extend to an increased speed of fracture healing. On the basis of biomechanical measurements of bone strength, GH therapy has been shown to accelerate the development of a calcified callus following a fracture when administered early in the healing process.[272-275] An acceleration of bone union was also observed in GH-treated humans in comparison to controls; however, the size of the study was small, and the statistical significance of this difference was not determined.[276] However, the conclusions of a study by Mosekilde and Bak[277] suggest that the improvement observed in earlier studies may be somewhat deceptive. They also observed accelerated formation of the bone callus in GH-treated rats, but the callus was histologically abnormal and resistant to bone remodeling. This effect may be partially due to stimulatory effects of GH on the immune system, as the callus was invaded with marrow cells. As a result, GH treatment actually delayed the recovery of the bone to its prefracture structure in comparison to untreated rats, as the callus was still present 80 days postfracture.

E. MECHANISM OF GH ACTION

GH-induced bone formation may be partially mediated by increased osteoblast activity, as 4-month GH therapy in GH-deficient patients induced a fivefold increase in plasma osteocalcin levels.[134,254,262,278,279] However, hydroxyproline, calcium, and pyridinoline/deoxypyridinoline excretion was also elevated to an equal or smaller extent (threefold), indicating that GH also stimulates osteoclast activity and bone resorption.[254,262,278,279] A similar increase in bone turnover is observed following short-term GH treatment of normal and aged volunteers.[266,280] The study of Franchimont[281] observed a more dramatic uncoupling of osteoblast and osteoclast activity, as plasma osteocalcin levels were elevated in postmenopausal women treated with GH for 1 week despite normal quantities of osteoclastic urinary markers. The difference in magnitude between osteoblast and osteoclast stimulation suggests that GH should induce a positive bone balance with every turnover cycle, which is opposite to the negative bone balance associated with increased turnover in other syndromes.[247] This conclusion is supported by the increased bone turnover and bone mass in acromegaly.[247,282] Inadequate doses of GH may account for the lack of effect on bone mass in the clinical studies described above. Bone turnover may also prevent excessive aging of osteocytes and thus contribute to the maintenance of healthy bone tissue.[283]

VII. WOUND HEALING

The anabolic effects of GH suggest that it may be of benefit in wound healing. Accordingly, increased GH induced by local or systemic GH injection or by exogenous GHRH stimulates collagen formation, increases wound strength, and accelerates wound healing.[284-291] A similar improvement in recovery from burns and chronic leg ulcers is also caused by topical or systemic GH.[292,293] This anabolic effect may be mediated by IGF-I, as IGF-I mRNA was increased in the wounds of GH-treated rats, and IGF-I can also promote wound healing.[294,295] GH can also reverse the inhibitory effects of cortisone[290] or malnutrition[296] on healing, suggesting that GH administration may be advantageous in wounded catabolic states. However, GH may be effective only in mild injury, as the coexistence of a second injury (burns) abolished the effectiveness of GH.[287]

VIII. THYROID DYSFUNCTION

Growth hormone is a crucial factor in the thyroid axis, as it stimulates conversion of thyroxine (T_4) into triiodothyronine (T_3), which is the more active form. The increased incidence of hypothyroidism in hypopituitary patients is therefore due in part to GH deficiency, as it can be corrected by GH therapy.[166,167] Indeed, GH therapy (6 months) in adult-onset or childhood-onset hypopituitary patients and normal patients increases total and free T_3 concentrations and transiently reduces T_4.[134,137,162,297] In GHD adults, the improvement in thyroid status persists after more than 16 months and is accompanied by an increase in total energy expenditure and basal metabolic rate.[134,298,300] In contrast, GH treatment of Turner's patients results in a transient depression in T_4 levels and no change in T_3 levels.[301]

IX. CARDIOVASCULAR DYSFUNCTION

The 100% increase in mortality rate in the GH-deficient population is largely due to cardiovascular dysfunction, including myocardial infarction, ischemic heart disease with congestive cardiac failure, and cerebrovascular disease.[302,303] These pathologies are due to the multiple GH effects, including muscle anabolism and lipid metabolism. For example, a reduced ventricular mass and ejection fraction was observed in patients that were GH deficient since childhood but did not receive GH replacement therapy as adults.[304] However, in the study by Johnston et al.,[305] these values were in the low-normal range for age- and sex-matched controls, and only the isovolumetric relaxation time (IRT) and the early atrial filling velocity ratio were abnormal in a proportion of their GH-deficient population. The reduced IRT is often indicative of early myocardial disease, thus structural alterations in the heart muscle cause a portion of the vascular problems in these patients. GH therapy may be effective in reducing myocardial dysfunction in GH deficiency, as GH administration for 6 months significantly improved exercise perfomance and several indices of cardiac performance, including LV diastolic dimension, LV mass, cardiac output, myocardial contraction, and stroke volume, in GH-deficient and/or normal patients.[261,306,307]

Growth hormone deficiency is also associated with an increased prevalence of arterial disease. A preliminary study by Markussis et al.[308] has observed an abnormally high incidence of asymptomatic carotid artery disease and of atheromatous plaques in the carotid and femoral arteries of GHD patients. Central obesity and subsequent insulin resistance, increased LDL and VLDL, decreased HDL, and glucose intolerance are all metabolic risk factors that are present in hypopituitary and GH-deficient patients[130,140,161,309] (see above). Body composition and lipid metabolism are favorably affected by GH therapy (see above); however, these alterations have not yet been linked to a decrease in arterial disease.[305] Curiously, the level of apolipoprotein A (a cardiac risk factor) in GHD and normal adults is increased by GH therapy.[143] Other risk factors in arterial disease such as increased levels of fibrinogen and von Willebrand factor are not improved by long-term treatment of GH-deficient adults.[299]

Hypertension is also commonly observed in GH-deficient patients. This observation is somewhat suprising, as the decrease in extracellular water volume and cardiac output also associated with GH deficiency[129,310] should provide some degree of hypotension and GH treatment is often associated with transient increases in blood pressure.[171] The abnormal cardiac compliance and distensibility identified in a group of hypopituitary patients[308] have been implicated,[140] but this is only a preliminary conclusion.

Curiously, acromegaly is also associated with a high incidence of cardiac disease, including hypertension, artery disease, congestive heart failure, and cardiac arrythmias.[282,310a] Some of these syndromes are due to the ventricular hypertrophy caused by the hypersecretion of GH, which can be reduced by normalization of GH levels with octreotide acetate, a somatostatin analog.[311] However, acromegalics also suffer from increased arterial disease, which may also be due to insulin resistance, noninsulin-dependent diabetes mellitus (NIDDM) and hypertension[312] and/or direct effects of GH on atherogenesis.[313-316]

X. THERMOREGULATION

GH therapy may also beneficially affect thermoregulation, as GHD patients have abnormally low sweating rates that are corrected by GH therapy.[317] Moreover, the increased core temperature and decreased sweating observed in GHD patients subjected to moderate heat stress (in comparison to normal controls) are also indicative of impaired thermoregulation.[318] GH may affect sweating directly at the skin, as GH receptors have been identified in sweat glands.[319]

XI. IMMUNE DYSFUNCTION

The immune system has been recognized as a bona fide GH target site (Chapter 24). Animal models of GH deficiency[320] have a reduction in immune function that is restored by exogenous GH administration. For example, the thymus atrophy and immature immune system observed in GH-deficient Snell-Bragg dwarf mice are partially reversed by

GH administration, which also significantly improves life expectancy.[321] The decrease in thymulin secretion (a marker of thymic activity) observed in aged rats, mice, and dogs can also be partially corrected by GH administration.[322,323]

However, studies with GH-deficient humans are less conclusive, as their immune function is generally reported to be normal[324-328] (reviewed by Wit et al.[329]). However, some indicators of decreased immune function have been observed in other studies, including decreased concentrations of interleukin 2 (IL-2)[330] and thymulin,[331] reduced natural killer (NK) cell activity,[332-334] and high numbers of B and T cells.[335] In children receiving GH therapy, reduced[327,328,332] or unchanged[324,336,337] B cell numbers, increased[334] or unchanged[333] NK cell numbers and activity, and increased IL-1 and IL-2 production have been detected. Although clinical reports on GH effects on T cell numbers appear inconsistent,[325,337,338] intramuscular,[325,338] but not subcutaneous,[327,328,336,337] GH administration decreased the ratio of helper T and suppressor T cells.

XII. RENAL DYSFUNCTION

There are indications that GH and/or IGF-I administration will have therapeutic applications in renal disease, both to increase renal function and to reverse the inhibition of growth in chronic renal failure (discussed above).

A. GLOMERULAR FUNCTION

It has long been known that glomerular filtration, renal blood flow, and renal size are decreased in GH-deficient states[133,339-343] and increased in acromegaly.[342-344] These changes in acromegaly or GHD are evident by the decreased or increased (respectively) levels or urinary albumin and creatine.[130,345] Indeed, a single GH injection in normal volunteers and GH-deficient patients induced a rise in renal plasma flow and glomerular filtration rate (GFR) after several hours[339,346,347] and GH induces renal and glomerular hypertrophy in rats.[348] The increased GFR is likely due to the hypertrophic effects of GH/IGF-I and the IGF-I mediated reduction in afferent and efferent arteriolar resistance.[339,347,349,350] IGF-I alone can induce a similar increase in GFR in GH-deficient rats and normal humans with a shorter lag time than GH,[349,351] and thus IGF-I therapy may be the optimal therapy to increase the GFR in patients with chronic renal failure. IGF-I administration also accelerates recovery from experimental ischemic tubular necrosis in the rat.[352] Initial studies show an initial favorable increase in GFR in GH-treated patients with chronic renal failure;[353] however, it remains to be determined if this improvement can be sustained. Moreover, in contrast to the rat,[354] IGF-II (not IGF-I) is the primary IGF in the human kidney.[355] The possible use of GH in improving the function of transplanted kidneys is presently under investigation.[356]

However, the use of GH as a therapeutic agent in chronic renal failure must be carefully considered, as pharmacological amounts GH may accelerate end-stage renal disease. GH administration to rats with puromyuclin aminonucleoside-induced nephropathy worsened the detrimental effects of this toxin on renal structure and function.[357] GH therapy in conjunction with calcitriol must also be carefully monitored, as calcitriol abolished the beneficial effect of GH on growth in uremic rats and greatly increased calcium excretion.[358]

B. FLUID VOLUME

Altered body water distribution is associated with GHD, as extracellular water volumes in GHD patients are reduced by up to 15%.[129] GH administration to normal or GHD individuals increases retention of sodium, potassium, and other electrolytes,[359,360] and increases extracellular water, primarily in the interstitial compartment.[134,360] This increase in interstitial water is also mediated by increased hydrophilic proteoglycans in the interstitial space. Body weight is transiently increased and urine volume is decreased as a result of the increased water retention.[359] However, this effect may be somewhat transient, as water volumes in some GH patients were increased significantly after 6 weeks of GH treatment but declined slightly between 6 and 26 weeks.[156] The antinatriuretic effects of GH are thought to be related in part by atrial natriutretic peptide (ANP), as ANP levels are decreased and extracellular water volume is increased in GH-treated normal men.[361] GH may also increase sodium retention directly at the sodium pumps in the distal tubule[359,362] and by activation of the renin-angiotensin-aldosterone system.[307,359]

The stimulatory effect of GH on fluid volume results in minor side effects in most GH-treated patients, resulting only in slight edema and minor increases in blood pressure.[359] Water balance is disrupted to a lesser extent in children, in which aldosterone secretion and the renin-angiotensin system are unaltered by GH therapy.[56] However, owing to this minor hypervolemic effect of GH, pharmacological doses must be administered with care to patients with Turner's syndrome and cardiopulmonary disease.

XIII. REPRODUCTIVE DYSFUNCTION

GHRs are present throughout the male and female reproductive systems[363] and GH is a permissive factor in a multitude of gonadotropin actions (see Chapter 23), suggesting the possible use of GH in male and female infertility.

A. MALE INFERTILITY

Although few clinical trials have investigated the possibility of GH as an adjunct to male infertility therapy, GH may play a secondary but important role in male sexual development, steroidogenesis, and perhaps spermatogenesis.

GHD is often associated with phallic underdevelopment and delayed puberty, and GH therapy can accelerate the initiation of puberty and correct hypogonadism.[364,365] The effect of GH on sexual development may be mediated by gonadal steroids. GH deficiency in men and animals is associated with a poor testicular response to human chorionic gonadotropin (hCG) and GH enhances the effect of hCG on testicular testosterone synthesis.[366-369] In contrast, spermatogenesis in men with isolated GH deficiency is relatively normal, and no direct effects of GH on spermatogenesis have been demonstrated.[366] However, GH enhanced the spermatogenic effect of luteinizing hormone/follicle-stimulating hormone (LH/FSH) in hypogonadotropic patients[366] and thus may be a permissive factor.

B. FEMALE INFERTILITY

The GH-IGF-I axis is thought to be involved in ovarian function, as GHRs and IGF-I are found in the ovarian follicle. Ovarian IGF-I synthesis is responsive to both GH and FSH, and IGF-I is thought to synergize with FSH to stimulate granulosa cell proliferation and steroidogenesis. Thus, GH may enhance the actions of gonatotropins on IGF-I formation and consequently stimulate follicular development and steroidogenesis.[370] Neither GH nor IGF-I is absolutely essential for follicular function, as Laron dwarfs are fertile and can menstruate normally despite very low follicular fluid IGF-I levels and the lack of ovarian GHRs. However, delayed puberty is common in GH-deficient and GH-resistant females and GH treatment is often followed closely (within 5 months) by sexual maturation, suggesting a synergistic role for GH.[371,372]

The possible use of GH as an adjunct to human menopausal gonadotropin (hMG) in reproductive therapies such as ovulation stimulation and *in vitro* fertilization is presently the focus of extensive research (for reviews see Shoham et al.[373] and Hillensjo and Bergh[374]). Clinical trials have demonstrated that GH may be a useful addition to hMG therapy in certain populations of infertile women. For instance, GH administration to anovulatory, hypogonadotropic women significantly reduced the dose of hMG and the duration of treatment required.[375-383] Importantly, a number of these anovulatory women were previously resistant to gonadotropin therapy. Normogonadotropic women with polycystic ovary syndrome showed a similar GH-induced improvement in gonadotropin response.[384,385] However, as female infertility is not a homogeneous disorder, not all women respond to GH therapy. Blumenfeld et al.[386,387] have observed that beneficial effects of GH on the pregnancy rate were generally confined to "clonidine-negative" anovulatory women, in which secretogogue-induced GH secretion was impaired. Thus, the infertility in these women is due in part to GH deficiency, whereas the infertility in nonresponding women may be due to an unrelated cause that is unaffected by GH treatment.

In these trials GH is usually administered in very high doses, six times the dose used in GH replacement therapy. However, very different doses of GH (24, 48, and 120 IU) are equally effective at enhancing gonadotropin sensitivity,[380] and thus the dose required may be considerably lower than previously thought. Moreover, a single GH injection at the beginning of one cycle effectively enhances hMG responsiveness for the four subsequent cycles.[376,380,381]

GH therapy may also be an effective adjunct in the induction of hyperovulation required for *in vitro* fertilization. Coadministration of GH with hMG to anovulatory or some normally cycling women can increase the number of oocytes collected and improve the fertilization and conception rate as compared with hMG treatment alone.[375,383,388,389] GH therapy is more effective in younger than in older women[389] and often requires previous pituitary suppression.[385,390] These caveats may account for the ineffectiveness of GH therapy in other studies of normally cycling, infertile women with an adequate number of follicles.[386,391-393]

XIV. GH AND LACTATION

The ability of GH to increase milk production in dairy cows (reviewed by Daughaday and Barbano[394]) has fostered research into the possible use of GH to correct lactational failure in women. The ability of hGH to stimulate lactation in primates has been demonstrated in rhesus monkeys, as the nursing infants of hGH-treated mothers show improved weight gain.[395] Milk production is also increased by rhGH treatment in normally lactating women, and the nutritional composition of the milk is unchanged.[396] The galactopoietic effect of GH may be mediated by circulating IGF-I and IGFBPs and increases in mammary blood flow.[397] However, it is unlikely that this use of hGH will be of much clinical importance, as the stimulatory effect of GH is minimal, and artificial feeding of infants provides a much simpler and more economic alternative than GH treatment of lactational insufficiency.

XV. THERAPEUTIC DISADVANTAGES

Although some concern has been raised as to the safety of GH, the European Society for Paediatric Endocrinology concluded that "human recombinant GH seems to be a remarkably safe drug when used in conventional substitution doses."[398] Some ethical and medical considerations that must be considered in the use of GH therapy are described below.

A. CREUTZFELDT-JAKOB DISEASE

The administration of pituitary-extracted GH has been implicated in over 50 cases of Creutzfeldt-Jakob disease (a rare spongiform encephalopathy) worldwide, with a curiously high prevalence in French patients (1.5/1000 in the United States, 4.7/1000 in the United Kingdom, and 10.9/1000 in France).[399,401] This disease is carried by a prion that does not contain nucleic acids, and this prion has been found in pituitary extracts used for the preparation of native GH. Although the possibility of future GHD patients developing this disease is minimal due to the widespread use of recombinantly derived hGH, the protocol developed by Pocchiari et al.[402] (ultrafiltration and 6 *M* urea treatment), which effectively decontaminates pituitary extracts from the scrapie virus, may still be used in some countries or for other pituitary hormones.

B. CANCER

The mitogenic effect of GH has led to the concern that GH treatment might be carcinogenic. This possibility is supported by the higher incidence of cancers (especially colonic neoplasias) in acromegalics[403-406] and the ability of supraphysiological doses of GH to induce tumors in experimental animals.[407,408] Tumor development in response to chemical carcinogens is significantly decreased in hypophysectomized rats, and this trend is reversed by injections of "purified" pituitary extract.[409] Moreover, the pulsatile nature of GH secretion in male rats is thought to be responsible for the increased male sensitivity to a chemical carcinogen, as hepatic tumors in males could be reduced to female levels by continuous GH administration in the female pattern.[410] *In vitro*, GH stimulates the proliferation and transformation of normal and leukemic human lymphocytes and can enhance the expression of protooncogenes c-*fos,* c-*myc,* and c-*jun* (see Chapters 17, 18, and 24). In addition, GH may act at lactogenic receptors to promote the growth of breast and prostatic tumors,[408,411-413] and octreotide (a somatostatin [SRIF] analog that inhibits GH secretion) is currently being investigated as a possible cancer treatment (reviewed in Ref. 414). Superficially, Japanese epidemiological data appear to support the premise of GH as a carcinogen, as the prevalence of leukemia is slightly higher in Japanese GH-treated children than in the total population. However, this increase is not observed when children with other risk factors for leukemia (such as previous leukemia) are excluded,[120,398,400,413] and GH therapy did not increase the incidence of tumor recurrence in children rendered GH deficient by irradiation for craniopharyngioma, brain tumor, or leukemia.[415,416] GH therapy has resulted in increased growth of pigmented nevi in the skin; however, there has been no documented case of skin cancer resulting from GH.[417] However, more GH-induced cancers may be observed as treatment with higher doses of GH becomes more prevalent, as GH increased the promotion of bladder carcinogenesis in rats.[407]

C. ANTIBODIES

Low concentrations of antibodies directed against GH are observed in a significant proportion of patients treated with synthetic or pituitary GH. Although these concentrations rarely reach a titer that interferes with the response to GH,[418,419] GH antibodies occasionally impair the growth reponse in patients totally lacking the *GH* gene.[420,421] Authentic recombinant human GH is the least immunogenic, whereas methionine-linked rhGH and extracted GH induce significantly greater immune responses.[32,399,418,422,423] Antibody titers seldom reach levels that affect response to GH.

D. SKELETAL EFFECTS

GH treatment appears to advance the onset of puberty, and may limit final height attainment by reducing the duration of the pubertal growth spurt.[424-426] To prolong the growth period, the concurrent administration of GnRH analogs has been proposed. GH therapy may also increase the incidence of slipped capital femoral epiphysis by weakening the epiphyseal plate.[427]

E. METABOLIC EFFECTS

High-dose GH therapy can also result in edema, atrial fibrillation, carpal tunnel syndrome, arthralgia, muscular aches, and hypertension.[134,155] Although the prevalence of these side effects of GH-induced anabolism and fluid retention is currently low, it may increase as chronic, high-dose treatment regimens become more widespread.

F. ETHICAL CONCERNS

The utility and necessity of GH treatment in GHD children are well accepted, as GH administration is restricted to a limited time period, involves restoration of a physiological state, and is of unquestionable benefit to the child. However, the known therapeutic benefits of GH in GH-deficient adults are partially offset by the high cost of lifelong GH treatment. Growth hormone treatment in the elderly poses a similar cost-benefit dilemma. It is, however, the use of GH therapy in idiopathic short stature that has raised the most furor, as "being short is not a medically recognized disease or affliction".[428] Importantly, it is very difficult to determine the cutoff point between treated and untreated short children, and any arbitrary cutoff will not be acceptable to many parents. Moreover, the cost of treatment of short children is in excess of $10,000 per year in the United States, which is beyond the means of most parents and not necessarily approved by third-party payers. At present, prolonged GH treatment should be provided only to children with the most severe growth retardation (SDS score less than –3).

The ethics of the study protocols themselves have also been challenged by lobby groups, who have focused their attention on long-term (up to 10 years) trials in idiopathic short stature in Turner's syndrome.[428] These studies require frequent examinations, injections, and X-rays during the crucial growing years, thus children in the control groups are subjected to the inconveniences of therapy but are irrevocably growth impaired. Thus, it is clear that the rapid advances in biotechnology have resulted in a plethora of novel therapeutic uses of GH; however, such advances must now be reconciled with medical ethics.

REFERENCES

1. **Christiansen, J. S., Jorgensen, J. O., Pedersen, S. A., et al.,** *Horm. Res.*, 36, 66, 1991.
2. **Vance, M. L.,** *Trends Endocrinol. Metab.*, 3, 46, 1992.
3. **Cuneo, R. C., Salomon, F., Mcgauley, G. A., et al.,** *Clin. Endocrinol.*, 37, 387, 1992.
4. **Baumann, G. and Silverman, B. L.,** *Growth Regul.*, 1, 43, 1991.
5. **Lippe, B. M. and Nakamoto, J. M.,** *Recent Prog. Horm. Res.*, 48, 179, 1993.
6. **Hindmarsh, P. C., Bridges, N. A., and Brook, C. G. D.,** *Clin. Endocrinol.*, 34, 417, 1991.
7. **Jorgensen, J. O. L.,** *Endocr.Res.*, 12, 189, 1991.
8. **Fryklund, L. M., Bierich, J. R., and Ranke, M. B.,** *J. Clin. Endocrinol. Metab.*, 15, 511, 1986.
9. **Lamberts, S. W. J., Valk, N. K. and Binnerts, A.,** *Clin. Endocrinol.*, 37, 111, 1992.
10. **Beshyah, S. A., Anyaoku, V., Niththyananthan, R., et al.,** *Clin. Endocrinol.*, 35, 409, 1991.
11. **Jorgensen, J. O. L., Moller, N., Lauritzen, T., et al.,** *Horm. Res.*, 3(Suppl. 4), 77, 1990.
12. **Zadik, Z., Lieberman, E., Altman, Y., et al.,** *Growth Regul.*, 39, 188, 1993.
13. **Chanoine, J., Vanderschueren-Lodeweyckx, M., Meas, M., et al.,** *J. Clin. Endocrinol. Metab.*, 73, 1269, 1991.
14. **Carlsson, L. and Jansson, J. O.,** *Endocrinology (Baltimore)*, 126, 6, 1990.
15. **Muller, E. E.,** *Horm. Res.*, 33(Suppl. 2), 90, 1990.
16. **Vance, M. L., Kaiser, D. L., Martha, P. M., Jr., et al.,** *J. Clin. Endocrinol. Metab.*, 68, 22, 1989.
17. **Barkan, A. L., Stred, S. E., Reno, K., et al.,** *J. Clin. Endocrinol. Metab.*, 69, 1225, 1989.
18. **Ross, R. J. M., Grossman, A., Besser, G. M., et al.,** *Acta Endocrinol. (Copenh.)*, 113(Suppl. 279), 123, 1986.
19. **Thorner, M. O., Reschke, J., Chitwood, J., et al.,** *N. Engl. J. Med.*, 312, 4, 1985.
20. **Thorner, M. O., Vance, M. L., Evans, W. S., et al.,** *Horm. Res.*, 24, 91, 1986.
21. **Bozzola, M., Biscaldi, I., Cisternino, M., et al.,** *J. Endocrinol. Invest.*, 13, 235, 1990.

22. **Su, C. M., Jensen, L. R., Egan, T. J., et al.,** *72nd Ann. Meet. Endocr. Soc.*, Atlanta, GA, 1990, p. 54.
23. **Deslauriers, N., Reeves, I., Gaudreau, P., et al.,** *72nd Ann. Meet. Endocr. Soc.*, Atlanta, GA, 1990, p. 54.
24. **Evans, W. S., Borges, J. L. C., Kaiser, D. L., et al.,** *J. Clin. Endocrinol. Metab.*, 57, 1081, 1983.
25. **Evans, W. S., Vance, M. L., Kaiser, D. L., et al.,** *J. Clin. Endocrinol. Metab.*, 61, 846, 1985.
26. **Khalfallah, Y., Vanailles, B., Cohen, R., et al.,** *J. Clin. Endocrinol. Metab.*, 71, 512, 1990.
27. **Bowers, C. Y. and Frentz, J. M.,** *J. Clin. Endocrinol. Metab.*, 74, 292, 1992.
28. **Cheng, K., Chan, W. W. S., Butler, B., et al.,** *Horm. Res.*, 40, 109, 1993.
29. **Froesch, E. R., Guler, H. P., Schid, C., et al.,** *Trends Endocrinol. Metab.*, 1, 254, 1990.
30. **Guler, H. P., Eckardt, K. U., Zapf, J., et al.,** *Acta Endocrinol. (Copenh.)*, 121, 101, 1989.
31. **Takano, K., Hizuka, N., Asakawa, K., et al.,** *Endocrinol. Jpn.*, 37, 309, 1990.
32. **Kaplan, S. L., Underwood, L. E., August, G. P., et al.,** *Lancet*, i, 697, 1986.
33. **Stubbe, P., Frasier, S. D., Stahnke, N., et al.,** *Horm. Res.*, 37, 28, 1992.
34. **Libber, S. M., Plotnick, L. P., Johanson, A. J., et al.,** *Medicine*, 69, 46, 1990.
35. **Laron, Z. and Galatzer, A.,** *Early Hum. Dev.*, 5, 211, 1981.
36. **Stanhope, R., Albanese, A., Hindmarsh, P., et al.,** *Horm. Res.*, 38, 9, 1992.
37. **Stanhope, R., Uruena, M., Hindmarsh, P., et al.,** *Acta Pediatr. Scand.*, 371, 47, 1991.
38. **Frasier, S. D., Rudlin, C. R., Zeisel, H. J., et al.,** *Am. J. Dis. Child.*, 146, 582, 1992.
39. **Duck, S. C., Schwarz, H. P., Costin, G., et al.,** *J. Clin. Endocrinol. Metab.*, 75, 1115, 1992.
40. **Hernandez, M., Nieto, J. A., Sobradillo, B., et al.,** *Horm. Res.*, 35, 13, 1991.
41. **Bercu, B. B., Shulman, D., Root, A. W., et al.,** *J. Clin. Endocrinol. Metab.*, 63, 709, 1986.
42. **Zadik, Z., Chalew, S. A., Raiti, S., et al.,** *Pediatrics*, 63, 709, 1085.
43. **Zadik, Z., Chalew, S. A., and Kowarski, A.,** *J. Clin. Endocrinol. Metab.*, 74, 801, 1992.
44. **Rose, S. R., Ross, J. L., Uriarte, M., et al.,** *N. Engl. J. Med.*, 319, 201, 1988.
45. **Rudman, D., Kutner, H., Blackston, R. D., et al.,** *N. Engl. J. Med.*, 305, 123, 1981.
46. **Blum, W. J., Ranke, M. B., Kietzmann, D., et al.,** *J. Clin. Endocrinol. Metab.*, 70, 1292, 1990.
47. **Cara, J. F. and Johnson, A. J.,** *Acta Pediatr. Scand.*, 109, 1990.
48. **Hopwood, N. J., Hintz, R. L., Gertner, J. M., et al.,** *J. Pediatr.*, 123, 215, 1993.
49. **Moore, W. V., Moore, K. C., Gifford, R., et al.,** *J. Pediatr.*, 120, 702, 1992.
50. **Hintz, R., Hopwood, N., and Genentech Study Group,** *Horm. Res.*, 35(Suppl. 2), 17, 1991.
51. **Wit, M. J., Kuilboer, M. M., De Muinch Keizer-Schrama, S. M. P. F., et al.,** *Horm. Res.*, 35, 31A, 1991.
52. **Albertsson-Wikland, K. and Karlberg, J.,** *Horm. Res.*, 35(Suppl. 2), 32, 1991.
53. **Lesage, C., Walker, J., Landler, F., et al.,** *J. Pediatr.*, 119, 29, 1991.
54. **Gertner, J. M., Tamborlane, W. M., Gianfredi, S. P., et al.,** *J. Pediatr.*, 110, 425, 1986.
55. **Zantleifer, D., Awadalla, S., and Brauner, R.,** *Horm. Res.*, 40, 123, 1993.
56. **Barton, J. S., Hindmarsh, P. C., Preece, M. A., et al.,** *Clin. Endocrinol.*, 38, 245, 1993.
57. **Fontoura, M., Mugnier, E., Brauner, R., et al.,** *Clin. Endocrinol. (Oxford)*, 37, 249, 1992.
58. **Hindemarsh, P. C. and Brook, C. G. D.,** *Br. Med. J.*, 295, 573, 1987.
59. **Whitehead, E. M., Shalet, S. M., Davies, D., et al.,** *Clin. Endocrinol. (Oxford)*, 17, 271, 1982.
60. **Wanke, R., Wolf, E., Hermanns, W., et al.,** *Horm. Res.*, 37, 74, 1992.
61. **Kaplan, S. A.,** *Clinical Pediatric Endocrinology* (Ed. Kaplan, S. A.). W. B. Saunders, Philadelphia, p. 1, 1989.
62. **Ranke, M. B., Pflugers, H., Rosendahl, W., et al.,** *Eur. J. Pediatr.*, 141, 81, 1983.
63. **Lyon, A. J., Preece, M. A., and Grant, D. B.,** *Arch. Dis. Child.*, 60, 932, 1985.
64. **Wilton, P.,** *Acta Pediatr.Scand.*, 76, 193, 1987.
65. **Rosenfeld, R. G., Attie, K. M., Johanson, A. J., et al.,** *Pediatr. Res.*, 29, 85A, 1991.
66. **Rosenfeld, R. G., Frane, J., Attie, K. M., et al.,** *J. Pediatr.*, 121, 49, 1992.
67. **Price, D. A., Clayton, P. E. C., Crowne, E. H., et al.,** *Horm. Res.*, 39(Suppl. 2), 44, 1993.
68. **Lenko, H. L., Hakulinen, A., Kaar, M. L., et al.,** *Horm. Res.*, 39(Suppl. 2), 3, 1997.
69. **Knudtzon, J., Aarskog, D., and the Norwegian Turner's Study Group,** *Horm. Res.*, 39(Suppl. 2), 7, 1993.
70. **Nienhuis, H. E., Rongen-Westerlaken, C., Wit, J. M., et al.,** *Horm. Res.*, 39(Suppl. 2), 31, 1993.
71. **Meyer, H. C. and Ranke, M. B.,** *Horm. Res.*, 35, 109, 1991.
72. **Anneren, G., Sara, V. R., Hall, K., et al.,** *Arch. Dis. Child.*, 61, 48, 1986.
73. **Torrada, C., Bastian, W., Wisniewski, K. E., et al.,** *J. Pediatr.*, 119, 478, 1991.
74. **Allen, D. B., Frasier, S. D., Foley, T. P., et al.,** *J. Pediatr.*, 123, 742, 1993.
75. **Butler, M. G., Meaney, F. J., and Palmer, C. G.,** *Am. J. Med. Genet.*, 23, 790, 1986.

76. **Bray, G. A., Dahms, W. T., Sverdloff, R. S., et al.,** *Medicine*, 62, 59, 1983.
77. **Lee, P. D., Wilson, D. M., Rountree, L., et al.,** *Am. J. Med. Genet.*, 28, 865, 1987.
78. **Angulo, M., Castro-Magana, M., Uy, J., et al.,** *Pediatr. Res.*, 29, 126A, 1991.
79. **Ahmed, L., Foot, A. B. M., Edge, J. A., et al.,** *Acta Pediatr. Scand.*, 80, 446, 1991.
80. **Cianfarani, S., Spadoni, G. L., Finocchi, G., et al.,** *Minerva Pediatr.*, 39, 281, 1987.
81. **Binkin, N. J., Yip, R., Fleshood, L., et al.,** *Pediatrics*, 82, 828, 1988.
82. **Albertsson-Wikland, K.,** *Acta Pediatr. Scand.*, 349, 35, 1989.
83. **Thieriot-Prevost, G., Boccara, J. F., Francuoal, C., et al.,** *Pediatr. Res.*, 24, 380, 1988.
84. **Albanese, A. and Stanhope, R.,** *Horm. Res.*, 39, 8, 1993.
85. **Rochiccioli, P., Tauber, M., Moisan, V., et al.,** *Acta Pediatr. Scand.*, 349, 42, 1989.
86. **Masala, A., Meloni, T., Gallisai, D., et al.,** *J. Clin. Endocrinol. Metab.*, 58, 667, 1984.
87. **Colombo, A., Larizza, D., Garibaldi, E., et al.,** *Minerva Pediatr.*, 29, 1235, 1977.
88. **Canale, V. C., Steinherz, P., New, M. I., et al.,** *Ann. N.Y. Acad. Sci.*, 232, 333, 1974.
89. **Scacchi, M., Danesi, L., De Martin, M., et al.,** *Clin. Endocrinol. (Oxford)*, 35, 335, 1991.
90. **Herington, A. C., Werther, G. A., Matthews, R. N., et al.,** *J. Clin. Endocrinol. Metab.*, 52, 393, 1981.
91. **Postel-Vinay, M. C., Girot, R., Leger, J., et al.,** *J. Clin. Endocrinol. Metab.*, 68, 94, 1989.
92. **Mullis, P. E., Patel, M. S., Brickell, P. M., et al.,** *Clin. Endocrinol. (Oxford)*, 34, 265, 1991.
93. **Appan, S., Laurent, S., Chapman, M., et al.,** *Acta Pediatr. Scand.*, 79, 796, 1990.
94. **Tick, D., Shohat, M., Baraket, S., et al.,** *Clin. Res.*, 39, 54A, 1991.
95. **Lanes, R. and Harrison, H. E.,** *J. Endocrinol. Invest.*, 13, 833, 1990.
96. **Wilson, D. M., Lee, P. D. K., Morris, A. H., et al.,** *Am. J. Dis. Child.*, 145, 1165, 1991.
97. **Rotenstien, D., Reige, D. H., and Flom, L. L.,** *J. Pediatr.*, 115, 417, 1989.
98. **Korkor, A. B.,** *N. Engl. J. Med.*, 316, 1573, 1987.
99. **Hodson, E. M., Shaw, P. F., Evans, R. A., et al.,** *J. Pediatr.*, 103, 735, 1983.
100. **Ramirez, G., O'Neill, W. M., Bloomer, A., et al.,** *Arch. Intern. Med.*, 138, 267, 1978.
101. **Hokken-Koelega, A. C. S., Hackeng, W. H. L., Stijnen, T., et al.,** *J. Clin. Endocrinol. Metab.*, 71, 688, 1990.
102. **Lee, P. K. D., Hintz, R. L., Sperry, J. B., et al.,** *Pediatr. Res.*, 26, 308, 1989.
103. **Lippe, B., Fine, R. N., Koch, V. H., et al.,** *Acta Pediatr. (Suppl.)*, 343, 127, 1988.
104. **Tonshoff, B., Mehls, O., Schauer, A., et al.,** *Kidney Int.*, 36(Suppl. 27), S201, 1989.
105. **Wilson, D. P., Jelley, D., Stratton, R., et al.,** *J. Pediatr.*, 115, 758, 1989.
106. **Hokken-Koelege, A. C. S., Stijnen, T., de Muinck Keiser-Schrama, S. M. P. F., et al.,** *Lancet*, 338, 585, 1991.
107. **Johansson, G., Sietnieks, A., Jannsens, F., et al.,** *Acta Pediatr. (Suppl.)*, 1990, 370, 1990.
108. **Fine, R. N., Yadin, O., Moulten, L., et al.,** *J. Am. Soc. Nephrol.*, 2(Suppl.), 274, 1992.
109. **Tonshoff, B., Mehls, O., Heinrich, U., et al.,** *J. Pediatr.*, 116, 561, 1990.
110. **Lippe, B., Yadin, O., Fine, R. N., et al.,** *Horm. Res.*, 40, 102, 1993.
111. **Ward, D. J., Harton, M., and Ansell, B. M.,** *Ann. Rheum. Dis.*, 26, 416, 1966.
112. **Morris, H. G., Jorgensen, J. R., Elrick, H., et al.,** *J. Clin. Invest.*, 47, 436, 1968.
113. **McCaffery, T. D., Nasr, K., Lawrence, A. M., et al.,** *Digest. Dis.*, 19, 411, 1974.
114. **Miell, J. P., Corder, R., Pralong, F. P., et al.,** *J. Clin. Invest.*, 72, 675, 1991.
115. **Unterman, T. G. and Phillips, L. S.,** *J. Clin. Endocrinol. Metab.*, 61, 618, 1985.
116. **Kovacs, G., Fine, R. N., Worgall, S., et al.,** *Kidney Int.*, 40, 1032, 1991.
117. **Boulton, T. J. C., Smith, R., and Single, T.,** *Acta Pediatr. Scand.*, 81, 322, 1992.
118. **Cicognani, A., Rosito, P., Mancini, A. F., et al.,** *33rd Annu. Meet. ESPE*, Masstricht, June 1994, p. 78.
119. **Bozzola, M., Giorgiani, G., Locatelli, F., et al.,** *Horm. Res.*, 39, 122, 1993.
120. **Sanders, J. E., Buckner, C. D., and Sullivan, K. M. E.,** *Horm. Res.*, 30, 92, 1988.
121. **Sanders, J. E., Pritchard, S., and Mahony, P. E.,** *Blood*, 68, 1129, 1986.
122. **Sanders, J., Sullivan, K., Witherspoon, R., et al.,** *Bone Marrow Transplant.*, 4(Suppl. 4), 27, 1989.
123. **Willi, S. M., Cooke, K., Goldwein, J., et al.,** *J. Pediatr.*, 120, 726, 1992.
124. **Olshan, J. S., Willi, S. M., Gruccio, D., et al.,** *Bone Marrow Transplant.*, 12, 381, 1993.
125. **Papdimitriou, A., Uruena, M., Hamil, G., et al.,** *Arch. Dis. Child.*, 66, 689, 1991.
126. **Christiansen, J. S. and Jorgensen, J. O. L.,** *Acta Endocrinol. (Copenh.)*, 125, 7, 1991.
127. **Moller, J., Jorgensen, J. O. L., Lauersen, T., et al.,** *Clin. Endocrinol. (Oxford)*, 39, 403, 1993.
128. **Vandeweghe, M., Taelman, P., and Kaufman, J.,** *Clin. Endocrinol. (Oxford)*, 39, 409, 1993.
129. **Rosen, T., Bosaeus, I., Tolli, J., et al.,** *Clin. Endocrinol. (Oxford)*, 38, 63, 1993.
130. **Salomon, F., Cuneo, R. C., Hesp, R., et al.,** *N. Engl. J. Med.*, 321, 1797, 1989.

131. **Jorgensen, J. O. L., Pedersen, S. A., Thuesen, L., et al.,** *Acta Endocrinol. (Copenh.)*, 125, 449, 1991.
132. **Whitehead, H. M., Boreham, C., Mcllrath, E. M., et al.,** *Clin. Endocrinol. (Oxford)*, 36, 45, 1992.
133. **Jorgensen, J. O. L., Thuesen, L., and Ingemann-Hansen, T. E.,** *Lancet*, 2, 1221, 1989.
134. **Bengtsson, B. A., Eden, S., Lonn, L., et al.,** *J. Clin. Endocrinol. Metab.*, 76, 309, 1993.
135. **Jorgensen, J. O. L., Moller, J., Alberti, G. M. M., et al.,** *J. Clin. Endocrinol. Metab.*, 77, 1589, 1993.
136. **Cuneo, R. C., Salomon, F., Watts, G. F., et al.,** *Metabolism*, 42, 519, 1993.
137. **Binnerts, A., Swart, G. R., Wilson, J. H. P., et al.,** *Clin. Endocrinol. (Oxford)*, 37, 79, 1992.
138. **Hintz, R. L., Wilson, D. M., Finno, J., et al.,** *Lancet*, i, 1276, 1982.
139. **Moller, A., Rasmussen, L. M., Thuesen, L., et al.,** *Horm. Metab. Res.*, 21, 207, 1989.
140. **Rosen, T., Eden, S., Wilhelmsen, L., et al.,** *Acta Endocrinol. (Copenh.)*, 129, 195, 1993.
141. **Keller, U. and Miles, J. M.,** *Horm. Res.*, 36, 36, 1991.
142. **Eden, S., Bengtsson, B., Loguluso, F., et al,** *Molecular and Clinical Advances in Pituitary Disorders* (Ed. Melmed, S.). Endocrine Research and Education, Los Angeles, CA, p. 173, 1993.
143. **Eden, S., Wiklund, O., Oscarsson, J., et al.,** *Arterioscler. Thromb.*, 13, 296, 1993.
144. **Asayama, K., Amemiya, S., Kusano, S., et al.,** *Metabolism*, 33, 129, 1984.
145. **Moller, A., Rasmussen, L. M., Thuesen, L., et al.,** *Horm. Metab. Res.*, 21, 207, 1989.
146. **White, R. M., Schaefer, E. J., and Papadopoulos, N. M.,** *Proc. Soc. Exp. Biol. Med.*, 173, 63, 1983.
147. **Friedman, M., Byers, S. O., Rosenman, R. H., et al.,** *Metabolism*, 23, 905, 1974.
148. **Nikkila, E. A. and Pelkonen, R.,** *Metabolism*, 24, 829, 1975.
149. **Murase, T., Yamada, N., Oshawa, N., et al.,** *Metabolism*, 29, 666, 1980.
150. **Cuneo, R. C., Salomon, F., Wiles, C. M., et al.,** *J. Appl. Physiol.*, 70, 695, 1991.
151. **Russel-Jones, D. L., Weissberger, A. J., Bowes, S. B., et al.,** *Clin. Endocrinol.*, 38, 427, 1993.
152. **Orme, S. M., Sebastian, J. P., Oldroyd, B., et al.,** *Clin. Endocrinol.*, 37, 453, 1992.
153. **Beshyah, S. A., Sharp, P. S., Gelding, S. V., et al.,** *Acta Endocrinol. (Copenh.)*, 129, 158, 1993.
154. **Kostyo, J. L. and Nutting, D. F.,** *Handbook of Physiology* (Eds. Knobil, E. and Sawyer, W. H.). American Physiological Society, Washington, D.C., p. 187, 1974.
155. **Sonksen, P. H., Cuneo, R. C., Salomon, F., et al.,** *Acta Pediatr. Scand.*, 139, 1991.
156. **Bengtsson, B. A., Rosen, T., Eden, S., Johansson, G., Bosaeus, I., and Sjostrom, L.,** *Molecular and Clinical Advances in Pituitary Disorders* (Ed. Melmed, S.). Endocrine Research and Education, Los Angeles, CA, p. 191, 1993.
157. **Rutherford, O. M., Jones, D. A., Round, J. M., et al.,** *Clin. Endocrinol. (Oxford)*, 34, 469, 1991.
158. **Preece, M. A., Round, J. M., and Jones, D. A.,** *Acta Pediatr. Scand.*, 331(Suppl.), 76, 1987.
159. **Sonksen, P. H., Salomon, F., and Cuneo, R.,** *Horm. Res.*, 36, 27, 1991.
160. **Merimee, T. J., Felig, P., Marliss, E., et al.,** *J. Clin. Invest.*, 50, 574, 1971.
161. **Salomon, F., Cuneo, R., and Sonksen, P. H.,** *Horm. Res.*, 40, 34, 1993.
162. **Fowelin, J., Attvall, S., Lager, I., et al.,** *Metabolism*, 42, 1443, 1993.
163. **Rosen, T., Wiren, L., Wilhelmsen, L., et al.,** *Clin. Endocrinol. (Oxford)*, 40, 111, 1994.
164. **Mcgauley, G. A.,** *Acta Pediatr. Scand. (Suppl.)*, 356, 70, 1989.
165. **Bjork, S., Jonsson, B., Westphal, O., et al.,** *Acta Pediatr. Scand. (Suppl.)*, 343, 3, 1989.
166. **Dean, H. J., McTaggart, T. T., Fish, D. G., et al.,** *Am. J. Dis. Child.*, 139, 1105, 1985.
167. **Blizzard, R. M.,** *Human Growth Hormone* (Eds. Raiti, S. and Tolman, R. A.). Plenum Press, New York, p. 93, 1986.
168. **O'Neill, P. A.,** *Med. Lab. Sci.*, 49, 283, 1992.
169. **Blackman, M. R., Bellantoni, M. F., Busby-Whitehead, J., Stevens, T., Vittone, J., and Harman, S. M.,** *Molecular and Clinical Advances in Pituitary Disorders* (Ed. Melmed, S.). Endocrine Research and Education, Los Angeles, CA, p. 179, 1993.
170. **Corpas, E., Hartman, S. M., and Blackman, M. R.,** *Endocr. Rev.*, 14, 20, 1993.
171. **Rudman, D., Feller, A. G., Nagraj, H. S., et al.,** *N. Engl. J. Med.*, 323, 1, 1990.
172. **Novack, L. P.,** *J. Gerontol.*, 27, 438, 1972.
173. **Chon, S., Vartsky, D., and Yasamura, S.,** *Am. J. Physiol.*, 239, E5524, 1980.
174. **Shimokata, H., Tobin, J. D., Muller, D. C., et al.,** *J. Gerontol.*, 44, M67, 1989.
175. **Enzi, G., Gasparo, M., Biondetti, P., et al.,** *Am. J. Clin. Nutr.*, 44, 739, 1986.
176. **Forbes, G. B. and Reina, J. C.,** *Metabolism*, 19, 653, 1970.
177. **Marcus, R., Butterfield, G., Holloway, L., et al.,** *J. Clin. Endocrinol. Metab.*, 70, 519, 1990.
178. **Rudman, D., Feller, A. G., Cohn, L., et al.,** *Horm. Res.*, 36, 73, 1991.

179. **Coh, L., Feller, A. G., Draper, M. W., et al.,** *Clin. Endocrinol. (Oxford)*, 39, 417, 1993.
180. **Corpas, E., Harman, S. M., Pineyro, M. A., et al.,** *J. Clin. Endocrinol. Metab.*, 75, 530, 1992.
181. **Wabitsch, M. and Heinze, E.,** *Horm. Res.*, 40, 5, 1993.
182. **Collipp, P. J., Curti, V., Thomas, J., et al.,** *Metabolism*, 22, 589, 1973.
183. **Dahms, W. T., Owens, R. P., Kalhan, S. C., et al.,** *Metabolism*, 38, 197, 1989.
184. **Tanner, J. M., Whitehouse, R. H., Hughes, P. C. R., et al.,** *Arch. Dis. Child.*, 84, 745, 1971.
185. **Bonnet, F., Vanderschueren-Lodeweyckx, M., Eeckels, R., et al.,** *Pediatr. Res.*, 8, 800, 1974.
186. **Vaisman, N., Zadik, Z., Shamai, Y., et al.,** *Metabolism*, 41, 483, 1992.
187. **Walker, J. M., Bond, S. A., Voss, L. D., et al.,** *Lancet*, 330, 1331, 1990.
188. **Rosenbaum, M., Gertner, J. M., Gidfar, N., et al.,** *J. Clin. Endocrinol. Metab.*, 75, 151, 1992.
189. **Leiberaman, L., Pilpej, D., Carel, C. A., et al.,** *Horm. Res.*, 40, 128, 1993.
190. **Press, M., Notarfrancesco, A., and Genel, M.,** *Lancet*, 1, 1002, 1987.
191. **Bougneres, P.,** *Horm. Res.*, 40, 31, 1993.
192. **Underwood, L. E., Van Den Brande, J. L., Antony, G. J., et al.,** *J. Pediatr.*, 82, 128, 1973.
193. **Ranke, M. B.,** *Horm. Res.*, 39, 104, 1993.
194. **Gregory, J. W., Green, S. A., Jung, R. T., et al.,** *Arch. Dis. Child.*, 66, 598, 1991.
195. **Winter, R. J., Thompson, R. G., and Green, O. C.,** *Metabolism*, 28, 1244, 1979.
196. **Caprio, S., Boulware, D., and Tamborlane, V.,** *Horm. Res.*, 38, 47, 1992.
197. **Wilson, D. M., France, J. W., Sherman, B., et al.,** *J. Pediatr.*, 112, 210, 1988.
198. **Weise, M., James, D., Leitner, C. H., et al.,** *Horm. Res.*, 39, 36, 1993.
199. **Takano, K., Hizuka, N., and Shizume, K.,** *Acta Pediatr. Scand.*, 325, 58, 1986.
200. **Vanderschueren-Loedeweyckx, M., Masa, G., Maes, M., et al.,** *J. Clin. Endocrinol. Metab.*, 70, 122, 1990.
201. **McCauley, E., Kay, Y., and Ito, J.,** *Child Dev.*, 58, 464, 1986.
202. **Siegel, P. T., Clopper, R., and Stabler, B.,** *Acta Pediatr. Scand.*, 377, 14, 1991.
203. **Rovet, J., Holland, J., and the Canadian Growth Hormone Advisory Group,** *Horm. Res.*, 39(Suppl. 2), 60, 1993.
204. **Huisman, J., Slijper, F. M. E., Sinnema, G., et al.,** *Horm. Res.*, 39(Suppl. 2), 56, 1993.
205. **Gertner, J. M.,** *Horm. Res.*, 40, 10, 1993.
206. **Moller, N., Porksen, N., Ovesen, P., et al.,** *Horm. Metab. Res.*, 25, 175, 1993.
207. **Moller, N., Moller, J., Jorgensen, J. O. L., et al.,** *Clin. Endocrinol. (Oxford)*, 39, 577, 1993.
208. **Clemmons, D. R., Snyder, D. K., and Williams, R.,** *J. Clin. Endocrinol. Metab.*, 64, 878, 1987.
209. **Snyder, D. K., Clemmons, D. R., and Underwood, L. E.,** *J. Clin. Endocrinol. Metab.*, 67, 53, 1988.
210. **Snyder, D. K., Clemmon, D. R., and Underwood, L. E.,** *J. Clin. Endocrinol. Metab.*, 69, 745, 1989.
211. **Snyder, D. K., Underwood, L. E., and Clemmons, D. R.,** *Am. J. Clin. Nutr.*, 52, 431, 1990.
212. **Skaggs, S. R. and Crist, D. M.,** *Horm. Res.*, 35, 19, 1991.
213. **MacIntyre, J. G.,** *Sports Med.*, 4, 129, 1987.
214. **Council on Scientific Affairs,** *JAMA*, 259, 1703, 1988.
215. **Crist, D. M., Peake, G. T., Egan, P. A., et al.,** *J. Appl. Physiol.*, 65, 579, 1988.
216. **Yarasheski, K. E., Campbell, J. A., Smith, K., et al.,** *Am. J. Physiol.*, 262, E261, 1992.
217. **Deyssig, R., Frisch, H., Blum, W. F., et al.,** *Acta Endocrinol. (Copenh.)*, 128, 313, 1993.
218. **Pape, G. S., Freidman, M., and Underwood, L. E.,** *Chest*, 99, 1495, 1991.
219. **Suchner, U., Rothkopf, M. M., Stanislaus, G., et al.,** *Arch. Intern. Med.*, 150, 1225, 1990.
220. **Wilmore, D. W., Moylan, J. A., Bristow, B. F., et al.,** *Surg. Gynecol. Obstet.*, 138, 875, 1974.
221. **Liljedahl, S. O., Gemzell, C. A., Plantin, L. O., et al.,** *Acta Chir. Scand.*, 122, 1, 1961.
222. **Roe, C. F. and Kinney, J.,** *Surg. Forum*, 13, 369, 1962.
223. **Soroff, H. S., Ronzin, R. R., Mooty, J., et al.,** *Ann. Surg.*, 166, 739, 1967.
224. **Fleming, R. Y. D., Rutan, R. L., Jahoor, F., et al.,** *J. Trauma*, 32, 698, 1992.
225. **Clemmons, D. R. and Underwood, L. E.,** *Horm. Res.*, 38, 37, 1992.
226. **Bennerts, A., Wilson, J. H. P., and Lamberts, S. W. J.,** *J. Clin. Endocrinol. Metab.*, 67, 1312, 1988.
227. **Kaiser, F. E., Silver, A. J., and Morley, J. E.,** *J. Am. Geriatr. Soc.*, 39, 235, 1991.
228. **Wolf, R. F., Heslin, M. J., Newman, E., et al.,** *Surgery*, 112, 284, 1992.
229. **Kotler, D. P., Tierney, A. R., Wang, J., et al.,** *Am. J. Clin. Nutr.*, 50, 444, 1989.
230. **Krentz, A. J., Koster, F. T., and Crist, D. M., et al.,** *J. Acquir. Immune Defic. Syndr.*, 6, 245, 1993.
231. **Mulligan, K., Grunfeld, C., Hellerstein, M. K., et al.,** *J. Clin. Endocrinol. Metab.*, 77, 956, 1993.

232. **Gottardis, M., Benzer, A., Koller, W., et al.,** *J. Trauma*, 31, 81, 1991.
233. **Hill, G. L.,** *J. Parenter. Enteral. Nutr.*, 16, 197, 1992.
234. **Warnold, I., Eden, E., and Lundholm, K.,** *Ann. Surg.*, 208, 143, 1988.
235. **Ward, H. C., Halliday, D., and Sim,** *Ann. Surg.*, 206, 56, 1987.
236. **Ponting, G. A., Halliday, D., Teale, J. D., et al.,** *Lancet*, 1, 438, 1988.
237. **Jiang, Z. M., He, G. Z., Zhang, S. Y., et al.,** *Ann. Surg.*, 210, 513, 1989.
238. **Byrne, T. A., Morrissey, T. B., Gatzen, C., et al.,** *Ann. Surg.*, 218, 400, 1993.
239. **Ziegler, T. R., Rombeau, J. L., Young, L. S., et al.,** *J. Clin. Endocrinol. Metab.*, 74, 865, 1992.
240. **Horber, F. F., Zurcher, R. M., Herren, H., et al.,** *Am. J. Clin. Nutr.*, 43, 758, 1986.
241. **Haymond, M. W., Horber, F. F., and Mauras, N.,** *Horm. Res.*, 38, 73, 1992.
242. **Haymond, M. W. and Horber, F. F.,** *Horm. Res.*, 38, 44, 1992.
243. **Bennet, W. M. and Haymond, M. W.,** *Clin. Endocrinol.*, 36, 161, 1992.
244. **Bentham, J., Rodriguez-Arnao, J., and Ross, R. J. M.,** *Horm. Res.*, 40, 87, 1993.
245. **Clemmons, D. R., Smith-Banks, A., and Celniker, A. C., et al.,** *J. Clin. Endocrinol. Metab.*, 75, 234, 1992.
246. **Clemmons, D. R. and Underwood, L. E.,** *Molecular and Clinical Advances in Pituitary Disorders* (Ed., Melmed, S.). Endocrine Research and Education, Los Angeles, CA, p. 166, 1993.
247. **Bouillon, R.,** *Horm. Res.*, 36(Suppl. 2), 49, 1991.
248. **Kassem, M., Blum, W., Ristelli, J., et al.,** *Calcif. Tiss. Int.*, 52, 222, 1993.
249. **Beck, J. C., McGarry, E. E., Dyenforth, D., et al.,** *Science*, 125, 884, 1957.
250. **Brixen, K., Nielsen, H. K., Bouillon, R., et al.,** *Acta Endocrinol. (Copenh.)*, 127, 331, 1992.
251. **Spencer, E. M. and Tobiassen, D.,** *Endocrinology (Baltimore)*, 108, 1064, 1981.
252. **Johansson, A. G., Burman, P., Westermark, K., et al.,** *J. Intern. Med.*, 232, 447, 1992.
253. **Rosen, T., Hansson, T., Granhed, G., et al.,** *Acta Endocrinol. (Copenh.)*, 129, 201, 1993.
254. **Degerblad, M., Elgindy, N., Hall, K., et al.,** *Acta Endocrinol. (Copenh.)*, 126, 387, 1992.
255. **Guevera-Aguirre, J., Rosenbloom, A. L., Fielder, P. J., et al.,** *J. Clin. Endocrinol. Metab.*, 76, 417, 1993.
256. **Shore, R. M., Chesney, R. B., Mazess, P. H., et al.,** *Calcif. Tiss. Int.*, 34, 519, 1982.
257. **Bergmann, P. J., Valsamis, J., Van Perborgh, J., et al.,** *J. Clin. Endocrinol. Metab.*, 71, 1461, 1990.
258. **Ross, J. L., Meyersenlong, P., Feuillan, F., et al.,** *J. Clin. Endocrinol. Metab.*, 73, 355, 1991.
259. **O'Halloran, D. J., Tsatsoulis, A., Whitehouse, R. W., et al.,** *J. Clin. Endocrinol. Metab.*, 76, 1344, 1993.
260. **Salomon, F., Cuneo, R. C., Hesp, R., et al.,** *New Engl. J. Med.*, 321, 1797, 1989.
261. **Cuneo, R. C., Salomon, F., Wiles, C. M., et al.,** *J. Appl. Physiol.*, 70, 695, 1991.
262. **van der Veen, E. A. and Netelenbos, J. C.,** *Horm. Res.*, 33(Suppl. 4), 65, 1990.
263. **Jensen, L. T., Jorgensen, J. O. L., Risteli, J., et al.,** *Acta Endocrinol. (Copenh.)*, 124, 278, 1991.
264. **Degerblad, M., Almkvist, O., Grunditz, R., et al.,** *Acta Endocrinol. (Copenh.)*, 123, 185, 1990.
265. **Antoniazzi, F., Radetti, G., Zamboni, G., et al.,** *Bone Miner.*, 21, 151, 1993.
266. **Marcus, R. G., Butterfield, L., Holloway, L., et al.,** *J. Clin. Endocrinol. Metab.*, 70, 519, 1990.
267. **Burstein, S., Chen, I. W., and Tsang, R. C.,** *J. Clin. Endocrinol. Metab.*, 56, 1246, 1983.
268. **Gertner, J. M., Tamborlane, W. V., Hintz, R. L., et al.,** *J. Clin. Endocrinol. Metab.*, 53, 818, 1981.
269. **Bagge, E., Eden, S., Rosen, T., et al.,** *Acta Endocrinol. (Copenh.)*, 129, 296, 1993.
270. **Aloia, J. F., Vaswani, A., Kapoor, A., et al.,** *Metabolism*, 34, 124, 1985.
271. **Bridges, N. A., Hindmarsh, P. C., and Brook, C. G. D.,** *Horm. Res.*, 36, 56, 1991.
272. **Bak, B., Jorgensen, P. H., and Andreassen, T. T.,** *Acta Orthoped. Scand.*, 61, 54, 1990.
273. **Bak, B., Jorgensen, P. H., and Andreassen, T. T.,** *Bone*, 11, 233, 1990.
274. **Bak, B., Jorgensen, P. H., and Andreassen, T. T.,** *Clin. Orthoped.*, 264, 295, 1991.
275. **Bak, B. and Andreassen, T. T.,** *Bone*, 12, 151, 1991.
276. **Lindholm, R. V., Koskinen, E. V. S., Puranen, J., et al.,** *Horm. Metab. Res.*, 9, 245, 1977.
277. **Mosekilde, L. and Bak, B.,** *Bone*, 14, 19, 1993.
278. **Johansen, J. S., Pedersen, S. A., Jorgensen, J. O. L., et al.,** *J. Clin. Endocrinol. Metab.*, 70, 916, 1990.
279. **Schlemmer, A., Johansen, J. S., Pedersen, S. A., et al.,** *Clin. Endocrinol. (Oxford)*, 35, 471, 1991.
280. **Brixen, K. H., Nielsen, L., Mosekilde, A., et al.,** *J. Bone Min. Res.*, 5, 609, 1990.
281. **Franchimont, P.,** *Acta Endocrinol. (Copenh.)*, 120, 121, 1989.

282. **Nabarro, J. D. N.,** *Clin. Endocrinol. (Oxford)*, 26, 481, 1987.
283. **Parfitt, A. M.,** *Progress in Basic and Clinical Pharmacology* (Ed. Kanis, J. A.). S. Karger, Basel, p. 1, 1990.
284. **Revhaug, A. and Mjaaland, M.,** *Horm. Res.*, 40, 99, 1993.
285. **Prudden, J. F., Nishihara, G., and Ocampo, L.,** *Surg. Gynecol. Obstet.*, 107, 481, 1958.
286. **Barbul, A., Rettura, G., Prior, E., et al.,** *Surg. Forum*, 29, 93, 1978.
287. **Belcher, H. J. and Ellis, H.,** *J. Clin. Endocrinol. Metab.*, 70, 939, 1990.
288. **Waago, H.,** *Lancet*, 1, 1485, 1987.
289. **Steenfos, H. H. and Jansson, J.-O.,** *J. Endocrinol.*, 132, 293, 1992.
290. **Kelley, S. F., Felix, A. M., and Ehrlich, H. P.,** *Proc. Soc. Exp. Biol. Med.*, 194, 320, 1990.
291. **Jorgensen, P. H. and Andreassen, T. T.,** *Acta Chir. Scand.*, 154, 623, 1988.
292. **Herndon, D. N., Barrow, R. E., Kunkel, K. R., et al.,** *Ann. Surg.*, 212, 424, 1990.
293. **Rasmussen, L. H., Karlsmark, T., Avnstorp, C., et al.,** *Phlebology*, 6, 23, 1991.
294. **Lynch, S. E., Colvin, R. B., and Antoniades, H. N.,** *J. Clin. Invest.*, 84, 640, 1989.
295. **Lynch, S. E., Nixon, J. C., Colvin, R. B., et al.,** *Proc. Natl. Acad. Sci. U.S.A.*, 84, 7696, 1987.
296. **Zeizen, Y., Ford, E. G., Costin, G., et al.,** *J. Pediatr. Surg.*, 25, 70, 1990.
297. **Jorgensen, J. O. L., Moller, J., Skakkeboek, N. E., et al.,** *Horm. Res.*, 38(Suppl. 1), 63, 1992.
298. **Chong, P. K. K., Jung, R. T., Scrimgeour, C. M., et al.,** *Clin. Endocrinol. (Oxford)*, 40, 103, 1994.
299. **Jorgensen, J. O. L., Pedersen, S. A., Ingerslev, J., et al.,** *Scand. J. Clin. Lab. Invest.*, 50, 417, 1990.
300. **Salomon, F., Cuneo, R. C., Hesp, R., et al.,** *Clin. Sci.*, 83, 325, 1992.
301. **Massa, G., de Zegher, F., and Vanderschueren-Lodeweyckx, M.,** *Clin. Endocrinol.*, 34, 205, 1991.
302. **Rosen, T. and Bengtsson, B. A.,** *Lancet*, 336, 285, 1990.
303. **Wuster, C., Slenczka, E., and Ziegler, R.,** *Klin. Wochenschr.*, 69, 769, 1991.
304. **Merela, B., Cittadini, A., Colao, A., et al.,** *Endocrinology (Baltimore)*, 77, 1658, 1993.
305. **Johnston, D. G., Beshyah, S. A., Markussis, V., et al.,** *Horm. Res.*, 38, 68, 1992.
306. **Thuesen, L., Christiansen, J. S., Sorensen, K. E., et al.,** *Dan. Med. Bull.*, 35, 193, 1988.
307. **Cuneo, R. C., Salomon, F., Wilmshurst, P., et al.,** *Clin. Sci.*, 81, 587, 1991.
308. **Markussis, V., Beshyah, S. A., Fisher, C., et al.,** *Lancet*, 340, 1188, 1992.
309. **Garrow, J.,** *Br. Med. J.*, 303, 1152, 1991.
310. **Falkheden, T. and Sjorgren, B.,** *Acta Endocrinol. (Copenh.)*, 46, 80, 1964.
310a. **Molitch, M. E.,** *Endocrinol. Metab. Clin. N. Am.*, 21, 597, 1992.
311. **Lim, M. J., Barkan, A. L., and Buda, A. J.,** *Ann. Intern. Med.*, 117, 719, 1992.
312. **Jorgensen, J. O. L., Moller, A., Vahl, N., et al.,** *Molecular and Clinical Advances in Pituitary Disorders — 1993* (Ed. Melmed, S.). Endocrine Education, Los Angeles, CA, p. 185, 1993.
313. **Lundbaek, K., Jensen, V. A., Olsen, T. S., et al.,** *Lancet*, i, 131, 1970.
314. **Ledet, T.,** *Diabetes*, 25, 1011, 1976.
315. **Ledet, T.,** *Diabetes*, 26, 798, 1989.
316. **Lundergan, C., Foegh, M. L., Vargas, R., et al.,** *Atherosclerosis*, 80, 49, 1989.
317. **Pedersen, S. A., Welling, K., Michaelsen, K. F., et al.,** *Lancet*, ii, 681, 1989.
318. **Juul, A., Behrenscheer, A., Tims, T., et al.,** *Clin. Endocrinol.*, 38, 237, 1993.
319. **Lobie, P. E., Breipohl, W., Lincoln, D. T., et al.,** *J. Endocrinol.*, 126, 467, 1990.
320. **Villanua, M. A., Szary, A., Bartke, A., et al.,** *J. Endocrinol. Invest.*, 15, 587, 1992.
321. **Murphy, W. J., Durum, S. K., and Longo, D. L.,** *J. Immunol.*, 149, 3851, 1992.
322. **Goya, R. G., Gagnerault, M. C., Demoraes, M. C. L., et al.,** *Brain Behav. Immun.*, 6, 341, 1992.
323. **Goff, B. L., Roth, J. A., Arp, L. H., et al.,** *Clin. Exp. Immunol.*, 68, 580, 1987.
324. **Abassi, V. and Bellanti, J. A.,** *Pediatr. Res.*, 19, 299, 1985.
325. **Rapaport, R., Oleske, J., Ahdieh, H., et al.,** *J. Pediatr.*, 109, 434, 1986.
326. **Bozzola, M. M., Maccario, R., Cisternino, M., et al.,** *Acta Pediatr. Scand.*, 77, 675, 1988.
327. **Petersen, B. H., Rapaport, R., Henry, D. P., et al.,** *J. Clin. Endocrinol. Metab.*, 70, 1756, 1990.
328. **Spandoni, G. L., Rossi, P., Ragno, W., et al.,** *Acta Pediatr. Scand.*, 80, 75, 1991.
329. **Wit, J. M., Kooijman, R., Rijkers, G. T., et al.,** *Horm. Res.*, 39, 107, 1993.
330. **Casanova, S., Repellin, A. M., and Schimpff, R. M.,** *Horm. Res.*, 34, 209, 1990.
331. **Mocchegiani, E., Paolucci, P., Balsamo, A., et al.,** *Horm. Res.*, 33, 248, 1990.
332. **Kiess, W., Doerr, H., Butenandt, O., et al.,** *N. Engl. J. Med.*, 314, 321, 1986.
333. **Kiess, W., Malozawsksi, S., Gelato, M., et al.,** *Clin. Immunol. Immunopathol.*, 48, 85, 1988.
334. **Bozzola, M., Valtorta, A., Moretta, A., et al.,** *J. Pediatr.*, 117, 596, 1990.

335. **Gupta, D., Fikrig, S. M., and Noval, M. S.,** *Clin. Exp. Immunol.*, 54, 87, 1983.
336. **Etzioni, A., Pollack, S., and Hochberg, Z.,** *Acta Pediatr. Scand.*, 77, 169, 1988.
337. **Church, J. A., Costin, G., and Brooks, J.,** *J. Pediatr.*, 115, 420, 1989.
338. **Bozzola, M. M., Cisternino, M., Valtorta, A., et al.,** *Horm. Res.*, 31, 153, 1989.
339. **Hirschberg, R. and Kopple, J.,** *Am. J. Nephrol.*, 8, 249, 1988.
340. **Falkheden, T.,** *Acta Endocrinol. (Copenh.)*, 48, 354, 1965.
341. **Falkheden, T. and Wickborn, I.,** *Acta Endocrinol. (Copenh.)*, 48, 348, 1965.
342. **Gerschberg, H., Heinemann, H., and Stumpf, H.,** *J. Clin. Endocrinol. Metab.*, 17, 377, 1957.
343. **Falkheden, T. and Sjogren, B.,** *Acta Endocrinol. (Copenh.)*, 46, 80, 1964.
344. **Ikkos, D., Ljunggren, H., and Luft, R.,** *Acta Endocrinol. (Copenh.)*, 21, 226, 1956.
345. **Hoogenberg, K., Sluiter, W. J., and Dullaart, P. F.,** *Acta Endocrinol. (Copenh.)*, 129, 151, 1993.
346. **Christiansen, J. S., Gammelgaard, J., Orskov, H., et al.,** *Eur. J. Clin. Invest.*, 11, 487, 1981.
347. **Hirschberg, R., Rabb, H., Bergamo, R., et al.,** *Kidney Int.*, 35, 865, 1989.
348. **Arnold, W. C., Shirkey, B., Frindik, P., et al.,** *Pediatr. Nephrol.*, 5, 529, 1991.
349. **Hirschberg, R. and Kopple, J.,** *J. Clin. Invest.*, 83, 326, 1989.
350. **Guler, H., Eckardt, K., Zapf, J., et al.,** *J. Clin. Invest.*, 87, 1200, 1991.
351. **Hirschberg, R.,** *Regul. Peptides*, 48, 241, 1993.
352. **Miller, S. B., Martin, D. R., Kissane, J., et al.,** *Proc. Natl. Acad. Sci. U.S.A.*, 89, 11876, 1992.
353. **O'Shea, M. H., Miller, S. B., and Hammerman, M. R.,** *Am. J. Physiol.*, 264, F917, 1993.
354. **Hammerman, M. R., Ryan, G., and Miller, S. B.,** *Miner. Electrolyte Metab.*, 18, 253, 1992.
355. **Chin, E. and Bondy, C.,** *J. Clin. Endocrinol. Metab.*, 75, 962, 1992.
356. **Janssen, F., Van Damme, L. R., Van Dyck, M., et al.,** *Transplant Proc.*, 25, 1049, 1993.
357. **Trachtman, H., Futterweit, S., Schwob, N., et al.,** *Kidney Int.*, 44, 1281, 1993.
358. **Kainer, G., Nakano, M., Massie, F. S., et al.,** *Physiol. Rev.*, 30, 528, 1991.
359. **Ho, K. Y. and Kelly, J. J.,** *Horm. Res.*, 39(Suppl. 1), 44, 1991.
360. **Ikkos, D., Luft, R., and Gemzell, C. A.,** *Lancet*, i, 720, 1958.
361. **Moller, J., Jorgensen, J. O. L., Moller, N., et al.,** *J. Clin. Endocrinol. Metab.*, 72, 768, 1991.
362. **Herlitz, H., Jonsson, O., and Bengtsson, B. A.,** *Acta Endocrinol. (Copenh.)*, 127, 38, 1992.
363. **Lobie, P. E., Breipohl, W., Garcia-Aragon, J., et al.,** *Endocrinology (Baltimore)*, 126, 2214, 1990.
364. **Cacciari, E., Cicognani, A., Pirazzoli, P., et al.,** *Helv. Paediatr. Acta*, 31, 481, 1976.
365. **Laron, Z. and Sarel, R.,** *Acta Endocrinol. (Copenh.)*, 63, 625, 1970.
366. **Shoham, Z., Conway, G. S., Ostergaard, H., et al.,** *Fertil. Steril.*, 57, 1044, 1992.
367. **Uy, J. S., Castro-Magana, M., and Angulo, M. A.,** *J. Pediatr. Endocrinol.*, 4, 185, 1991.
368. **Chatelain, P. G., Sanchez, P., and Saez, J. M.,** *Endocrinology (Baltimore)*, 128, 1857, 1991.
369. **Balducci, R., Toscano, V., Mangiantini, A., et al.,** *Acta Endocrinol. (Copenh.)*, 128, 19, 1993.
370. **Katz, E., Ricciarelli, E., and Adashi, E. Y.,** *Fertil. Steril.*, 59, 8, 1993.
371. **Sheikholislam, B. M. and Stempfel, R. S.,** *Pediatrics*, 49, 362, 1972.
372. **Menashe, Y., Sack, J., and Mashiach, S.,** *Hum. Reprod.*, 6, 670, 1991.
373. **Shoham, Z., Homburg, R., Owen, E. J., et al.,** *Baillieres Clin. Obstet. Gynaecol.*, 6, 267, 1992.
374. **Hillensjo, T. and Bergh, C.,** *Acta Endocrinol. (Copenh.)*, 128(Suppl. 2), 23, 1993.
375. **Stone, B. A. and Marrs, R. P.,** *Fertil. Steril.*, 58, 32, 1992.
376. **Homburg, R., Eshel, A., Adalla, H. I., et al.,** *Clin. Endocrinol. (Oxford)*, 29, 113, 1990.
377. **Homburg, R., West, C., Torressani, T., et al.,** *Fertil. Steril.*, 53, 254, 1990.
378. **Homburg, R., West, C., Torresani, T., et al.,** *Clin. Endocrinol. (Oxford)*, 32, 781, 1990.
379. **Volpe, A., Coukos, G., Artini, P. G., et al.,** *Hum. Reprod.*, 3, 345, 1990.
380. **Burger, H. G., Kovacs, G. T., Polson, D. M., et al.,** *Clin. Endocrinol. (Oxford)*, 35, 119, 1991.
381. **Fowler, P. A. and Templeton, A.,** *Clin. Endocrinol. (Oxford)*, 35, 117, 1991.
382. **Homburg, R., West, C., Torresani, T., et al.,** *Clin. Endocrinol. (Oxford)*, 32, 781, 1990.
383. **Jacobs, H. S.,** *Horm. Res.*, 38, 14, 1992.
384. **Owen, E. J., Shoham, Z., Mason, B. A., et al.,** *Fertil. Steril.*, 56, 1104, 1991.
385. **Owen, E. J., West, C., Mason, B. A., et al.,** *Hum. Reprod.*, 6, 524, 1991.
386. **Blumenfeld, Z. and Amit, T.,** *Basic and Clinical Advances in Pituitary Disease* (Ed. Melmed, S.). Endocrine Research and Education, Los Angeles, CA, p. 205, 1993.
387. **Blumenfeld, Z., Amit, T., Barkley, R. J., et al.,** *Ann. N.Y. Acad. Sci.*, 626, 250, 1991.
388. **Ibrahim, Z. H. Z., Matson, P. L., Buck, P., et al.,** *Fertil. Steril.*, 55, 202, 1991.
389. **Volpe, A., Coukos, G., Barreca, A., et al.,** *Gynecol. Endocrinol.*, 3, 125, 1989.
390. **Owen, E. J., Shoham, Z., Mason, B. A., et al.,** *Fertil. Steril.*, 56, 1104, 1991.
391. **Shaker, A. G., Yates, R. W. S., Fleming, R., et al.,** *Fertil. Steril.*, 58, 919, 1992.

392. **Younis, J. S., Dorembus, D., Simon, A., et al.,** *Fertil. Steril.*, 58, 575, 1992.
393. **Tapanainen, J., Martikainen, H., Voutilainen, R., et al.,** *Fertil. Steril.*, 58, 726, 1992.
394. **Daughaday, W. H. and Barbano, D. M.,** *JAMA*, 264, 1003, 1990.
395. **Wilson, M. E., Gordon, T. P., Cnikazawa, L., et al.,** *J. Clin. Endocrinol. Metab.*, 72, 1302, 1991.
396. **Milsom, S. R., Breier, B. H., Gallaher, B. W., et al.,** *Acta Endocrinol. (Copenh.)*, 127, 337, 1992.
397. **Breier, B. H., Milsom, S. R., Blum, W. F., et al.,** *Acta Endocrinol. (Copenh.)*, 129, 427, 1993.
398. **Ritzen, E. M., Czernichow, P., Preece, M., et al.,** *Horm. Res.*, 39, 92, 1993.
399. **Hintz, R. L.,** *Horm. Res.*, 38, 44, 1992.
400. **Job, J. C., Maillard, F., and Goujard, J.,** *Horm. Res.*, 38, 35, 1992.
401. **Preece, M.,** *Horm. Res.*, 39, 95, 1993.
402. **Pocchiari, M., Peano, S., Conz, A., et al.,** *Horm. Res.*, 35, 161, 1991.
403. **Ezzat, S., Strom, C., and Melmed, S.,** *Ann. Intern. Med.*, 114, 754, 1991.
404. **Ron, E., Gridley, G., Hrubec, Z., et al.,** *Cancer*, 68, 1673, 1991.
405. **Brunner, J. F. and Mellinger, R. C.,** *Clin. Endocrinol. (Oxford)*, 32, 65, 1990.
406. **Pines, A., Rosen, P., Ron, E., et al.,** *Am. J. Gastroenterol.*, 80, 266, 1985.
407. **Akaza, H., Matsuki, K., Matsushima, H., et al.,** *Cancer*, 68, 2418, 1991.
408. **Manni, A., Wright, C., David, G., et al.,** *Cancer Res.*, 46, 1669, 1986.
409. **Moon, H. D., Li, C. H., and Simpson, M. E.,** *Cancer Res.*, 16, 111, 1956.
410. **Liao, D., Porsch-Hallstrom, I., Gustafsson, J., et al.,** *Carcinogenesis*, 14, 2045, 1993.
411. **Murphy, L. J., Vrhovsek, E., Sutherland, R. L., et al.,** *J. Clin. Endocrinol. Metab.*, 58, 149, 1984.
412. **Malarkey, W. B., Kennedy, M., Alfred, L. E., et al.,** *J. Clin. Endocrinol. Metab.*, 56, 673, 1983.
413. **Chatelain, P. G. and Wilton, P.,** *KIGS Biannu. Rep. No. 8*, 1, 42, 1992.
414. **Manni, A.,** *Biotherapy*, 4, 31, 1992.
415. **Ogilvystuart, A. L., Ryder, W. D. J., Gattamaneni, H. R., et al.,** *Br. Med. J.*, 304, 1601, 1992.
416. **Wilton, P. and Price, D. A.,** *KIGS Biannu. Rep. No. 8*, 1, 48, 1992.
417. **Bourgignon, J. P., Pierard, E., Ernould, C., et al.,** *Lancet*, 341, 1505, 1993.
418. **Massa, G., Vanderschueren-Lodeweyckx, M., and Bouillon, R.,** *Clin. Endocrinol.*, 38, 137, 1993.
419. **Rougeot, C., Marchand, P., Dray, F., et al.,** *Horm. Res.*, 35, 76, 1991.
420. **Zeisel, H. J., Lutz, A., and Petrykowski, W. V.,** *Horm. Res.*, 37, 47, 1992.
421. **Parks, J. S., Meacham, L. R., McKean, M. C., et al.,** *Pediatr. Res.*, 25, 90A, 1989.
422. **Cardoso, A. I., Llera, A. S., Iacono, R. F., et al.,** *Acta Endocrinol. (Copenh.)*, 129, 20, 1993.
423. **Underwood, L. E., Voina, S. J., and Van Wyk, J. J.,** *J. Clin. Endocrinol. Metab.*, 38, 288, 1974.
424. **Manfredi, R., Zucchini, A., Azzaroli, L., et al.,** *J. Endocrinol. Invest.*, 16, 709, 1993.
425. **Hibi, I., Tanaka, T., Tane, A., et al.,** *J. Clin. Endocrinol. Metab.*, 69, 221, 1989.
426. **Darendeliler, F., Hindmarsh, P. K. C., Preece, M. A., et al.,** *Acta Endocrinol. (Copenh.)*, 122, 414, 1990.
427. **Rappaport, E. B. and Fife, D.,** *Am. J. Dis. Child.*, 136, 901, 1985.
428. **Anderson, C.,** *Nature (Lond.)*, 358, 4, 1992.

INDEX

A

B

C

D

E

F

G

H

I

N

O

P

Q

R

S

T

U

V

W

X

Z